Mastering
AutoCAD® Release 14

Mastering AutoCAD® Release 14

Ron K. C. Cheng
The Hong Kong Polytechnic University

PWS Publishing
An imprint of Brooks/Cole Publishing Company
I(T)P® An International Thomson Publishing Company

Pacific Grove • Albany • Belmont • Bonn • Boston • Cincinnati • Detroit • Johannesburg • London
Madrid • Melbourne • Mexico City • New York • Paris • Singapore • Tokyo • Toronto • Washington

Sponsoring Editor: *Bill Stenquist*
Project Development Editor: *Suzanne Jeans*
Production Editor: *Kelsey McGee*
Marketing Manager: *Nathan Wilbur*
Editorial Assistant: *Shelley Gesicki*
Manuscript Editor: *Connie Day*
Composition/Interior Design/
Cover Image: *Ron C.K. Cheng*
Cover Design: *Denise Davidson*
Printing and Binding: *Patterson*

For more information, contact:

Autodesk Press
3 Columbia Circle, Box 15-015
Albany, New York 12212-5015

BROOKS/COLE PUBLISHING COMPANY
511 Forest Lodge Road
Pacific Grove, CA 93950
USA

International Thomson Editores
Seneca, 53
Colonia Polanco
11560 México D. F., México

International Thomson Publishing Europe
Berkshire House 168-173
High Holborn
London WC1V 7AA
England

International Thomson Publishing GmbH
Königswinterer Strasse 418
53227 Bonn
Germany

Thomas Nelson Australia
102 Dodds Street
South Melbourne, 3205
Victoria, Australia

International Thomson Publishing Asia
60 Albert Street
#15-01 Albert Complex
Singapore 189969

Nelson Canada
1120 Birchmount Road
Scarborough, Ontario
Canada M1K 5G4

International Thomson Publishing Japan
Hirakawacho Kyowa Building, 3F
2-2-1 Hirakawacho
Chiyoda-ku, Tokyo 102
Japan

Printed in the United States of America

10 9 8 7 6 5 4 3 2 1

Library of Congress Cataloging-in-Publication Data

Cheng, Ron.
Mastering AutoCAD release 14/Ron K. C. Cheng.
p. cm.
Includes index.
ISBN 0-534-95761-7
1. Computer graphics. 2. AutoCAD (Computer file) I. Title.
II. Title: Mastering AutoCad release fourteen
T385.C4761998
604.2'0285'5369—dc21

98-46364

8.8 3D Model for the Roofs 543
8.9 Complete Model 547
8.10 Key Points and Exercises 556

9 Documentation and Plotting 565
9.1 Engineering Drawing Title Block 565
9.2 Documentation for a 2D Drawing 569
9.3 Documentation for a 3D Solid Model 587
9.4 Advanced Drawing Setup 608
9.5 Plot a Document 610
9.6 Batch Plotting 615
9.7 Key Points and Exercises 659

10 Visualization and RP Application 621
10.1 Perspective View 621
10.2 Hide and Shade 625
10.3 Photo-realistic Rendering 627
10.4 Introduction to RP and STL 657
10.5 Key Points and Exercises

Appendix A Application Window Configuration 659

Appendix B Quick Command Reference 675

Appendix C Export, Import, Windows Clipboard, and Raster Image 685

Appendix D Script File 693

Index 697

Preface

This book is intended for all those who want to gain a thorough understanding of how to use AutoCAD Release 14 to create CAD drawings and 3D solid objects. It is also developed for use as a textbook for university and college students of all disciplines and to meet the needs of industrial AutoCAD users who want to gain hands-on practical experience in using AutoCAD as a tool in engineering design. This book is practically oriented, and it has been carefully written in such a way that as many commands as possible are applied in solving practical problems in technical drawing. There are ten chapters in this book.

AutoCAD R14 is installed in the Windows environment. It is 100% compatible with other Windows software. You can import from, export to, and link objects with other Windows applications. In Chapter 1, you will learn how to use AutoCAD interface, execute AutoCAD commands, and use various tools.

Before starting a CAD drawing, you need to do some preparatory work. Chapter 2 tells you what you should do to start a new drawing.

Simply speaking, producing a CAD drawing involves two basic tasks: entity creation and entity editing. Chapter 3 offers guidance in completing a number of engineering drawings. You will gain an appreciation of the basic entity creation and editing commands, together with other tools in engineering application, through hands-on practical work. Although the majority of the exercises in this chapter are mechanical engineering drawings, they are appropriate to all other engineering disciplines as well, because the objective of practicing these exercises is to gain the skills and master the techniques used in AutoCAD rather than merely to produce the drawings themselves. Therefore, you may find that some of the drawings can be created by using a different approach that takes less time. In this case, you should be congratulated, because it means that you have already mastered some skills in using the CAD system. You may also notice that in the exercises, only those dimensions necessary for executing the specific command at a certain stage are given. This is done deliberately in order to guide you through all the steps as delineated.

After you know how to create and edit a drawing, you should learn some techniques of entity manipulation. Chapter 4 discusses the use of blocks and external references in managing data in a drawing and across several drawings.

It is absolutely essential that you create a CAD drawing in its exact size. Naturally, dimensional information is already an integrated part of the drawing database, and you do not have to add further dimensions to a drawing if the drawing is to be transmitted to the end user electronically. In a drawing document, the hard copy would normally have to be scaled up or down to suit the plotter and the paper. Therefore, you need to include dimensions in a drawing on a hard copy. AutoCAD provides a very comprehensive set of commands and variables for dimensioning. Study Chapter 5 carefully to learn how to put annotations and dimensions on a drawing and make inquiries about a drawing.

There are three major types of 3D models in a computer—wireframe, surface, and solid. Solid models are further divided into explicit and parametric solids. In AutoCAD terms, explicit solids are called native solids. AutoCAD is a 3D CAD system. You can use AutoCAD R14 to construct native solids and use Mechanical Desktop to construct parametric solids. If you are interested in learning about Mechanical Desktop, please refer to *Mastering Mechanical Desktop R3, Mastering Mechanical Desktop: Parametric Modeling,* and *Mastering Mechanical Desktop: Surface Modeling.*

Chapter 6 begins with an overview of solid objects, explains how to work in 3D space in a CAD system, and delineates how solid models can be built in a computer and how AutoCAD native solids are compatible with Mechanical Desktop parametric solids. Apart from learning how to create primitive solids, extruded solids, and revolved solids and to create a complex solid by using Boolean operations, you will also learn how to use the utility commands on a solid.

To enhance your knowledge of 3D solid modeling, Chapter 7 gives you an opportunity to work on a complex thin-shell solid with internal bosses and webs, and in Chapter 8 you will explore the use of constructive solid geometry technique on an architectural project. After working through the projects, you should have a good understanding of solid modeling techniques and be able to apply the techniques to projects of your own.

The drawing space of a CAD drawing is virtually unlimited in size. You may draw an aircraft carrier, or even larger objects, in a single drawing file and see the entire object in the display unit. When you come to presentation using hard copies, you need to use the plotters. Chapter 9 introduces the two conceptual working environments in AutoCAD—paper space and model space. In the model space environment, you create all the entities necessary to depict an object in a drawing. In paper space, you prepare a document for plotting. This chapter shows you how to prepare a document from both 2D and 3D objects and to plot it on a hard copy.

In the final chapter of this book, you will learn how to manipulate 3D viewing and outputting photo-realistic rendered images. In addition, the use of AutoCAD 3D solids in making rapid prototypes is introduced.

To conclude, this book has ten chapters. It starts from the basic level and progresses to the advanced level of AutoCAD drawing creation. Chapter 1 serves as an introduction. Chapter 2 concerns drawing preparation. Chapter 3 deals with 2D draw and edit. Chapter 4 details block and external references. Chapter 5 discusses dimensioning and annotation. Chapter 6 is about 3D solid modeling techniques. Chapter 7 and Chapter 8 depict two solid modeling projects. Chapter 9 delineates outputting of engineering documents. Chapter 10 focuses on visualization and output to rapid prototyping systems. To learn more, you should proceed to 3D NURBS surface modeling and 3D parametric solid modeling by using Mechanical Desktop. (See the separate book on Mastering Mechanical Desktop R3.) Good luck in your work. Studying this book and following the instructions in it should enable you to master the technique of creating CAD drawings and 3D solids with AutoCAD.

Acknowledgments

This book never would have been realized without the contributions of many individuals.

I am grateful to the following reviewers for their thoughtful suggestions and help: Vera Anand, Clemson University; Robert A. Chin, East Carolina University; Hollis Driskell, Trinity Valley Community College; Arvid Myklebust, Virgina Polytechnic Institute and State University; William Ross, Purdue University; Ed Wheeler, University of Tennessee at Martin; and Tricia A. Zaremba, Ball State University.

Several people at Brooks/Cole Publishing also deserve special mention, particularly Suzanne Jeans, the project development editor who worked closely with me on this and previous books; Shelley Gesicki, the editorial assistant; Kelsey McGee, the production editor; and Vernon Boes, the art director.

Ron K. C. Cheng

Mastering AutoCAD® Release 14

Chapter 1

Introduction

1.1 Start Up
1.2 AutoCAD Application Window
1.3 Command Interaction
1.4 Function Keys
1.5 File Handling
1.6 Utilities
1.7 Key Points and Exercises

Aims and Objectives

The aim of this chapter is to introduce the use of AutoCAD R14 as an engineering communication tool. After studying this chapter, you should be able to:

- start a drawing session in AutoCAD,
- describe and explain the use of various areas of the AutoCAD Application Window,
- run AutoCAD commands,
- use various function keys,
- manage AutoCAD drawing files, and
- describe the use of various utilities.

Overview

AutoCAD R14 is a 3D computer-aided design tool. By default, the AutoCAD program is placed in the directory "Program files\AutoCAD R14." For details regarding application installation, please consult the Installation Guide published by Autodesk Inc. Here, we presume that AutoCAD has been properly installed in your computer, and you will learn how to use AutoCAD as an engineering communication and design tool.

1.1 Start Up

Start AutoCAD by selecting the AutoCAD icon from the Windows main menu. Figure 1.1 shows the AutoCAD start up dialog box. See Figure 1.1. There are four ways to start a drawing session: Use a Wizard, Use a Template, Start from Scratch, and Open a Drawing.

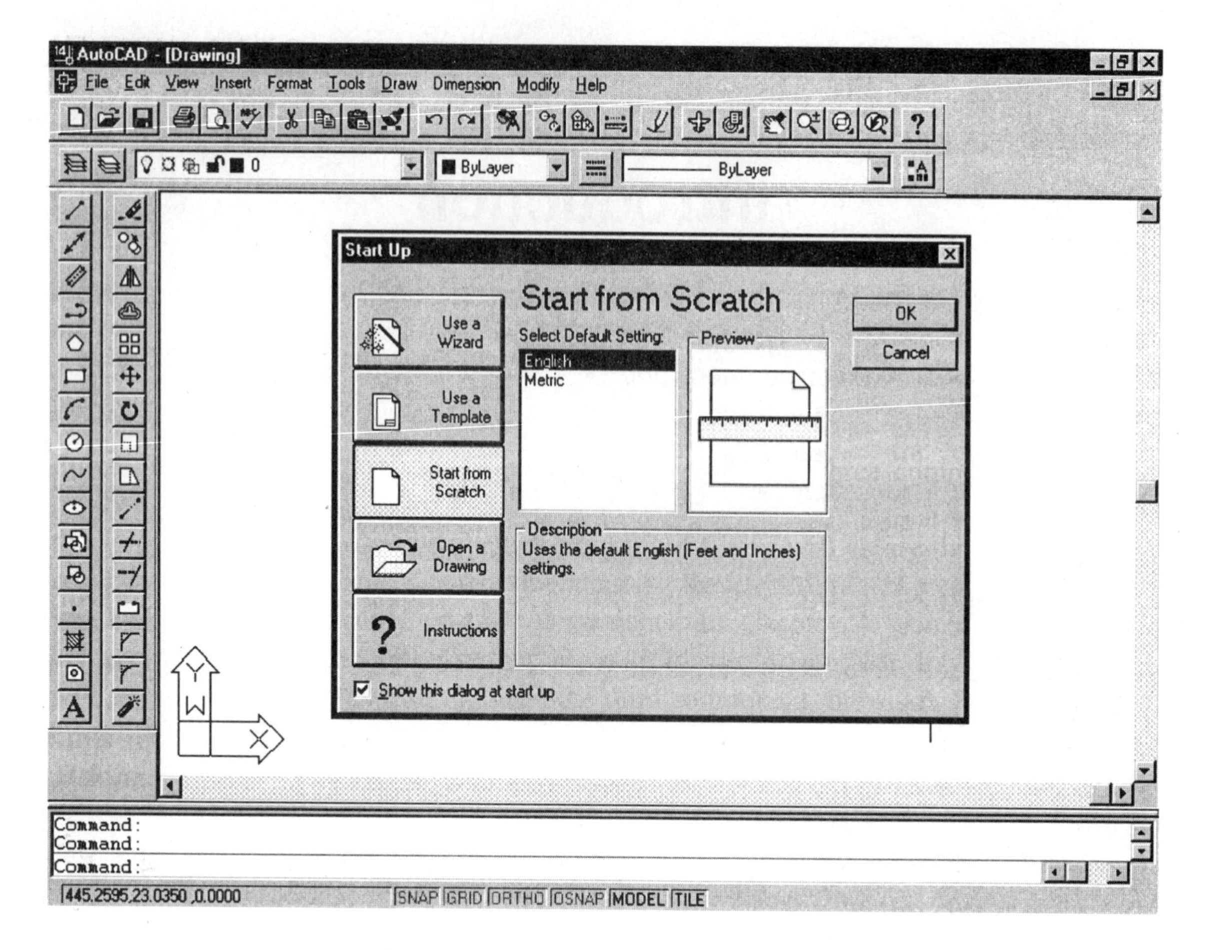

Figure 1.1 AutoCAD Start Up dialog box

If you do not want to display this dialog box the next time you start AutoCAD, you can unselect the box next to the item Show this dialog at start up. However, if you wish to display this dialog box again, you have to use the PREFERENCES command.

Refer to Figure 1.2. Use your pointing device to select the Tools pull-down menu and then select the Preferences... item to run the PREFERENCES command.

<Tools> **<Preferences...>**

Note:
In the delineation that follows, <AAA> <BBB> will mean selecting the <AAA> pull-down menu and then selecting the <BBB> item.

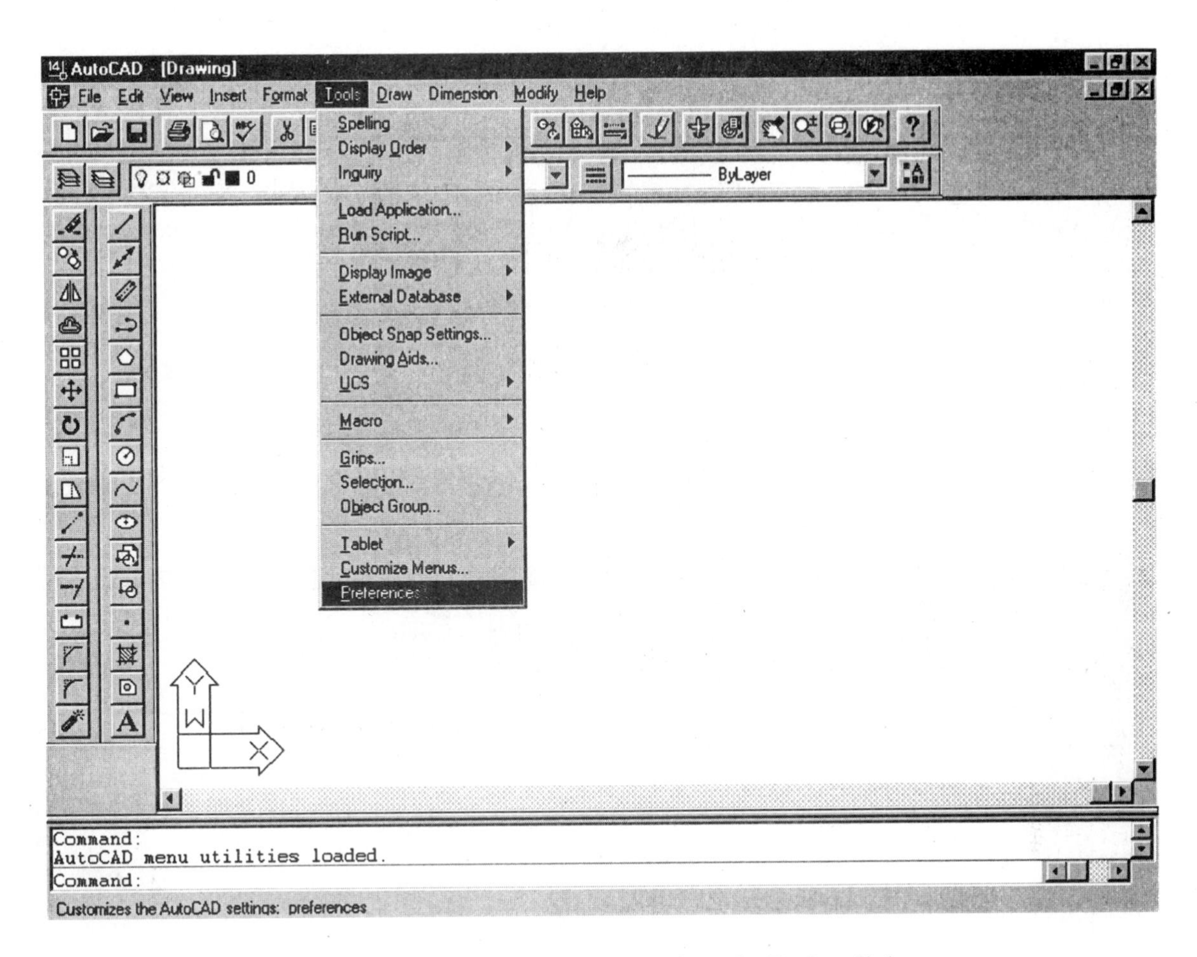

Figure 1.2 Selecting the Preferences... item from the Tools pull-down menu

In the PREFERENCES command dialog box, there are eight tabs. Select the Compatibility tab and select the box next to Show the Start Up dialog box. See Figure 1.3.

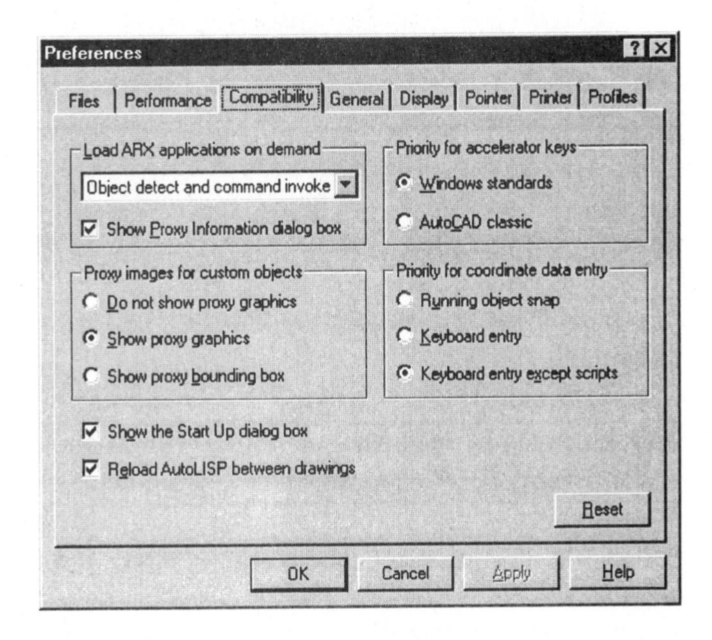

Figure 1.3 Compatibility tab of the Preferences dialog box

1.2 AutoCAD Application Window

You may have taken one of the options from the Start Up dialog box. To start a new drawing, select the New... item from File pull-down menu to apply the NEW command. Then select the [Start from Scratch] button. As can be seen in the dialog box, there are two settings, English and metric. Here, you will use metric settings.

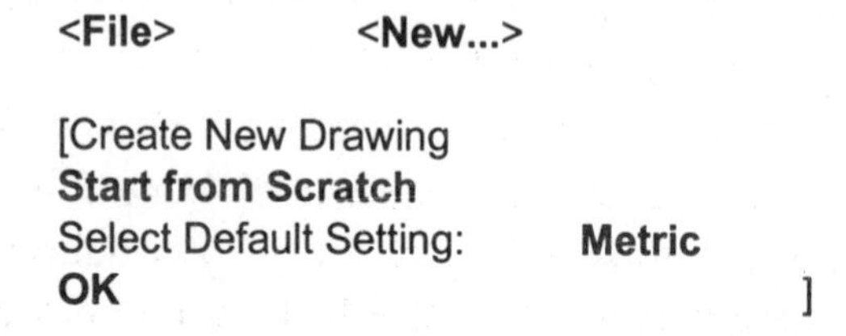

<File> **<New...>**

[Create New Drawing
Start from Scratch
Select Default Setting: **Metric**
OK]

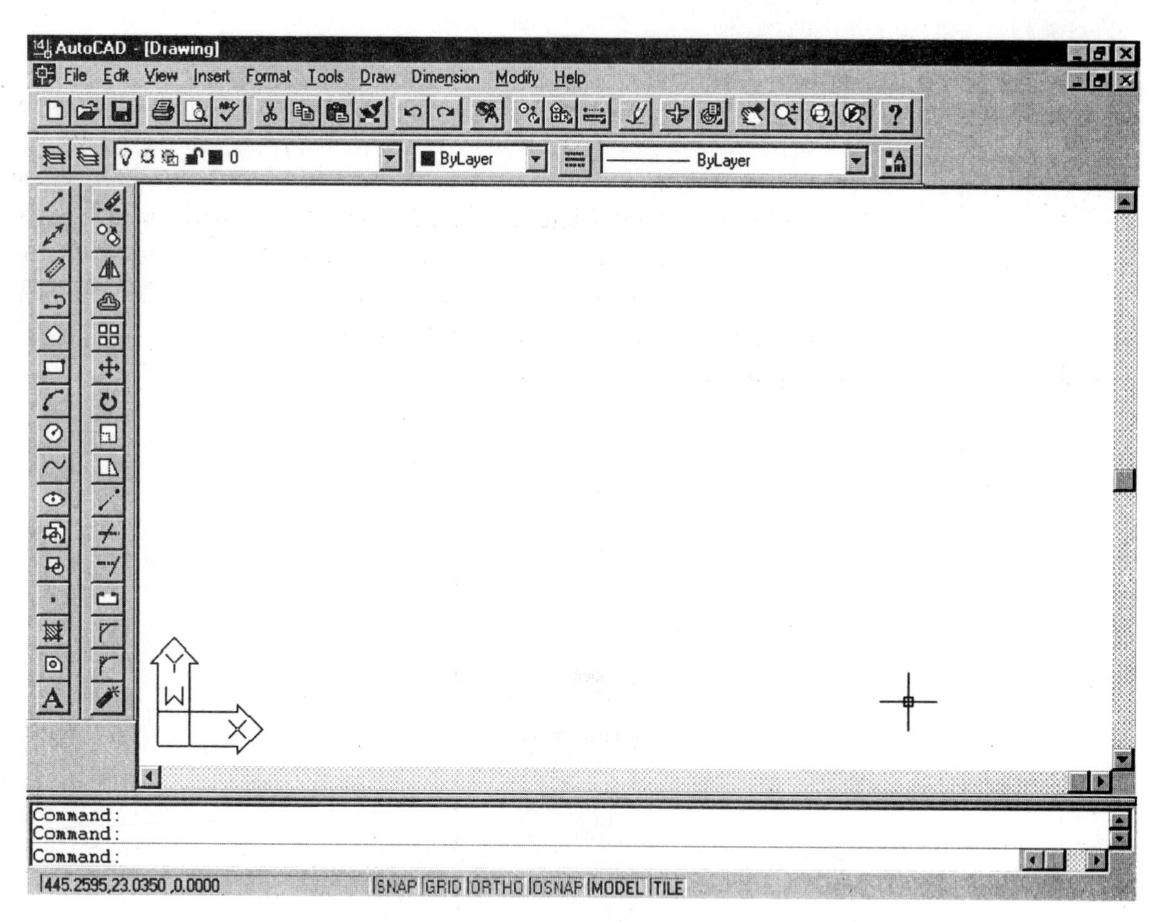

Figure 1.4 AutoCAD application window

Figure 1.4 shows the AutoCAD application window. Note that this figure may not be the same as your screen display because there might be minor differences in toolbar and command window configuration. In this application window, there are a number of window areas. At the top of the screen, there is a standard Windows heading bar showing the name of the current application, AutoCAD, and the current file name. Down below, there is a Windows menu bar. Right below the menu bar, there are a number of Windows toolbars. Farther down the screen, there is the graphic window where you create your drawing. Within the graphic window, you will find a pair of crosshairs. Below the graphic window, there is a command line interface window. At the bottom of the screen, there is the status bar.

Pointing Device and Cursor

To transit around in the screen areas, you need a pointing device. The most commonly used pointing device is the mouse. Moving it around on the screen, you will find that a pair of crosshairs in the graphic window move around as well. If you move the crosshairs outside the graphic window, the crosshairs change to a pointer.

A mouse normally has two or three buttons. For a mouse with three buttons, the left button is the first button, the right button is the second button, and the central button is the third button. If your mouse has only two buttons, selecting the second button while holding down the [Shift] key of your keyboard is equivalent to selecting the third button. The button assignments are:

1st button — is used for selecting a point or selecting an object.

2nd button — performs a number of tasks. If you place the cursor inside the graphic window, this button is equivalent to the [Enter] key. If you place the cursor on an icon, this button brings up the Toolbar dialog box of the TOOLBAR command. The TOOLBAR command will be explained later in this chapter. If you place the cursor in the command line interface or in a dialog box such as the LAYER or LINETYPE command, this button brings up a pop-up menu.

3rd button — brings up the cursor menu. See Figure 1.5.

Cursor Menu

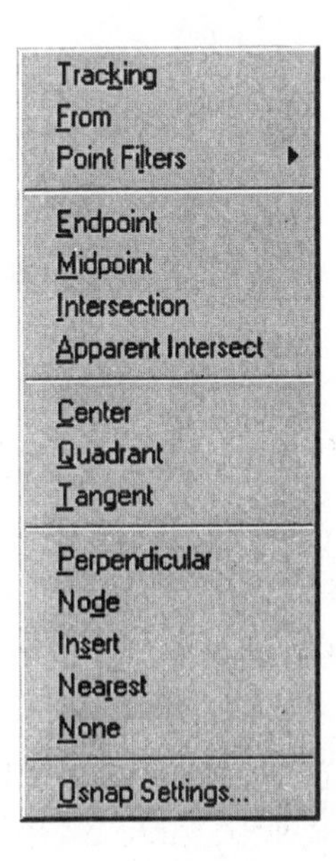

Figure 1.5 Cursor menu

Use the pointing device to move the crosshairs to the central part of the screen. Then press the third mouse button, or press the [Shift] key in conjunction with the second button. A very handy cursor menu appears. This menu provides frequently used menu items. You will use these menu items in Chapter 3.

Graphic Window

Let us start from the largest screen area of the AutoCAD application window — the graphic window. This is the main working area. You may regard it as a view window of an imaginary camera. Through this imaginary camera, you can see the 3D working space where you create and edit drawing entities. You can use the ZOOM command to zoom in and out, and you can use the PAN command to pan across the view.

Here, the working space is not limited by what you currently see through this window. In fact, the working area is unlimited in size, and you can place an entity anywhere you like. You do not have to worry about the size of the object. If you want to draw a line that is 10,000 units long, you do not have to bother about the scale of the drawing so that the line can be shown on a drawing or in the screen. You simply create a line 10,000 units long.

In this 3D working space, you may consider that there is an imaginary construction plane. When you select a point on the screen with the pointing device, you are selecting a point in this plane. To help you visualize this construction plane, there is an icon displayed at the lower left corner. It is called the UCS icon. See Figure 1.6. UCS stands for User Coordinate System. This icon is useful when you create three-dimensional objects. The default coordinate system is called the world coordinate system (WCS); Another user- defined system is called the user coordinate system (UCS). You will learn more about UCS in Chapter 6.

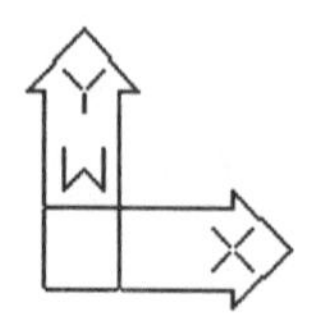

Figure 1.6 UCS icon

Status Bar

At the bottom of the screen, you will find a status bar. See Figure 1.7. At the left, it displays the coordinates of the crosshairs position. Move the crosshairs across the graphics area. See whether the coordinate display is updating all the time. If it is not, press the [F6] key or the [^D] key. These are the toggle keys for turning the coordinate update on or off.

Apart from the coordinate display, there are buttons for controlling the snap mode, the grid mode, the ortho mode, the running osnap mode, the working space (model or paper), and the tilemode.

445.2595,23.0350 ,0.0000 SNAP GRID ORTHO OSNAP MODEL TILE

Figure 1.7 Status bar

You will learn snap mode, grid mode, and ortho mode in Chapters 2 and 3. Working space and tilemode involve engineering documentation, and we will discuss them in Chapter 9.

Command Line Interface (CLI)

Above the status bar, you will see a command line interface window. See Figure 1.8. Here, you can execute any command by typing the command name and then pressing the [Enter] key or the [Space bar.] After a command is run, further prompts or instructions will appear in this area or any pop-out dialog boxes. After you complete a command and the prompt Command: appears again, you can issue another command.

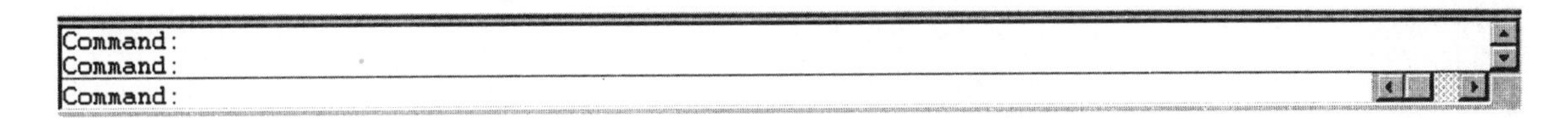

Figure 1.8 Command line interface

At the right end of the command line interface window, there is a pair of scrolling arrows. By selecting these arrows, you can scroll up or down to browse all the commands that you run and their corresponding prompts. This window can be docked or placed elsewhere. When docked, it is called the docked command window. When placed elsewhere in the screen, it is called the floating command window.

Windows Menu Bar

Let us turn to the top of the screen to look at the Windows menu bar. This area appears below the standard Windows heading bar. It provides you with a pull-down menu and cascading menus. When you pick an item from this menu bar, the pull-down menu appears. See Figure 1.9.

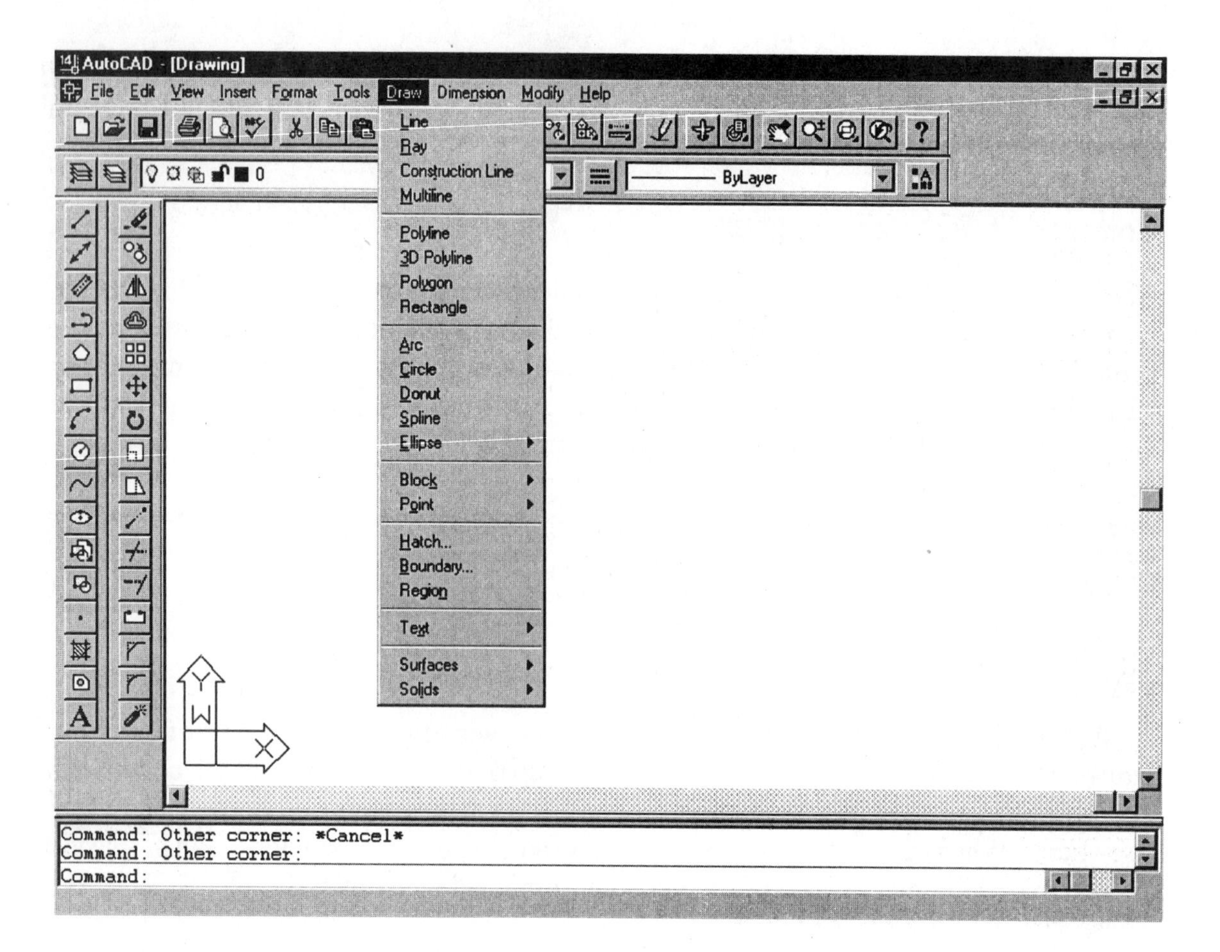

Figure 1.9 Pull-down menu activated from Windows menu bar

There is a small triangle in some of the items of the pull-down menu. Selecting these triangles brings out a cascading menu. See Figure 1.10. To run a command, you can select any menu item without a small triangle in the pull-down menu or the cascading menu.

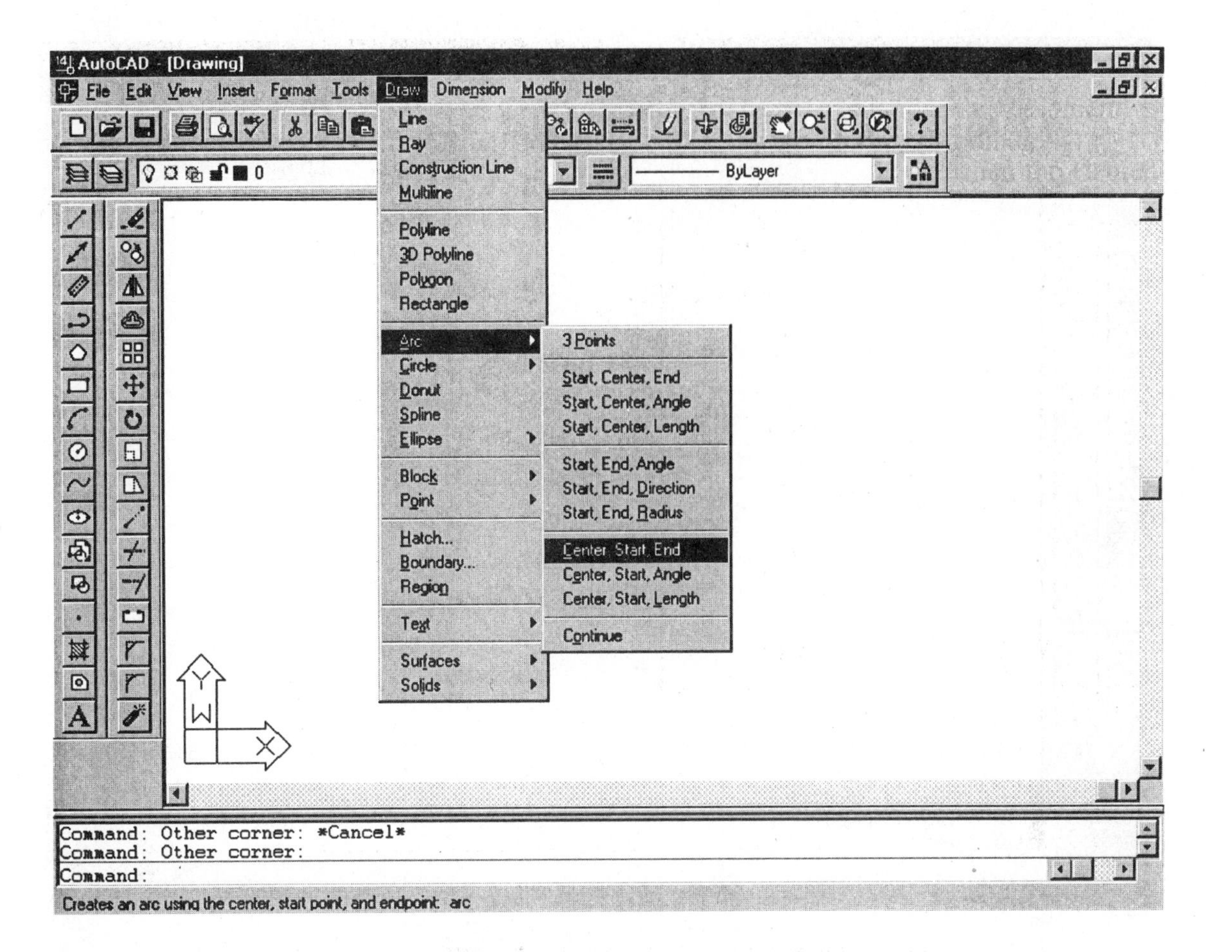

Figure 1.10 Cascading menu

The Windows menu bar is configurable. See Appendix A for details.

Windows Toolbars and Flyouts

The third way to run a command is to select an icon from the toolbars. AutoCAD toolbars, like other Windows toolbars, can be docked to the edge of the screen or can be placed anywhere. You can select and drag them to dock to the edge to become docked toolbars. Or you can select and drag them to anywhere on the screen to become floating toolbars. In some icons of the toolbars, you will find a small triangle at the lower right corner. Selecting an icon with a small triangle will bring out an extension of the toolbar button, flyout. See Figure 1.11.

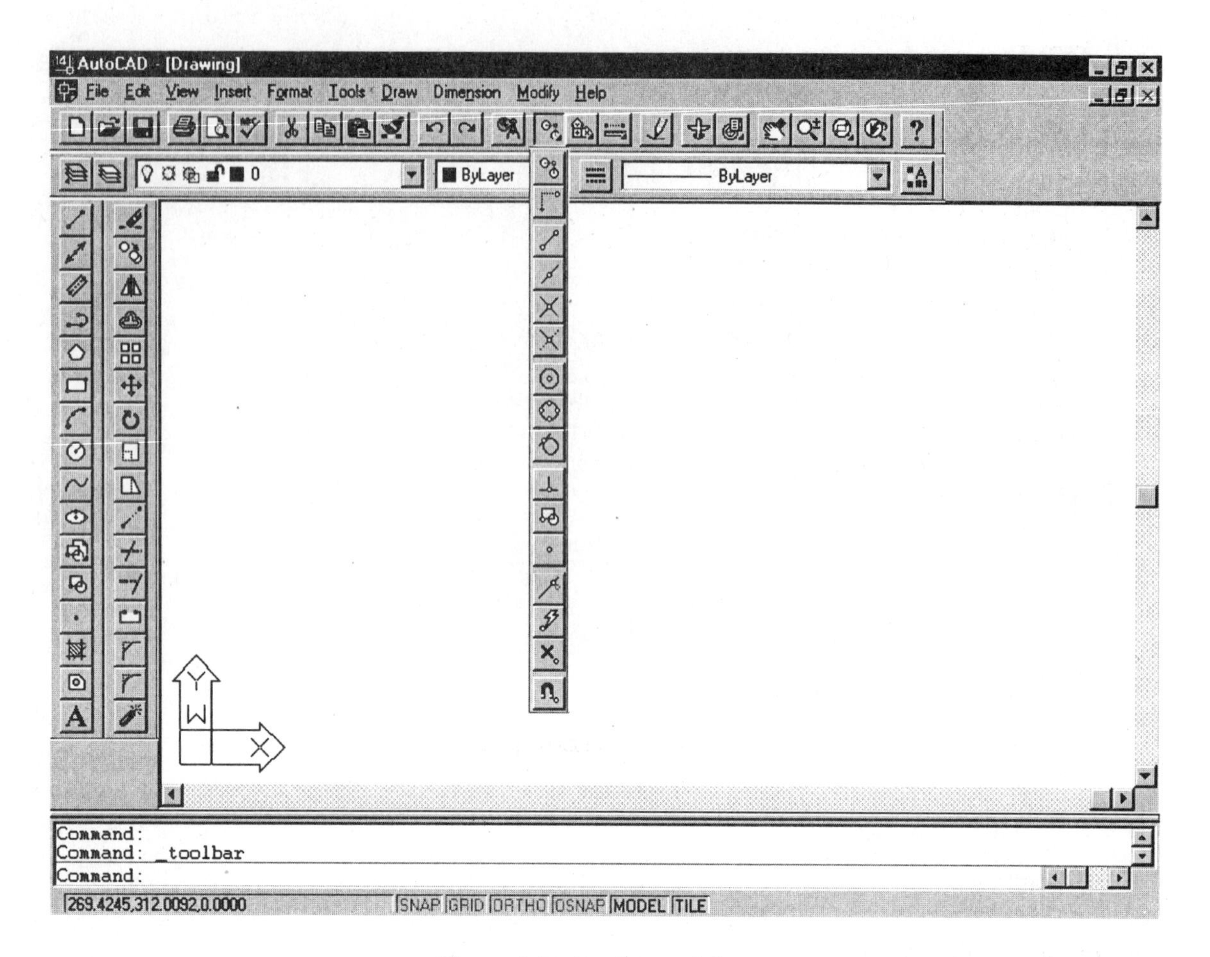

Figure 1.11 Toolbar and flyout

Windows toolbars are configurable. See Appendix A for details.

To sum up, AutoCAD has a number of window areas: standard Windows heading bar, Windows menu bar, Windows toolbars, graphic window, command line interface window, and status bar.

1.3 Command Interaction

As we have said, you can run a command by selecting an item from the cascading menu or the pull-down menu from the Windows menu bar, selecting the appropriate icon from the Windows toolbar, or typing the command name directly at the command window and pressing the [Enter] key or [Space bar.]

Recalling a Command

To recall a previously issued command, you can press the [↑] key continuously, select the command, and press the [Enter] key.

Repeating a Command

In many circumstances, you may have to repeat a command. To do so, you can press the [Enter] key or the [Space bar.] Another way to repeat a command is to issue the

MULTIPLE command prior to running the required command. For example, entering MULTIPLE and CIRCLE at the command prompt will repeat the CIRCLE command infinitely until you call for a stop.

Command: **MULTIPLE**
Multiple command: **CIRCLE**
3P/2P/TTR<Center point>: **[Select any point on the screen.]**
Diameter<Radius>: **[Select another point.]**
3P/2P/TTR<Center point>: **[Select any point on the screen.]**
Diameter<Radius>: **[Select another point.]**
3P/2P/TTR<Center point>: **[Select any point on the screen.]**
Diameter<Radius>: **[Select another point.]**
3P/2P/TTR<Center point>: **[Select any point on the screen.]**
Diameter<Radius>: **[Select another point.]**
[ESC]

Here [ESC] means pressing the [ESC] key in the keyboard.

[ESC] Key

You can cancel the command by pressing the [ESC] key.

[ESC]

Command: *Cancel*

Transparent Commands

You can suspend a command that is in operation and run a second command. The second command is called a transparent command. To run a transparent command, you can key in an apostrophe and then the command name of the second command. After the transparent command is complete, the original command resumes automatically.

Run the CIRCLE command by selecting the Circle icon of the Draw toolbar. See Figure 1.12.

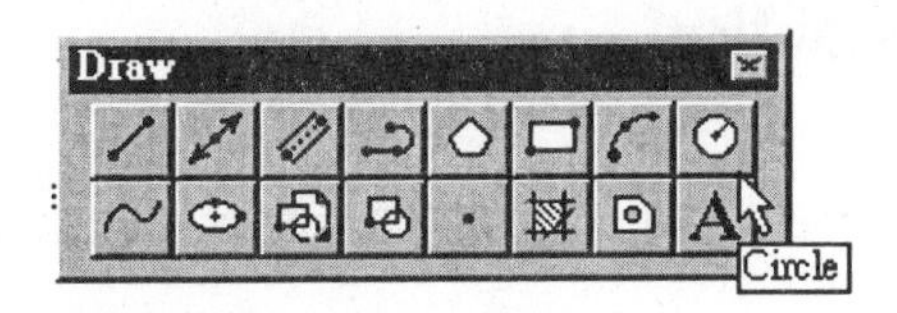

Figure 1.12 Circle icon of Draw toolbar

[Draw] **[Circle]**

Note:
In the delineation that follows, [AAA] [BBB] will mean selecting the [AAA] Windows toolbar and then the [BBB] icon.

Command: **CIRCLE**

Suppose that you want to get help regarding this command. Issue the HELP command transparently. See Figure 1.13.

[Standard Toolbar] **[Help]**

3P/2P/TTR<Center point>: **'HELP** [or **'?**]

Now the CIRCLE command suspends temporarily, and a Help dialog box pops up to provide help on the CIRCLE command. See Figure 1.13.

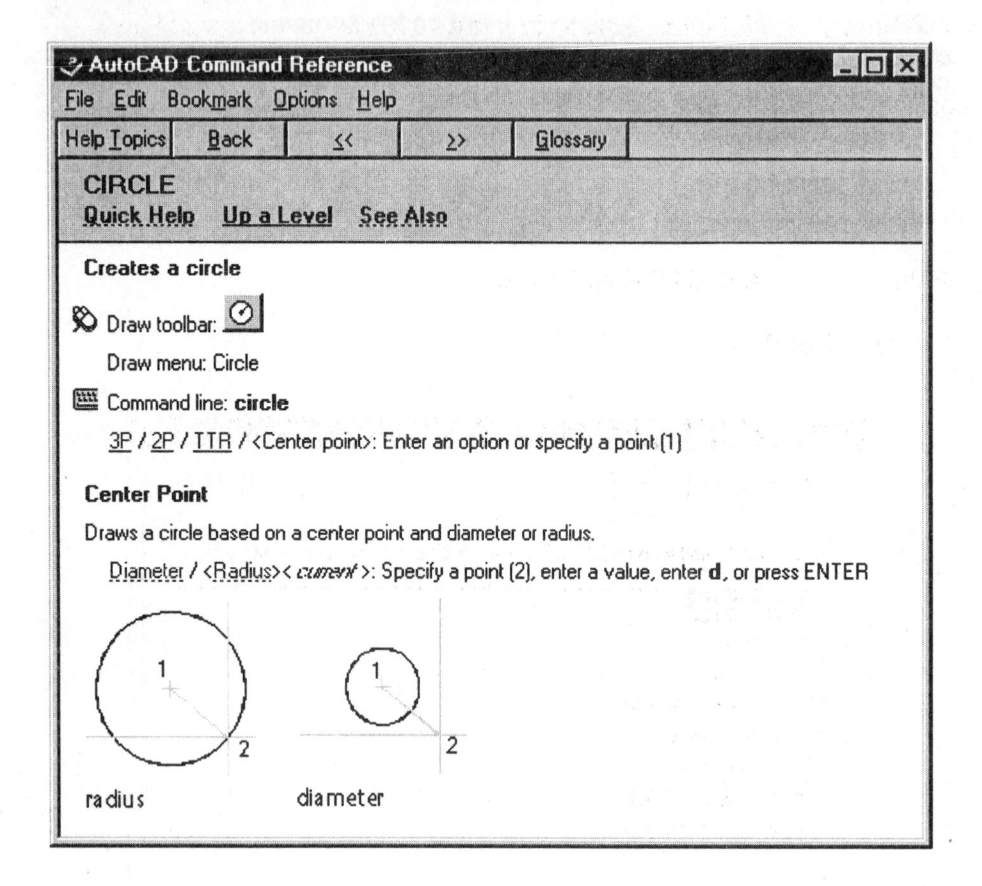

Figure 1.13 Help dialog box for CIRCLE command

After reading the help screen, select the [X] button of the dialog box to resume the CIRCLE command.

Resuming the CIRCLE command.
3P/2P/TTR<Center point>: **10,40**

Suppose that the radius is derived from an equation of 9+10/3. Instead of doing the calculation beforehand, you can call out the calculator transparently.

Diameter/<Radius>: **'CAL**
Initializing...>> Expression: **9+10/3**

In response to the request for diameter/radius, the 'CAL command works transparently to calculate an expression and returns the value to the CIRCLE command. The value is taken as the radius of the circle.

Not all the commands can run transparently. For example, you cannot run the LINE command in the middle of the CIRCLE command.

Dialog Boxes

Some commands bring up a user-friendly dialog box. With a dialog box, you can select items, pick buttons, check boxes, or type in a text string at the designated area.

Select the AutoCAD Help Topics item from the Help pull-down menu to run the HELP command to bring up a dialog box. See Figure 1.14. You can press the short cut key [F1].

[F1]

<Help> **<AutoCAD Help Topics>**

Command: **HELP**

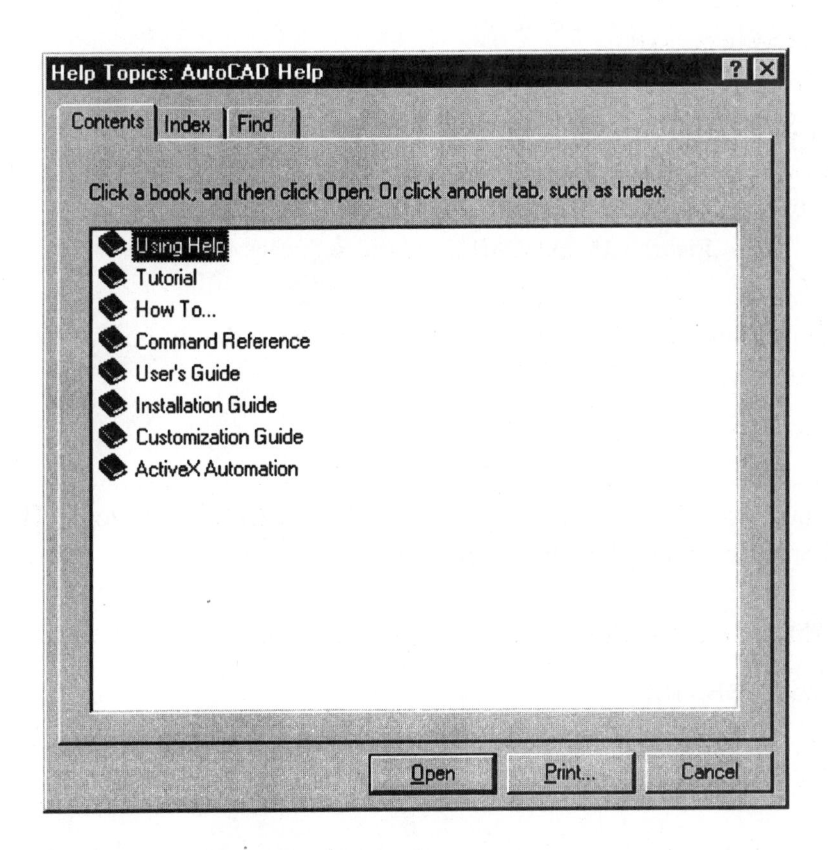

Figure 1.14 AutoCAD Help dialog box

From the Contents tab of the AutoCAD Help dialog box, select the Command Reference item and then the Commands item. See Figure 1.15.

[Help Topics: AutoCAD Help
Contents
Command Reference
Commands]

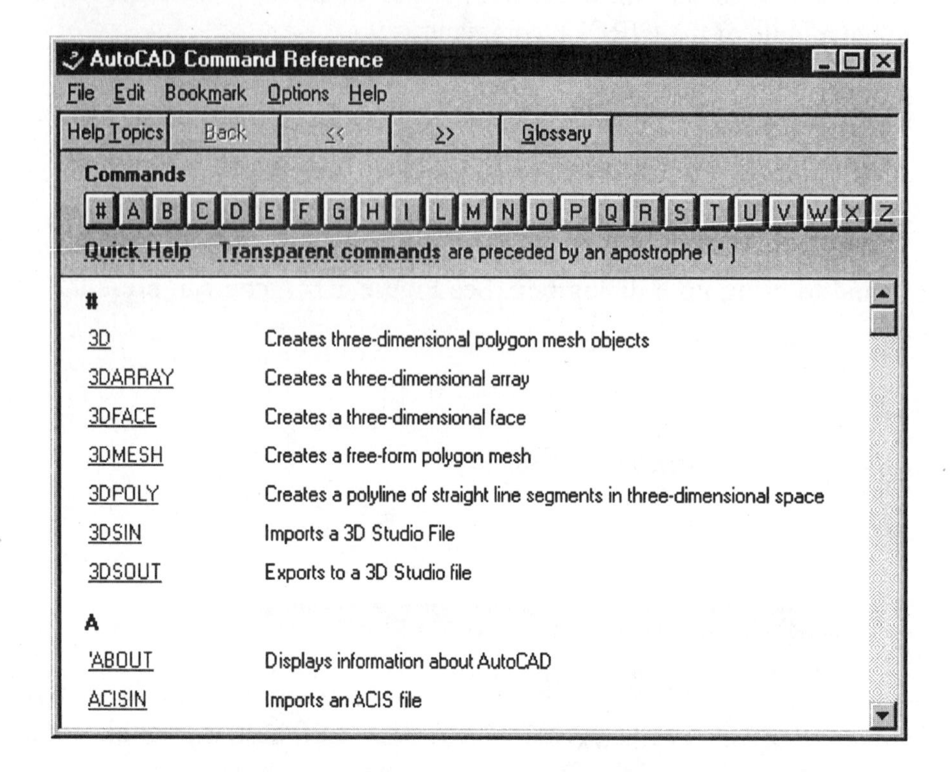

Figure 1.15 AutoCAD Command Reference dialog box

Select the [X] button of the dialog box to exit the Help dialog box. A quick command reference is shown in Appendix B.

Image Tiles

Some commands provide an image tile to display how a feature will look. The BHATCH command is one such example. See Figure 1.16.

<Draw> **<Hatch...>**

Command: **BHATCH**

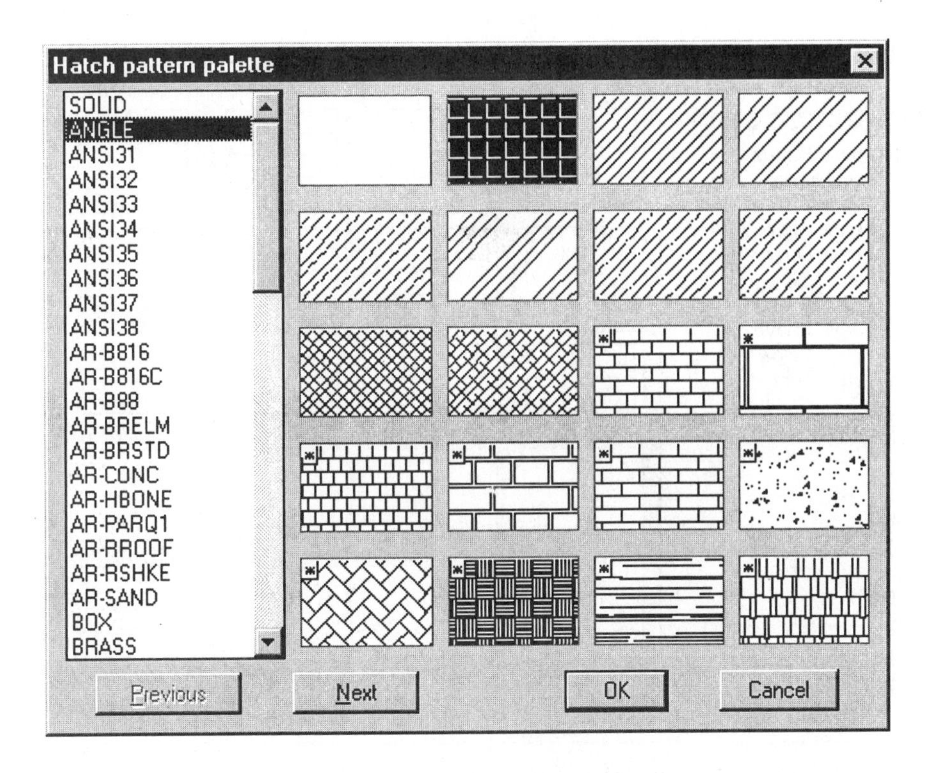

Figure 1.16 Image tiles of BHATCH command

The image tiles of the BHATCH command indicate how the hatch pattern looks. Select the image to toggle to other patterns. You will learn how to use this command in Chapter 3. Now select the [Cancel] button to terminate the command.

Icon Menu

In an icon menu, the available menu choices are shown in pictorial form. This makes the selection of commands easier. To display a typical icon menu, select the View item of the menu bar, then the Tiled Viewports item, and finally the Layout... item. To exit from this command, select the [Cancel] button. See Figure 1.17.

<View> **<Tiled Viewports>** **<Layout...>**

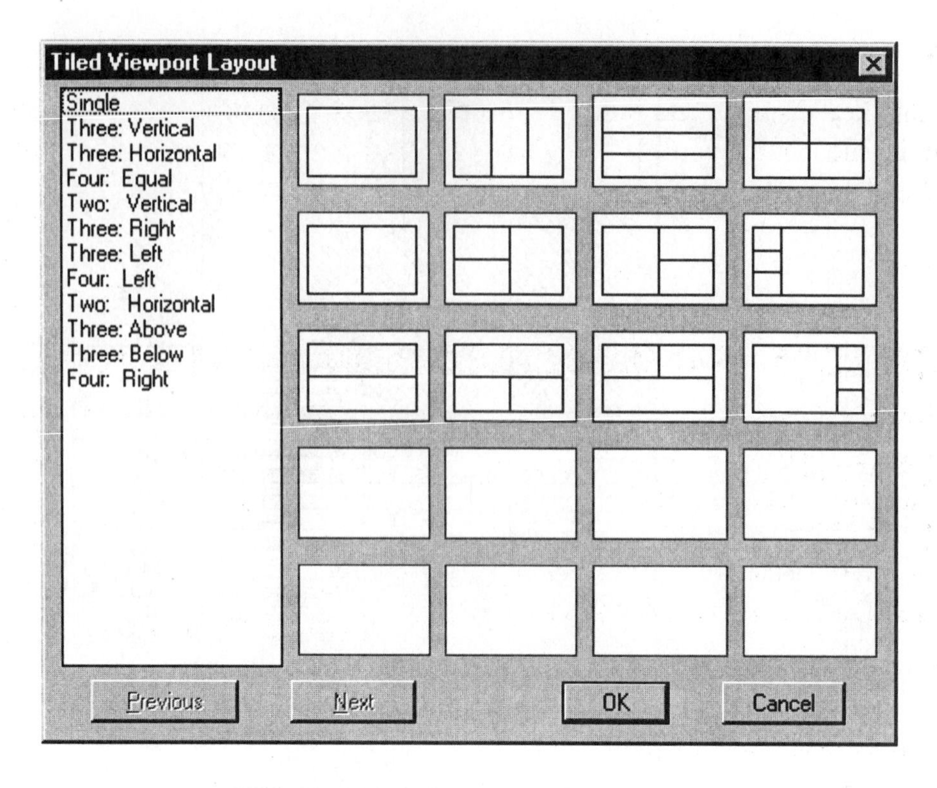

Figure 1.17 Icon menu for setting viewports

To sum up, you can run a command in three ways — using a pointing device to select an icon from the toolbar, using a pointing device to select a menu item from the pull-down menu, and typing the command name at the command line interface and pressing the [Space bar] or the [Enter] key.

You can select a previously executed command by pressing the [↑] key and the [Enter] key or the [Space bar.] You can repeat the last command by pressing the [Enter] key or the [Space bar.] You can repeatedly run a command by running the MULTIPLE command in conjunction with the said command. To terminate or cancel a command, you can use the [ESC] key. In the middle of a command, you can suspend execution and run another command such as ZOOM, HELP, or CAL. These are called transparent commands. Some commands may bring up a dialog box, an image tile, or an icon menu.

1.4 Function Keys

The keys [F1] to [F12] in your keyboard are called function keys. In AutoCAD, the [F1] to [F10] keys serve the following functions:

[F1] Key

Pressing this key brings out the Help dialog box. You can also use the [?] key.

[F2] Key

Pressing this key displays and hides the AutoCAD text window, which shows the text displayed in the command line interface window. See Figure 1.18. The equivalent command to display the text screen is the TEXTSCR command.

Command: **TEXTSCR**

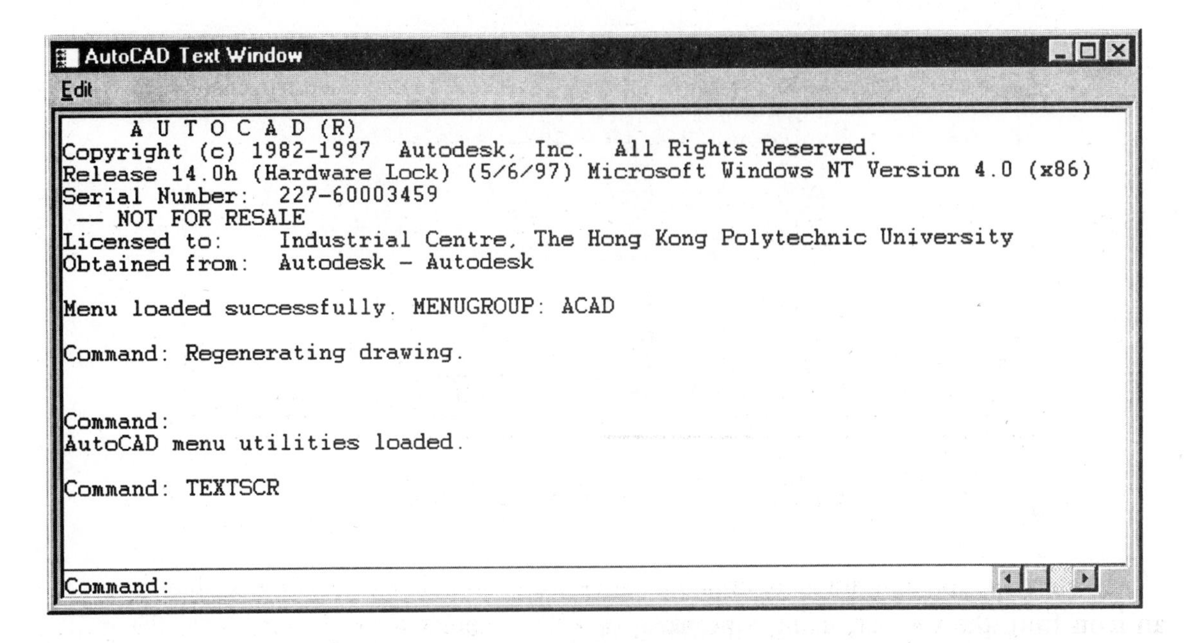

Figure 1.18 AutoCAD Text Window dialog box

The equivalent command to hide the text screen is the GRAPHSCR command.

Command: **GRAPHSCR**

[F3] Key

Pressing this key (or selecting the OSNAP item of the status bar) enables or disables the running object snap mode. If you have not set any running object snap mode, this key calls out the OSNAP command. See Figure 1.19. You will learn object snap in Chapter 3.

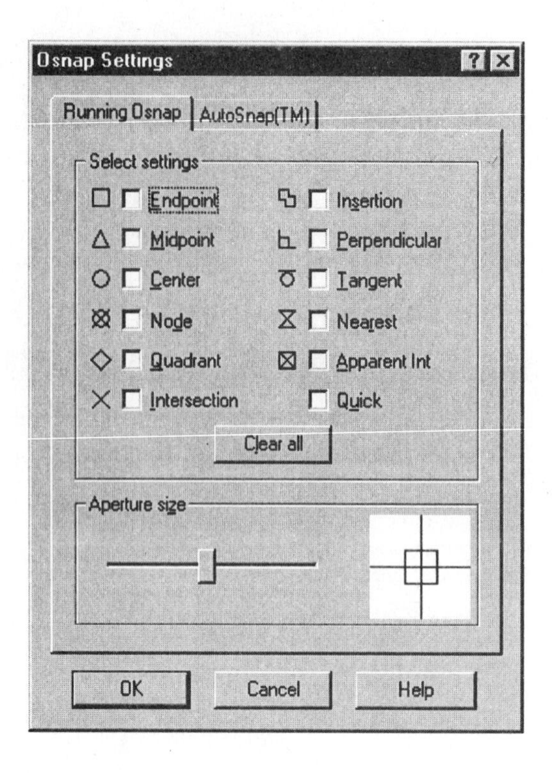

Figure 1.19 Osnap Settings dialog box

[F4] Key

Pressing this key turns on and off the tablet. You can also use the [^T] key.

[F5] Key

Pressing this key toggles among the three isoplanes when you produce a "2D" isometric drawing. You can also use the [^E] key. See Chapter 3.

[F6] Key

Pressing this key (or selecting the coordinate display box of the status bar) turns on and off the coordinate display. You can also use the [^D] key. See Chapter 2.

[F7] Key

Pressing this key (or selecting the GRID item of the status bar) turns on and off the grid display. You can also use the [^G] key. See Chapter 2.

[F8] Key

Pressing this key (or selecting the ORTHO item of the status bar) turns on and off the ortho mode. You can also use the [^L] key. See Chapter 3.

[F9] Key

Pressing this key (or selecting the SNAP item of the status bar) turns on and off the snap mode. You can also use the [^B] key. See Chapter 3.

[F10] Key

This key turns on and off the display of the status bar.

1.5 File Handling

Once you have started the AutoCAD program, you are in the drawing editor environment. Therefore, you can start creating and editing entities. If you start right away, you then work on an unnamed drawing with default settings. If you prefer to start a new drawing session with various set up options, use the NEW command. See Figure 1.20.

<File> **<New...>**

Command: **New**

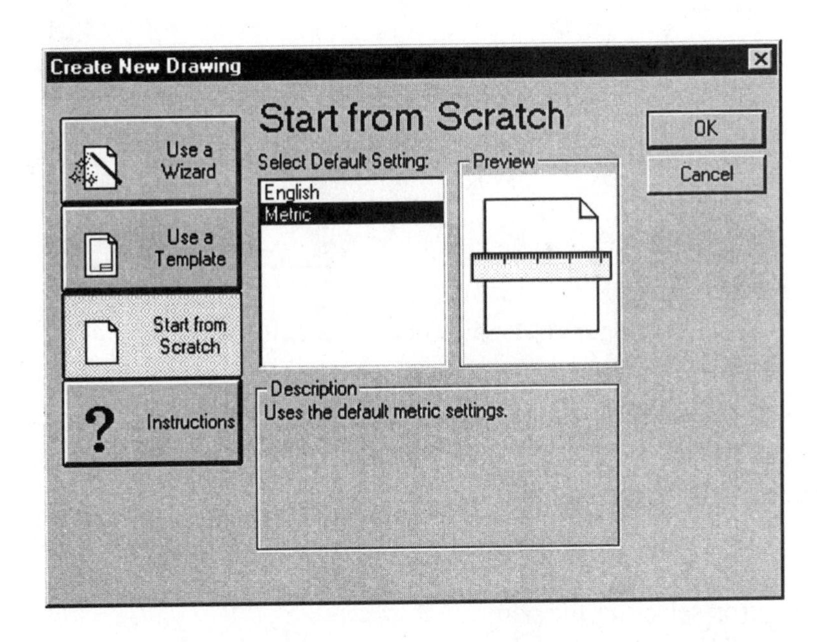

Figure 1.20 Create New Drawing dialog box

For details on how to start a new drawing, see Chapter 2.

Saving a File

You should make a habit of saving your file from time to time to prevent accidental data loss. If you have already specified a drawing name, or if you are working on an old drawing, you can simply use the QSAVE command.

<File> **<Save...>**

Command: **QSAVE**

This command quickly saves the drawing to a file name that you have specified earlier. If you have not yet specified a drawing name, the QSAVE command will bring up a dialog box and will set the current drawing to the saved file name. See Figure 1.21.

Another way to save a file is to use the SAVE command. This command also brings up a similar dialog box, saves the drawing to another file name, and keeps you on the current working drawing file name.

Command: **SAVE**

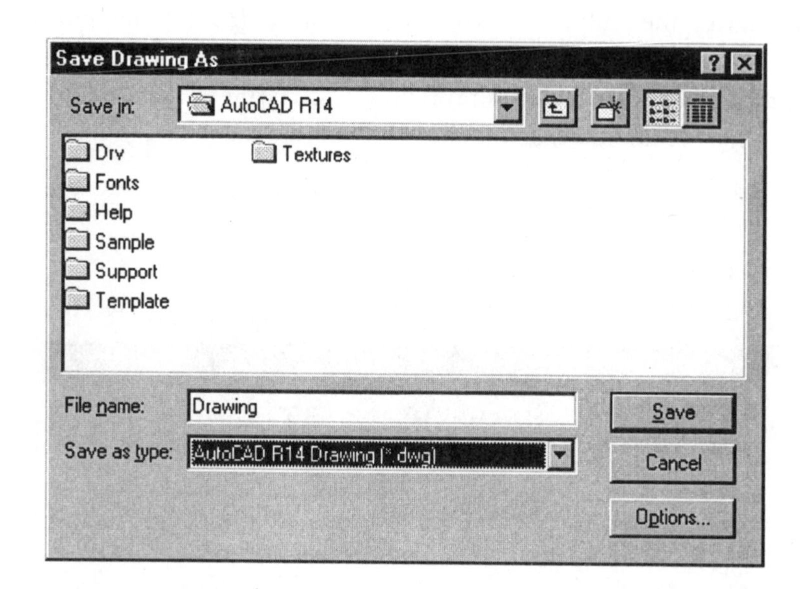

Figure 1.21 Save Drawing As dialog box

If you open a drawing but want to save it to another drawing file name, you can use the SAVEAS command. Apart from saving to a new file name, this command also sets the current working file to the new file.

<File> **<Save As...>**

Command: **SAVEAS**

You can configure AutoCAD to do automatic saving at intervals. Use the PREFERENCES command to set the automatic saving interval to 60 minutes.

<Tools> **<Preferences...>**

Command: **PREFERENCES**

[Preferences
General
Drawing session safety precautions
Automatic save
Minutes between saves: **60**
OK]

Compatibility with Older AutoCAD Releases

AutoCAD is upward compatible. You can open a file saved in an earlier version with R14 AutoCAD. But once a file of old format is saved to R14 format, you may have difficulty opening the file again with a R13 or older version of AutoCAD.

If you wish to edit a drawing of older releases in AutoCAD R14, but wish to open it again in AutoCAD R13 or R12, then you have to specify the file type when you use the SAVE, QSAVE, or SAVEAS command. See Figure 1.22.

<File> **<Save...>**

Save as Type: **[Select R13 or R12 (*.dwg).]**

<File> **<Save As...>**

Save as Type: **[Select R13 or R12 (*.dwg).]**

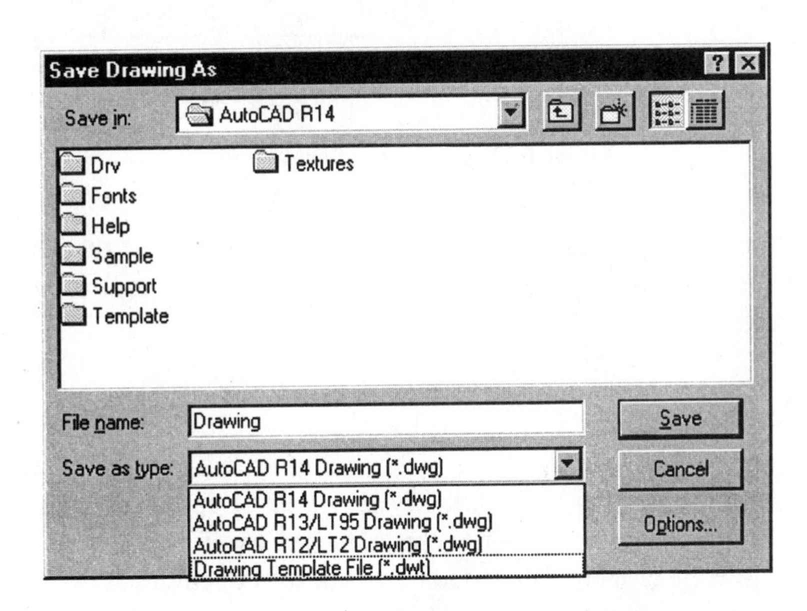

Figure 1.22 Save to R13, R12, or DWT files

Save a Template

If you want to save a file to become a template, you can use the DWT option. For details about a template file, see Chapter 2.

<File> **<Save...>**

Save as Type: **[Select Drawing Template File (*.dwt).]**

Opening an Existing File

When you want to work on an old drawing file, you can use the OPEN command. See Figure 1.23.

<File> **<Open...>**

Command: **OPEN**

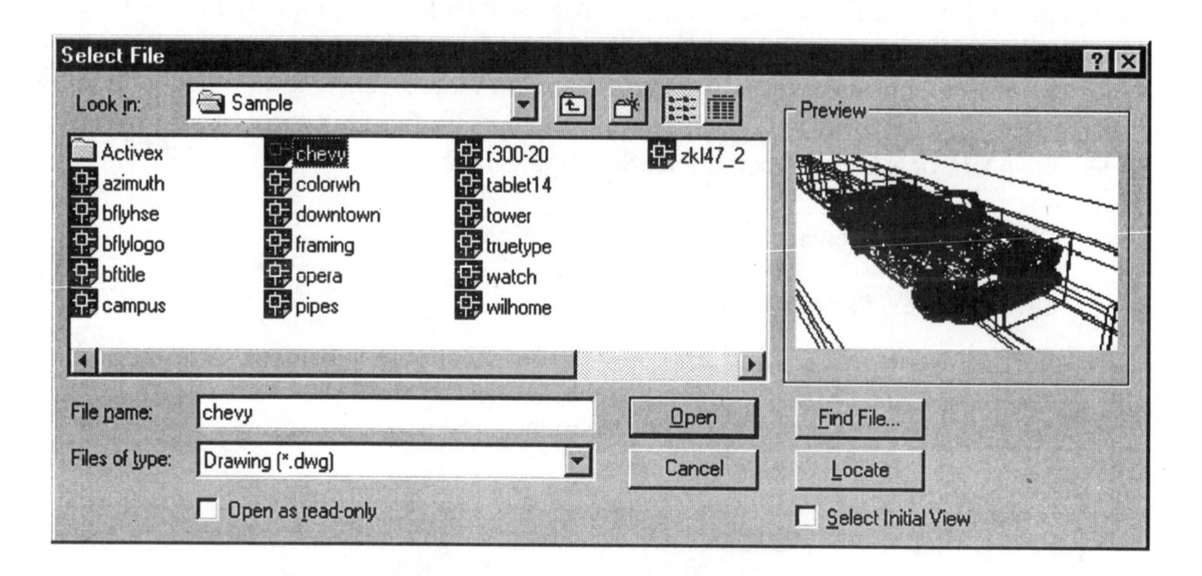

Figure 1.23 Circle icon of Draw toolbar

There is an image tile in the dialog box. It shows a preview of the selected drawing. By looking at the preview images, you can scan through the content of the drawings without having to open them one by one.

File Preview

Drawing file preview image is available for drawings saved in R14 and R13 formats. If the drawing that you are going to open is saved in R12 or earlier versions of AutoCAD, the image tile will show nothing. After the older release file is saved to R14 or R13 format, you can see the preview image the next time you open the drawing. If you do not want to have a preview image in your R14 or R13 drawings, use the PREFERENCES command to disable preview making.

<Tools> **<Preferences...>**

Command: **PREFERENCES**

[Preferences
General
Save thumbnail preview image
OK]

Recovering a Damaged File

When you encounter any problem in opening a file, you can use the RECOVER command to try to repair any damage.

<File> <Drawing Utilities> <Recover...>

Command: **RECOVER**

Auditing the Current Drawing

To evaluate the integrity of a drawing, you can use the AUDIT command. It is a diagnostic tool for examining and correcting errors of the current drawing.

<Files> <Drawing Utilities> <Audit>

Command: **AUDIT**

Ending a Drawing Session

To end a drawing session, you can use the EXIT command. After you have issued this command, an alert dialog box appears. See Figure 1.24.

<File> <Exit>

Command: **EXIT**

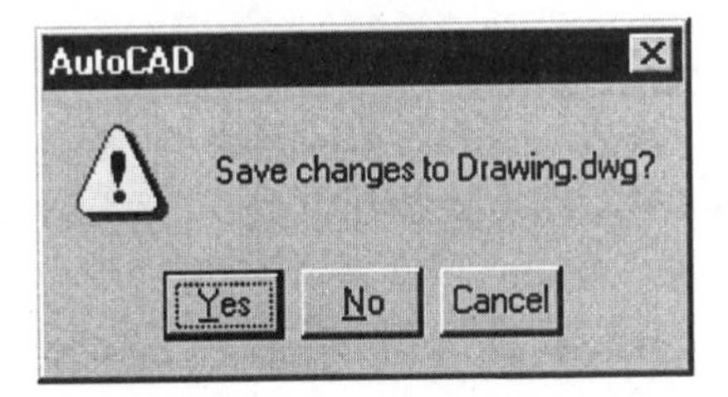

Figure 1.24 Warning dialog box

If you select the [Yes] button, AutoCAD will end the drawing session and will save the drawing if you have already assigned a name to the current drawing or will bring out a warning dialog box if you have not assigned a name.

Export and Import

If you have to use the data from other applications or have to use AutoCAD drawing data in other applications, you need to transform data. See Appendix C for details.

1.6 Utilities

The following are miscellaneous utilities.

Re-initialization

To re-initialize input and output ports without exiting AutoCAD, you can use the REINIT command. See Figure 1.25.

Command: **REINIT**

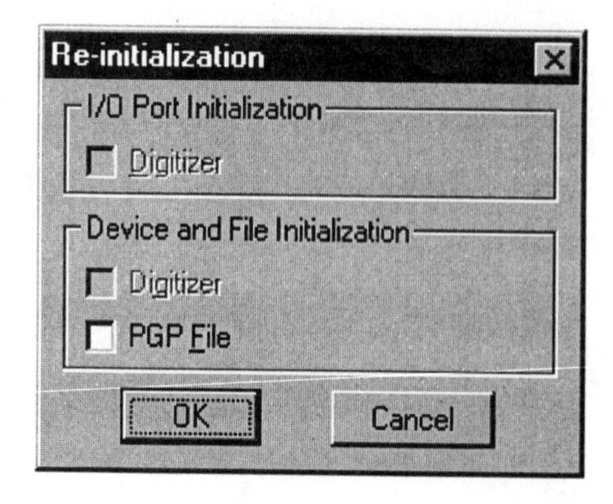

Figure 1.25 Warning dialog box

Log File

AutoCAD enables you to record the content of the text window to a log file for later reference. If you want to do so, you can use the LOGFILEON command to open a log file for recording. The log file name is ACAD.LOG.

Command: **LOGFILEON**

The log file will grow with each drawing session. Thus, the file can become very large. If you want to close the log file and stop recording, use the LOGFILEOFF command.

Command: **LOGFILEOFF**

Command History

Apart from using the log file to record the command line history, you can copy the history from the command window to the clipboard. Apply the COPYHIST command.

Command: **COPYHIST**

After copying the command line history to the Windows clipboard, you can paste it to other Windows software.

Calculation

To perform mathematical calculation, you can use the CAL command. This command can run transparently. See the example on page 12.

Command: **CAL**
>> Expression: **(2.0+4.5)/2**
3.25

Load Applications

To run AutoLISP, ADS, and ARX applications that are not loaded into memory automatically, you can use the APPLOAD command.

<Tools> **<Load Applications...>**

Command: **APPLOAD**

System Variable

An AutoCAD system variable can be set by directly typing the variable name at the command prompt and then keying in the new setting. The SETVAR command enables you set system variables and to check the current system variable settings.

<Tools> **<Inquiry>** **<Set Variable>**

Command: **SETVAR**

Command Modification

You may write a program to define a new command. If you wish to use an existing command name as your newly defined command name, you have to undefine the existing command.

Apply the UNDEFINE command to undefine the LINE command.

Command: **UNDEFINE**
Command name: **LINE**

Try the LINE command after it is undefined.

Command: **LINE**
Unknown command "LINE". Type ? for list of commands.

If you wish to use the original LINE command, run the REDEFINE command.

Command: **REDEFINE**
Command name: **LINE**

1.7 Key Points and Exercises

When you start a drawing session by using AutoCAD, a Start Up dialog box appears. Using this dialog box, you can start a new drawing by using one of the three methods, or you can open an existing drawing.

After you start a drawing session, you work in the AutoCAD application window in which there are several window areas: standard Windows heading bar, Windows menu bar, Windows toolbar, graphic window, command interface window, and status window.

To run a command, you can select an item from the pull-down menu or the cascading menu of the menu bar, select an icon from the toolbars, or type the command name in the command interface window. There are twelve function keys, and they serve different purposes in AutoCAD. You should make a the habit of saving your drawing file at regular intervals. You can quickly save a file to the current working file, save a file to a new file name and set to that file, and save to a new file but continue working on the current file. Besides saving a file in DWG format, you can save it as a template file or save the file to earlier versions of AutoCAD.

In this chapter, you became familiar with the AutoCAD user interface. You learned the various methods of command execution. In addition, you learned how to use some file utilities, how to input and output to other file formats, and how to use miscellaneous drawing tools.

Exercise 1.1

How many ways can you start an AutoCAD drawing session? What are they? How can you disable and reactivate the display of the Start Up dialog box?

Exercise 1.2

Describe the AutoCAD application window and explain each window area.

Exercise 1.3

What are the three ways to run AutoCAD commands? What is a transparent command? How can you run a transparent command?

Exercise 1.4

State the functions of the function keys [F1] to [F12].

Exercise 1.5

Explain how the SAVE, SAVEAS, and QSAVE commands differ.

Exercise 1.6

How can you retrieve a record of your work in AutoCAD?

Chapter 2

Getting Started

2.1 Start a New Drawing
2.2 Set Up a New Drawing
2.3 Linetype and Linetype Scale
2.4 Layer Management
2.5 Entity Properties
2.6 Key Points and Exercises

Aims and Objectives

The aim of this chapter is to introduce the basic steps in setting up a new drawing and to explain the properties of geometric entities in a drawing. After studying this chapter, you should be able to:

- load linetypes and set linetype scale,
- create and manage layers, and
- explain the various properties of geometric entities in a drawing.

Overview

To create a new drawing, it is sensible to set the drawing limits to a size that covers your planned work. You may have to set grid and snap meshes to facilitate drafting. There are different types of lines in a drawing, so you need to load the relevant linetypes. The scale of linetype is important because it affects the visual appearance of your drawing. For better management of data, you should create layers and organize the entities into layers. Because entities possess properties, you should know how to change these properties.

2.1 Start a New Drawing

As we noted in Chapter 1, there are three ways to start a new drawing — Use a Wizard, Use a Template, and Start from Scratch. If you are a beginner, you should use the Start from Scratch option and then follow the steps delineated in Sections 2.2, 2.3, and 2.4 to set up a drawing. But first, let's get an overview of the three methods.

Use a Wizard

A wizard displays dialog boxes to guide you through the steps of setting up a new drawing. There are two wizards, Quick Setup and Advanced Setup. See Figure 2.1.

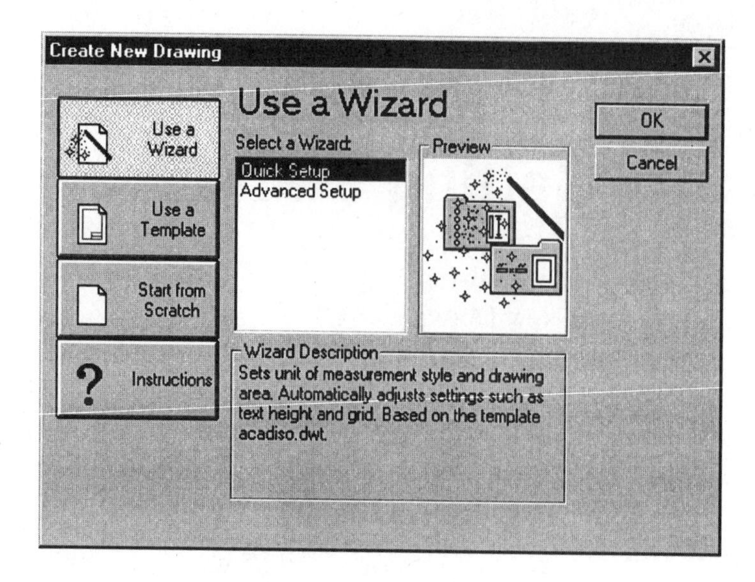

Figure 2.1 Use a Wizard

In the Quick Setup Wizard (Figure 2.2), you go through two steps. First, you select the units of measurement of the drawing. There are five kinds of measurement — Decimal, Engineering, Architectural, Fractional, and Scientific. The Decimal option displays the measurements in decimal notation (in either English or metric units). The Engineering option displays the measurements in feet and decimal inches. The Architectural option displays the measurements in feet, inches, and fractional inches. The Fractional option displays the measurements in mixed-number (integer and fractional) notation. The Scientific option displays the measurements in scientific notation with numbers expressed in the form of the product of a decimal number between 0 and 10 and a power of 10. See the UNITS and DDUNITS commands explained in Section 2.2.

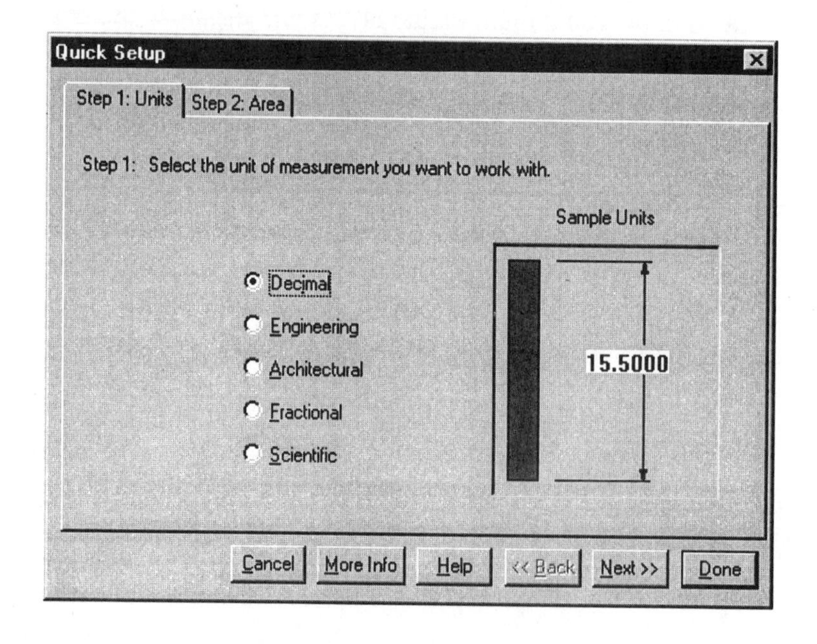

Figure 2.2 Quick Setup wizard

Second, you set the initial working area you want to display in full screen. If you turn grid on, this area is covered by grid dots. See the LIMITS command explained in Section 2.2.

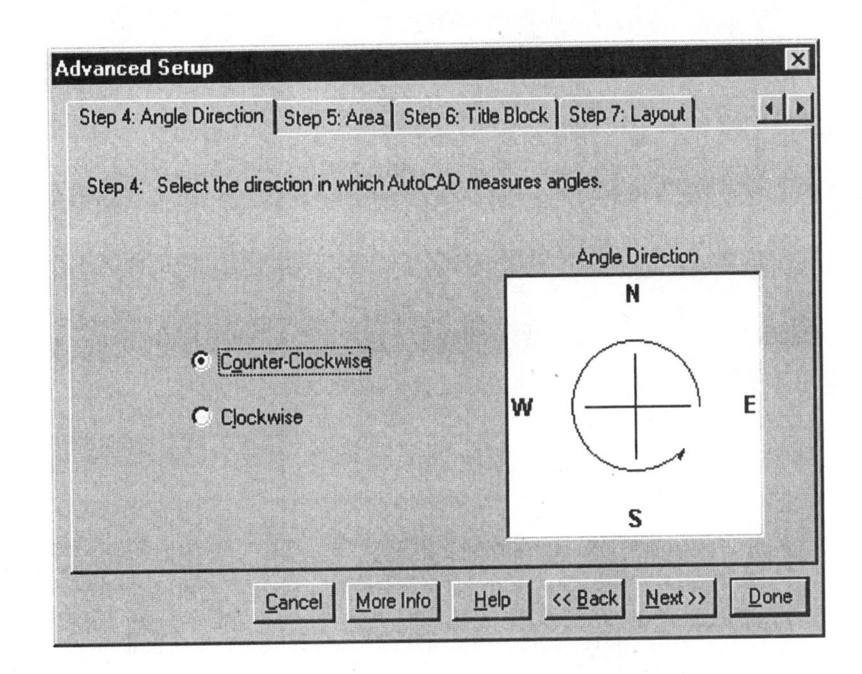

Figure 2.3 Advanced Setup wizard

In the Advanced Setup Wizard (Figure 2.3), you go through seven steps. First, you set up the units and the precision of units in which you want the linear measurement to be displayed. Second, you set up the angle measurement method and the precision of angular display. Third, you decide the direction of the zero angle for the entry of angles. Fourth, you decide the direction of angular measurement: counterclockwise or clockwise. Fifth, you set the area of initial screen display and the area covered by grid dots. Sixth, you can choose to insert a title block. Seventh, you select whether you want the title block to be inserted in paper space or in model space. You also select where you want to start if the title block is inserted in paper space. You will learn more about paper space, model space, and tilemode in Chapter 9.

Use a Template

A template is equivalent to the prototype drawing in the previous releases of AutoCAD. By default, the file extension of a template file is DWT. Although a DWT template file is the same as a DWG drawing file in all respects, the difference in file extension is to distinguish it from other drawing files so that you will not accidentally edit a template file. You can also use a DWG file as a template. After a template file is chosen, a preview appears in the Preview box.

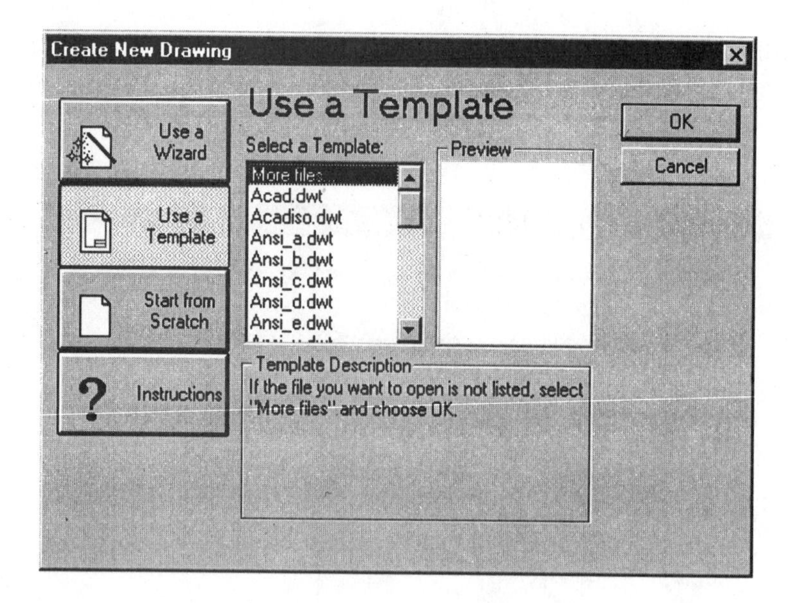

Figure 2.4 Use a Template

Refer to Figure 2.4. You can choose a template from the Select a Template box. If the template that you want does not appear in the box, you can select the More files... item in the box. To use a DWG template, select the More files... item and select Drawing (*.dwg) in the Files of type box. Then you can browse to select an AutoCAD drawing file to become the template drawing. Whether you use a DWT or a DWG file, the new drawing you are going to create will have all the entities and settings of the template.

Start from Scratch

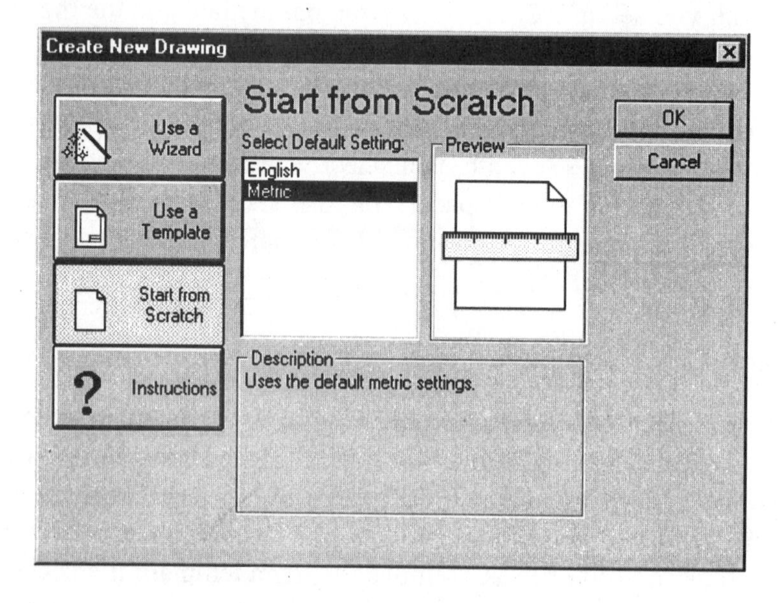

Figure 2.5 Start from Scratch

This is the simplest way of starting a new drawing. You can use default English (feet and inches) settings or default metric settings. See Figure 2.5.

2.2 Set Up a New Drawing

After using the Start from Scratch option to start a new drawing, you can follow the steps below to set up your drawing.

Units of Measurement

By using the UNITS or the DDUNITS command, you can set the units, display precision, and direction of angle measurement of a drawing. Both commands serve the same purpose. The UNITS command works at the command line interface, and the DDUNITS command brings out dialog boxes. Select the Units... item from the Format pull-down menu to use the DDUNITS command. See Figure 2.6.

<Format> **<Units...>**

Command: **DDUNITS**

[Units Control
Units **Decimal**
Angles **Decimal Degrees**
Direction...]

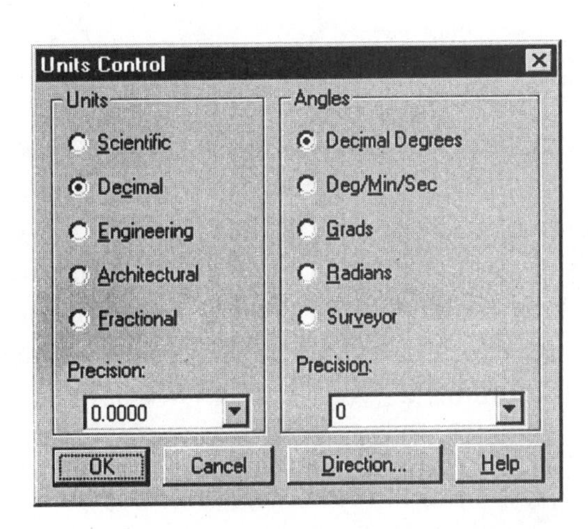

Figure 2.6 Units Control dialog box

From the Units Control dialog box, select the [Direction...] button to set the direction of angular measurement. See Figure 2.7.

[Direction Control
Angle 0 Direction **East** **Counter-Clockwise**
OK]

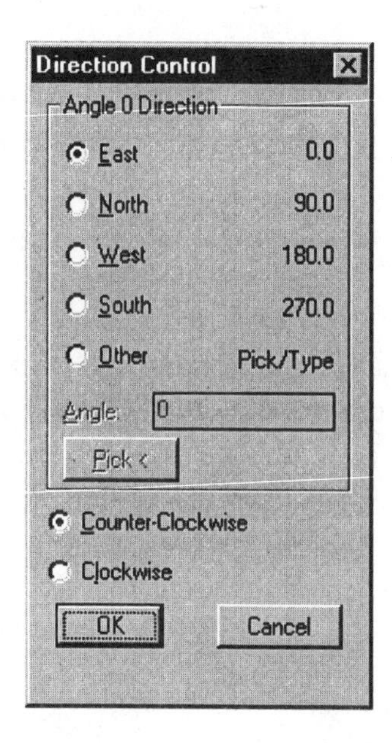

Figure 2.7 Direction Control dialog box

After making appropriate settings in accordance with Figure 2.6 and Figure 2.7, select the [OK] buttons to exit.

Drawing Limits and Coordinate Display

When you start a new drawing, the size of the working area displayed in the graphic window depends on the settings of the template drawing or the area you have selected in the Setup Wizard. Assuming that you want to have an initial working area of 420 units times 297 units (this is a typical metric A3 size setting), you can use the LIMITS command to set the working limits and use the ZOOM command to zoom to ALL. The area of the view window returned by the ALL option of the ZOOM command is determined by the LIMITS command.

<Format> **<Drawing Limits>**

Command: **LIMITS**
ON/OFF/<Lower left corner>: **0,0**
Upper right corner: **420,297**

<View> **<Zoom>** **<All>**

The LIMITS command sets up the initial working space. Make sure that it is set to OFF. If you set it on, it prevents any entity being created outside the limits.

<Format> **<Drawing Limits>**

Command: **LIMITS**
Reset Model space limits:
ON/OFF/<Lower left corner>: **OFF**

Move the crosshairs around, and watch to see whether the coordinate display in the status bar is updating all the time. If not, press the [F6] key or the [^D] key to turn it on. You can also select the coordinate display box in the status bar.

Command: <Coords off> <Coords on>

Move the crosshairs to the top right corner of the graphic window. Take down the reading of the crosshairs position from the status line. Move the crosshairs to the lower left corner of the graphic window and take down the reading again. These two readings give you some idea about the current window size.

Grid Mesh

Grid points of known spacing on the screen give you a sense of the actual size of the current view window. Because you have set the limits to 420 units x 297 units, use the GRID command to display a grid mesh of 10-units spacing.

Command: **GRID**
Grid spacing(X) or ON/OFF/Snap/Aspect: **10**

This command turns on the grid points within an area defined by the LIMITS command. These points appear only on the screen. They will not appear on the printed copy of your drawing. The toggle key for turning on and off the grid mesh is the [F7] key or the [^G] key. Press the key twice. The grid goes off and on again. You can also select the Grid item of the status bar.

Command: <Grid off> <Grid on>

By using the GRID command, you can set a different vertical and horizontal spacing between grids. Use the Aspect option if you want to do so.

Issue the ZOOM command and zoom to a scale of 0.5. The scale is relative to the window size defined by the LIMITS command. See Figure 2.8.

<View> **<Zoom>** **<Scale>**

Command: **ZOOM**
All/Center/Dynamic/Extents/Previous/Scale(X/XP)/Window/<Realtime>: **S**
Enter scale factor: **0.5**

If you suffix the letter X to the zoom scale, you will get a screen display of 0.5 relative to the previous window size. You will learn about the XP suffix in Chapter 9.

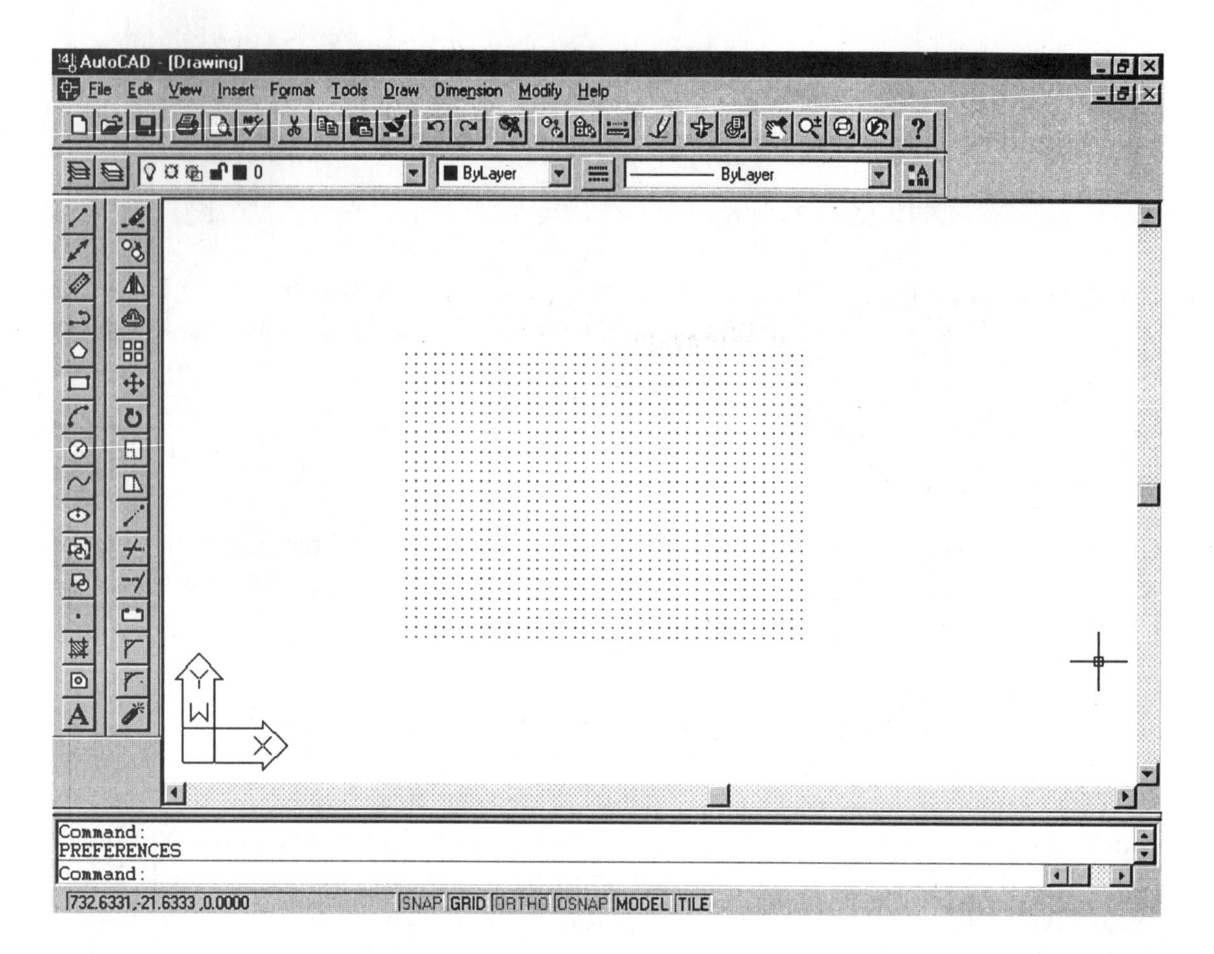

Figure 2.8 Grid points displayed within an area defined by the LIMITS command

From the screen, you can see that the grid points cover only part of the screen. Run the GRID command again, and set both the vertical and the horizontal spacing to 1 unit.

Command: **GRID**
Grid spacing(X) or ON/OFF/Snap/Aspect: **1**
Grid too dense to display

With grid spacing set to 1 unit, the grid mesh becomes very dense with respect to the current view window size. If the grids were to be displayed, you would see only an area filled up with grids closely packed together. Therefore, AutoCAD refuses to display the grid points and issues the message that the grid is too dense to display.

Undo and Redo

We all make mistakes. Sometimes, after issuing a command, we wish we had not done so. The UNDO or U command returns you to where you were before the last command was executed. Issue this command now.

[Standard Toolbar] **[Undo]**

Command: **U**

GRID

This command undoes the last GRID command. The grid spacing returns to 10 units x 10 units again. You can see the grid mesh again. Repeat the U command.

Command: **[Enter]**
U
ZOOM

This time, the U command undoes the ZOOM command. The opposite of the U command is the REDO command. Run this command.

[Standard Toolbar] **[Redo]**

Command: **REDO**

Repeat the REDO command.

Command: **[Enter]**
REDO
Previous command did not undo things

The REDO command does not work the second time. Remember, you can undo as many times as you like, but you can redo only once.

Snap Intervals

Grid points are for visual reference only. Without the use of any further aids, it is virtually impossible to select these points precisely using the pointing device. Apply the SNAP command to set snap intervals to 10 units.

Command: **SNAP**
Snap spacing or ON/OFF/Aspect/Rotate/Style: **10**

After you have set snap mesh to 10-units intervals, the crosshairs are locked to the nearest snap mesh point throughout the entire drawing. Move the pointing device around. You will find that the crosshairs can stop only at a snap point. Unlike the GRID command, the coverage of the SNAP command is not restricted by the LIMITS command.

The toggle key for turning on and off snap is the [F9] key or the [^B] key. Press the key twice. You can also select the Snap item from the status bar.

Command: <Snap off> <Snap on>

Snap Rotation

Repeat the SNAP command to rotate the snap mesh about a base point at (0,0) for an angle of 15 degrees.

Command: **SNAP**
Snap spacing or ON/OFF/Aspect/Rotate/Style: **ROTATE**
Base point: **0,0**
Rotation angle: **15**

The command rotates the crosshairs as well as the grid mesh. To verify where the origin (0,0) is, you can use the UCSICON command. See Figure 2.9.

<View> **<Display>** **<UCS Icon>** **<Origin>**

Command: **UCSICON**
ON/OFF/All/Noorigin/ORigin <ON>: **OR**

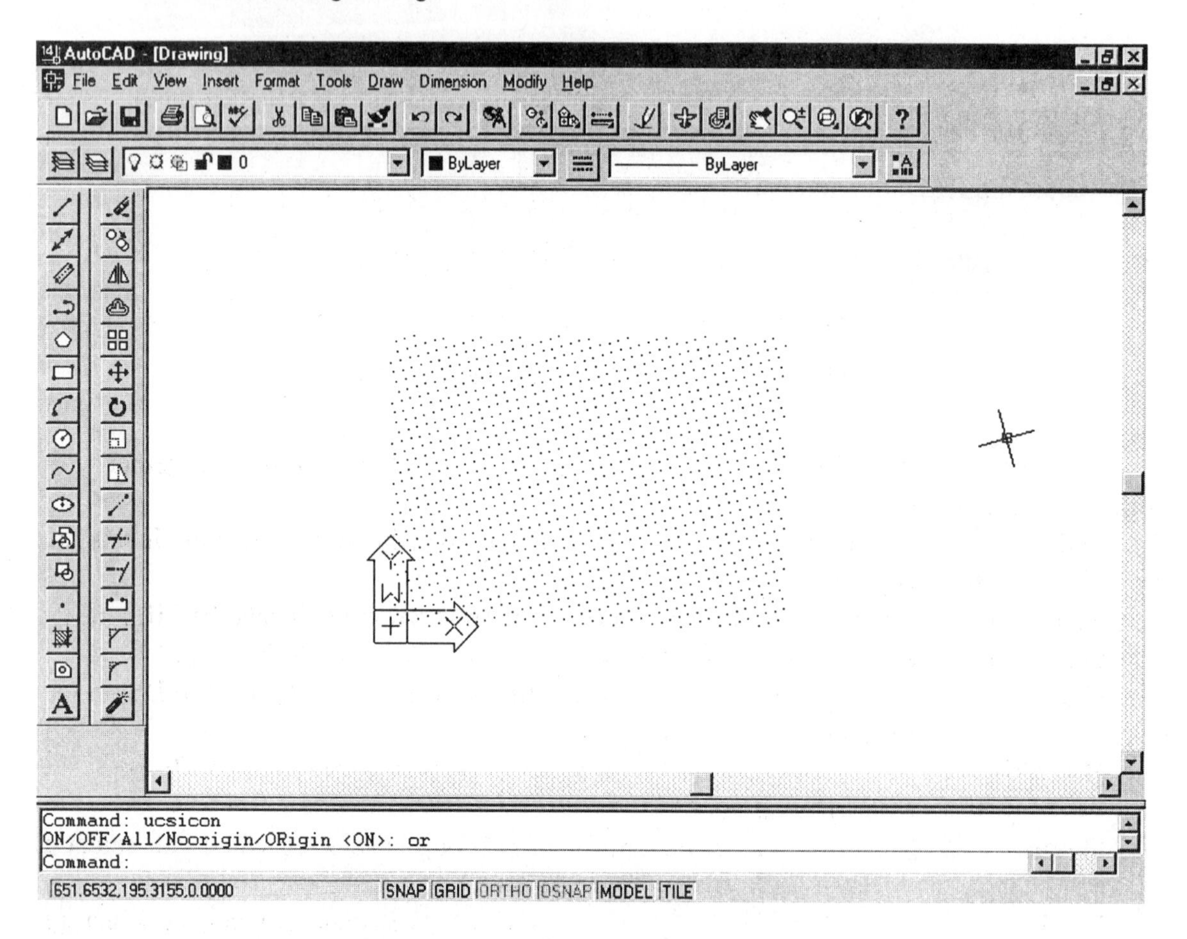

Figure 2.9 Crosshairs and grid points rotated

The UCS icon now displays at the origin point. Use the U command twice to undo the UCSICON and SNAP commands.

[Standard Toolbar] **[Undo]**

Command: **U** UCSICON

Command: **[Enter]**
U SNAP

DDRMODES

To set up the snap and grid collectively, you can use the DDRMODES command. See Figure 2.10.

<Tools> **<Drawing Aids...>**

Command: **DDRMODES**

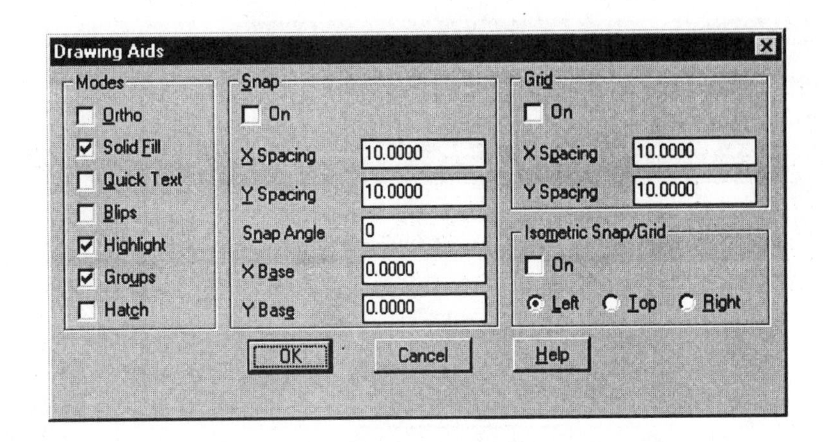

Figure 2.10 DDRMODES command dialog box

With the DDRMODES command, you can set the grid and snap modes, together with other drawing modes:

Ortho	Constrains the crosshairs movement to orthogonal positions. It sets the Ortho mode.
Solid Fill	Controls the filling of wide polylines and donuts. It sets the Fill mode.
Quick Text	Controls the display of text on screen. It sets the Qtext mode.
Blips	Turns on and off the marker blips on the screen. It sets the Blipmode variable.
Highlight	Controls whether the selected objects are highlighted. It sets the Highlight variable.
Groups	Controls whether the entire entity group is selected when you select a member. It concerns the GROUP command.
Hatch	Controls whether the boundary object is selected when you select an associative hatching. It concerns the BHATCH command.

2.3 Linetype and Linetype Scale

You need different linetypes in a drawing. For example, you need a long chain line for the center line and a dashed line for hidden details. To load the required linetypes to your drawing, you can use the LINETYPE command. This command enables you to load linetypes from a file of LIN extension to your drawing.

Linetype

Select the Linetype... item from the Format pull-down menu to apply the LINETYPE command to load linetype Center and Dashed from the file ACAD.LIN.

<Format> **<Linetype...>**

Command: **LINETYPE**

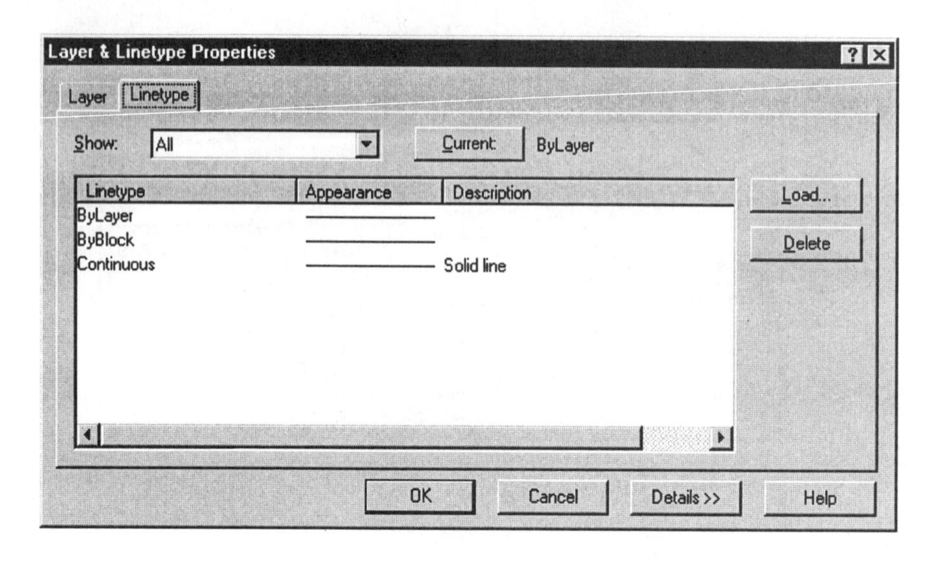

Figure 2.11 Layer and Linetype Properties dialog box

Refer to Figure 2.11. There are two tabs in this dialog box — Linetype and Layer. The LINETYPE and LAYER commands share the same dialog box. You will work on the LAYER command in the next paragraph. Select the [Load...] button of the Linetype tab. See Figure 2.12.

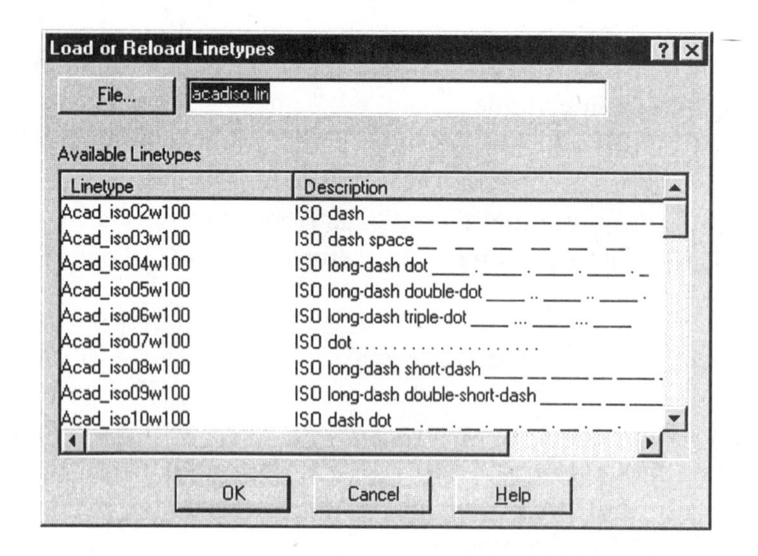

Figure 2.12 Load or Reload Linetypes dialog box

To load the linetype CENTER and DASHED, select these two items in the Available Linetypes box. Remember to hold down the [Control] key while selecting. After selecting the linetypes, select the [OK] button twice to exit from the Load or Reload Linetype dialog box and the Layer & Linetype Properties dialog box.

Linetype Scale

Depending on the units of measurement of the drawing and the actual size of the objects that you are going to create in the drawing, you may have to alter the linetype scale. Linetype scale concerns the relative length of the dash-dot linetype per drawing unit. To change the linetype scale, you can use the LINETYPE command. Key in the value 25 in the Global scale factor box. If you cannot find this box, select the [Details>>] button. To return to the previous tab, select the [Details<<] button.

<Format> **<Linetype...>**

Command: **LINETYPE**

[Linetype
Global scale factor: **25**
OK]

In previous releases of AutoCAD, there are two commands for linetype management, LINETYPE command and DDLTYPE command. In AutoCAD R14, these commands become identical and display dialog boxes for interaction. If you want to work at command line interface, you can use the -LINETYPE command.

Command: **-LINETYPE**

2.4 Layer Management

The concept of layer is analogous to having a series of transparent overlays put on your drawing. You need to create layers in your drawing and to organize your entities into different layers for ease of manipulation. For example, you can put all the center lines on one layer and the outlines on another.

Layer Creation

You can use the LAYER command to create and manage layers in your drawing. Select the Layer... item of the Format pull-down menu to apply the LAYER command to create a new layer called Center with linetype Center and color red. As we have noted, the LINETYPE and LAYER commands share the same dialog box. See Figure 2.13.

<Format> **<Layers...>**

Command: **LAYER**

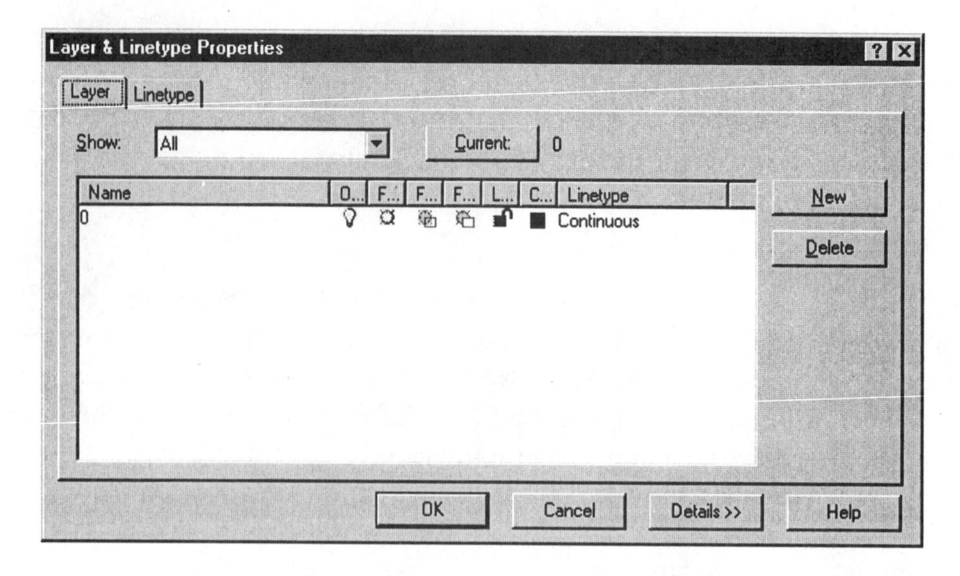

Figure 2.13 Layer tab of the Layer & Linetype Properties dialog box

To create a new layer, select the [New] button. You will find that a new layer called Layer1 displays in the layer list. See Figure 2.14.

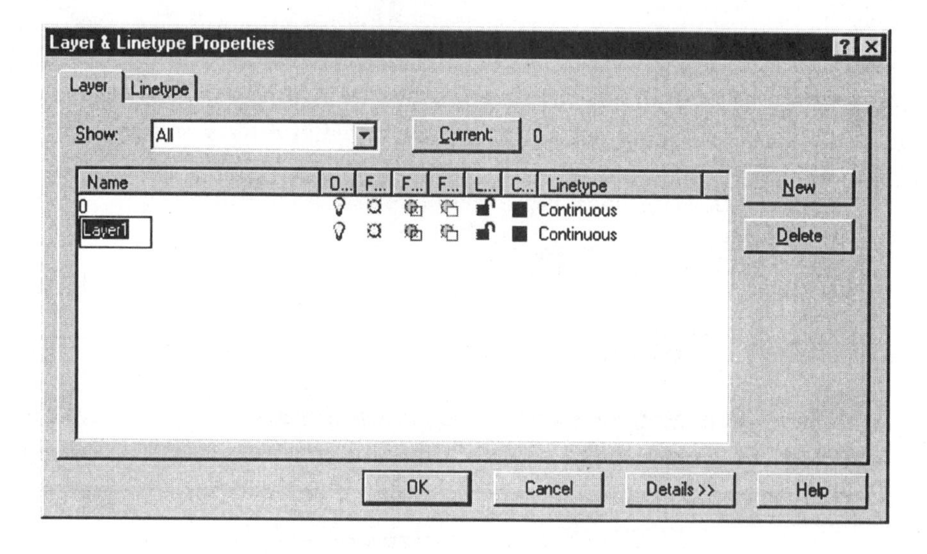

Figure 2.14 Layer1 created

Select Layer1 from the layer list box and type the name CENTER to change the layer name. To change the linetype assignment, select the linetype of this layer to display the Select Linetype dialog box. See Figure 2.15.

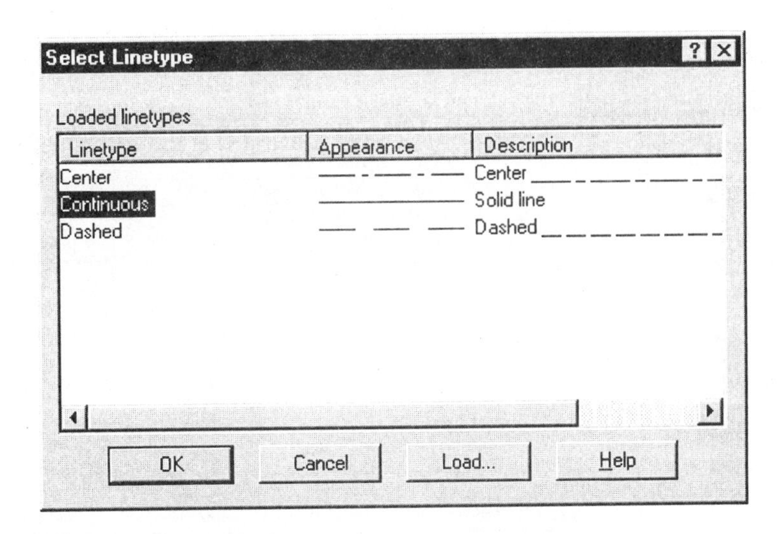

Figure 2.15 Select Linetype dialog box

From the Select Linetype dialog box, select the linetype Center. If the linetype that you want to select does not display in this list, you can load it by selecting the [Load...] button. Now, you can argue that you do not need to use the LINETYPE command to load the linetypes beforehand.

After selecting the appropriate linetype, select the [OK] button. To change the color setting of a layer, select the square box of this item under the C... column. C... stands for Color. A Select Color dialog box appears. See Figure 2.16.

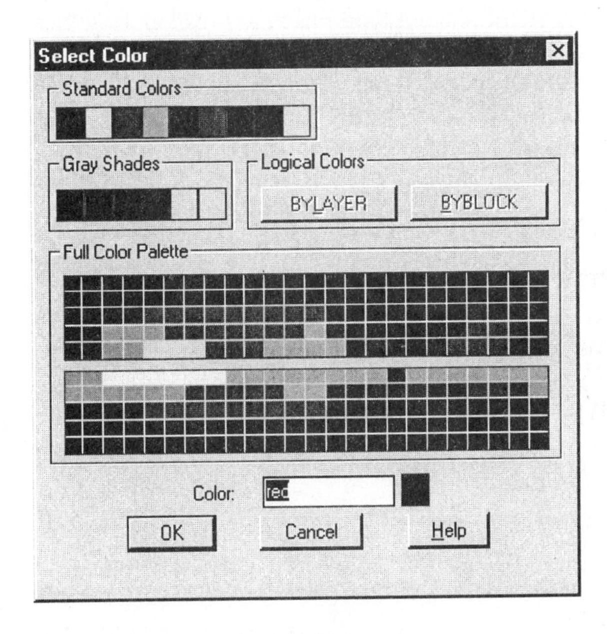

Figure 2.16 Select Color dialog box

In the Select Color dialog box, select the color Red and then select the [OK] button. In the Layer & Linetype Properties dialog box, select the [OK] button to exit. A new layer called Center with linetype Center and color red is created. See Figure 2.17.

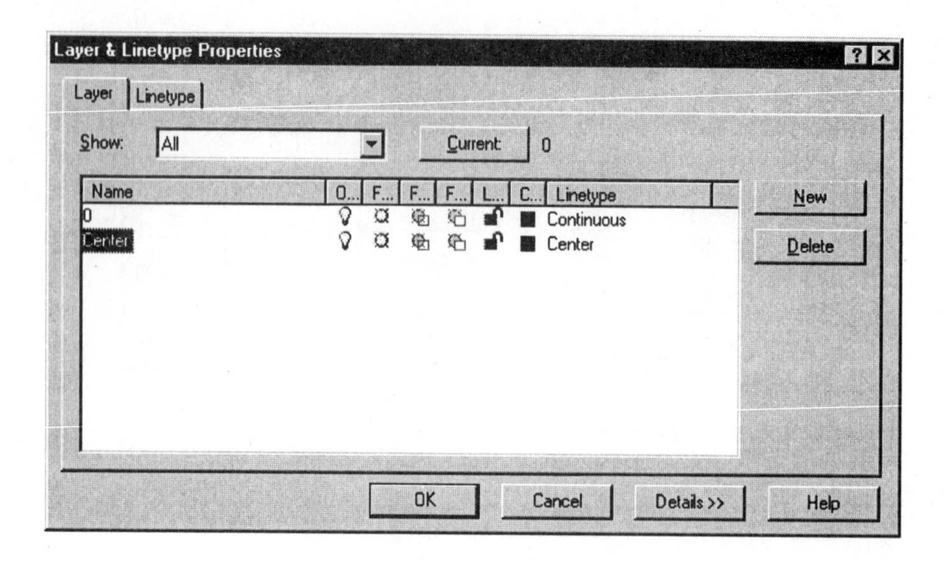

Figure 2.17 New layer created

In previous releases of AutoCAD, there are two commands to manipulate layers, LAYER command and DDLMODES command. The LAYER command works in command line and the DDLMODES command displays dialog boxes. In AutoCAD R14, these commands become identical. If you wish to have command line input, you can key in -LAYER at the command line interface.

Command: **-LAYER**
?/Make/Set/New/ON/OFF/Color/Ltype/Freeze/Thaw/LOck/Unlock: **NEW**
New layer name(s): **CENTER**
?/Make/Set/New/ON/OFF/Color/Ltype/Freeze/Thaw/LOck/Unlock: **LT**
Linetype (or ?) <CONTINUOUS>: **CENTER**
Layer name(s) for linetype CENTER <0>: **CENTER**
?/Make/Set/New/ON/OFF/Color/Ltype/Freeze/Thaw/LOck/Unlock: **C**
Color: **RED**
Layer name(s) for color 1 (red) <0>: **CENTER**
?/Make/Set/New/ON/OFF/Color/Ltype/Freeze/Thaw/LOck/Unlock: **[Enter]**

Changing Color and Linetype Assignment

To change the linetype or color assignments of a layer, you can use the LAYER command and follow a route similar to the layer creation procedure we have just described.

Renaming a Layer

You can rename any layer except layer 0. To rename a layer, you can use the LAYER command, highlight the layer name, and key in a new name. Another way of renaming a layer is to use the RENAME command. See Figure 2.18. Select the layer name in the Items box, key in a new name in the Rename To: box, and select the [Rename To:] button.

<Format> **<Rename...>**

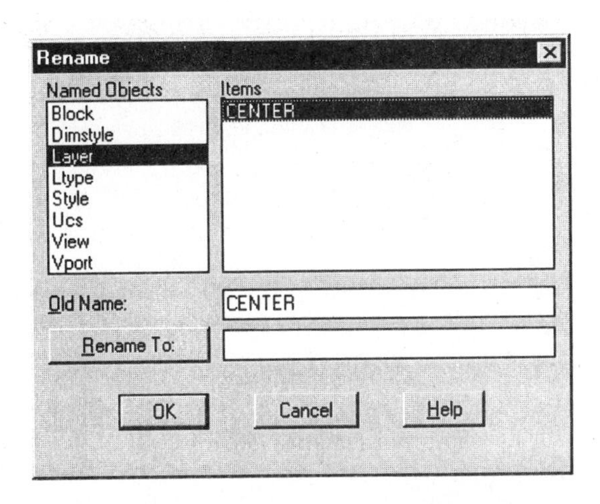

Figure 2.18 Rename dialog box

Current Layer

With more than one layer in your drawing, you have to decide which is the current layer. The entity that you create is always placed in the current layer. Suppose that you want to work on layer 0. With the LAYER command, select the layer 0, then the [Current:] button, and then the [OK] button.

<Format> **<Layers...>**

Command: **LAYER**

[Layer
Current: **0**
OK]

A quicker way to select a layer to become the current layer is to use the Layer Control box of the Object Properties toolbar. Select the Layer Control box and then layer 0. See Figure 2.19.

[Object Properties] **[Layer Control]**

Figure 2.19 Layer Control of the Object Properties toolbar

Layer Removal

Only empty layers can be removed. To remove an empty layer, apply the LAYER command, select the empty layer, and then select the [Delete] button.

<Format> **<Layers...>**

Command: **LAYER**

[Layer
Select an empty layer.
Delete
OK]

Besides using the LAYER command, you can remove an empty layer by using the PURGE command.

<File> **<Drawing Utilities>** **<Purge>** **<Layers>**

Command: **PURGE**
Blocks/Dimstyles/LAyers/LTypes/SHapes/STyles/APpids/Mlinestyles/All: **LAYER**

Layer Management

Referring to the LAYER command dialog box, there are eight columns in the list of layers box. They are, from left to right, Name, On, Freeze in All Viewports, Freeze in Current Viewport, Freeze in New Viewports, Lock, Color, and Linetype. In normal view, you will see them display as Name, O..., F..., F..., F..., L..., C..., and Linetype. See Figure 2.13.

To see their full names, you can select the partition line between the columns and drag the line to a new position. See Figure 2.20.

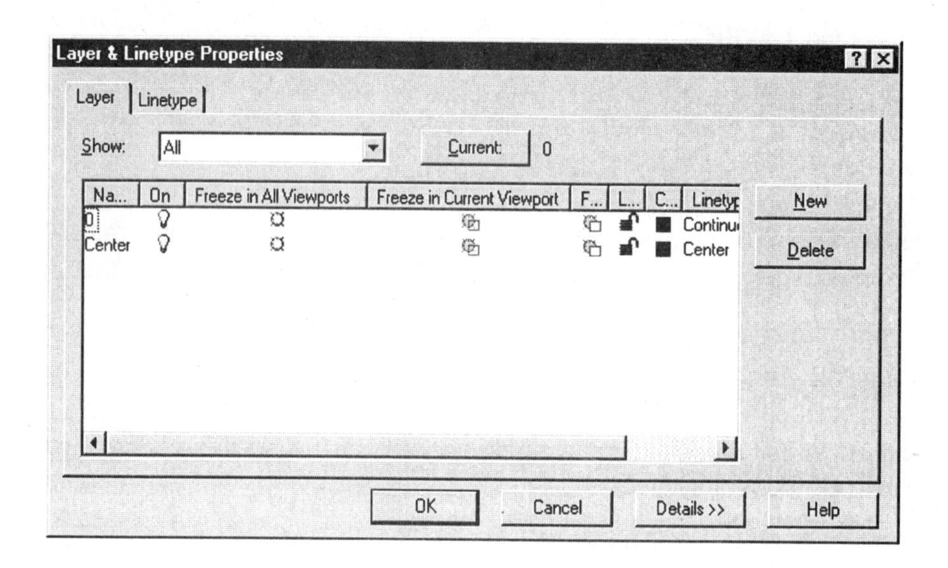

Figure 2.20 Column names of list of layers box

You can select and unselect the symbol under these columns to perform the following tasks:

Name	Sets the name of a layer. You can change the layer name by selecting the old name and key in a new name.
On	Turns on or off a layer. Entities on a layer that is turned off cannot be seen.

Freeze in All Viewports	Freezes or thaws a layer in all viewports. A frozen layer is ignored by the program in addition to being invisible. This speeds up program execution. Thaw is the opposite of freeze.
Freeze in Current Viewport	Freezes or thaws a layer in the current floating viewport that is created in paper space. You will learn about paper space in Chapter 9.
Freeze in New Viewports	Freezes or thaws a layer in all subsequently created new viewports in paper space.
Lock	Locks or unlocks a layer. A locked layer remains visible, but entities on it cannot be selected for editing.
Color	Sets the color of the entities residing on a layer.
Linetype	Sets the linetype of the entities residing on a layer.

2.5 Entity Properties

A geometric entity possesses at least three properties — color, layer, and linetype. Some have a fourth, thickness, and a fifth, linetype scale. By default, color and linetype are BYLAYER. This means that the color and linetype of an entity depend on the color and linetype assignment of the layer where the entity is placed.

If you change the color or linetype assignment of a layer, the color and linetype of an entity that is BYLAYER change as well. If the color or linetype of an entity is not BYLAYER, then any change to the color or linetype assignment of the layer will have no effect on the entity.

Change Properties

You can use the DDMODIFY command, the CHPROP command, or the DDCHPROP command to change the properties of an entity. The CHPROP command works at the command line interface.

Command: **CHPROP**
Select objects: [**Select an object.**]
Select objects: [**Enter**]
Change what property (Color/LAyer/LType/ltScale/Thickness) ? **C**
New color <BYLAYER>: **YELLOW**
Change what property (Color/LAyer/LType/ltScale/Thickness) ? [**Enter**]

The DDCHPROP command displays a dialog box. See Figure 2.21.

Command: **DDCHPROP**
Select objects: [**Select an object.**]
Select objects: [**Enter**]

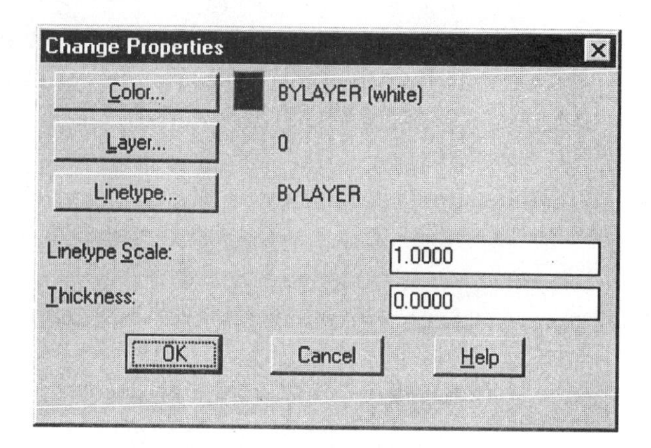

Figure 2.21 DDCHPROP command

The DDMODIFY command is entity-sensitive. Depending on what type of entity is selected, this command displays an appropriate dialog box to change relevant aspects of the selected object. Select the Properties... item of the Modify pull-down menu or select the Properties icon of the Object Properties toolbar. See Figure 2.22.

<Modify> **<Properties...>**

[Object Properties] **[Properties...]**

Select objects: **[Select an object.]**
Select objects: **[Enter]**

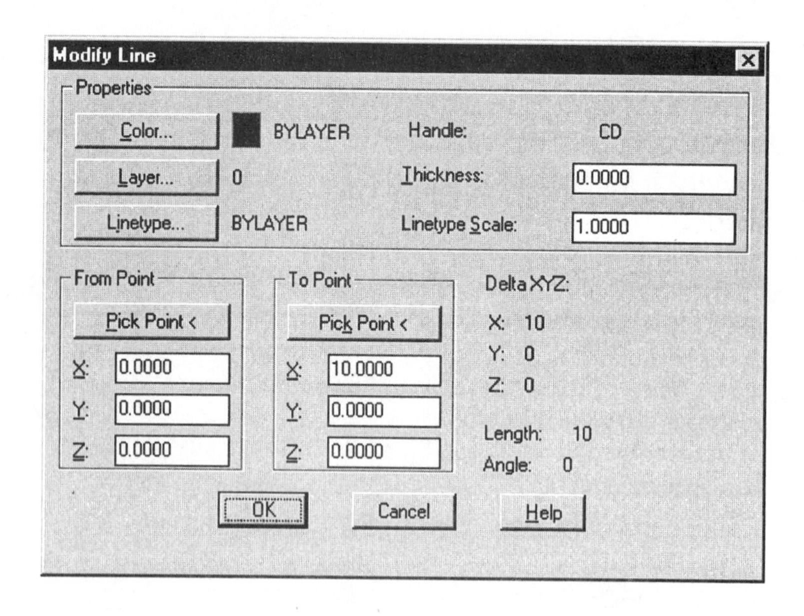

Figure 2.22 DDMODIFY command

Match Properties

To match properties is to change the properties of a selected entity to match those of another selected entity. Select the Match Properties item of the Modify pull-down menu or the Match Properties icon of the Standard toolbar.

<Modify> **<Match Properties>**

[Standard Toolbar] **[Match Properties]**

Command: **MATCHPROP**
Select Source Object: **[Select the source object.]**
Settings/<Select Destination Object(s)>: **[Select the target object.]**

Color

If you want to set the color of new entities not to follow the color assignment of layers, use the COLOR command or the DDCOLOR command. Use the COLOR command to set the color of new entities to red. After that, the color of all new entities will be red regardless of the color assignment of the layer on which they are created.

Command: **COLOR**
New object color <BYLAYER>: **RED**

After you set the entity color to red, all subsequently created entities will be red regardless of which layer they reside in.

Select the Color... item of the Format pull-down menu to apply the DDCOLOR command to set the entity color to BYLAYER. See Figure 2.16.

<Format> **<Color...>**

Command: **DDCOLOR**

[Select Color
Logical Colors **BYLAYER**
OK]

After you set the color to BYLAYER, the color of the entities that you create will follow the color assignment of the current layer.

Thickness

The thickness of an entity is measured in the Z-direction. The default thickness of an entity is zero. If you change the thickness of a line to a non-zero value and then view it from an angle, you can see a rectangular object. This is useful in creating 2-1/2 D effect.

<Format> **<Thickness>**

The THICKNESS command is an old command. You should not use it. To construct 3D objects, you should use the solid modeling commands that are explained in Chapter 6.

Object Properties Toolbar

To control the properties of entities collectively, you can use the Object Properties toolbar. See Figure 2.23.

Figure 2.23 Object Properties toolbar

From left to right, the Object Properties toolbar performs the following tasks:

Make Object's Layer Current	Selects objects whose layer will become the current layer.
Layers	Brings up the LAYER command.
Layer Control	Selects a layer to be the current layer, sets the visibility of a layer, freezes/thaws a layer in all viewports, freezes/thaws a layer in the current viewport, or locks/unlocks a layer.
Color Control	Sets the color of the entities that are created subsequently.
Linetype	Brings up the LINETYPE command.
Linetype Control	Sets the linetype of the entities that are created subsequently.
Properties	Changes the properties of the selected entities.

2.6 Key Points and Exercises

There are three ways to start a new drawing — Use a Wizard, Use a Template, and Start from Scratch. If you are a beginner, you should use the Start from Scratch option. Then you should set the units of measurement, set the drawing limits, zoom the display to the drawing limits, load the required linetypes, set the linetype scale, and create layers.

In this chapter, you learned how to set up drawing limits, display grid meshes, set snap intervals, load linetypes, and create layers. You also examined the properties of entities.

Exercise 2.1

Explain briefly the three options for starting a new drawing by using the NEW command. If you start a drawing from scratch, what should you do to set up a drawing?

Exercise 2.2

Why is it necessary to load linetype to a drawing and why should the linetype scale be set? List the steps required to create a layer in a drawing.

Exercise 2.4

Name three properties of a geometric entity. Explain the term "BYLAYER" with respect to linetype and color of an entity.

Chapter 3

Draw and Modify

3.1 Pipe Support Assembly
3.2 Top View of the First Component
3.3 Sectional Front View of the First Component
3.4 Side View of the First Component
3.5 Top View of the Second Component
3.6 Sectional Front View of the Second Component
3.7 Side View of the Second Component
3.8 Floor Plan
3.9 Regions
3.10 Sketching
3.11 2D Isometric Drawing
3.12 Grips
3.13 Pick First
3.14 Entity Selection
3.15 Geometric Snap Tools
3.16 Coordinates Systems
3.17 Key Points and Exercises

Aims and Objectives

The aim of this chapter is to let you practice producing orthographic engineering drawings by using various methods and techniques. The drawings have been designed in such a way that as many commands as possible are applied. After studying this chapter, you should be able to:

- use various draw and modify commands to construct mechanical engineering drawings and architectural drawings,
- use absolute/relative coordinates systems to locate points in a drawing,
- select points by filtering coordinates of existing points, by tracking, and by entering direct distance,
- use geometricl snap tools to select exact locations of existing geometric objects,
- use various methods to select objects in a drawing, and
- create and refine a block in a drawing.

Overview

To help you gain an appreciation of applying AutoCAD commands in engineering drawing construction, this chapter guides you in creating a series of engineering drawings.

First, you will produce orthographic engineering drawings of a pipe support assembly. Second, you will create a floor plan by using multiple lines and wide objects. Third, you will work on regions. Fourth, you will learn about the concept of sketching. Fifth, you will make a 2D isometric view. Finally, you will practice the use of grip points to edit a drawing. In practicing the creation of engineering drawings, you will learn how to draw, modify, create a block, and refine a block. You will also learn how to input coordinates and become familiar with various drawing utilities — pick first, entity selection methods, and geometric snaps.

3.1 Pipe Support Assembly

Figure 3.1 shows the pipe support assembly. It consists of three components — base, roller, and shaft. In this chapter, you will start work on the top view, the front view, and the side view of the first component, the base. Next, you will create three views of the second component, the roller, in the same drawing but on different layers. In this chapter, you will have an assembly of two components. In Chapter 4, you will export two components from this assembly to two separate files to create part drawings. Then you will start another new drawing to create the third component, the shaft. Finally, you will open the assembly drawing and import the third component.

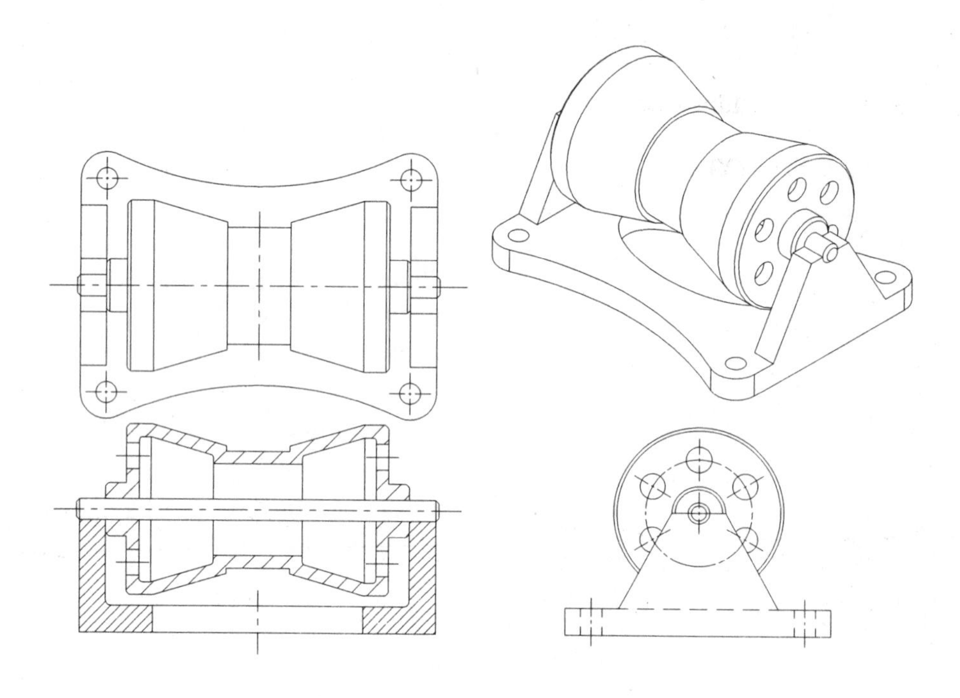

Figure 3.1 Pipe support assembly

Start a New Drawing

Select the New... item of the File pull-down menu to use the NEW command to start a new drawing. Select the [Start from Scratch] button and use metric default setting.

<File> **<New...>**

Command: **NEW**

[**Start from Scratch** **Metric**
OK]

Note:
In the following delineation, [AAA] means selecting the [AAA] toolbar. <BBB> means selecting the <BBB> pull-down menu. If you have difficulty in locating the respective toolbar or pull-down menu, you can simply enter the command name at the command line interface.

Load Linetype and Create New Layers

In this drawing, there are outlines, hidden lines, center lines, and hatching lines. You need the linetypes Center, Center2, and Dashed. For the purpose of better entity management, you need to create three more layers in addition to the default layer 0. If you do not know how to load linetypes and create layers, please refer to Chapter 2.

Select the Linetype... item from the Format pull-down menu to use the LINETYPE command to load the aforementioned linetypes to your drawing. Select the [Load...] button. Then hold down the [Control] key to select the linetypes Center, Center2, and Dashed in the Available Linetypes box.

<Format> **<Linetype...>**

Command: **LINETYPE**

[Linetype
Load...]

[Available Linetypes
Center
Center2
Dashed
OK]

The linetypes, Center, Center2, and Dashed are now loaded to your drawing. Because the LINETYPE command and the LAYER command share the same dialog box, the Layer & Linetype Properties dialog box, select the Layer tab to apply the LAYER command.

[Layer & Linetype Properties
Layer]

On the Layer tab of the Layer & Linetype Properties dialog box, create three additional layers and set their color and linetype assignments as follows:

Name	Color	Linetype
0	White	Continuous
Center	Red	Center
Hatch	Blue	Continuous
Hidden	Green	Dashed

After creating three additional layers, select layer 0 and then the [Current:] button to set it as the current layer. See Figure 3.2.

[Current: 0
OK]

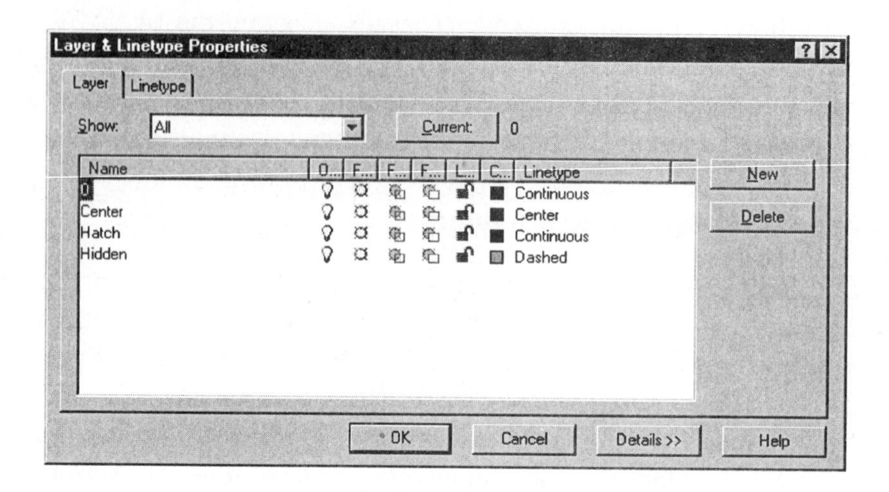

Figure 3.2 Three additional layers created

Set Limits

For the sake of obtaining a view window that is large enough for creating the top view, select the Drawing Limits item from the Format pull-down menu to run the LIMITS command to set the drawing limits to 420 units x 297 units. Then select the All item of the Zoom cascading menu of the View pull-down menu to use the ZOOM command to zoom the display to an area of approximately 420 units x 297 units.

<Format> **<Drawing Limits>**

Command: **LIMITS**
ON/OFF/<Lower left corner>: **0,0**
Upper right corner: **420,297**

<View> **<Zoom>** **<All>**

Set Grid and Snap Meshes

Now the graphic window displays an area of approximately 420 units x 297 units.

Grid points are a useful visual aid. Select the Drawing Aids... item of the Tools pull-down menu to use the DDRMODES command to turn on the grid mesh, set grid spacing to 10 units x 10 units, turn on snap, and set snap intervals to 10 units x 10 units. See Figure 3.3.

<Tools> **<Drawing Aids...>**

Command: **DDRMODES**

[Drawing Aids

Snap		Grid	
On		**On**	
X Spacing	**10**	X Spacing	**10**
Y Spacing	**10**	Y Spacing	**10**
OK			]

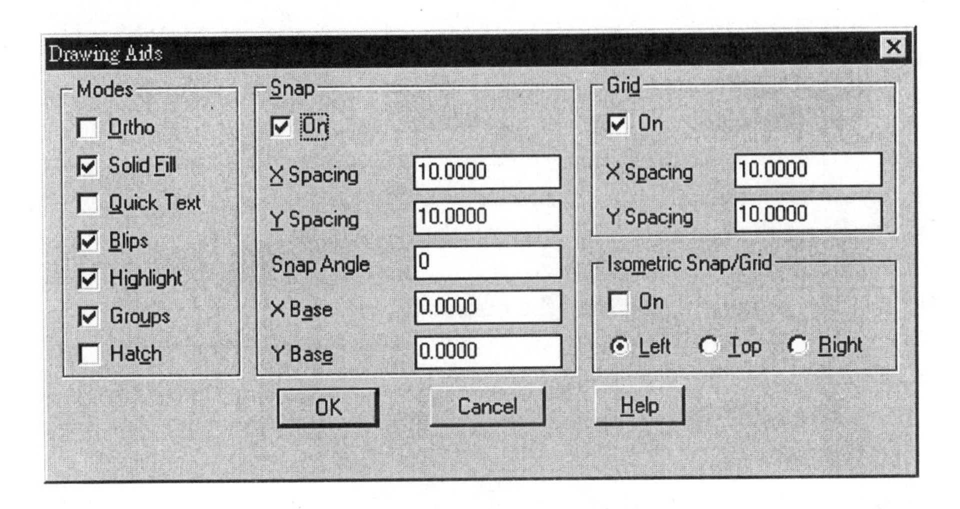

Figure 3.3 Set grid and snap

Save as Template

Drawing preparation takes some time. The foregoing are routines that you may have to do every time you start a new drawing from scratch. If you find these settings useful to you, you should save them to a template file with DWT extension and use it as a template in the future.

3.2 Top View of the First Component

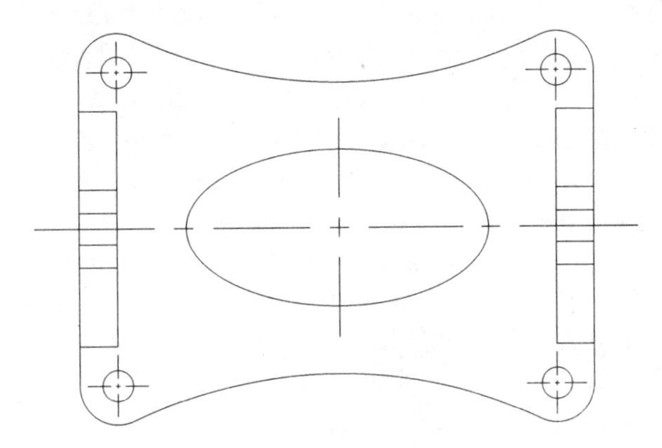

Figure 3.4 Top view of first component

The top view of the first component is shown in Figure 3.4. You will create the outlines on layer 0 and the center lines on layer Center.

Create a Circle

Although you can start your first entity anywhere you like, the position of the first entity is fixed for the sake of better communication between this book and you.

Select the Center, Radius item from the Circle cascading menu of the Draw pull-down menu to use the CIRCLE command to create a circle at (60,50) with a radius of 10 units.

<Draw> **<Circle>** **<Center, Radius>**

Command: **CIRCLE**

Because you have turned on snap mode, you can move the cursor to select the point (60,50) by checking the coordinate display on the status bar.

3P/2P/TTR<Center point>: [**Snap the cursor to a snap point at (60,50).**]

After selecting the center point, you can see the image of a circle varying in size with a radius defined by the latest position of the crosshairs in the screen. This is called dragging.

Diameter/<Radius>: **10**

The DRAGMODE system variable controls this dragging effect. When DRAGMODE is turned off, there is no dragging, and the computer runs faster. The default setting is Auto. Dragging is performed wherever possible.

Command: **DRAGMODE**
ON/OFF/Auto <Auto>: **AUTO**

Array the Circle

In this drawing view, there are four circles. To create the other three circles, use the ARRAY command instead of repeating the CIRCLE command. The ARRAY command provides two options — rectangular array and polar array. Rectangular array creates an array of rows and columns. Polar array creates an array about a center point.

From the Modify pull-down menu, select the Array item to run the ARRAY command. After running the command, you need to select objects.

There are many ways to select objects. See Section 3.14 of this chapter for details. Now respond with LAST (means the last entity that you create) to the prompt for entity selection. Then press the [Enter] key to end the selection. Next, key in [R] to take the rectangular array option and supply the necessary data for the array. See Figure 3.5.

<Modify> **<Array>**

Command: **ARRAY**
Select objects: **LAST**
Select objects: **[Enter]**
Rectangular or Polar array (R/P): **R**
Number of rows (---): **2**

Number of columns (|||): **2**
Unit cell or distance between rows (---): **200**
Distance between columns (|||): **290**

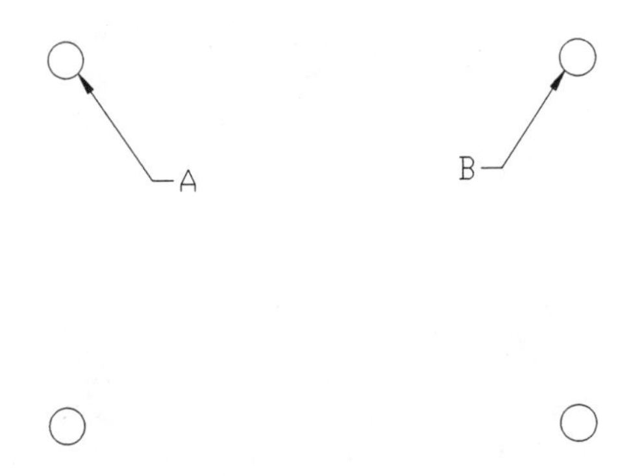

Figure 3.5 Rectangular array of circles

Pickbox Size

Notice that the crosshairs change to a small square box when you issue the ARRAY command. This is called the selection box. It appears whenever you are required to select objects. The size of this selection box is controlled by the PICKBOX system variable. Set the selection box size to 10 pixels.

Command: **PICKBOX**
New value for PICKBOX: **10**
Object snap target height (1-50 pixels): **10**

To set the Pickbox size graphically, you can use the DDSELECT command. From the Tools pull-down menu, select the Selection... item.

<Tools> **<Selection...>**

Command: **DDSELECT**

In the Pickbox Size box, move the slider bar to adjust the pickbox size. Then select the [OK] button. See Figure 3.6.

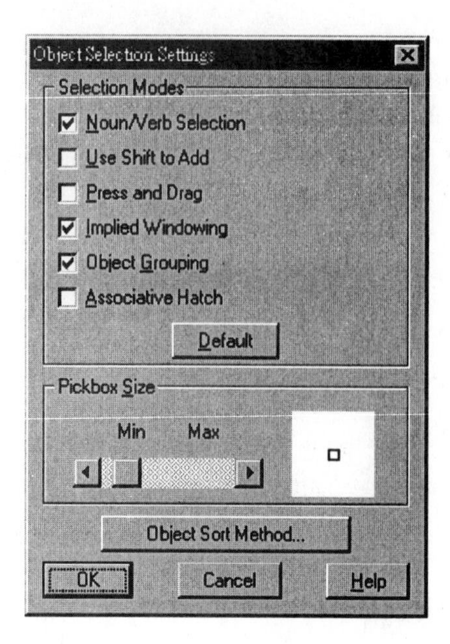

Figure 3.6 Object Selection Settings dialog box

Marker Blips

When you select a point on the screen, a marker blip remains on the screen. This marker serves the purpose of echoing to your selection. To remove the markers, select the Redraw item from the View pull-down menu to apply the REDRAW command.

<View> **<Redraw>**

To disable or enable marker blips, you can also use the DDRMODES command by selecting the Drawing Aids... item from the Tools pull-down menu. Refer to Figure 3.3. In the Modes box of the dialog box, unselect Blips to disable marker blips.

Arcs and Tangential Circle

Refer to Figure 3.4. The upper part of the drawing has three tangential arcs. To draw these arcs, you will first draw two arcs that are concentric to the two upper circles. Then you will draw a circle tangential to the arcs. Finally, you will trim off the excess parts.

You can draw an arc in several ways. Here you will specify the center point, the start point, and the end point. In order to select the center of the upper left center precisely, you have to use an object snap tool — CEN. You can enter CEN at the command line interface or select the Center item from the cursor menu (Figure 1.5). To bring up the cursor menu, hold down the [Shift] key and select the second mouse button. Details of geometric snap tools are explained in Section 3.15.

Two system variables are related to the object snap tools, APERTURE and APBOX. APERTURE sets the aperture size that controls the affected area of the object snap tool. It determines how close your cursor has to move to the entity for the related geometry to be chosen. APBOX constrols the display of the target box at the cross hairs. The target box displays the affected area of the object snap tool. Set the APBOX variable to 1.

Command: **APBOX**
New value for APBOX: **1**

From the Draw pull-down menu, select the Arc cascading menu and then the Center, Start, End item to run the ARC command.

<Draw> **<Arc>** **<Center, Start, End>**

Command: **ARC**
Center/<Start point>: **C**

Specify the geometric object snap CEN and select any part of the upper left circle. Notice the target box at the cross hairs and the object snap symbol at the center of the circle.

Center: **CEN** of [**Select A (Figure 3.5).**]

Set the start point as a relative polar coordinate 25 units in the 0° direction from the center and the second point 25 units in the 180° direction from the center. When inputting a point, you can use polar coordinates or Cartesian coordinates, absolute or relative. See Section 3.16 for details.

Start point: **@25<0**
Angle/Length of chord/<End point>: **@25<180**

Press the [Enter] key to repeat the ARC command to draw a similar arc at the upper right circle. See Figure 3.7.

Command: [**Enter**]
ARC
Center/<Start point>: **C**
Center: **CEN** of [**Select B (Figure 3.5).**]
Start point: **@25<0**
Angle/Length of chord/<End point>: **@25<180**

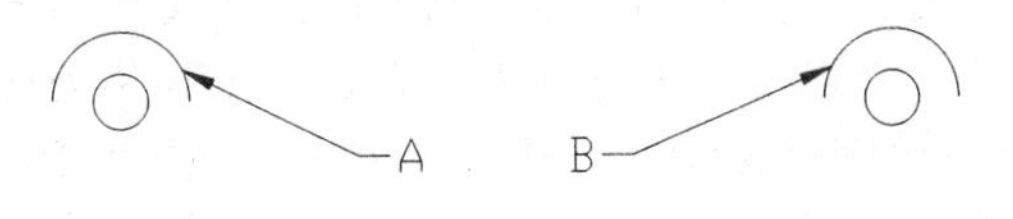

Figure 3.7 Arcs drawn

Turning on the snap mode enables you to move the cursor to a fixed interval quickly and precisely, but it prevents you from selecting any point not at the intervals. Press the

[F9] key or select the Snap item of the status bar to turn off snap mode.

[F9]

Select the Tan, Tan, Radius item of the Circle cascading menu of the Draw pull-down menu to use the CIRCLE command to create a circle of radius 320 units that is tangential to the two arcs. After you have taken the TTR option, select the arcs one by one. Notice that an object snap symbol appears at the cross hairs. The system will locate the tangential point for you automatically. See Figure 3.8.

<Draw> <Circle> <Tan, Tan, Radius>

Command: **CIRCLE**
3P/2P/TTR<Center point>: **TTR**
Enter Tangent spec: **[Select A (Figure 3.7).]**
Enter second Tangent spec: **[Select B (Figure 3.7).]**
Radius: **320**

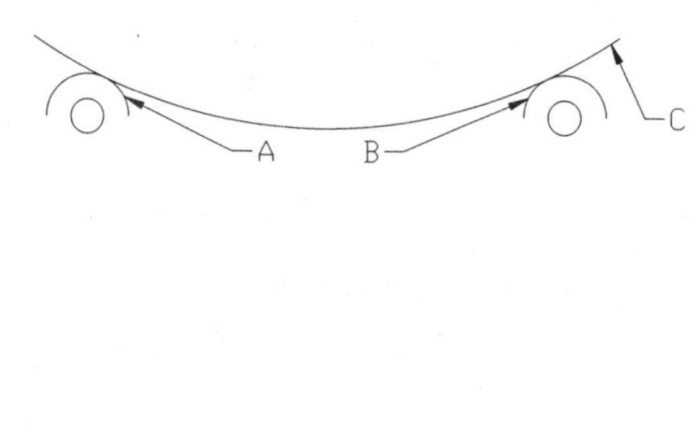

Figure 3.8 Tangential circle drawn

If you are a beginner, you may ask why your screen apparently displays three circular arcs instead of two arcs and one circle. The answer is that the graphic window displays only an area of 420 units x 297 units. Part of the circle that you created is outside the display window. To view the entire drawing, you can select the All item from the Zoom cascading menu of the View pull-down menu to zoom to all.

<View> <Zoom> <All>

If you have zoomed to all, select the Previous item of the Zoom cascading menu to return to the previous screen display.

<View> <Zoom> <Previous>

Trim Entities

Entities that are too long can be trimmed away. From the Modify pull-down menu, select the Trim item to apply the TRIM command to trim the circle and the two arcs at their tangential points. Because the two arcs are the cutting objects for the circle, and the circle is the cutting object for the two arcs, select all of them as cutting edges. After selecting the cutting edges, select the part that you want to remove. If you select the wrong part, use the U option to undo. See Figure 3.9.

<Modify> **<Trim>**

Command: **TRIM**
Select objects: **[Select A, B, and C (Figure 3.8).]**
Select objects: **[Enter]**
<Select object to trim>/Project/Edge/Undo: **[Select A, B, and C (Figure 3.8).]**
<Select object to trim>/Project/Edge/Undo: **[Enter]**

Figure 3.9 Arcs and circle trimmed

Create a Line

From the Draw pull-down menu, select the Line item to apply the LINE command to draw a line from the end point of the left arc to a distance of 200 units in the 270° direction.

Hold down the [Shift] key while selecting the second mouse button to bring up the Cursor menu. Then use the object snap mode END to locate the start point of the line precisely. See Figure 3.10.

<Draw> **<Line>**

Command: **LINE**
From point: **END** of **[Select A (Figure 3.9).]**
To point: **@200<270**
To point: **[Enter]**

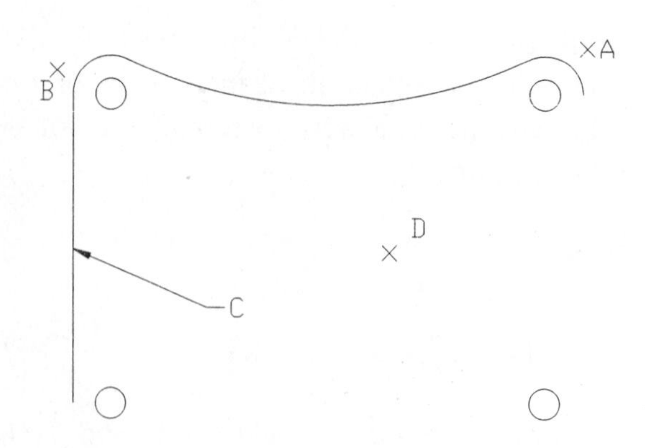

Figure 3.10 Line drawn

Set Ortho Mode

Refer to Figure 3.4. You can find that there is some repetition of entities. Repeated entities can be created by mirroring. Before doing so, issue the ORTHO command to turn on ortho mode. The toggle key for turning on and off this mode is the [F8] key.

Command: **ORTHO**
ON/OFF: **ON**

After you turn on the ortho mode, the second select point will be forced to be orthogonal to the first point.

Mirror Entities

Select the Mirror item of the Modify pull-down menu to apply the MIRROR command to mirror the three arcs. You need to define a mirror line by specifying two points. The first point of the mirror line is the midpoint of the vertical line. The second point can be any point to the right or left of the first point, because you have set ortho mode on, and the mirror line will always be orthogonally horizontal or vertical. See Figure 3.11.

<Modify> **<Mirror>**

Command: **MIRROR**
Select objects: **C**
First corner: **[Select A (Figure 3.10).]**
Other corner: **[Select B (Figure 3.10).]**
Select objects: **[Enter]**
First point of mirror line: **MID** of **[Select C (Figure 3.10).]**
Second point: **[Select D (Figure 3.10).]**
Delete old objects?<N>: **[Enter]**

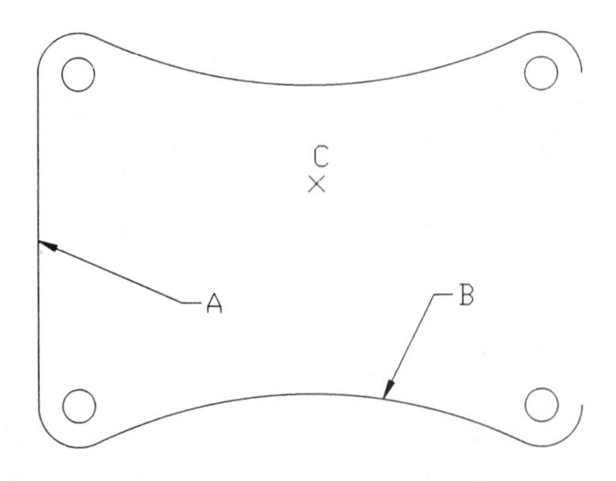

Figure 3.11 Arcs mirrored

Repeat the MIRROR command to mirror the vertical line. Select the center of the R320 arc as the first point, and select any point below or above the first selected point to define a vertical mirror line. See Figure 3.12.

Command: **[Enter]**
MIRROR
Select objects: **[Select A (Figure 3.11).]**
Select objects: **[Enter]**
First point of mirror line: **CEN** of **[Select B (Figure 3.11).]**
Second point: **[Select C (Figure 3.11).]**
Delete old objects?<N>: **[Enter]**

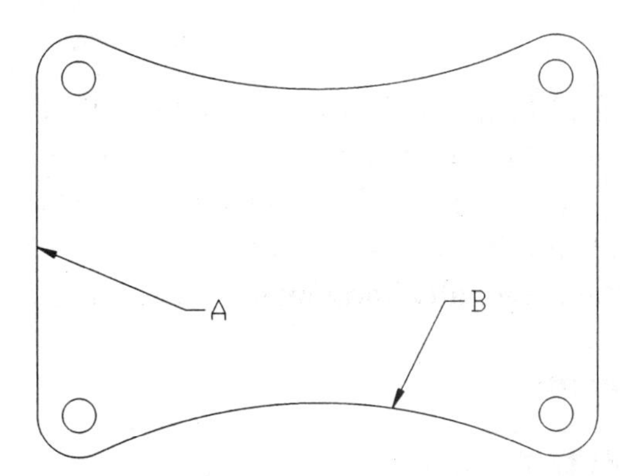

Figure 3.12 Line mirrored

Filter Points

The concept of filtering is to build up a 2D or 3D point from selected X, Y, and Z coordinates of intermediate points. For 2D points, you can use .X, .Y, or .Z. For 3D points, you can use .XY, .XZ, and .YZ. You can key in filters at the command line interface or use the Point filters cascading menu of the cursor menu. In the next paragraph, you will use this technique to compose a center point for drawing an ellipse at the central position of the component.

Create an Ellipse

From the Draw pull-down menu, select the Ellipse cascading menu and then the Center item to issue the ELLIPSE command with the C option to specify the center position.

<Draw> **<Ellipse>** **<Center>**

Command: **ELLIPSE**
Arc/Center/<Axis endpoint1>: **C**

In reply to the request for center position, hold down the [Shift] key, select the second mouse button to bring up the cursor menu, select the Point Filters cascading menu, and then select the .Y item (you can type .Y at the command line). Next, select the MID of the vertical line. The system then asks for the X value. Select the CEN of the R320 arc.

Center of ellipse: **.Y** of **MID** of **[Select A Figure 3.12).]**
(need x): **CEN** of **[Select B (Figure 3.12).]**

The Y value of the midpoint of the vertical line and the X value of the center of the arc build up a point for the center of the ellipse. See Figure 3.13.

Axis endpoint: **@100<0**
<Other axis distance>/Rotation: **50**

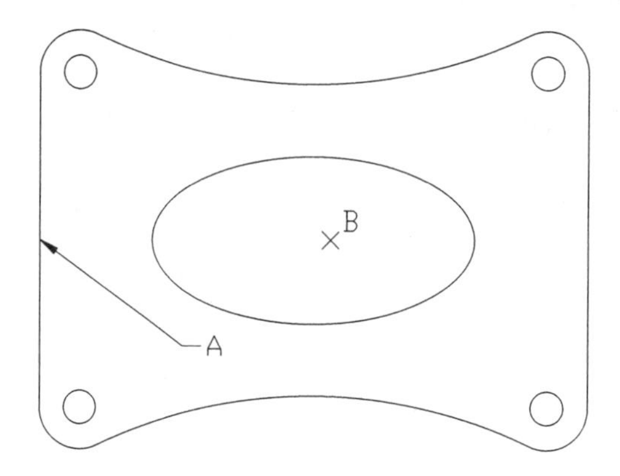

Figure 3.13 Ellipse drawn with the point filtering technique

Tracking

Tracking is a new technique available in AutoCAD Release 14. It combines both X and Y point filters. To illustrate how tracking can be done, erase the ellipse and create it again by tracking. See Figure 3.13.

<Edit> **<Clear>**

Command: **ERASE**
Select objects: **LAST**
Select objects: **[Enter]**

<Draw> **<Ellipse>** **<Center>**

Command: **ELLIPSE**
Arc/Center/<Axis endpoint1>: **C**

In reply to the request for center position, hold down the [Shift] key, select the second mouse button, and select the item Tracking (you can enter TK at the command line). The first tracking point is the center of the R320 arc, and the second tracking point is the midpoint of the left vertical line. Select the [Enter] key when tracking is complete.

Center of ellipse: **TK**
First tracking point: **CEN** of **[Select B (Figure 3.12).]**
Next point (Press ENTER to end tracking): **MID** of **[Select A (Figure 3.12).]**
Next point (Press ENTER to end tracking): **[Enter]**
Axis endpoint: **@100<0**
<Other axis distance>/Rotation: **50**

Direct Distance Entry

Direct distance entry is another new technique. You can specify a point by moving the cursor to indicate a direction and enter the distance from the last point. Use this technique to create a line 370 units distant from the midpoint of the left vertical line. See Figure 3.14.

<Draw> **<Line>**

Command: **LINE**
From point: **MID** of **[Select A (Figure 3.13).]**

Place the cursor on point B of Figure 3.13 to indicate a direction. Because ortho mode is on, the cursor points to a horizontal direction. In reply to the request for the next point, enter 370 to specify a distance.

To point: **370**
To point: **[Enter]**

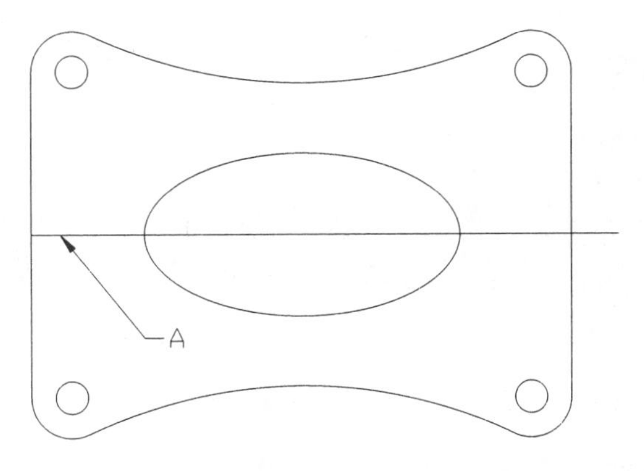

Figure 3.14 Line drawn by direct distance entry

Lengthening

Lines can be lengthened. Select the Lengthen item from the Modify pull-down menu to use the LENGTHEN command to find out the length of the last line and to increase its length. See Figure 3.15.

<Modify> **<Lengthen>**

Command: **LENGTHEN**
DElta/Percent/Total/DYnamic/<Select object>: **[Select A (Figure 3.14).]**
Current length: **370**
DElta/Percent/Total/DYnamic/<Select object>: **DE**
Angle/<Enter delta length>: **30**
<Select object to change>/Undo: **[Select A (Figure 3.14).]**
<Select object to change>/Undo: **[Enter]**

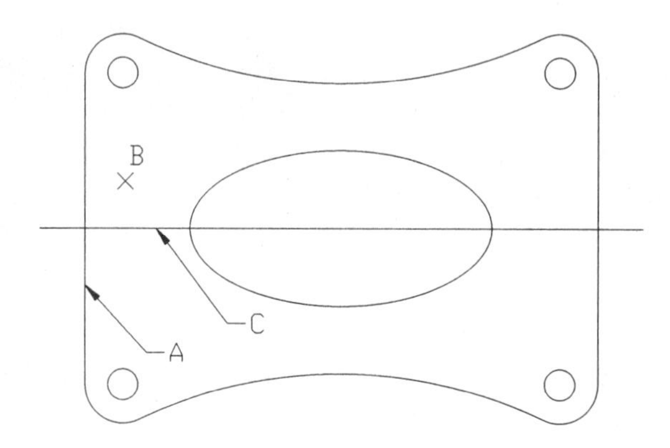

Figure 3.15 Line lengthened

Offset

Offset creates an entity that is similar in shape to the original object. From the Modify pull-down menu, select the Offset item to use the OFFSET command to create an offset line from the left vertical line and the horizontal line. See Figure 3.16. Note that if you offset a circle, you will have a concentric circle.

<Modify> **<Offset>**

Command: **OFFSET**
Offset distance or Through: **25**
Select object to offset: **[Select A (Figure 3.15).]**
Side to offset? **[Select B (Figure 3.15).]**
Select object to offset: **[Enter]**

Command: **[Enter]**
OFFSET
Offset distance or Through: **10**
Select object to offset: **[Select C (Figure 3.15).]**
Side to offset? **[Select B (Figure 3.15).]**
Select object to offset: **[Enter]**

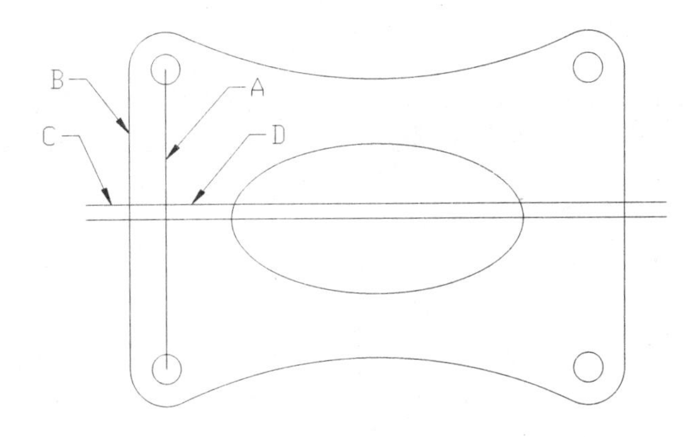

Figure 3.16 Lines offset

Further Work

Apply the TRIM command to trim away the two ends of the newly offset horizontal line. See Figure 3.17.

<Modify> **<Trim>**

Command: **TRIM**
Select objects: [**Select A and B (Figure 3.16).**]
Select objects: [**Enter**]
<Select object to trim>/Project/Edge/Undo: [**Select C and D (Figure 3.16).**]
<Select object to trim>/Project/Edge/Undo: [**Enter**]

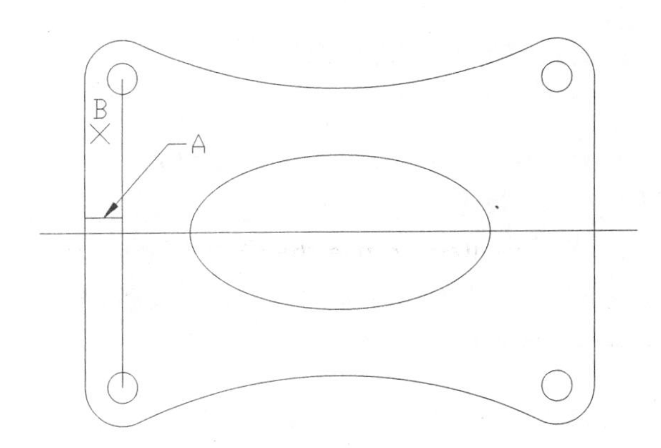

Figure 3.17 Horizontal offset line trimmed

Use the OFFSET command to offset the trimmed horizontal line twice. See Figure 3.18.

<Modify> **<Offset>**

Command: **OFFSET**
Offset distance or Through: **15**
Select object to offset: [**Select A (Figure 3.17).**]
Side to offset? [**Select B (Figure 3.17).**]
Select object to offset: [**Enter**]

Command: **[Enter]**
OFFSET
Offset distance or Through: **65**
Select object to offset: **[Select A (Figure 3.17).]**
Side to offset? **[Select B (Figure 3.17).]**
Select object to offset: **[Enter]**

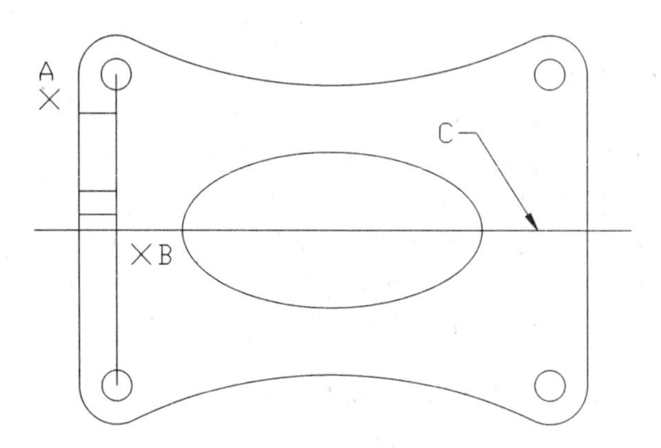

Figure 3.18 Two more offset lines

Apply the MIRROR command to mirror the three offset lines about a horizontal mirror line. Make sure that ortho mode is still on. See Figure 3.19.

<Modify> **<Mirror>**

Command: **MIRROR**
Select objects: **[Select A (Figure 3.18).]**
Other corner: **[Select B (Figure 3.18).]**
Select objects: **[Enter]**
First point of mirror line: **END** of **[Select C (Figure 3.18).]**
Second point: **[Select B (Figure 3.18).]**
Delete old objects?<N>: **[Enter]**

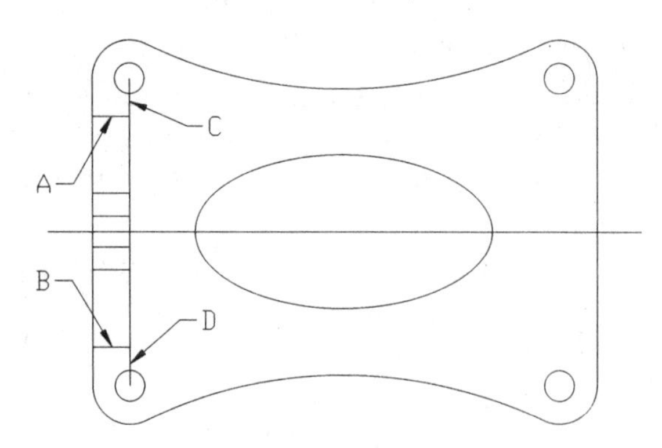

Figure 3.19 Lines mirrored

Use the Trim command to trim away both ends of the offset vertical line. See Figure 3.20.

<Modify> **<Trim>**

Command: **TRIM**
Select objects: [**Select A and B (Figure 3.19).**]
Select objects: [**Enter**]
<Select object to trim>/Project/Edge/Undo: [**Select C and D (Figure 3.19).**]
<Select object to trim>/Project/Edge/Undo: [**Enter**]

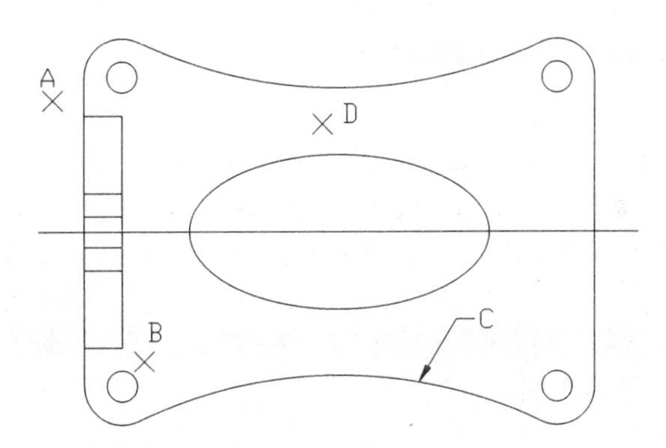

Figure 3.20 Line trimmed

Both sides of the drawing are symmetrical. Run the MIRROR command to mirror a group of entities to the right side of the component. See Figure 3.21.

<Modify> **<Mirror>**

Command: **MIRROR**
Select objects: [**Select A (Figure 3.20).**]
Other corner: [**Select B (Figure 3.20).**]
Select objects: [**Enter**]
First point of mirror line: **CEN** of [**Select C (Figure 3.20).**]
Second point: [**Select D (Figure 3.20).**]
Delete old objects?<N>: [**Enter**]

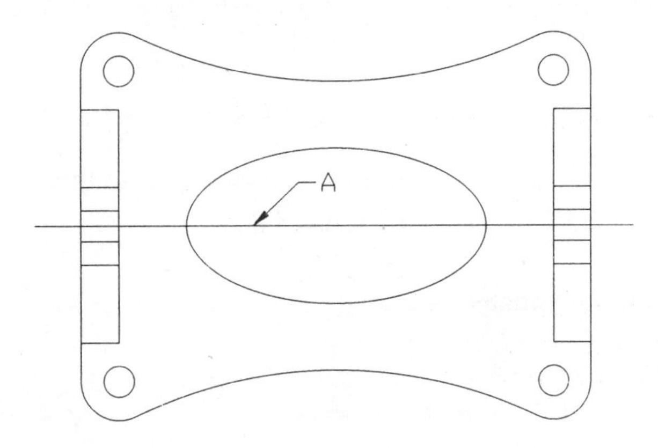

Figure 3.21 Lines mirrored

Change Entity Properties

Entities possess a number of properties. See Chapter 2. The central horizontal line should be a center line instead of a continuous line. Use the DDMODIFY command to change its property from layer 0 to layer CENTER. See Figure 3.22.

<Modify> **<Properties...>**

Select objects: **[Select A (Figure 3.21).]**
Select objects: **[Enter]**

[Properties
Layer... **Center**
OK]

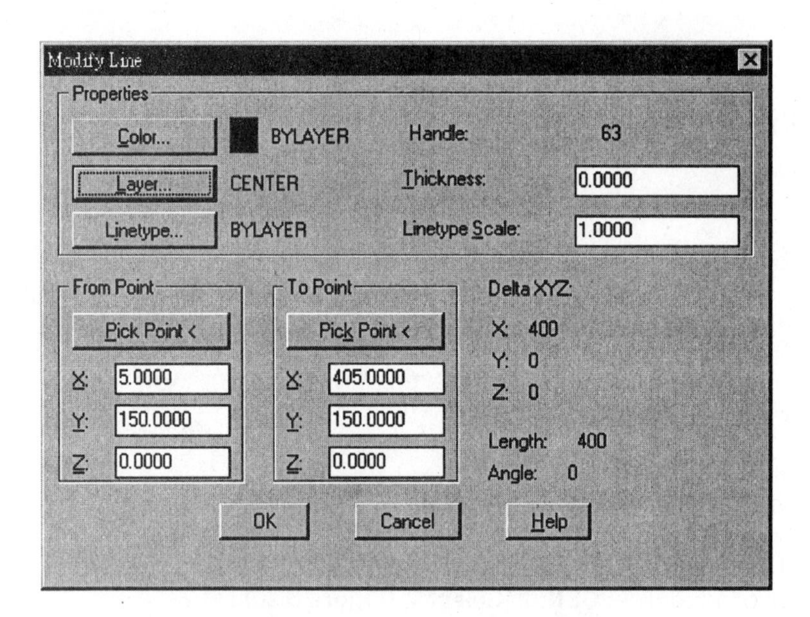

Figure 3.22 Modify Line dialog box

Now the line resides on layer Center instead of layer 0.

Set Linetype Scale

The proportion of the center line may not look very good. Use the LINETYPE command to set the global linetype scale to 0.8. See Figure 3.23.

<Format> **<Linetypes>**

[Linetype
Details>>
Global scale factor: **0.8**
OK]

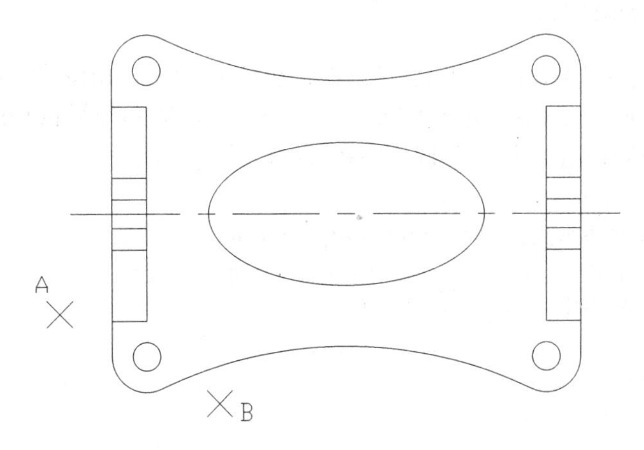

Figure 3.23 Linetype scale changed

Zoom into the Lower Left Corner

Use the ZOOM command to zoom into the lower left circle. See Figure 3.24.

<View> **<Zoom>** **<Window>**

Command: **ZOOM**
All/Center/Dynamic/Extents/Previous/Scale(X/XP)/Window/<Realtime>: **W**
First corner: [**Select A (Figure 3.23).**]
Other corner: [**Select B (Figure 3.23).**]

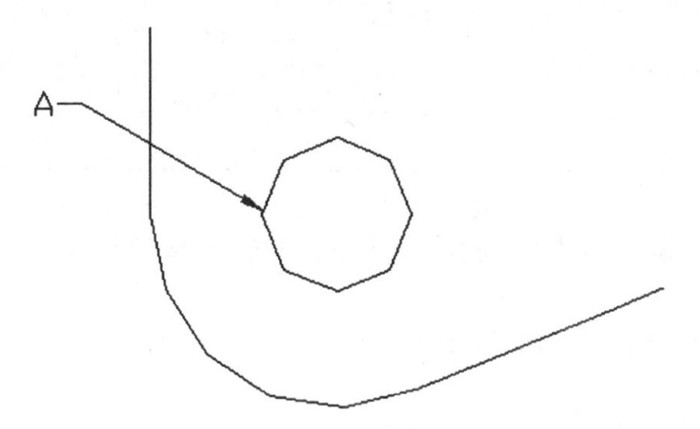

Figure 3.24 Circle and arc represented by small vectors

You can find from your screen that both the circle and the arc appear as polygons rather than smooth curves. Do not worry; they are still circle and arc. In fact, AutoCAD represents circles and arcs on screen with many short vectors in order to save time. Normally, there are more vectors for larger circles and fewer vectors for smaller circles.

By default, small circles that are zoomed to larger circles on the screen are still represented by the same number of vectors. As a result, the circle and arc appear as polygons. If you wish to have smooth circles and arcs, you can use the REGEN command.

<View> **<Regen>**

Redraw and Regeneration

The REDRAW command and the REGEN command are very similar. The REGEN command regenerates the entire drawing by recomputing the screen coordinates from the drawing database. The REDRAW command simply refreshes the current screen and removes any marker blips.

Regeneration takes longer than redrawing. Some commands call for a regeneration automatically. If you do not want automatic regeneration, issue the REGENAUTO command to suppress such regeneration.

Command: **REGENAUTO**
ON/OFF: **OFF**

Set REGENAUTO on again.

Command: **REGENAUTO**
ON/OFF: **ON**

Regeneration Speed

Another way to display smooth circles and arcs all the time is to turn off fast zooming mode. Issue the VIEWRES command.

Command: **VIEWRES**
Do you want fast zooms? **N**
Enter circle zoom percent (1-20000): **100**

In the foregoing command prompts, you can regard the zoom percent as an index to control the number of vectors to represent a circle or an arc on the screen. The default value is 100. If this value is smaller than 100, AutoCAD uses fewer vectors, and the regeneration speed is faster. If your computer runs very slowly during regeneration, turn on fast zoom, and set the zoom percent value to a smaller number.

Add Center Lines

From the Object Properties toolbar, select the Layer Control box and the layer Center to set the current layer to Center.

[Object Properties] **[Layer Control]**

Current layer: **Center**

Apply the LINE command to draw a vertical center line and a horizontal center line to pass through the center of the lower left circle. Regarding center lines, you can also use the DIMCENTER command. See Chapter 5. However, you will use the LINE command here to learn how to make use of the From option.

To draw the line, use the From object snap, and make use of the center of the circle as the Base point.

<Draw> **<Line>**

Command: **LINE**

The line should start at a point 20 units in the 90° direction from the center of the circle. To select the start point, hold down the [Shift] key and select the second mouse button to bring up the cursor menu, and then select the From item.

From point: **FROM**
Base point: **CEN** of **[Select A (Figure 3.24).]**
<offset>: **@20<90**
To point: **@40<270**
To point: **[Enter]**

In the foregoing command lines, the start point of the line is at 20 units in the 90° direction offset from the center of the circle. It then runs to 40 units in the 270° direction. Run the LINE command again to draw the horizontal center line.

Command: **[Enter]**
LINE
From point: **FROM**
Base point: **CEN** of **[Select A (Figure 3.24).]**
<offset>: **@20<180**
To point: **@40<0**
To point: **[Enter]**

Apply the ZOOM command with P option to zoom out, and go back to the previous screen. See Figure 3.25.

<View> **<Zoom>** **<Previous>**

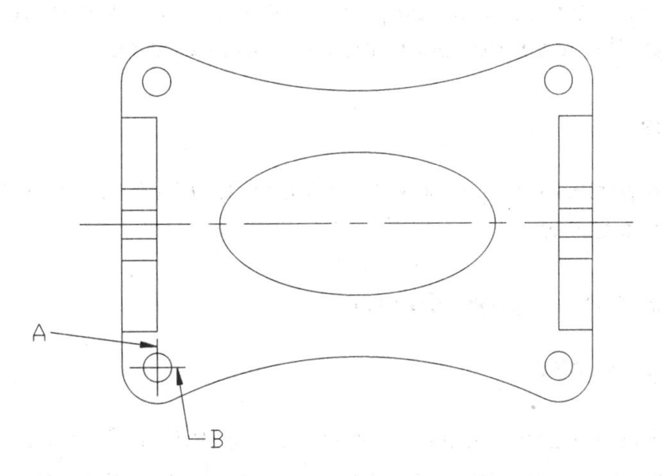

Figure 3.25 Two center lines added

Because the two lines are comparatively short, the chain line feature may not be obvious visually. The lines may appear as continuous lines. If you change the global linetype scale, all the entities are affected. Therefore, you should change only the properties of these two lines. Change the linetype from BYLAYER to Center2, color from BYLAYER to yellow, and relative linetype scale to 0.8. See Figure 3.26.

<Modify> **<Properties...>**

Select objects: **[Select A and B (Figure 3.25).]**
Select objects: **[Enter]**

[Change Properties
Color... **Yellow**
Linetype... **Center2**
Linetype Scale... **0.8**
OK]

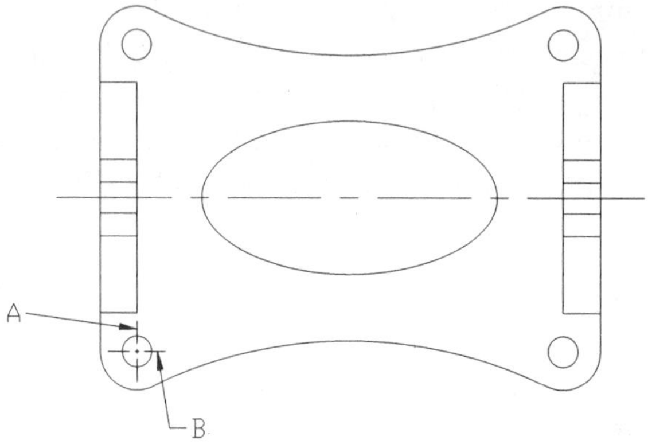

Figure 3.26 Linetype and color of the two center lines changed

The default linetype and color of an entity are BYLAYER. BYLAYER means that the linetype and color are in accordance with the linetype and color assignment of the layer. If the properties of an entity are not BYLAYER, any subsequent change to the color or linetype setting of the layer will have no effect on that particular entity.

Global and Non-global Linetype Scale

In the last command, you changed the linetype scale of two lines to 0.8. The linetype scale of the lines then becomes 0.8 times that of the global linetype scale.

Complete the Top View

Use the ARRAY command to array the two center lines. See Figure 3.27.

<Modify> **<Array>**

Command: **ARRAY**
Select objects: **P**
Select objects: **[Enter]**
Rectangular or Polar array (R/P): **R**
Number of rows (---): **2**
Number of columns (|||): **2**

Unit cell or distance between rows (---): **200**
Distance between columns (|||): **290**

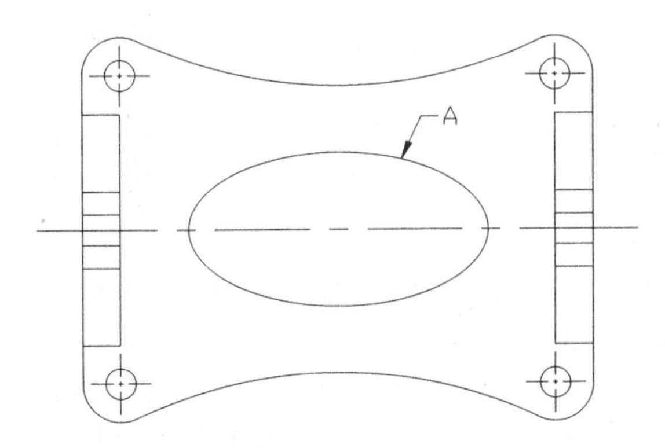

Figure 3.27 Center lines arrayed

To complete the top view, issue the LINE command to add a vertical center line to the ellipse. Apply the FROM object snap. Then use the CEN of the ellipse as the Base point. See Figure 3.28.

<Draw> **<Line>**

Command: **LINE**
From point: **FROM**
Base point: **CEN** of [**Select A (Figure 3.27).**]
<offset>: **@70<90**
To point: **@140<270**
To point: [**Enter**]

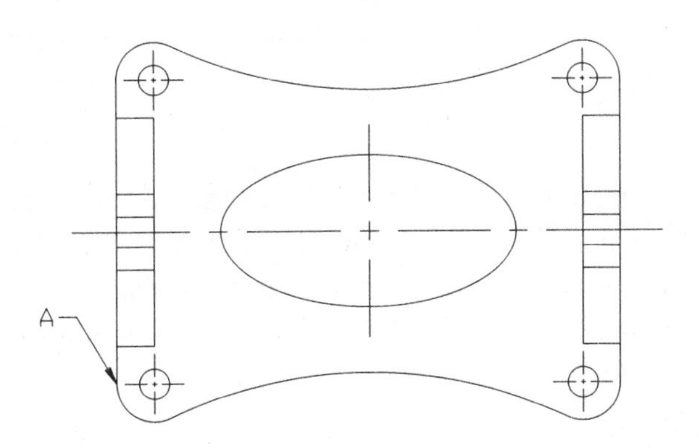

Figure 3.28 Complete top view of the first component

The top view is complete. In making this drawing view, you created circles, arcs, lines, and an ellipse. You edited by arraying, lengthening, mirroring, offsetting, and changing properties of entities. You also learned dragmode, pickbox size, aperture size, blipmode, direct distance entry, REDRAW, REGEN, REGENAUTO, geometric object snap, coordinate input, trim, ortho mode, filter points, tracking, linetype scale, VIEWRES, and global and local linetype scales.

3.3 Sectional Front View of the First Component

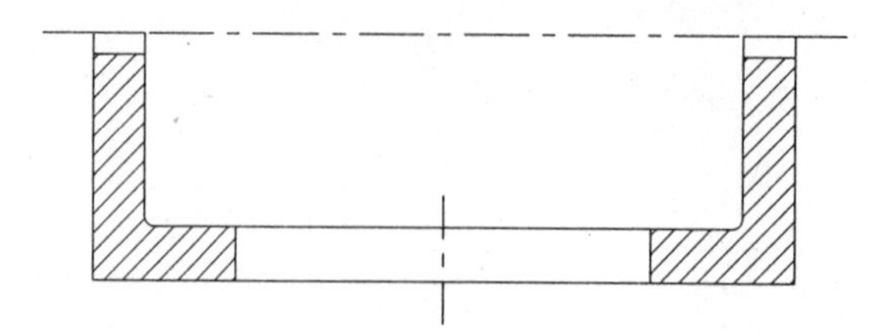

Figure 3.29 Sectional front view of the first component

Figure 3.29 shows the sectional front view of the first component. You will create outlines on layer 0, center lines on layer Center, and hatching lines on layer Hatch.

In an orthographic projection, the front view has to align with the top view. Apply the LINE command to draw two lines. Make use of the FROM object snap mode to locate the starting point of the first line. Then use the lower end point of the left vertical line of the top view as the Base point.

<Draw> **<Line>**

Command: **LINE**
From point: **FROM**
Base point: **END** of **[Select A (Figure 3.28).]**
<offset>: **@130<270**
To point: **@115<270**
To point: **@340<0**
To point: **[Enter]**

Although you have drawn two lines, you may not see them because they are probably outside the current view window. Apply the ZOOM command to zoom to All.

<View> **<Zoom>** **<All>**

In order to make the entities on the screen more readily selectable, repeat the ZOOM command to zoom to a scale of 0.9X. A zoom scale suffixed by the letter X means that the zoom scale is relative to the previous screen size. See Figure 3.30.

<View> **<Zoom>** **<Scale>**

Command: **ZOOM**
All/Center/Dynamic/Extents/Previous/Scale(X/XP)/Window/<Realtime>: **S**
Enter scale factor: **0.9X**

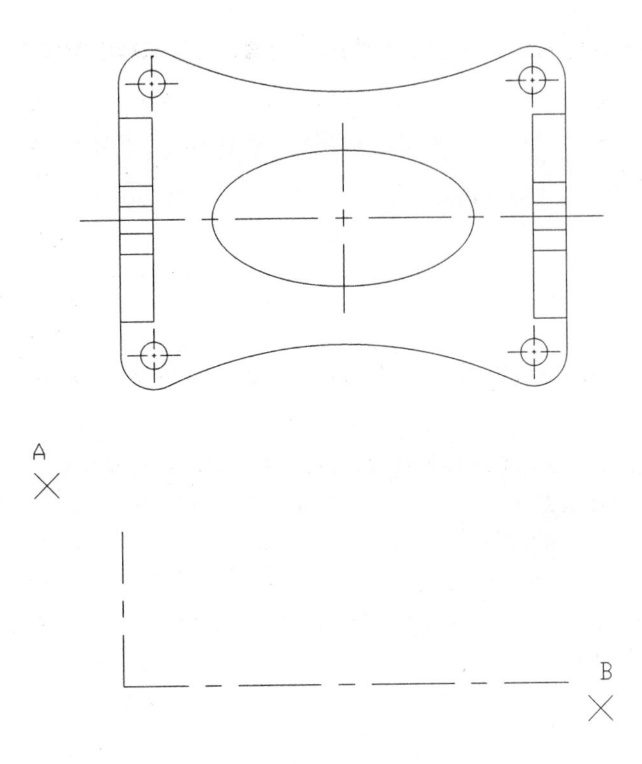

Figure 3.30 Zoom to ALL and then to a scale of 0.9X

Because you will be working on the front view, use the ZOOM command to zoom to the front view. See Figure 3.35.

<View> **<Zoom>** **<Window>**

Command: **ZOOM**
All/Center/Dynamic/Extents/Previous/Scale(X/XP)/Window/<Realtime>: **W**
First corner: [**Select A (Figure 3.30).**]
Other corner: [**Select B (Figure 3.30).**]

Save a View

Before proceeding to the sectional front view, you will practice dynamic zoom and display an aerial view. In order to restore the current view after the practice, you will use the VIEW command or the DDVIEW command to save the setting of the current view. The VIEW command works at the command line interface. The DDVIEW command brings up dialog boxes. Select the Named Views... item from the View pull-down menu to use the DDVIEW command to save the current display view. See Figure 3.31.

<View> **<Named Views...>**

Command: **DDVIEW**

[View Control **New...**]
[Define New View
New Name: **VIEW1**
Current Display

Save View]
[OK]

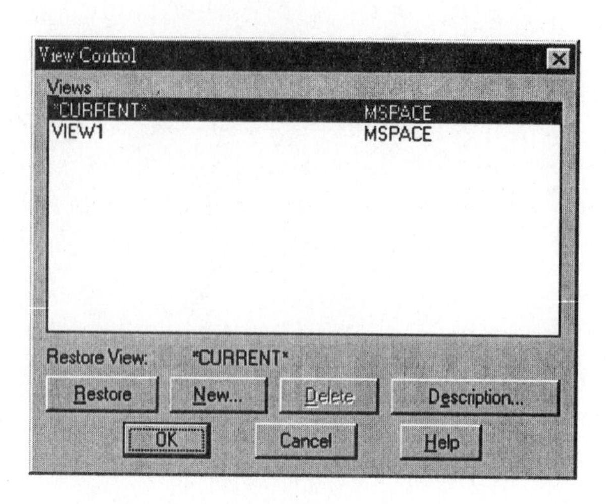

Figure 3.31 View Control dialog box

Dynamic Zoom

You have saved the current display view. Now you will work on dynamic zooming. Later, you will restore the saved view and continue to work on the front view.

Set VIEWRES to Yes. Then select the Dynamic item from the Zoom cascading menu of the View pull-down menu to apply dynamic zoom. See Figure 3.32.

Command: **VIEWRES**
Do you want fast zooms? **Y**
Enter circle zoom percent (1-20000): **100**

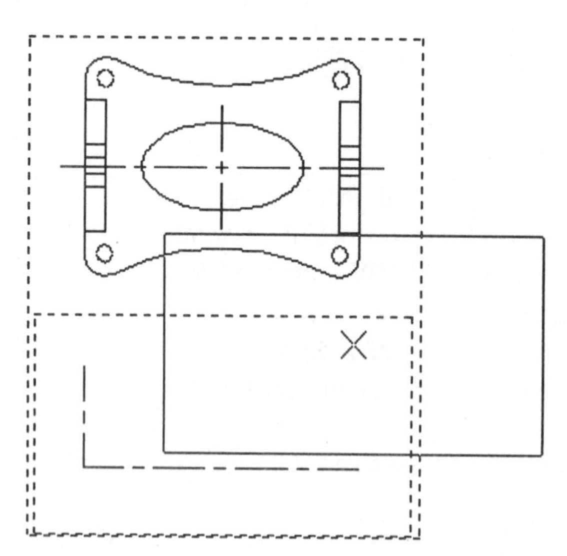

Figure 3.32 Dynamic zoom with panning view box

Now you will find three boxes in your screen display:

A dashed blue box	Shows the drawing extent. The area shown will be the drawing limit or the actual area, whichever is larger.
A dashed red box	Shows the area covered by the current view.
A panning view box	Features a small cross at the middle. This is the view area of what you saw before you executed the ZOOM command. You may move it to pan to the required position and then press the [Enter] key. If you press the first mouse button, this box will be replaced by the zooming view box. See Figure 3.33.

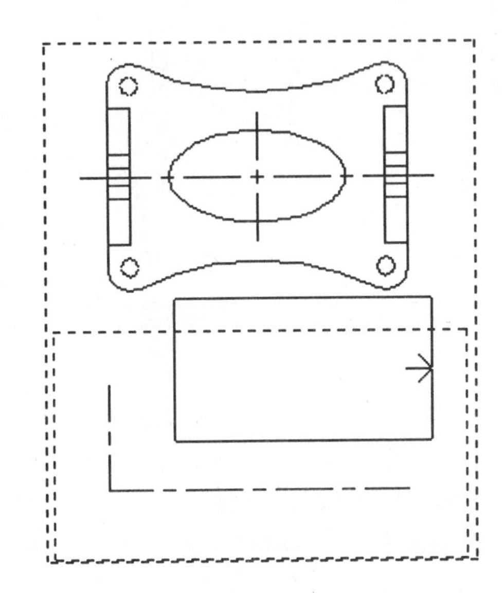

Figure 3.33 Dynamic zoom with zooming view box

Figure 3.33 shows a zooming view box. It features a small arrow on the right side. You may move the pointing device to increase or decrease this view box to zoom out and zoom in the drawing.

Realtime Zoom

Realtime zooming is the default zoom option. From the View pull-down menu, select the Zoom cascading menu and then the Realtime item.

<View> **<Zoom>** **<Realtime>**

Hold down the left button of your mouse and drag the cursor along the screen. The view will zoom in or out accordingly. While you are performing realtime zooming, you may select the second mouse button to bring up a menu. There are six items in the menu: Exit, Pan, Zoom, Zoom Window, Zoom Previous, and Zoom Extents. Try each of them to appreciate the effect.

Aerial view

After practicing how to zoom dynamically and realtime zooming, select the Aerial View item from the View pull-down menu to display an aerial view dialog box. An aerial view is a bird's-eye view of the entire drawing. See Figure 3.34.

<View> **<Aerial View>**

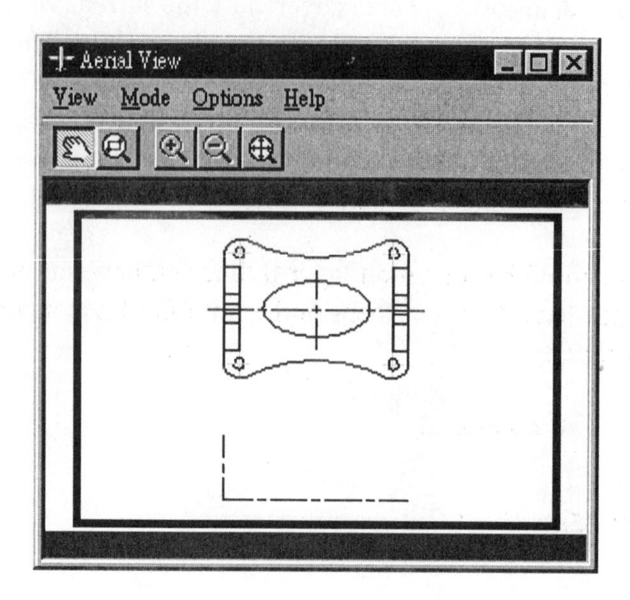

Figure 3.34 Aerial View dialog box

When you work on a large drawing, the aerial view window is a very useful navigation tool. Because it displays the entire drawing in a window, you can quickly locate and zoom to any specific area of the drawing. To close the aerial window, select the [X] button of the Aerial View dialog box.

Zoom and Edit

After playing around with various zoom and display tools, use the DDVIEW command to restore the saved view. See Figure 3.35.

<View> **<Named Views...>**

Command: **DDVIEW**

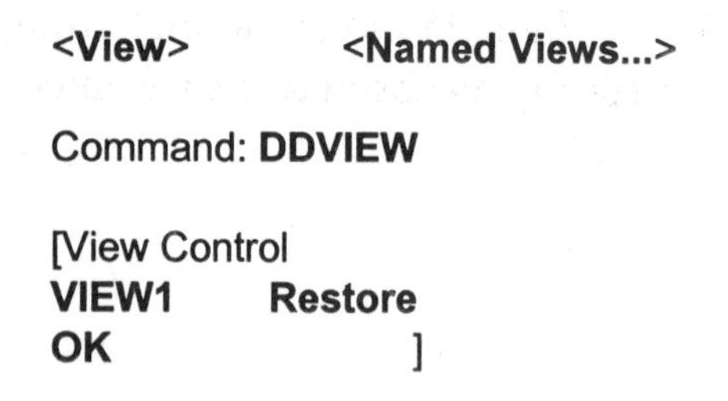
[View Control
VIEW1 **Restore**
OK]

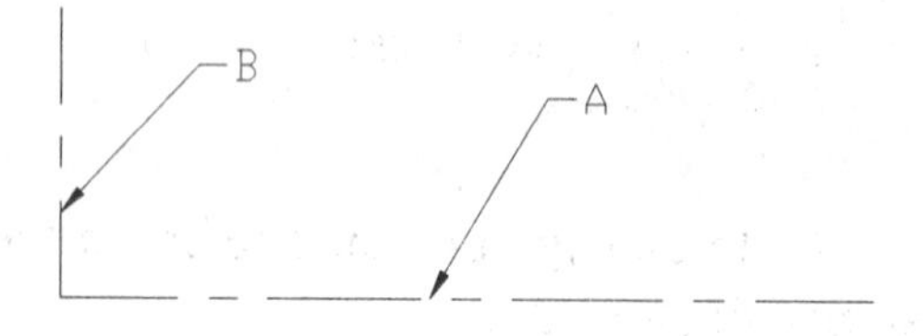

Figure 3.35 Restored view

The positions of entities can be moved. Use the MOVE command to move the last two lines 20 units in the 90° direction.

<Modify> **<Move>**

Command: **MOVE**
Select objects: **[Select A and B (Figure 3.35).]**
Select objects: **[Enter]**
Base point or displacement: **20<90**
Second point of displacement: **[Enter]**

These two lines should be drawn on layer 0, but you have put them on layer Center. However, you do not have to erase them and create the lines again. Instead, you may change their layer property.

<Modify> **<Properties...>**

Select objects: **P**
Select objects: **[Enter]**

[Properties
Layer... **0**
OK]

The P (Previous) object in the foregoing command refers to the horizontal line and the vertical line because they are selected in the previous command.

To proceed, set the current layer to 0.

[Object Properties] **[Layer Control]**

Current layer: **0**

Add More Lines

Use the COPY command to copy the horizontal line. Enter 25<90 in the first prompt and press [Enter] to the prompt for the second point to specify an absolute displacement of 25 units in the 90° direction.

<Modify> **<Copy>**

Command: **COPY**
Select objects: **L**
Select objects: **[Enter]**
<Base point or displacement>/Multiple: **25<90**
Second point of displacement: **[Enter]**

The L (Last) object in the foregoing command refers to the horizontal line because this is the line created last. See Figure 3.36.

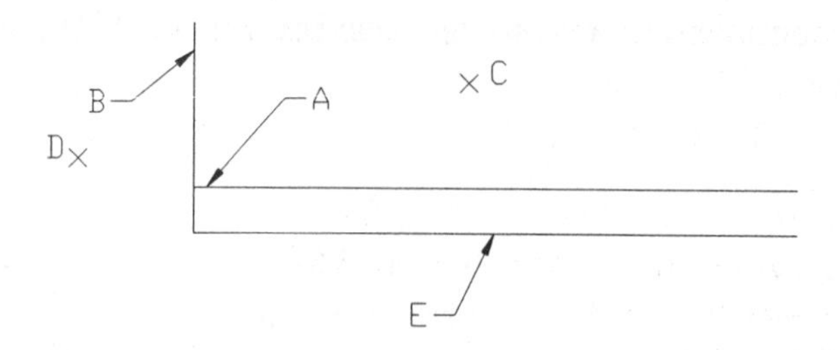

Figure 3.36 Line copied

Repeat the COPY command to copy another line.

Command: **[Enter]**
COPY
Select objects: **L [Or select A (Figure 3.36).]**
Select objects: **[Enter]**
<Base point or displacement>/Multiple: **END** of **[Select A (Figure 3.36).]**
Second point of displacement: **END** of **[Select B (Figure 3.36).]**

Apply the OFFSET command to offset the left vertical line. Then use the MIRROR command to mirror two vertical lines. See Figure 3.37.

<Modify> **<Offset>**

Command: **OFFSET**
Offset distance or Through: **25**
Select object to offset: **[Select B (Figure 3.36).]**
Side to offset? **[Select C (Figure 3.36).]**
Select object to offset: **[Enter]**

<Modify> **<Mirror>**

Command: **MIRROR**
Select objects: **C**
First corner: **[Select C (Figure 3.36).]**
Other corner: **[Select D (Figure 3.36).]**
Select objects: **[Enter]**
First point of mirror line: **MID** of **[Select E (Figure 3.36).]**
Second point: **[Select C (Figure 3.36).]**
Delete old objects?<N>: **[Enter]**

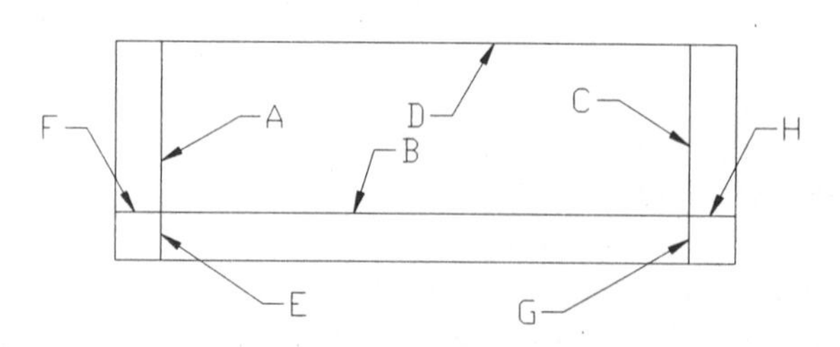

Figure 3.37 Lines offset and mirrored

Issue the TRIM command to trim some lines. See Figure 3.38.

<Modify> <Trim>

Command: **TRIM**
Select objects: **[Select A, B, and C (Figure 3.37).]**
Select objects: **[Enter]**
<Select object to trim>/Project/Edge/Undo: **[Select D, E, F, G, and H (Figure 3.37).]**
<Select object to trim>/Project/Edge/Undo: **[Enter]**

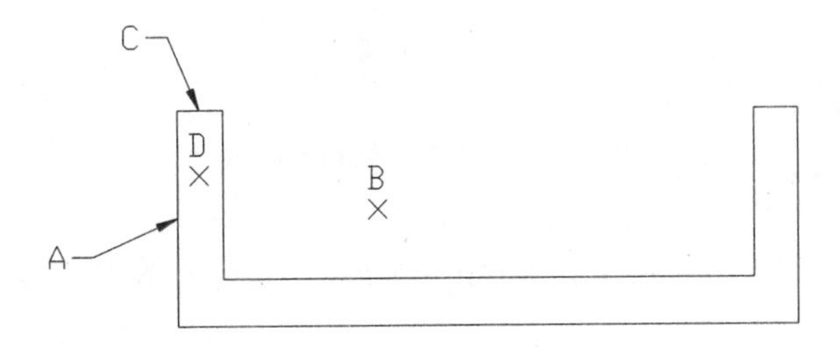

Figure 3.38 Lines trimmed

Use the OFFSET command to create two offset lines, one from the upper left horizontal line and the other from the left vertical line. See Figure 3.39.

<Modify> **<Offset>**

Command: **OFFSET**
Offset distance or Through: **70**
Select object to offset: **[Select A (Figure 3.38).]**
Side to offset? **[Select B (Figure 3.38).]**
Select object to offset: **[Enter]**

Command: **[Enter]**
OFFSET
Offset distance or Through: **10**
Select object to offset: **[Select C (Figure 3.38).]**
Side to offset? **[Select D (Figure 3.38).]**
Select object to offset: **[Enter]**

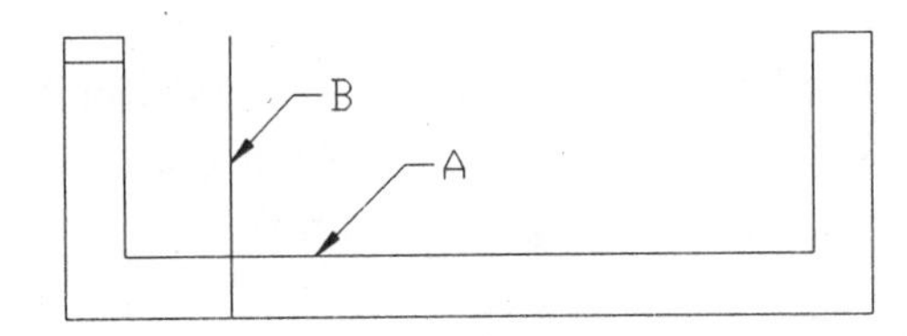

Figure 3.39 Offset lines created

Apply the TRIM command to trim the newly offset vertical line. Keep the lower portion. See Figure 3.40.

<Modify> <Trim>

Command: **TRIM**
Select objects: **[Select A (Figure 3.39).]**
Select objects: **[Enter]**
<Select object to trim>/Project/Edge/Undo: **[Select B (Figure 3.39).]**
<Select object to trim>/Project/Edge/Undo: **[Enter]**

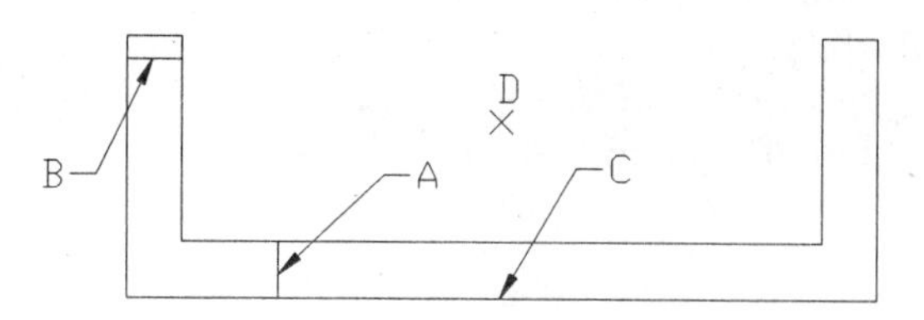

Figure 3.40 Offset line trimmed

The two sides of the drawing are symmetrical. Run the MIRROR command to mirror two lines. See Figure 3.41.

<Modify> **<Mirror>**

Command: **MIRROR**
Select objects: **[Select A (Figure 3.40).]**
Select objects: **[Select B (Figure 3.40).]**
Select objects: **[Enter]**
First point of mirror line: **MID** of **[Select C (Figure 3.40).]**
Second point: **[Select D (Figure 3.40).]**
Delete old objects?<N>: **[Enter]**

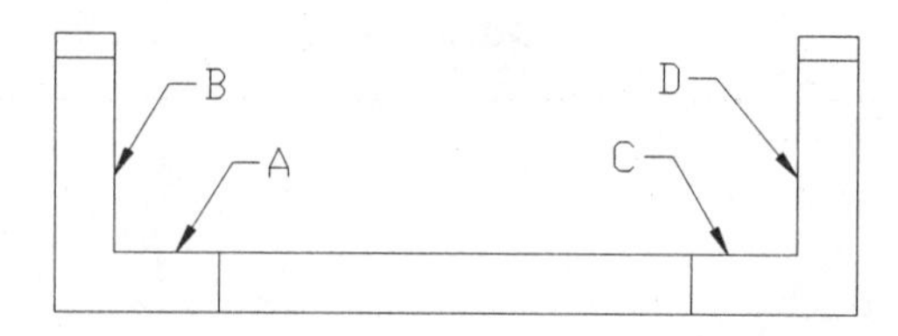

Figure 3.41 Two offset lines mirrored

Create Rounded Corners (Fillets)

You need to round off two corners. Rounding off corners takes two steps, setting a fillet radius and then filleting. From the Modify pull-down menu, select the Fillet item to run the FILLET command to set the fillet radius to 5 units.

<Modify> **<Fillet>**

Command: **FILLET**
Polyline/Radius/Trim/<Select first object>: **RADIUS**
Enter fillet radius: **5**

Repeat the FILLET command twice to create two rounded corners. See Figure 3.42.

Command: **[Enter]**
FILLET
Polyline/Radius/Trim/<Select first object>: **[Select A (Figure 3.41).]**
Select second object: **[Select B (Figure 3.41).]**

Command: **[Enter]**
FILLET
Polyline/Radius/Trim/<Select first object>: **[Select C (Figure 3.41).]**
Select second object: **[Select D (Figure 3.41).]**

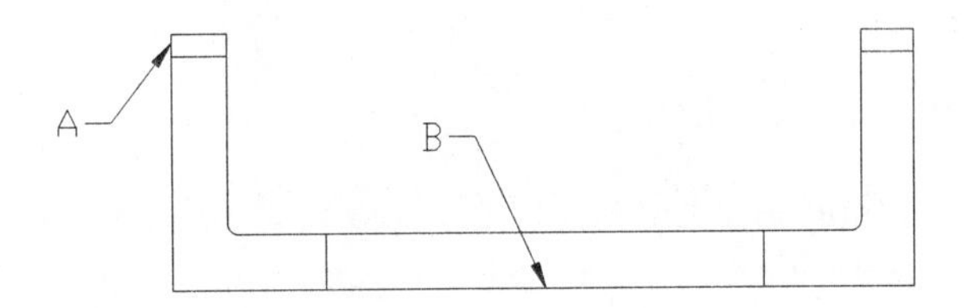

Figure 3.42 Two corners filleted

Add Two Center Lines

You need to add two center lines to the sectional front view. From the Layer Control box of the Object Properties toolbar, select the layer Center to set it as the current layer.

[Object Properties] **[Layer Control]**

Current layer: **Center**

On the layer Center, apply the LINE command with the FROM object snap to create two center lines. The first line uses the upper left corner as the Base point. The second line uses the midpoint of the lower horizontal line as the Base point. See Figure 3.43.

<Draw> **<Line>**

Command: **LINE**
From point: **FROM**
Base point: **END** of **[Select A (Figure 3.42).]**
<offset>: **@25<180**
To point: **@390<0**
To point: **[Enter]**

Command: **[Enter]**
LINE
From point: **FROM**
Base point: **MID** of **[Select B (Figure 3.42).]**
<offset>: **@40<90**
To point: **@60<270**
To point: **[Enter]**

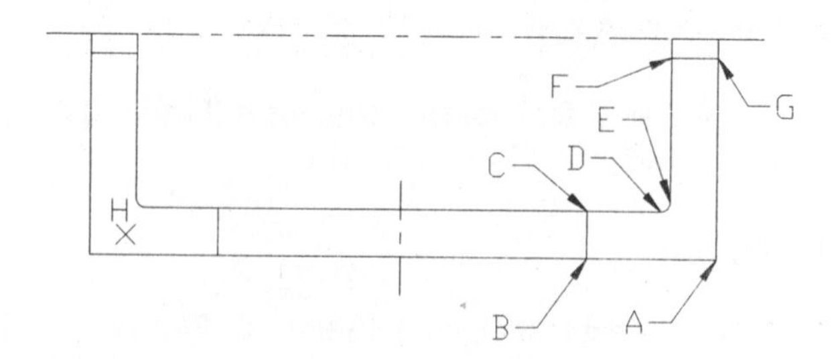

Figure 3.43 Two center lines created

Create Boundaries for Hatching

To create hatching, you may use the HATCH command or the BHATCH command. You will use the HATCH command for the first component and the BHATCH command for the second component. To use the HATCH command, you have to create closed boundaries. Here you will learn two ways of making boundaries for hatching. First, you will use the PLINE command. Second, you will use the BOUNDARY command.

Because the hatching is going to be placed on layer Hatch, set this layer as the current layer.

[Object Properties] **[Layer Control]**

Current layer: **Hatch**

Ortho mode is not required in making the hatch boundaries. Turn off ortho mode by pressing the [F8] key or selecting the ORTHO item from the status bar.

[F8]

The default object snap mode is none. You have to supply the object snap mode during each selecting. In using the PLINE command, you will use the intersection object snap repeatedly, it is more convenient to set it as the default snap mode, running object snap.

From the Tools pull-down menu, select the Object Snap Settings... item to use the OSNAP command to set the default object snap mode to intersection. See Figure 3.44.

<Tools> **<Object Snap Settings...>**

Command: **OSNAP**

[Osnap Settings
Running Osnap
Select settings
Intersection
OK]

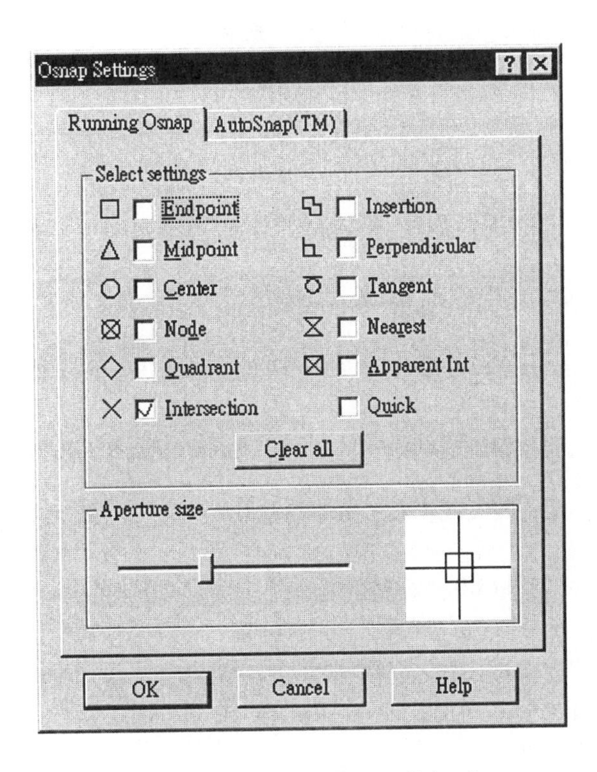

Figure 3.44 Osnap Settings dialog box

Refer to Figure 3.44. Select the slider bar in the Aperture size box to adjust the aperture size. The aperture size is measured in absolute screen pixel size. It affects the selection target box size. Set the aperture size (target box size) to 8 pixels and enable the display of the aperture target box.

Command: **APERTURE**
Object snap target height (1-50 pixels): **8**

Command: **APBOX**
New value for APBOX: **1**

A polyline is a set of line and arc segments that are connected together. They are treated as a single object. From the Draw pull-down menu, select the Polyline item to apply the PLINE command to create a polyline. Because the default object snap mode is set to intersection, the target box appears automatically for each prompt in the following command. You simply have to select a point near the specified intersection point.

<Draw> **<Polyline>**

Command: **PLINE**
From point: **[Select A (Figure 3.43).]**
Arc/Close/Halfwidth/Length/Undo/Width/<Endpoint of line>: **[Select B (Figure 3.43).]**
Arc/Close/Halfwidth/Length/Undo/Width/<Endpoint of line>: **[Select C (Figure 3.43).]**
Arc/Close/Halfwidth/Length/Undo/Width/<Endpoint of line>: **[Select D (Figure 3.43).]**

After creating three line segments, key in A to draw the next arc.

Arc/Close/Halfwidth/Length/Undo/Width/<Endpoint of line>: **A**

Angle/CEnter/CLose/Direction/Halfwidth/Line/Radius/Second pt/Undo/Width/
<Endpoint of arc>: **[Select E (Figure 3.43).]**

To create the next three line segments, key in L to return to line mode.

Angle/CEnter/CLose/Direction/Halfwidth/Line/Radius/Second pt/Undo/Width/
<Endpoint of arc>: **L**
Arc/Close/Halfwidth/Length/Undo/Width/<Endpoint of line>: **[Select F (Figure 3.43).]**
Arc/Close/Halfwidth/Length/Undo/Width/<Endpoint of line>: **[Select G (Figure 3.43).]**
Arc/Close/Halfwidth/Length/Undo/Width/<Endpoint of line>: **[Select A (Figure 3.43).]**
Arc/Close/Halfwidth/Length/Undo/Width/<Endpoint of line>: **[Enter]**

After completing the PLINE command, use the OSNAP command to clear all the running snap mode.

<Tools> **<Object Snap Settings...>**

Command: **OSNAP**

[Osnap Settings
Running Osnap
Select settings
Clear all
OK]

Because the polyline overlaps with the other line segments, it may be hard to find on the display. To change the display order to bring it to the front, select the Bring to Front item from the Display Order cascading menu of the Tools pull-down menu.

<Tools> **<Display Order>** **<Bring to Front>**

Command: **DRAWORDER**
Select objects: **LAST**
Select objects: **[Enter]**
Above object/Under object/Front/<Back>: **FRONT**

Boundary Line

The second way to create a closed polyline from a closed boundary is to use the BOUNDARY command. See Figure 3.45.

<Draw> **<Boundary...>**

Command: **BOUNDARY**

[Boundary Creation
Define Boundary Set
From Everything on Screen
Island Detection
Pick Points<]

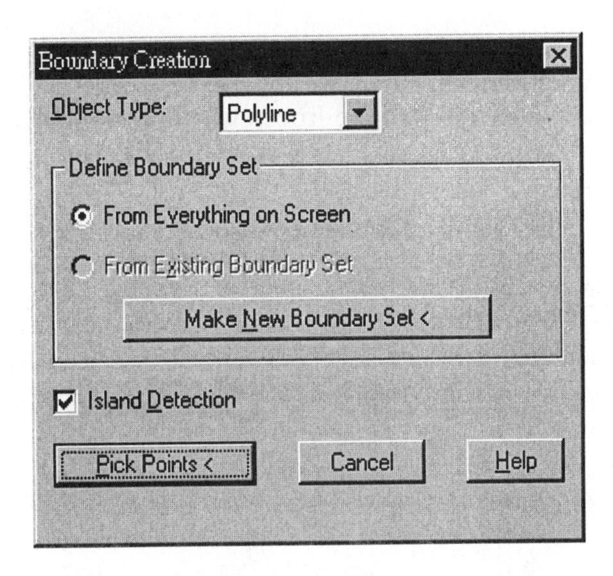

Figure 3.45 Boundary Creation dialog box

Select internal point: [**Select H (Figure 3.43).**]
Select internal point: [**Enter**]
BOUNDARY created 1 polyline

In order for you to see clearly the polylines created by the PLINE command and the BOUNDARY command, freeze all layers except the current layer Hatch. See Figure 3.46.

[**Object Properties**] [**Layer Control**]

Freeze	0
Freeze	Center
On	Hatch
Freeze	Hidden

Current layer: **Hatch**

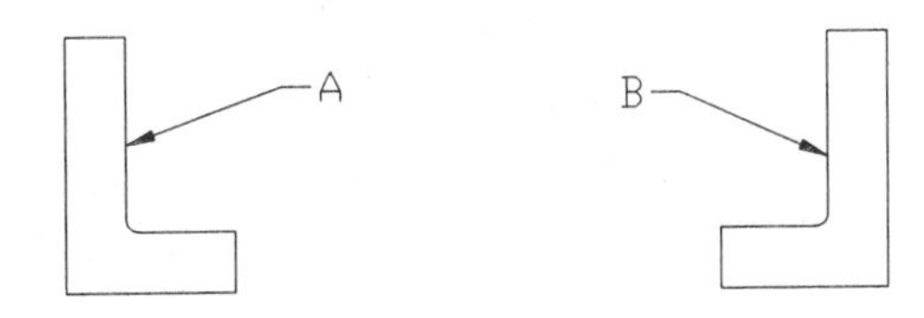

Figure 3.46 Polyline and boundary line

Hatching

Issue the HATCH command to create hatching in the areas enclosed by the two polylines.

Command: **HATCH**
Enter pattern name or [?/Solid/User defined]: **LINE**
Scale for pattern: **2**
Angle for pattern: **45**
Select hatch boundaries or press ENTER for direct hatch option,
Select objects: [**Select A and B (Figure 3.46).**]

Select objects: **[Enter]**

After hatching, thaw (unfreeze) all the frozen layers. See Figure 3.47.

[Object Properties] **[Layer Control]**

On/Thaw	0
On/Thaw	Center
On	Hatch
On/Thaw	Hidden

Current layer: **Hatch**

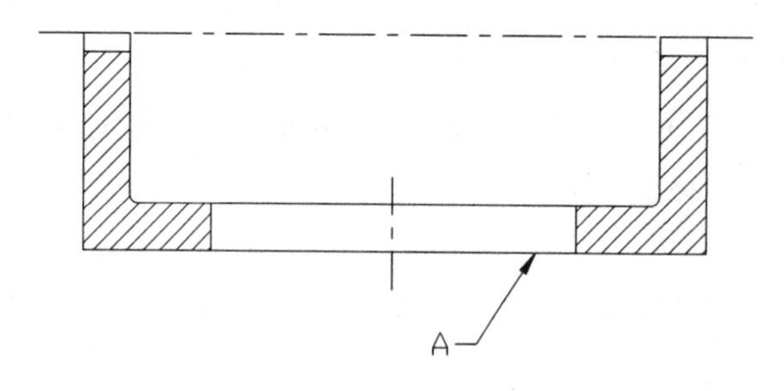

Figure 3.47 Hatching created

The sectional front view is complete. In making this drawing view, you created lines, boundary lines, polylines, and hatching. You edited with moving, copying, offsetting, mirroring, trimming, and filleting. You also learned saving a view, dynamic zoom, aerial view, and running object snap.

3.4 Side View of the First Component

Figure 3.48 shows the side view of the first component. As in the top view, you will place the outlines on layer 0, hidden lines on layer Hidden, and center lines on layer Center.

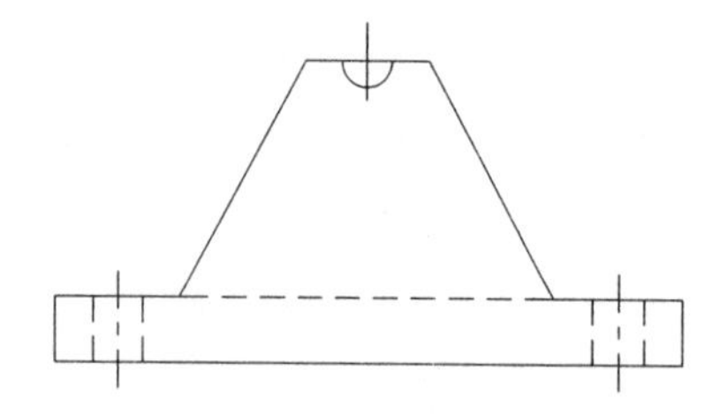

Figure 3.48 Side view of first component

Create Lines

Select layer 0 in the Layer control box of the Object Properties toolbar to set layer 0 as the current layer. Then use the LINE command to draw a line. Because the side view should align with the front view, apply the FROM object snap, and use the lower right corner of the front view as the Base point to set the start point of this line.

[Object Properties] **[Layer Control]**

Current layer: **0**

<Draw> **<Line>**

Command: **LINE**
From point: **FROM**
Base point: **END** of **[Select A (Figure 3.47).]**
<offset>: **@125<0**
To point: **@250<0**
To point: **[Enter]**

After drawing the first line of the side view, use the ZOOM command to zoom to All.

<View> **<Zoom>** **<All>**

Repeat the LINE command to draw a vertical line from the left end point of the last line. See Figure 3.49.

<Draw> **<Line>**

Command: **LINE**
From point: **END** of **[Select A (Figure 3.49).]**
To point: **@115<90**
To point: **[Enter]**

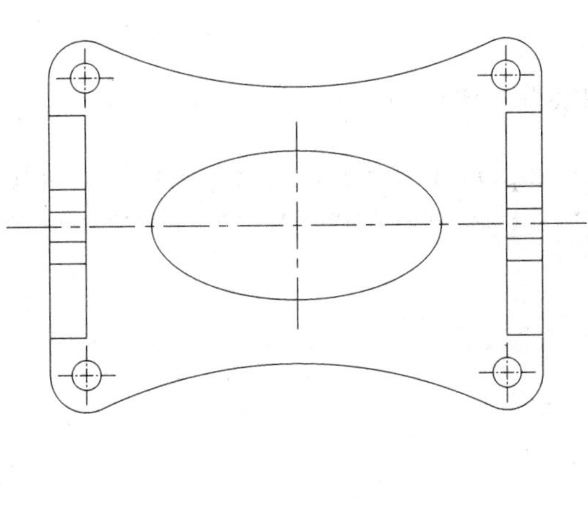

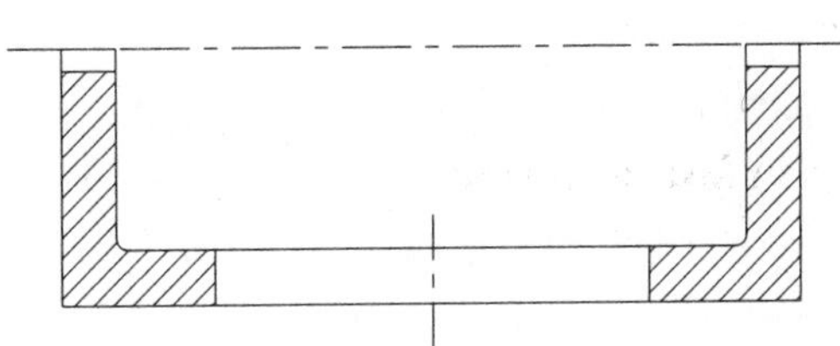

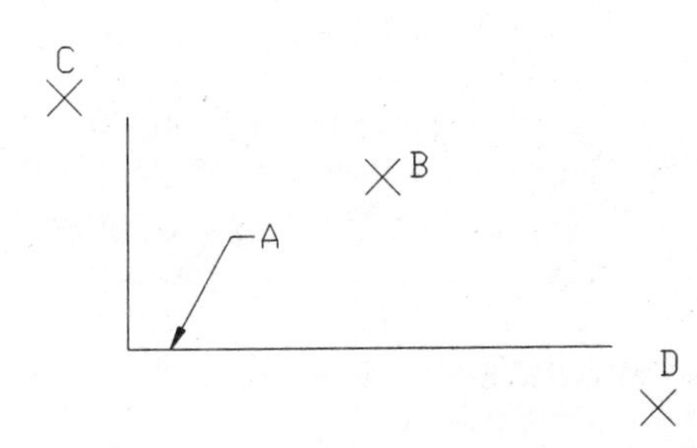

Figure 3.49 Two lines created

From the View pull-down menu, select the Zoom cascading menu and then the Window item to zoom the display to the side view.

<View> **<Zoom>** **<Window>**

Command: **ZOOM**
All/Center/Dynamic/Extents/Previous/Scale(X/XP)/Window/<Realtime>: **W**
First corner: **[Select C (Figure 3.49).]**
Other corner: **[Select D (Figure 3.49).]**

Apply the OFFSET command to offset the lower horizontal line twice, 25 units and 115 units upward. See Figure 3.50.

<Modify> **<Offset>**

Command: **OFFSET**
Offset distance or Through: **25**
Select object to offset: **[Select A (Figure 3.49).]**
Side to offset? **[Select B (Figure 3.49).]**
Select object to offset: **[Enter]**

Command: **[Enter]**
OFFSET
Offset distance or Through: **115**
Select object to offset: **[Select A (Figure 3.49).]**
Side to offset? **[Select B (Figure 3.49).]**
Select object to offset: **[Enter]**

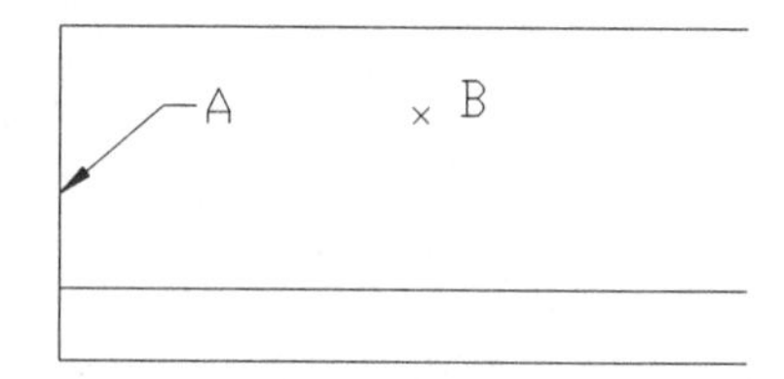

Figure 3.50 Two offset lines created

Repeat the OFFSET command twice to offset the left vertical line, 50 units and 100 units to the right. See Figure 3.51.

Command: **[Enter]**
OFFSET
Offset distance or Through: **50**
Select object to offset: **[Select A (Figure 3.50).]**
Side to offset? **[Select B (Figure 3.50).]**
Select object to offset: **[Enter]**

Command: **[Enter]**
OFFSET
Offset distance or Through: **100**
Select object to offset: **[Select A (Figure 3.50).]**
Side to offset? **[Select B (Figure 3.50).]**
Select object to offset: **[Enter]**

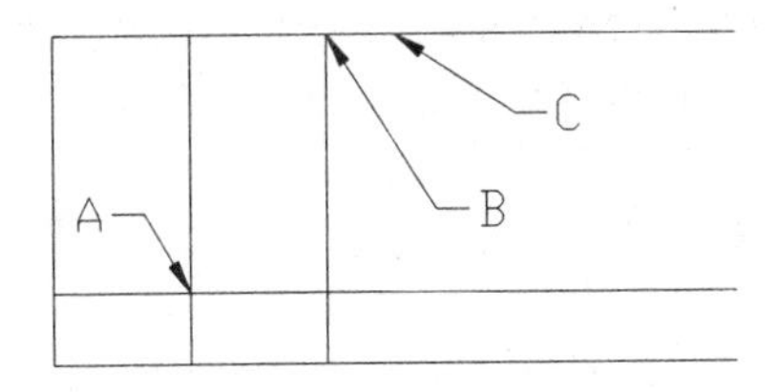

Figure 3.51 Two more offset lines created

Use the LINE command to draw an inclined line and the CIRCLE command to draw a circle. See Figure 3.52.

<Draw> **<Line>**

Command: **LINE**
From point: **INTER** of **[Select A (Figure 3.51).]**
To point: **INTER** of **[Select B (Figure 3.51).]**
To point: **[Enter]**

<Draw> **<Circle>** **<Center, Radius>**

Command: **CIRCLE**
3P/2P/TTR<Center point>: **MID** of **[Select C (Figure 3.51).]**
Diameter/<Radius>: **10**

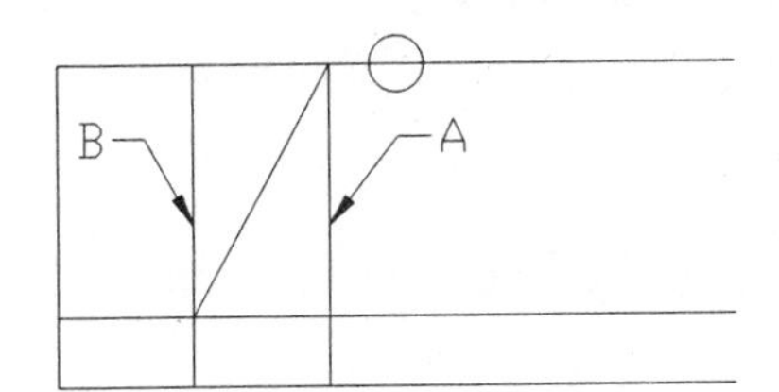

Figure 3.52 Inclined line and circle drawn

Erase and Trim

If you want to shorten an entity, you should trim it. However, if you do not want an entity any more, you should erase it. Apply the ERASE command. See Figure 3.53.

<Modify> **<Erase>**

Command: **ERASE**
Select objects: **[Select A and B (Figure 3.52).]**
Select objects: **[Enter]**

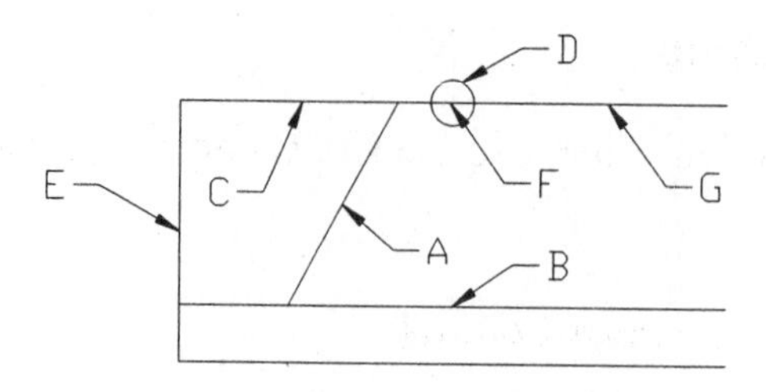

Figure 3.53 Two vertical lines erased

Issue the TRIM command to trim the lines and the circle. See Figure 3.54.

<Modify> **<Trim>**

Command: **TRIM**
Select objects: **[Select A, B, C, and D (Figure 3.53).]**
Select objects: **[Enter]**
<Select object to trim>/Project/Edge/Undo: **[Select B, C, D, E, G, and F (Figure 3.53).]**
<Select object to trim>/Project/Edge/Undo: **[Enter]**

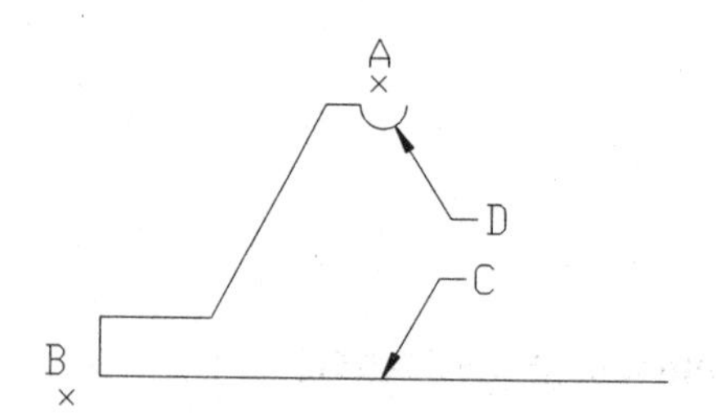

Figure 3.54 Lines and circle trimmed

Run the MIRROR command to mirror four lines. See Figure 3.55.

<Modify> **<Mirror>**

Command: **MIRROR**
Select objects: **[Select A (Figure 3.54).]**
Other corner: **[Select B (Figure 3.54).]**
Select objects: **[Enter]**
First point of mirror line: **MID** of **[Select C (Figure 3.54).]**
Second point: **CEN** of **[Select D (Figure 3.54).]**
Delete old objects?<N>: **[Enter]**

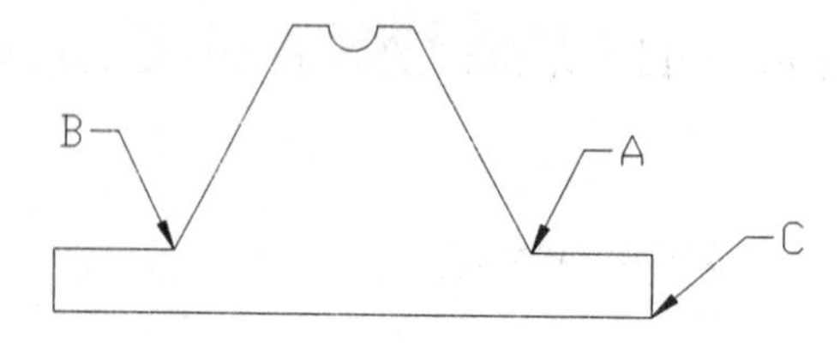

Figure 3.55 Four lines mirrored

Create Three Hidden Lines

Set the current layer to Hidden. Then use the LINE command to create three hidden lines. See Figure 3.56.

[Object Properties] **[Layer Control]**

Current layer: **Hidden**

<Draw> **<Line>**

Command: **LINE**
From point: **INT** of **[Select A (Figure 3.55).]**
To point: **INT** of **[Select B (Figure 3.55).]**
To point: **[Enter]**

Command: **[Enter]**
LINE
From point: **FROM**
Base point: **END** of **[Select C (Figure 3.55).]**
<offset>: **@15<180**
To point: **@25<90**
To point: **[Enter]**

Command: **[Enter]**
LINE
From point: **FROM**
Base point: **END** of **[Select C (Figure 3.55).]**
<offset>: **@35<180**
To point: **@25<90**
To point: **[Enter]**

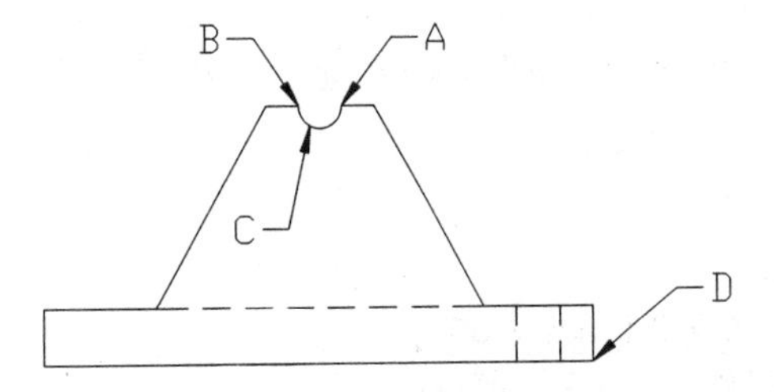

Figure 3.56 Three hidden lines created

Create Three Center Lines

Set the current layer to Center and use the LINE command to create three center lines. See Figure 3.57.

[Object Properties] **[Layer Control]**

Current layer: **Center**

<Draw> **<Line>**

Command: **LINE**

From point: **INT** of **[Select A (Figure 3.56).]**
To point: **INT** of **[Select B (Figure 3.56).]**
To point: **[Enter]**

Command: **[Enter]**
LINE
From point: **FROM**
Base point: **CEN** of **[Select C (Figure 3.56).]**
<offset>: **@15<90**
To point: **@30<270**
To point: **[Enter]**

Command: **[Enter]**
LINE
From point: **FROM**
Base point: **END** of **[Select D (Figure 3.56).]**
<offset>: **@-25,-10**
To point: **@45<90**
To point: **[Enter]**

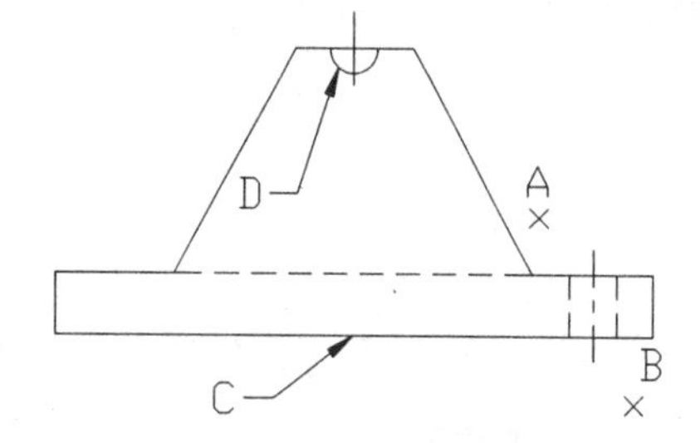

Figure 3.57 Three center lines created

Apply the MIRROR command to mirror two hidden lines and a center line about the center of the side view. See Figure 3.58. The side view is complete.

<Modify> **<Mirror>**

Command: **MIRROR**
Select objects: **[Select A (Figure 3.57).]**
Other corner: **[Select B (Figure 3.57).]**
Select objects: **[Enter]**
First point of mirror line: **MID** of **[Select C (Figure 3.57).]**
Second point: **CEN** of **[Select D (Figure 3.57).]**
Delete old objects?<N>: **[Enter]**

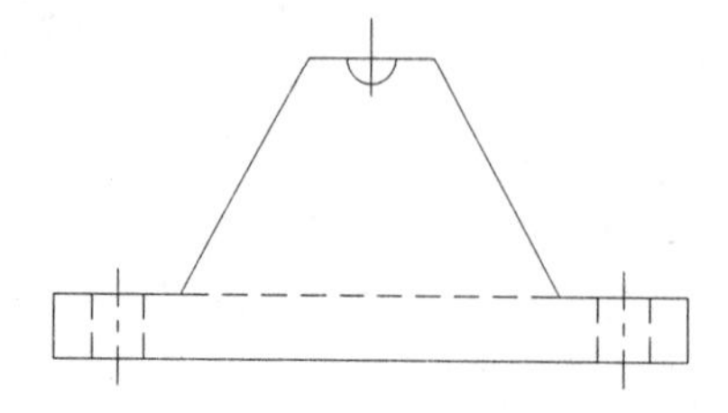

Figure 3.58 Two hidden lines and a center lines mirrored

3.5 Top View of the Second Component

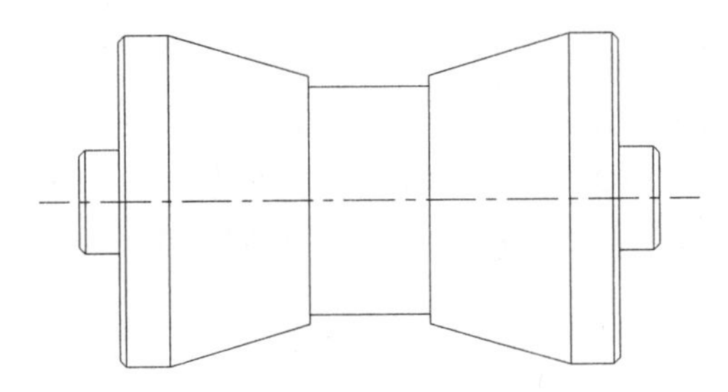

Figure 3.59 Top view of second component

Figure 3.59 shows the top view of the second component. To separate the second component from the first component, organize three additional layers — Point, Roller, and Hatch1.

Add Layers

Run the LAYER command to create three more layers. See Figure 3.60. The color of layer Point is blue. The color of layer Roller is cyan. The color of layer Hatch1 is green. The linetype for the three layers is Continuous. Set layer Roller as the current layer.

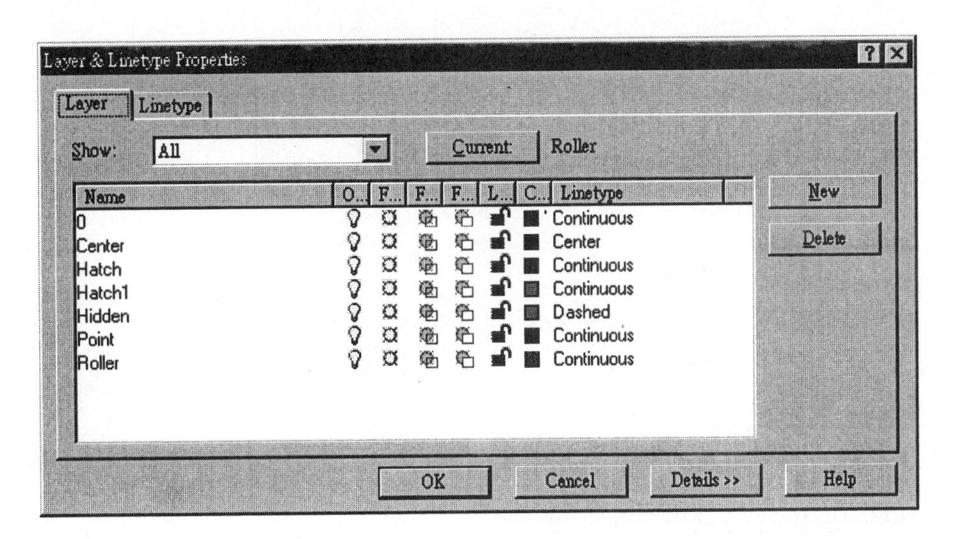

Figure 3.60 Three more layers created

<Format> **<Layers...>**

Name	Color	Linetype
0	White	Continuous
Center	Red	Center
Hatch	Blue	Continuous
Hatch1	Green	Continuous
Hidden	Green	Dashed
Point	Blue	Continuous
Roller	Cyan	Continuous

Current layer: **Roller**

Create Lines

Use the ZOOM command to zoom to the top view. Then apply the LINE command to create three lines. Use the FROM object snap to locate the first point, and use the midpoint of the central horizontal center line of the top view of the first component as the Base point. See Figure 3.61.

<Draw> **<Line>**

Command: **LINE**
From point: **FROM**
Base point: **MID** of **[Select A (Figure 3.61).]**
<offset>: **@125<180**
To point: **@250<180**
To point: **@80<90**
To point: **@250<0**
To point: **[Enter]**

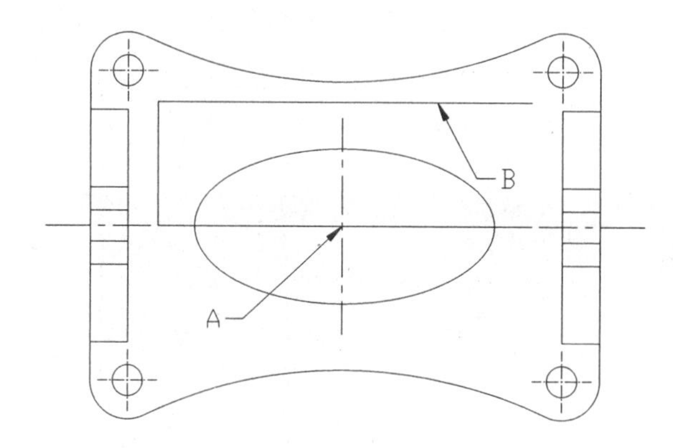

Figure 3.61 Three lines drawn

Create Points

Set the current layer to Point and freeze all other layers except layer Roller. Then create a series of points on the two horizontal lines.

Use the Layer Control box of the Object Properties toolbar to manage layers.

[Object Properties] **[Layer Control]**

Freeze	0
Freeze	Center
Freeze	Hatch
Freeze	Hatch1
Freeze	Hidden
On	Point
On	Roller

Current layer: **Point**

A point in a drawing can be displayed in many forms. The way that it is shown depends on the setting of the PDMODE. Enter PDMODE at the command prompt to set it to 3. This sets the appearance to a cross.

Command: **PDMODE**
New value for PDMODE: **3**

The corresponding dialog box command to set point display mode is the DDPTYPE command. See Figure 3.62.

<Format> **<Point Style...>**

Command: **DDPTYPE**

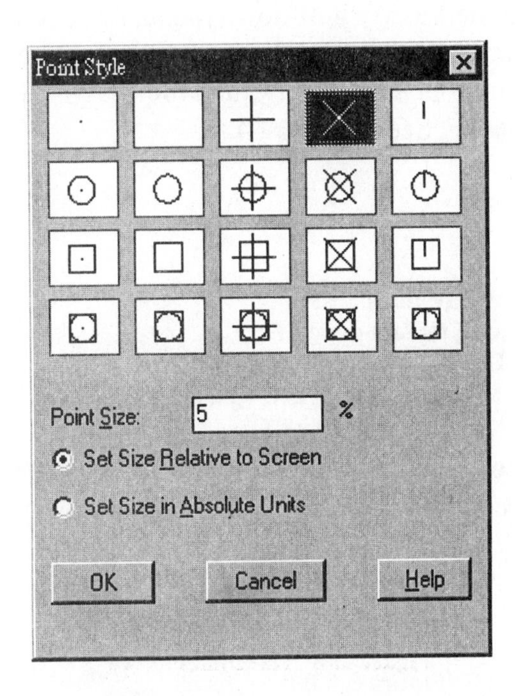

Figure 3.62 Point Style dialog box

Points can be created by three commands. They are the POINT command, the DIVIDE command, and the MEASURE command. If you wish to create a point, you may issue the POINT command and select a location. You can snap to a point with the NODE object snap.

From the Draw pull-down menu, select the Point cascading menu and then the Divide item to apply the DIVIDE command to place four points equally spaced on the upper horizontal line. Select the upper line. The DIVIDE command does not actually cut the line into segments; it merely places points on it. See Figure 3.63.

<Draw> **<Point>** **<Divide>**

Command: **DIVIDE**
Select object to divide: **[Select B (Figure 3.62).]**
<Number of segments>/Block: **5**

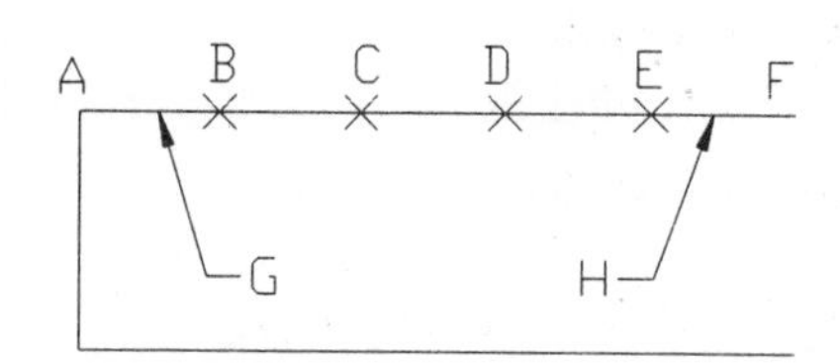

Figure 3.63 Four points created

Break an Object

As we have said, the DIVIDE command merely places points on a line. It does not break a line into a number of parts. To break a line into segments, you have to issue the BREAK command. From the Modify pull-down menu, select the Break item to apply the command on the upper horizontal line. After selecting the line, enter F to override the select point. Then select the NODE of point C and the NODE of point D to break the line into two segments AC and DF. See Figure 3.64.

<Modify> **<Break>**

Command: **BREAK**
Select object: [**Select G (Figure 3.63).**]
Enter second point (or F for first point): **F**
Enter first point: **NODE** of [**Select C (Figure 3.63).**]
Enter second point: **NODE** of [**Select D (Figure 3.63).**]

Repeat the BREAK command on the line segment AC. After entering [F], select the NODE of point B; the second point is at (@) the first point. The line AC is now broken into lines AB and BC.

Command: [**Enter**]
BREAK
Select object: [**Select G (Figure 3.63).**]
Enter second point (or F for first point): **F**
Enter first point: **NODE** of [**Select B (Figure 3.63).**]
Enter second point: **@**

Similarly, apply the BREAK command on line segment DF to break it into segments DE and EF.

Command: [**Enter**]
BREAK
Select object: [**Select H (Figure 3.63).**]
Enter second point (or F for first point): **F**
Enter first point: **NODE** of [**Select E (Figure 3.63).**]
Enter second point: **@**

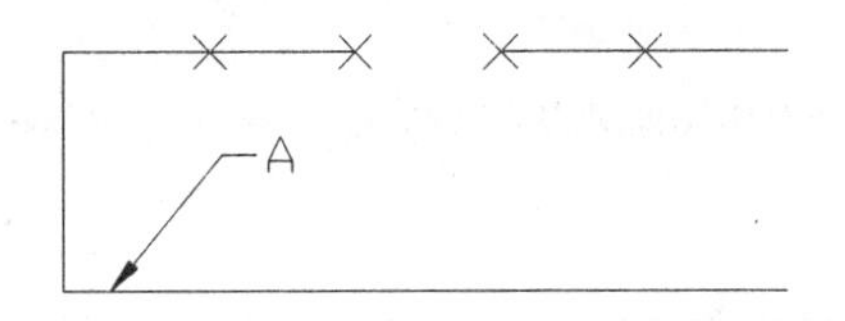

Figure 3.64 Line AF broken into AB, BC, DE, and EF

As mentioned, the MEASURE command also places points on a line. From the Draw pull-down menu, select the Point cascading menu and then the Measure item to use the MEASURE command to place points on the lower horizontal line. Note the difference between the DIVIDE command and the MEASURE command. See Figure 3.65.

<Draw> **<Point>** **<Measure>**

Command: **MEASURE**
Select object to measure: **[Select A (Figure 3.64).]**
<Segment length>/Block: **45**

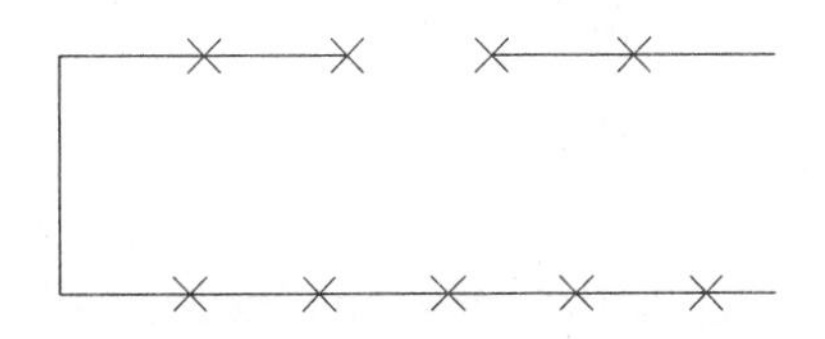

Figure 3.65 Five points placed on the lower horizontal line

Filter

The nine points are not required in the final drawing. In order to select all the points collectively, you can use the FILTER command transparently while using the ERASE command. See Figure 3.66. The FILTER command builds a list of properties required of an object for it to be selected. From the Edit pull-down menu, select the Clear item to apply the ERASE command to erase all the points. See Figure 3.67.

<Edit> **<Clear>**

Command: **ERASE**

At the prompt to select objects, enter the FILTER command transparently.

Select objects: **'FILTER**

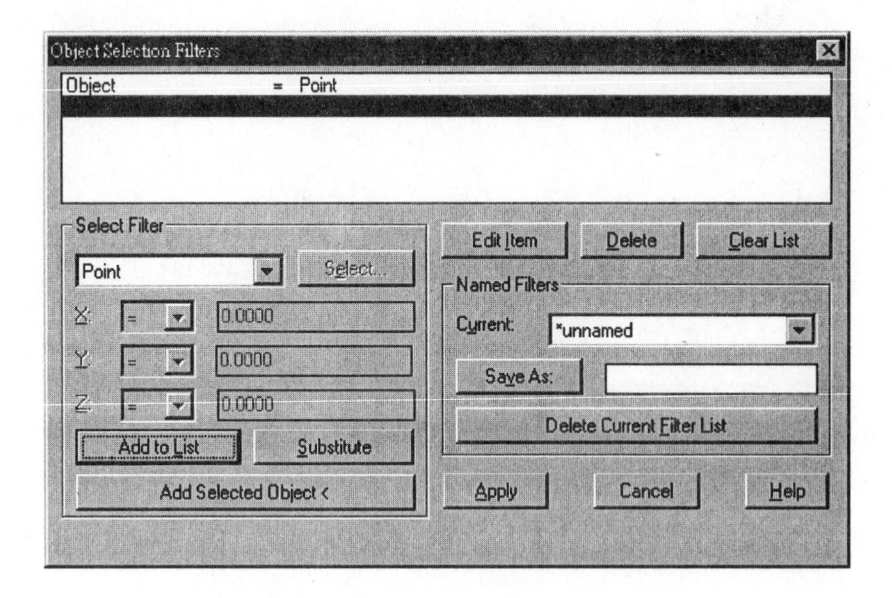

Figure 3.66 Object Selection Filters dialog box

Under the Select Filter column of the Object Selection Filters dialog box, select the object type, Point. Click the [Add to List] button and then the [Apply] button.

```
Applying filter to selection
```

At the command prompt, enter All to include all the objects in the filter list.

Select objects: **ALL**
15 found
6 were filtered out.
Select objects: [**Enter**]
Exiting filtered selection. 9 found
Select objects: [**Enter**]

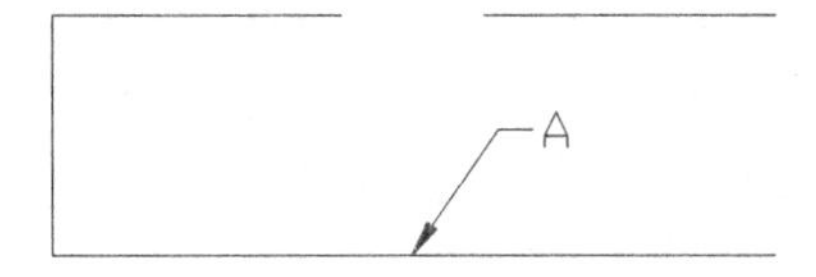

Figure 3.67 All points erased

Set Layer and Create Line

Set the layer Roller as the current layer. Then draw a line with the LINE command from the midpoint of the lower horizontal line. See Figure 3.68.

[Object Properties] **[Layer Control]**

Freeze	0
Freeze	Center
Freeze	Hatch

Freeze	Hatch1
Freeze	Hidden
On	Point
On	Roller

Current layer: **Roller**

<Draw> **<Line>**

Command: **LINE**
From point: **MID** of [**Select A (Figure 3.67).**]
To point: **@60<90**
To point: [**Enter**]

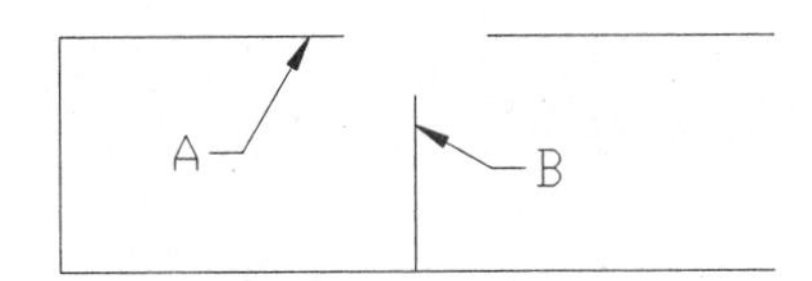

Figure 3.68 Vertical line drawn

Change End Point Location

Refer to Figure 3.68. The end point A should align with the end point B. To relocate the end point A, apply the CHANGE command. This command can be used to change entity properties or the end point position. See Figure 3.69.

Command: **CHANGE**
Select objects: [**Select A (Figure 3.68).**]
Select objects: [**Enter**]
Properties/<Change point>: **END** of [**Select B (Figure 3.68).**]

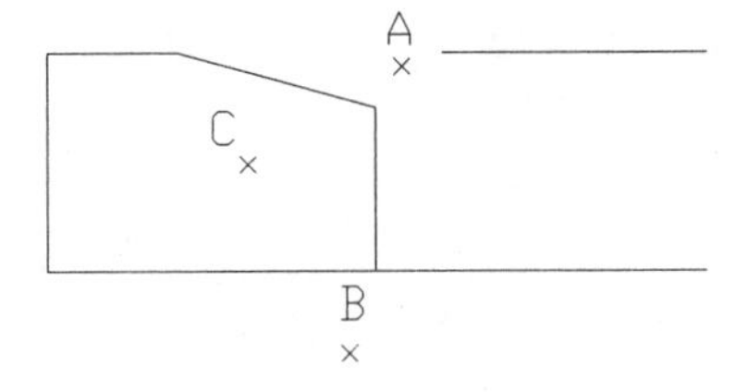

Figure 3.69 End point position changed

Stretch Entities

Lines can be stretched with the STRETCH command. To appreciate how lines can be stretched, select the Stretch item from the Modify pull-down menu to apply this command. Use the C object selection to describe a crossing window about (215,225) and (195,120). To stretch an object, you have to include the definition points of the entities being stretched in the selection. See Figure 3.70.

<Modify> **<Stretch>**

Command: **STRETCH**
Select objects: **C**
First corner: [**Select A (Figure 3.69).**]
Other corner: [**Select B (Figure 3.69).**]
Select objects: [**Enter**]
Base point or displacement: [**Select C (Figure 3.69).**]
Second point of displacement: **@30<180**

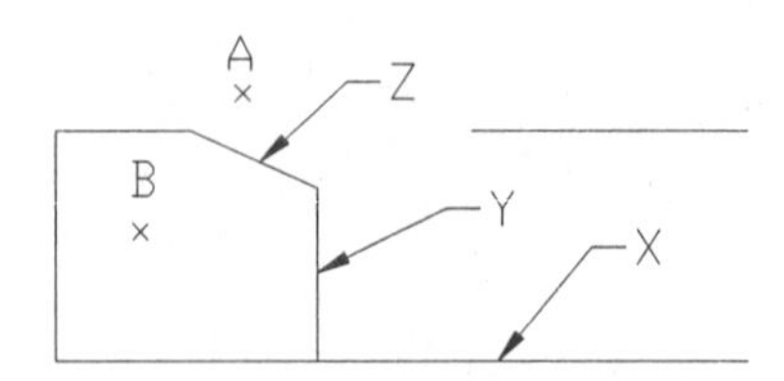

Figure 3.70 Lines stretched

In the foregoing command, three lines are selected. The lower line, X, does not change because neither end point is selected. The vertical line, Y, is moved because both end points are selected. The inclined line, Z, is stretched because only one end point is selected. In the Base point or displacement selection, you selected a random point on the screen. In the Second point of displacement, you keyed in @30<180 — a point relative to the randomly selected point. These two points established a vector of length 30 in the 180° direction. Repeat the STRETCH command again. Watch how the lines are stretched like flexible rubber bands. See Figure 3.71.

Command: [**Enter**]
STRETCH
Select objects: **C**
First corner: [**Select A (Figure 3.70).**]
Other corner: [**Select B (Figure 3.70).**]
Select objects: [**Enter**]
Base point or displacement: **25<180**
Second point of displacement: [**Enter**]

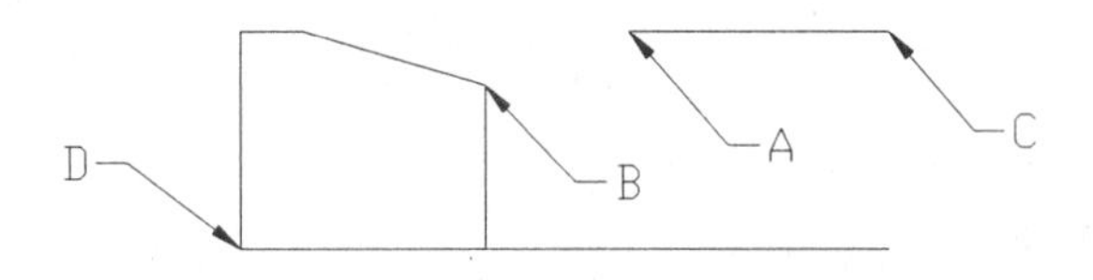

Figure 3.71 Two more lines stretched

In the foregoing command, the 25<180 reply to the request for a Base point or displacement is treated as a vector of displacement because you pressed the [Enter] key in response to the request for a Second point of displacement.

Move Entities

Apply the MOVE command to move two line segments. See Figure 3.72.

<Modify> **<Move>**

Command: **MOVE**
Select objects: [**Select A (Figure 3.71).**]
Select objects: [**Enter**]
<Base point or displacement>/Multiple: **END** of [**Select A (Figure 3.71).**]
Second point of displacement: **END** of [**Select B (Figure 3.71).**]

Command: [**Enter**]
MOVE
Select objects: [**Select C (Figure 3.71).**]
Select objects: [**Enter**]
<Base point or displacement>/Multiple: **END** of [**Select C (Figure 3.71).**]
Second point of displacement: **END** of [**Select D (Figure 3.71).**]

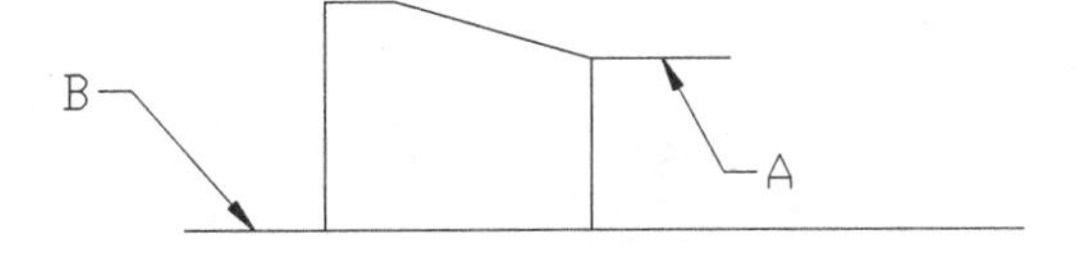

Figure 3.72 Lines moved

Repeat the MOVE command twice to move the two line segments again. See Figure 3.73.

Command: [**Enter**]
MOVE
Select objects: [**Select A (Figure 3.72).**]
Select objects: [**Enter**]
<Base point or displacement>/Multiple: **5<270**
Second point of displacement: [**Enter**]

Command: [**Enter**]
MOVE
Select objects: [**Select B (Figure 3.72).**]
Select objects: [**Enter**]
<Base point or displacement>/Multiple: **25<90**
Second point of displacement: [**Enter**]

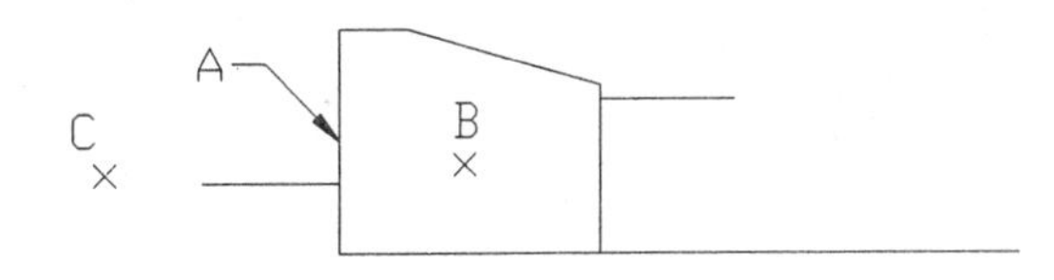

Figure 3.73 Lines moved again

Offset and Trim Entities

Use the OFFSET command to create three lines. See Figure 3.74.

<Modify> **<Offset>**

Command: **OFFSET**
Offset distance or Through: **25**
Select object to offset: [**Select A (Figure 3.73).**]
Side to offset? [**Select B (Figure 3.73).**]
Select object to offset: [**Enter**]

Command: [**Enter**]
OFFSET
Offset distance or Through: **3**
Select object to offset: [**Select A (Figure 3.73).**]
Side to offset? [**Select B (Figure 3.73).**]
Select object to offset: [**Enter**]

Command: [**Enter**]
OFFSET
Offset distance or Through: **20**
Select object to offset: [**Select A (Figure 3.73).**]
Side to offset? [**Select C (Figure 3.73).**]
Select object to offset: [**Enter**]

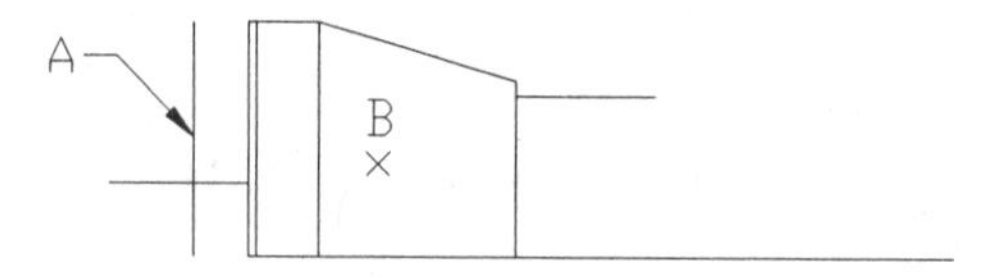

Figure 3.74 Three offset lines created

Repeat the OFFSET command to create one more line. See Figure 3.75.

Command: [**Enter**]
OFFSET
Offset distance or Through: **3**
Select object to offset: [**Select A (Figure 3.74).**]
Side to offset? [**Select B (Figure 3.74).**]
Select object to offset: [**Enter**]

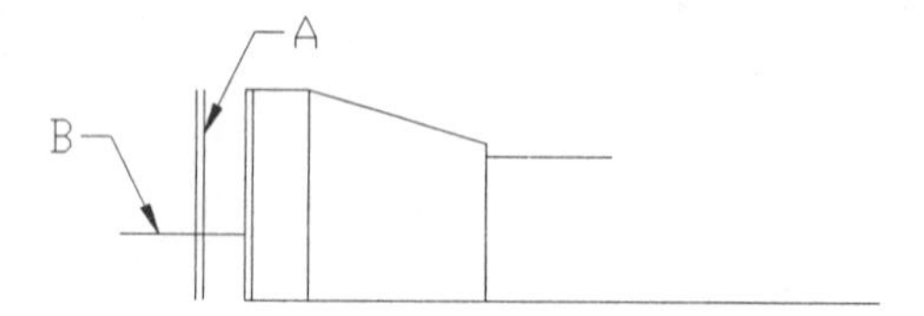

Figure 3.75 One more offset line created

Issue the TRIM command to trim a line. See Figure 3.76.

<Modify> **<Trim>**

Command: **TRIM**
Select objects: **[Select B (Figure 3.75).]**
Select objects: **[Enter]**
<Select object to trim>/Project/Edge/Undo: **[Select A (Figure 3.75).]**
<Select object to trim>/Project/Edge/Undo: **[Enter]**

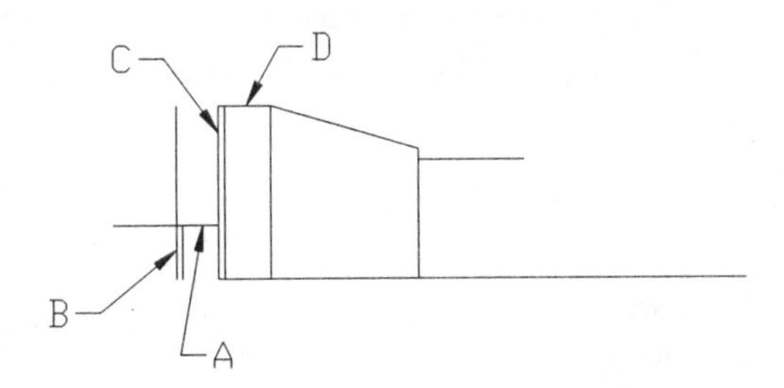

Figure 3.76 Lines trimmed

Chamfer Corners

To chamfer is to bevel two coplanar non-parallel lines. It takes two steps to do chamfering. Here you will first set the chamfer distance to 3 units x 3 units. Then you will bevel the edges. See Figure 3.77.

<Modify> **<Chamfer>**

Command: **CHAMFER**
Polyline/Distance/Angle/Trim/Method/<Select first line>: **D**
Enter first chamfer distance: **3**
Enter second chamfer distance: **3**

Command: **[Enter]**
CHAMFER
Polyline/Distance/Angle/Trim/Method/<Select first line>: **[Select A (Figure 3.76).]**
Select second line: **[Select B (Figure 3.76).]**

Command: **[Enter]**
CHAMFER
Polyline/Distance/Angle/Trim/Method/<Select first line>: **[Select C (Figure 3.76).]**
Select second line: **[Select D (Figure 3.76).]**

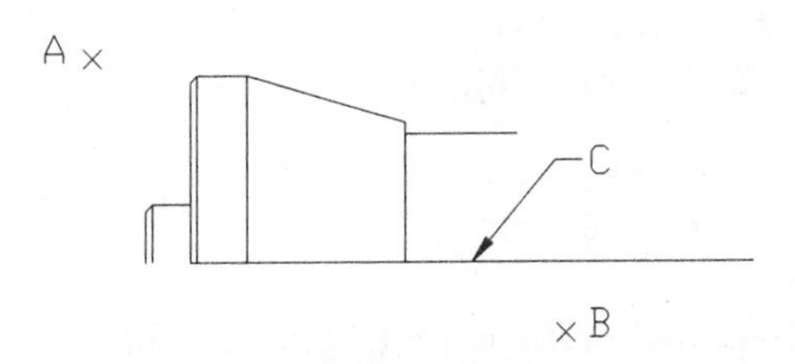

Figure 3.77 Corners chamfered

Array and Mirror Entities

As we have said, there are two kinds of arrays, rectangular and polar. A rectangular array creates a number of rows and columns of copies of the selected objects. A polar array replicates objects about a center point. In making the top view of the first component, you created rectangular array. Now, you will create a polar array.

From the Modify pull-down menu, select the Array item to use the ARRAY command to create a polar array of all the entities except the lower horizontal line. The center of the array is the midpoint of that horizontal line. See Figure 3.78.

<Modify> **<Array>**

Command: **ARRAY**
Select objects: **[Select A (Figure 3.77).]**
Other corner: **[Select B (Figure 3.77).]**
Select objects: **[Enter]**
Rectangular or Polar array (R/P): **P**
Base/<Specify center point of array>: **MID** of **[Select C (Figure 3.77).]**
Number of items: **2**
Angle to fill (+ =CCW,- =CW): **360**
Rotate objects as they are copied? <Y> **[Enter]**

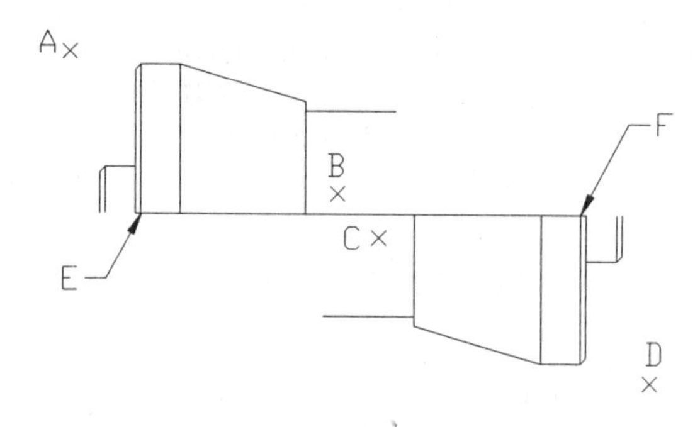

Figure 3.78 Twelve lines arrayed

Apply the MIRROR command to mirror 10 line segments. See Figure 3.79.

<Modify> **<Mirror>**

Command: **MIRROR**
Select objects: **W**
First corner: **[Select A (Figure 3.78).]**
Other corner: **[Select B (Figure 3.78).]**
Select objects: **W**
First corner: **[Select C (Figure 3.78).]**
Other corner: **[Select D (Figure 3.78).]**
Select objects: **[Enter]**
First point of mirror line: **END** of **[Select E (Figure 3.78).]**
Second point: **END** of **[Select F (Figure 3.78).]**
Delete old objects? <N> **[Enter]**

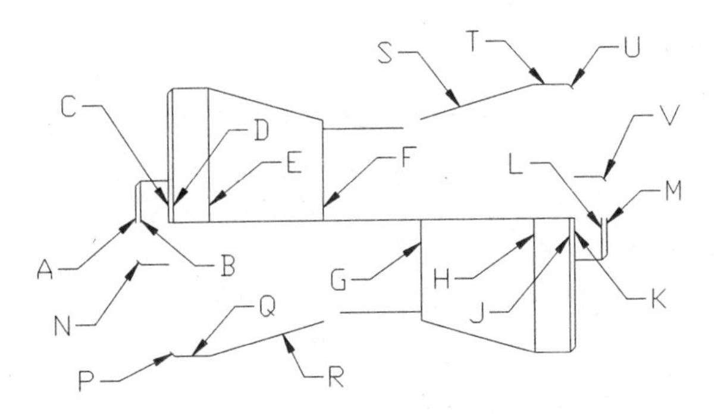

Figure 3.79 Ten lines mirrored

Extend and Lengthen Lines

The opposite of to trim is to extend. From the Modify pull-down menu, select the Extend item to use the EXTEND command to extend 12 vertical lines. See Figure 3.80.

<Modify> **<Extend>**

Command: **EXTEND**
Select objects: **[Select N, P, Q, R, S, T, U, and V (Figure 3.79).]**
Select objects: **[Enter]**
<Select object to extend>/Project/Edge/Undo: **[Select A, B, C, D, E, F, G, H, J, K, L, AND M (Figure 3.79).]**
<Select object to extend>/Project/Edge/Undo: **[Enter]**

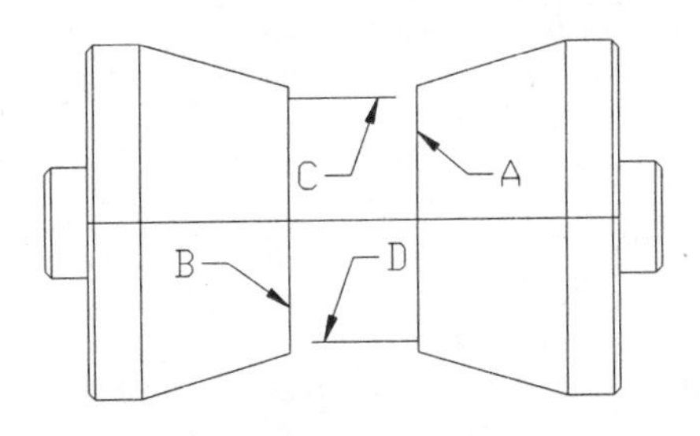

Figure 3.80 Twelve lines extended

Repeat the EXTEND command to extend two horizontal lines. See Figure 3.81.

Command: **[Enter]**
EXTEND
Select objects: **[Select A and B (Figure 3.80).]**
Select objects: **[Enter]**
<Select object to extend>/Project/Edge/Undo: **[Select C and D (Figure 3.80).]**
<Select object to extend>/Project/Edge/Undo: **[Enter]**

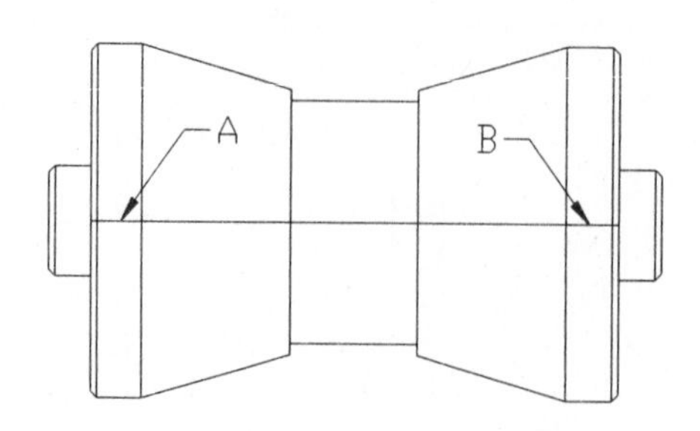

Figure 3.81 Two horizontal lines extended

Both ends of the central horizontal line are not long enough. Use the LENGTHEN command to lengthen this line. See Figure 3.82.

<Modify> **<Lengthen>**

Command: **LENGTHEN**
DElta/Percent/Total/DYnamic/<Select object>: **DE**
Angle/<Enter delta length>: **40**
<Select object to change>/Undo: **[Select A and B (Figure 3.81).]**
<Select object to change>/Undo: **[Enter]**

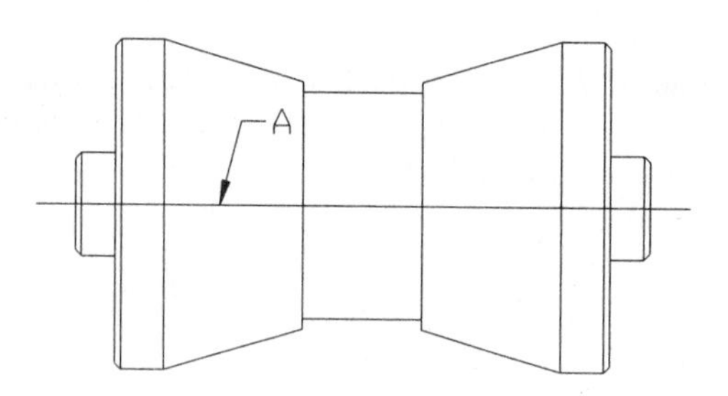

Figure 3.82 Line lengthened

Line A of Figure 3.82 should be a center line. To change it to a center line, you have two options, create a new layer with linetype assignment of Center and place this line on this layer, or change the linetype property of this line to Center.

Considering that there will be too many layers in your drawing if you create a unique layer for each object type in the drawing, you will change its linetype to center. To make it readily identifiable, change its color to yellow. To make the change, select the Properties... item from the Modify pull-down menu. See Figure 3.83.

<Modify> **<Properties...>**

Select objects: **[Select A (Figure 3.82).]**
Select objects: **[Enter]**

[Modify Line
Properties
Color... **Yellow**
Linetype... **Center**
OK]

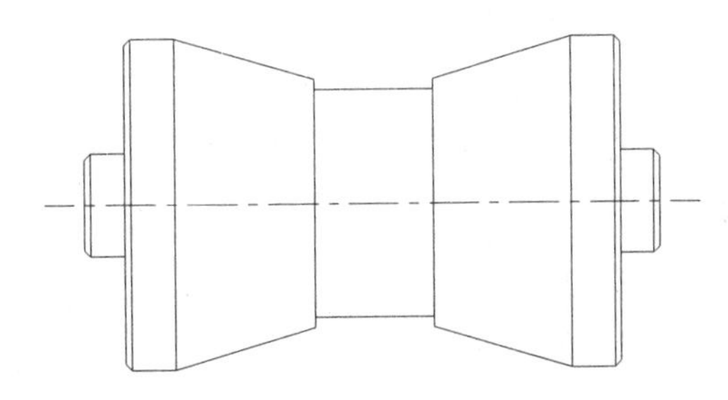

Figure 3.83 Complete top view of the second component

The top view of the second component is complete. In making this drawing view, you created points by using divide and measure, broke a line into segments, used filter in selecting objects, changed the end point position, stretched lines, chamfered edges, and polar arrayed.

3.6 Sectional Front View of the Second Component

Figure 3.84 shows the sectional front view of the second component of the assembly. You will create the outlines and center lines on layer Roller and hatching on layer Hatch1.

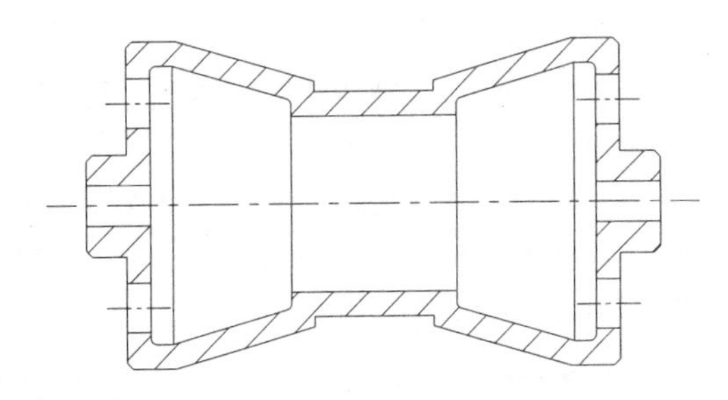

Figure 3.84 Sectional front view of the second component

Layer Management and Copy Entities

For this component, the front view is very similar to the top view. It is faster to copy the entire top view to the front view position and then do editing. In order to position the sectional front view properly relative to the top view, you will thaw all layers so that the first component is displayed. In order not to temper with the entities of the first component accidentally, you will lock all the layers related to the first component.

Use the Layer Control box of the Object Properties toolbar to thaw all layers, lock all layers except layer Roller, and set layer Roller as the current layer. See Figure 3.85.

[Object Properties] **[Layer Control]**

On, Thaw, & Lock	0
On, Thaw, & Lock	Center
On, Thaw, & Lock	Hatch
On, Thaw, & Lock	Hatch1

On, Thaw, & Lock	Hidden
On, Thaw, & Lock	Point
On	Roller

Current layer: **Roller**

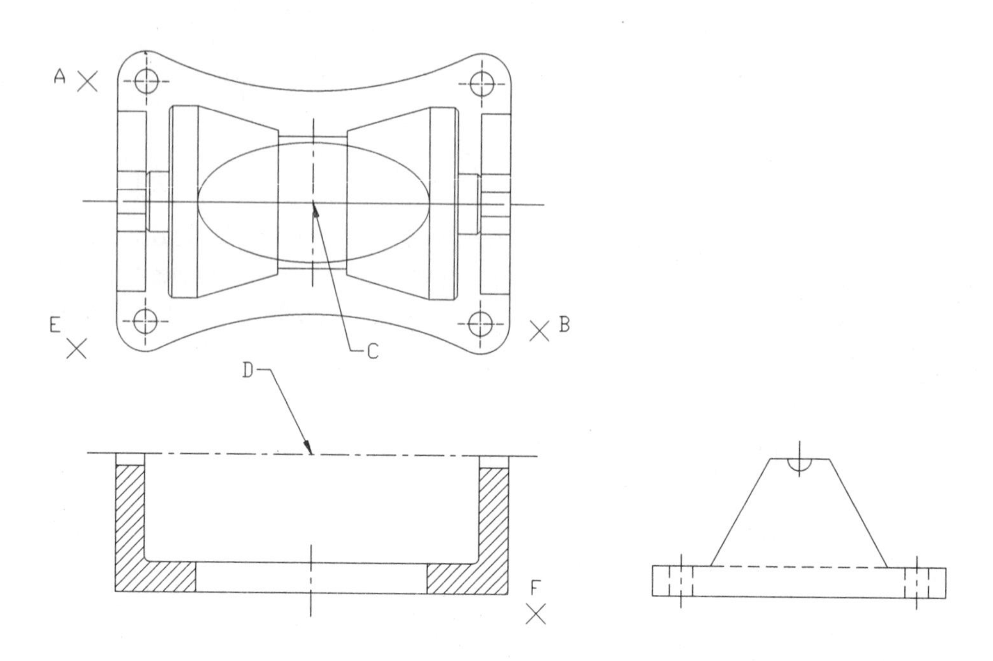

Figure 3.85 Layers thawed and locked

Use the COPY command to copy the entire top view of the second component. Use a window selection to include all entities of the top view. Because other layers are locked, only those entities on layer Roller are selected.

<Modify> **<Copy>**

Command: **COPY**
Select objects: **W**
First corner: **[Select A (Figure 3.85).]**
Other corner: **[Select B (Figure 3.85).]**
Select objects: **[Enter]**
<Base point or displacement>/Multiple: **MID** of **[Select C (Figure 3.85).]**
Second point of displacement: **PERP** of **[Select D (Figure 3.85).]**

Use the ZOOM command to zoom to the front view of the drawing. See Figure 3.86.

<View> **<Zoom>** **<Window>**

Command: **ZOOM**
All/Center/Dynamic/Extents/Previous/Scale(X/XP)/Window/<Realtime>: **W**
First corner: **[Select E (Figure 3.85).]**
Other corner: **[Select F (Figure 3.85).]**

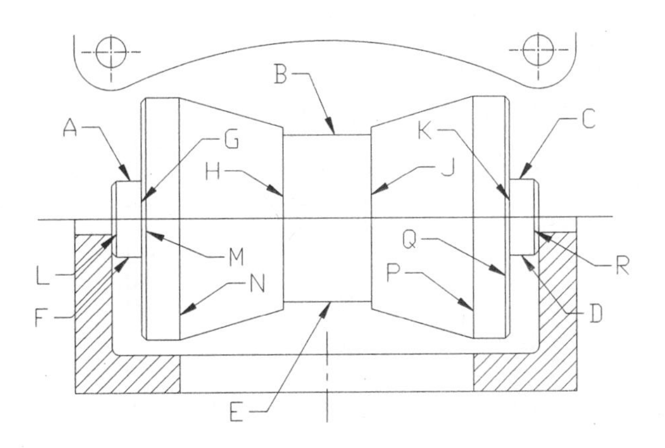

Figure 3.86 Zoomed to front view

In order to concentrate on the second component, freeze all the layers except the current layer.

[Object Properties] **[Layer Control]**

On, Freeze, & Lock	0
On, Freeze, & Lock	Center
On, Freeze, & Lock	Hatch
On, Freeze, & Lock	Hatch1
On, Freeze, & Lock	Hidden
On, Freeze, & Lock	Point
On	Roller

Current layer: **Roller**

Trim, Erase, and Offset Entities

Use the TRIM command and the ERASE command to trim and erase the redundant entities. See Figure 3.87.

<Modify> **<Trim>**

Command: **TRIM**
Select objects: [**Select A, B, C, D, E, and F (Figure 3.86).**]
Select objects: [**Enter**]
<Select object to trim>/Project/Edge/Undo: [**Select G, H, J, and K (Figure 3.86).**]
<Select object to trim>/Project/Edge/Undo: [**Enter**]

<Modify> **<Erase>**

Command: **ERASE**
Select objects: [**Select L, M, N, P, Q, and R (Figure 3.86).**]
Select objects: [**Enter**]

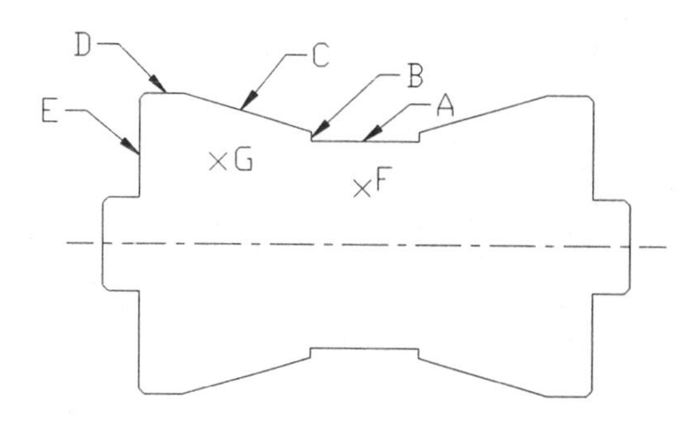

Figure 3.87 Redundant entities trimmed and erased

Issue the OFFSET command to create five offset lines. See Figure 3.88.

<Modify> **<Offset>**

Command: **OFFSET**
Offset distance or Through: **12**
Select object to offset: [**Select A (Figure 3.87).**]
Side to offset? [**Select F (Figure 3.87).**]
Select object to offset: [**Select B (Figure 3.87).**]
Side to offset? [**Select G (Figure 3.87).**]
Select object to offset: [**Select C (Figure 3.87).**]
Side to offset? [**Select G (Figure 3.87).**]
Select object to offset: [**Select D (Figure 3.87).**]
Side to offset? [**Select G (Figure 3.87).**]
Select object to offset: [**Select E (Figure 3.87).**]
Side to offset? [**Select G (Figure 3.87).**]
Select object to offset: [**Enter**]

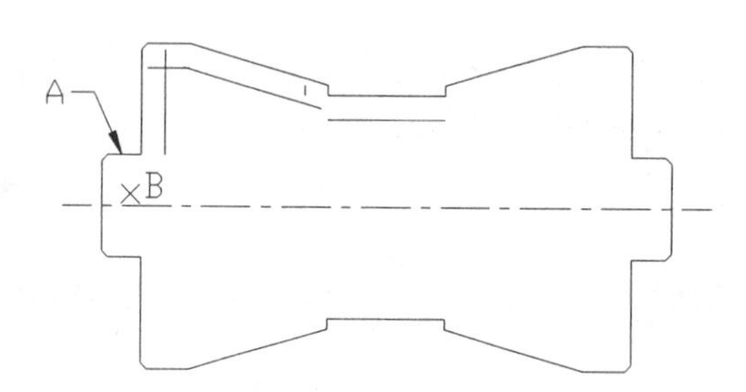

Figure 3.88 Five offset lines created

Repeat the OFFSET command. See Figure 3.89.

Command: [**Enter**]
OFFSET
Offset distance or Through: **15**
Select object to offset: [**Select A (Figure 3.88).**]
Side to offset? [**Select B (Figure 3.88).**]
Select object to offset: [**Enter**]

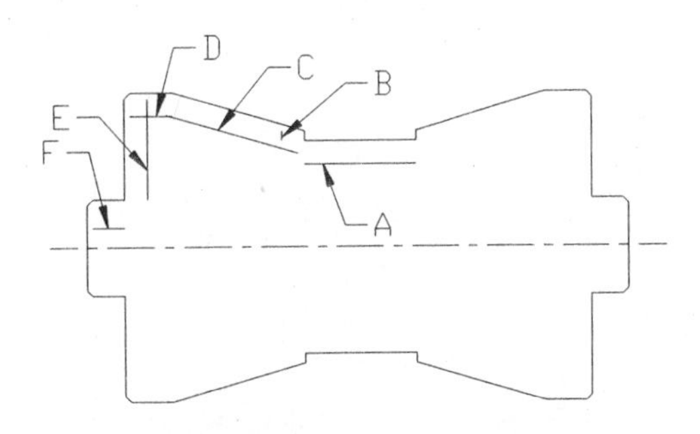

Figure 3.89 One more offset line created

Fillet Corners

Filleting to zero radius is a quick way to obtain a sharp edge between two non-parallel lines. Use the FILLET command to set the fillet radius to zero. Then repeat the FILLET command three times to create three corners of zero fillet radius. See Figure 3.90.

<Modify> **<Fillet>**

Command: **FILLET**
Polyline/Radius/Trim/<Select first object>: **R**
Enter fillet radius: **0**

Command: **[Enter]**
FILLET
Polyline/Radius/Trim/<Select first object>: **[Select A (Figure 3.89).]**
Select second object: **[Select B (Figure 3.89).]**

Command: **[Enter]**
FILLET
Polyline/Radius/Trim/<Select first object>: **[Select C (Figure 3.89).]**
Select second object: **[Select D (Figure 3.89).]**

Command: **[Enter]**
FILLET
Polyline/Radius/Trim/<Select first object>: **[Select E (Figure 3.89).]**
Select second object: **[Select F (Figure 3.89).]**

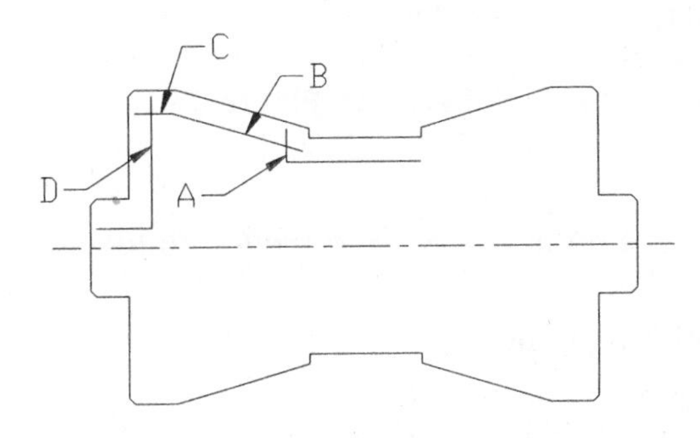

Figure 3.90 Three corners of zero fillet radius created

Repeat the FILLET command to set the fillet radius to 3 units and to round off two corners. See Figure 3.91.

Command: **[Enter]**
FILLET
Polyline/Radius/Trim/<Select first object>: **R**
Enter fillet radius: **3**

Command: **[Enter]**
FILLET
Polyline/Radius/Trim/<Select first object>: **[Select A (Figure 3.90).]**
Select second object: **[Select B (Figure 3.90).]**

Command: **[Enter]**
FILLET
Polyline/Radius/Trim/<Select first object>: **[Select C (Figure 3.90).]**
Select second object: **[Select D (Figure 3.90).]**

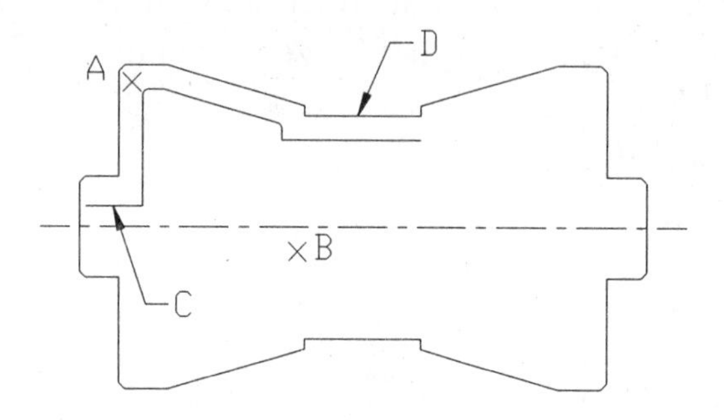

Figure 3.91 Two fillets created

Mirror and Extend

Use the MIRROR command to mirror the offset lines and the fillets. See Figure 3.92.

<Modify> **<Mirror>**

Command: **MIRROR**
Select objects: **W**
First corner: **[Select A (Figure 3.91).]**
Other corner: **[Select B (Figure 3.91).]**
Select objects: **[Select C (Figure 3.91).]**
Select objects: **[Enter]**
First point of mirror line: **MID** of **[Select D (Figure 3.91).]**
Second point: **@1<90**

The above "@1<90" is a point in the 90° direction to locate a direction.

Delete old objects? <N>: **[Enter]**

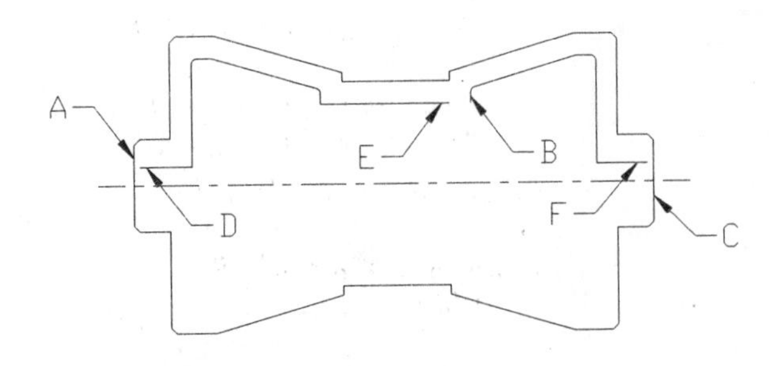

Figure 3.92 Entities mirrored

Run the EXTEND command to extend three horizontal lines. See Figure 3.93.

<Modify> **<Extend>**

Command: **EXTEND**
Select objects: [**Select A, B, and C (Figure 3.92).**]
Select objects: [**Enter**]
<Select object to extend>/Project/Edge/Undo: [**Select D, E, and F (Figure 3.92).**]
<Select object to extend>/Project/Edge/Undo: [**Enter**]

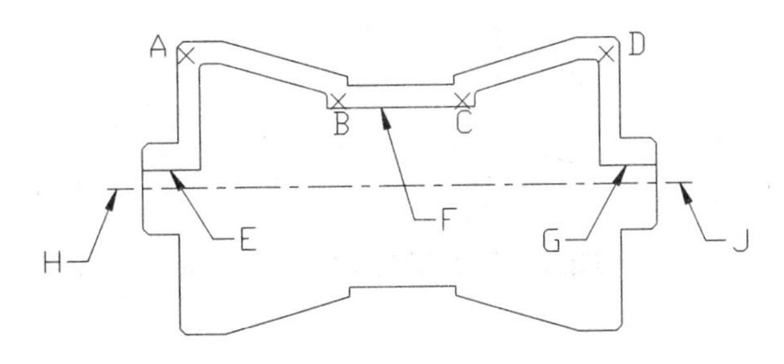

Figure 3.93 Three horizontal lines extended

Apply the MIRROR command to mirror entities about the central horizontal line. See Figure 3.94.

<Modify> **<Mirror>**

Command: **MIRROR**
Select objects: **W**
First corner: [**Select A (Figure 3.93).**]
Other corner: [**Select B (Figure 3.93).**]
Select objects: **W**
First corner: [**Select C (Figure 3.93).**]
Other corner: [**Select D (Figure 3.93).**]
Select objects: [**Select E, F, and G (Figure 3.93).**]
Select objects: [**Enter**]
First point of mirror line: **END** of [**Select H (Figure 3.93).**]
Second point: **END** of [**Select J (Figure 3.93).**]
Delete old objects? <N>: [**Enter**]

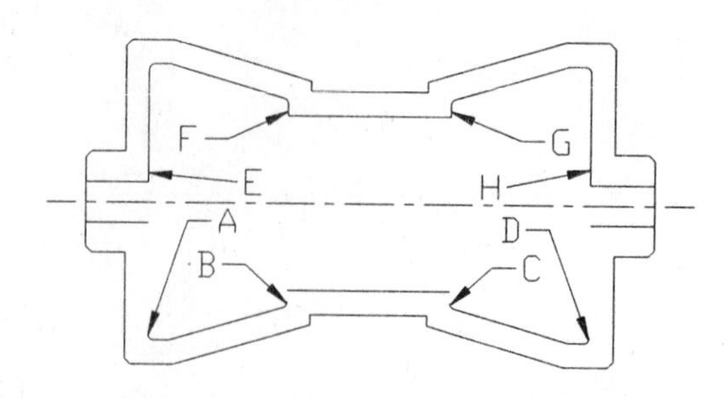

Figure 3.94 Entities mirrored

Issue the EXTEND command to extend four vertical lines. See Figure 3.95.

<Modify> **<Extend>**

Command: **EXTEND**
Select objects: [**Select A, B, C, and D (Figure 3.94).**]
Select objects: [**Enter**]
<Select object to extend>/Project/Edge/Undo: [**Select E, F, G, and H (Figure 3.94).**]
<Select object to extend>/Project/Edge/Undo: [**Enter**]

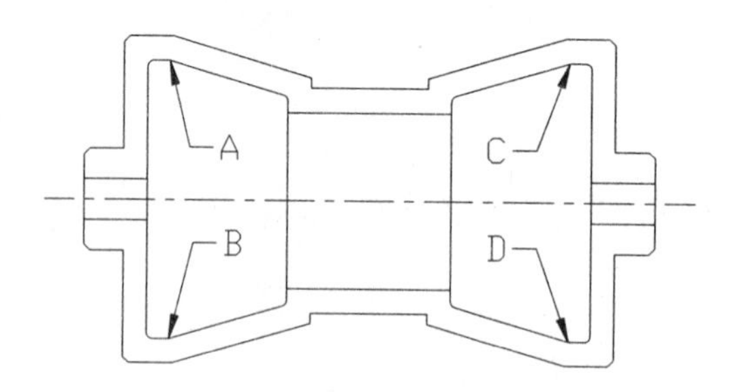

Figure 3.95 Four vertical lines extended

Create Lines

Apply the LINE command to draw two vertical lines. See Figure 3.96.

<Draw> **<Line>**

Command: **LINE**
From point: **INTER** of [**Select A (Figure 3.95).**]
To point: **INTER** of [**Select B (Figure 3.95).**]
To point: [**Enter**]

Command: [**Enter**]
LINE
From point: **INTER** of [**Select C (Figure 3.95).**]
To point: **INTER** of [**Select D (Figure 3.95).**]
To point: [**Enter**]

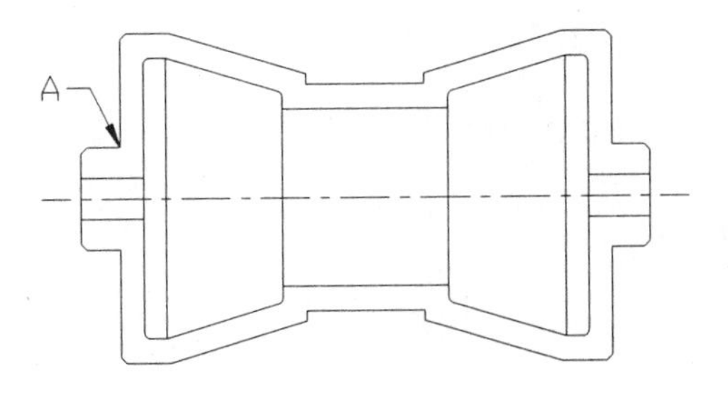

Figure 3.96 Two vertical lines drawn

Issue the LINE command three times. Apply the FROM object snap and use point A of Figure 3.96 as the Base point. See Figure 3.97.

Command: **[Enter]**
LINE
From point: **FROM**
Base point: **INT** of **[Select A (Figure 3.96).]**
<offset>: **@37<90**
To point: **@12<0**
To point: **[Enter]**

Command: **[Enter]**
LINE
From point: **FROM**
Base point: **INT** of **[Select A (Figure 3.96).]**
<offset>: **@13<90**
To point: **@12<0**
To point: **[Enter]**

Command: **[Enter]**
LINE
From point: **FROM**
Base point: **INT** of **[Select A (Figure 3.96).]**
<offset>: **@-10,25**
To point: **@32<180**
To point: **[Enter]**

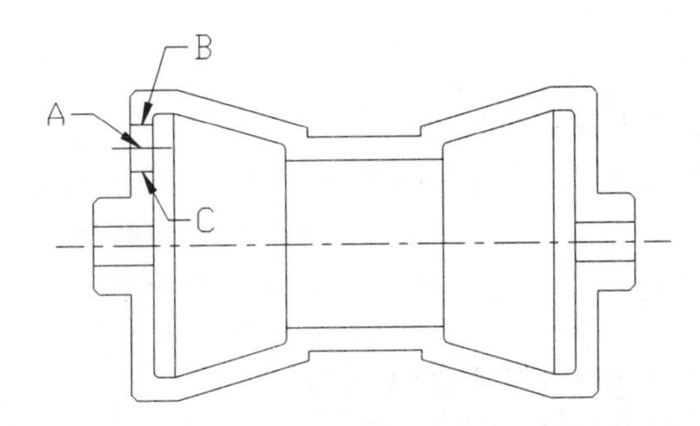

Figure 3.97 Three lines created

Change Entity Properties and Array

Change the properties of the last line to linetype Center2 and color yellow.

<Modify> **<Properties...>**

Select objects: **[Select A (Figure 3.97).]**
Select objects: **[Enter]**

[Modify Line
Properties
Color... **Yellow**
Linetype... **Center2**
OK]

Use the ARRAY command to array the three lines. See Figure 3.98.

<Modify> **<Array>**

Command: **ARRAY**
Select objects: **[Select A, B, and C (Figure 3.97).]**
Select objects: **[Enter]**
Rectangular or Polar array (R/P): **R**
Number of rows (---): **2**
Number of columns (|||): **2**
Unit cell or distance between rows (---): **-100**
Distance between columns (|||): **238**

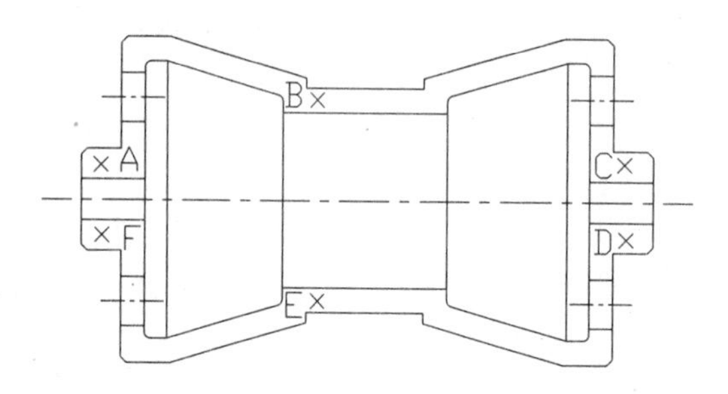

Figure 3.98 Lines arrayed

Set Layer and Create Boundary Hatching

To complete the sectional front view of the second component, you will add hatching lines. In creating the hatching for the first component, you used the HATCH command. Now you will use the BHATCH command to create an associative hatching. As its name implies, an associative hatching is associative with the boundaries. If you change the boundary, the hatching changes automatically.

Thaw all layers, unlock layer Hatch1, and set layer Hatch1 as the current layer.

[Object Properties] **[Layer Control]**

On, Thaw, & Lock	0
On, Thaw, & Lock	Center
On, Thaw, & Lock	Hatch
On, Thaw, & Unlock	Hatch1
On, Thaw, & Lock	Hidden
On, Thaw, & Lock	Point
On	Roller

Current layer: **Hatch1**

Create a hatching by using the BHATCH command. See Figure 3.99. This command automatically creates boundaries for hatching. See Figure 3.100.

<Draw> **<Hatch...>**

Command: **BHATCH**

[Boundary Hatch
Pattern...]

[Hatch pattern palette
ANSI31 **OK**]

[Scale: **3**
Angle: **0**
Boundary:
Pick Points<]

Select internal point: **[Select A, B, C, D, E, and F (Figure 3.98).]**
Select internal point: **[Enter]**

[Preview Hatch<]

Refer to Figure 3.100. Select the [Apply] button if the hatchings are correct.

[Apply]

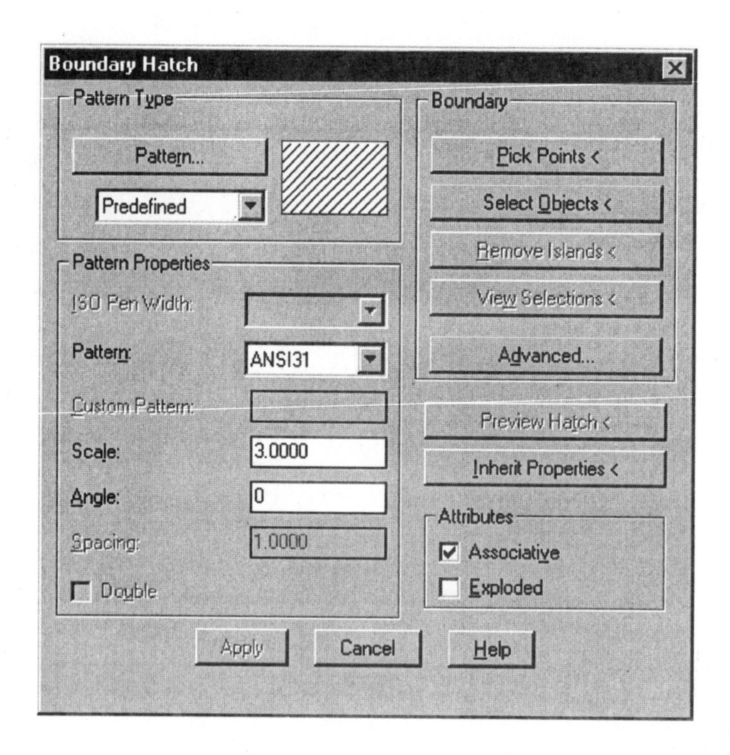

Figure 3.99 Boundary Hatch dialog box

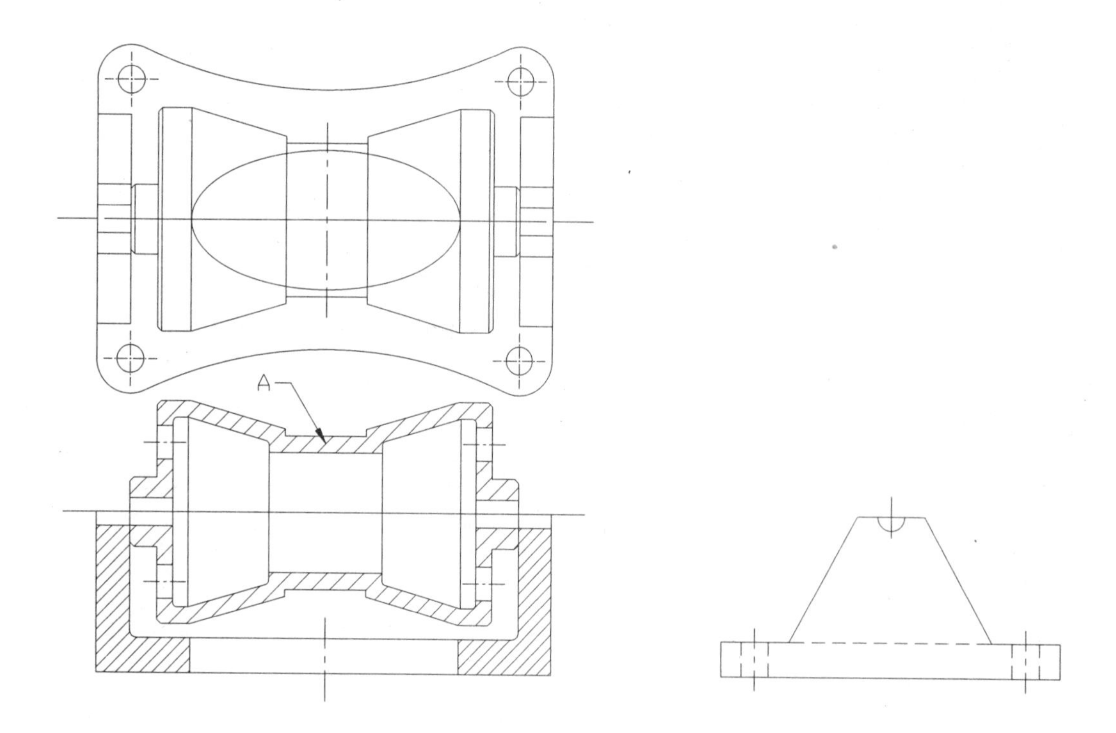

Figure 3.100 Hatching applied

The sectional front view for the second component is complete. To change the hatching, you may use the HATCHEDIT command. From the View pull-down menu, select the Toolbars... item to run the TOOLBAR command.

<View> **<Toolbars...>**

Select the Modify II item in the toolbars list to display the Modify II toolbar.

[Toolbars
Toolbars:
Modify II
Close]

From the Modify II toolbar, select the Edit Hatch icon to apply the HATCHEDIT command.

[Modify II] **[Edit Hatch]**

Command: **HATCHEDIT**
Select hatch object: **[Select A (Figure 3.100).]**

When the dialog box appears, change the hatch scale to 4, and then select the [Apply] button again. See Figure 3.101.

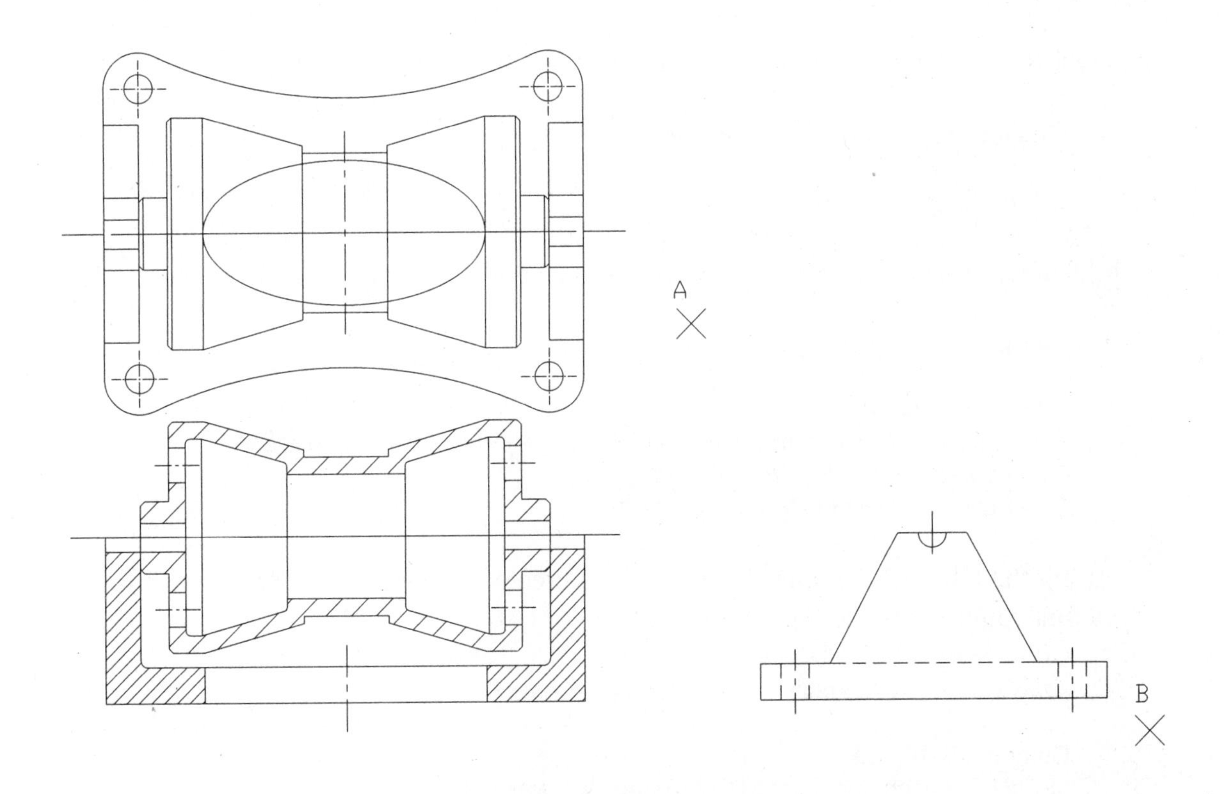

Figure 3.101 Hatching edited

The sectional front view of the second component is complete. In making this view, you copied and edited the entities from the top view and created a boundary hatching.

3.7 Side View of the Second Component

Figure 3.102 shows the completed side view of the second component. You will create all the entities on layer Roller. You will create and redefine a block and use the DIVIDE command to place the instances of the block at equal intervals along the circular center line.

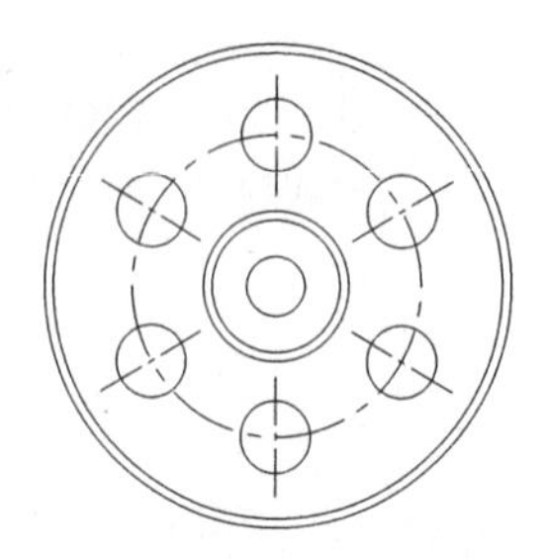

Figure 3.102 Side view of the second component

Set Layer and Create Circles

Set layer Roller as the current layer. You will create the side view on layer Roller.

[Object Properties] **[Layer Control]**

Current layer: **Roller**

Use the ZOOM command to zoom to the side view.

<View> **<Zoom>** **<Window>**

Command: **ZOOM**
All/Center/Dynamic/Extents/Previous/Scale(X/XP)/Window/<Realtime>: **W**
First corner: **[Select A (Figure 3.101).]**
Other corner: **[Select B (Figure 3.101).]**

Apply the CIRCLE command five times to create five circles concentric with the arc of the first component. See Figure 3.103.

<Draw> **<Circle>** **<Center, Radius>**

Command: **CIRCLE**
3P/2P/TTR<Center point>: **INT** of **[Select A (Figure 3.103).]**
Diameter/<Radius>: **10**

Command: **[Enter]**
CIRCLE
3P/2P/TTR<Center point>: **INT** of **[Select A (Figure 3.103).]**
Diameter/<Radius>: **22**

Command: **[Enter]**

CIRCLE
3P/2P/TTR<Center point>: **INT** of [**Select A (Figure 3.103).**]
Diameter/<Radius>: **25**

Command: **[Enter]**
CIRCLE
3P/2P/TTR<Center point>: **INT** of [**Select A (Figure 3.103).**]
Diameter/<Radius>: **77**

Command: **[Enter]**
CIRCLE
3P/2P/TTR<Center point>: **INT** of [**Select A (Figure 3.103).**]
Diameter/<Radius>: **80**

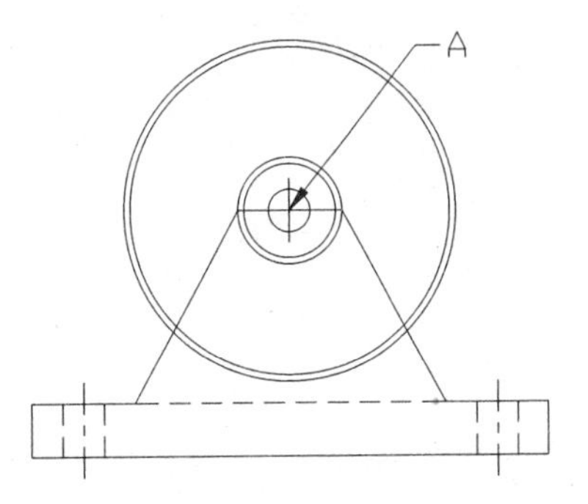

Figure 3.103 Five circles drawn

Repeat the CIRCLE command to draw one more concentric circle. Then use the CHPROP command to change its linetype to Center and its color to yellow. See Figure 3.104.

Command: **[Enter]**
CIRCLE
3P/2P/TTR<Center point>: **INT** of [**Select A (Figure 3.103).**]
Diameter/<Radius>: **50**

<Modify> **<Properties...>**

Select objects: **LAST**
Select objects: **[Enter]**

[Modify Line
Properties
Color... **Yellow**
Linetype... **Center**
OK]

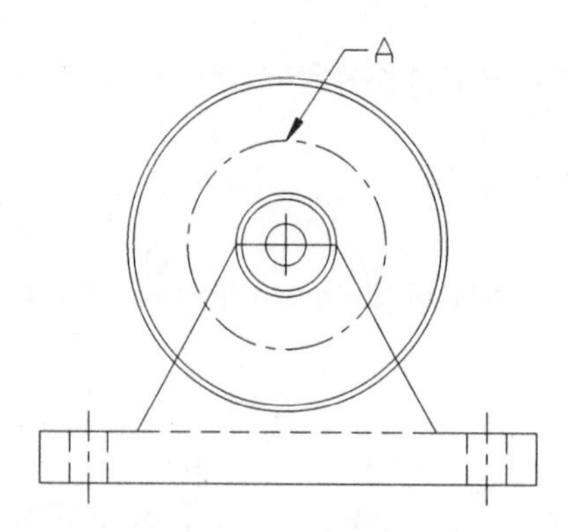

Figure 3.104 Circle drawn and properties changed

In order to concentrate on the side view of the second component, freeze all layers except the current layer.

[Object Properties] **[Layer Control]**

Freeze	0
Freeze	Center
Freeze	Hatch
Freeze	Hatch1
Freeze	Hidden
Freeze	Point
On	Roller

Current layer: **Roller**

Create and Scale Polygon

From the Draw pull-down menu, select the Polygon item to use the POLYGON command to create a polygon with five sides and a circumscribing radius of 15 units. See Figure 3.105.

<Modify> **<Polygon>**

Command: **POLYGON**
Number of sides: **5**
Edge/<Center of polygon>: **QUA** of **[Select A (Figure 3.104).]**
Inscribed in circle/Circumscribed about circle (I/C): **C**
Radius of circle: **15**

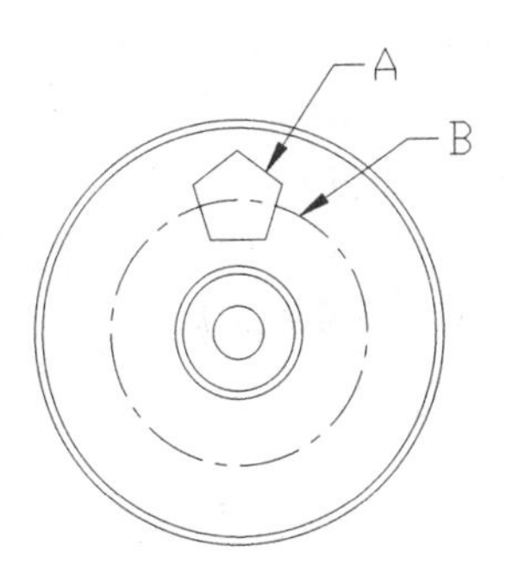

Figure 3.105 Polygon created

Suppose the polygon that you have created is too large. Select the Scale item of the Modify pull-down menu to run the SCALE command to scale it down.

<Modify> **<Scale>**

Command: **SCALE**
Select objects: [**Select A (Figure 3.105).**]
Select objects: [**Enter**]
Base point: **QUA** of [**Select B (Figure 3.105).**]
<Scale factor>/Reference: **12/15**

Polar Array and Undo

You need to create six polygons. To produce five more polygons, use the ARRAY command to make a polar array. See Figure 3.106.

<Modify> **<Array>**

Command: **ARRAY**
Select objects: **LAST**
Select objects: [**Enter**]
Rectangular or Polar array (R/P): **P**
Base/<Specify center point of array>: **CEN** of [**Select B (Figure 3.105).**]
Number of items: **6**
Angle to fill (+ =CCW,- =CW): **360**
Rotate objects as they are copied? <Y> [**Enter**]

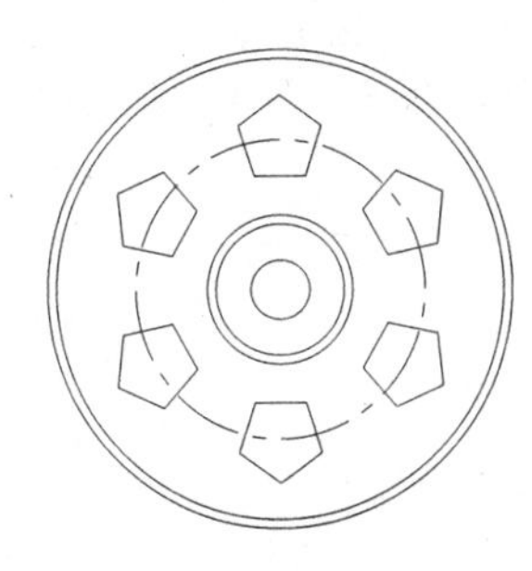

Figure 3.106 Polygon arrayed

The ARRAY command and the COPY command duplicate entities in your drawing database. There is a better way of handling identical or similar entities in the drawing. Use the U command to undo the array of polygons. Do not over-undo! Otherwise, use the REDO command. Remember that you may UNDO as many times as you like, but you can REDO only once! See Figure 3.105 again.

<Edit> **<Undo>**

Create Block

To handle repeating entities in a drawing, it is better to form a block and then save the block in the drawing database under a user-defined name. Later on, you may make use of the block by applying the INSERT, MINSERT, DIVIDE, and MEASURE commands.

To create a block, you can use the BLOCK command or the BMAKE command. The BLOCK command works at the command line interface. The BMAKE command brings out dialog boxes.

From the Draw pull-down menu, select the Block cascading menu and then the Make... item to use the BMAKE command to create a block from the polygon. See Figure 3.107.

<Draw> **<Block>** **<Make...>**

Command: **BMAKE**

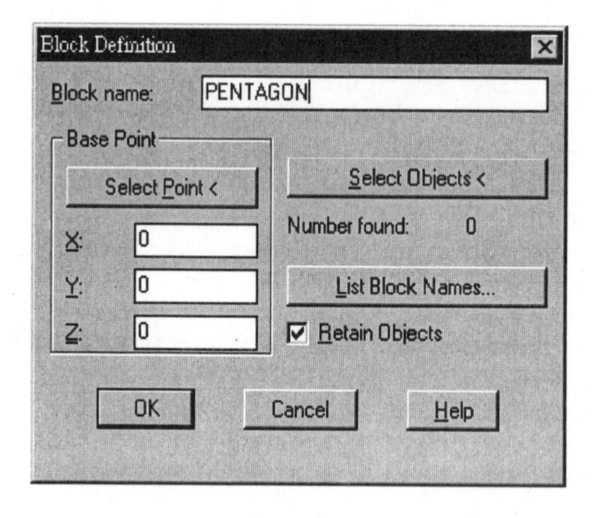

Figure 3.107 Block Definition dialog box

[Block Definition
Block name: **PENTAGON**
Select Objects<]

Select objects: [**Select the pentagon.**]
Select objects: [**Enter**]

[Base Point: **Select Point<**]

Insertion base point: **QUA** of [**Select B (Figure 3.105).**]

[Retain objects: **NO**
OK]

In the last command, AutoCAD deletes the polygon because you have chosen not to retain the object. If you want to retrieve an entity that is deleted after it is placed in a block, you must not use the UNDO command because this command undoes the effect of blocking as well. Instead, you should use the OOPS command. This command brings back the erased entities.

Command: **OOPS**

The polygon returns after the OOPS command. Because you do not need the polygon, run the U command to undo the OOPS command.

<Edit> **<Undo>**

Command: **U**

Now, you have defined a block called Pentagon. In order to know the block names defined in your drawing, run the BLOCK command and take the ? option.

Command: **BLOCK**
Block name (or ?): **?**
Block(s) to list <*>: **[Enter]**
Defined blocks.
PENTAGON

User Blocks	External References	Dependent Blocks	Unnamed Blocks
1	0	0	0

Insert Block by Dividing

As mentioned earlier, a block can be used by the DIVIDE command. From the Draw pull-down menu, select the Point cascading menu and then the Divide item to run the DIVIDE command. Use the BLOCK option. See Figure 3.108.

<Draw> **<Point>** **<Divide>**

Command: **DIVIDE**
Select object to divide: [**Select B (Figure 3.105).**]
<Number of segments>/Block: **BLOCK**
Block name to insert: **PENTAGON**
Align block with object? <Y>: **Y**
Number of segments: **6**

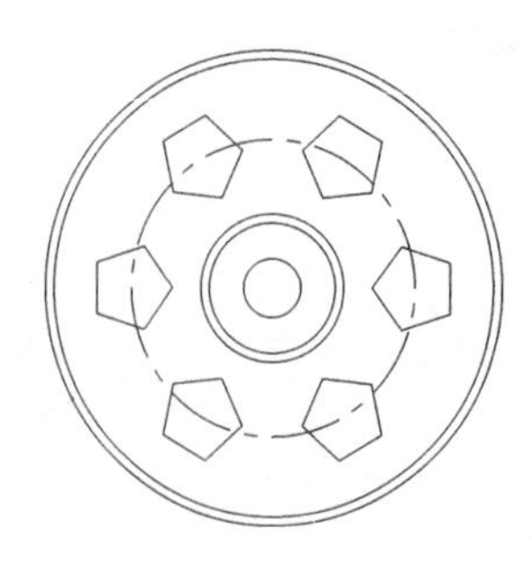

Figure 3.108 Insertion of blocks with the DIVIDE command

Refer to Figure 3.106 and Figure 3.108. Compare the results achieved by the DIVIDE command and the ARRAY command.

With the ARRAY command, the selected object is used as the starting point, and the orientations of the arrayed objects follow that of the selected object.

With the DIVIDE command, the first block is placed at the start point of the entity, in this case a circle. For a circle, the default starting point is the zero degree quadrant point. As for orientation, the upward direction points to the center of the circle. Therefore, the "top" of the polygon at 3 o'clock position points toward the center of the pitch circle.

In the drawing database, the ARRAY command duplicates the entities many times, but the DIVIDE command refers only to the entities described by the block. Generally speaking, the BLOCK command uses less memory storage.

Redefining a Block

There is a defined block called PENTAGON in your drawing. You will redefine this block by changing its content from a pentagon to a circle and a center line.

Execute the CIRCLE command to draw a circle at (500,-20) with a radius of 12 units.

<Draw> **<Circle>** **<Center, Radius>**

Command: **CIRCLE**
3P/2P/TTR<Center point>: **500,-20**
Diameter/<Radius>: **12**

Run the LINE command to create a line passing through the center of the circle. Use the FROM object snap. Then take the center of the last circle as the Base point.

<Draw> **<Line>**

Command: **LINE**
From point: **FROM**
Base point: **CEN** of [**Select the last circle.**]
<offset>: **@20<90**
To point: **@40<270**
To point: [**Enter**]

Change the last line to linetype Center2 and color yellow. See Figure 3.109.

<Modify> **<Properties...>**

Select objects: **LAST**
Select objects: **[Enter]**

[Modify Line
Properties
Layer... **Center2**
Color... **Yellow**
OK]

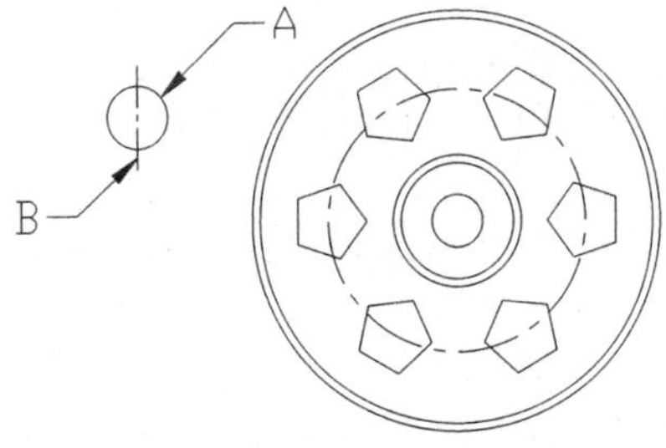

Figure 3.109 Circle and center line created

After creating a circle and a center line, apply the BLOCK command at the command line interface to redefine the block PENTAGON. See Figure 3.110.

Command: **BLOCK**
Block name (or ?): **PENTAGON**
Block PENTAGON already exists.
Redefine it? <N> **Y**
Insertion base point: **CEN** of **[Select A (Figure 3.109).]**
Select objects: **[Select A and B (Figure 3.109).]**
Select objects: **[Enter]**

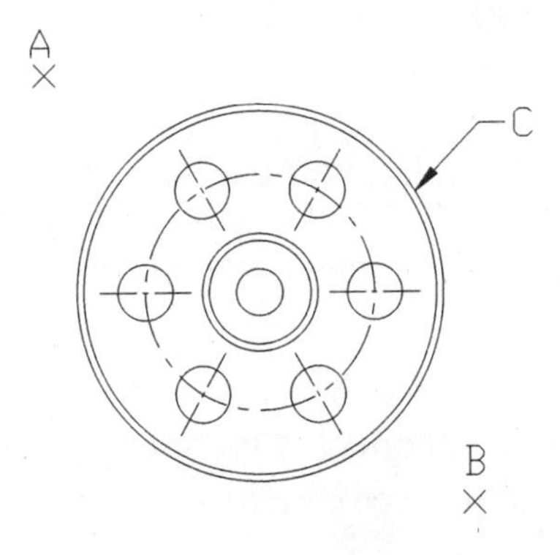

Figure 3.110 Blocks redefined

As you can see in Figure 3.110, all the inserted blocks change after the block is redefined. You will learn more about blocks in the next chapter.

Rotate

Because the starting point of the first block of the array is at the 0° position, use the ROTATE command to rotate the inserted blocks for 90°. See Figure 3.111.

<Modify> **<Rotate>**

Command: **ROTATE**
Select objects: **W**
First corner: [**Select A (Figure 3.110).**]
Other corner: [**Select B (Figure 3.110).**]
Select objects: [**Enter**]
Base point: **CEN** of [**Select C (Figure 3.110).**]
<Rotation angle>/Reference: **90**

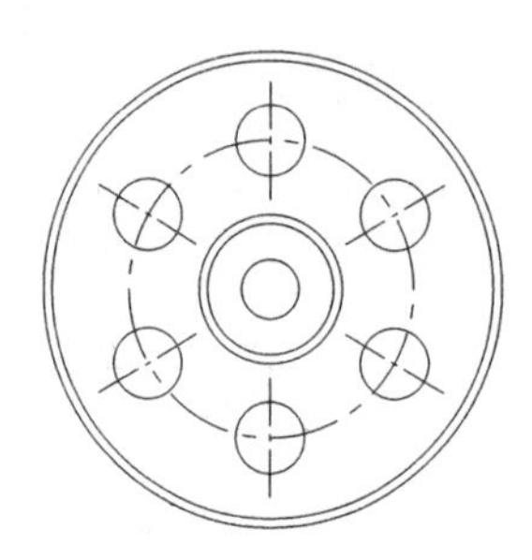

Figure 3.111 Entities rotated

Rename Objects

To rename a block or other named objects in a drawing, you may use the RENAME command or the DDRENAME command.

Command: **RENAME**
Block/Dimstyle/LAyer/LType/Style/Ucs/View/VPort: **BLOCK**
Old block name: **PENTAGON**
New block name: **HOLE**

To run the DDRENAME command, you may select Rename... item from the Format pull-down menu.

<Format> **<Rename...>**

The side view of the second component is complete. In making this view, you created blocks, used the block in dividing, redefined a block, renamed a block, and rotated objects.

Complete Drawing of Two Components

Unlock and thaw all layers. Two components in three orthographic views are complete. See Figure 3.112.

[Object Properties] **[Layer Control]**

Thaw & Unlock	0
Thaw & Unlock	Center
Thaw & Unlock	Hatch
Thaw & Unlock	Hatch1
Thaw & Unlock	Hidden
Thaw & Unlock	Point
On	Roller

Current layer: **Roller**

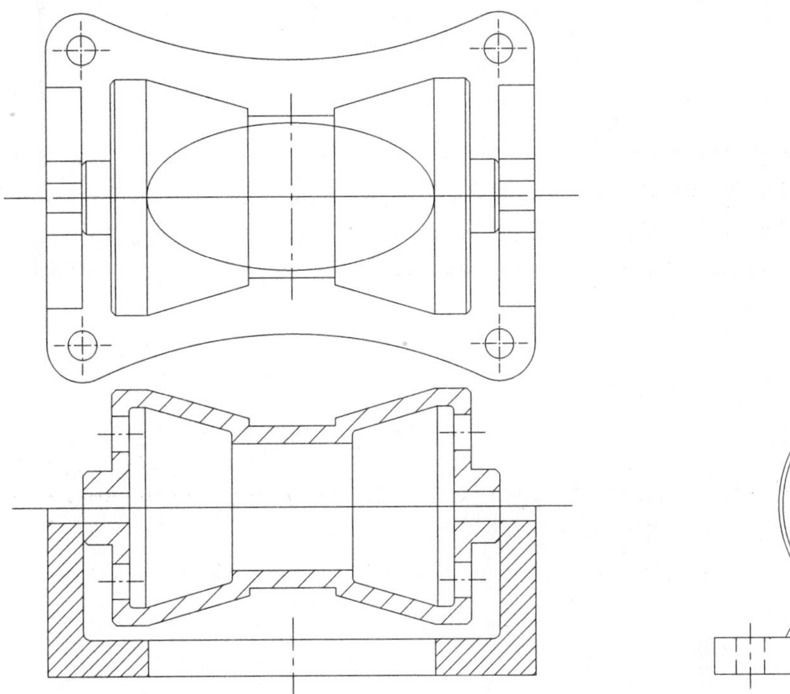

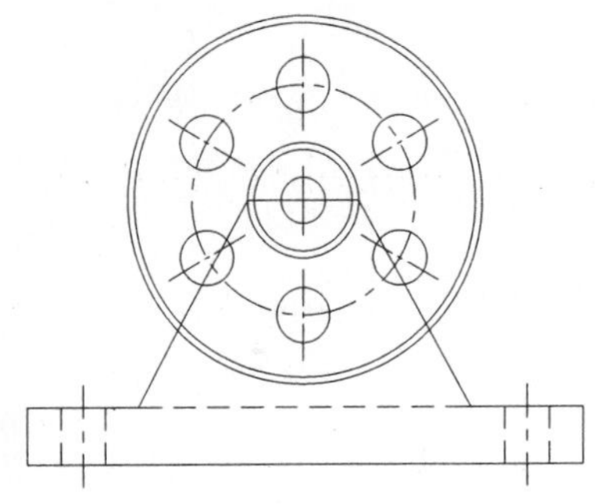

Figure 3.112 All layers unlocked and thawed

Save the File

Use the SAVE command or the SAVEAS command to save the drawing. You will continue to make the third component in the next chapter.

<File> **<Save>**

File name: **PIPE.DWG**

In constructing the orthographic views of the two engineering components, you learned most of the draw, modify, and utility commands.

3.8 Floor Plan

Figure 3.113 shows a floor plan. In constructing this drawing, you will use multiple lines, polyline, trace, donut, and 2D solid object. Start a new drawing with the NEW command. Use a wizard and quick setup procedure. Select decimal units and set an initial display area of 8,000 x 12,000.

<File> **<New...>**

Command: **NEW**

[Create New Drawing
Use a Wizard **Quick Setup**
OK]
[Step 1: Units
Decimal **Next>>**]
[Step 2: Area
Width: **8000** Length: **12000**
Done]

Use the LAYER command to create an additional layer called Plan, and set this layer as the current layer.

<Format> **<Layer...>**

Current layer: **Plan**

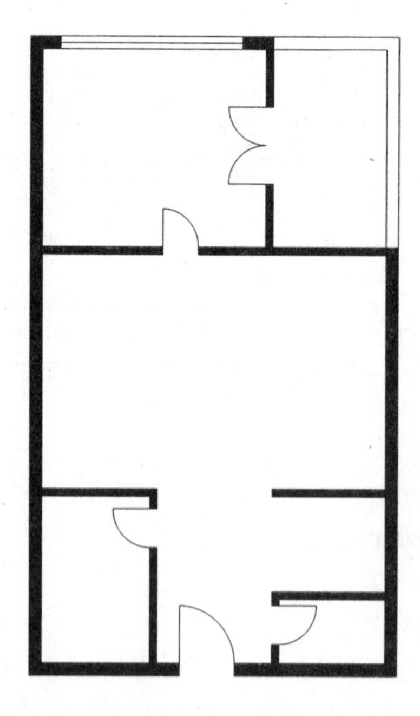

Figure 3.113 Floor plan

Multiple Lines

Multiple lines are sets of parallel lines. They are very useful in 2D architectural or interior design drawings. By using the MLINE command, you can create up to 16 parallel lines at the same time. The style of the multiple lines is set by the MLSTYLE command. To edit a multiple line, you can use the MLEDIT command.

From the Format pull-down menu, select the Multiline Style... item to use the MLSTYLE command to set up four multiline styles. See Figure 3.114.

<Format> **<Multiline Style...>**

Command: **MLSTYLE**

[Multiline Style
Name: **120WALL**
Description: **Internal Wall**
Add]

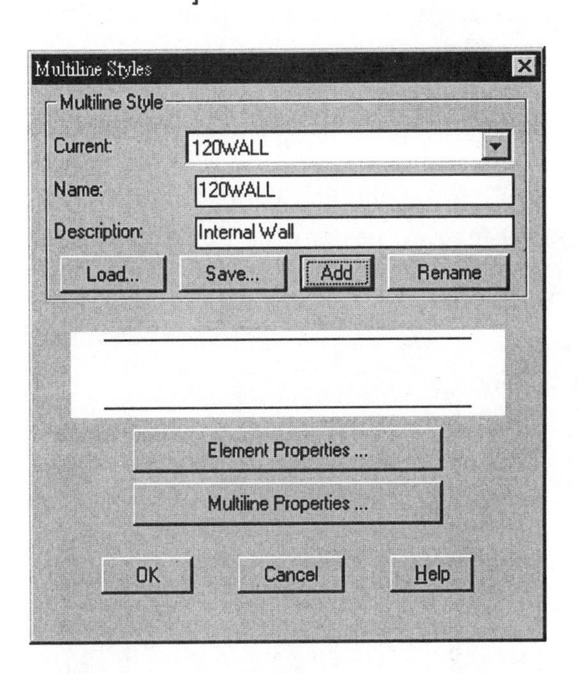

Figure 3.114 Multiline Style dialog box

Select the [Element Properties...] button to bring out another dialog box. See Figure 3.115.

[Element Properties...]

The second dialog box allows you to set the number, color, and linetypes of parallel lines. Change the offset of the first line to 60 units, and change the offset of the second line to -60 units. Then select the [OK] button.

[Element Properties
Offset Color Ltype
60 BYLAYER BYLAYER
-60 BYLAYER BYLAYER

OK]

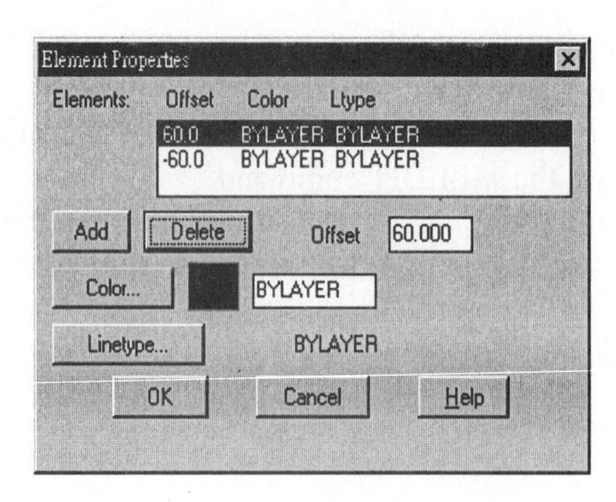

Figure 3.115 Element Properties dialog box

When the main dialog box returns, select the [Multiline Properties...] button. Another dialog box is displayed. See Figure 3.116.

[Multiline Properties...]

In this dialog box, you can set the end cap style and fill the multilines. Select the boxes under Start and End that are next to the item Line. Also turn Fill on with color cyan. Then select the [OK] button.

[Multiline Properties
Line **Start** **End**
Angle **90°** **90°**
Fill: **Yes**
Color: **Cyan**
OK]

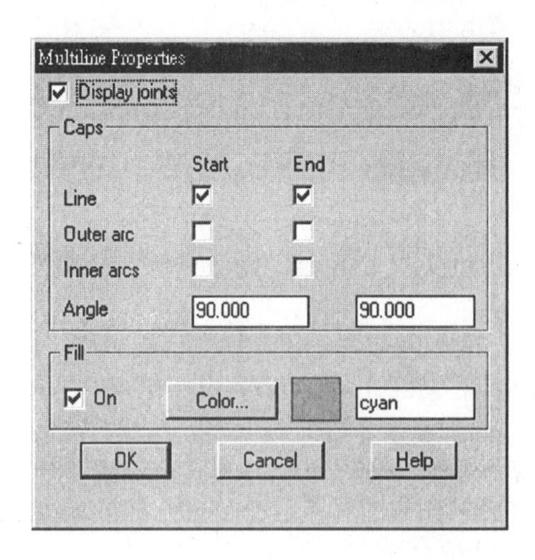

Figure 3.116 Multiline Properties dialog box

On returning to the main dialog box, select the [Save...] button. A multiline style is created.

[Save...]

Try on your own to create three more multiline styles. In total, you should have four styles. See the following table.

Name	Description	Element Properties			Multiline Properties
		Offset	Color	Ltype	
120WALL	Internal Wall	60	BYLAYER	BYLAYER	Line Start End
		-60	BYLAYER	BYLAYER	Angle 90° 90°
					Fill: Yes
					Color: Cyan
200WALL	External Wall	100	BYLAYER	BYLAYER	Line Start End
		100	BYLAYER	BYLAYER	Angle 90° 90°
					Fill: Yes
					Color: Cyan
200WINDOW	Window	100	BYLAYER	BYLAYER	Line Start End
		0	BYLAYER	BYLAYER	Angle 90° 90°
		-100	BYLAYER	BYLAYER	Fill: No
200BALCONY	Balcony	100	BYLAYER	BYLAYER	Line Start End
		-100	BYLAYER	BYLAYER	Angle 90° 90°
					Fill: No

After setting four multiline styles, you can start to create the floor plan. From the Draw pull-down menu, select the Multiline item to run the MLINE command to create three segments of the outer walls. See Figure 3.117.

<Draw> **<Multiline>**

Command: **MLINE**
Justification = Top, Scale = 1.00, Style = 200WINDOW
Justification/Scale/STyle/<From point>: **ST**

Choose the 200WALL style for the first part of the floor plan.

Mstyle name (or ?): **200WALL**
Justification = Top, Scale = 1.00, Style = 200WALL
Justification/Scale/STyle/<From point>: **J**

Justification determines how the multiline is drawn between the points that you select. Top justification means that all measurements will be taken from the topmost multiline.

Top/Zero/Bottom <top>: **TOP**
Justification = Top, Scale = 1.00, Style = 200WALL
Justification/Scale/STyle/<From point>: **2500,0**
<To point>: **@2500<180**
Undo/<To point>: **@10000<90**
Close/Undo/<To point>: **@500<0**
Close/Undo/<To point>: **[Enter]**

Figure 3.117 Multiline 200WALL created

Because the first multiline segment draws from right to left, AB of Figure 3.117 is the topmost multiline. Thus, the lengths of the walls AB, BC, and CD are 2500, 10000, and 500, respectively. Create another multiline for the window opening. Use style 200WINDOW. Again, justification is top. See Figure 3.118.

Command: **[Enter]**
MLINE
Justification = Top, Scale = 1.00, Style = 200WALL
Justification/Scale/STyle/<From point>: **ST**
Mstyle name (or ?): 200WINDOW
Justification = Top, Scale = 1.00, Style = 200WINDOW
Justification/Scale/STyle/<From point>: **J**
Top/Zero/Bottom <top>: **TOP**
Justification = Top, Scale = 1.00, Style = 200WINDOW
Justification/Scale/STyle/<From point>: **END** of **[Select E (Figure 3.117).]**
<To point>: **@3000<0**
Undo/<To point>: **[Enter]**

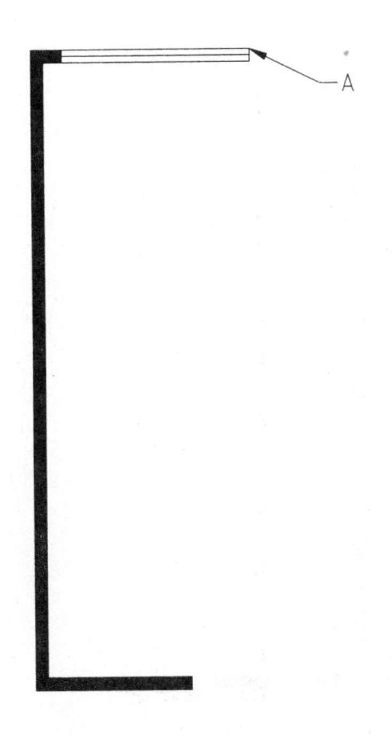

Figure 3.118 Multiline 200WINDOW created

Construct another multiline segment with 200WALL style. See Figure 3.119.

Command: **[Enter]**
MLINE
Justification = Top, Scale = 1.00, Style = 200WINDOW
Justification/Scale/STyle/<From point>: **ST**
Mstyle name (or ?): **200WALL**
Justification = Top, Scale = 1.00, Style = 200WALL
Justification/Scale/STyle/<From point>: **J**
Top/Zero/Bottom <top>: **TOP**
Justification = Top, Scale = 1.00, Style = 200WALL
Justification/Scale/STyle/<From point>: **END** of **[Select A (Figure 3.118).]**
<To point>: **@500<0**
Undo/<To point>: **[Enter]**

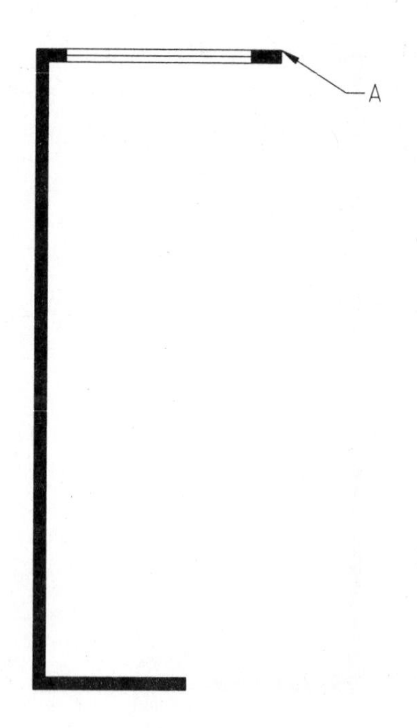

Figure 3.119 Multiline 200WALL created

Draw two segments of the balcony with 200BALCONY style. See Figure 3.120.

```
Command: [Enter]
MLINE
Justification = Top, Scale = 1.00, Style = 200WALL
Justification/Scale/STyle/<From point>: ST
Mstyle name (or ?): 200BALCONY
Justification = Top, Scale = 1.00, Style = 200BALCONY
Justification/Scale/STyle/<From point>: J
Top/Zero/Bottom <top>: TOP
Justification = Top, Scale = 1.00, Style = 200BALCONY
Justification/Scale/STyle/<From point>: END of [Select A (Figure 3.119).]
<To point>: @2000<0
Undo/<To point>: @3300<270
Close/Undo/<To point>: [Enter]
```

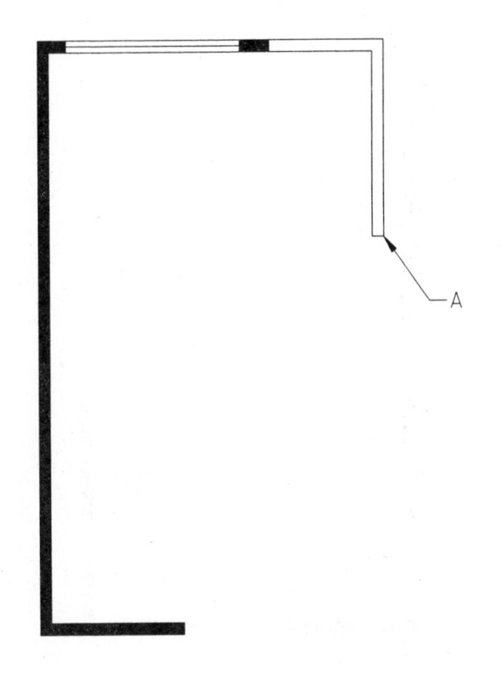

Figure 3.120 Multiline 200BALCONY created

Repeat the MLINE command to complete the outer wall. Use 200WALL style. See Figure 3.121.

```
Command: [Enter]
MLINE
Justification = Top, Scale = 1.00, Style = 200BALCONY
Justification/Scale/STyle/<From point>: ST
Mstyle name (or ?): 200WALL
Justification = Top, Scale = 1.00, Style = 200WALL
Justification/Scale/STyle/<From point>: J
Top/Zero/Bottom <top>: T
Justification = Top, Scale = 1.00, Style = 200WALL
Justification/Scale/STyle/<From point>: END of [Select A (Figure 3.120).]
<To point>: @6700<270
Undo/<To point>: @2600<180
Close/Undo/<To point>: [Enter]
```

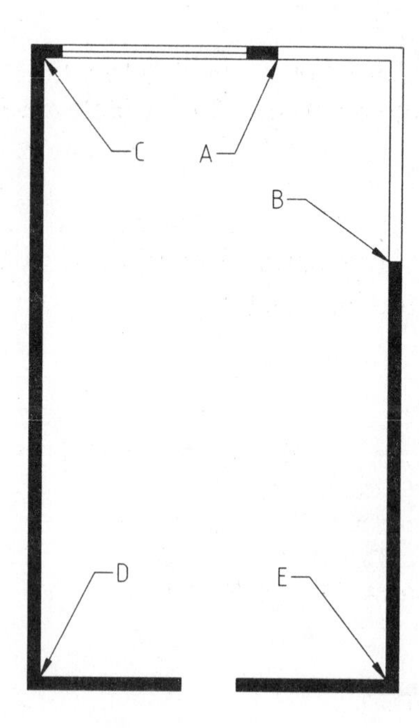

Figure 3.121 Outer walls created

The outer walls are complete. In making these walls, you have used three types of multiline styles: 200WALL, 200WINDOW, and 200BALCONY. Now repeat the MLINE command to create eight wall segments with 120WALL style. See Figure 3.122.

Command: **[Enter]**
MLINE
Justification = Top, Scale = 1.00, Style = 200WALL
Justification/Scale/STyle/<From point>: **ST**

The multiline style for the internal walls is 120WALL.

Mstyle name (or ?): **120WALL**
Justification = Top, Scale = 1.00, Style = 120WALL
Justification/Scale/STyle/<From point>: **J**
Top/Zero/Bottom <top>: **T**
Justification = Top, Scale = 1.00, Style = 120WALL
Justification/Scale/STyle/<From point>: **END** of **[Select A (Figure 3.121).]**
<To point>: **@900<270**
Undo/<To point>: **[Enter]**

With top justification and the second point of the multiline segment pointing in the 270° direction, the start point of the multiline aligns with point A of Figure 3.121.

The next line segment should align with point B of Figure 3.121. Because this line segment points in the 180° direction, we have bottom justification. See Figure 3.122.

Command: **[Enter]**
MLINE
Justification = Top, Scale = 1.00, Style = 120WALL
Justification/Scale/STyle/<From point>: **J**

Top/Zero/Bottom <top>: **B**
Justification = Bottom, Scale = 1.00, Style = 120WALL
Justification/Scale/STyle/<From point>: **END** of [**Select B (Figure 3.121).**]
<To point>: **@3000<180**
Undo/<To point>: [**Enter**]

The next segment is 3100 units offset from point C of Figure 3.121 and is in the 0° direction. Therefore, we have top justification.

Command: [**Enter**]
MLINE
Justification = Bottom, Scale = 1.00, Style = 120WALL
Justification/Scale/STyle/<From point>: **J**
Top/Zero/Bottom <bottom>: **T**
Justification/Scale/STyle/<From point>: **FROM**
Base point: **END** of [**Select C (Figure 3.121).**]
<Offset>: **@3100<270**
<To point>: **@2000<0**
Undo/<To point>: [**Enter**]

The next segment is 2600 units offset from point D of Figure 3.121 in the 90° direction.

Command: [**Enter**]
MLINE
Justification = Top, Scale = 1.00, Style = 120WALL
Justification/Scale/STyle/<From point>: **J**
Top/Zero/Bottom <top>: **B**
Justification = Bottom, Scale = 1.00, Style = 120WALL
Justification/Scale/STyle/<From point>: **FROM**
Base point: **END** of [**Select D (Figure 3.121).**]
<Offset>: **@2600<90**
<To point>: **@1800<0**
Undo/<To point>: **@200<270**
Close/Undo/<To point>: [**Enter**]

The next segment is 1800 units offset from point D of Figure 3.121 in the 0° direction.

Command: [**Enter**]
MLINE
Justification = Bottom, Scale = 1.00, Style = 120WALL
Justification/Scale/STyle/<From point>: **J**
Top/Zero/Bottom <bottom>: **T**
Justification = Top, Scale = 1.00, Style = 120WALL
Justification/Scale/STyle/<From point>: **FROM**
Base point: **END** of [**Select D (Figure 3.121).**]
<Offset>: **@1800<0**
<To point>: **@1800<90**
Undo/<To point>: [**Enter**]

The next segment is 2600 units offset from point E of Figure 3.121 in the 90° direction.

Command: **[Enter]**
MLINE
Justification = Top, Scale = 1.00, Style = 120WALL
Justification/Scale/STyle/<From point>: **J**
Top/Zero/Bottom <top>: **T**
Justification = Top, Scale = 1.00, Style = 120WALL
Justification/Scale/STyle/<From point>: **FROM**
Base point: **END** of **[Select E (Figure 3.121).]**
<Offset>: **@2600<90**
<To point>: **@1800<180**
Undo/<To point>: **[Enter]**

The next two segments are 1000 units offset from point E of Figure 3.121 in the 90° direction.

Command: **[Enter]**
MLINE
Justification = Top, Scale = 1.00, Style = 120WALL
Justification/Scale/STyle/<From point>: **J**
Top/Zero/Bottom <top>: **T**
Justification = Top, Scale = 1.00, Style = 120WALL
Justification/Scale/STyle/<From point>: **FROM**
Base point: **END** of **[Select E (Figure 3.121).]**
<Offset>: **@1000<90**
<To point>: **@1680<180**
Undo/<To point>: **@100<270**
Close/Undo/<To point>: **[Enter]**

The next segment are 1800 units offset from point E of Figure 3.121 in the 180° direction.

Command: **[Enter]**
MLINE
Justification = Top, Scale = 1.00, Style = 120WALL
Justification/Scale/STyle/<From point>: **J**
Top/Zero/Bottom <top>: **T**
Justification = Top, Scale = 1.00, Style = 120WALL
Justification/Scale/STyle/<From point>: **FROM**
Base point: **END** of **[Select E (Figure 3.121).]**
<Offset>: **@1800<180**
<To point>: **@100<90**
Undo/<To point>: **[Enter]**

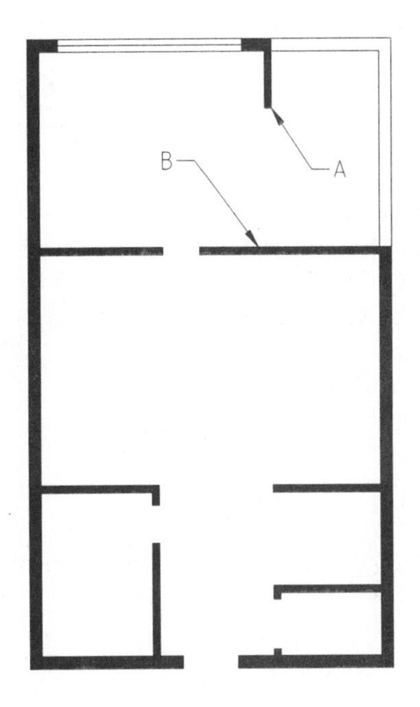

Figure 3.122 Internal walls created

Repeat the MLINE command to create another multiline segment.

Command: **[Enter]**
MLINE
Justification = Top, Scale = 1.00, Style = 120WALL
Justification/Scale/STyle/<From point>: **J**
Top/Zero/Bottom <top>: **T**
Justification = Top, Scale = 1.00, Style = 120WALL
Justification/Scale/STyle/<From point>: **FROM**
Base point: **END** of **[Select A (Figure 3.122).]**
<Offset>: **@1200<270**
<To point>: **PERP** to **[Select B (Figure 3.122).]**
Undo/<To point>: **[Enter]**

At present, the multilines created are filled. To speed up display regeneration, turn off the Fill mode by applying the FILL command. In order for the new Fill mode to take effect, issue the REGEN command. See Figure 3.123.

Command: **FILL**
ON/OFF <ON>: **OFF**

<View> **<Regen>**

Command: **REGEN**

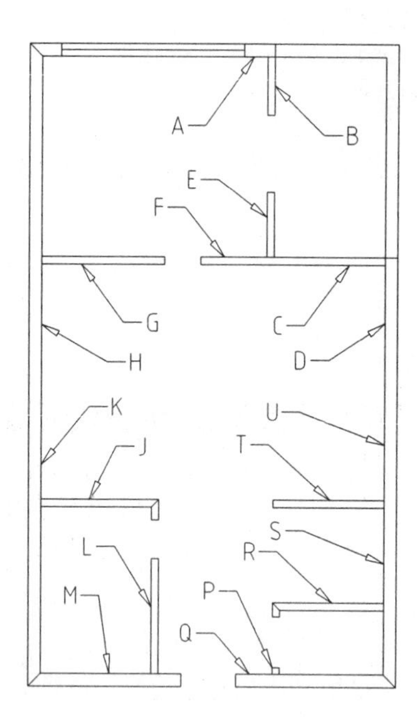

Figure 3.123 Fill mode turned off

You can join multiline line segments together by using the MLEDIT command. From the Modify pull-down menu, select the Object cascading menu and then the Multiline... item. See Figure 3.124.

<Modify> **<Object>** **<Multiline...>**

Command: **MLEDIT**

In the Multiline Edit Tools dialog box, select the first row of the third column, Corner joint. Then select the [OK] button.

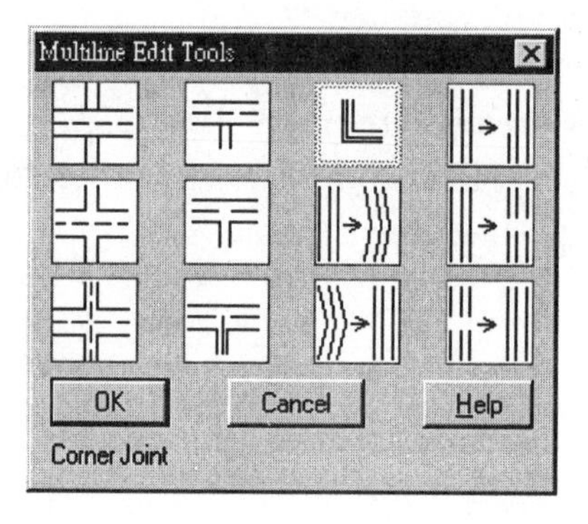

Figure 3.124 Corner joint selected

After selecting the corner joint, edit two joints.

Select first mline: **[Select A (Figure 3.123).]**
Select second mline: **[Select B (Figure 3.123).]**

Select first mline: [**Select C (Figure 3.123).**]
Select second mline: [**Select D (Figure 3.123).**]
Select first mline(or Undo): [**Enter**]

Repeat the MLEDIT command and select the second row of the second column, Open tee. See Figure 3.125.

Command: [**Enter**]
MLEDIT

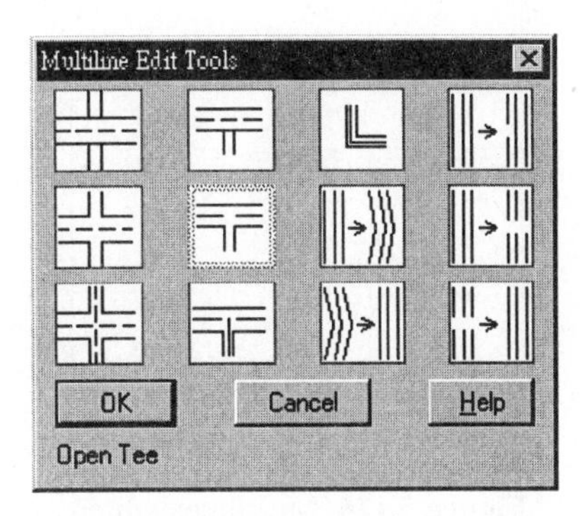

Figure 3.125 Open tee selected

Edit the other joints. The main walls and windows are complete. See Figure 3.126.

Select first mline: [**Select E (Figure 3.123).**]
Select second mline: [**Select F (Figure 3.123).**]
Select first mline: [**Select G (Figure 3.123).**]
Select second mline: [**Select H (Figure 3.123).**]
Select first mline: [**Select J (Figure 3.123).**]
Select second mline: [**Select K (Figure 3.123).**]
Select first mline: [**Select L (Figure 3.123).**]
Select second mline: [**Select M (Figure 3.123).**]
Select first mline: [**Select P (Figure 3.123).**]
Select second mline: [**Select Q (Figure 3.123).**]
Select first mline: [**Select R (Figure 3.123).**]
Select second mline: [**Select S (Figure 3.123).**]
Select first mline: [**Select T (Figure 3.123).**]
Select second mline: [**Select U (Figure 3.123).**]
Select first mline (or Undo): [**Enter**]

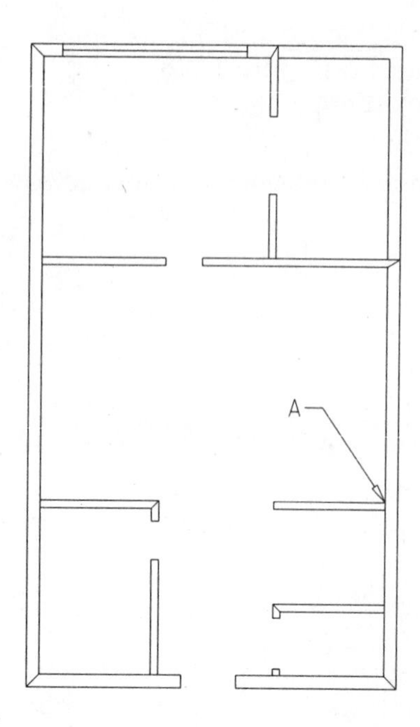

Figure 3.126 Multilines edited

Wide Objects

Besides multilines, there are other 2D entities with specific width. You can use the PLINE command, the TRACE command, the SOLID command, and the DONUT command to create different types of wide objects.

Use the PLINE command to create a wide polyline with a starting width of 100 and an ending width of 50. See Figure 3.127.

<Draw> **<Polyline>**

Command: **PLINE**
From point: **FROM**
Base point: **END** of **[Select A (Figure 3.126).]**
<Offset>: **@1000<90**

The default width of a polyline is zero. To create a wide polyline, you have to specify a width.

Arc/Close/Halfwidth/Length/Undo/Width/<Endpoint of line>: **WIDTH**
Starting width: **100**
Ending width: **50**
Arc/Close/Halfwidth/Length/Undo/Width/<Endpoint of line>: **@1000<180**
Arc/Close/Halfwidth/Length/Undo/Width/<Endpoint of line>: **[Enter]**

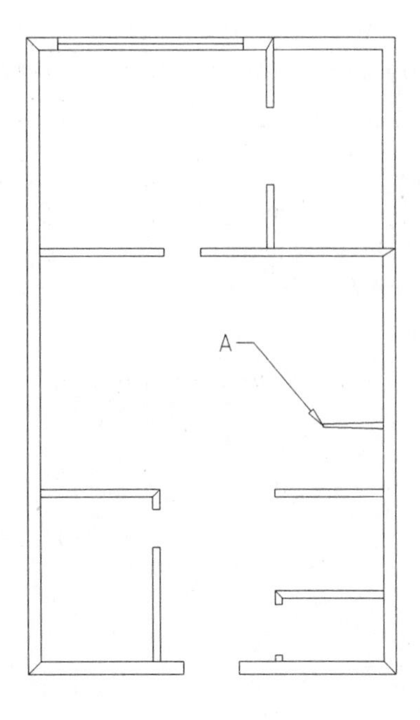

Figure 3.127 Wide polyline created

Create a 2D constant-width trace by using the TRACE command. See Figure 3.128.

Command: **TRACE**
Trace width: **50**
From point: **END** of [**Select A (Figure 3.127).**]
To point: **@1000<90**
To point: [**Enter**]

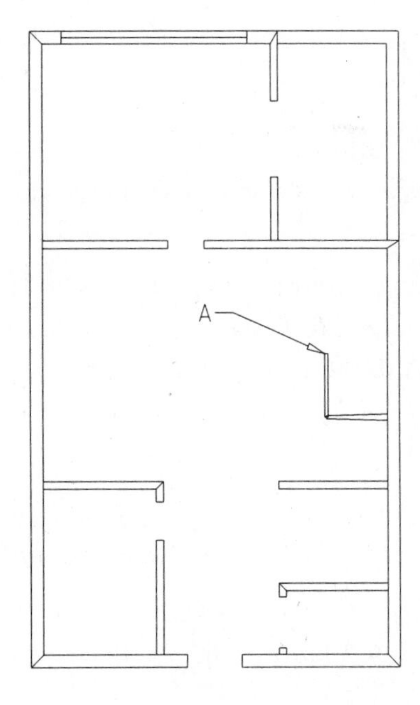

Figure 3.128 Trace created

A polyline can have variable width. A trace can have only constant width. If you want to have a quadrilateral object by specifying four corner positions, you can use the SOLID command. See Figure 3.129.

[Surfaces] **[2D Solid]**

Command: **SOLID**
First point: **END** of **[Select A (Figure 3.128).]**
Second point: **@500<180**
Third point: **@500,300**
Fourth point: **@500<180**
Third point: **@400,500**
Fourth point: **@200<180**
Third point: **[Enter]**

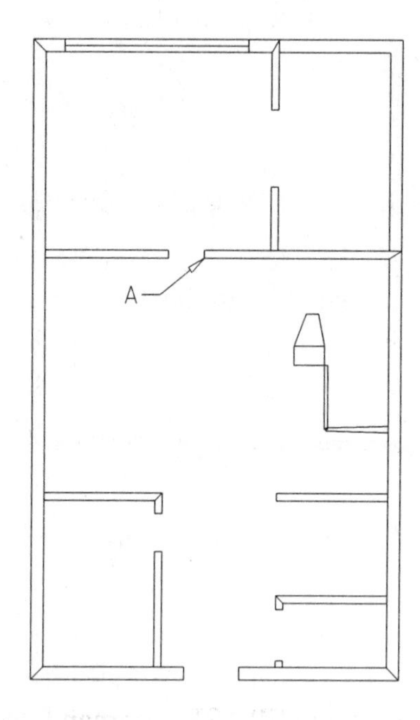

Figure 3.129 2D solid filled object created

So far, you have been dealing with linear wide objects. If you want to create a circular filled object, you can use the DONUT command. Execute the DONUT command to create a donut-shaped object. A donut is a circular object with width.

<Draw> **<Donut>**

Command: **DONUT**
Inside diameter: **300**
Outside diameter: **400**
Center of doughnut: **FROM**
Base point: **END** of **[Select A (Figure 129).]**
<Offset>: **@1000<270**
Center of doughnut: **[Enter]**

Set fill mode on with the FILL command and perform a regeneration with the REGEN command. See Figure 3.130.

Command: **FILL**
ON/OFF: **ON**

<View> **<Regen>**

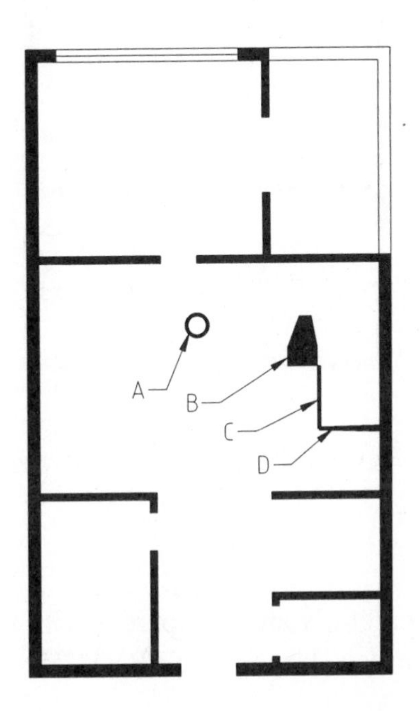

Figure 3.130 Donut created and fill mode turned on

In the foregoing, you made a polyline line with varying width, a trace, a 2D quadrilateral "solid" object, and a donut in your floor plan. They are not required in the final drawing. Erase them by using the ERASE command. See Figure 3.131.

<Modify> **<Erase>**

Command: **ERASE**
Select objects: [**Select A, B, C, and D (Figure 3.130).**]
Select objects: [**Enter**]

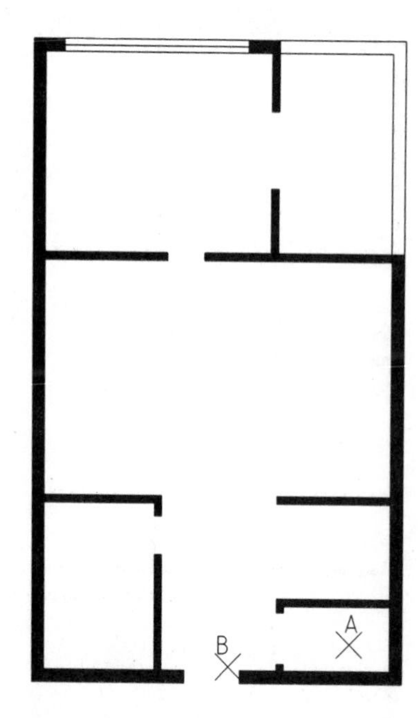

Figure 3.131 Floor plan completed

Edit by Stretching

If you examine Figure 3.131 carefully and compare it with Figure 3.113, you will see that the size of a wall is incorrect. Use the STRETCH command to correct it. See Figure 3.132.

<Modify> **<Stretch>**

Command: **STRETCH**
Select objects: **C**
First corner: **[Select A (Figure 3.131).]**
Other corner: **[Select B (Figure 3.131).]**
Select objects: **[Enter]**
Base point or displacement: **200<90**
Second point of displacement: **[Enter]**

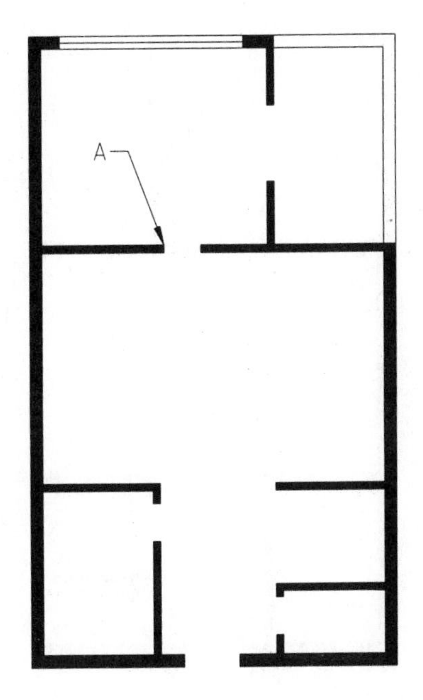

Figure 3.132 Wall stretched

Door

In the floor plan, there are six doors. You will draw one door, use it to create a block, and insert the block six times.

To draw the door, use the ARC command to draw an arc and the LINE command to draw a line. See Figure 3.133.

<Draw> **<Line>**

Command: **LINE**
From point: **END** of [**Select A (Figure 3.132).**]
To point: **@600<90**
To point: **[Enter]**

<Draw> **<Arc>** **<Center, Start, End>**

Command: **ARC**
Center/<Start point>: **C**
Center: **END** of [**Select A (Figure 3.132).**]
Start point: **@600<0**
Angle/Length of chord/<End point>: **@600<90**

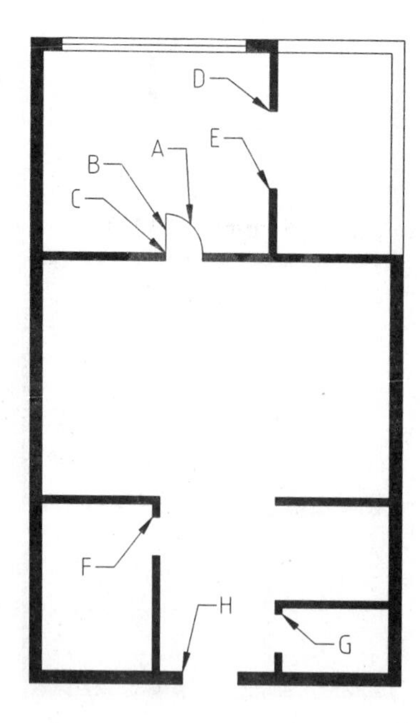

Figure 3.133 Door created

From the Draw pull-down menu, select the Block cascading menu and then the Make... item to run the BMAKE command to make a block called Door. Select the line and arc as objects for the block.

<Draw> **<Block>** **<Make...>**

Command: **BMAKE**

[Block Definition
Block name: **DOOR**
Base Point
Select Point<]

Insertion base point: **END** of [**Select C (Figure 3.133).**]

[**Select Objects<**]

Select objects: [**Select A and B (Figure 3.133).**]
Select objects: [**Enter**]

[Retain Objects **No**
OK]

Because the line and arc selected for making the block are not retained, they are deleted. Now you have a block called Door residing in memory. With this block, use the DDINSERT command to make six insertions.

Select the Block... item from the Insert pull-down menu to insert the block Door.

<Insert> **<Block...>**

Command: **DDINSERT**

[Insert
Block: **DOOR**
OK]

Insertion point: **END** of [**Select C (Figure 3.133).**]
X scale factor <1> / Corner / XYZ: **1**
Y scale factor (default=X): **1**
Rotation angle <0>: **0**

Repeat the DDINSERT command to insert five more doors. There are six doors in all. See Figure 3.134. Details of the remaining five insertions follow.

Insertion point	X scale factor	Y scale factor	Rotation angle
END of [Select D (Figure 3.133).]	1	-1	270
END of [Select E (Figure 3.133).]	1	1	90
END of [Select F (Figure 3.133).]	1	-1	270
END of [Select G (Figure 3.133).]	1	1	-90
END of [Select H (Figure 3.133).]	1.5	1.5	0

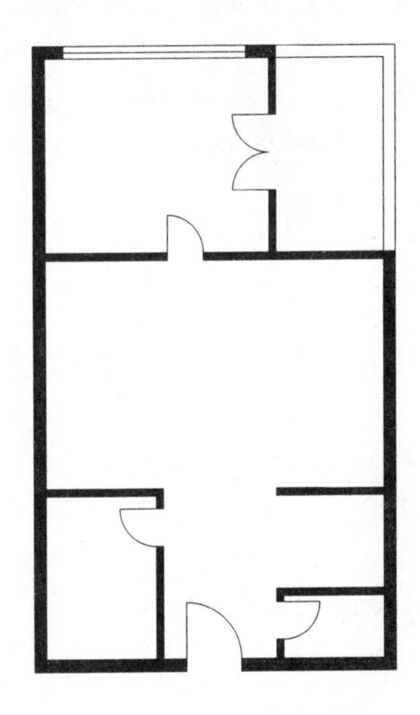

Figure 3.134 Doors inserted

The floor plan is complete. Save your drawing with the SAVE command. You will continue to work on this floor plan in the next chapter.

<File> **<Save>**

File name: **FLOOR.DWG**

3.9 Regions

A region is a special kind of entity. It is defined from a closed, 2D area. You can create a closed area using a combination of lines, arcs, polylines, planar 3D polylines, circles, arcs, ellipses, elliptical arcs, and splines. The entities that form a closed area must not intersect each other. With two or more regions, you can combine them together by joining (use the UNION command), cutting (use the SUBTRACT command), or intersecting (use the INTERSECT command). Figure 3.135 shows three washroom facilities for the floor plan that you created in Section 3.8. You will create them as 2D regions in this chapter and will use them as blocks in the next chapter. You will learn more about regions in Chapter 8.

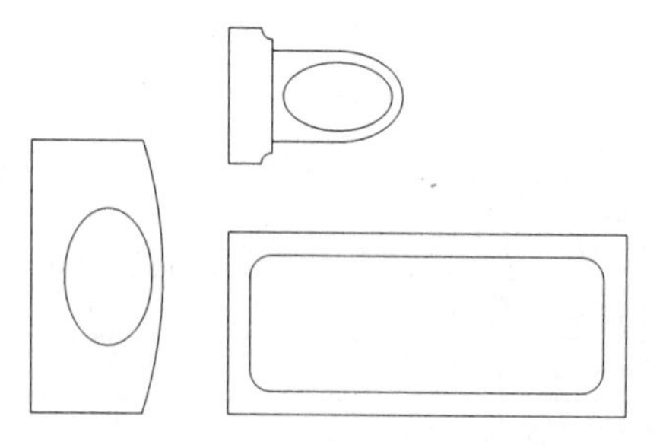

Figure 3.135 Washroom facilities

Start a new drawing. Use the quick wizard setup procedure with decimal units and an initial display area of 3000 units x 2000 units.

<File> **<New...>**

Command: **NEW**

[Create New Drawing
Use a Wizard
Quick Setup
OK]

[Quick Setup
Step 1: Units
Decimal
Next>>]

[Step 2: Area
Width: **3000**
Length: **2000**
Done]

Rectangle

First, you will draw a rectangle. A rectangle is a closed rectangular polyline. From the Draw pull-down menu, select the Rectangle item to use the RECTANG command. Set the fillet radius and chamfer distances of the four corners to zero.

<Draw> **<Rectangle>**

Command: **RECTANG**
Chamfer/Elevation/Fillet/Thickness/Width/<First corner>: **F**
Fillet radius for rectangles: **0**
Chamfer/Elevation/Fillet/Thickness/Width/<First corner>: **C**
First chamfer distance for rectangles: **0**
Second chamfer distance for rectangles<0>: **[Enter]**
Chamfer/Elevation/Fillet/Thickness/Width/<First corner>: **100,100**
Other corner: **@600,1200**

After making a rectangle, select the 2 Points item from the Circle cascading menu of the Draw menu to create a circle. See Figure 3.136.

<Draw> **<Circle>** **<2 Points>**

Command: **CIRCLE**
3P/2P/TTR/<Center point>: **2P**
First point on diameter: **MID** of [**Select A (Figure 3.136).**]
Second point on diameter: **@4000<180**

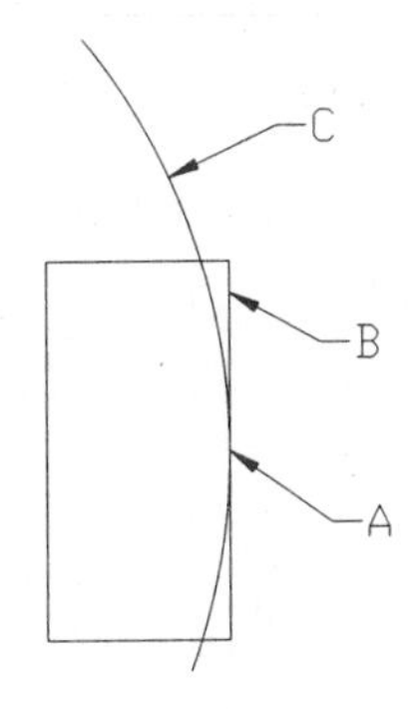

Figure 3.136 Rectangle and circle created

Use the REGION command to create two regions from the rectangle and the circle. When you form a region from a closed area, you can choose to retain or to delete the original objects. To delete the original objects after they have been used to form a region, set the DELOBJ system variable to 1.

Command: **DELOBJ**
New value for DELOBJ: **1**

From the Draw pull-down menu, select the Region item to use the REGION command. Because you have set DELOBJ to 1, the original rectangle and circle are deleted after the regions have been created.

<Draw> **<Region>**

Command: **REGION**
Select objects: [**Select B and C (Figure 3.136).**]
Select objects: **[Enter]**

Although there is no noticeable visual change in your screen display, the two objects are no longer rectangular polyline and a circle. They are regions.

Complex Region

You can unite, subtract, and intersect two or more regions to form a complex region by using the UNION, SUBTRACT, and INTERSECT commands. The union of two regions is a complex region whose area contains the area of the first region and that of the second region. The subtraction of two regions is a complex region whose area is found in the first region but not in the second region. The intersection of two regions is a complex region whose area is found in both the first region and the second region.

From the Modify pull-down menu, select the Boolean cascading menu and then the Intersect item to use the INTERSECT command to form an intersection of the rectangular and circular regions. See Figure 3.137.

<Modify> <Boolean> <Intersect>

Command: **INTERSECT**
Select objects: [**Select B and C (Figure 3.136).**]
Select objects: [**Enter**]

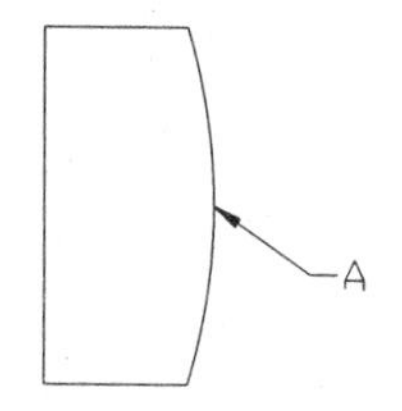

Figure 3.137 Common area of rectangular and circular regions

To complete the washbasin, add an ellipse. See Figure 3.138.

<Draw> <Ellipse> <Axis, End>

Command: **ELLIPSE**
Arc/Center/<Axis endpoint 1>: **FROM**
Base point: **QUA** of [**Select A (Figure 3.137).**]
<Offset>: **@50<180**
Axis endpoint 2: **@400<180**
<Other axis distance>/Rotation: **300**

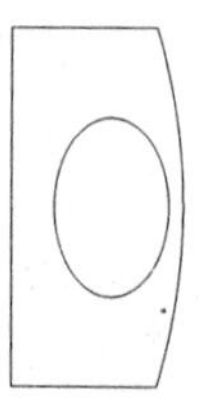

Figure 3.138 Ellipse created

The washbasin is complete. Now you will create the toilet. Use the RECTANG command to create two rectangles. See Figure 3.139.

<Draw> **<Rectangle>**

Command: **RECTANG**
Chamfer/Elevation/Fillet/Thickness/Width/<First corner>: **1000,1200**
Other corner: **@200,600**

Command: **[Enter]**
RECTANG
Chamfer/Elevation/Fillet/Thickness/Width/<First corner>: **FROM**
Base point: **END** of **[Select A (Figure 3.139).]**
<Offset>: **@100<90**
Other corner: **@300,400**

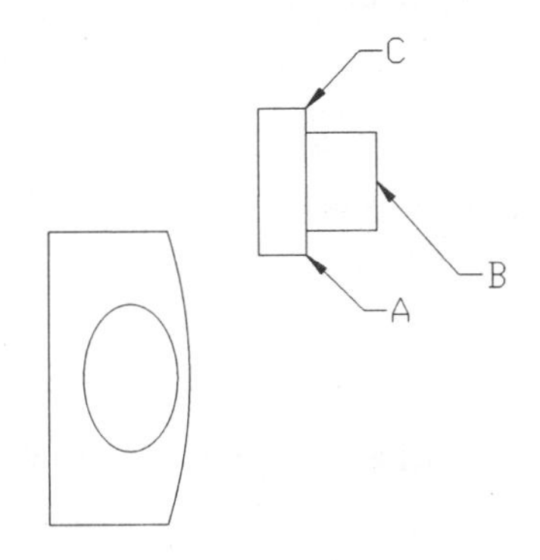

Figure 3.139 Rectangles created

Create two ellipses and two circles. See Figure 3.140.

<Draw> **<Ellipse>** **<Center>**

Command: **ELLIPSE**
Arc/Center/<Axis endpoint 1>: **C**
Center of ellipse: **MID** of **[Select B (Figure 3.139).]**
Axis endpoint: **@300<0**
<Other axis distance>/Rotation: **200**

Command: **[Enter]**
ELLIPSE
Arc/Center/<Axis endpoint 1>: **C**
Center of ellipse: **MID** of **[Select B (Figure 3.139).]**
Axis endpoint: **@250<0**
<Other axis distance>/Rotation: **150**

<Draw> **<Circle>** **<Center, Radius>**

Command: **CIRCLE**
3P/2P/TTR/<Center point>: **END** of **[Select A (Figure 3.139).]**
Diameter/<Radius>: **50**

Command: **[Enter]**

CIRCLE
3P/2P/TTR/<Center point>: **END** of **[Select C (Figure 3.139).]**
Diameter/<Radius>: **50**

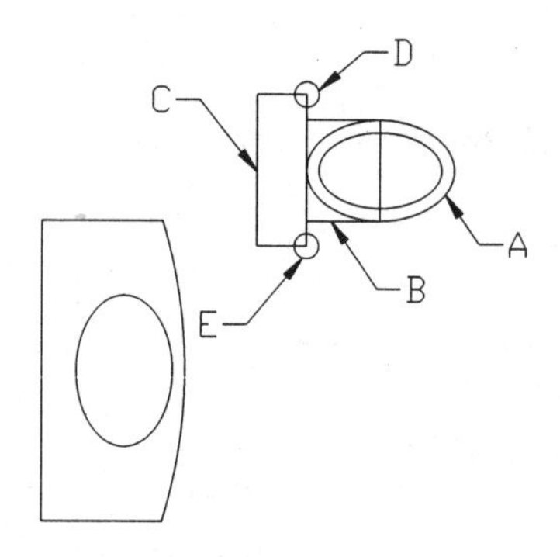

Figure 3.140 Ellipses and circles created

From the objects created, form five regions. Then use the UNION command and the SUBTRACT command to form complex regions. See Figure 3.141.

<Draw> **<Region>**

Command: **REGION**
Select objects: **[Select A, B, C, D, and E (Figure 3.140).]**
Select objects: **[Enter]**

<Modify> **<Boolean>** **<Union>**

Command: **UNION**
Select objects: **[Select A and B (Figure 3.140).]**
Select objects: **[Enter]**

<Modify> **<Boolean>** **<Subtract>**

Command: **SUBTRACT**
Select solids and regions to subtract from...
Select objects: **[Select C (Figure 3.140).]**
Select objects: **[Enter]**
Select solids and regions to subtract...
Select objects: **[Select D and E (Figure 3.140).]**
Select objects: **[Enter]**

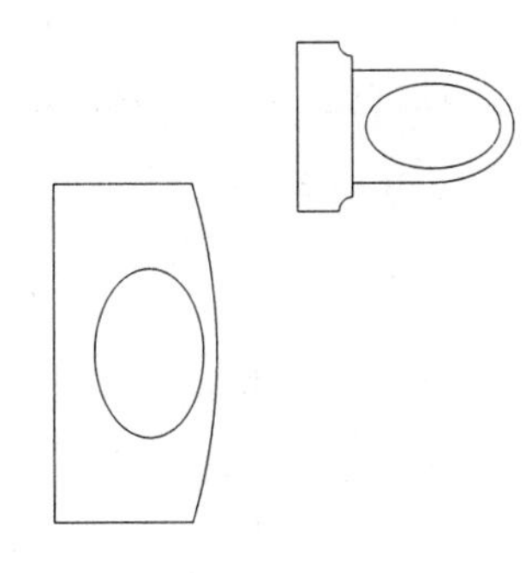

Figure 3.141 Regions united and subtracted

The toilet is complete. The next item to create is the bath. Create two rectangles. One rectangle has square corners. The other one has rounded corners. See Figure 3.142.

<Draw> **<Rectangle>**

Command: **RECTANG**
Chamfer/Elevation/Fillet/Thickness/Width/<First corner>: **1000,100**
Other corner: **@1800,800**

Command: [**Enter**]
RECTANG
Chamfer/Elevation/Fillet/Thickness/Width/<First corner>: **F**
Fillet radius for rectangles: **100**
Chamfer/Elevation/Fillet/Thickness/Width/<First corner>: **FROM**
Base point: **END** of [**Select A (Figure 3.142).**]
<Offset>: **@-100,-100**
Other corner: **FROM**
Base point: **END** of [**Select B (Figure 3.142).**]
<Offset>: **@100,100**

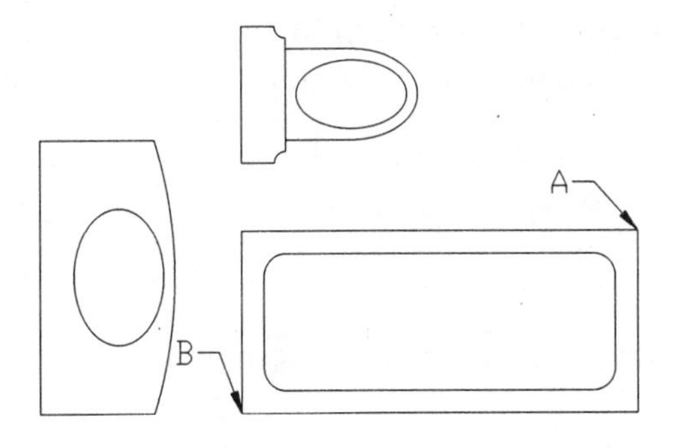

Figure 3.142 Rectangles created

The drawing is complete. Save your drawing. You will insert these washroom facilities in the floor plan in the next chapter.

<File> **<Save>**

File name: **WASH.DWG**

As you can see, converting entities into regions and applying Boolean operations to them is a quick way to produce a complex outline. If you want to reduce a region to an outline, you can use the EXPLODE command.

<Modify> **<Explode>**

Command: **EXPLODE**
Select objects: **[Select the region that you wanted to revert to an outline.]**
Select objects: **[Enter]**

3.10 Sketching

You can do free-hand sketching by using the SKETCH command. Sketching creates a set of short line segments that form a profile defined by the movement of the pointing device. You can use this command to draw contour lines of a map or signatures. Start a new drawing. Then apply the SKETCH command at the command line interface.

Command: **SKETCH**
Record increment: **5**

The Record increment is the length of the short line segments. If you set a very short increment, there will be a lot of line segments. Consequently, the file size will be very large. On the other hand, if you set a very long increment, any smooth curve you intend to draw will be approximated with too few line segments and will not appear smooth.

Sketch. Pen eXit Quit Record Erase Connect . <Pen down> <Pen up>
6 lines recorded.

Press the first button of the pointing device, and then move the device to create a series of free-hand line segments. When you finish sketching, press the first button again to raise the pen. Then press the second button to record the line segments.

Because the free-hand sketched lines are not required, use the ERASE command to delete them.

<Modify> **<Erase>**

Command: **ERASE**
Select objects: **[Select the sketched line segments.]**
Select objects: **[Enter]**

3.11 2D Isometric Drawing

A 2D isometric drawing is a 2D drawing that looks like a 3D drawing. It is useful in making simple presentations. Figure 3.143 shows an isometric drawing that you will construct.

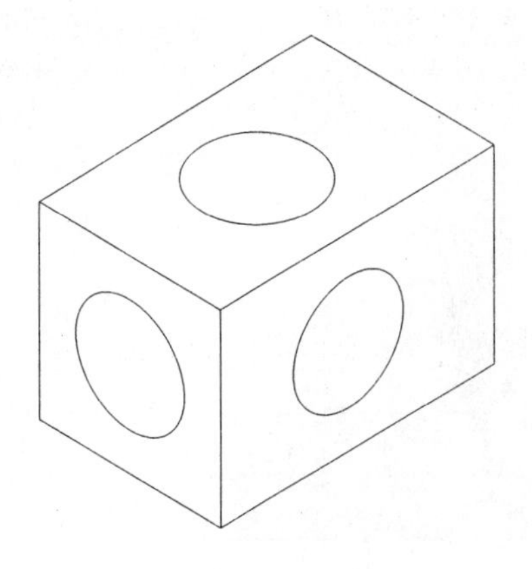

Figure 3.143 2D isometric drawing

Start a new drawing and use default metric settings.

<File> **<New...>**

Command: **NEW**

[Create New Drawing
Start from Scratch
Metric
OK]

To facilitate isometric drafting, select the Drawing Aids... item from the Tools pull-down menu to use the DDRMODES command to set the grid and snap spacing to 10 units, and turn on isometric grid and snap modes. See Figure 3.144.

<Tools> **<Drawing Aids...>**

Command: **DDRMODES**

[Drawing Aids
Snap Grid
On **On**
Y Spacing **10** Y Spacing **10**
Isometric Snap/Grid
On
Left
OK]

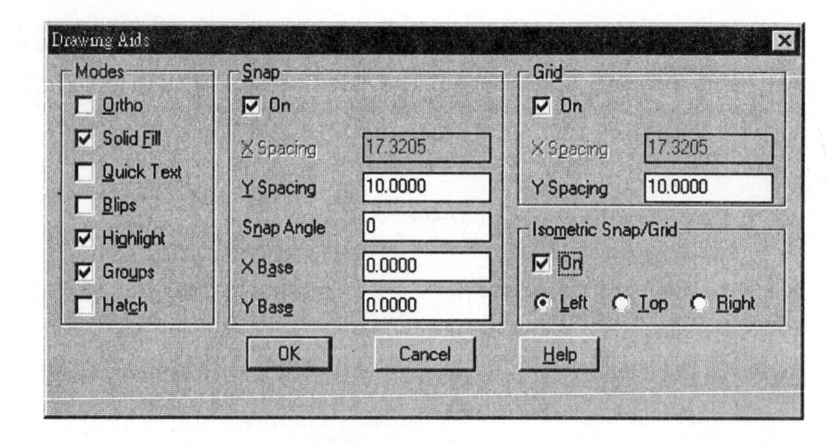

Figure 3.144 Isometric grid and snap spacing set with Drawing Aids dialog box

Construction Lines

There are two kinds of construction lines — XLINE and RAY. An xline is a line of infinite length. A ray is a semi-infinite line. It has a finite starting point and extends to infinity. Select the Construction line item from the Draw pull-down menu to use the XLINE command to create two lines of infinite length that are inclined at 30° to the horizontal. See Figure 3.145.

<Draw> **<Construction Line>**

Command: **XLINE**
Hor/Ver/Ang/Bisect/Offset/<From point>: **A**
Reference/<Enter angle (150)>: **30**
Through point: **0,0**
Through point: **0,100**
Through point: **[Enter]**

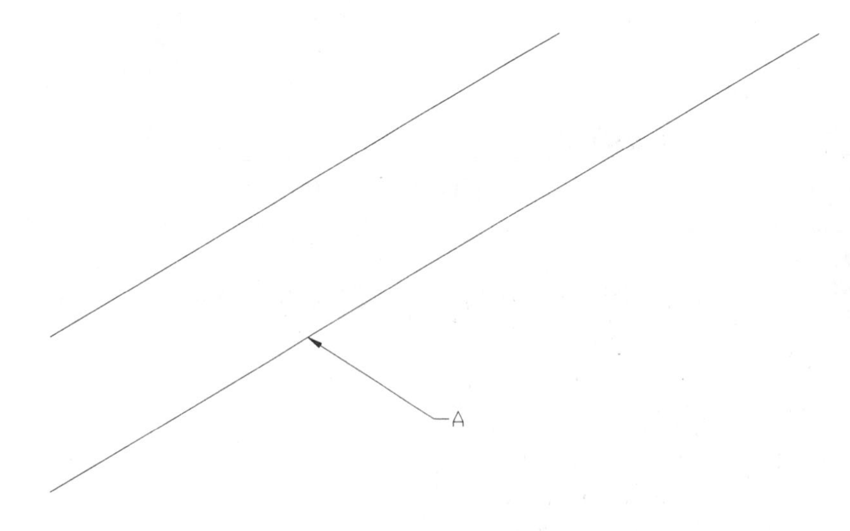

Figure 3.145 Two construction lines drawn

Repeat the XLINE command to create a vertical line. See Figure 3.146.

Command: **[Enter]**
XLINE
Hor/Ver/Ang/Bisect/Offset/<From point>: **V**
Through point: **[Select a snap point near A (Figure 3.145).]**
Through point: **[Enter]**

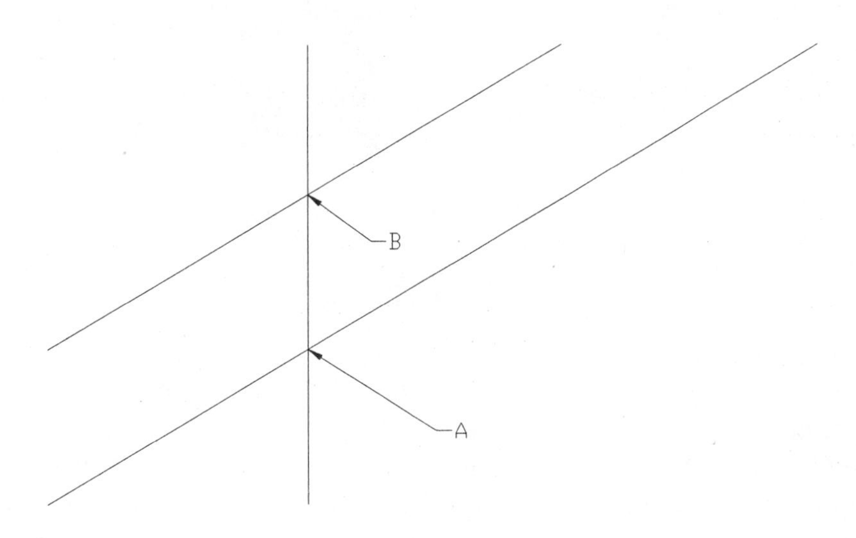

Figure 3.146 Vertical construction line drawn

From the Draw pull-down menu, select the Ray item to use the RAY command to create two semi-infinite lines. See Figure 3.147.

<Draw> **<Ray>**

Command: **RAY**
From point: **[Select A (Figure 3.146).]**
Through point: **@1<150**

In the foregoing command line input, @1<150 indicates a direction. The magnitude is not important because the object to draw is a ray of semi-infinite length.

Through point: **[Enter]**

Command: **[Enter]**
RAY
From point: **[Select B (Figure 3.146).]**
Through point: **@1<150**
Through point: **[Enter]**

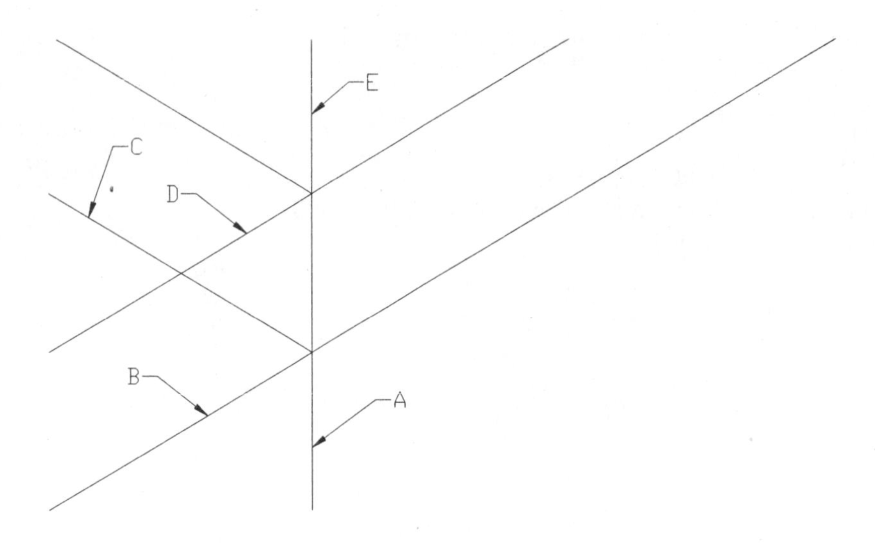

Figure 3.147 Rays drawn

An xline line is a line of infinite length. A ray is a line from a specific point that extends to infinity. Both xlines and rays are construction lines. They are very useful when you want to draw a line but have not decided how long it will be. Naturally, you will trim them down to specific lengths. If you trim one end of an xline, it becomes a ray. If you trim both ends of an xline or the infinite end of a ray, it becomes a line. To bring about the change, select the List item from the Inquiry cascading menu of the Tools pull-down menu to use the LIST command to list the data of the xlines and rays created.

<Tools> **<Inquiry>** **<List>**

```
Command: LIST
Select objects: [Select A, B, and C (Figure 3.147).]
Select objects: [Enter]
        XLINE    Layer: 0
            Space: Model space      Handle = F0
            base point, X= 173.2051  Y= 100.0000  Z=  0.0000
            unit direction, X=  0.0000  Y=   1.0000  Z=  0.0000
            angle in XY plane = 90
        XLINE    Layer: 0
            Space: Model space      Handle = E4
            base point, X=  0.0000  Y=   0.0000  Z=  0.0000
            unit direction, X=  0.8660  Y=   0.5000  Z=  0.0000
            angle in XY plane = 30
        RAY      Layer: 0
            Space: Model space      Handle = F3
            base point, X= 173.2051  Y= 100.0000  Z=  0.0000
            unit direction, X= -0.8660  Y=   0.5000  Z=  0.0000
            angle in XY plane = 150
```

Use the TRIM command to trim three xlines and one ray. See Figure 3.148.

<Modify> **<Trim>**

Command: **TRIM**
Select objects: [**Select A, B, and D (Figure 3.147).**]
Select objects: [**Enter**]
<Select object to trim>/Project/Edge/Undo: [**Select A, B, C, D, and E (Figure 3.147).**]
<Select object to trim>/Project/Edge/Undo: [**Enter**]

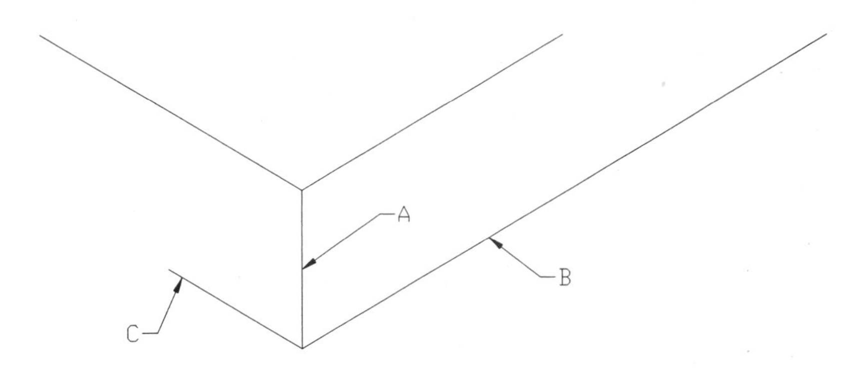

Figure 3.148 Lines trimmed

To verify that the xline changes to a ray and the ray to a line after trimming, use the LIST command to check again.

<Tools> **<Inquiry>** **<List>**

Command: **LIST**
Select objects: [**Select A, B, and C (Figure 3.148).**]
Select objects: [**Enter**]
LINE Layer: 0
Space: Model space Handle = F0
from point, X= 173.2051 Y= 100.0000 Z= 0.0000
to point, X= 173.2051 Y= 200.0000 Z= 0.0000
Length = 100.0000, Angle in XY Plane = 90
Delta X = 0.0000, Delta Y = 100.0000, Delta Z = 0.0000
RAY Layer: 0
Space: Model space Handle = E4
base point, X= 173.2051 Y= 100.0000 Z= 0.0000
unit direction, X= 0.8660 Y= 0.5000 Z= 0.0000
angle in XY plane = 30
LINE Layer: 0
Space: Model space Handle = F3
from point, X= 173.2051 Y= 100.0000 Z= 0.0000
to point, X= 86.6025 Y= 150.0000 Z= 0.0000
Length = 100.0000, Angle in XY Plane = 150
Delta X = -86.6025, Delta Y = 50.0000, Delta Z = 0.0000

As can be seen, an xline line trimmed at both ends becomes a line, an xline line trimmed at one end becomes a ray, and a ray trimmed at its infinite end becomes a line.

To continue, use the COPY command to copy two lines. See Figure 3.149.

<Modify> **<Copy>**

Command: **COPY**

Select objects: **[Select A (Figure 3.148).]**
Select objects: **[Enter]**
<Base point or displacement>/Multiple: **150<30**
Second point of displacement: **[Enter]**

Command: **[Enter]**
COPY
Select objects: **[Select A (Figure 3.148).]**
Select objects: **[Enter]**
<Base point or displacement>/Multiple: **100<150**
Second point of displacement: **[Enter]**

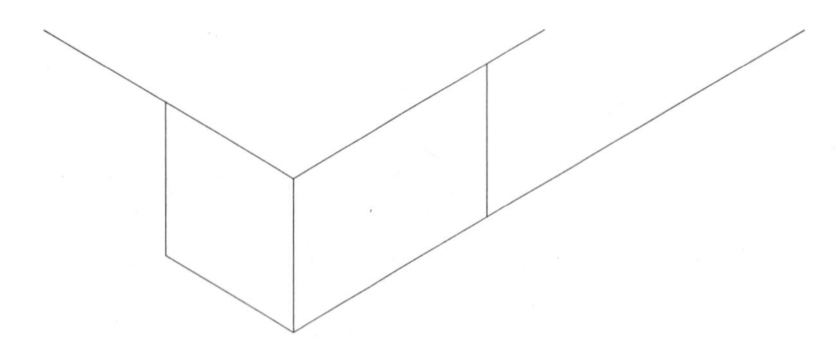

Figure 3.149 Lines copied

Complete the isometric box according to Figure 3.150.

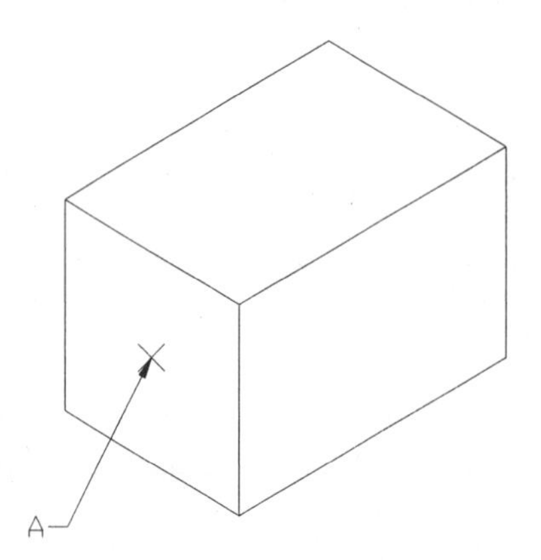

Figure 3.150 Isometric box created

Note the XY orientation as depicted by the UCS icon at the lower left corner of your screen. Although the isometric drawing looks like a box, it is, in fact, a 2D representation of a 3D object. Circles on the isometric planes will look like ellipses. To draw a circle, you have to select an isoplane and then use the ELLIPSE command. Refer to Figure 3.150: the current isometric plane is the left plane. Use the ELLIPSE command to create an isometric ellipse on this plane. See Figure 3.151.

<Draw> **<Ellipse>** **<Axis, End>**

Command: **ELLIPSE**
Arc/Center/Isocircle/<Axis endpoint 1>: **ISOCIRCLE**
Center of circle: [**Snap to the central point of the left isoplane, A (Figure 3.150).**]
<Circle radius>/Diameter: **30**

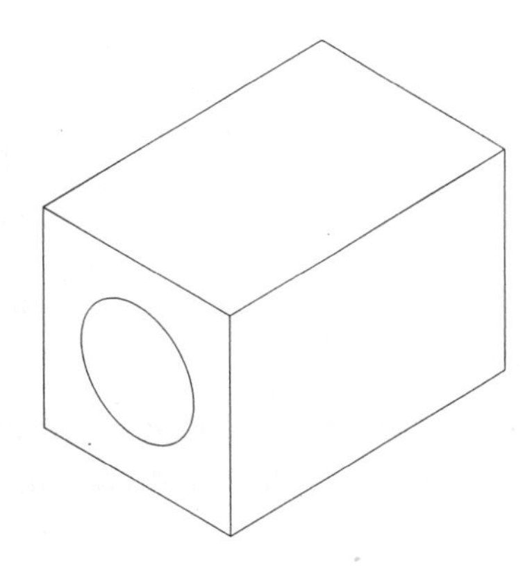

Figure 3.151 Isocircle created on left isoplane

Select the Drawing Aids... item from the Tools pull-down menu to use the DDRMODES command to select the top plane to be the isoplane.

<Tools> **<Drawing Aids...>**

Command: **DDRMODES**

[Drawing Aids
Snap Grid
On **On**
Y Spacing **10** Y Spacing **10**
Isometric Snap/Grid
On
Top
OK]

After setting the isoplane to the top plane, use the ELLIPSE command to create an isocircle. See Figure 3.152.

<Draw> **<Ellipse>** **<Axis, End>**

Command: **ELLIPSE**
Arc/Center/Isocircle/<Axis endpoint 1>: **ISOCIRCLE**
Center of circle: [**Snap to the top isoplane.**]
<Circle radius>/Diameter: **30**

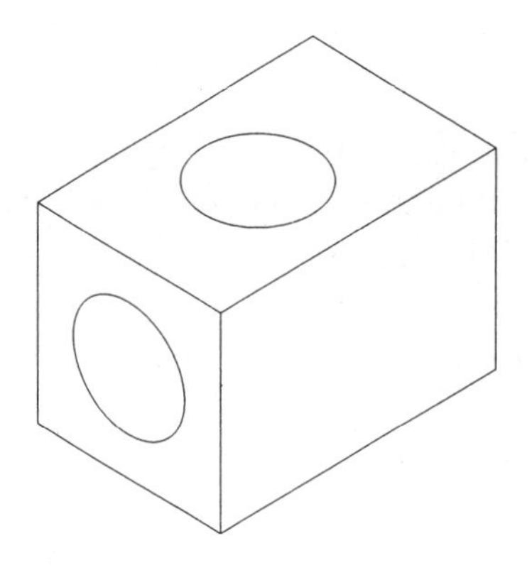

Figure 3.152 Isocircle created on top isoplane

Apart from using the DDRMODES command, you can use the ISOPLANE command to set the isoplane. Set the isoplane to the right and create an ellipse. See Figure 3.153.

Command: **ISOPLANE**
Left/Top/Right/<Toggle>: **RIGHT**
Current Isometric plane is: Right

<Draw> **<Ellipse>** **<Axis, End>**

Command: **ELLIPSE**
Arc/Center/Isocircle/<Axis endpoint 1>: **ISOCIRCLE**
Center of circle: [**Snap to the right isoplane.**]
<Circle radius>/Diameter: **30**

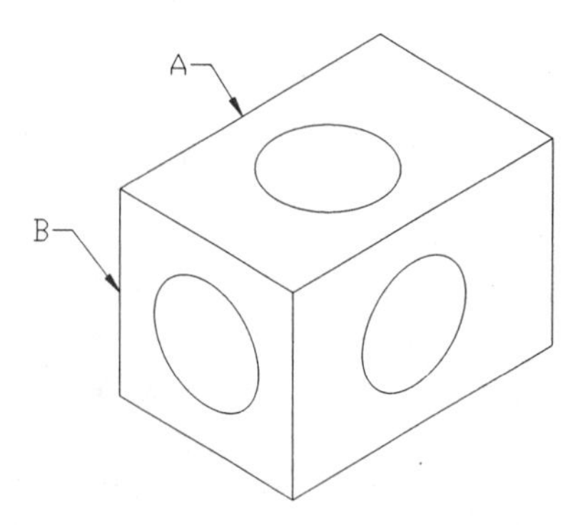

Figure 3.153 Isocircle created on right isoplane

The 2D isometric drawing is complete. Save your drawing with the SAVE command.

<File> **<Save>**

File name: **ISO_BOX.DWG**

3.12 Grips

Objects and entities have defining points. They are called grip points. For example, the defining points of a line are the midpoint and the two end points; the defining points of a circle are the center point and the four quadrant points. Use the OPEN command to open the 2D isometric drawing that you created in Section 3.11. See Figure 3.153. You will enable grip points and edit the drawing by manipulating the grip points.

<File> **<Open...>**

There are two ways to enable grips: setting the GRIPS system variable to 1 and using the DDGRIPS command.

Command: **GRIPS**
New value for GRIPS: **1**

From the Tools pull-down menu, select the Grips... item to apply the DDGRIPS command. See Figure 3.154.

<Tools> **<Grips...>**

Command: **DDGRIPS**

In the Grips dialog box, select the box next to Enable Grips to enable grips, select the [Unselected...] button to bring up a standard color dialog box to set the unselected grips to color green, and select the [Selected...] button to change the selected grips to color yellow. To adjust the size of the grip boxes, move the box in the sliding bar to an approximately central position. Select the [OK] button.

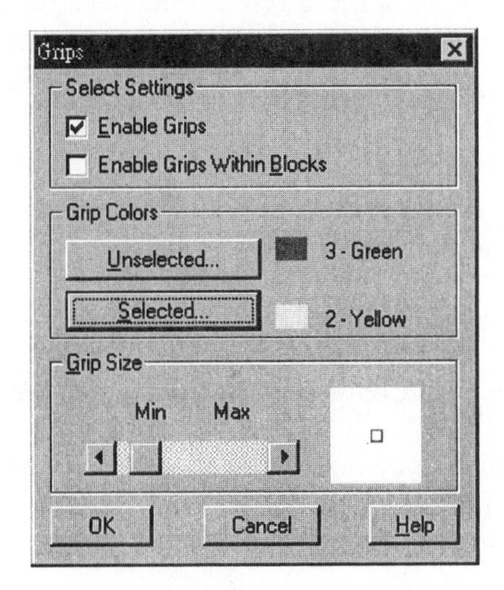

Figure 3.154 Grips dialog box

After you enable GRIPS, some small square boxes will display at the defining points as grips if you select the object without issuing a command.

Edit Entities

Ensure that snap mode is on. If not, select the Snap item from the status bar or press the [F9] key. Do not apply any command. Simply use the pointing device to select A and B (Figure 3.153). After you select the objects, hollow green boxes appear at the defining points of the lines. These are unselected grips. Select A (Figure 3.155). The green hollow box at A changes to a yellow solid box. It is selected. See Figure 3.155.

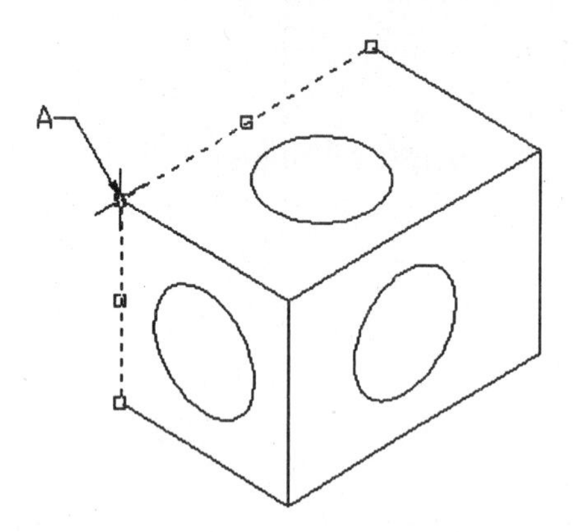

Figure 3.155 Grips appearing at defining points

At the command line interface, you will see the following prompts:

Command: **[Select A and B (Figure 3.153).]**
[Select A (Figure 3.154).]
** STRETCH **
<Stretch to point>/Base point/Copy/Undo/eXit:

The initial grips editing mode is the Stretch mode. All together, there are five editing modes. The other four modes are Move, Rotate, Scale, and Mirror. To cycle through the various modes, press the [Enter] key. Continue to press the [Enter] key until you cycle back to the Stretch mode.

** STRETCH **
<Stretch to point>/Base point/Copy/Undo/eXit: **[Enter]**
** MOVE **
<Move to point>/Base point/Copy/Undo/eXit: **[Enter]**
** ROTATE **
<Rotation angle>/Base point/Copy/Undo/Reference/eXit: **[Enter]**
** SCALE **
<Scale factor>/Base point/Copy/Undo/Reference/eXit: **[Enter]**
** MIRROR **
<Second point>/Base point/Copy/Undo/eXit: **[Enter]**

Apart from using the [Enter] key to cycle through the five modes, you can select the second mouse button to bring up a menu. See Figure 3.156.

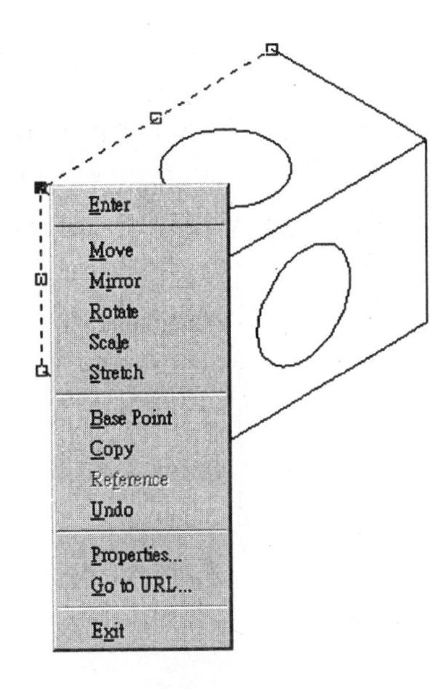

Figure 3.156 Grips editing menu

With this menu, you can select one of the editing mode quickly without the need to cycle through all the modes. Select the Move item to use the Move mode. Then select a point 20 units in the 150° direction. See Figure 3.157.

** STRETCH **
<Stretch to point>/Base point/Copy/Undo/eXit: **MOVE**
** MOVE **
<Move to point>/Base point/Copy/Undo/eXit: **[Select E (Figure 3.157).]**

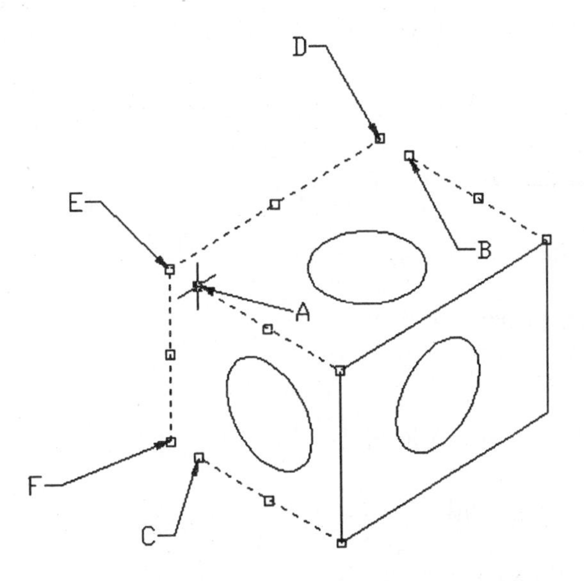

Figure 3.157 Two lines moved

Select the lines A, B, and C (Figure 3.157) to enable grips of these lines. Then select A. The current editing mode is Stretch. Select point E to stretch point A to point E. Do the same for points B and C. See Figure 3.158.

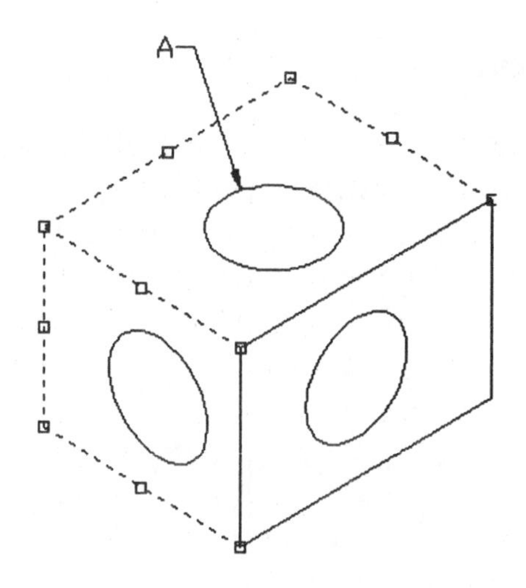

Figure 3.158 Lines stretched

To exit grips mode, press the [ESC] key twice. To disable grips, change the value of the system variable GRIPS to 0.

Command: **GRIPS**
New value for GRIPS: **0**

3.13 Pick First

As its name implies, the PICKFIRST facility enables you to select objects prior to running any editing command. By default, PICKFIRST is enabled. To disable this facility, enter PICKFIRST at the command prompt. Then key in 0 to set this system variable to zero.

Command: **PICKFIRST**
New value for PICKFIRST <1>: **0**

Select A (Figure 3.158). The isocircle is not highlighted. Now set PICKFIRST to 1.

Command: **PICKFIRST**
New value for PICKFIRST <1>: **1**

Select the isocircle again. It is highlighted and is ready for use in the next editing command.

Issue the ERASE command.

<Modify> **<Erase>**

Command: **ERASE**
1 found

The preselected isocircle is erased immediately, without warning! Therefore, you should be very careful in editing while PICKFIRST is enabled.

3.14 Entity Selection

Whenever you need to select objects, you can use the following methods:

Select	Select the object with the pointing device.
Window	Enter W, and then select two points to describe a rectangular window. All objects totally enclosed within the window are selected.
Crossing	Enter C, and then select two points to describe a rectangular crossing window. All objects totally or partly enclosed within the crossing window are selected.
Box	Enter BOX, and then select two points to describe a box. If the second such point is to the right side of the first point, the box is a WINDOW box. If the second point is to the left side of the first point, the box is a CROSSING box.
Auto	Enter AU, and select on the screen. If you select an object, then the object is selected. If you select a blank space, then it becomes a BOX option that requires a second point to form either a WINDOW box or a CROSSING box. This is the default mode.
Multiple	Enter M, and then continue the selection process. M suppresses AutoCAD from scanning through the entire database before you end the selection.
WPolygon	Enter WP, and then select a series of points to describe a polygon. All objects totally enclosed within the window polygon are selected.
CPolygon	Enter CP, and then select a series of points to describe a polygon. All objects totally or partly enclosed within the crossing polygon are selected.
Fence	Enter F, and then select a series of points to define a series of lines to form a fence. All objects crossed by the fence are selected.
Last	Enter L to select the last object created.
Previous	Enter P to take the most recent selection set. A selection set is a set of entities that you have selected in the previous command or a set of entities chosen with the SELECT command. This option will not work if you do not make a selection set beforehand.
Group	Enter G and then the group name. You have to use the GROUP command to create entity groups before you can use this option.
Single	Enter SI to have the single mode. Selection ends when an object or a selection set is selected.
All	Enter ALL to select all entities except those on the frozen layers.
Add	Enter A, and then use other selection modes to select objects for adding to the selection set. This is the default mode.
Remove	Enter R, and then use other selection modes to select objects for removal from the selection set.
Undo	Enter U to undo the last selection made.
Enter	Press ENTER to terminate the selection set.

Pickbox Size

When a command requires you to select an object, the crosshairs change to a pickbox.

The size of a pickbox is controlled by the PICKBOX system variable.

Command: **PICKBOX**
New value for PICKBOX: **10**

Selection Set

After you have selected the entities for editing, a selection set is created in memory. You can explicitly make a selection set with the SELECT command.

Command: **SELECT**
Select objects: [**Use the above methods to select objects.**]
Select objects: [**Enter**]

Object Selection Setting

The DDSELECT command concerns object selection settings. Select the Selection... item of the Tools pull-down menu to run this command. See Figure 3.159.

<Tools> **<Selection...>**

Command: **DDSELECT**

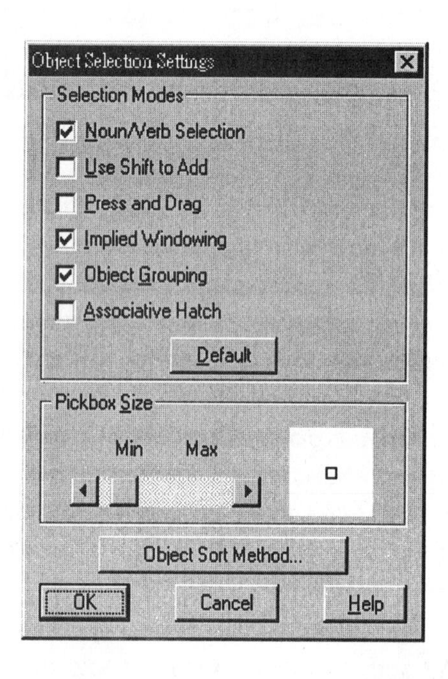

Figure 3.159 Object Selection Settings dialog box

Entity Group

With the SELECT command, you can create only one selection set. Moreover, the set is overwritten every time you make a new selection for an editing command. If you wish to maintain one or more selection sets in a drawing, you should use the GROUP command.

From the Tools pull-down menu, select the Object Group... item to apply the GROUP command. See Figure 3.160.

<Tools> **<Object Group...>**

Command: **GROUP**

In the dialog box, you can specify the group name and a brief description of the group for reference, decide whether a group is selectable by simply selecting its member, create a group, and change a group.

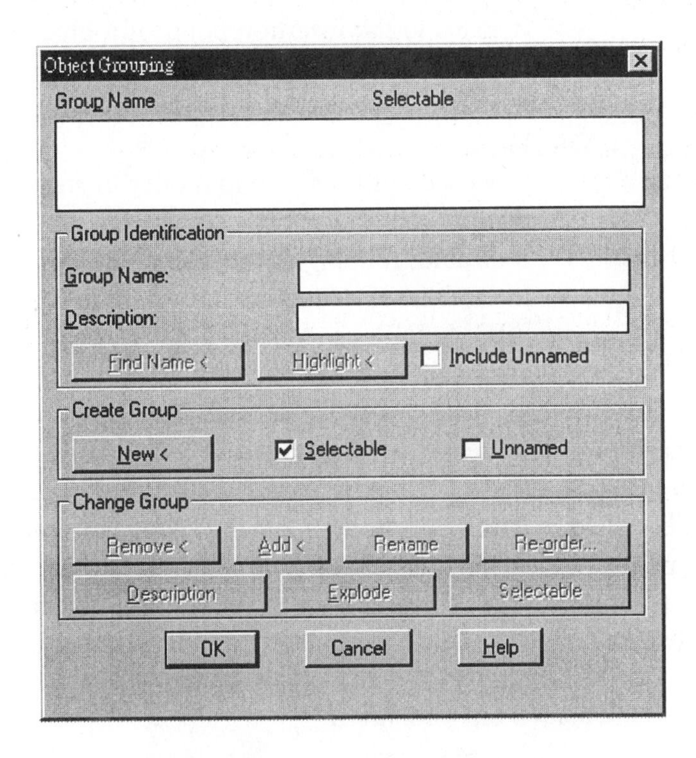

Figure 3.160 Object Grouping dialog box

3.15 Geometric Snap Tools

In order to locate the position of a feature quickly, you can use the object snap facility. You can specify these modes in response to the request for point input to instruct the system to locate a point for you. The various modes are as follows:

ENDpoint	It snaps to the nearest endpoint of an object. The short form is END.
MIDpoint	It snaps to the midpoint of an object. The short form is MID.
INTersection	It snaps to the intersection of two objects. The short form is INT.
APParent intersection	It snaps to the apparent intersection of two objects. Two objects that do not intersect spatially may be seen to be intersecting from a certain direction of viewing. The apparent intersection depends on the viewing direction. The short form is APP.
CENter	It snaps to the center of an arc or circle. The short form is CEN.

QUAdrant — It snaps to the quadrant points of an arc or circle. The short form is QUA.

PERpendicular — It snaps to a point that is perpendicular to the last point. The short form is PER.

TANgent — It snaps to a tangential point on an arc or circle from a specified point. The short form is TAN.

NODe — It snaps to a point object. The short form is NOD.

INSertion — It snaps to the insertion point of a block. The short form is INS.

NEArest — It snaps to a point closest to an object. The short form is NEA.

QUIck — It is used with other snap modes to snap to the first eligible point found. The short form is QUI.

None — It disables running object snaps.

From — It establishes a temporary reference point. For example, FROM END establishes a reference from the endpoint of an object.

To use these geometric snap tools, you can enter their short form at the command prompt or click the appropriate item from the cursor menu. If a certain geometric snap tool is going to be used repeatedly, it is more convenient to set it as the default object snap, running object snap. To set the running object snap, you can use the OSNAP command. You can select the OSNAP item of the status bar, press the [F3] to enable running object snap, or disable it. If you have not yet set any object snap, the OSNAP item of the status bar or the [F3] key brings out the OSNAP command. See Figure 3.161.

<Tools> **<Object Snap Settings...>**

Command: **OSNAP**

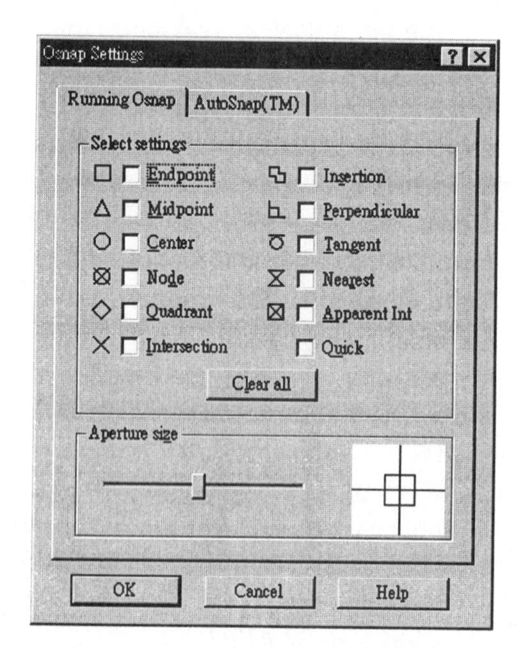

Figure 3.161 Object Snap toolbar

The related system variable and command are APBOX and APERTURE, respectively. The APBOX system variable controls the display of this target box. The APERTURE command controls the size of the object snap target box.

3.16 Coordinates Systems

In a drawing, there is an origin (0,0) defined by the world coordinate system (WCS). The imaginary construction plane is set on the XY plane of the WCS. You can translate the origin and the imaginary construction plane with the UCS command. The translated coordinate system is called the user coordinate system (UCS). For some 3D entities, there is also an object coordinate system (OCS).

No matter how many coordinate systems you have in a drawing, there is one system that is current. If you specify or select a point, you are selecting with reference to the current coordinate system.

Absolute Coordinates

To specify a point, you can use either the Cartesian coordinates or the polar coordinates.

A point (10,20) is specified in Cartesian coordinates. It is 10 units in the X direction and 20 units in the Y direction. Measurement is from the current origin point.

A point (10<0) is specified in polar coordinates. It is 10 units in the 0° direction. The 0° direction is controlled by the UNITS command. The default is the positive X direction. Measurement is, again, from the current origin point.

@, Relative Coordinates

A point (@10,20) is specified in relative Cartesian coordinates. It is 10 units in the X direction and 20 units in the Y direction from the last selected point.

A point (@10<0) is specified in relative polar coordinates. It is 10 units in the 0° direction from the last selected point.

3.17 Key Points and Exercises

To create a new AutoCAD drawing, you have to decide which method you use to start: Use a Wizard, Use a Template, or Start from Scratch. Being a beginner, you should choose the Start from Scratch method. Then you should set working limits, set grid meshes, set snap intervals, load linetype, set linetype scale, and create layers. After you have set up a drawing, you can save it as a template file for future use. Basically, Use a Wizard should be reserved for advanced users. After you have studied all the chapters in this book, you may consider using this option.

In brief, making a CAD drawing involves drawing and modifying. You can use appropriate drawing commands to construct lines, arcs, circles, ellipses, polylines, and points. You can translate objects by arraying, copying, moving, and mirroring. You can modify objects by offsetting, erasing, lengthening, extending, trimming, breaking, and manipulating the grip points. To modify objects, you have to select them by using the selection methods provided. In drawing and editing, you need to locate points. To select a point on the screen, you can use absolute/relative Cartesian coordinates and polar coordinates. To locate existing geometric entities, you can use object snap tools.

In this chapter, you created a number of engineering drawings. Basically, the drawings are very simple. Any skillful AutoCAD user can complete them very quickly by using a few commands and many shortcuts. However, you followed an indirect touring route and spent much time. In fact, the main aim of producing the drawings was not simply to make the drawings themselves but to practice the use of as many commands as possible. As a result, you practiced most AutoCAD entity draw and modify commands and learned how to make and redefine a block. To discover how to create a script file to work in AutoCAD, you can refer to Appendix D. There are two entity manipulation facilities, Grips and Pick First. Entities have grip points that are used for selecting. The Pick First facility allows you to select an object first and then issue the editing command afterwards. To select entities for editing, you can use one of the many ways to create a selection set. To do preliminary design work, you can create construction lines and rays.

In order to enhance your learning, do the following exercises by referring to the guidelines offered in this chapter.

Exercise 3.1

Outline absolute/relative Cartesian and polar coordinates systems and explain filtering, tracking, and direct distance entry. State the various methods used to select objects and various geometric snap tools.

Exercise 3.2

What is the difference between the DIVIDE and MEASURE commands? What commands will you use to increase or to decrease the length of a line? What are the three commands needed to construct multilines? What command will you use in conjunction with an editing command to filter selection?

Exercise 3.3

Cite the advantages of creating and using blocks in a drawing file.

Exercise 3.4

Start a new drawing from scratch with default metric settings. Load linetypes Center and Hidden. Create three more layers in addition to the default layer 0 — one layer for putting in the center lines, one layer for putting in the hatching lines, and one layer for putting in the hidden lines. The color and linetype assignments of the layers are as follows:

Layer Name	Color	Linetype	Remark
0	White	Continuous	For outlines
Center	Red	Center	For center lines
Hatch	Blue	Continuous	For hatching lines
Hidden	Green	Hidden	For hidden lines

Refer to Figure Figure 3.162. Estimate the working space required, set up drawing limits accordingly, zoom to such limits, and create the top view, the front view, and the sectional side view. You do not have to add dimensions to your drawing now. You will learn how in Chapter 5. Save your drawing named EX3_1.DWG.

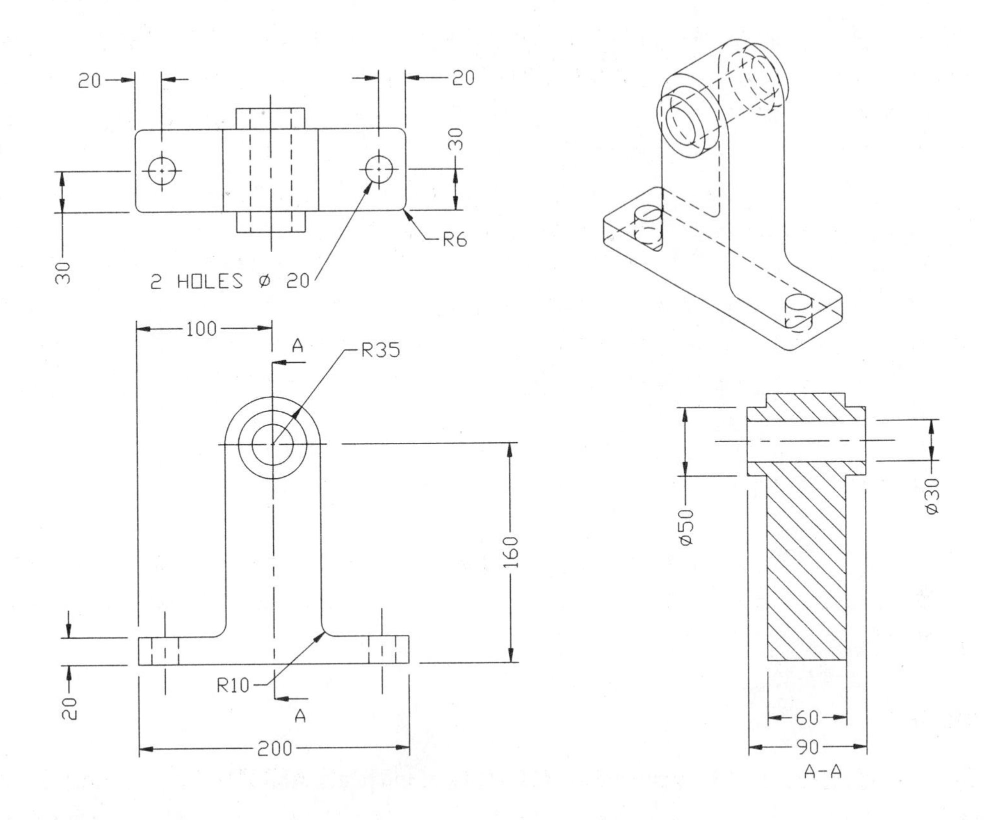

Figure 3.162 Support column

Exercise 3.5

Open the drawing EX3_1.DWG. Use the SAVEAS command to save it to a new drawing file – EX3_2.DWG. In addition to the existing layers, create one more layer Pendulum with color blue and linetype continuous. Set this layer current, and lock all other layers.

Layer Name	Color	Linetype	Status
0	White	Continuous	Locked
Center	Red	Center	Locked
Hatch	Blue	Continuous	Locked
Hidden	Green	Hidden	Locked
Pendulum	Blue	Continuous	Current layer

According to the dimensions shown in Figure 3.163, construct the outlines, center lines, and hatchings of the pendulum on layer Pendulum. Color and linetype assignments of the entities are as follows:

Entities of the pendulum	Layer	Color	Linetype
Outlines	Pendulum	Bylayer	Bylayer

Center lines	Pendulum	Yellow	Center
Hidden lines	Pendulum	Green	Hidden
Hatching lines	Pendulum	Magenta	Bylayer

The completed drawing is shown in Figure 3.164.

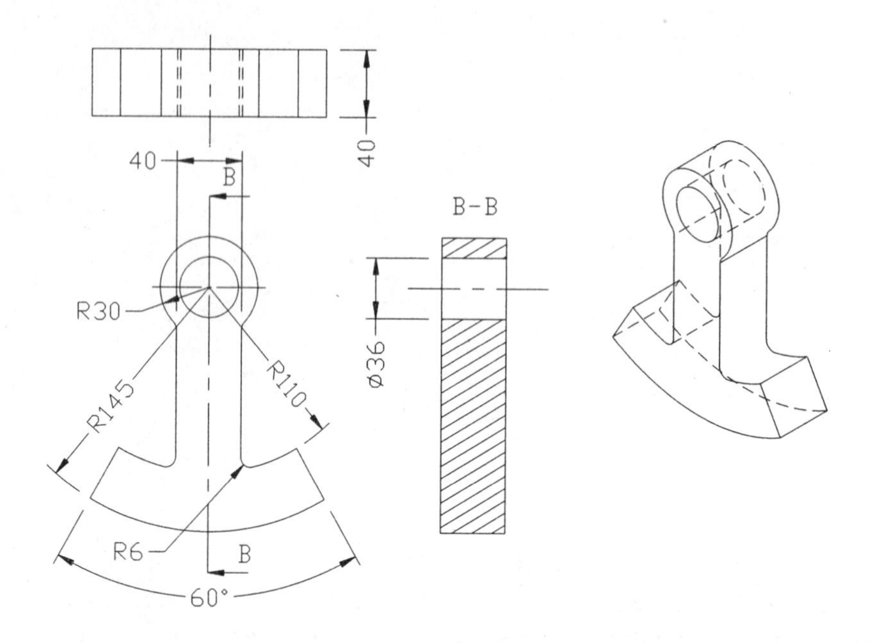

Figure 3.163 Dimensions of the pendulum

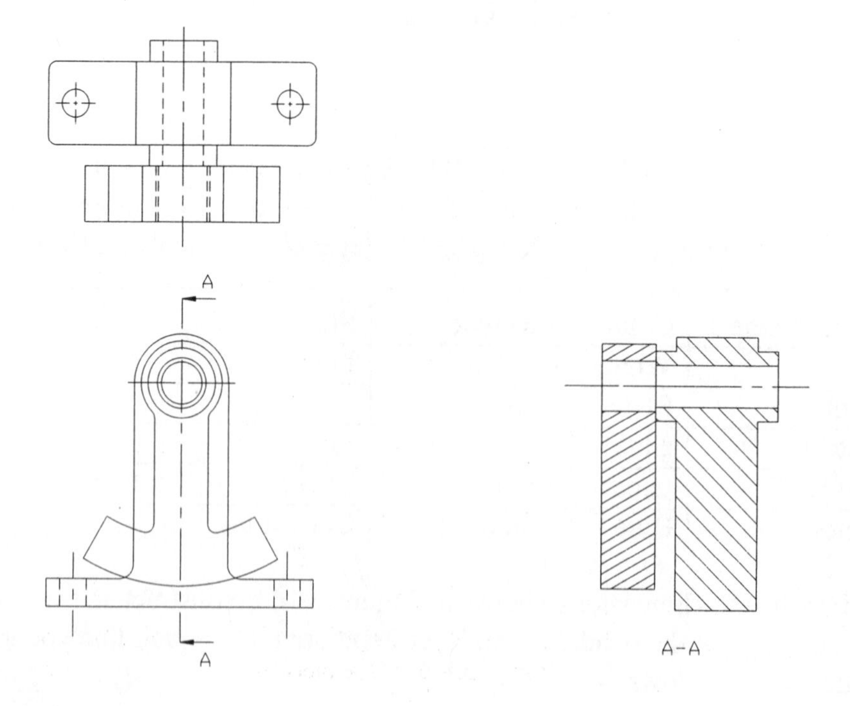

Figure 3.164 Two components created

Exercise 3.6

Figure 3.165 shows the orthographic projection of a shaft for inserting to the pendulum assembly. Start a new drawing from scratch. Use metric defaults. Create a new layer called Shaft with color cyan. Create two orthographic views on this layer. Save your drawing named EX3_3.DWG.

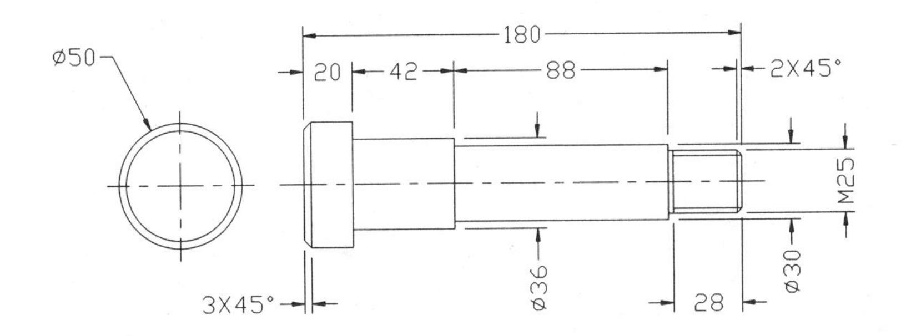

Figure 3.165 Shaft

Exercise 3.7

Referring to Figure 3.166, construct the orthographic projection of an angle block. Start a new drawing from scratch with metric defaults. Set up two additional layers: layer Center with linetype center and color red for center lines, and layer Hidden with linetype hidden and color green for hidden lines. Save your drawing named EX3_4.DWG.

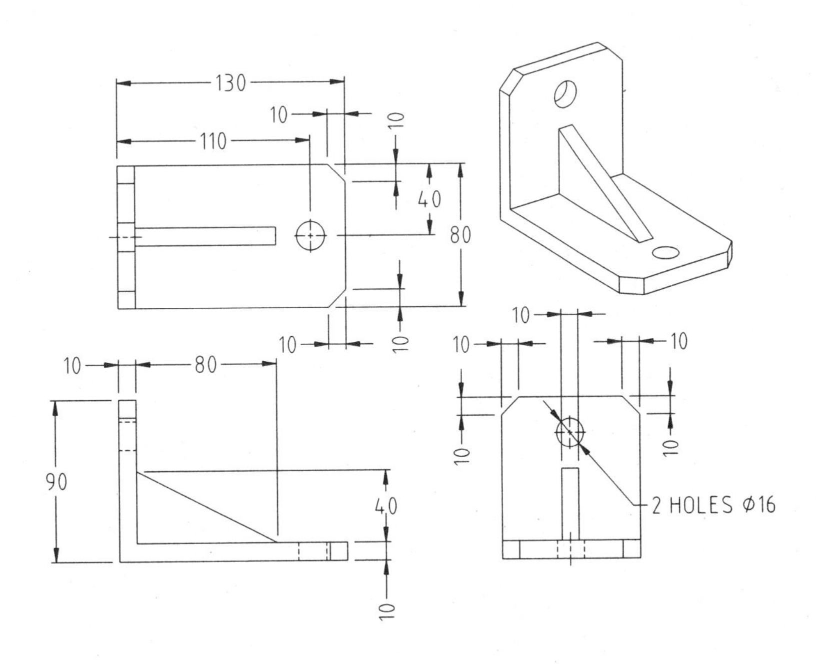

Figure 3.166 Angle block

Exercise 3.8

Figure 3.167 shows the orthographic projection of a mounting plate. Create the top view and the front view. You should put the hidden lines and the center lines on separate layers. Save your drawing named EX3_5.DWG.

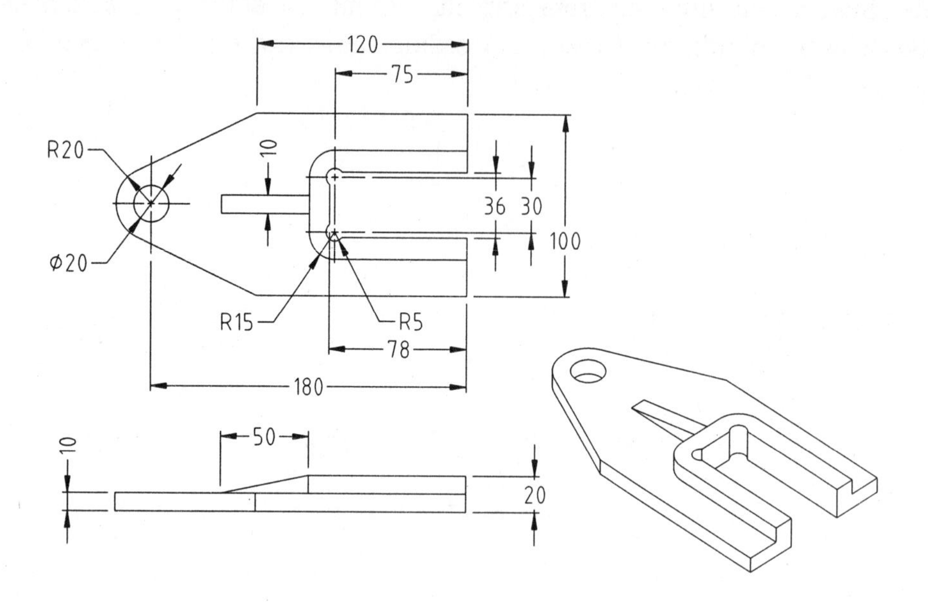

Figure 3.167 Mounting plate

Exercise 3.9

A bell cover is shown in Figure 3.168. Load linetypes. Set up three additional layers for the hidden lines, center lines, and hatching lines. Create the top view, the front view, and the sectional view accordingly. Save your drawing named EX3_6.DWG.

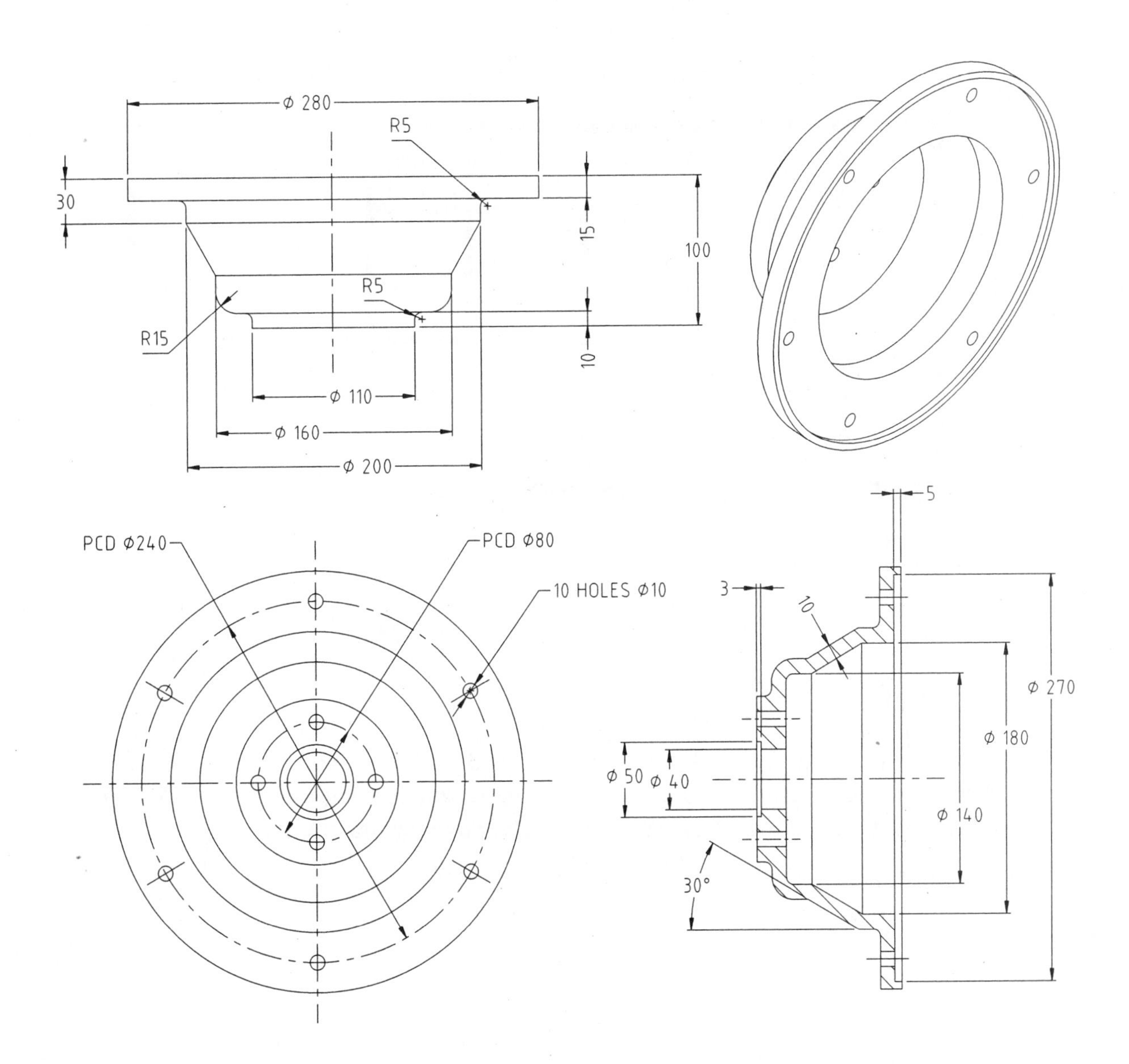

Figure 3.168 Bell cover

Exercise 3.10

Figure 3.169 shows the layout of a condominium. Set up a multiline style with two lines that are 200 units apart, have closed ends, and are filled. Create the drawing and save your drawing named EX3_7.DWG.

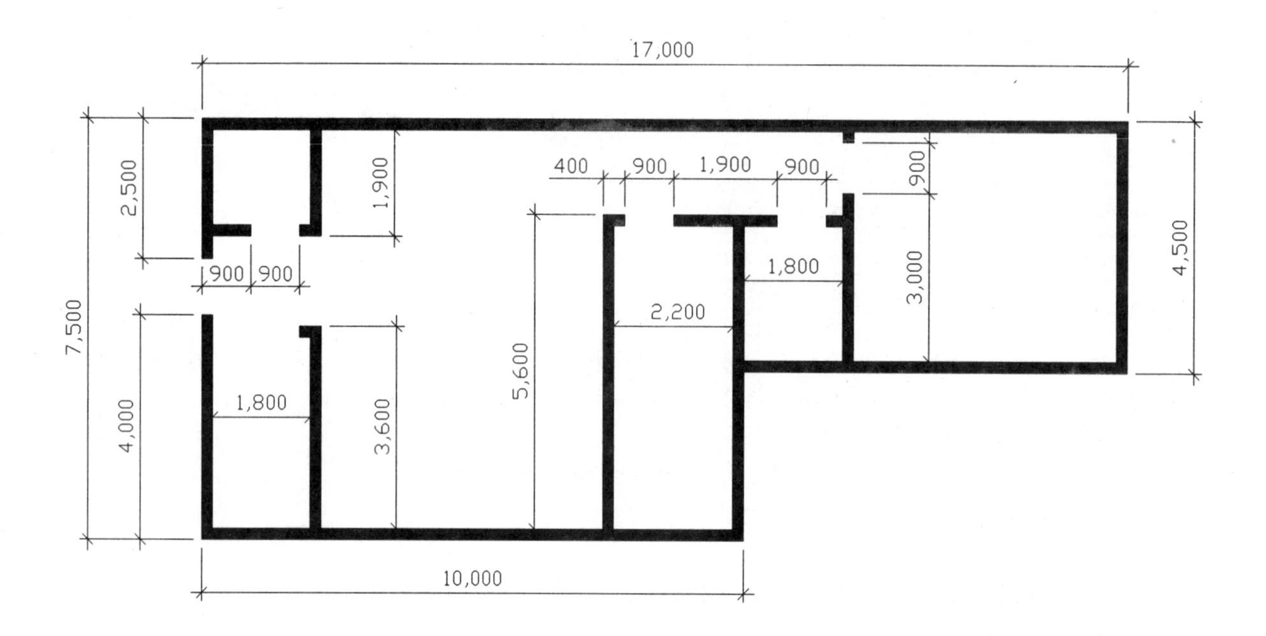

Figure 3.169 Layout of a condominium

Exercise 3.11

Open the drawings EX3_1.DWG, EX3_4.DWG, EX3_5.DWG, and EX3_6.DWG one by one. Add an isometric drawing to each of them in accordance with Figure 3.162, Figure 3.166, Figure 3.167, and Figure 3.168, respectively. Save the drawings.

Chapter 4

Blocks and External References

4.1 Export Entities to Form New Drawing Files
4.2 Prepare a Drawing for Insertion
4.3 Insert and Edit
4.4 Block and External References
4.5 Block Attributes
4.6 Key Points and Exercises

Aims and Objectives

The aim of this chapter is to give you an in-depth understanding of blocks and external references. After studying this chapter, you should be able to:

- export selected blocks and entities to form new drawing files,
- insert an external drawing to the current working drawing,
- edit inserted entities,
- attach an external drawing,
- distinguish between insertions and external references, and
- manipulate block attributes.

Overview

In the last chapter, you created a number of engineering drawings. In making these drawings, you learned how to use various drawing tools to draw, to modify, and to create and redefine a block.

In the pipe support assembly drawing that you created in the last chapter, you created a block and used the DIVIDE command to place multiple instances at regular intervals along a circle.

In the floor plan of the last chapter, you created a block and used the DDINSERT command to create instances at various locations.

With a block, you can make multiple instances of a group of entities by referring to a block definition saved in memory. The advantages of using block rather than copying and arraying are that file size is smaller (because block insertion does not actually duplicate the entities) and that change to a block can cause global change to all the instances of the blocks (because instances refer to block definitions).

There are seven commands to create block instances — BLOCK, BMAKE, INSERT, DDINSERT, MINSERT, DIVIDE, and MEASURE commands. The BLOCK and BMAKE commands create block definitions from selected entities within a drawing file. The INSERT command, the DDINSERT command, and the MINSERT command create instances of a block, or form a block from an external drawing file and create instances from the defined block. The DIVIDE command and the MEASURE command create instances from a defined block. Regarding instances, the DDINSERT and INSERT commands create single instance and the MINSERT, DIVIDE, and MEASURE commands create multiple instances.

In this chapter, you will continue to work on the assembly drawing and the floor plan to learn the use of blocks and external references in drawing object management. You will also learn how to include textual attributes in an inserted block.

From the pipe support assembly drawing, you will use the WBLOCK command to export a block and selected entities to create three external drawing files. The exported files then become AutoCAD drawing files that you can open and edit in a normal way. This is a top down approach to making a set of engineering drawings. You create all the necessary parts in a single drawing and export the individual parts to external drawing files.

Besides exporting, you will start a new drawing file, prepare the third component, and use the DDINSERT command to import it to the assembly drawing. This is down top approach in engineering drawing production. You create the individual parts and import them to the assembly. After insertion, the inserted drawing forms an internal block. To edit the inserted block instance, you use the EXPLODE command or the XPLODE command to break it into separate entities.

After you insert an external file to a drawing, file size increases accordingly. Also, the inserted block has no relationship to the original file. Therefore, changes in the original file will not have any effect on the inserted block. To overcome these two drawbacks, you can use the XREF command, instead of insertion, to attach external files. An attached file does not form a part of the drawing. Only the attachment data are stored. Consequently, the drawing file size is much smaller than with insertion. When you open a file with external references, AutoCAD loads the attachment. As a result, the latest change in the external file is reflected in the drawing. In making an external reference, you can choose to attach or to overlay. You will appreciate the difference between attaching and overlaying in a nested situation.

Attributes are textual information that you can include when inserting a block. In the final part of this chapter, you will learn how to define attribute templates in a source file and how the attribute templates are used while the source file is inserted to another file. In addition, you will learn how to redefine a block that has attributes.

4.1 Export Entities to Form New Drawing Files

In the pipe support assembly that you created in Chapter 3, there are two components residing on different layers. For the first component, the outlines are placed on layer 0, the center lines on layer Center, the hatching lines on layer Hatch, and the hidden lines on layer Hidden. For the second component, the outlines and the center lines are placed on layer Roller and the hatching lines are placed on layer Hatch1. In addition, you have defined a block called Hole.

Use the OPEN command to open the pipe support assembly that was saved in the last chapter. See Figure 4.1.

<File> **<Open...>**

Command: **OPEN**

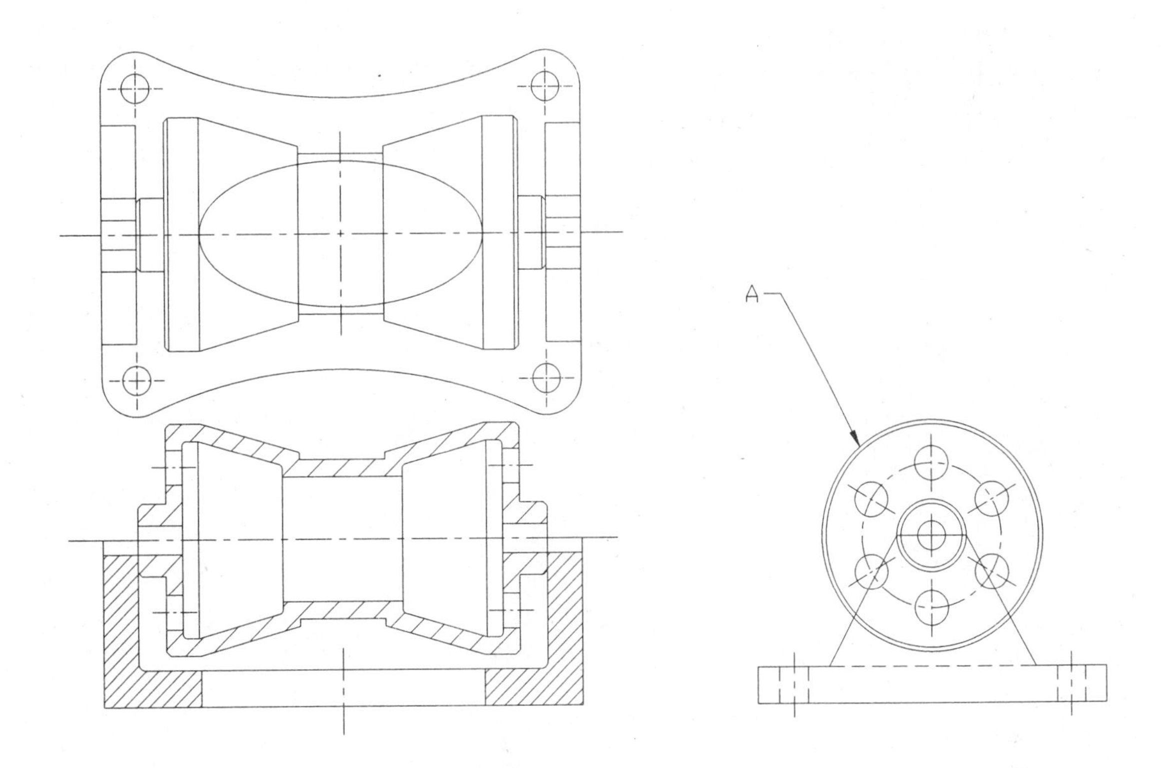

Figure 4.1 Pipe support assembly with two components

From this drawing, you will export the block and export selected entities of the two components to become individual drawing files that you can open and edit subsequently.

Write Block to a File

This drawing has a block called Hole. You have used the DIVIDE command to make multiple instances of it along a circular center line. If you want to make this block available to other drawings, you can export this block to an external drawing file. Apply the WBLOCK command at the command line interface to export this block.

```
Command: WBLOCK
```

When the dialog box appears, select the drive, and key in a valid file name.

```
Block name: HOLE
```

After you specify a file name and the block name, AutoCAD creates a file that consists of the entities defined by the block. The command completes.

Write Entities to a File

Apart from defined blocks, you can export selected entities to form a separate drawing file. There are two components in this assembly drawing. You will create two drawing files from them. You will export the first component to a file and the second component to another file.

To facilitate selection of the entities of the first component, use the Layer Control box of the Object Properties toolbar to freeze the layers Roller, Point, and Hatch1 such that only those entities that belong to the first component are left on the screen. Set layer 0 as the current layer. See Figure 4.2.

[Object Properties] [Layer Control]

On	0
On	Center
On	Hatch
Freeze	Hatch1
On	Hidden
Freeze	Point
Freeze	Roller

Current layer: **0**

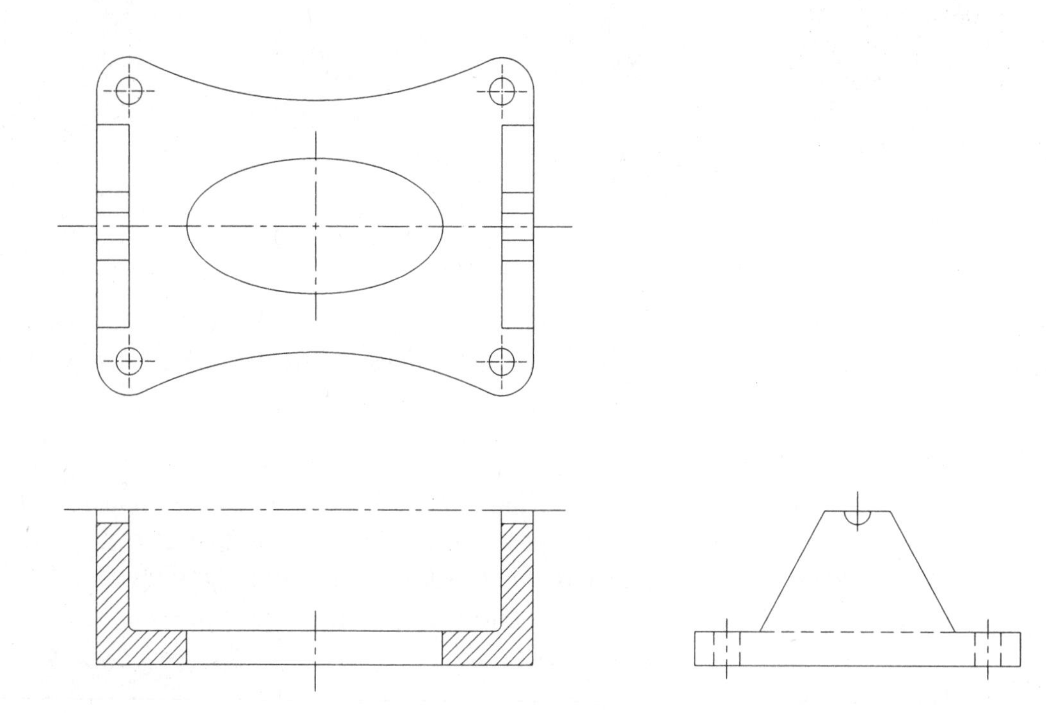

Figure 4.2 Layers Roller, Point and Hatch1 frozen

Use the WBLOCK command to select all the entities related to the first component and write them to a file.

Command: **WBLOCK**
File name: **[Specify a valid file name.]**
Block name: **[Enter]**
Insertion base point: **0,0**
Select objects: **[Select all the entities shown in Figure 4.2.]**
Select objects: **[Enter]**

Because AutoCAD assumes that you no longer need the entities chosen for writing to an external file, they are erased. If you wish to retain them, apply the OOPS command to retrieve the erased entities.

Command: **OOPS**

To leave only the entities of the second component on the display screen, use the Layer Control box of the Object Properties toolbar to thaw all the frozen layers, and freeze the layers 0, Center, Hidden and Hatch. Set layer Roller as the current layer. See Figure 4.3.

[Object Properties] **[Layer Control]**

Freeze	0
Freeze	Center
Freeze	Hatch
On	Hatch1
Freeze	Hidden
On	Point
On	Roller

Current layer: **Roller**

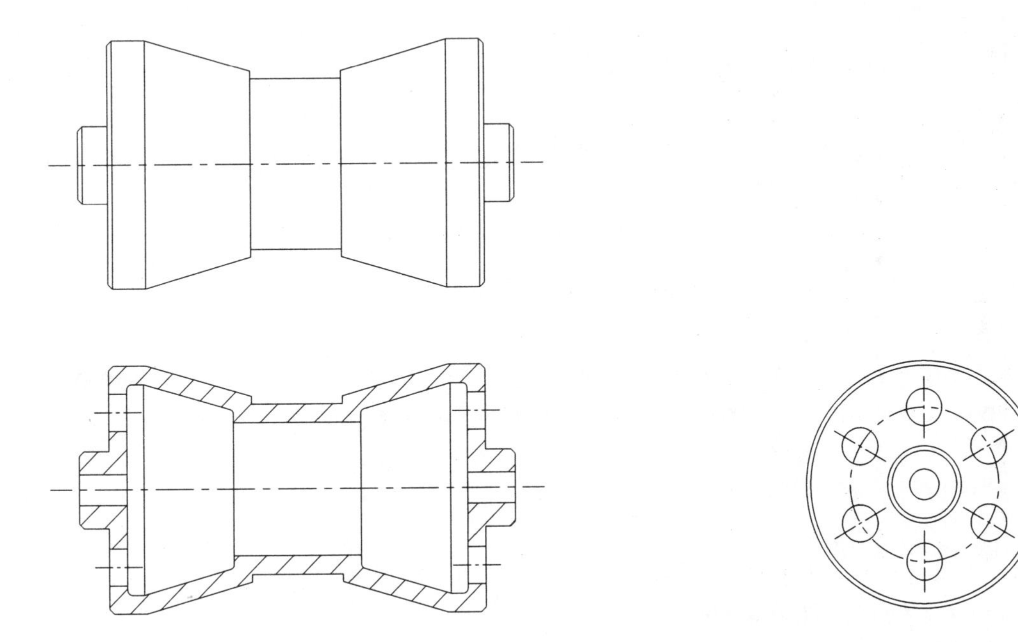

Figure 4.3 Layers 0, Center, Hidden, and Hatch frozen

Similarly, apply the WBLOCK command to export the entities related to the second component to another file. After that, issue the OOPS command to retrieve the erased objects.

Command: **WBLOCK**
File name: **[Specify a valid file name.]**
Block name: **[Enter]**

Insertion base point: **0,0**
Select objects: **[Select all the entities shown in Figure 4.3.]**
Select objects: **[Enter]**

Command: **OOPS**

Now you have three additional drawings derived from the assembly drawing — the block Hole, the first component, and the second component.

4.2 Prepare a Drawing for Insertion

As we have said, there are three components in the pipe support assembly. Here, you will create the third component and insert it to the assembly drawing.

Use the NEW command to start a new drawing from scratch and use metric default settings. The finished drawing is shown in Figure 4.4.

<File> **<New...>**

Command: **NEW**

[Create New Drawing
Start from Scratch
Metric
OK]

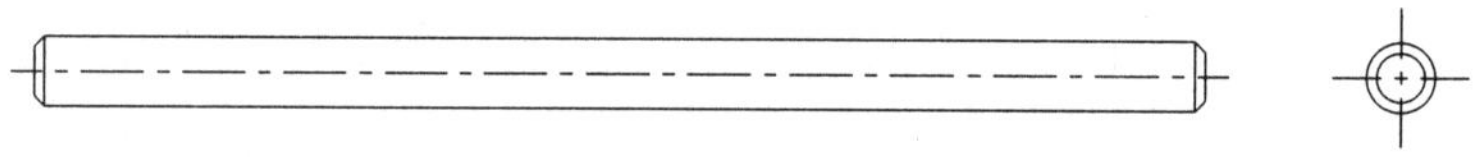

Figure 4.4 Third component

In the drawing, there are center lines and outlines. Use the LINETYPE command to load linetype Center to your drawing and set the global linetype scale to 15.

<Format> **<Linetype...>**

Command: **LINETYPE**

[Layer & Linetype Properties
Linetype
Load...]

[Load or Reload Linetypes
Available Linetypes
Center
OK]

[Global scale factor: **15**
OK]

Select the Layer tab to run the LAYER command to create a new layer Shaft and set it as the current layer. You will put all the entities related to the third component on this layer.

[Layer]

Name	Color	Linetype
0	White	Continuous
Shaft	Magenta	Continuous

[OK]

Side View

Use the CIRCLE command to create two circles at center (410,30) with a radii of 10 units and 7 units, respectively. See Figure 4.5.

<Draw> **<Circle>** **<Center,Radius>**

Command: **CIRCLE**
3P/2P/TTR/<Center point>: **410,30**
Diameter/<Radius>: **10**

Command: **[Enter]**
CIRCLE
3P/2P/TTR/<Center point>: **CEN** of **[Select the last circle.]**
Diameter/<Radius>: **7**

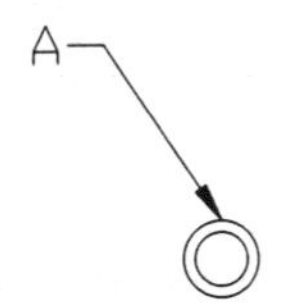

Figure 4.5 Two circles drawn

Front View

Apply the LINE command to create lines for the front view. Because the front view has to align with the side view, use the FROM object snap. Then select the 90° quadrant position of the circle as the Base point. See Figure 4.6.

<Draw> **<Line>**

Command: **LINE**
From point: **FROM**
Base point: **QUA** of **[Select A (Figure 4.5).]**
<Offset>: **@60<180**
To point: **@340<180**
To point: **@20<270**
To point: **@340<0**
To point: **C**

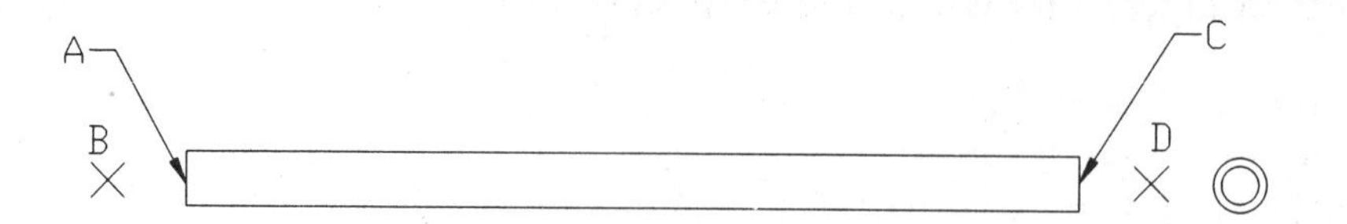

Figure 4.6 Four lines drawn

Apply the OFFSET command to offset two vertical lines. See Figure 4.7.

<Modify> **<Offset>**

Command: **OFFSET**
Offset distance or Through: **3**
Select object to offset: **[Select A (Figure 4.6).]**
Side to offset? **[Select B (Figure 4.6).]**
Select object to offset: **[Select C (Figure 4.6).]**
Side to offset? **[Select D (Figure 4.6).]**
Select object to offset: **[Enter]**

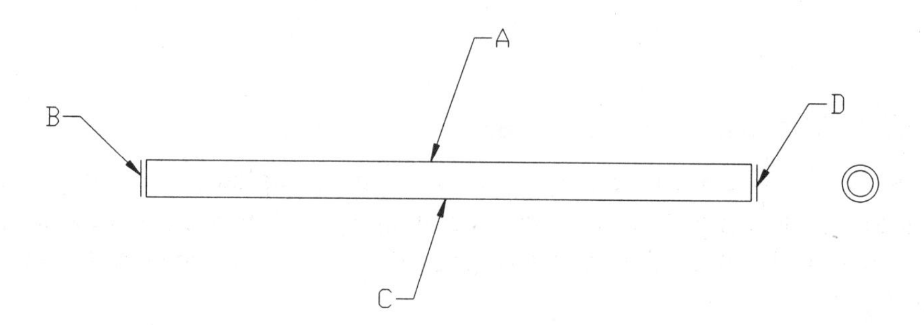

Figure 4.7 Two offset lines added

Run the CHAMFER command to set the chamfer distance to 3 units x 3 units and create bevel edges at the four corners. See Figure 4.8.

<Modify> **<Chamfer>**

Command: **CHAMFER**
Polyline/Distance/Angle/Trim/Method/<Select first line>: **D**
Enter first chamfer distance: **3**
Enter second chamfer distance: **3**

Command: **[Enter]**
CHAMFER
(TRIM mode) Current chamfer Dist1 = 3.0000, Dist2 = 3.0000
Polyline/Distance/Angle/Trim/Method/<Select first line>: **[Select A (Figure 4.7).]**
Select second line: **[Select B (Figure 4.7).]**

Command: **[Enter]**
CHAMFER
(TRIM mode) Current chamfer Dist1 = 3.0000, Dist2 = 3.0000
Polyline/Distance/Angle/Trim/Method/<Select first line>: **[Select B (Figure 4.7).]**

Select second line: **[Select C (Figure 4.7).]**

Command: **[Enter]**
CHAMFER
(TRIM mode) Current chamfer Dist1 = 3.0000, Dist2 = 3.0000
Polyline/Distance/Angle/Trim/Method/<Select first line>: **[Select C (Figure 4.7).]**
Select second line: **[Select D (Figure 4.7).]**

Command: **[Enter]**
CHAMFER
(TRIM mode) Current chamfer Dist1 = 3.0000, Dist2 = 3.0000
Polyline/Distance/Angle/Trim/Method/<Select first line>: **[Select D (Figure 4.7).]**
Select second line: **[Select A (Figure 4.7).]**

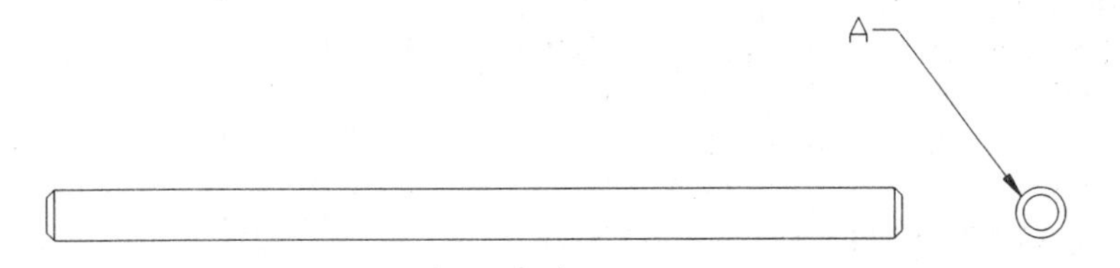

Figure 4.8 Four corners chamfered

Add Center Lines

The outlines for the side view and the front view are complete. You need to add three center lines to the drawing. Use the LINE command to draw two lines passing through the center of the circle. With the FROM object snap, select the center of the circle as the Base point. See Figure 4.9.

<Draw> **<Line>**

Command: **LINE**
From point: **FROM**
Base point: **CEN** of **[Select A (Figure 4.8).]**
<Offset>: **@20<90**
To point: **@40<270**
To point: **[Enter]**

Command: **[Enter]**
LINE
From point: **FROM**
Base point: **CEN** of **[Select A (Figure 4.8).]**
<Offset>: **@20<0**
To point: **@40<180**
To point: **[Enter]**

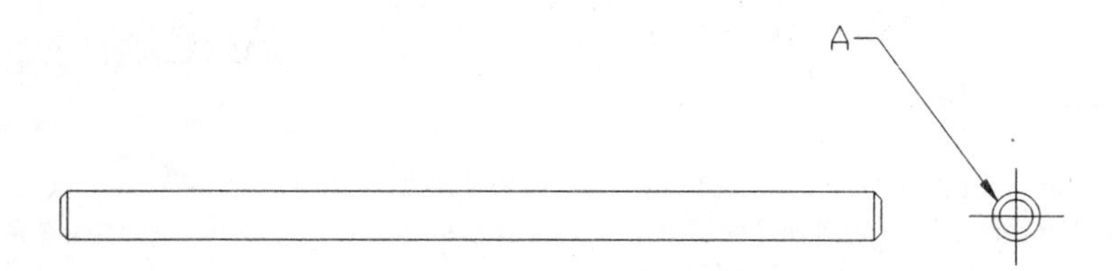

Figure 4.9 Two lines drawn

Run the LINE command again to draw another line. Use the FROM object snap, and take the center of the circle as the Base point. See Figure 4.10.

<Draw> **<Line>**

Command: **[Enter]**
LINE
From point: **FROM**
Base point: **CEN** of **[Select A (Figure 4.9).]**
<Offset>: **@50<180**
To point: **@360<180**
To point: **[Enter]**

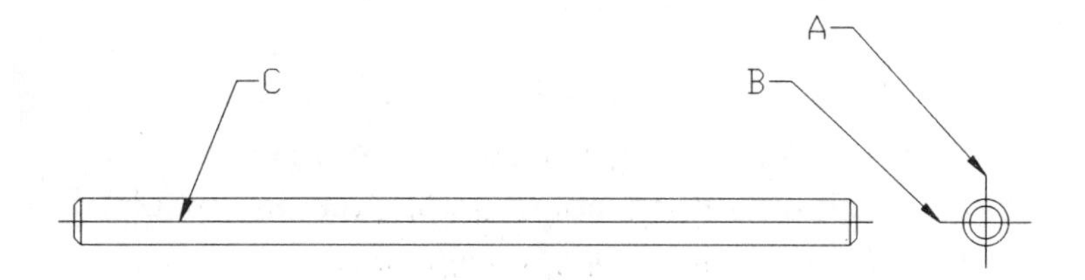

Figure 4.10 Another line drawn

The linetype for the last three lines is Continuous. Because they should be center lines, use the DDCHPROP command to change their linetype property to Center. From the Modify pull-down menu, select the Properties... item to run the DDCHPROP command. See Figure 4.11.

<Modify> **<Properties...>**

Select objects: **[Select A, B, and C (Figure 4.10).]**
Select objects: **[Enter]**

[Properties
Linetype... **Center**
OK]

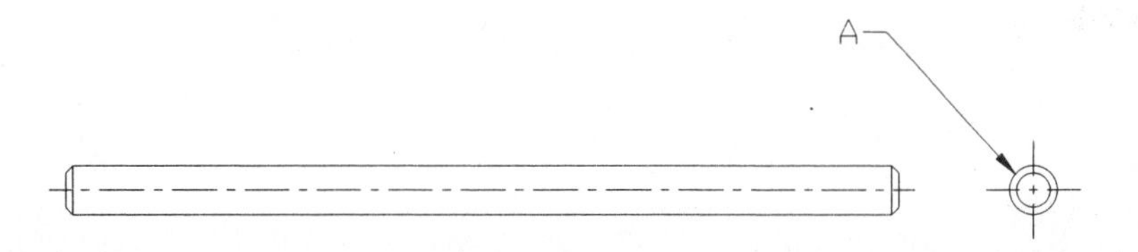

Figure 4.11 Three lines changed to CENTER linetype

Insertion Base Point

The front view and side view of the third component are complete. In Section 4.3, you will insert this file to the pipe support assembly drawing. Because the default insertion base point is (0,0) unless otherwise specified, run the BASE command to set the insertion base point to the center of the circle.

<Draw> **<Block>** **<Base>**

Command: **BASE**
Base point: **CEN** of **[Select A (Figure 4.11).]**

The drawing is complete. Use the QSAVE command to save the file.

<File> **<Save>**

File name: **SHAFT**

4.3 Insert and Edit

To insert an external drawing file to the current drawing, you can use the INSERT command, the DDINSERT command, or the MINSERT command. These commands create instances from a block definition (if there is a block definition) or form a block definition from an external drawing file and then create instances from the defined block.

Open the pipe support assembly drawing with the OPEN command.

<File> **<Open...>**

Command: **OPEN**

To display all the entities related to the two drawn components, select the Layer Control box of the Object Properties toolbar to check that all layers are unlocked, thawed, and turned on.

[Object Properties] **[Layer Control]**

Thaw & Unlock	0
Thaw & Unlock	Center
Thaw & Unlock	Hatch
Thaw & Unlock	Hatch1
Thaw & Unlock	Hidden

Thaw & Unlock	Point
On	Roller

Current layer: **Roller**

Issue the INSERT command to insert the drawing of the third component to your drawing. See Figure 4.12.

<Insert> **<Block...>**

Command: **DDINSERT**

[Insert
File...]

When the file selection dialog box appears, select the file that you saved in Section 4.2.

[OK]

Insertion point: **CEN** of **[Select A (Figure 4.1).]**
X scale factor <1> / Corner / XYZ: **1**
Y scale factor (default=X): **1**
Rotation angle: **0**

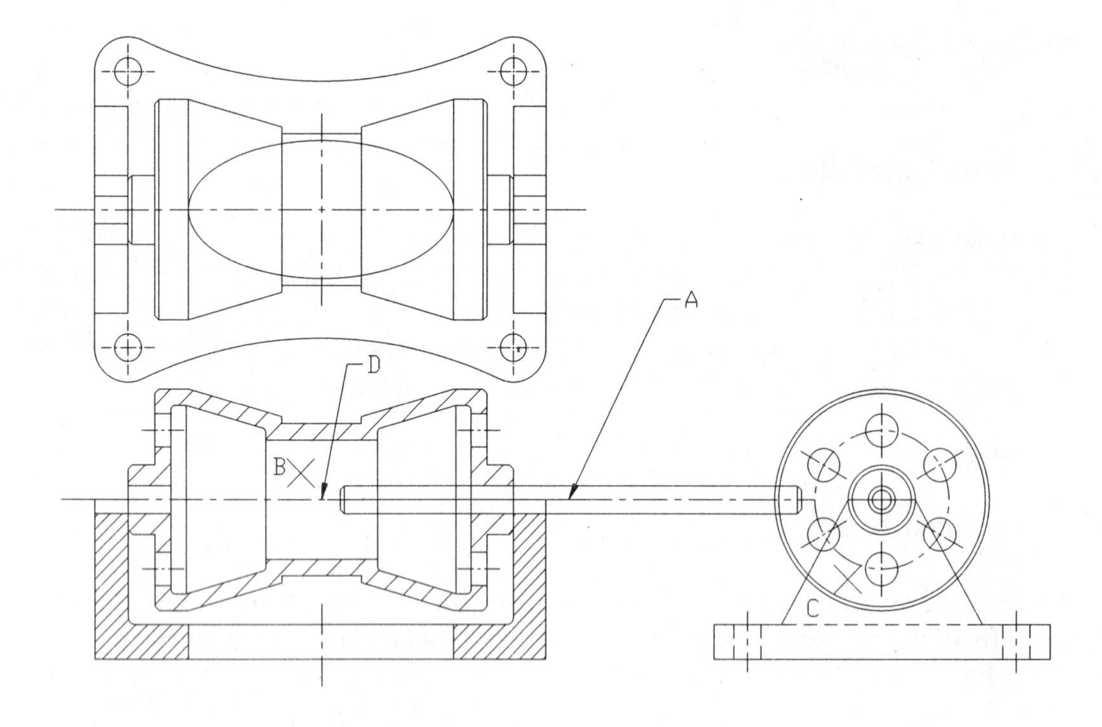

Figure 4.12 Third component inserted

When you insert an external drawing file to a drawing, the imported file forms an internal block in the drawing. Refer to Figure 4.12. The objects that are displayed at the insertion point are an instance of the newly defined block. It is a single compound entity.

Explode a Block

To modify the instance of a block, you can redefine the block in accordance with what you have learned in Chapter 3. However, if you wish to edit the individual entities of an instance without redefining the block, you have to change the instance into a copy of the block as a separate set of entities at the insertion point and remove the link to the defined block. To do so, apply the EXPLODE command.

<Modify> **<Explode>**

Command: **EXPLODE**
Select objects: [**Select A (Figure 4.12).**]
Select objects: [**Enter**]

After using the EXPLODE command on an instance of a block, you will get a copy of the set of entities of the defined block in addition to the defined block stored in the file memory. The file size increases.

Apart from exploding the instances of a block, the EXPLODE command also works on any compound objects. It breaks a compound object into its component objects. For example, you can use this command to explode a polyline into individual line and arc segments. Another command similar to the EXPLODE command is the XPLODE command. The XPLODE command has additional options, allowing you to explode multiple compound objects and at the same time change the color, layer, and linetype.

Purge Unreferenced Objects

As we have said, there is a copy of the defined block at the insertion point in addition to the defined block after you exploded the inserted block. In this drawing, you will not need the defined block anymore. This block becomes an unreferenced block because there are no instances referenced to it.

In this drawing, there is a layer called Point. You have erased all the objects created on this layer. Because there is no entity residing on this layer, it is an unreferenced layer.

To save memory space, you should remove the unreferenced block definition and the unreferenced layer by using the PURGE command.

<File> **<Drawing Utilities>** **<Purge>** **<All>**

Command: **PURGE**
Purge unused Blocks/Dimstyles/LAyers/LTypes/SHapes/STyles/Mlinestyles/All: **ALL**
Names to purge <*>: [**Enter**]
Verify each name to be purged? <Y> [**Enter**]
Purge block Name_of_the_third_component ? <N> **Y**
Purge layer POINT? <N> **Y**

Besides removing the unreferenced layer and block, the PURGE command also removes other unreferenced objects such as dimstyles, linetypes, shapes, and styles.

When you insert a drawing file to another drawing file, a block forms. When you insert a drawing file with a block to another drawing file, a nested block forms. In some complex drawings, there are nested blocks and references. The PURGE command purges only one level of reference at one time. To purge the unreferenced nested objects, you may have to use the PURGE command repeatedly. You may wish to check the file size each time after purging.

Insert as Separate Entities

When you insert a block or a file to a drawing, you can choose to treat the inserted objects as separate entities or as an instance of a block. To treat the objects as separate entities means to explode them. If you do not want to define a block when inserting a file, select the box next to the item Explode. See Figure 4.13.

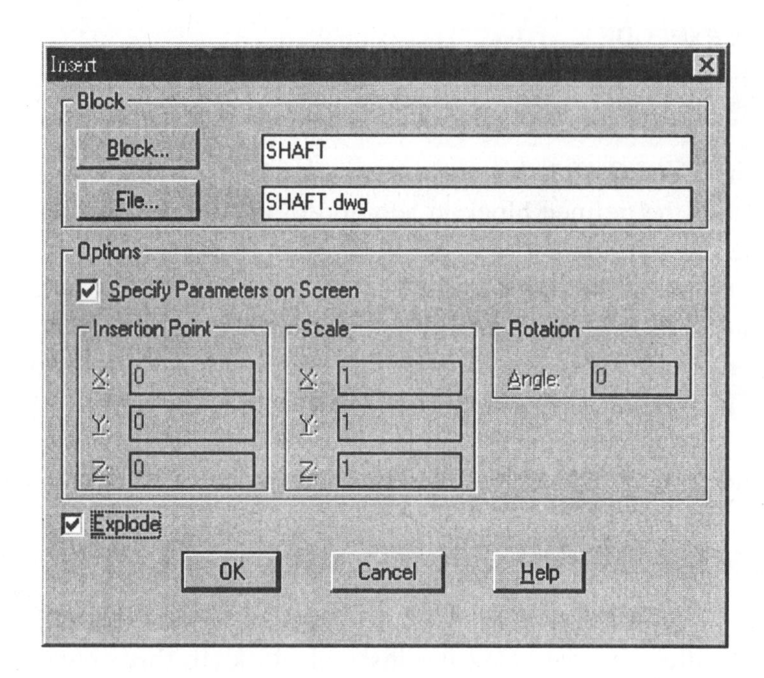

Figure 4.13 Explode while inserting

Remove Unreferenced Layer

As we have noted, you can use the PURGE command to remove unreferenced layers. In the LAYER command dialog box shown in Figure 4.14, there is a [Delete] button. You can also use this button to remove any unreferenced layer.

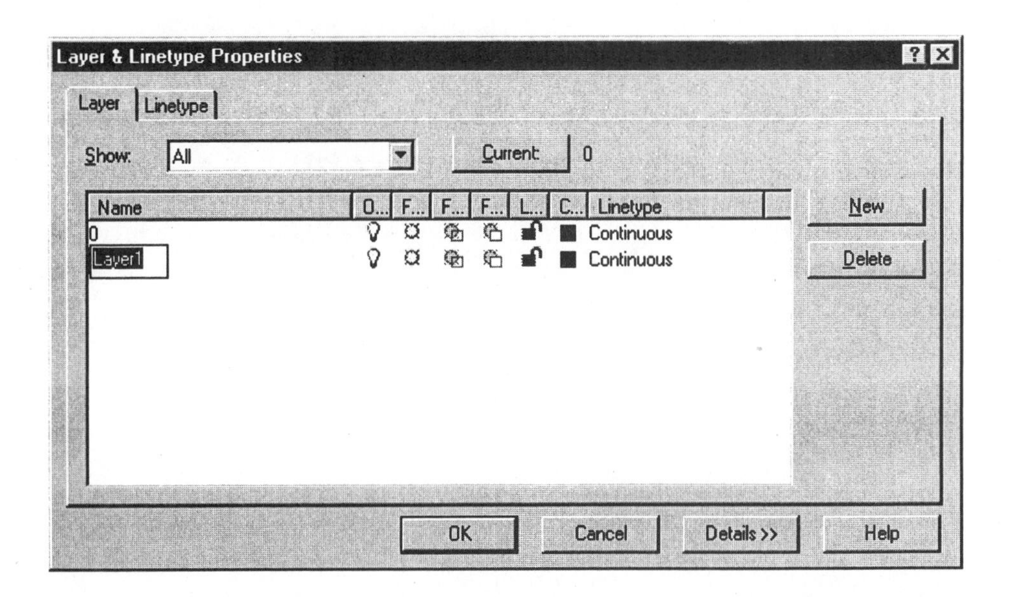

Figure 4.14 Deleting unused layers by using the Layer & Linetype Properties dialog box

Move Entities

The exploded instance of the inserted block is placed at the side view of the assembly drawing. Because some of the entities of this exploded block belong to the front view, you should use the MOVE command to translate these entities to align with the front view of the assembly.

In order that you will not accidentally edit the entities on other layers, use the Layer Control box of the Object Properties toolbar to lock all layers except layer Shaft.

[Object Properties] **[Layer Control]**

Lock	0
Lock	Center
Lock	Hatch
Lock	Hatch1
Lock	Hidden
On	Shaft
Lock	Roller

Current layer: **Shaft**

Apply the MOVE command to move the entities related to the front view of the assembly. See Figure 4.15.

<Modify> **<Move>**

Command: **MOVE**
Select objects: **W**
First corner: [**Select B (Figure 4.12).**]
Other corner: [**Select C (Figure 4.12).**]
Select objects: [**Enter**]

Here, entities other than those on the layer Shaft are not selected because they are locked.

Base point or displacement: **MID** of **[Select A (Figure 4.12).]**
Second point of displacement: **MID** of **[Select D (Figure 4.12).]**

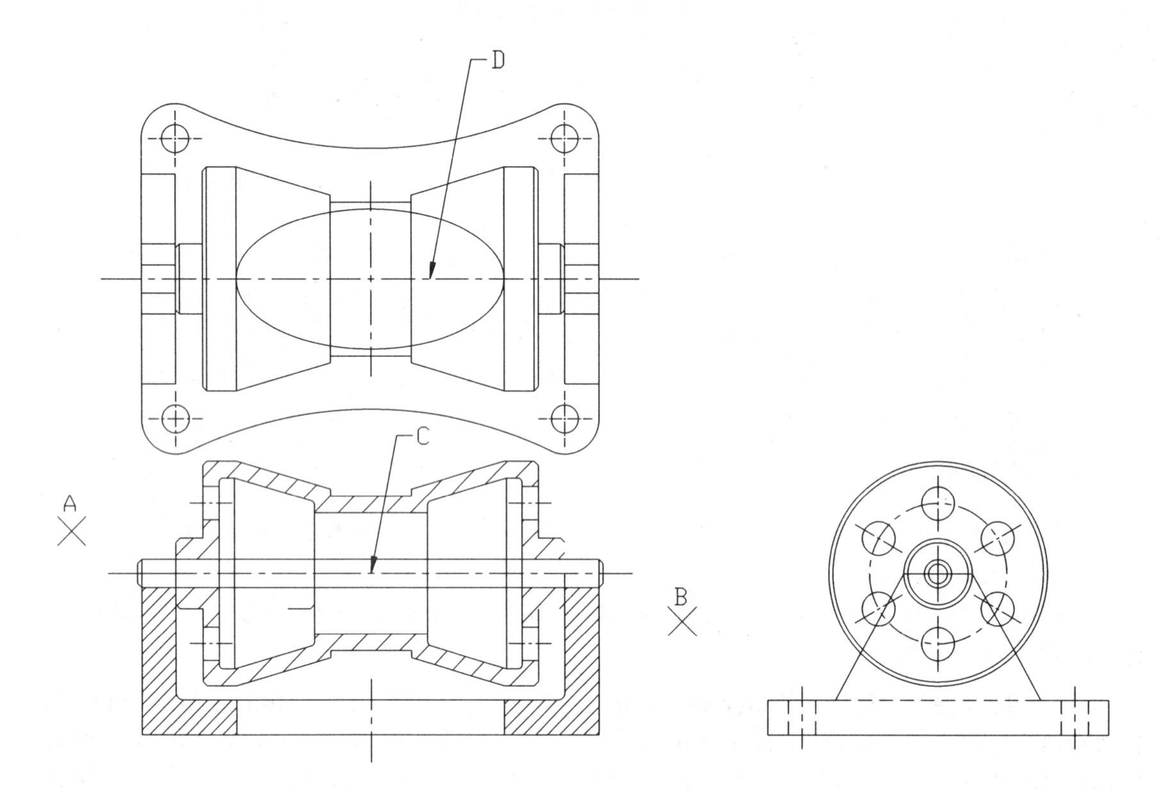

Figure 4.15 Front view of the third component moved

Refer to Figure 4.15. You may find in your screen display that the line C appears to be a continuous line instead of a center line. This is the result of having more than one center line and placing one on top of the other.

Create and Edit the Top View

The top view of the shaft is similar to the front view. Use the COPY command to copy the entities related to the front view of the shaft to the top. See Figure 4.16.

<Modify> **<Copy>**

Command: **COPY**
Select objects: **W**
First corner: **[Select A (Figure 4.15).]**
Other corner: **[Select B (Figure 4.15).]**
Select objects: **[Enter]**
<Base point or displacement>/Multiple: **MID** of **[Select C (Figure 4.15).]**
Second point of displacement: **PER** to **[Select D (Figure 4.15).]**

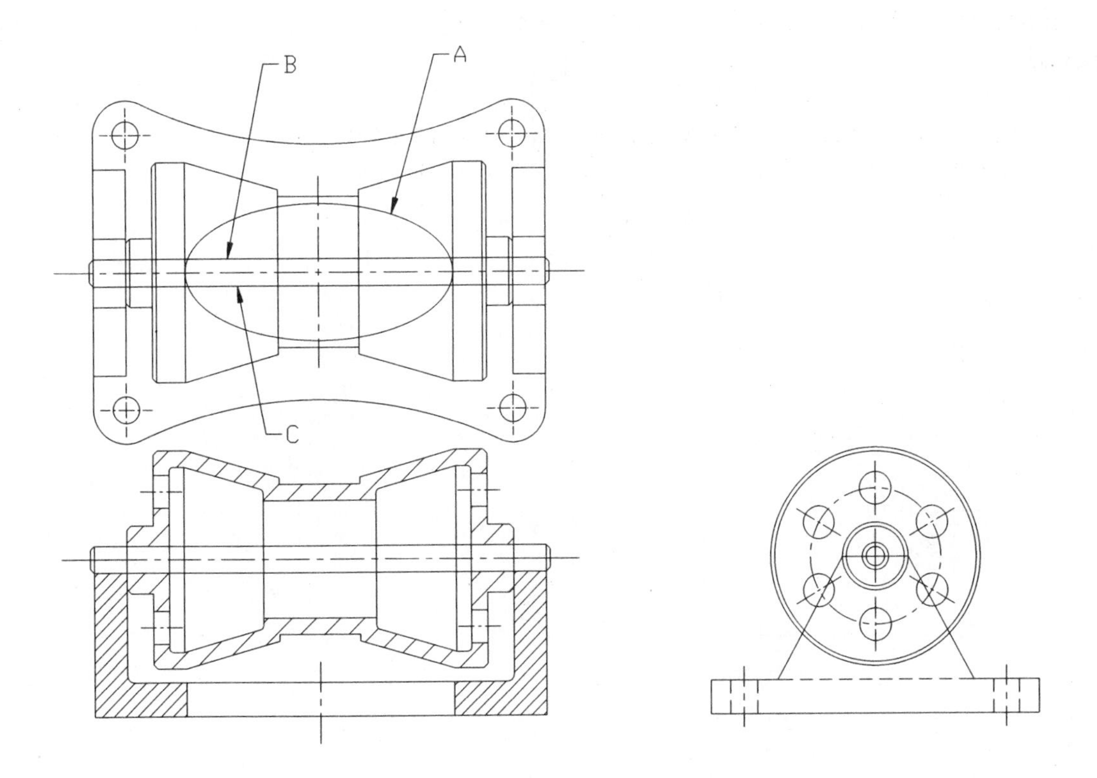

Figure 4.16 Top view of the third component copied

As in Figure 4.15, the center lines of the top view and the sectional front view in Figure 4.16 also appear to be continuous lines. To get a better effect, you can erase one of the overlapping lines. To continue with editing, use the Layer Control box of the Object Properties toolbar to unlock all layers.

[Object Properties] **[Layer Control]**

Unlock	0
Unlock	Center
Unlock	Hatch
Unlock	Hatch1
Unlock	Hidden
On	Shaft
Unlock	Roller

Current layer: **Shaft**

Apply the ERASE command to erase the two horizontal lines and ellipse that are hidden by the inserted third component. The top view of the assembly is complete. See Figure 4.17.

<Modify> **<Erase>**

Command: **ERASE**
Select objects: **[Select A, B, and C (Figure 4.16).]**
Select objects: **[Enter]**

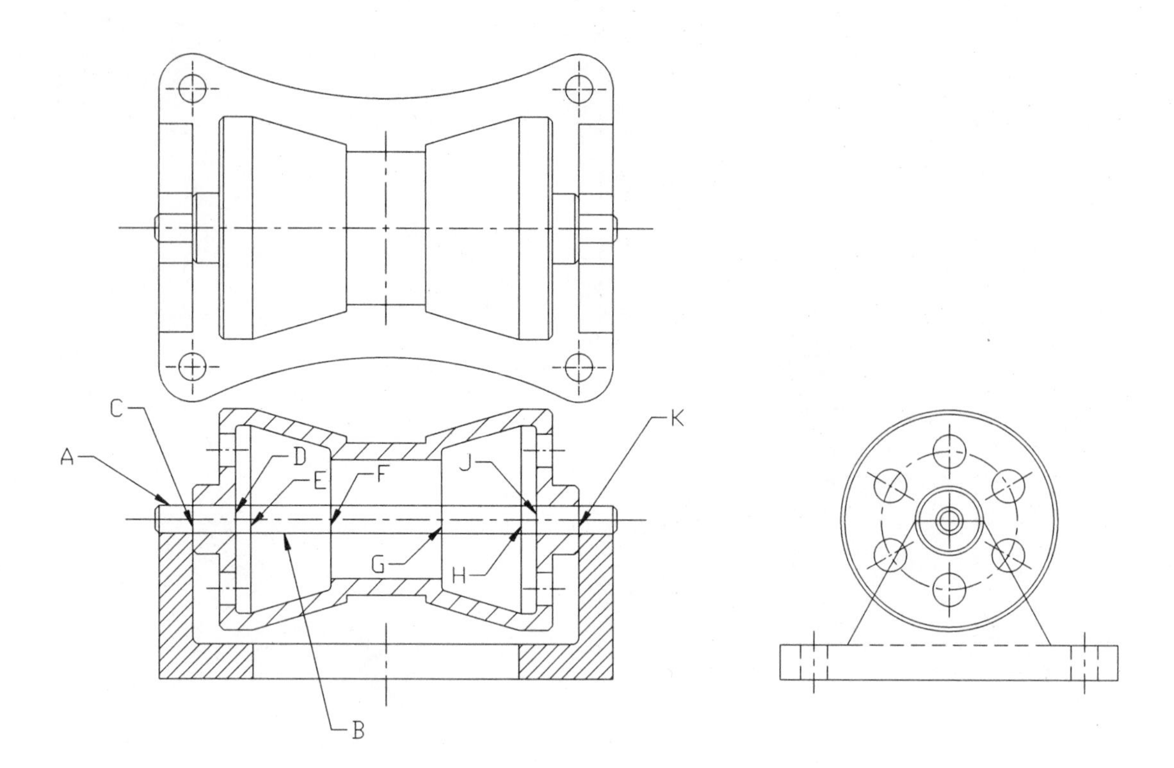

Figure 4.17 Top view edited

Edit the Front View

To complete the front view, trim the vertical lines of the first and second components that are partly hidden by the third component. See Figure 4.18.

<Modify> **<Trim>**

Command: **TRIM**
Select cutting edges: (Projmode = UCS, Edgemode = No extend)
Select objects: **[Select A and B (Figure 4.17).]**
Select objects: **[Enter]**
<Select object to trim>/Project/Edge/Undo: **[Select C, D, E, F, G, H, J, and K (Figure 4.17).]**
<Select object to trim>/Project/Edge/Undo: **[Enter]**

<View> **<Redraw>**

Command: **REDRAW**

<Modify> **<Trim>**

Command: **TRIM**
Select cutting edges: (Projmode = UCS, Edgemode = No extend)
Select objects: **[Select B (Figure 4.17).]**
Select objects: **[Enter]**
<Select object to trim>/Project/Edge/Undo: **[Select C and K (Figure 4.17).]**
<Select object to trim>/Project/Edge/Undo: **[Enter]**

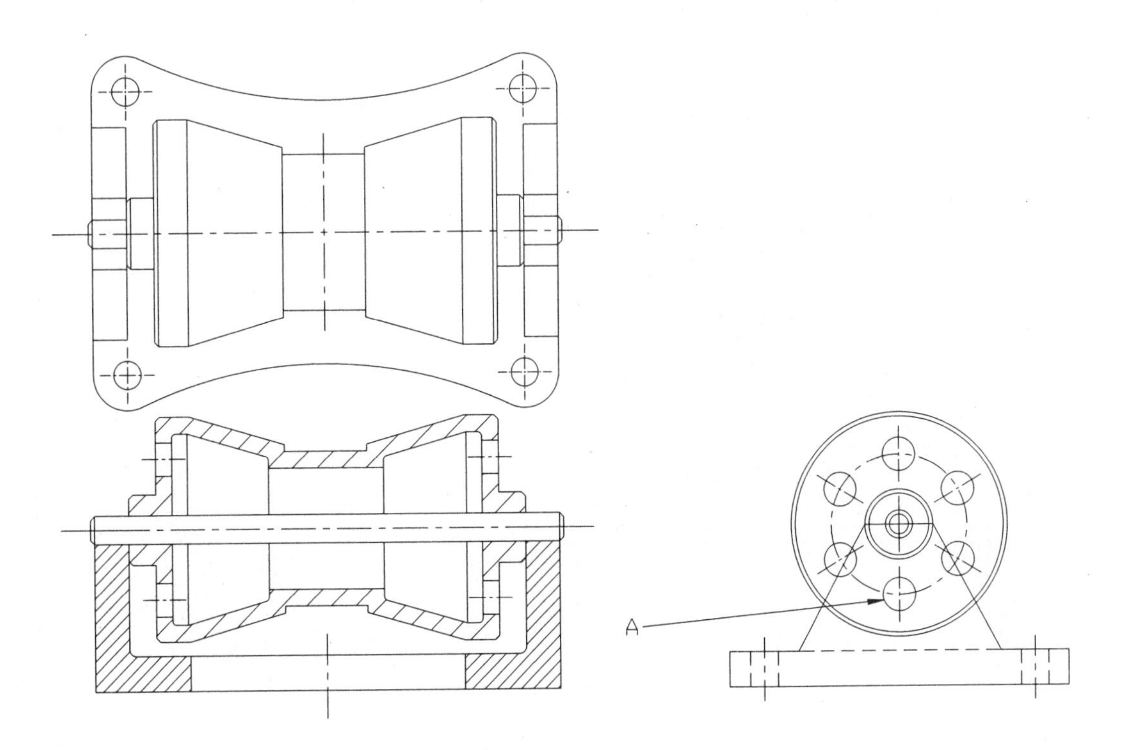

Figure 4.18 Front view edited

The front view of the assembly is complete.

Edit the Side View

Run the ERASE command to erase an instance of the block Hole in the side view. See Figure 4.19.

<Modify> **<Erase>**

Command: **ERASE**
Select objects: **[Select A (Figure 4.18).]**
Select objects: **[Enter]**

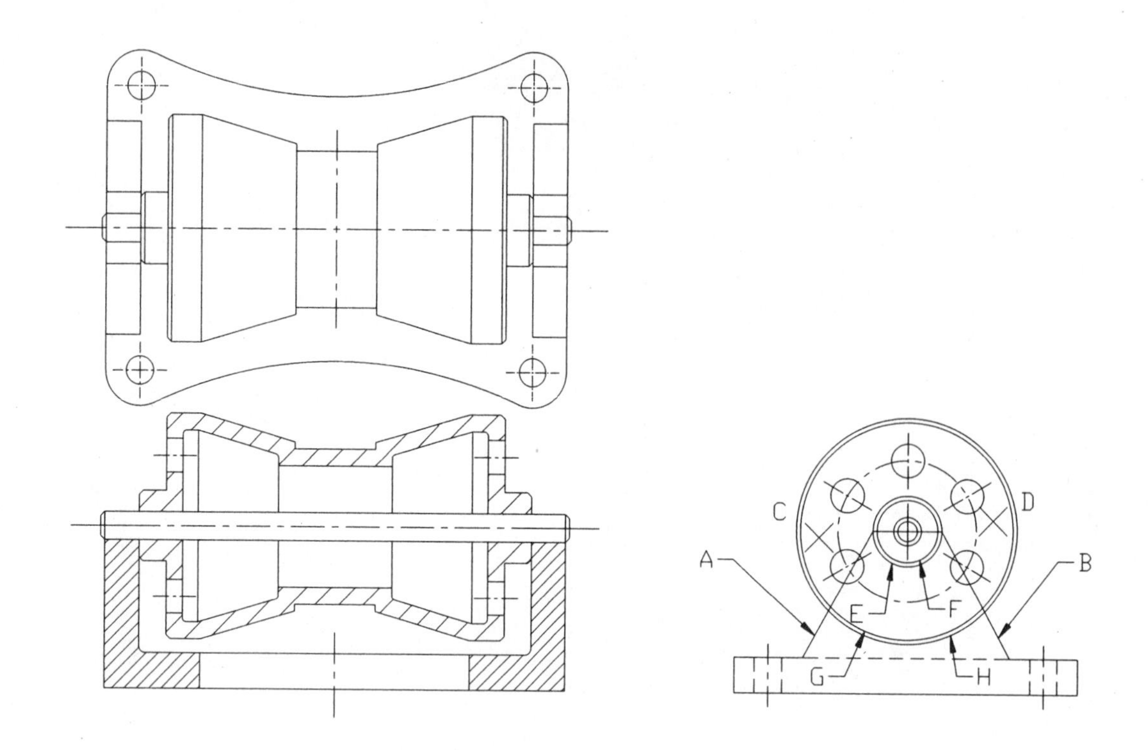

Figure 4.19 Block instance erased

Four circles of the second component need to be trimmed. Execute the TRIM command. See Figure 4.20.

<Modify> **<Trim>**

Command: **TRIM**
Select cutting edges: (Projmode = UCS, Edgemode = No extend)
Select objects: [**Select A and B (Figure 4.19).**]
Select objects: **W**
First corner: [**Select C (Figure 4.19).**]
Other corner: [**Select D (Figure 4.19).**]
Select objects: [**Enter**]
<Select object to trim>/Project/Edge/Undo: [**Select E, F, G, and H (Figure 4.19).**]
<Select object to trim>/Project/Edge/Undo: [**Enter**]

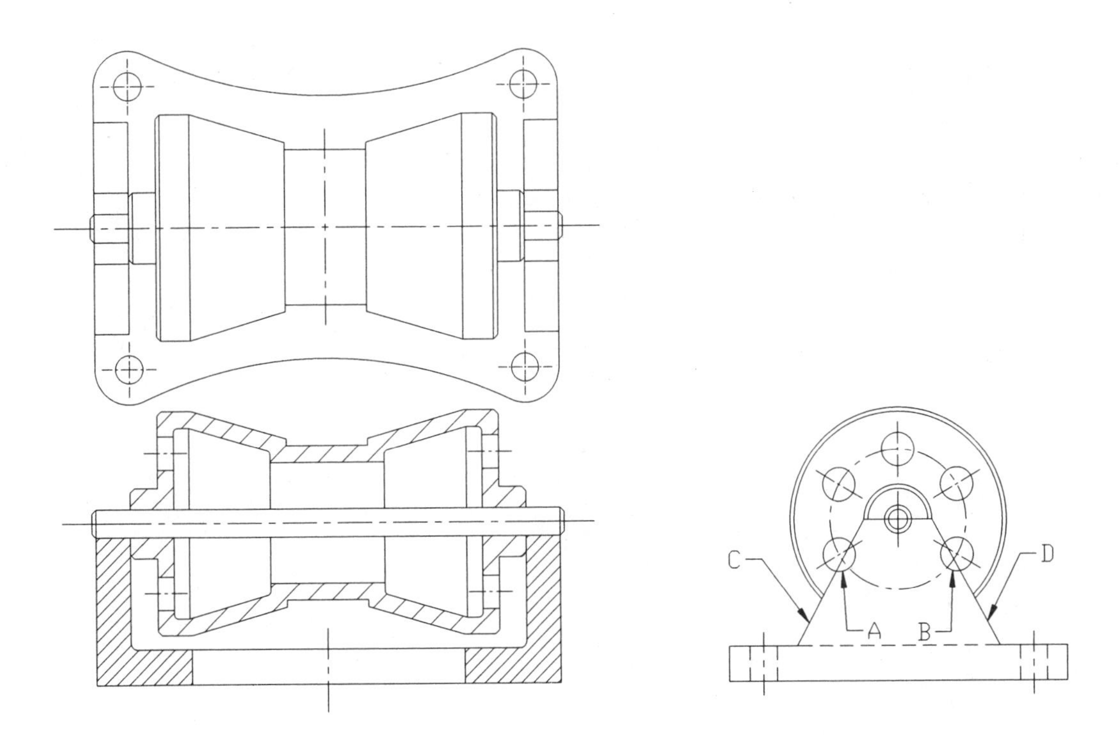

Figure 4.20 Circles trimmed

Entities A and B of Figure 4.20 are instances of the block Hole. Because you cannot trim the individual entities of an instance, you have to use the EXPLODE command to break the block reference before you can edit them.

<Modify> **<Explode>**

Command: **EXPLODE**
Select objects: [**Select A and B (Figure 4.20).**]
Select objects: [**Enter**]

After exploding, the two instances become individual entities. You can trim them with the TRIM command. See Figure 4.21.

<Modify> **<Trim>**

Command: **TRIM**
Select cutting edges: (Projmode = UCS, Edgemode = No extend)
Select objects: [**Select C and D (Figure 4.20).**]
Select objects: [**Enter**]
<Select object to trim>/Project/Edge/Undo: [**Select A and B (Figure 4.20).**]
<Select object to trim>/Project/Edge/Undo: [**Enter**]

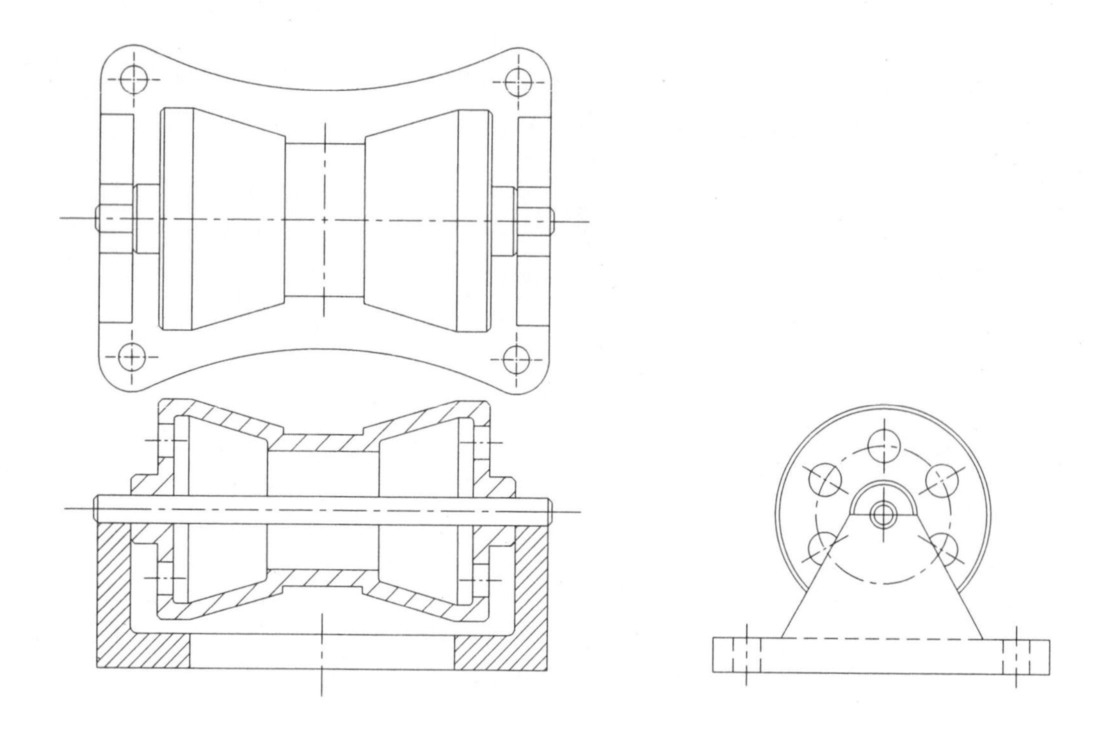

Figure 4.21 Side view edited

The assembly of three components is complete. Save the drawing.

<File> **<Save As...>**

Filename: **PIPESP.DWG**

In Chapter 3, you created two components of the assembly on different layers. In this chapter, you exported the entities related to the first and second components and a block to three external files, created the third component, imported the third component, and edited the assembly drawing. In making this set of assembly drawings, you have used two approaches — top down approach and down top approach. In making the first and second components, the approach is top down. You created everything on a single drawing, in which you can check how they fit together. In making the third component, the approach is down top. You created the component drawing and inserted it to the assembly.

Write Entire File to Another File

Earlier, you learned that you can remove unreferenced objects by using the PURGE command and you can remove unreferenced layers by using the LAYER command. If you want to remove all unreferenced objects quickly, you can use the WBLOCK command.

In Section 4.1 of this chapter, you used this command to export defined blocks and selected objects to form external drawing files. Apart from making these exports, you can export the entire drawing file to another drawing file. In the process of writing the entire block, any unreferenced objects are excluded.

Run the WBLOCK command at the command line interface. When asked for block name, respond with * to write everything to the external file. You can specify a different file name for this output file.

Command: **WBLOCK**

When the dialog box appears, specify a file name.

Block name: *

4.4 Blocks and External References

To reiterate, the INSERT command, the DDINSERT command, and the MINSERT command create instances of a block or create block definitions from an external drawing file and then create instances of the defined block. The INSERT command and the MINSERT command work at the command line interface. The DDINSERT command brings up a dialog box.

Both the INSERT command and the MINSERT command ask for a block name. After you supply a name, AutoCAD searches for a defined block with the specified block name. If there is one, AutoCAD uses that block to perform block insertion. If there is none, AutoCAD searches through the current paths for a file of the specified name. When it finds one, it copies the entire file to the current drawing to form a block. Then insertion proceeds. For the DDINSERT command, you can specify a block name to create an instance from an internal block or a file name to create a block definition and an instance from the block definition.

When you insert an external file, the file size of the current drawing increases. If you are handling a very large project with a lot of insertions, the resulting drawing file size can be extremely large. Furthermore, any change to the original drawing that is made after the insertion is not reflected in the current drawing.

An alternative to inserting a file is to attach an external file with the XREF command. Instead of copying the file to form an internal block, this command establishes a data link between the external drawing and the current drawing file. As a result, the master drawing file size is much smaller than with insertion, because the external drawing is not copied. When you save the drawing, only the link data are stored. Naturally, the next time you open the file, the external drawing has to be loaded again. Because of this reloading, the latest change in the external file is reflected in the master file. However, you should be careful in handling files when you use a lot of external references. It can be confusing to other people who do not know the organization of your drawing and reference files.

When you externally reference a file, you can choose to attach the file or overlay the file. An attached file will be displayed in subsequently nested references, but the overlay will not.

Prepare Three Drawings for Comparison

In order to appreciate how the XREF command works and how it compares with the DDINSERT command (or the INSERT command and the MINSERT command), prepare three identical drawings — one for insertion by using the DDINSERT command, one for attachment by using the XREF command, and one for overlay by using the XREF command.

Select the New... item from the File pull-down menu to start a new drawing by using the NEW command. Start from scratch and use metric default settings.

<File> **<New...>**

Command: **NEW**

[Create New Drawing
Start from Scratch
Metric
OK]

From the Format pull-down menu, select the Linetype... item to apply the LINETYPE command to load linetype Center to the drawing.

<Format> **<Linetype...>**

Command: **LINETYPE**

[Linetype
Load...]

[Load or Reload Linetypes
Available Linetypes
Center
OK]

[Global scale factor: **2**
OK]

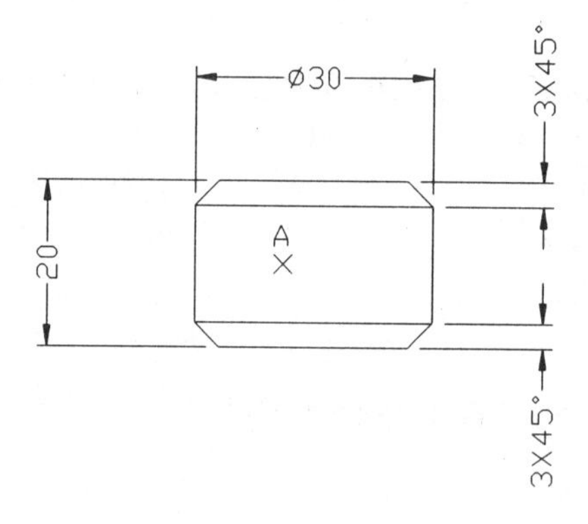

Figure 4.22 Dimensioned drawing

According to the dimensions given in Figure 4.22, create a set of line segments. Do not add the dimensions to your drawing. You will learn how to include dimensions in the next chapter.

Run the BHATCH command to create hatching. See Figure 4.23.

<Draw> **<Hatch...>**

Command: **BHATCH**

[Boundary Hatch
Pattern...]

[Hatch pattern palette
ANSI37 OK]

[Scale: **1**
Angle: **0**
Boundary:
Pick Points<]

Select internal point: **[Select A (Figure 4.22).]**
Select internal point: **[Enter]**

[Preview Hatch<]

Refer to Figure 4.23. Select the [Apply] button if the hatching is correct.

[Apply]

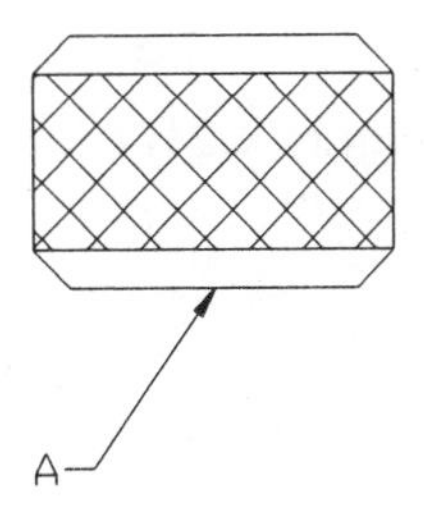

Figure 4.23 Hatching created

Issue the LINE command to create a line, and then change its linetype property to Center. See Figure 4.24.

<Draw> <Line>

Command: **LINE**
From point: **FROM**
Base point: **MID** of **[Select A (Figure 4.23).]**
<offset>: **@10<270**
To point: **@40<90**
To point: **[Enter]**

<Modify> <Properties...>

Select objects: **LAST**
Select objects: **[Enter]**

[Modify Line
Properties
Linetype... Center
OK]

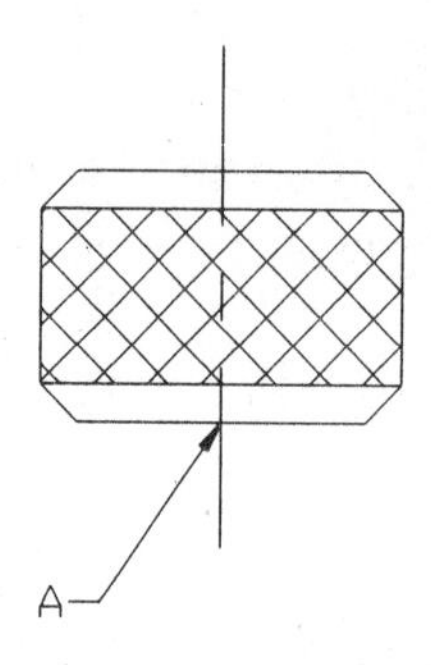

Figure 4.24 Completed drawing

Insertion Base Point

Because this drawing is going to be inserted to another drawing, use the BASE command to set the base point for insertion at the midpoint of the lower horizontal line.

<Draw> **<Block>** **<Base>**

Command: **BASE**
Base point : **INT** of **[Select A (Figure 4.24).]**

The drawing is complete.

Save Three Files

In order to make a comparison between inserting and externally referencing a drawing file, apply the SAVE command to save this file under a name SCREWHD, and use the SAVEAS command to save the same set of entities to another file called SCREWHD1. In using the XREF command to reference a file externally, you can choose to attach or to overlay the file. To compare the effect of attach and overlay, you need another drawing file. Use the SAVEAS command to save to a third file called SCREWHD2.

<File> **<Save>**

Command: **QSAVE**
Current drawing name set to C:\SCREWHD.

<File> **<Saveas...>**

Command: **SAVEAS**
Current drawing name set to C:\SCREWHD1.

<File> **<Saveas...>**

Command: **SAVEAS**
Current drawing name set to C:\SCREWHD2.

Now you have three identical drawings. You will insert SCREWHD.DWG, externally reference and attach SCREWHD1.DWG, and externally reference and overlay SCREWHD2.DWG to the pipe support assembly.

Insert a File

With three identical drawings ready, open the assembly drawing again with the OPEN command. Then use the ZOOM command to zoom to the side view. See Figure 4.25.

<File> **<Open...>**

Command: **OPEN**

File name: [**Select the pipe support assembly drawing file.**]

<View> **<Zoom>**

Command: **ZOOM**
All/Center/Dynamic/Extents/Previous/Scale(X/XP)/Window/<Realtime>: [**Enter**]

Use realtime zooming mode to obtain a display similar to Figure 4.25.

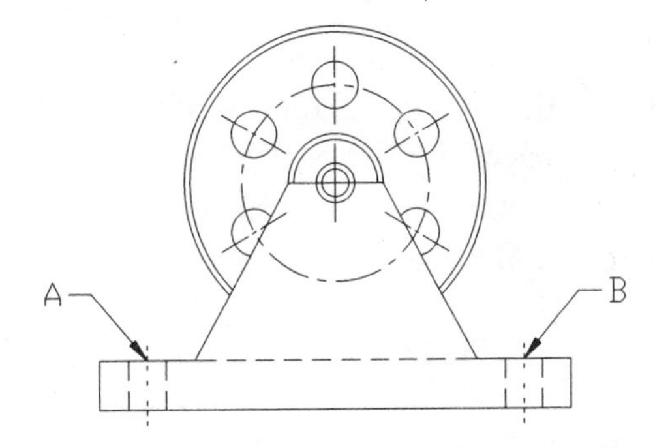

Figure 4.25 Side view of assembly drawing

Run the INSERT command to insert the file SCREWHD to location A. The insertion scale for both the X and the Y axis is 1, and the insertion rotation angle is 0. See Figure 4.27.

<Insert> **<Block...>**

Command: **DDINSERT**

[Insert
File...]

In the file selection dialog box, select the file SCREWHD and then the [Open] button. If you saved the SCREWHD.DWG under a different name, select the file name that you used to save the file.

[**OK**]

Insertion point: **INT** of [**Select A (Figure 4.25).**]
X scale factor <1> / Corner / XYZ: **1**
Y scale factor (default=X): **1**
Rotation angle <0>: **0**

The DDINSERT, INSERT, and MINSERT commands perform two tasks while inserting a file. They copy the external file to form a block definition and create an instance of the block definition at the insertion point. Thus the file SCREWHD.DWG is copied to the current drawing to create a block called SCREWHD, and the display shows an instance of this block.

Attach a File

After inserting a file, apply the XREF command to attach an externally referenced file called SCREWHD1 at location B of the assembly drawing. See Figure 4.27.

<Insert> **<External Reference...>**

Command: **XREF**

[External Reference
Attach...]

The first time you use the XREF command on a drawing file, a file selection dialog box displays when you select the [Attach...] button of the XREF command dialog box. In the file selection dialog box, select the file SCREWHD1 (or the file name that you used to save the second file) and then select the [Open] button. The XATTACH command dialog box displays. See Figure 4.26. Use the Attachment option.

[Attach Xref
Reference Type
Attachment
Parameters
X Scale Factor: **1**
Y Scale Factor: **1**
Z Scale Factor: **1**
Rotation Angle: **0**
OK]

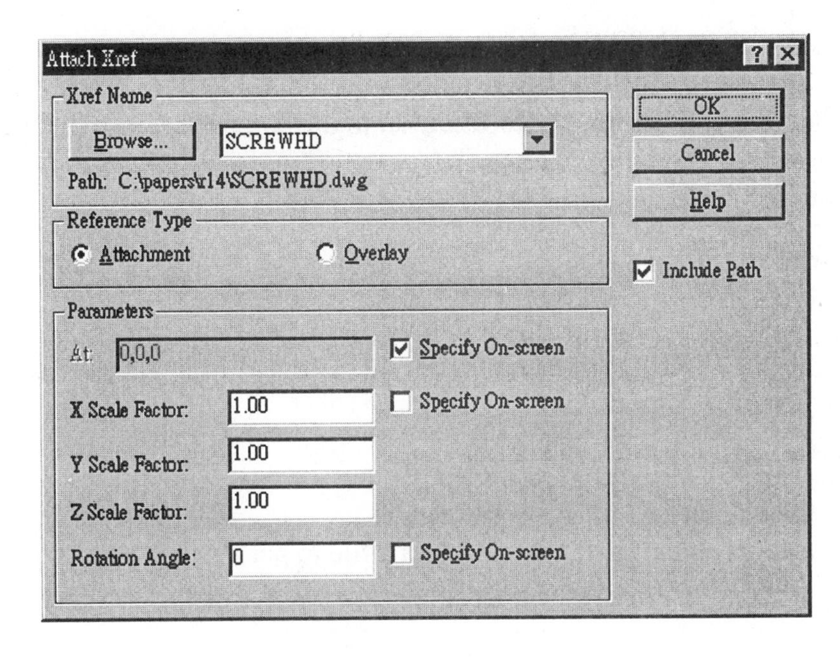

Figure 4.26 Attach Xref dialog box

Insertion point: **INT** of [**Select B (Figure 4.25).**]

The file SCREWHD1.DWG is externally referenced and attached. See Figure 4.27. The display reveals no difference between the insertion (left side) and the external reference (right side). In fact, however, they are not the same. The content of the attached file does not constitute a part of the current drawing file. After saving, only the attachment data are stored.

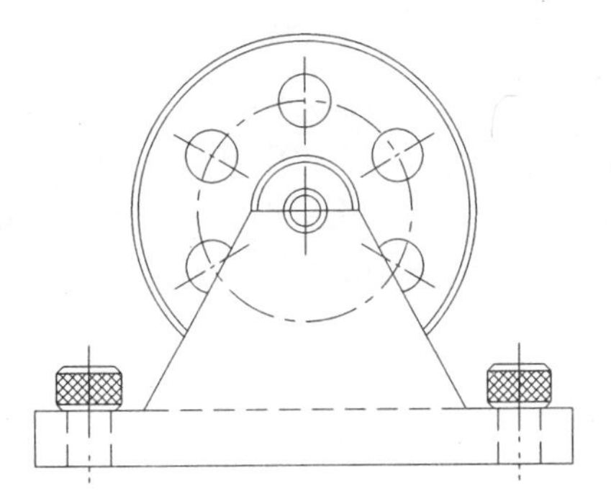

Figure 4.27 SCREW inserted and SCREWHD1 attached

Refer to Figure 4.26. It is the dialog box of the XATTACH command. The XREF command and the XATTACH command are quite similar. The XATTACH command attaches an external reference to the drawing. The XREF command controls external references. To run the XATTACH command, you can enter the command name at the command line interface or select the External Reference Attach icon of the Reference toolbar. To display the Reference toolbar, select the Toolbar... item of the View pull-down menu to apply the TOOLBAR command.

<View> <Toolbars...>

Select the Reference item in the toolbars list to display the Reference toolbar.

[Toolbars
Toolbars:
Reference
Close]

To run the XATTACH command, select the External Reference Attach icon of the Reference toolbar. See Figure 4.26.

[Reference] [External Reference Attach]

Select the [Cancel] button to exit, or select a file to perform external referencing.

[Attach Xref
Cancel]

Use the QSAVE command to save the pipe support assembly drawing.

<File> <Save>

Command: **QSAVE**

To reiterate, the pipe support assembly now has an insertion of the file SCREWHD.DWG and an external reference of the file SCREWHD1.DWG.

Edit the Source Files

In order to appreciate the differences between inserting and external referencing, open the source files and modify them. Apply the OPEN command to open the files SCREWHD and SCREWHD1 one by one, and issue the ERASE command to erase the hatching. Use the QSAVE command to quick-save them.

<File> <Open...>

Command: **OPEN**

When the dialog box appears, specify the drawing SCREWHD. See Figure 4.28.

<Modify> <Erase>

Command: **ERASE**
Select objects: **[Select the hatching.]**
Select objects: **[Enter]**

<File> <Save>

Command: **QSAVE**

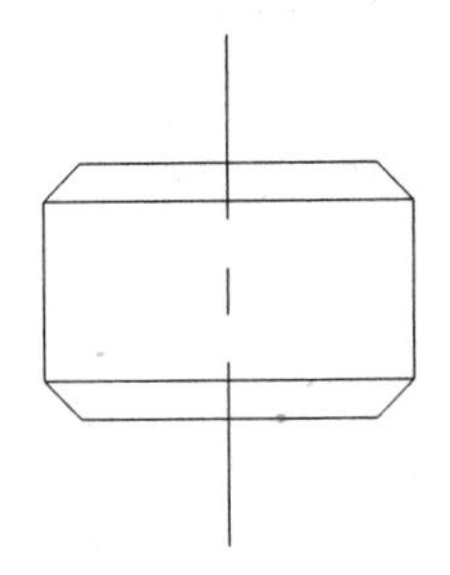

Figure 4.28 Hatching erased

<File> **<Open...>**

Command: **OPEN**

When the dialog box appears, specify the drawing SCREWHD1.

<Modify> **<Erase>**

Command: **ERASE**
Select objects: [**Select the hatching.**]
Select objects: [**Enter**]

<File> **<Save>**

Command: **QSAVE**

The source files SCREWHD.DWG and SCREWHD1.DWG are modified.

Effect of Changes

To appreciate the effect of changes, run the OPEN command to open the pipe support assembly drawing again. See Figure 4.29.

<File> **<Open...>**

Command: **OPEN**

Notice the following messages as you open the drawing:

Resolve Xref SCREWHD1: SCREWHD1.dwg
SCREWHD1 loaded. Regenerating drawing.

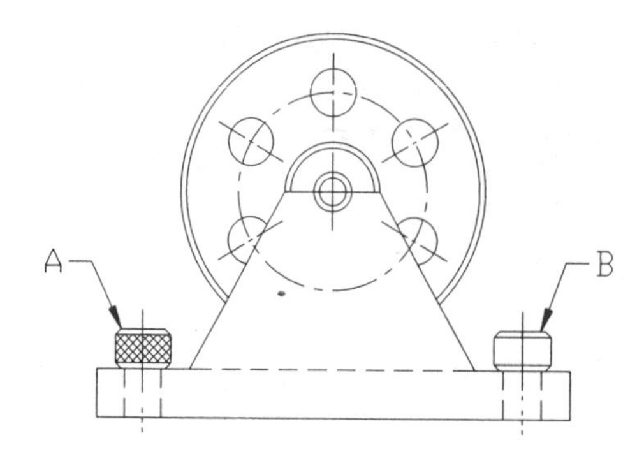

Figure 4.29 Only the attached block shows the change

When you open a file, AutoCAD loads all attached externally referenced drawings. Because the most recently saved drawing is loaded, you will see the change in the source file being reflected in the attached block. As for insertion, a block has already been formed, and there is no linkage to the external file. Change is not reflected.

Update an Inserted Block

If you want to update a block definition with an external drawing file, you have to re-insert the file. To re-insert a file, you can use the DDINSERT command or the INSERT command.

From the Insert pull-down menu, select the Block... item to run the DDINSERT command.

<Insert> **<Block...>**

Command: **DDINSERT**

[Insert
File...]

In the file selection dialog box, select the file SCREWHD and then the [Open] button.

[OK]

A warning dialog box appears. See Figure 4.30. Because you want to redefine the block, select the [OK] button.

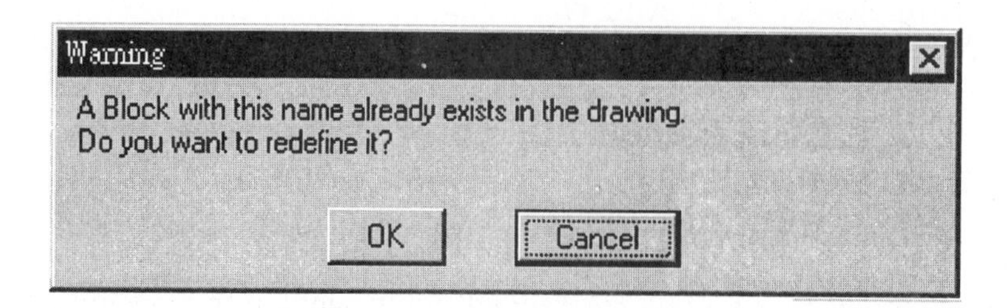

Figure 4.30 Warning dialog box

After you select the [OK] button, the normal prompt asking for an insertion point displays at the command line interface. Because the purpose of running this command is to redefine the block and not to create another instance, press the [ESC] key to terminate the command.

```
Insertion point: [ESC]*Cancel*
```

If you prefer to use the INSERT command, key in the command name at the command line interface. Then key in the file name, and suffix it with the = symbol. This instructs AutoCAD to ignore the internal block and to look for an external file of the same specified name. When it finds one, it copies the file to the current drawing to update the existing block.

```
Command: INSERT
Block name (or ?): SCREWHD=SCREWHD
```

In the foregoing input, you should replace the file name with the name that you used to save the file SCREWHD, and you may have to specify the full path where you saved your source file.

```
Block SCREWHD redefined
Regenerating drawing.
Insertion point: [ESC] *Cancel*
```

Copying of the external file to update the internal block happens before the prompt for insertion point appears. Because the purpose of this re-insertion is to update the block, press the [ESC] key to terminate the command. See Figure 4.31.

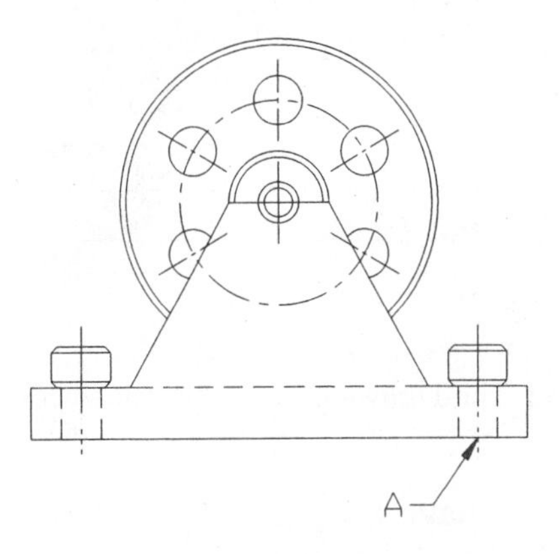

Figure 4.31 Inserted block updated

Attach and Overlay

There are two options to reference a file externally — attach and overlay. The difference between them is how the externally referenced file is handled in a nested situation.

Consider three drawing files A, B, and C. If you externally reference A to B and then externally reference B to C, then file A is said to be nested. If file A is attached to B, it is displayed in file C. If file A is overlaid in B, it is not displayed in file C.

Select the External Reference... item of the Insert pull-down menu to use the XREF command to overlay the file SCREWHD2.DWG to the pipe support assembly. See Figure 4.32.

<Insert> **<External Reference...>**

Command: **XREF**

[External Reference
Attach...]

Select the file SCREWHD2.DWG from the file selection dialog box. Then select the [OK] button.

[Attach Xref
Reference Type
Overlay
Parameters X Scale Factor: **1**
Y Scale Factor: **1**
Z Scale Factor: **1**
Rotation Angle: **0**
OK]

Insertion point: **INT** of [**Select A (Figure 4.31).**]

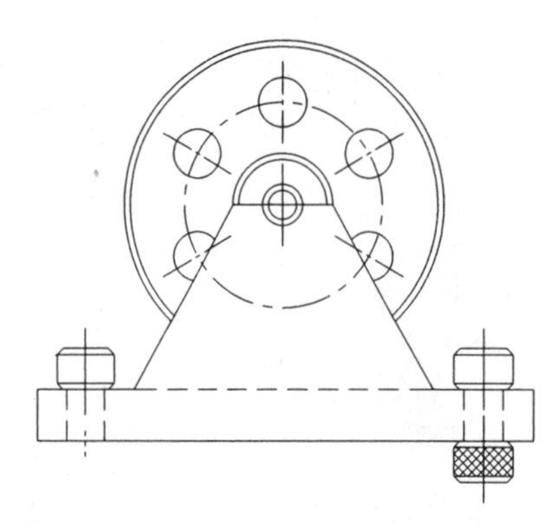

Figure 4.32 Third drawing externally referenced as an overlay

Refer to Figure 4.32. You have an insertion, an attached external reference, and an overlaid external reference in the pipe support assembly drawing.

Use the SAVE command to save the drawing.

<File> **<Save>**

Command: **QSAVE**

To appreciate the difference between attach and overlay, start a new drawing and use the XREF command to attach the pipe support assembly. See Figure 4.33.

<File> **<New...>**

Command: **NEW**

[Create New Drawing
Start from Scratch
Metric
OK]

<Insert> **<External Reference...>**

Command: **XREF**

[External Reference
Attach...]

Select the file pipe support assembly drawing from the file selection dialog box. Then select the [OK] button.

[Attach Xref
Reference Type
Attach
OK]

Insertion point: **0,0**

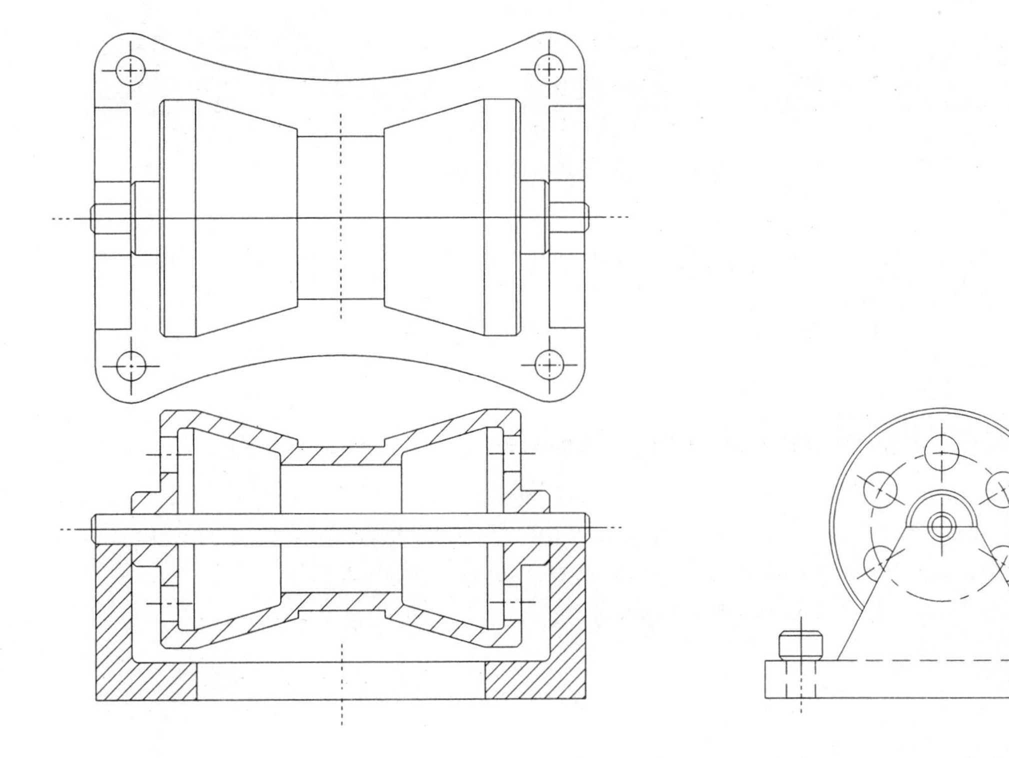

Figure 4.33 Nested overlay does not display

Refer to Figure 4.33. Both the attached and the overlaid external references are nested. The attached is displayed and the overlaid is not.

XBind

Attached external files are not part of the current drawing file. You cannot use the dependent symbols and linetypes that belong to the external file. If you want to use them, you need to use the XBIND command to bind the type of object you want. See Figure 4.34. After XBINDING, the relevant symbols become readily accessible.

<Modify> <Object> <External Reference> <Bind...>

Command: **XBIND**

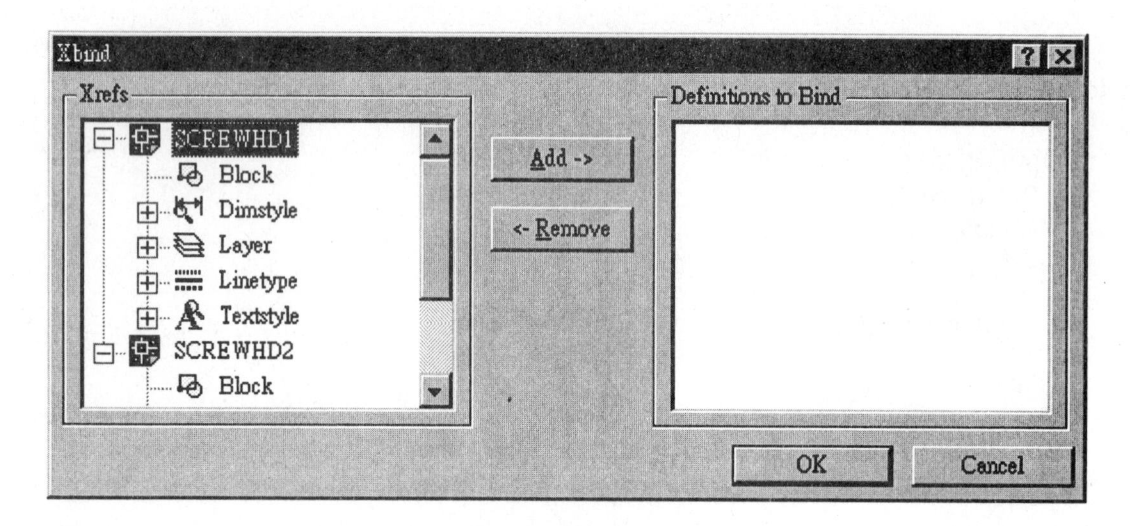

Figure 4.34 Xbind dialog box

Select the item that you want to bind to the current drawing. Then select the [Add->] button and the [OK] button.

Xref Options

Figure 4.35 shows the External Reference dialog box.

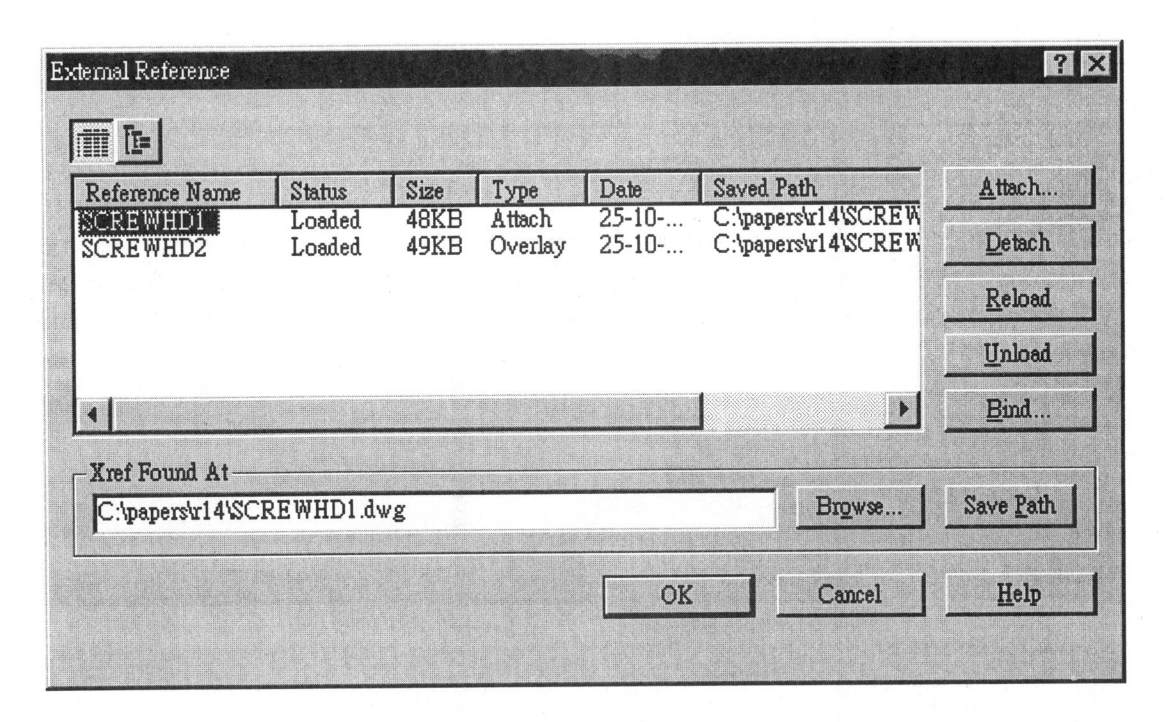

Figure 4.35 External Reference dialog box

There are several options available.

1. If you want to get rid of an externally referenced drawing, use the DETACH option.
2. If you want to reload the externally referenced drawing during the drawing session, use the RELOAD option.
3. If you want to suppress the display and regeneration of the external references, use the UNLOAD option. Unlike detaching, unloading does not remove the external references. You can reload them quickly.
4. If you wish to copy the external file to the master file with a result that is similar to insertion, use the BIND option.

Do not save the current drawing.

Clipping External References

Clipping is cutting out a portion of an instance of an external reference. To appreciate how clipping works, externally reference the washroom facilities to the floor plan and then clip the instances.

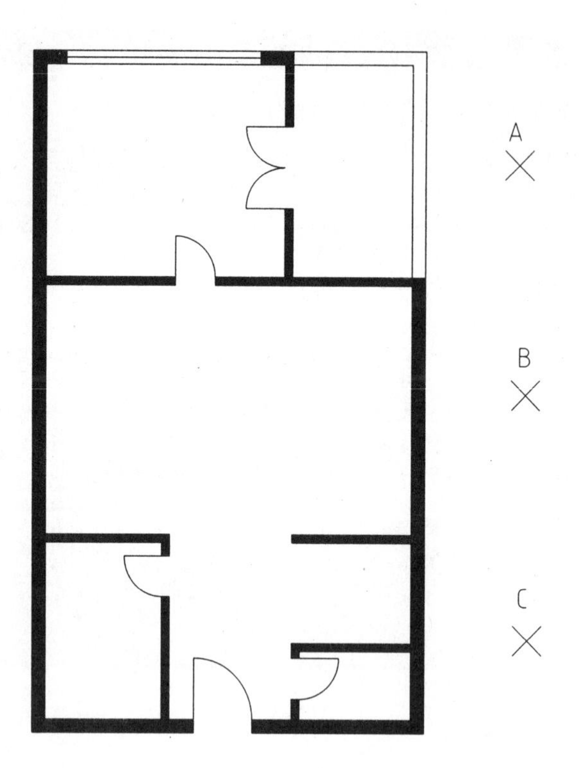

Figure 4.36 Floor plan created in Chapter 3

Open the floor plan that you created in Chapter 3. See Figure 4.36. You will use the XREF command to externally reference the washroom facilities to the floor plan and the XCLIP command to clip the instances of the reference.

<File> **<Open...>**

Command: **OPEN**
File name: **FLOOR.DWG**

Refer to Figure 3.135 of Chapter 3. There are three pieces of washroom facilities in a single file.

Select the External Reference... item from the Insert pull-down menu to use the XREF command to externally reference the washroom facilities to the floor plan. See Figure 4.37.

<Insert> **<External Reference...>**

Command: **XREF**

[External Reference
Attach...]

Select the file that you saved in Chapter 3. Refer to Figure 3.135.

[Attach Xref
Reference Type

Attachment
Parameters
X Scale Factor: **1**
Y Scale Factor: **1**
Z Scale Factor: **1**
Rotation Angle: **0**
OK]

Insertion point: [**Select A (Figure 4.36).**]

Repeat the XREF command twice to create two more external references. See Figure 4.37.

Command: [**Enter**]
XREF

[External Reference
Reference Name
WASH
Attach...]

[Attach Xref
Reference Type
Attachment
OK]

Insertion point: [**Select B (Figure 4.36).**]

Command: [**Enter**]
XREF

[External Reference
Reference Name
WASH
Attach...]

[Attach Xref
Reference Type
Attachment
OK]

Insertion point: [**Select C (Figure 4.36).**]

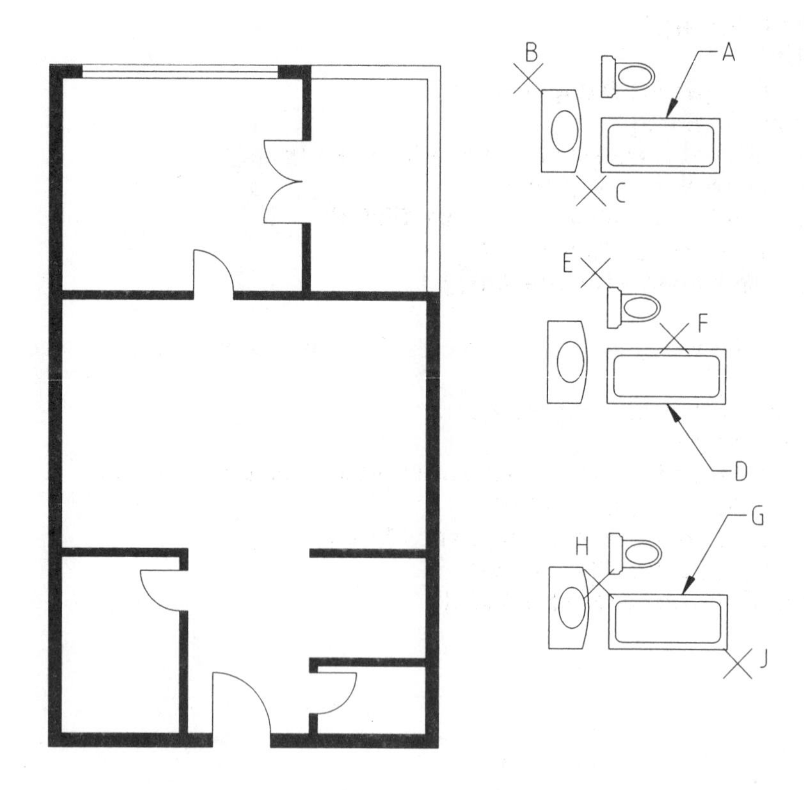

Figure 4.37 Washroom facilities externally referenced three times

Select the Toolbar... item of the View pull-down menu to use the TOOLBAR command to display the Reference toolbar.

<View> **<Toolbars...>**

Select the Reference item in the toolbars list to display the Reference toolbar.

[Toolbars
Toolbars:
Reference
Close]

From the Reference toolbar, select the External Reference Clip icon to run the XCLIP command to clip the three instances of the external reference. See Figure 4.38.

[Reference] **[External Reference Clip]**

Command: **XCLIP**
Select objects: **[Select A (Figure 4.37).]**
Select objects: **[Enter]**
ON/OFF/Clipdepth/Delete/generate Polyline/<New boundary>: **[Enter]**
Specify clipping boundary: **[Enter]**
Select polyline/Polygonal/<Rectangular>: **[Enter]**
First corner: **[Select B (Figure 4.37).]**
Other corner: **[Select C (Figure 4.37).]**

Command: **[Enter]**
XCLIP
Select objects: **[Select D (Figure 4.37).]**
Select objects: **[Enter]**
ON/OFF/Clipdepth/Delete/generate Polyline/<New boundary>: **[Enter]**
Specify clipping boundary: **[Enter]**
Select polyline/Polygonal/<Rectangular>: **[Enter]**
First corner: **[Select E (Figure 4.37).]**
Other corner: **[Select F (Figure 4.37).]**

Command: **[Enter]**
XCLIP
Select objects: **[Select G(Figure 4.37).]**
Select objects: **[Enter]**
ON/OFF/Clipdepth/Delete/generate Polyline/<New boundary>: **[Enter]**
Specify clipping boundary: **[Enter]**
Select polyline/Polygonal/<Rectangular>: **[Enter]**
First corner: **[Select H (Figure 4.37).]**
Other corner: **[Select J (Figure 4.37).]**

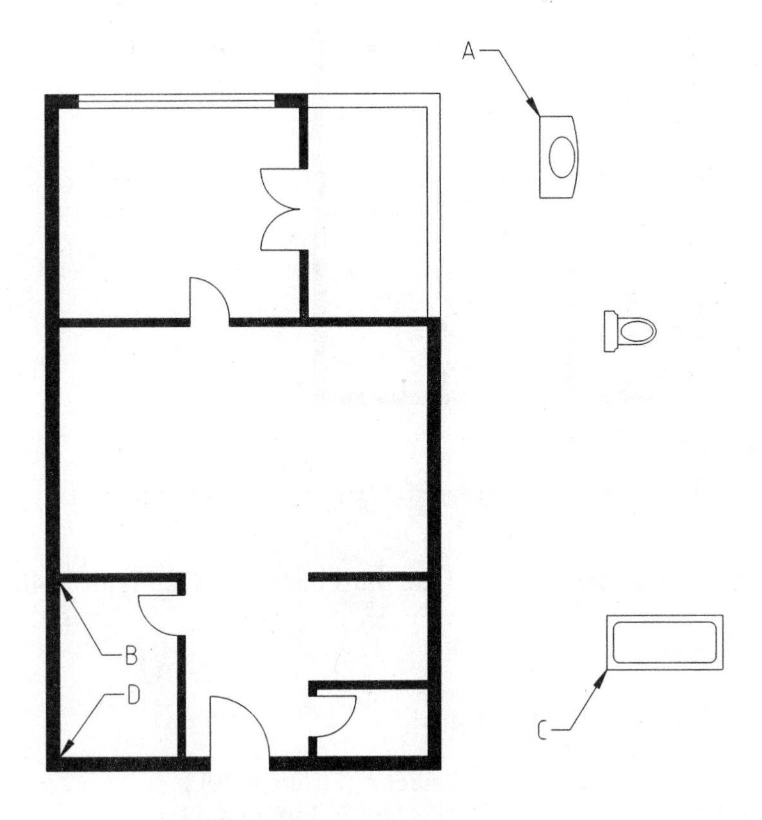

Figure 4.38 External references clipped

After clipping, use the MOVE command to move two instances. See Figure 4.39.

<Modify> **<Move>**

Command: **MOVE**
Select objects: **[Select A (Figure 4.38).]**
Select objects: **[Enter]**

Base point or displacement: **INT** of [**Select A (Figure 4.38).**]
Second point of displacement: **INT** of [**Select B (Figure 4.38).**]

Command: [**Enter**]
MOVE
Select objects: [**Select C (Figure 4.38).**]
Select objects: [**Enter**]
Base point or displacement: **INT** of [**Select C (Figure 4.38).**]
Second point of displacement: **INT** of [**Select D (Figure 4.38).**]

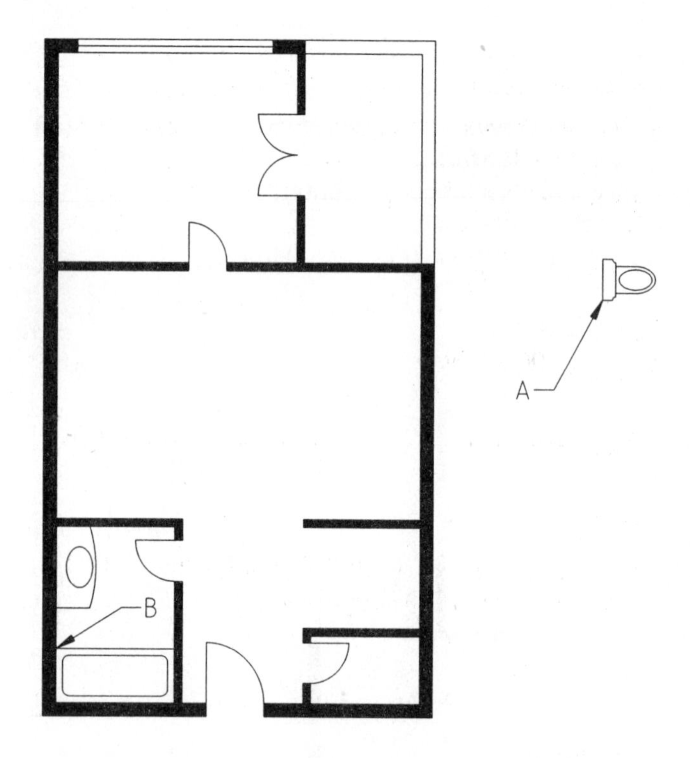

Figure 4.39 Two clipped external references moved

Repeat the MOVE command to move the third instance. See Figure 4.40.

Command: [**Enter**]
MOVE
Select objects: [**Select A (Figure 4.39).**]
Select objects: [**Enter**]
Base point or displacement: **INT** of [**Select A (Figure 4.39).**]
Second point of displacement: **INT** of [**Select B (Figure 4.39).**]

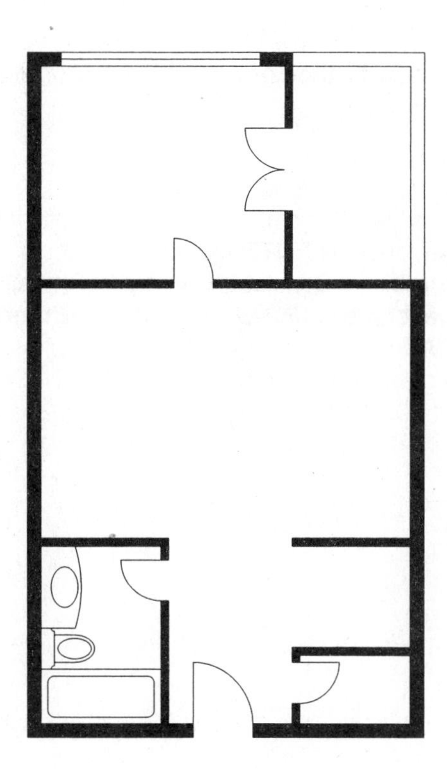

Figure 4.40 Clipped external references moved

The drawing is complete. There are three instances of the external reference WASH in this drawing. You clipped and moved them into position. Use the SAVEAS command to save your drawing to a new file name.

<File> **<Save As...>**

Command: **SAVEAS**
File name: **FLOOR1.DWG**

4.5 Block Attributes

Attributes are informational text that you can include during insertion of a block. Prior to insertion of a block, you have to define an attribute template. You will open the drawing SCREWHD.DWG and define a set of attribute templates. Then, you will insert this set to another drawing.

Select the Open... item from the File pull-down menu to apply the OPEN command.

<File> **<Open...>**

Command: **OPEN**
File name: **SCREWHD**

Define Attribute Templates

To define an attribute template, you can use either the ATTDEF command or the

DDATTDEF command.

Run the ATTDEF command at the command line interface to define a template. See Figure 4.41.

Command: **ATTDEF**
Attribute modes -- Invisible:N Constant:N Verify:N Preset:N
Enter (ICVP) to change, or press ENTER when done: **V**
Attribute modes -- Invisible:N Constant:N Verify:Y Preset:N
Enter (ICVP) to change, or press ENTER when done: **[Enter]**
Attribute tag: **MATERIAL**
Attribute prompt: **Please enter material...**
Default attribute value: **MILD STEEL**
Justify/Style/<Start point>: **[Select A (Figure 4.41).]**
Height: **5**
Rotation angle: **0**

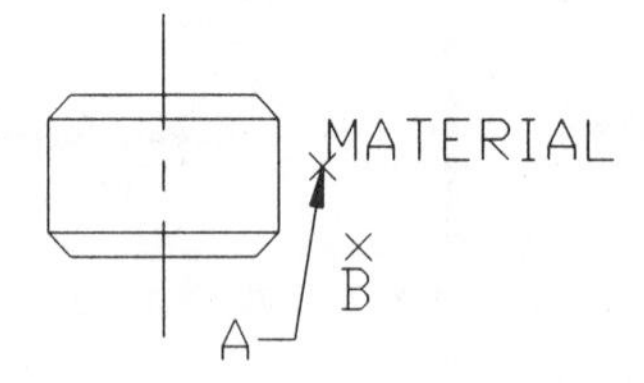

Figure 4.41 Attribute template defined by the ATTDEF command

Select the Define Attributes... item of the Block cascading menu of the Draw pull-down menu to use the DDATTDEF command to define another template. See Figure 4.42 and Figure 4.43.

<Draw> **<Block>** **<Define Attributes...>**

Command: **DDATTDEF**

[Attribute Definition
Attribute
Tag: **SUPPLER**
Prompt: **Supplier name please...**
Value: **SLL Inc.**
Text Options
Justification: **Left**
Text Style: **Standard**
Height< **5**
Rotation< **0**
Insertion Point
Pick Point<]

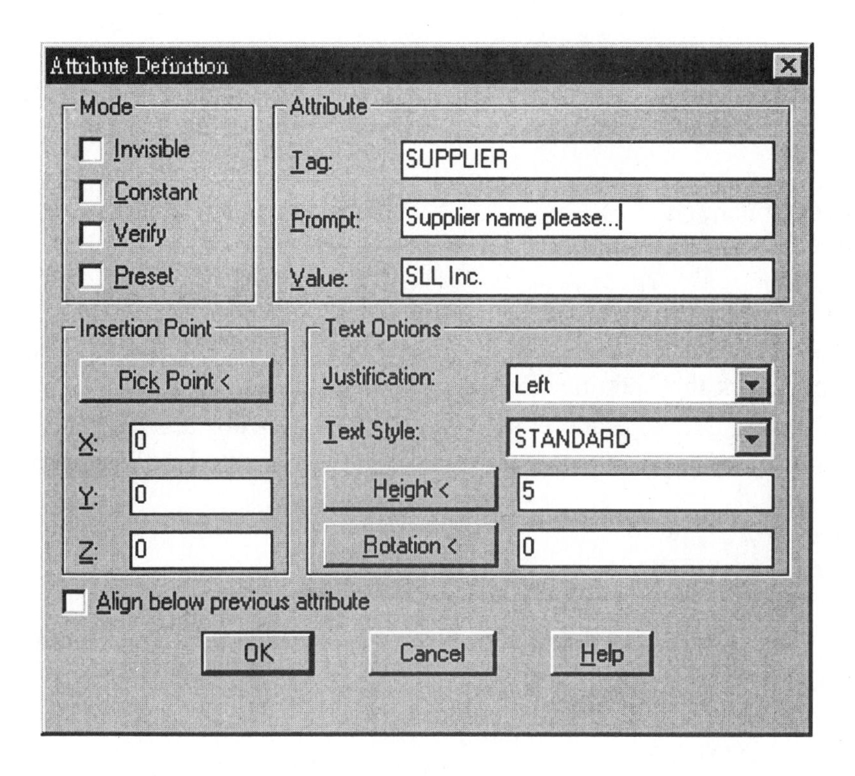

Figure 4.42 Attribute Definition dialog box

Start point: **[Select A (Figure 4.41).]**

[OK]

Figure 4.43 Attribute template defined by the DDATTDEF command

The second attribute template is defined. Refer to Figure 4.43. The tags of the attribute templates are displayed. The prompt and default value will be displayed during insertion to another file.

The drawing is complete. Quick-save it with the QSAVE command.

<File> **<Save>**

Command: **QSAVE**

Insert a File with Attribute Templates

To appreciate how a file with attribute templates is inserted to another file, start a new drawing. Select the New... item of the File pull-down menu to use the NEW command. Start from scratch and use metric default settings.

<File> **<New...>**

Command: **NEW**

[Create New Drawing
Start from Scratch
Metric
OK]

Because the ATTDIA and ATTREQ system variables affect the insertion of a block with attributes, check their settings before you insert the file SCREWHD into the current drawing. If the ATTDIA variable is set to 1, a dialog box for attribute entry will appear during insertion. If it is set to 0, attribute entry will be requested at the command prompt.

Command: **ATTDIA**
New value for ATTDIA: **1**

If the ATTREQ variable is set to 1, entry of the attribute value is requested. If it is set to 0, there will be no attribute request. Default values will be used.

Command: **ATTREQ**
New value for ATTREQ: **1**

Select the Block... item of the Insert pull-down menu to run the DDINSERT command to insert the file SCREWHD to the current drawing.

<Insert> **<Block...>**

Command: **DDINSERT**

[Insert
File...]

When the file selection dialog box appears, select the file that you saved in the last paragraph.

[OK]

Insertion point: **[Select a point near the central part of the drawing window.]**
X scale factor <1> / Corner / XYZ: **1**
Y scale factor (default=X): **1**
Rotation angle: **0**

After you select an insertion point and specify the insertion scale and rotation, an Enter Attribute dialog box displays. See Figure 4.44. Select the [OK] button to accept the default attributes values. See Figure 4.45.

Enter Attributes

Block Name: SCREWHD

Please enter material... MILD STEEL

Supplier name please... SLL Inc.

OK Cancel Previous Next Help

Figure 4.44 Enter Attributes dialog box

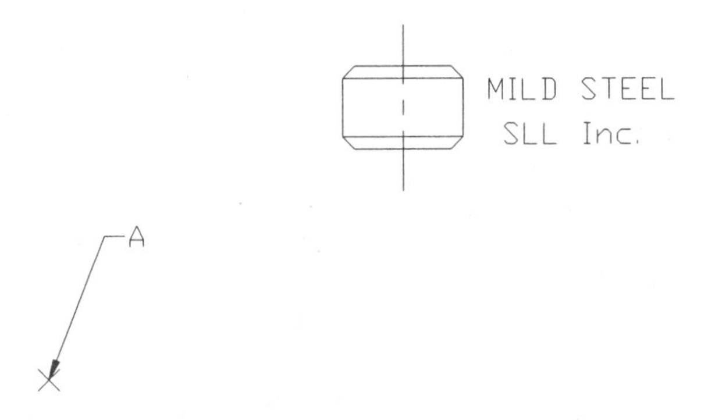

Figure 4.45 Instance inserted with attributes

Set the system variable ATTDIA to 0 and run the DDINSERT command again. See Figure 4.46.

Command: **ATTDIA**
New value for ATTDIA: **0**

<Insert> **<Block...>**

Command: **DDINSERT**

[Insert
Block...SCREWHD
OK]

Insertion point: [**Select A (Figure 4.45).**]
X scale factor <1> / Corner / XYZ: [**Enter**]
Y scale factor (default=X): [**Enter**]

Rotation angle <0>: **[Enter]**
Enter attribute values
Please enter material... <MILD STEEL>: **COPPER**
Supplier name please... <SLL Inc.>: **ALL Inc.**
Verify attribute values

Because you specified V (verify) when you defined the Material attribute, you are now prompted to verify your input.

Please enter material... <COPPER>: **[Enter]**

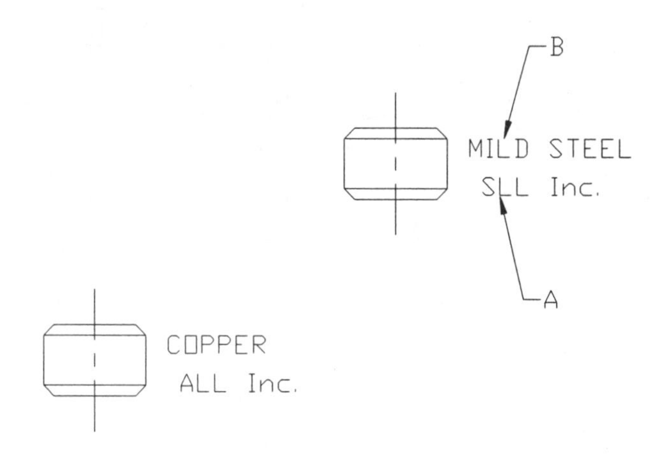

Figure 4.46 File inserted again

Manipulating Attributes

Attributes can be set to be visible or invisible when the template is defined. To override visibility, you can use the ATTDISP command. In normal mode, the attributes remain visible or invisible according to the settings of the template.

Command: **ATTDISP**
Normal/ON/OFF <Normal>: **NORMAL**
Regenerating drawing.

After the attribute values are entered, you can edit them with either the ATTEDIT command or the DDATTE command. From the Modify pull-down menu, select the Object and the Attribute cascading menu and then the Global item to apply the ATTEDIT command. See Figure 4.47.

<Modify> **<Object>** **<Attribute><Global>**

Command: **ATTEDIT**
Edit attributes one at a time? <Y> **[Enter]**
Block name specification <*>: **[Enter]**
Attribute tag specification <*>: **[Enter]**
Attribute value specification <*>: **[Enter]**
Select Attributes: **[Select A (Figure 4.46).]**
Select Attributes: **[Select B (Figure 4.46).]**
Select Attributes: **[Enter]**

Value/Position/Height/Angle/Style/Layer/Color/Next <N>: **COLOR**
New color: **RED**
Value/Position/Height/Angle/Style/Layer/Color/Next <N>: **HEIGHT**
New height: **6**
Value/Position/Height/Angle/Style/Layer/Color/Next <N>: **[Enter]**
Value/Position/Height/Angle/Style/Layer/Color/Next <N>: **VALUE**
Change or Replace? <R>: **R**
New attribute value: **BRONZE**
Value/Position/Height/Angle/Style/Layer/Color/Next <N>: **[Enter]**

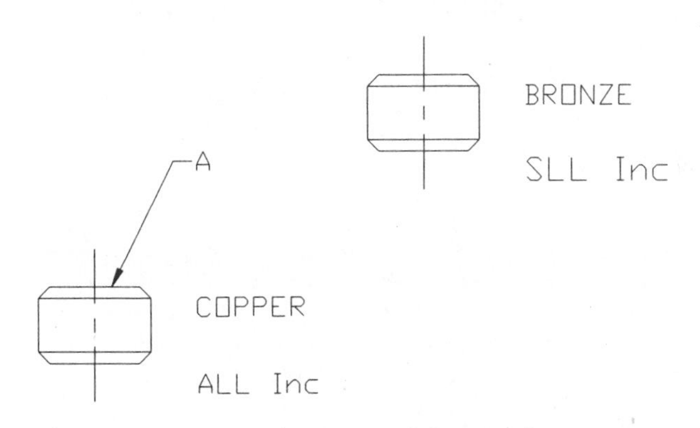

Figure 4.47 Attribute values edited

From the Modify pull-down menu, select the Object and the Attribute cascading menu and then the Single item to run the DDATTE command. See Figure 4.48.

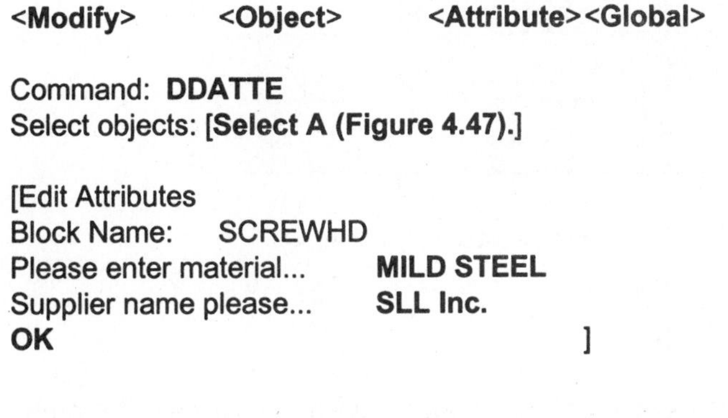

<Modify> **<Object>** **<Attribute><Global>**

Command: **DDATTE**
Select objects: **[Select A (Figure 4.47).]**

[Edit Attributes
Block Name: SCREWHD
Please enter material... **MILD STEEL**
Supplier name please... **SLL Inc.**
OK]

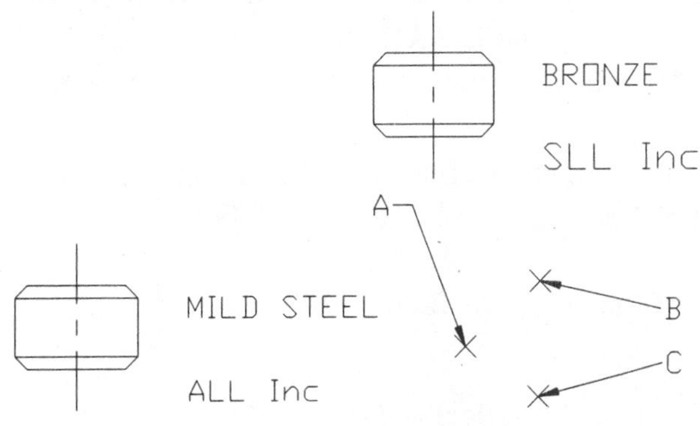

Figure 4.48 Attribute edited

You can export all the attributes of a drawing to an external text file with the ATTEXT command or the DDATTEXT command. The ATTEXT command works at the command line interface. The DDATTEXT command brings out a dialog box. There are three export formats, the CDF, SDF, and DXF formats. For the CDF and SDF formats, you need to predefined a template file.

Run the ATTEXT command.

Command: **ATTEXT**
CDF, SDF, or DXF Attribute extract (or Objects)? <C>: **DXF**

Run the DDATTEXT command. See Figure 4.49.

Command: **DDATTEXT**

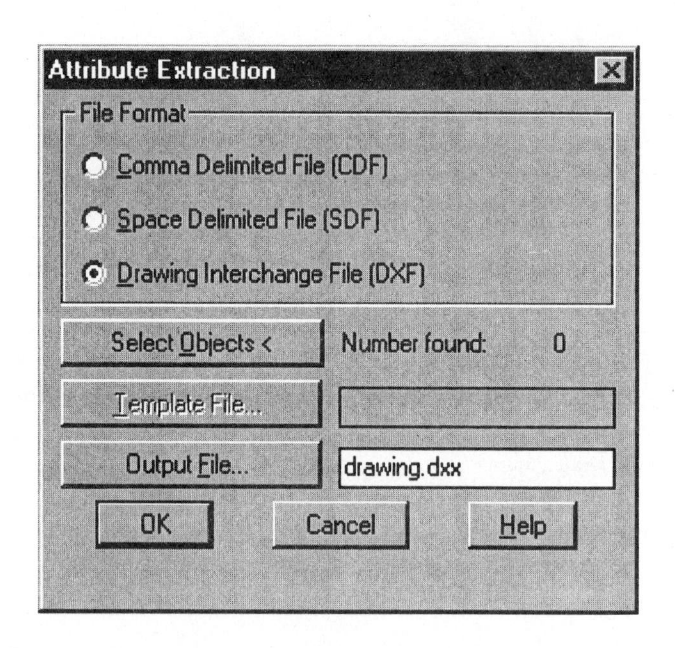

Figure 4.49 Attribute Extraction dialog box

Redefine a Block with Attributes

In the last chapter, you learned how to redefine a block with the BMAKE command. If a block contains attributes, using the BMAKE command for this purpose may give rise to unpredictable results. To redefine a block with attributes, you should use the ATTREDEF command.

Use the CIRCLE command to create a circle to replace the content of the block SCREWHD. See Figure 4.50.

<Draw> **<Circle>** **<Center,Radius>**

Command: **CIRCLE**
3P/2P/TTR/<Center point>: **[Select A (Figure 4.48).]**
Diameter/<Radius>: **10**

From the Block cascading menu of the Draw pull-down menu, select the Define

Attributes... to apply the DDATTDEF command twice to define two attribute templates for the new block. See Figure 4.50.

<Draw> **<Block>** **<Define Attributes...>**

Command: **DDATTDEF**

[Attribute Definition
Attribute
Tag: **MATERIAL**
Prompt: **Please enter material...**
Value: **COPPER**
Text Options Justification: **Left** Text Style: **Standard**
Height< **5** Rotation< **0**
Insertion Point **Pick Point<**]

Start point: [**Select B (Figure 4.48).**]

[OK]

Command: **[Enter]**
DDATTDEF

[Attribute
Tag: **PRICE**
Prompt: **Price please...**
Value: **200**
Text Options Justification: **Left** Text Style: **Standard**
Height< **5** Rotation< **0**
Insertion Point **Pick Point<**]

Start point: [**Select C (Figure 4.48).**]

[OK]

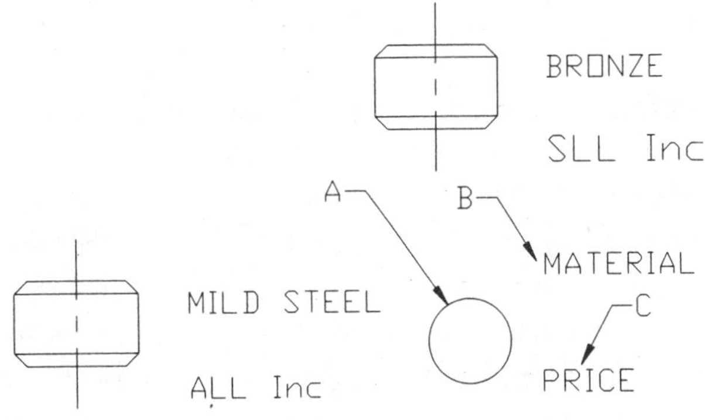

Figure 4.50 Circle and attribute templates created

To redefine the block SCREWHD, apply the ATTREDEF command at the command line interface. Specify the block SCREWHD. Then select the circle and the two attribute templates. See Figure 4.51.

Command: **ATTREDEF**
Name of Block you wish to redefine: **SCREWHD**
Select objects for new Block...
Select objects: **[Select A, B, and C (Figure 4.50).]**
Select objects: **[Enter]**
Insertion base point of new Block: **CEN** of **[Select A (Figure 4.50).]**

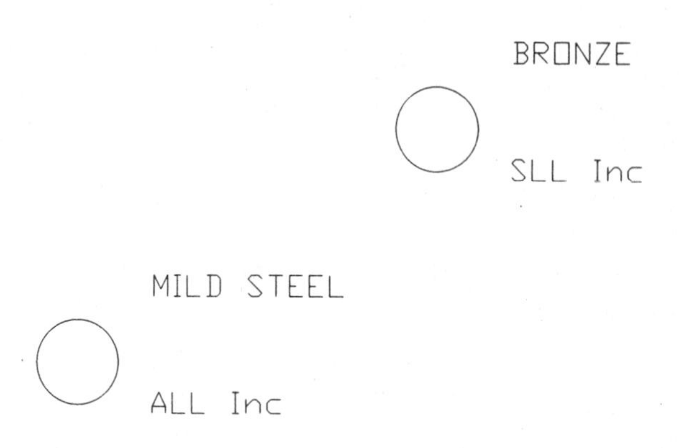

Figure 4.51 Block together with its attributes redefined

The block SCREWHD and its instances are redefined. The attribute MATERIAL that exists in both the new block and the old block is retained. The attribute SUPPLIER that does not exist in the new block is lost. The attribute PRICE that exists in the new block but not in the old block is given the default value.

Save the drawing.

<File> **<Save>**

4.6 Key Points and Exercises

There are two ways to make a block: construct a block definition from entities within the current drawing by using the BLOCK command or the BMAKE command and insert an external drawing file by using the INSERT command, MINSERT command, or DDINSERT command. To control the insertion location of a drawing when it is inserted into another drawing, you can use the BASE command.

An instance is a reference to the block definition which is a collection of data. With a block defined, you can make instances by using the INSERT command, the DDINSERT command, the MINSERT command, the DIVIDE command, or the MEASURE command. The INSERT command and the DDINSERT command make single instances of a block. The MINSERT command, the DIVIDE command, and the MEASURE command make multiple instances. In order to edit the individual objects of an instance, you have to use the EXPLODE or XPLODE command to remove the link to the block definition and to create a copy of the block definition at the insertion point. Both commands break a compound object into its component objects. The XPLODE command has additional options, allowing you to explode multiple compound objects and at the same time change the color, layer, and linetype.

The INSERT command imports a file to form a block definition. After insertion, the file size of the current drawing increases and any change in the original drawing is not reflected. To overcome these two drawbacks, you can use the XREF command to attach a file. With attachment instead of insertion, the external file is loaded to the drawing every time the drawing is opened. As a result, changes in the attached file are reflected and the file size is much less than the insertion. The XREF command controls external references to a drawing file, and the XATTACH command attaches external references. To control the display content of the instances of an external reference, you can use the XCLIP command. To use the symbols and block definitions of an external reference, you need to use the XBIND command. When a drawing with an attached drawing is attached to another drawing, the original attached file is nested. To suppress the display of the nested file, you can use the overlay option during attachment.

The first part of this chapter is a continuation of Chapter 3. Here you learned how to use the WBLOCK command to export a block and selected entities in a drawing so that they become independent drawing files. Then you prepared a drawing and inserted it into the drawing that was created in Chapter 3. You also learned how to explode a block in order to do editing. In completing the pipe support assembly, you learned the top down approach and the down top approach in engineering drawing production. You created two components in a drawing and then exported them to become individual drawing files — the top down approach. You also created a component and imported it to the assembly drawing — the down top approach.

In the second part of the chapter, you discovered the differences between inserted blocks and externally referenced objects. You also discovered the difference between attaching and overlaying an external reference. In the final part of this chapter, you worked on attributes of a block. You defined attribute templates with the ATTDEF command and the DDATTDEF command. After insertion of a block that has attributes, you can edit the attribute values by using the ATTEDIT command or the DDATTE command. If you redefined a block that has attributes, you should use the ATTREDEF command. Working the following exercises will reinforce your learning.

Exercise 4.1

There are two ways of making a block in a drawing — form a block from selected entities and import a drawing file. How can you edit the content of such blocks? State the commands used for making block instances and explain their uses.

Exercise 4.2

Describe how you would export selected entities or blocks of a drawing file to become an individual drawing file.

Exercise 4.3

Delineate the steps in defining attributes in a drawing file for subsequent insertion into another drawing file. From your personal experience, name some uses of attributes.

Exercise 4.4

Distinguish between block insertion and external reference. What are the advantages and disadvantages of each method?

Exercise 4.5

Open the drawing EX3_2.DWG that you completed in Chapter 3. In this drawing, there are two components residing on different layers. Set layer Pendulum as the current layer, and freeze all other layers. See Figure 4.52.

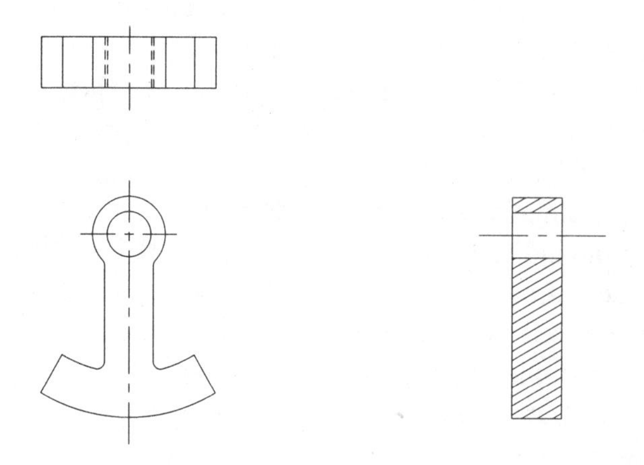

Figure 4.52 Pendulum

Use the WBLOCK command to write all the entities related to the pendulum to an external drawing called EX4_1A.DWG. After writing to the external file, remember to use the OOPS command to retrieve the erased entities.

Thaw all layers, set layer 0 current, and freeze layer Pendulum. See Figure 4.53.

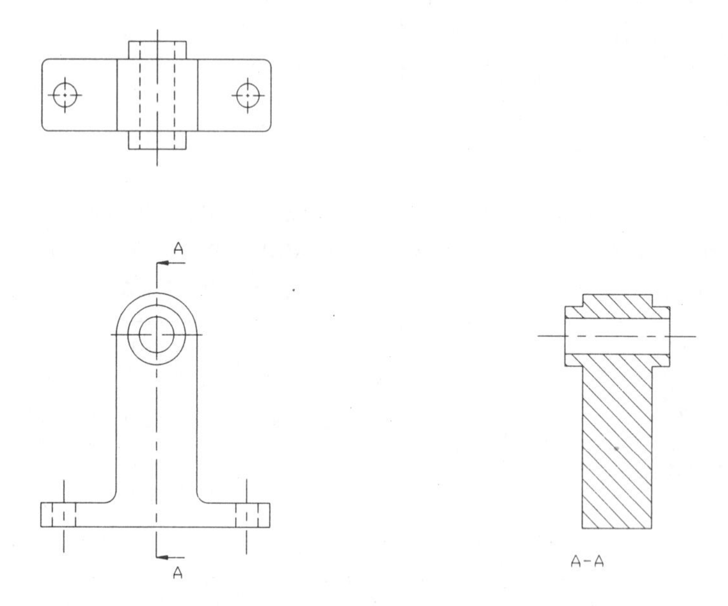

Figure 4.53 Support column

As in the pendulum, use the WBLOCK command to write all the displayed entities in Figure 4.53 to an external drawing called EX4_1B.DWG. Then, use the OOPS command to retrieve the erased entities.

Now you should have an assembly of two components (EX3_2.DWG), a drawing for the first component (EX4_1A.DWG), and a drawing for the second component (EX4_1B.DWG).

Thaw all layers. Insert the drawing file EX3_3.DWG to the drawing EX3_2.DWG. Edit the drawing according to Figure 4.54. Save the drawing file as EX4_1C.DWG.

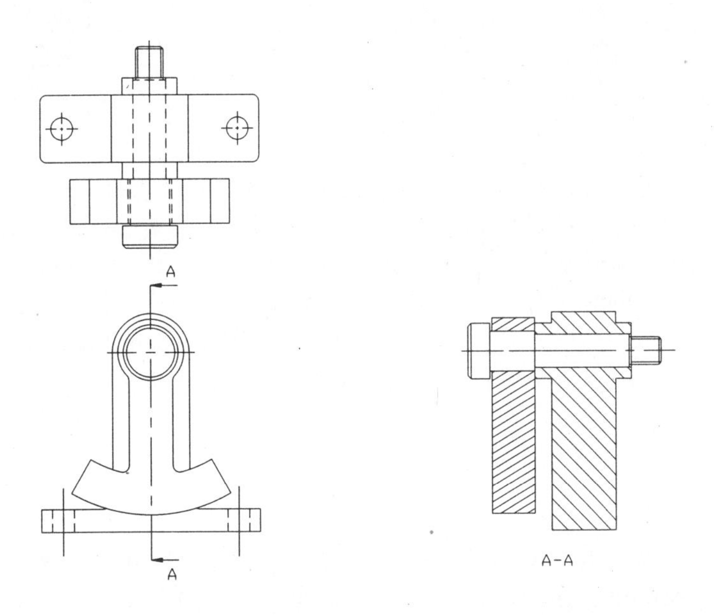

Figure 4.54 Pendulum assembly

Exercise 4.6

Open the file WASH.DWG that you created in Chapter 3. See Figure 4.55.

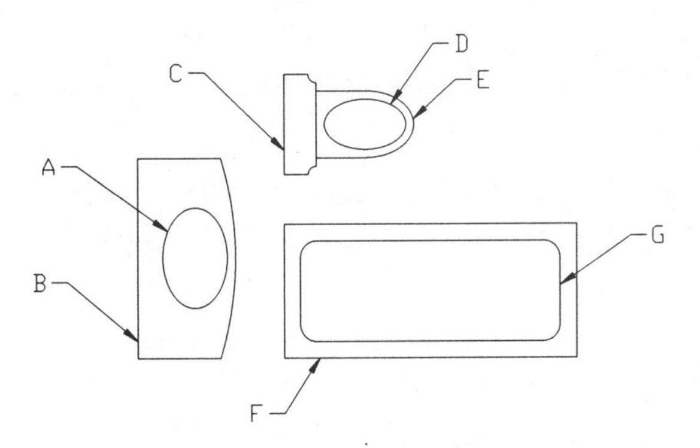

Figure 4.55 Bathroom facilities

According to the following table, use the WBLOCK command to write three external files: EX4-2A, EX4-2B, and EX4-2C. After each WBLOCK command, apply the OOPS command to retrieve the erased entities.

File name	WBLOCK Entities	Insertion base point
EX4-2A	Select A and B	END point of B

EX4-2B	Select C, D, and E	END point of C
EX4-2C	Select F and G	END point of F

Open the drawings EX4-2A, EX4-2B, and EX4-2C one by one, and use the DDATTDEF command to add attribute templates as follows:

File	Tag	Prompt	Value	Mode
EX4-2A	PRICE	Please enter price...	400	Invisible
EX4-2B	PRICE	Please enter price...	300	Invisible
EX4-2C	PRICE	Please enter price...	500	Invisible

Save the drawings and open the drawing FLOOR.DWG. See Figure 4.36. According to Figure 4.40, use the INSERT command to insert the files EX4-2A, EX4-2B, and EX4-2C. During insertion, you will be prompted to input the price value of the wash room objects. Because the display mode of the attributes is invisible, they are not displayed. However, you can retrieve them by using the DDATTEXT command and can make them visible by using the ATTDISP command. Save the drawing as EX4_2D.DWG.

Exercise 4.7

Refer to Figure 4.56. Start a new drawing. Set up a layer called Bath and create the drawing on this layer. With the BASE command, set the insertion base point at the lower left corner of the rectangle. Save the drawing named BATH.DWG.

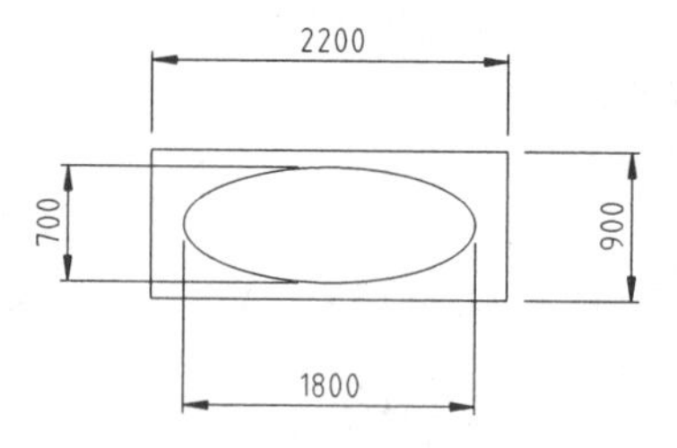

Figure 4.56 Bath

Open the drawing EX3_7.DWG (Figure 3.169). Insert the drawing BATH in accordance with Figure 4.57. Save the drawing as EX4_3.DWG.

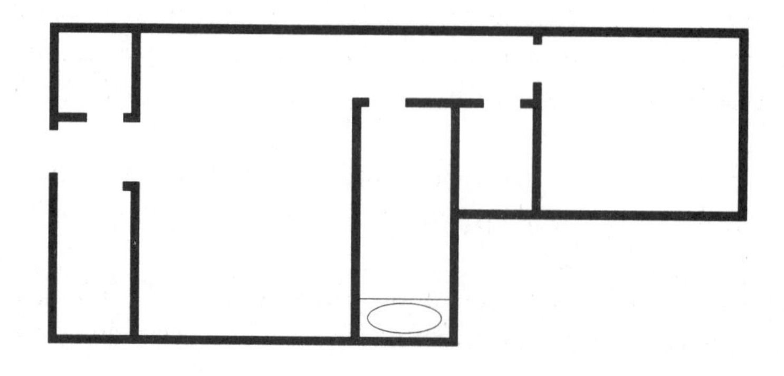

Figure 4.57 Condominium with an inserted bath

Chapter 5

Annotation, Inquiry, and Dimension

5.1 Annotation
5.2 Inquiry
5.3 Dimension
5.4 Dimension Style
5.5 Dimension Variables
5.6 Dimensioning Practice 1
5.7 Dimensioning Practice 2
5.8 Edit Dimension
5.9 Tolerance
5.10 Key Points and Exercises

Aims and Objectives

The aim of this chapter is to explain how annotations, dimensions, and tolerances can be added to a drawing, to introduce various inquiry commands, to explain how to set dimension styles, to examine the dimension variables, to give you practice adding and modifying dimensions in a drawing, and to introduce lateral and geometric tolerances. After studying this chapter, you should be able to:

- add annotations to a drawing,
- modify annotations in a drawing,
- set dimension styles,
- explain the use of the dimension variables,
- include and edit dimensions in a drawing, and
- add lateral and geometric tolerances to a drawing.

Overview

Along with graphical data, AutoCAD enables you to add textual information to a drawing. You can include single-line or multiline text. AutoCAD can even check your spelling!

In construction of a CAD drawing, entities are drawn to exact size. Therefore, dimensions are implicitly included in the data file. When you plot a drawing, the output hardcopy may have to be scaled up or down. Therefore, dimensional information is necessary.

To produce or manufacture an actual part from a drawing, you have to specify tolerances in addition to dimensions. In an AutoCAD drawing, you can add linear tolerance and geometric tolerance.

In this chapter, you will add annotation, dimensions, and tolerances to drawings created in the previous chapters. You will also learn how to use inquiry commands.

5.1 Annotation

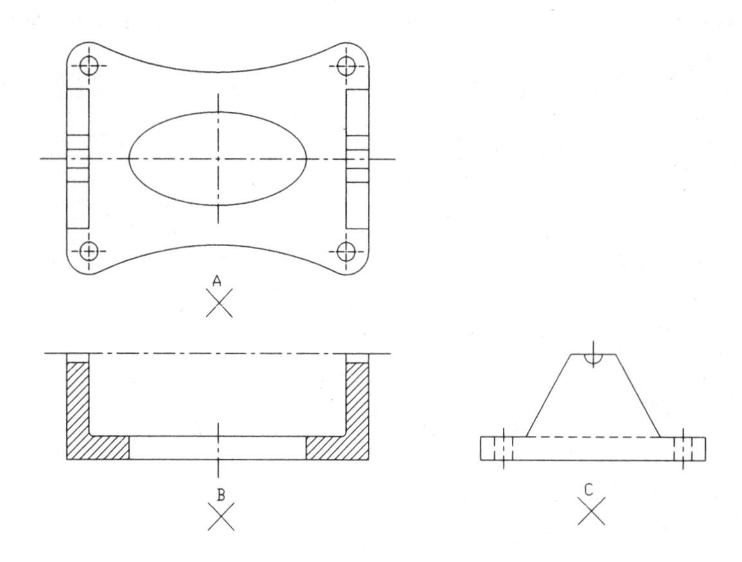

Figure 5.1 First component of the pipe support assembly

Open the file of the first component of the pipe support assembly that you saved in Chapter 4. See Figure 5.1. You will add single-line text and multiline text to this drawing.

<File> **<Open...>**

Command: **OPEN**

To add text to your drawing, you can use the TEXT command, the DTEXT command, or the MTEXT command. The TEXT command and the DTEXT command enable you to add a single line of text. The MTEXT command brings up a text editor and enables you to add multiline text.

Text Style

Before adding textual information to your drawing, you need to set up a text style to set the type and style of the text fonts. The default text style name is Standard. You may use the default text style or a user-defined new text style. Apply the STYLE command to define a new text style called AA.

<Format> **<Text Style...>**

Command: **STYLE**

From the dialog box, select the [New...] button and key in the new style name AA. Under the Font table, select TXT.SHX as the Font Name and set the Height to zero. Setting the text height to zero allows you to specify various text heights during text creation. If you specify a non-zero value, the text height for this style is fixed. In the Effects table, set the Width Factor to 1 and Oblique Angle to 0. See Figure 5.2

Select the [Apply] button to set this style as the default style, and select the [Close] button to exit the command.

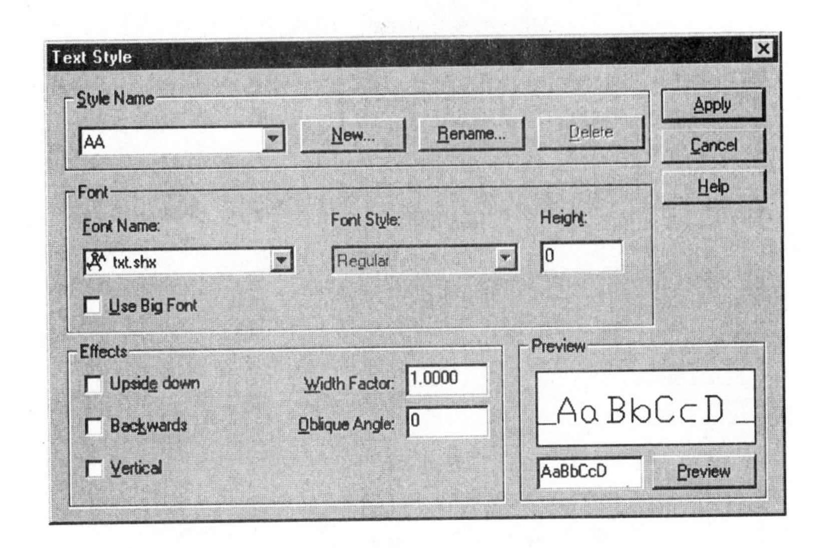

Figure 5.2 Text Style dialog box

If you want to create a text style by working at the command line, you can use the STYLE command and prefix the command name with a hyphen.

Command: **-STYLE**
Text style name (or ?) <AA>:**[Enter]**
....

Single-Line Text

The TEXT command and the DTEXT command create single-line text. They are similar. The TEXT command prints the line of text after the return key. The DTEXT command prints the text on the screen as you type it. Run the TEXT command to create a single line of text. Use the text style AA that you defined. See Figure 5.3.

Command: **TEXT**
Justify/Style/<Start point>: **STYLE**
Style name (or ?) <STANDARD>: **AA**
Justify/Style/<Start point>: **[Select A (Figure 5.1).]**
Height: **12**
Rotation angle: **0**
Text: **TOP VIEW. [Enter]**

Execute the DTEXT command to create another line of text.

<Draw> **<Text>** **<Single Line Text>**

Command: **DTEXT**
Justify/Style/<Start point>: **[Select B (Figure 5.1).]**
Height: **12**

Rotation angle: **0**
Text: **FRONT VIEW**
Text: **[Select C (Figure 5.1).]**
Text: **SIDE VIEW**
Text: **[Enter]**

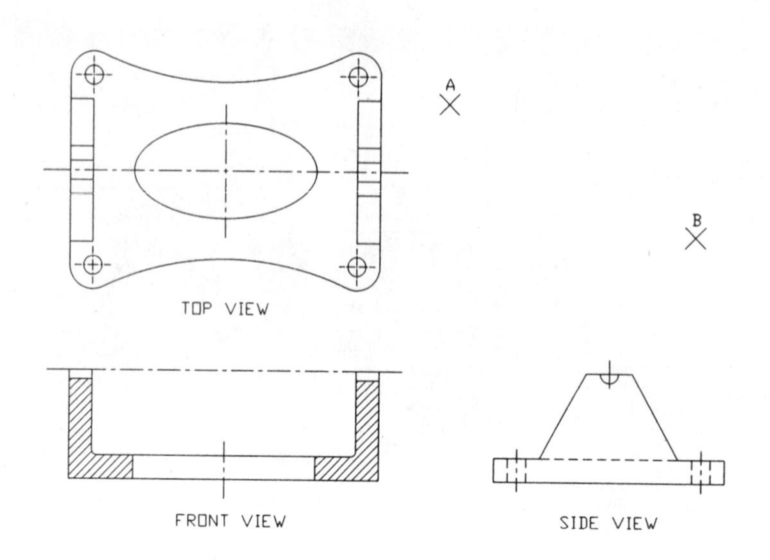

Figure 5.3 Single-line text created

Multiline Text

Issue the MTEXT command to add multiline text to your drawing. Use the text style AA.

<Draw> <Text> <Multiline Text...>

Command: **MTEXT**
Current text style: AA. Text height: 12
Specify first corner: **[Select A (Figure 5.3).]**
Specify opposite corner or [Height/Justify/Rotation/Style/Width]: **[Select B (Figure 5.3).]**

After you define the insertion point and the other corner, a text editor appears. See Figure 5.4. Enter the following text.

ALONG WITH GRAPHICAL DATA, YOU MAY ADD TEXTUAL INFORMATION TO A DRAWING. YOU MAY INCLUDE A SINGLE LINE OR PARAGRAPH OF TEXT. IN ADDITION, YOU MAY CHCEK SPELLING.

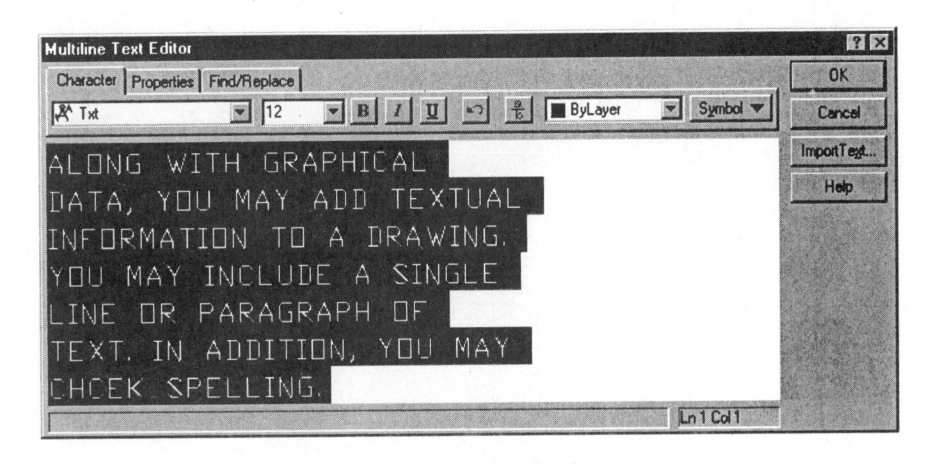

Figure 5.4 Multiline Text Editor dialog box

Select the typed text, change the text height to 12, and then select the [OK] button. See Figure 5.5.

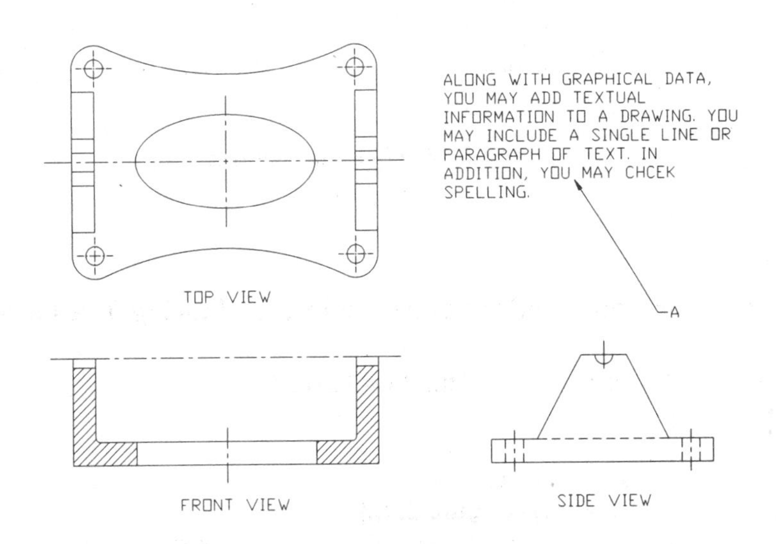

Figure 5.5 Multiline text created

Check Spelling

Obviously, the next-to-last word is spelled wrong. This is done intentionally. Apply the SPELL command to check the spelling of the last paragraph of text.

<Tools> **<Spelling>**

Command: **SPELL**
Select objects: [**Select A (Figure 5.5).**]
Select objects: [**Enter**]

A dialog box appears. See Figure 5.6. The word "chcek" is misspelled. The suggestion is to change to "check." Select the [Change] button. Because this is the only mistake, checking is finished. See Figure 5.7.

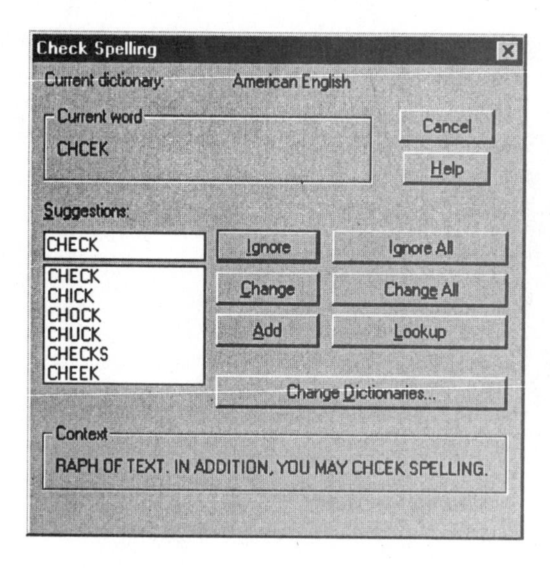

Figure 5.6 Check Spelling dialog box

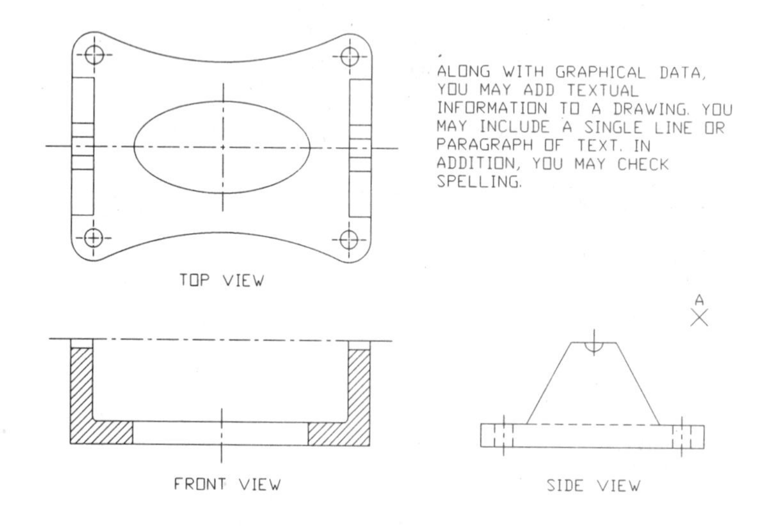

Figure 5.7 Wrong spelling corrected

Edit Text

You may edit single-line text and multiline text by using the DDMODIFY command or the DDEDIT command.

Select the Properties... item of the Modify pull-down menu to run the DDMODIFY command to edit the multiline text.

<Modify> **<Properties...>**

Command: **DDMODIFY**
Select one object to modify: [**Select the multiline text.**]

After you select the multiline text, the Modify M Text dialog box appears. Refer to Figure 5.9. From this dialog box, select the [Full Editor...] button to bring out the

Multiline Text Editor. See Figure 5.8.

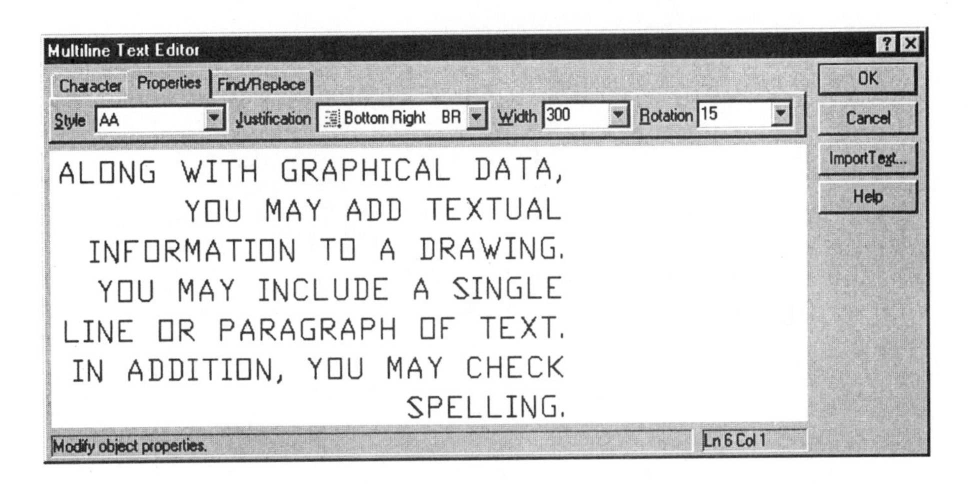

Figure 5.8 Properties tab of Multiline Text Editor dialog box

To change the properties of the multiline text, select the Properties tab. Change Justification to Bottom Right, Width to 300, and Rotation to 15°. Select the [OK] button to return to the Modify M Text dialog box. See Figure 5.9.

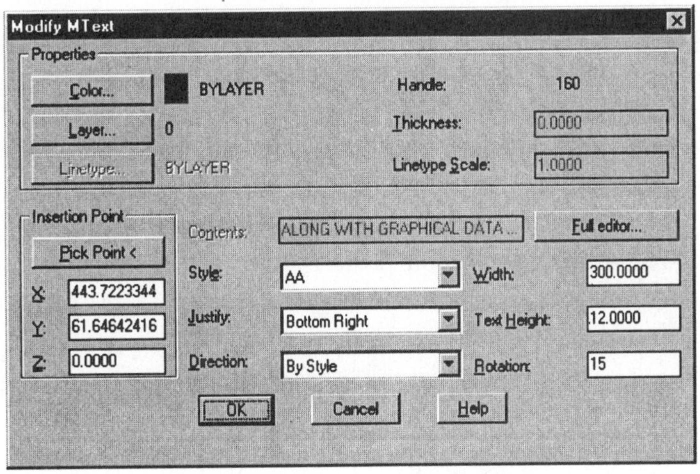

Figure 5.9 Modify M Text dialog box

To change the multiline text location, select the [Pick Point<] button and select a new insertion point for the text.

New Insertion Point: **[Select A (Figure 5.7).]**

On returning to the Modify M Text dialog box, select the [OK] button. See Figure 5.10.

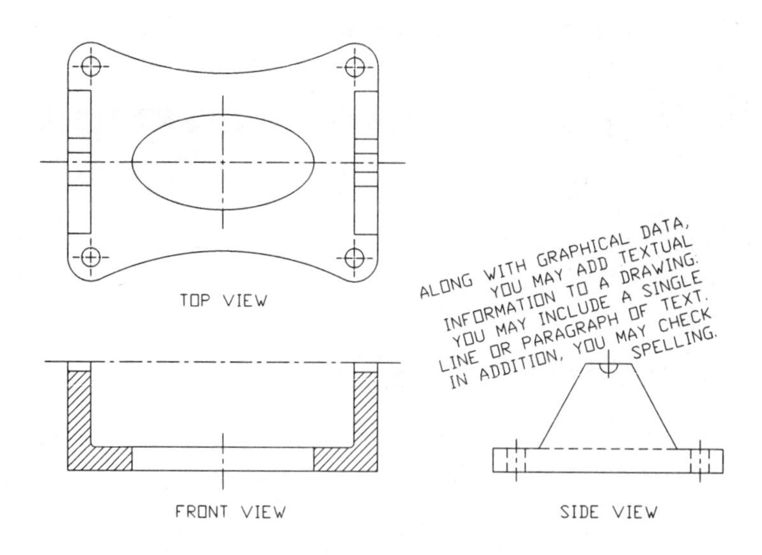

Figure 5.10 Multiline text insertion point changed

By selecting the Edit Text icon of the Modify II toolbar, run the DDEDIT command to edit the multiline text again. See Figure 5.11.

[Modify II] **[Edit Text]**

Command: **DDEDIT**
<Select an annotation object>/Undo: **[Select the multiline text.]**

The Multiline text editor appears. See Figure 5.8 again. Select the Properties tab, set the rotation angle to 0°, and then select the [OK] button. See Figure 5.11.

<Select an annotation object>/Undo: **[Enter]**

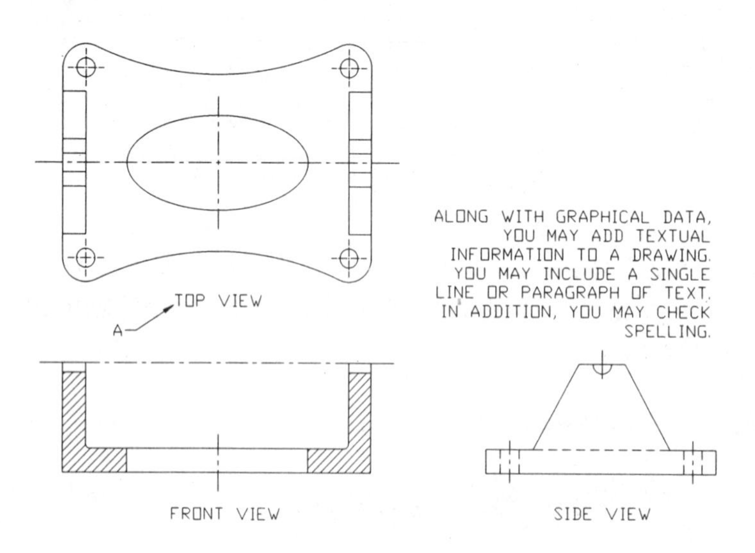

Figure 5.11 Text insertion point changed

Repeat the DDEDIT command on a single-line text.

Command: **[Enter]**
DDEDIT
<Select an annotation object>/Undo: **[Select A (Figure 5.11).]**

The Edit Text dialog box appears. See Figure 5.12. Change the text content and select the [OK] button. See Figure 5.13.

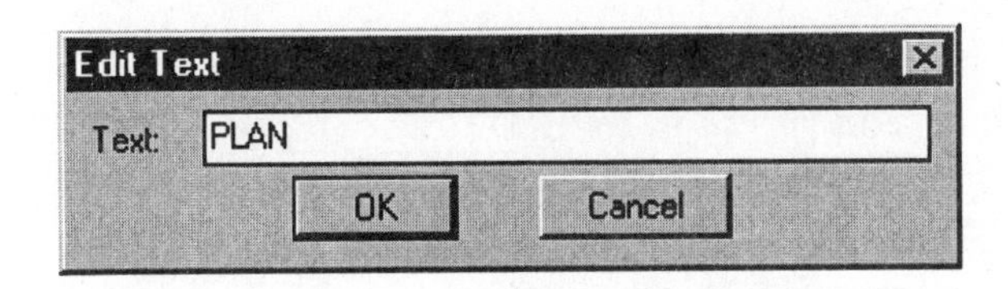

Figure 5.12 Edit Text dialog box

<Select an annotation object>/Undo: **[Enter]**

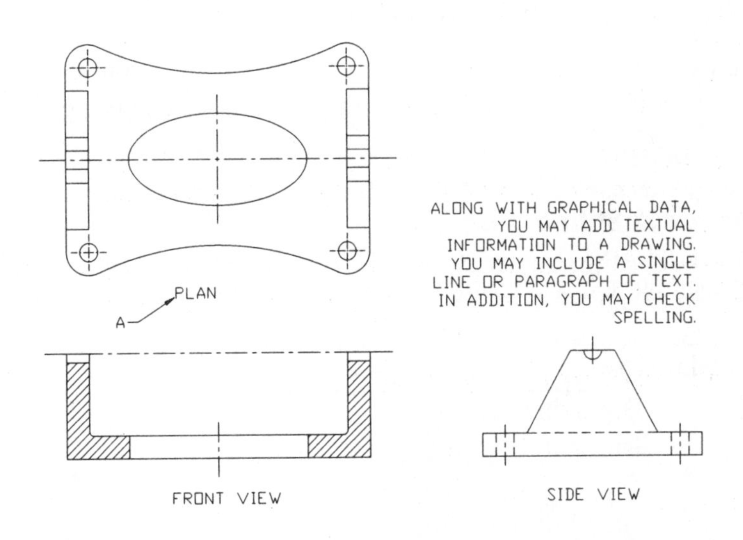

Figure 5.13 Single-line text edited

When you apply the DDEDIT command on a single-line text, you can edit only its text content. To edit other attributes of a single line text, you have to use the DDMODIFY command. See Figure 5.14.

<Modify> **<Properties...>**

Command: **DDMODIFY**
Select one object to modify: **[Select A (Figure 5.13).]**

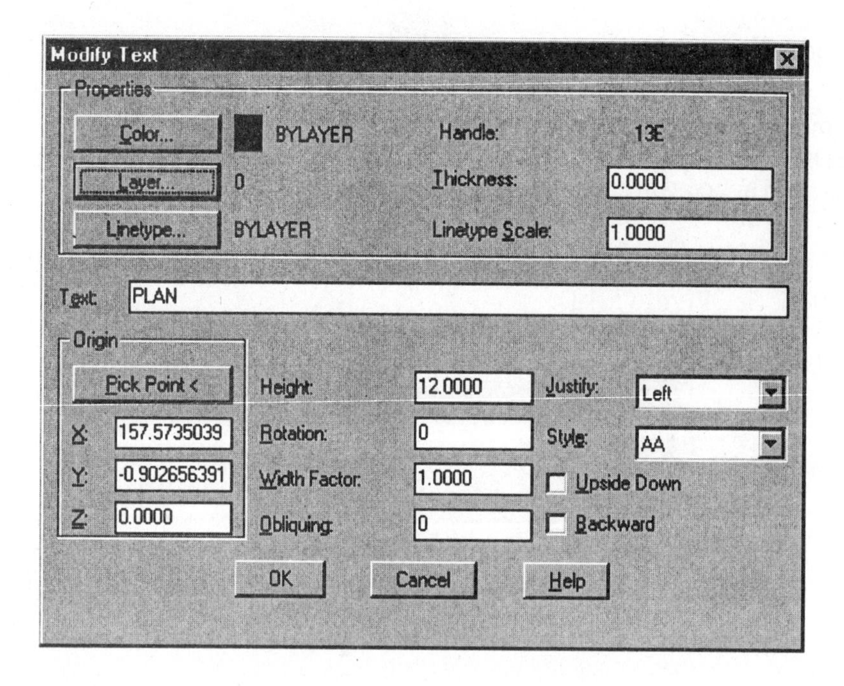

Figure 5.14 Modify Text dialog box

Refer to Figure 5.14. The DDMODIFY command enables you to change the attribute as well as the text content. To exit the command without changing the text, select the [OK] button.

Format Codes for Multiline Text

You can format a paragraph text by including format codes.

Format code	Use
\O	Turns on overline.
\o	Turns off overline.
\L	Turns on underline.
\l	Turns off underline.
\~	Keeps two words together.
\\	Adds the \ character.
\{	Adds the { character.
\}	Adds the } character.
\Cxx;	Changes the text that follows to the color number xx.
\Fxx;	Changes the text that follows to the font type xx.
\Hxx;	Changes the text that follows to a height of xx points.
\Saa^bb	Stacks the text "aa" upon the text "bb".
\Txx;	Changes the text that follows to a spacing of xx times.
\Qxx;	Changes the text that follows to an oblique angle of xx degrees.
\Wxx;	Changes the text that follows to a width factor of xx.
\A	Sets text alignment: 0 for bottom, 1 for center, and 2 for top.

\P	Causes the text that follows to start a new paragraph.
%%d	Displays the degree ° symbol.
%%c	Displays the diameter ∅ symbol.

Quick Text Mode

In a drawing that consists of a lot of text entities, each regeneration may take considerable time. You can speed up regeneration by turning on Quick Text mode. When this mode is on, rectangular boxes appear on the screen to represent the actual text.

```
Command: QTEXT
ON/OFF: ON
```

The change will be effective in the next regeneration. Apply the REGEN command. See Figure 5.15.

```
Command: REGEN
```

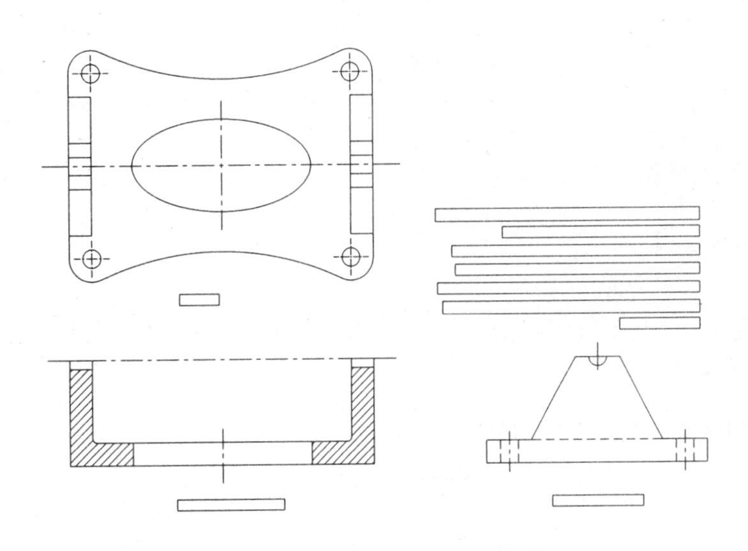

Figure 5.15 Quick Text mode turned on

Reset QTEXT mode to 1 and perform a REGEN. See Figure 5.13.

```
Command: QTEXT
ON/OFF: OFF

Command: REGEN
```

Text Fill and Quality

The display of TrueType fonts and Type 1 fonts also affects the drawing regeneration speed. You can adjust the text quality and the text fill by manipulating two system variables, TEXTQLTY and TEXTFILL.

From the Inquiry cascading menu of the Tools pull-down menu, select the Set Variable item to run the SETVAR command.

<Tools> **<Inquiry>** **<Set Variable>**

Command: **SETVAR**
Variable name or ?: **TEXTFILL**
New value for TEXTFILL: **0**

When the TEXTFILL variable is set to zero, TrueType fonts and Type 1 fonts are not filled. TEXTFILL controls the filling of TrueType fonts while plotting, exporting with PSOUT, and rendering.

Repeat the SETVAR command.

Command: **[Enter]**
SETVAR
Variable name or ?: **TEXTQLTY**
New value for TEXTQLTY: **0**

The TEXTQLTY variable can be set within a range of 0 to 100. 100 is the highest quality and 0 is the lowest quality. TEXTQLTY sets the resolution of TrueType fonts while plotting, exporting with PSOUT, and rendering. Values represent dots per inch. Lower values decrease resolution and increase plotting speed. Higher values increase resolution and decrease plotting speed.

Save your drawing with the QSAVE command.

<File> **<Save...>**

Command: **QSAVE**

Annotation to the drawing is complete.

5.2 Inquiry

Now you will work on inquiry commands. Select the About AutoCAD item of the Help pull-down menu to run the ABOUT command. This command displays the AutoCAD version number, serial number, license information, dealer name, and copyright information. See Figure 5.16. You may find this information useful.

<Help> **<About AutoCAD>**

Command: **ABOUT**

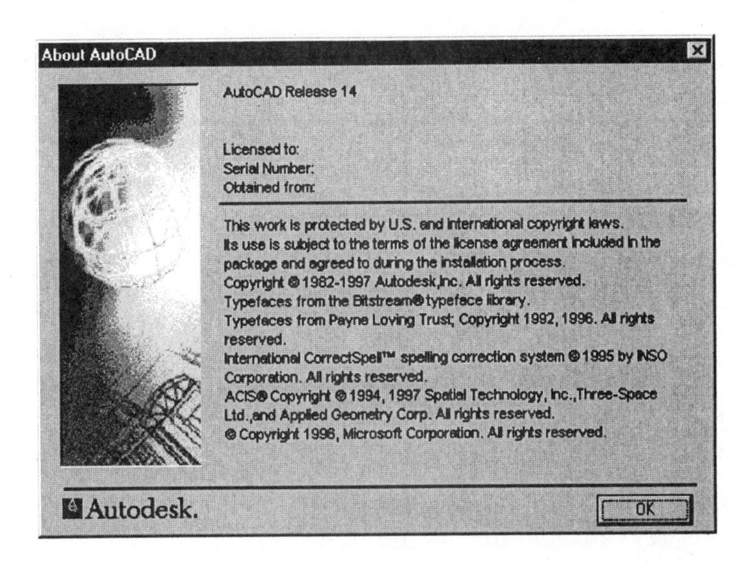

Figure 5.16 About AutoCAD dialog box

Open the file WASH that you created in Chapter 3. This is a washroom facility. See Figure 5.17. You will use inquiry commands on this drawing.

<File> **<Open...>**

File name: **WASH.DWG**

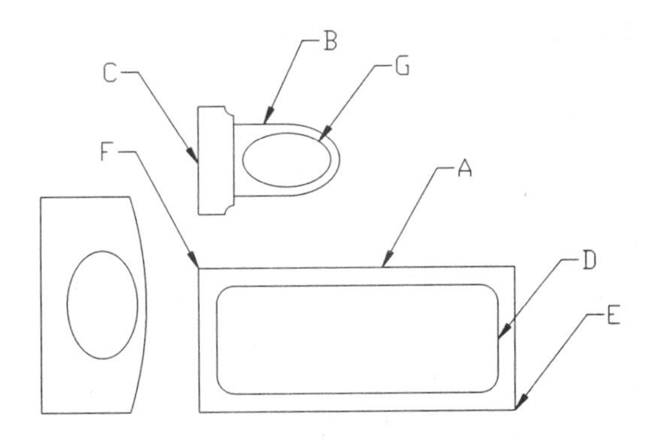

Figure 5.17 Washroom facility

General Information

You can obtain general information on a drawing by using the LIST command, the DBLIST command, the TIME command, and the STATUS command.

The LIST command lists the database information of an entity. Select the List item of the Inquiry cascading menu of the Tools pull-down menu to apply the LIST command on a rectangle. The information displayed on your screen might be slightly different from the delineation that follows.

<Tools> **<Inquiry>** **<List>**

Command: **LIST**
Select objects: [**Select A (Figure 5.17).**]

Select objects: **[Enter]**

LWPOLYLINE Layer: 0
...

The DBLIST command lists the entire database of the current drawing. Execute the DBLIST command.

Command: **DBLIST**
REGION Layer: 0
...
Press ENTER to continue: **[ESC]**

After a full page of listing, the command pauses. You can continue the command by selecting the [Enter] key or can terminate the command by selecting the [ESC] key.

The TIME command gives you some statistics on the timing of the current drawing. Run this command.

<Tools> **<Inquiry>** **<Time>**

Command: **TIME**
Current time: ...
...
Display/ON/OFF/Reset: **[Enter]**

The STATUS command provides a report on the current drawing. Run this command.

<Tools> **<Inquiry>** **<Status>**

Command: **STATUS**
230 objects ...
...

Area

The AREA command takes polylines, circles, or polygons that are defined by a series of selected points. If you want to determine the area enclosed by a set of entities shown in the screen, you have to convert the entities into a polyline with the PEDIT command.

To find the difference or the sum of two or more areas, you have to select the Add mode first. If the second object is to be subtracted from the first, switch to Subtract mode and pick the second object.

Issue the AREA command to find the sum of the two areas.

<Tools> **<Inquiry>** **<Area>**

Command: **AREA**
<First point>/Object/Add/Subtract: **ADD**
<First point>/Object/Subtract: **OBJECT**
(ADD mode) Select objects: **[Select B (Figure 5.17).]**
Area = 214247.7796, Perimeter = 1793.2717
Total area = 214247.7796

```
(ADD mode) Select objects: [Select C (Figure 5.17).]
Area = 116073.0092, Perimeter = 1557.0796
Total area = 330320.7888
(ADD mode) Select objects: [Enter]
<First point>/Object/Subtract: [Enter]
```

The area and perimeter of the first selected object are 214247.7796 and 1793.2717, respectively. The area and perimeter of the second selected object are 116073.0092 and 1557.0796, respectively. The total area of the two objects is 330320.7888.

Repeat the AREA command to find the difference between the two areas.

```
Command: [Enter]
AREA
<First point>/Object/Add/Subtract: A
<First point>/Object/Subtract: O
(ADD mode) Select objects: [Select A (Figure 5.17).]
Area = 1440000.0000, Perimeter = 5200.0000
Total area = 1440000.0000
(ADD mode) Select objects: [Enter]
<First point>/Object/Subtract: S
<First point>/Object/Add: O
(SUBTRACT mode) Select objects: [Select D (Figure 5.17).]
Area = 951415.9265, Perimeter = 4228.3185
Total area = 488584.0735
(SUBTRACT mode) Select objects: [Enter]
<First point>/Object/Add: [Enter]
```

The area and perimeter of the first object are 1440000.0000 and 5200.0000, respectively. The area and perimeter of the second selected object are 951415.9265 and 4228.3185, respectively. The difference in area is 488584.0735.

Distance

The DISTANCE command tells you the distance between two selected points. Run this command.

<Tools> **<Inquiry>** **<Distance>**

```
Command: DIST
First point: END of [Select E (Figure 5.17).]
Second point: END of [Select F (Figure 5.17).]
Distance = 1969.7716,  Angle in XY Plane = 156,  Angle from XY Plane = 0
Delta X = -1800.0000,  Delta Y = 800.0000,  Delta Z = 0.0000
```

Identity

The ID command tells you the coordinates of a point.

<Tools> **<Inquiry>** **<ID Point>**

```
Command: ID
Point: CEN of [Select G (Figure 5.17).]
X = 1500.0000     Y = 1500.0000     Z = 0.0000
```

5.3 Dimension

Although dimensional information is an integral part of the drawing database, the dimensions of an entity are not readily perceivable if they are not explicitly displayed on the drawing. This is why you have to add dimensions to your drawing.

Dimensioning Principles

There are two basic principles to follow when you add dimensions to a drawing:

1. Each dimension required for the accurate definition of a feature should appear only once in the drawing. Assignment of more than one dimension to a feature should be avoided.
2. As far as possible, the reader of the drawing should not be required to do calculation in order to obtain the dimension of a feature.

Components of a Dimension

Refer to Figure 5.18. A dimension consists of four components:

1. A dimension value.
2. A dimension line that is parallel to the direction of the described feature.
3. A pair of arrowheads.
4. A pair of extension lines projecting from the feature to which the dimension refers. There should be a small gap between the end of the extension line and the feature. The extension line should project a short distance away from the intersection with the dimension line.

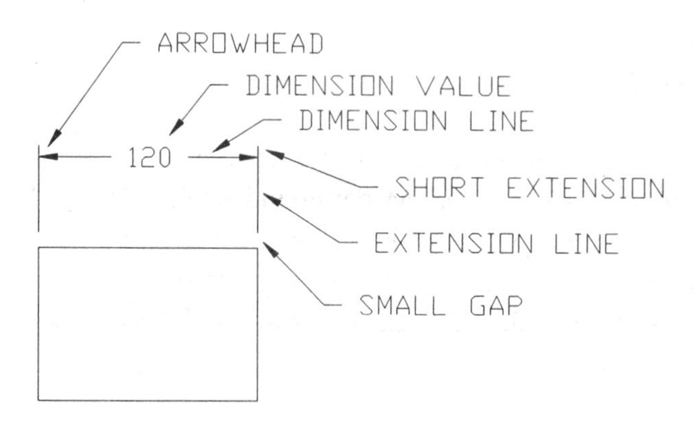

Figure 5.18 Components of a dimension

To create a dimension in AutoCAD, you do not have to create the above-mentioned four components of a dimension all by yourself. Instead, you need only issue a relevant command and then specify a location for the dimension. Then AutoCAD creates the four components of a dimension automatically. The way the components of a dimension will look is determined by the current dimension style and the setting of the dimension variables.

Definition Points

In addition to the four general components of a dimension, AutoCAD includes definition points in a dimension block for reference. These points are stored in a special layer called DEFPOINTS. You may leave this layer ON all the time because the entities on this layer will not be plotted. When you edit an associative dimension, AutoCAD refers to these points. See Figure 5.19.

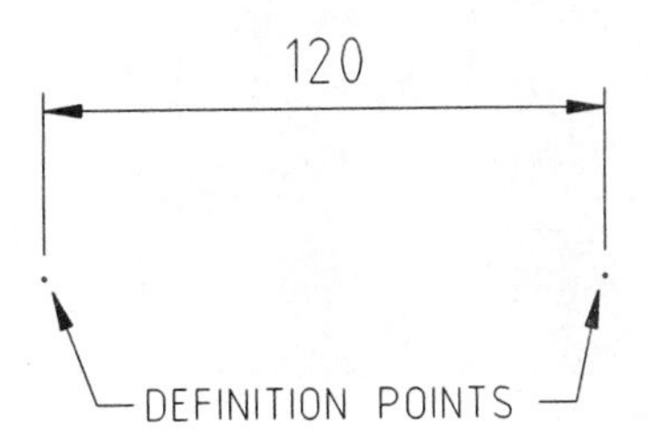

Figure 5.19 Definition points

Associative Dimension

The components of a dimension that are listed above can be grouped together as a block or treated as separate entities. If they are grouped as a block, the dimension is called an associative dimension. You can edit or modify the appearance of an associative dimension as a single unit. The dimensions will update automatically after you edit a related feature. On the other hand, if they are created as separate entities, you can edit them independently using general editing commands.

5.4 Dimension Style

The way a dimension appears depends on the current dimension style. To set up a dimension style, you may use the DDIM command or the DIMSTYLE command. The DDIM command works graphically by providing interactive dialog boxes. The DIMSTYLE command works at the command line interface.

From the Dimension pull-down menu, select the Style... item to run the DDIM command. See Figure 5.20.

<Dimension> **<Style...>**

Command: **DDIM**

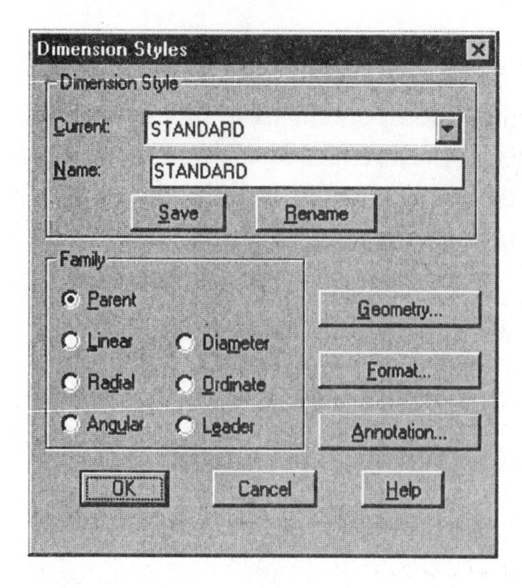

Figure 5.20 Dimension Styles dialog box

Dimension Style

At the upper part of the dialog box, the current dimension style name is displayed. You can create a new dimension style, choose a saved style, save changes to a style name, and rename a dimension style.

Family

For each dimension style, there is a family of seven members. These consist of a parent and six other family members — linear, diameter, radial, ordinate, angular, and leader. The parent is a set of variables that applies to all types of dimensions. The other family members are subsets of variable settings that are specific to the types of dimensions that they refer to.

Parent	enables you to set the overall dimension style.
Linear	enables you to set the linear dimension style.
Diameter	enables you to set the diameter dimension style.
Radial	enables you to set the radial dimension style.
Ordinate	enables you to set the ordinate dimension style.
Angular	enables you to set the angular dimension style.
Leader	enables you to set the leader style.

To create a new style, or to modify an existing style, select the [Parent] button, and then change the settings through the sub-dialog boxes by selecting the buttons [Geometry], [Format], and [Annotation]. If you wish to have minor variations on the appearance of different kinds of dimensions, select the buttons [Linear], [Diameter], [Radial], [Ordinate], [Angular], and [Leader]. Then proceed, changing the settings in a similar way to alter the parent settings.

Geometry

The geometry of a dimension includes the dimension line, the arrowheads, the extension lines, the center mark, and the overall geometry scale. Select the [Geometry...] button to bring out the Geometry sub-dialog box. See Figure 5.21.

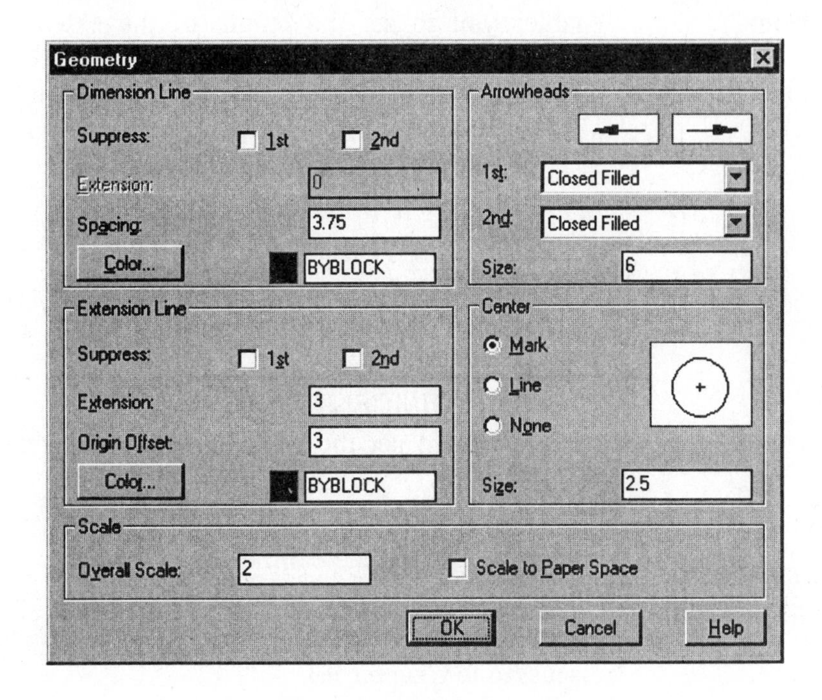

Figure 5.21 Geometry dialog box

There are five areas in the Geometry dialog box.

1. Dimension Line

   Suppress: 1st — enables you to suppress the display of the part of the dimension line that starts from the first selected end point of a dimension.

   Suppress: 2nd — enables you to suppress the display of the part of the dimension line that starts from the second selected end point of a dimension.

   Extension: — enables you to set the length of extension of the dimension line beyond the extension line. It works only with the architectural tick arrowhead.

   Spacing: — enables you to set the incremental distance between consecutive baseline dimensions.

   Color... — enables you to set dimension line color.

2. Arrowheads

   1st: — enables you to choose an arrowhead at the first selected end point of a dimension.

   2nd: — enables you to choose an arrowhead at the second selected end point of a dimension.

   Size: — sets the size of an arrowhead, the size of an architectural tick, or the overall scale of a user-defined arrowhead block.

3. Extension Line

Suppress: 1st	enables you to suppress the display of the extension line at the first selected end point of a dimension.
Suppress: 2nd	enables you to suppress the display of the extension line at the second selected end point of a dimension.
Extension:	enables you to set the length of the extension line that extends beyond the dimension line.
Origin Offset:	enables you to set the distance of the starting point of the extension line from the selected end points of an object.
Color...	enables you to set extension line color.

4. Center

Mark	enables you to cause the DIMCENTER command to produce a pair of center marks.
Line	enables you to cause the DIMCENTER command to produce a pair of center lines.
None	disables the DIMCENTER command.
Size:	enables you to set the size of the center marks or center lines.

5. Scale

Overall Scale:	enables you to scale the dimension geometry.
Scale to Paper Space	enables you to scale the dimension geometry relative to paper space environment. (Paper space environment will be discussed in Chapter 9.)

Format

Dimension format includes the placement of dimension text, arrowheads, and leader lines. Select the [Format...] button to bring out the Format sub-dialog box. See Figure 5.22.

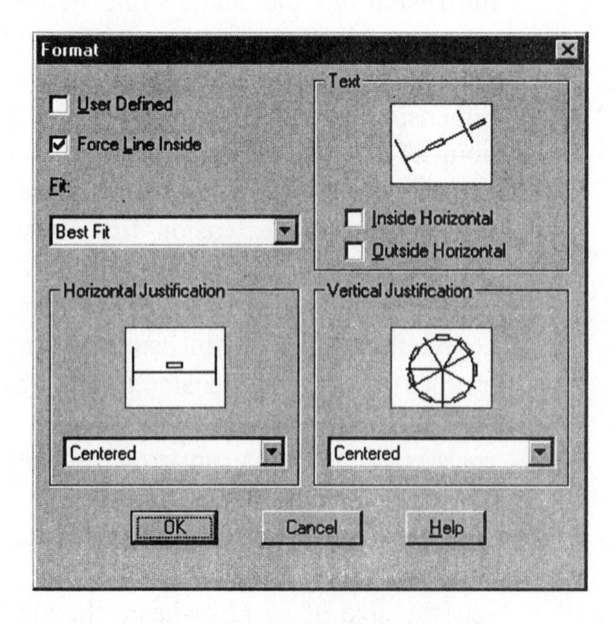

Figure 5.22 Format dialog box

There are four areas in the Format dialog box.

1. Unnamed

User Defined	enables you to select a dimension text location.
Force Line Inside	enables you to place the dimension line within the pair of extension lines in case the dimension text is placed outside the extension lines.
Fit:	enables you to decide how the dimension text and arrowheads are placed within the pair of extension lines in case there is not enough space for both.

2. Text

Inside Horizontal	enables you to force the dimension text placed within the extension lines to be horizontal.
Outside Horizontal	enables you to force the dimension text placed outside the extension line to be horizontal.

3. Horizontal
   Justification — enables you to set the dimension text horizontal location.
4. Vertical
   Justification — enables you to set the dimension text vertical location.

Annotation

Annotation concerns the way the dimension text is displayed. Select the [Annotation...] button to bring out the Annotation sub-dialog box. See Figure 5.23.

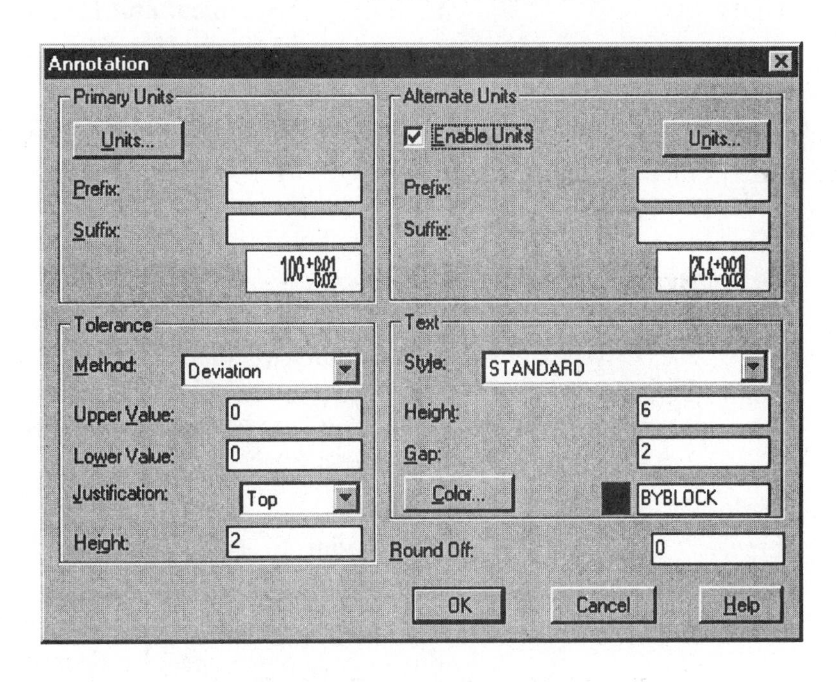

Figure 5.23 Annotation dialog box

There are five areas in the Annotation dialog box.

1. Primary Units

Units...	enables you to bring out the Primary Units dialog box. See Figure 5.24.

Prefix: enables you to set the primary annotation prefix.
Suffix: enables you to set the primary annotation suffix.
Display box displays the primary annotation format.

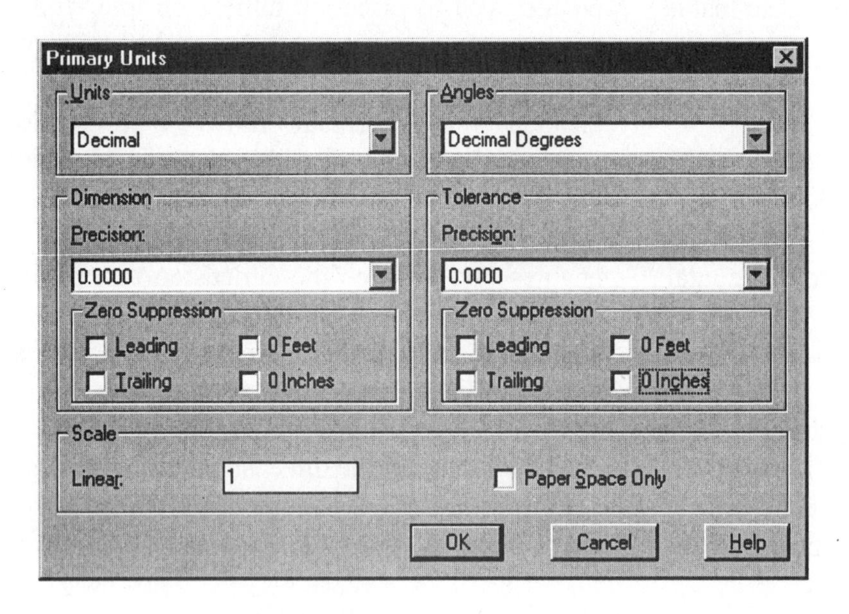

Figure 5.24 Primary Units dialog box

In the Primary Units dialog box, there are five areas.

a. Units — enables you to set linear unit format.
b. Angles — enables you to set angular unit format.
c. Dimension
   Precision: enables you to set the number of decimal places of the primary dimension text.
   Zero Suppression enables you to decide whether zero inch or zero feet is displayed if English units are used.
d. Tolerance
   Precision: enables you to set the number of decimal places of the primary tolerance text.
   Zero Suppression enables you to decide whether zero inch or zero feet is displayed in the tolerance text if English units are used.
e. Scale
   Linear: enables you to scale the primary dimension value.
   Paper Space Only enables you to scale the primary dimension value only in paper space environment.

2. Alternate Units
   Enables Units enables you to control whether alternate units are enabled.
   Units... enables you to bring out the Alternate Units dialog box. See Figure 5.25.
   Prefix: enables you to set the alternate annotation prefix.

Suffix — enables you to set the alternate annotation suffix.
Display box — displays the alternate annotation format.

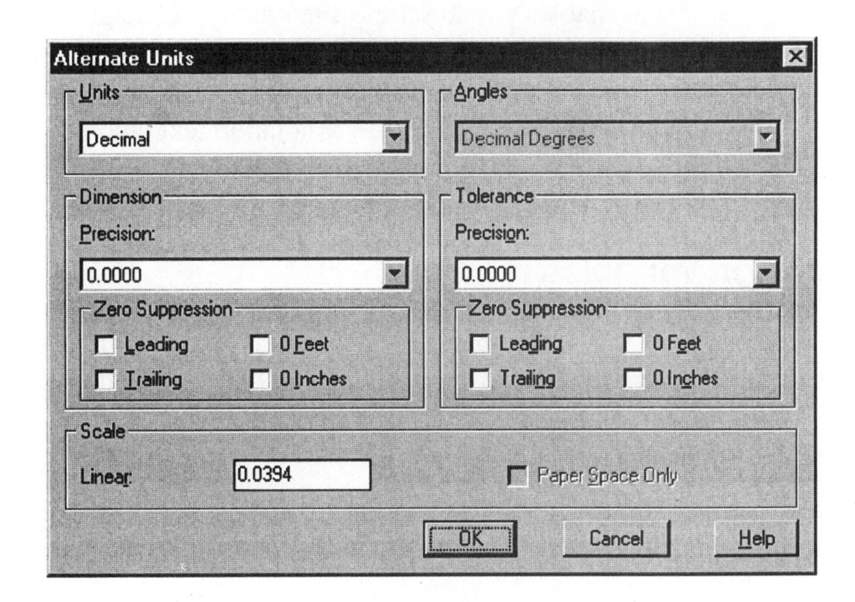

Figure 5.25 Alternate Units dialog box

There are four useable areas in the Alternate Units dialog box.

a. Units — enables you to set linear unit format.

b. Dimension
   - Precision: — enables you to set the number of decimal places of the alternate dimension text.
   - Zero Suppression — enables you to decide whether zero inch or zero feet is displayed if English units are used.

c. Tolerance
   - Precision: — enables you to set the number of decimal places of the alternate tolerance text.
   - Zero Suppression — enables you to decide whether zero inch or zero feet is displayed in the tolerance text if English units are used.

d. Scale
   - Linear: — enables you to scale the alternate dimension value.
   - Paper Space Only — enables you to scale the alternate dimension value only in paper space environment.

3. Tolerance
   - Method: — enables you to select the tolerance method: None, Symmetrical, Deviation, Limits, or Basic.
   - Upper Value: — enables you to set the positive deviation of size.
   - Lower Value: — enables you to set the negative deviation of size.
   - Justification: — enables you to decide how the tolerance text is justified with the dimension text.
   - Height: — enables you to set the tolerance text height relative to the dimension text.

4. Text

Style:	enables you to select the annotation text style.
Height:	enables you to set the annotation text height.
Gap:	enables you to set the distance between the annotation text and other adjacent geometry.
Color...	enables you to set the annotation text color.

5. Round Off — sets the annotation text rounding-off value.

5.5 Dimension Variables

Apart from using the DDIM command to set the style of a dimension, you can manipulate the dimension variables by using the DIMSTYLE command.

Command: **DIMSTYLE**

dimension style: STANDARD
Dimension Style Edit (Save/Restore/STatus/Variables/Apply/?) <Restore>: **ST**

DIMALT Off	Alternate units selected
DIMALTD 4	Alternate unit decimal places
DIMALTF 0.0394	Alternate unit scale factor
DIMALTTD 4	Alternate tolerance decimal places
DIMALTTZ 0	Alternate tolerance zero suppression
DIMALTU 8	Alternate units
DIMALTZ 0	Alternate unit zero suppression
DIMAPOST	Prefix and suffix for alternate text
DIMASO On	Create associative dimensions
DIMASZ 6.0000	Arrow size
DIMAUNIT 0	Angular unit format
DIMBLK	Arrow block name
DIMBLK1	First arrow block name
DIMBLK2	Second arrow block name
DIMCEN 2.5000	Center mark size
DIMCLRD BYBLOCK	Dimension line and leader color
DIMCLRE BYBLOCK	Extension line color
DIMCLRT BYBLOCK	Dimension text color
DIMDEC 4	Decimal places
DIMDLE 0.0000	Dimension line extension
DIMDLI 3.7500	Dimension line spacing
DIMEXE 3.0000	Extension above dimension line
DIMEXO 3.0000	Extension line origin offset
DIMFIT 3	Fit text
DIMGAP 2.0000	Gap from dimension line to text
DIMJUST 0	Justification of text on dimension line
DIMLFAC 1.0000	Linear unit scale factor
DIMLIM Off	Generate dimension limits
DIMPOST	Prefix and suffix for dimension text
DIMRND 0.0000	Rounding value
DIMSAH Off	Separate arrow blocks
DIMSCALE 2.0000	Overall scale factor
DIMSD1 Off	Suppress the first dimension line
DIMSD2 Off	Suppress the second dimension line

DIMSE1 Off	Suppress the first extension line
DIMSE2 Off	Suppress the second extension line
DIMSHO On	Update dimensions while dragging
DIMSOXD Off	Suppress outside dimension lines
DIMSTYLE STANDARD	Current dimension style (read-only)
DIMTAD 0	Place text above the dimension line
DIMTDEC 4	Tolerance decimal places
DIMTFAC 2.0000	Tolerance text height scaling factor
DIMTIH Off	Text inside extensions is horizontal
DIMTIX Off	Place text inside extensions
DIMTM 0.0000	Minus tolerance
DIMTOFL On	Force line inside extension lines
DIMTOH Off	Text outside horizontal
DIMTOL Off	Tolerance dimensioning
DIMTOLJ 1	Tolerance vertical justification
DIMTP 0.0000	Plus tolerance
DIMTSZ 0.0000	Tick size
DIMTVP 0.0000	Text vertical position
DIMTXSTY STANDARD	Text style
DIMTXT 6.0000	Text height
DIMTZIN 0	Tolerance zero suppression
DIMUNIT 8	Unit format
DIMUPT Off	User positioned text
DIMZIN 8	Zero suppression

This list of variables by the DIMSTYLE command is in alphabetical order. The same variables are regrouped in logical order, and briefly explained, below.

Associativeness Variable

DIMASO — It turns on and off associative dimensioning. When it is set to OFF, the components of a dimension are broken into separate entities. Keep this variable ON all the time.

Dimension Line Variables

DIMCLRD — It sets the color of the dimension line and the arrowheads.

DIMDLE — It sets the length of extension of the dimension line beyond the extension line, when DIMTSZ is non-zero. See Figure 5.26.

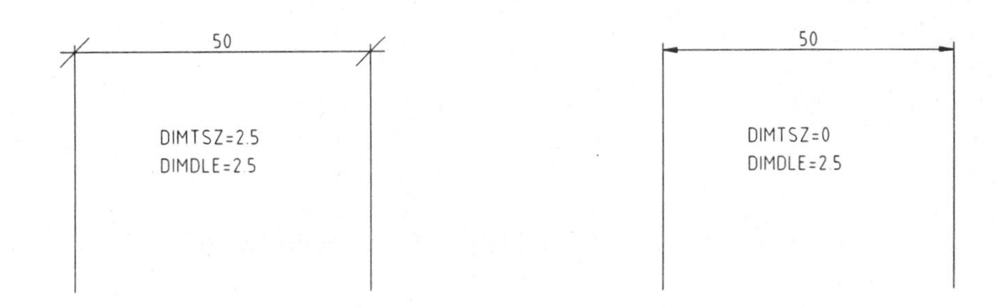

Figure 5.26 DIMDLE and DIMTSZ variables

DIMDLI — It sets the incremental distances between dimension lines for base dimension or continuous dimension. See Figure 5.27.

Figure 5.27 DIMDLI variable

DIMSD1 — It suppresses the display of the first dimension line. You should apply DIMSD1 in conjunction with DIMSE1. See Figure 5.30. When you create a linear dimension, you normally specify two points. The first dimension line points to the first pick point.

DIMSD2 — It suppresses the display of the second dimension line. It should be applied in conjunction with DIMSE2.

DIMSOXD — It suppresses the display of the part of the dimension line that is outside the extension lines. It works on dimensions of short distances when DIMTIX is ON. See Figure 5.28.

Figure 5.28 DIMSOXD and DIMTIX variables

DIMTOFL — It works on dimensions of short distances. It forces the dimension line to be placed inside the extension lines. See Figure 5.29.

Figure 5.29 DIMTOFL, DIMSOXD, and DIMTIX variables

Arrowhead/Arrow-Block Variables

DIMASZ — It sets the arrowhead size. Arrowhead appears if DIMTSZ is 0, DIMBLK is nil, and DIMSAH is OFF. If you use arrow-block

	instead of arrowhead, DIMASZ sets the scale factor of the arrow-blocks.
DIMTSZ	It sets the tick size of the oblique stroke arrowheads. Oblique stroke arrowhead appears if DIMTSZ is non-zero, DIMBLK is nil, and DIMSAH is OFF. You should apply DIMTSZ in conjunction with DIMDLE. See Figure 5.26.
DIMBLK	It sets the name of arrow-block. Arrow-block appears when DIMBLK is set to a valid block name and DIMSAH is OFF. To reset DIMBLK to nil, enter ".". See custom arrow-blocks later in this chapter.
DIMSAH	It turns on and off separate arrow-blocks.
DIMBLK1	It sets the block name of the first separate arrowhead. It works together with DIMSAH and DIMBLK2.
DIMBLK2	It sets the block name of the second separate arrow-block.

Extension Line Variables

DIMCLRE	It sets the color of the extension lines.
DIMSE1	It suppresses the display of the first extension line. See Figure 5.30.
DIMSE2	It suppresses the display of the second extension line.

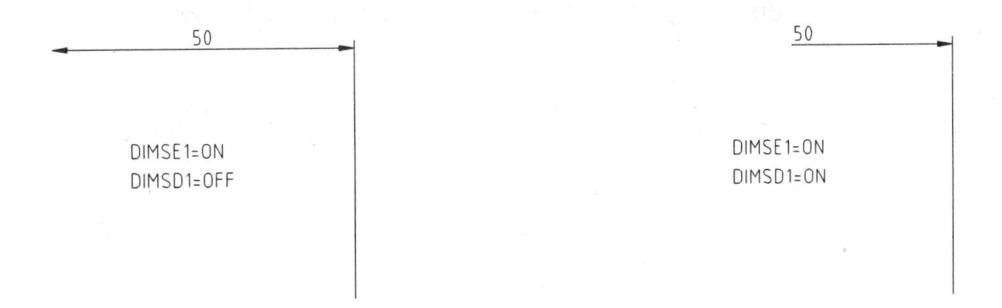

Figure 5.30 DIMSD1 and DIMSE1 variables

DIMEXE	It sets the length of extension of the extension line beyond the dimension line. See Figure 5.31.

Figure 5.31 DIMEXE variable

DIMEXO	It sets the offset distance of the extension line from the definition points of the dimension. See Figure 5.32.

Figure 5.32 DIMEXO variable

Text Format Variables

DIMCLRT	It sets the color of the dimension text.
DIMTXT	It sets the dimension text height.
DIMTXSTY	It sets the text style.
DIMGAP	It sets the size of the gap between the dimension text and its surrounding features. A negative value creates a box around the dimension text. See Figure 5.33.

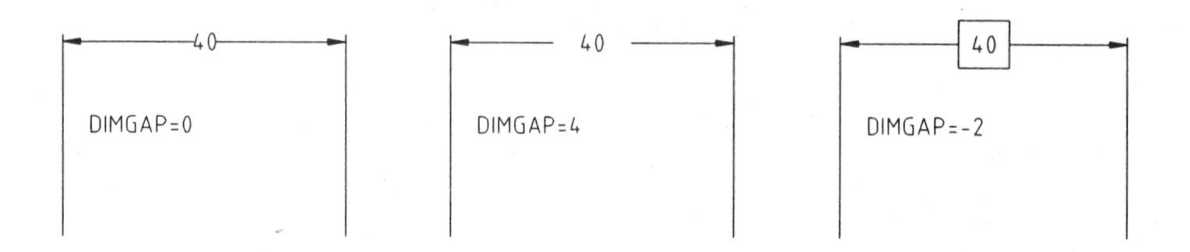

Figure 5.33 DIMGAP variable

Text Location Variables

DIMUPT	It allows the user to specify the position of the dimension text. See Figure 5.34.

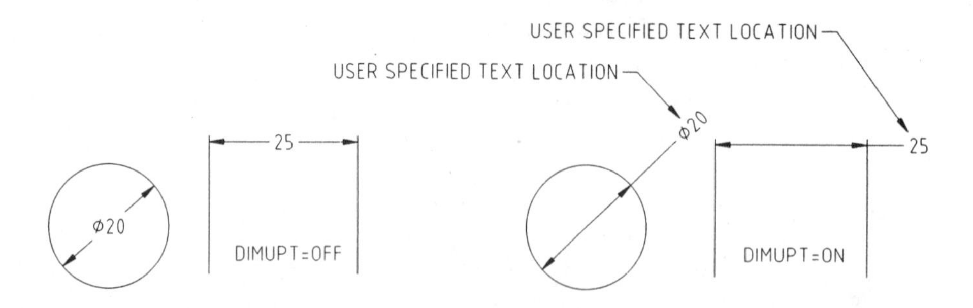

Figure 5.34 DIMUPT variable

DIMFIT It controls how the dimension text and the arrowheads are fitted within the extension lines when space within the pair of extension lines is limited. See Figure 5.35.

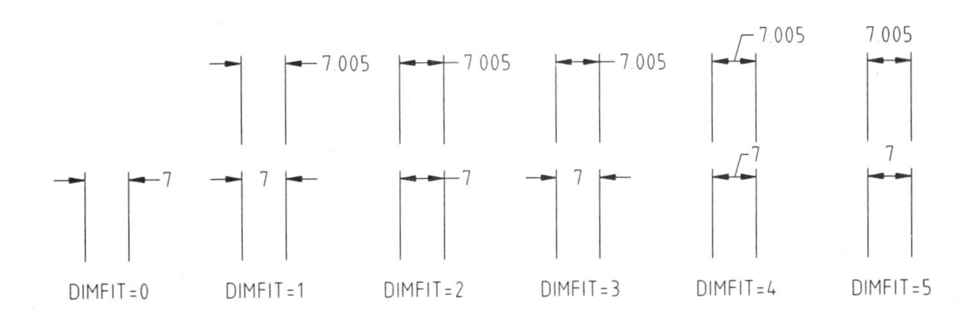

Figure 5.35 DIMFIT variable

DIMJUST It sets the horizontal justification of the text on the dimension line. See Figure 5.36.

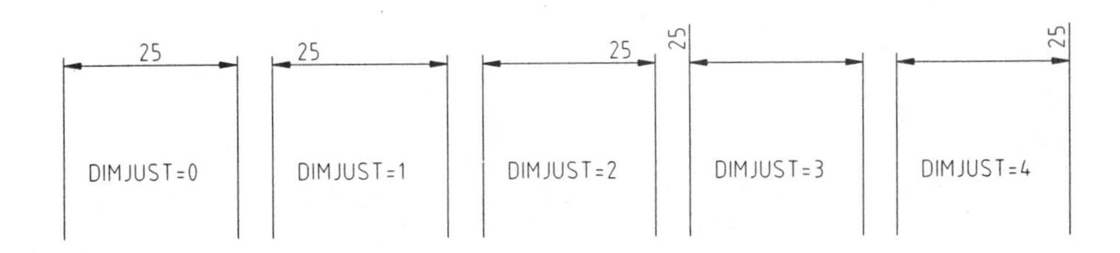

Figure 5.36 DIMJUST variable

DIMTAD It places the text above the dimension line.

DIMTVP It sets the vertical text position relative to the dimension line when DIMTAD is 0. See Figure 5.37.

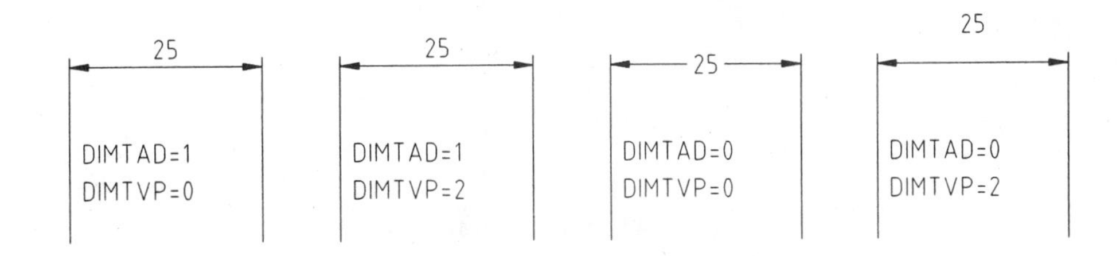

Figure 5.37 DIMTAD and DIMTVP variables

DIMTIX For dimensions of short distances, the arrowheads and the dimension text value are normally placed outside the extension lines. DIMTIX forces the dimension text to be placed inside the extension lines. See Figure 5.38.

Figure 5.38 DIMTIX variable

Text Orientation Variables

DIMTIH	It sets the text inside the extension line to be horizontal. See Figure 6.39.
DIMTOH	It sets the text outside the extension line to be horizontal. See Figure 5.39.

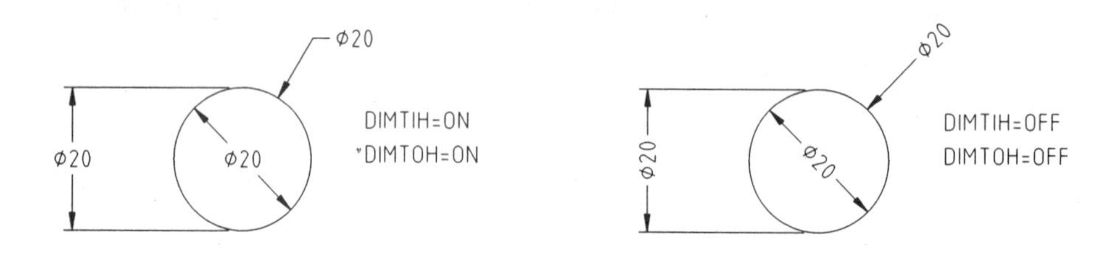

Figure 5.39 DIMTIH and DIMTOH variables

Text Primary Unit Variables

DIMAUNIT	It sets the angular unit format.
DIMDEC	It sets the number of decimal places of the dimension text.
DIMPOST	It sets the suffix of the dimension text. To reset DIMPOST to nil, enter ".". See Figure 5.40.

Figure 5.40 DIMPOST variable

DIMUNIT	It sets the unit format.
DIMZIN	It suppresses the display of zero inch or zero feet of the dimension text if English units are used.
DIMLFAC	It sets the linear unit scale factor.

Text Alternate Unit Variables

DIMALT It turns on and off the display of dimensions of the alternate unit. See Figure 5.41.

DIMALTD It sets the number of decimal places of the alternate dimensions. See Figure 5.41.

DIMALTF It sets the scale factor of the alternate dimensions. See Figure 5.41.

Figure 5.41 DIMALT, DIMALTD, and DIMALTF variables

DIMALTU It sets the units of the alternate dimensions.

DIMALTZ It suppresses the display of zero inch or zero feet of the alternate dimensions text if English units are used.

DIMAPOST It sets the suffix of the alternate dimensions. To reset DIMAPOST to nil, enter ".".

Text Tolerance and Limits Variables

DIMLIM It turns on and off the dimension limits. See Figure 5.42.

DIMTOL It turns on and off the dimension tolerances. See Figure 5.42.

DIMTP It sets the tolerance plus value. See Figure 5.42.

DIMTM It sets the tolerance minus value. See Figure 5.42.

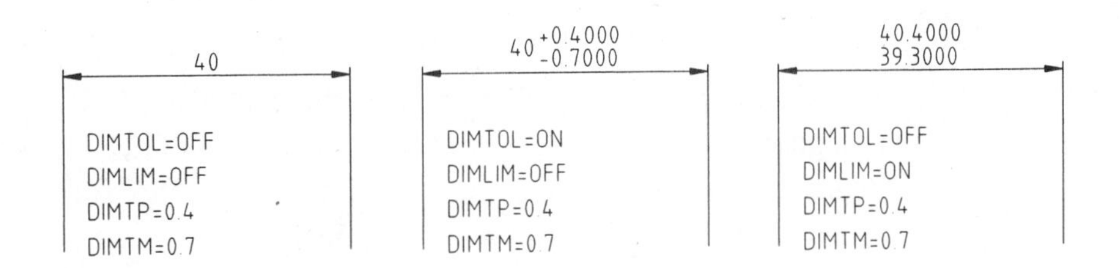

Figure 5.42 Tolerance and limits variables

DIMTOLJ It sets the tolerance vertical justification. See Figure 5.43.

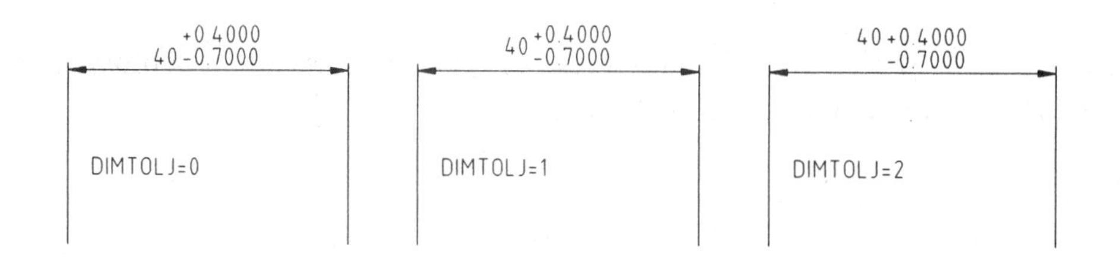

Figure 5.43 DIMTOLJ variable

DIMTFAC It sets the tolerance text height scale factor. See Figure 5.44.

Figure 5.44 DIMTFAC variable

DIMTDEC It sets the decimal places for the dimension tolerance.

DIMTZIN It suppresses the display of zero inch or zero feet of dimension tolerances if English units are used.

DIMALTTD It sets the number of decimal places of the alternate tolerances.

DIMALTTZ It suppresses the display of zero inch or zero feet of the alternate tolerances if English units are used.

Utilities Variables

DIMCEN It sets the appearance and the size of the center mark. If you supply a negative value, it will cause the DIMCENTER command to create a pair of "center lines" that are in the form of three horizontal and three vertical dot and dashes.

DIMRND It sets the rounding-off value for the display of dimension value.

DIMSCALE It sets the overall scale factor of the arrowheads, text height, and other dimension components. Set DIMSCALE to 0 when TILEMODE is 0 and dimension is placed in MSPACE.

DIMSHO It controls whether the dimension text value is updating on the screen while dragging.

DIMSTYLE It stores the current dimension style. It is read-only. Do not confuse this variable with the command of the same name. Issuing DIMSTYLE at the command prompt calls out the DIMSTYLE command. To read this variable, use the SETVAR command.

5.6 Dimensioning Practice 1

Open the drawing of the second component of the pipe support assembly. See Figure 5.45. You will add dimensions to this drawing.

<File> **<Open>**

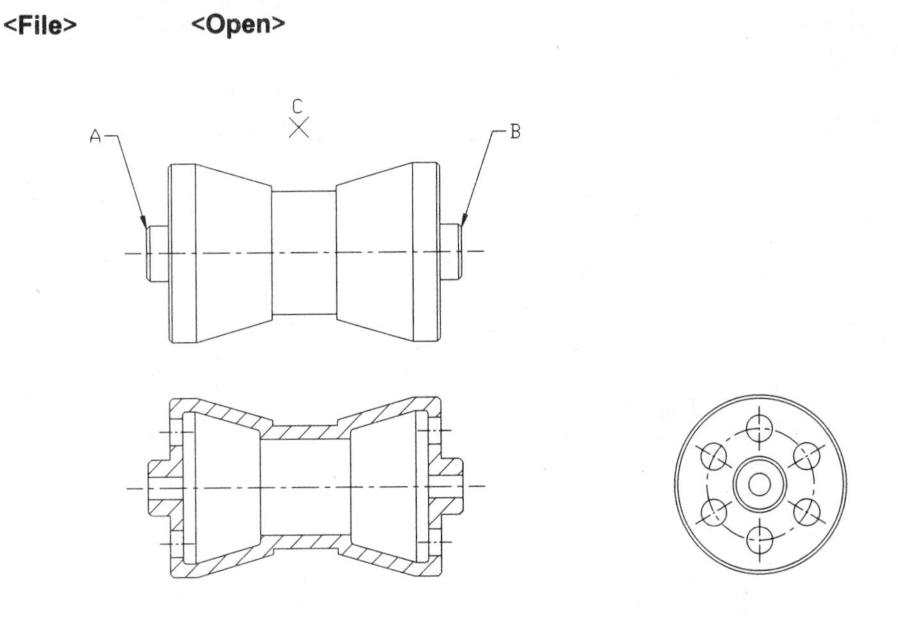

Figure 5.45 Second component of the pipe support assembly

Before adding dimensions to your drawing, make sure that the dimension style settings are appropriate. Select the Style... item of the Dimension pull-down menu to run the DDIM command.

<Dimension> **<Style...>**

Command: **DDIM**

Select the Parent item of the Family box to modify the parent dimension style. Then select the [Geometry...] button to change the geometry setting.

[Family **Parent**
Geometry...]

[Dimension Line Spacing: **12**
Arrowheads 1st **Closed Filled**
2nd **Closed Filled**
Size: **6**
Extension Line Extension: **3**
Origin Offset: **3**
Center **None**
OK]

Select the [Format...] button to change the format setting.

[Format...]

[User Defined **No**
Force Line Inside **No**
Fit: **Best Fit**
Text Inside Horizontal **No**
Outside Horizontal **No**
Horizontal Justification **Centered**
Vertical Justification **Centered**
OK]

Select the [Annotation...] button to change the annotation setting.

[Annotation...]

Select the [Units...] button of the Primary Units box to set the primary units.

[Primary Units **Units...**]

[Units **Decimal**
Angles **Decimal Degrees**
Dimension Precision **0.0**
OK]

[Alternate Units Enable Units **No**
Tolerance Method: **None**
Text Style: **STANDARD**
Height: **6**
Gap: **2**
OK]

Select the [Save] button to save the setting. Then select the Diameter item of the Family box to change the dimension style setting for diameter dimensions.

**[Save
Diameter
Format...]**

[Outside Horizontal **Yes**
OK]

Save the setting and then change the radial dimension setting.

**[Save
Radial
Format...]**

[Outside Horizontal **Yes**
OK]

Save the setting and exit the command.

**[Save
OK]**

Linear Dimensions

Select the Linear item of the Dimension pull-down menu to use the DIMLINEAR command to create a horizontal dimension. See Figure 5.46.

<Dimension> **<Linear>**

Command: **DIMLINEAR**
First extension line origin or press ENTER to select: **END** of **[Select A (Figure 5.45).]**
Second extension line origin: **END** of **[Select B (Figure 5.45).]**
Dimension line location (Mtext/Text/Angle/Horizontal/Vertical/Rotated): **[Select C (Figure 5.45).]**
Dimension text = 290

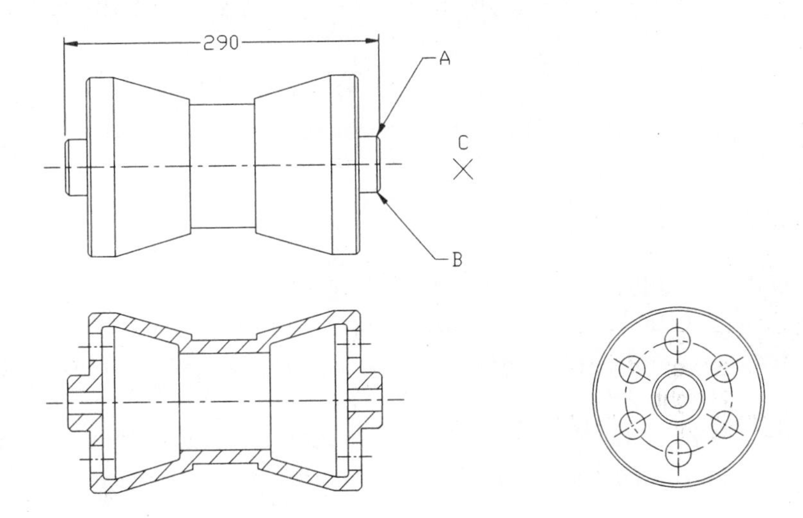

Figure 5.46 Horizontal linear dimension created

Depending on where you specify the dimension line location, the DIMLINEAR command creates either a vertical or a horizontal dimension. In the above command, the dimension location determines a horizontal dimension.

Repeat the DIMLINEAR command to create a vertical dimension. See Figure 5.47.

Command: **[Enter]**
DIMLINEAR
First extension line origin or press ENTER to select: **END** of **[Select A (Figure 5.46).]**
Second extension line origin: **END** of **[Select B (Figure 5.46).]**

According to the drawing, this dimension should be a diameter dimension. Because a linear dimension command is used, the ∅ sign is not supplied. To prefix the dimension text with the ∅ sign, select the Text option and change the dimension text to %%C<>. %%C means the ∅ sign. <> means using the default dimension value.

Dimension line location (Mtext/Text/Angle/Horizontal/Vertical/Rotated): **TEXT**
Dimension text <50>: **%%c<>**
Dimension line location (Mtext/Text/Angle/Horizontal/Vertical/Rotated): **[Select C (Figure 5.46).]**
Dimension text = 50

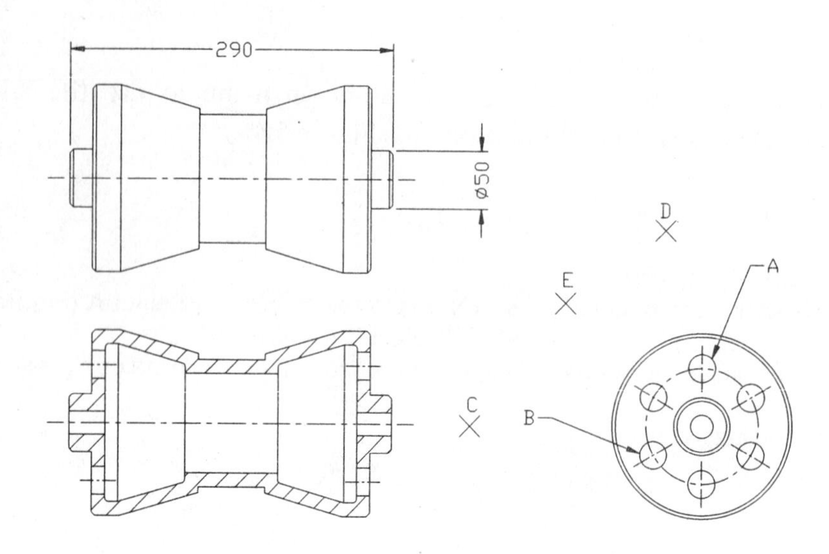

Figure 5.47 Vertical dimension prefixed with Ø sign created

Diameter, Radius, and Degree Symbols

The Ø, R, and ° symbols are automatically generated by the relevant commands: DIMDIAMETER, DIMRADIUS, and DIMANGULAR. If you use a command that does not generate these symbols in the dimension text, you can use the code %%D for ° and %%C for Ø.

Repeat the DIMLINEAR command to create a rotated dimension. A rotated dimension is a linear dimension measured in a specified rotation angle. See Figure 5.48.

Command: **[Enter]**
DIMLINEAR
First extension line origin or press ENTER to select: **CEN** of **[Select A (Figure 5.47).]**
Second extension line origin: **CEN** of **[Select B (Figure 5.47).]**

Move the cursor to point C of Figure 5.47 and note that the dimension becomes a vertical dimension. Then move the cursor to D of Figure 5.47 and note that the dimension becomes a horizontal dimension. As we have said, the dimension location determines whether it is a horizontal dimension or a vertical dimension. To create a rotated dimension, select the R option.

Dimension line location (Mtext/Text/Angle/Horizontal/Vertical/Rotated): **R**
Dimension line angle: **45**
Dimension line location (Mtext/Text/Angle/Horizontal/Vertical/Rotated): **[Select E (Figure 5.47).]**
Dimension text = 83.7

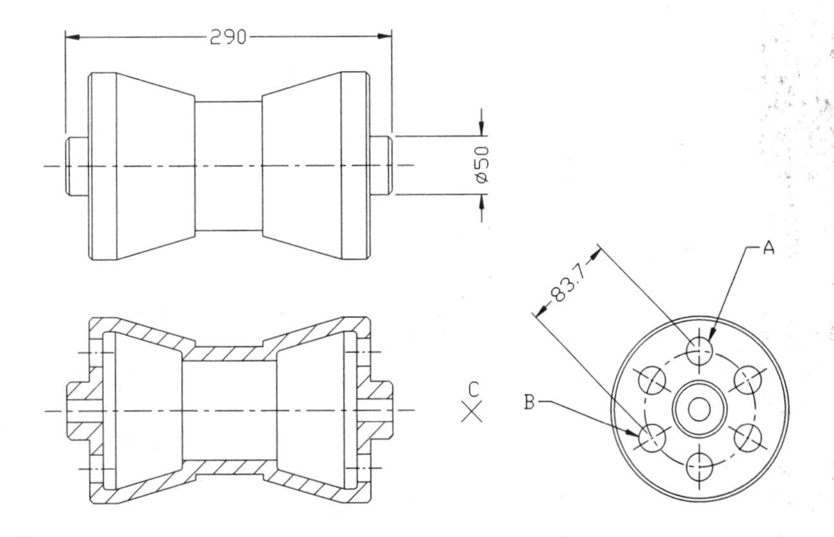

Figure 5.48 Rotated dimension created

Refer to Figure 5.48. The rotated dimension you create is measured in a direction that rotates 45°. Repeat the DIMLINEAR command to produce another horizontal dimension. To reiterate, the dimension location determines whether a horizontal or a vertical dimension is created. To force a horizontal dimension regardless of dimension location, you have to use the H option. See Figure 5.49.

Command: **[Enter]**
DIMLINEAR
First extension line origin or press ENTER to select: **CEN** of **[Select A (Figure 5.48).]**
Second extension line origin: **CEN** of **[Select B (Figure 5.48).]**

As you move the cursor to C of Figure 5.48, a vertical dimension is displayed. Select the H option to force a horizontal dimension.

Dimension line location (Mtext/Text/Angle/Horizontal/Vertical/Rotated): **H**
Dimension line location (Mtext/Text/Angle): **[Select C (Figure 5.48).]**
Dimension text = 43.3

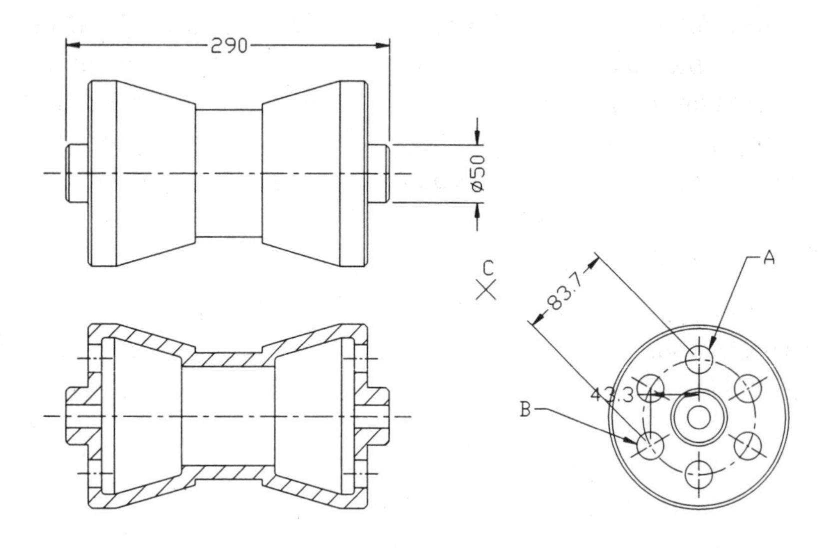

Figure 5.49 Horizontal dimension forced

Aligned Dimension

Select the Aligned item of the Dimension pull-down menu to run the DIMALIGNED dimension. An aligned dimension creates a dimension that measures the distance from two selected points. See Figure 5.50.

<Dimension> **<Aligned>**

Command: **DIMALIGNED**
First extension line origin or press ENTER to select: **CEN** of [**Select A (Figure 5.49).**]
Second extension line origin: **CEN** of [**Select B (Figure 5.49).**]
Dimension line location (Mtext/Text/Angle): [**Select C (Figure 5.49).**]
Dimension text = 86.6

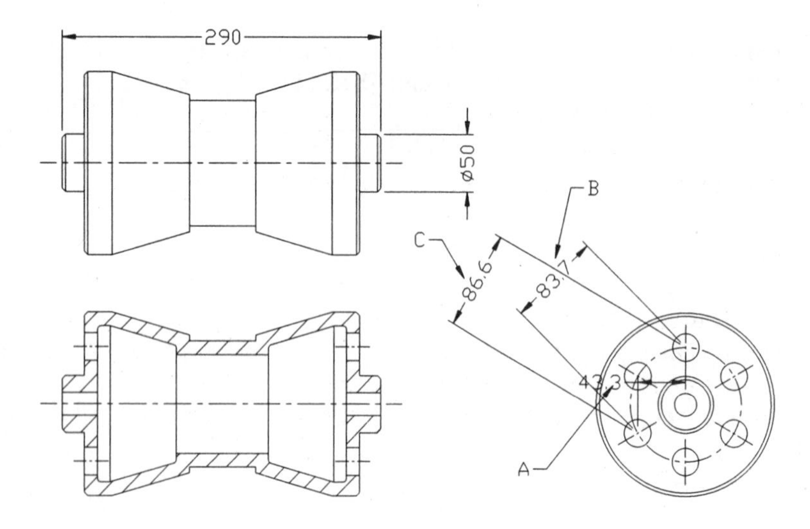

Figure 5.50 Aligned dimension created

Refer to Figure 5.50. Compare dimension B with dimension C. Dimension B is a rotated dimension; it measures the distance in a specified, rotated direction. Dimension C is an aligned dimension; it measures the shortest distance between two selected points.

These two dimensions, together with the last horizontal dimension, are not required in the final drawing. Erase them. See Figure 5.51.

<Modify> **<Erase>**

Command: **ERASE**
Select objects: **[Select A, B, and C (Figure 5.50).]**
Select objects: **[Enter]**

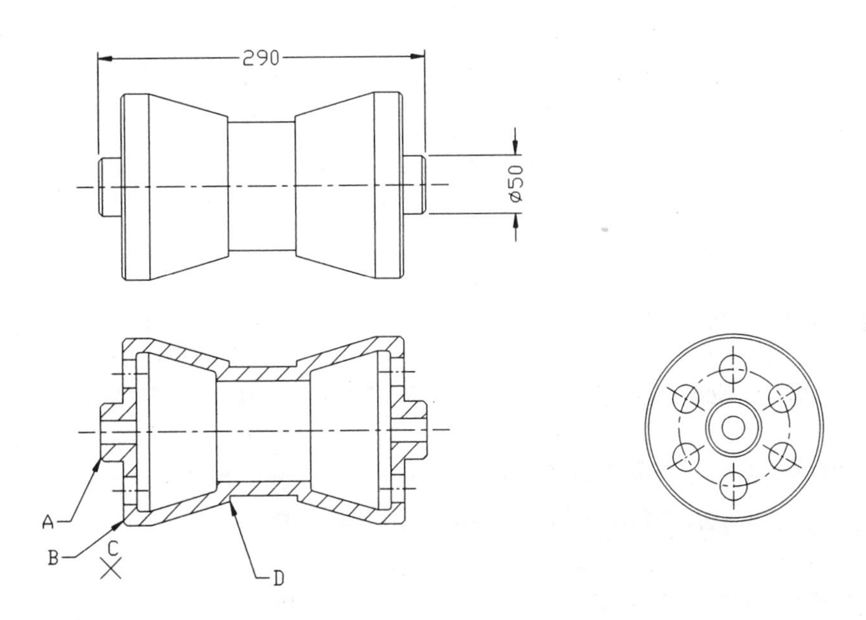

Figure 5.51 Three dimensions erased

Baseline Dimension

Baseline dimensions are dimensions that share a common extension line. Issue the DIMLINEAR command to produce a horizontal dimension. Then select the Baseline item of the Dimension pull-down menu to create a baseline dimension. See Figure 5.52.

<Dimension> **<Linear>**

Command: **DIMLINEAR**
First extension line origin or press ENTER to select: **END** of **[Select A (Figure 5.51).]**
Second extension line origin: **END** of **[Select B (Figure 5.51).]**
Dimension line location (Mtext/Text/Angle/Horizontal/Vertical/Rotated): **[Select C (Figure 5.51).]**
Dimension text = 20

<Dimension> **<Baseline>**

Command: **DIMBASELINE**
Specify a second extension line origin or (Undo/<Select>): **END** of **[Select D (Figure 5.51).]**

Dimension text = 115
Specify a second extension line origin or (Undo/<Select>): **[Enter]**
Select base dimension: **[Enter]**

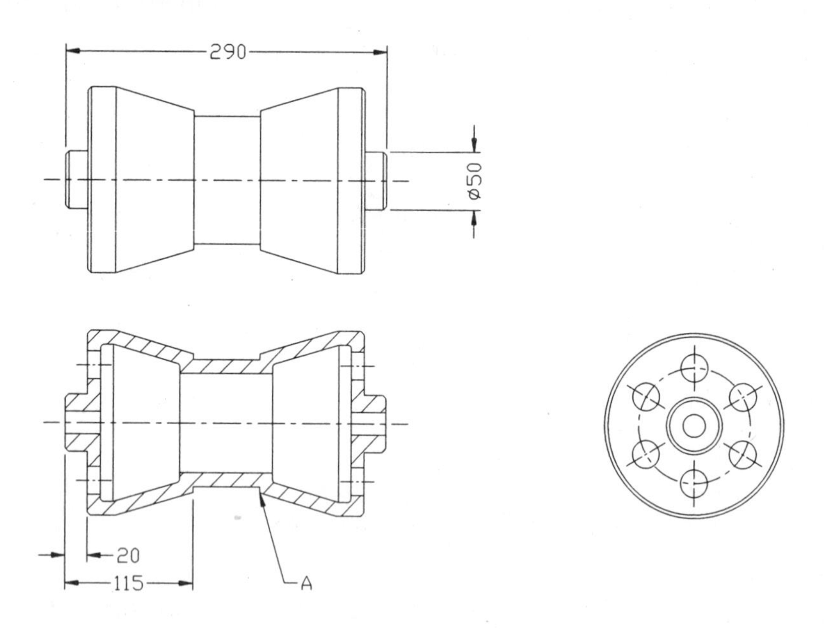

Figure 5.52 Linear dimension and baseline dimension created

Refer to Figure 5.52. The incremental distance between the horizontal dimension and the baseline dimension is determined by the DIMDLI variable.

Continue Dimension

The next dimension is a continue dimension. A continue dimension continues from an existing dimension. Its dimension line aligns with that of the selected dimension. Select the Continue item of the Dimension pull-down menu to apply the DIMCONTINUE command to produce a continue dimension. See Figure 5.53.

<Dimension> **<Continue>**

Command: **DIMCONTINUE**
Specify a second extension line origin or (Undo/<Select>): **END** of **[Select A (Figure 5.52).]**
Dimension text = 60
Specify a second extension line origin or (Undo/<Select>): **[Enter]**
Select continued dimension: **[Enter]**

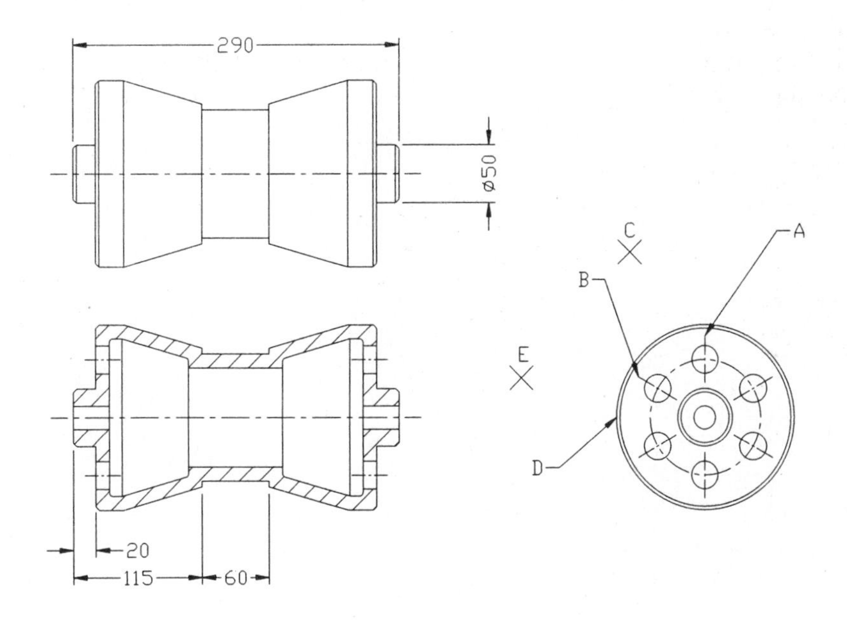

Figure 5.53 Continue dimension created

Refer to Figure 5.53. Note that the dimension line of the continue dimension aligns with that of the previous dimension.

Angular Dimension

From the Dimension pull-down menu, select the Angular item to apply the DIMANGULAR command to create two angular dimensions. See Figure 5.54. Note that the DIMANGULAR command adds the ° symbol to the dimension text automatically.

<Dimension> **<Angular>**

Command: **DIMANGULAR**
Select arc, circle, line, or press ENTER: **[Select A (Figure 5.53).]**
Second line: **[Select B (Figure 5.53).]**
Dimension arc line location (Mtext/Text/Angle): **[Select C (Figure 5.53).]**
Dimension text = 60

In the last command, you select two non-parallel lines. In repeating the command below, you will select an angle vertex and two points to define an angle.

Command: **[Enter]**
DIMANGULAR
Select arc, circle, line, or press ENTER: **[Enter]**
Angle vertex: **CEN** of **[Select D (Figure 5.53).]**
First angle endpoint: **QUA** of **[Select D (Figure 5.53).]**
Second angle endpoint: **END** of **[Select B (Figure 5.53).]**
Dimension arc line location (Mtext/Text/Angle): **[Select E (Figure 5.53).]**
Dimension text = 30

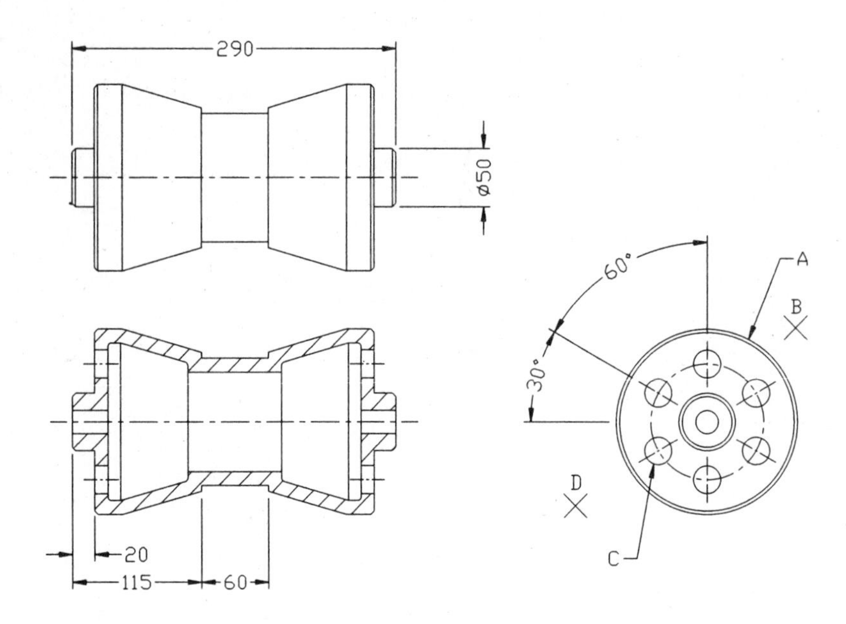

Figure 5.54 Angular dimensions created

Diameter Dimension

From the Dimension pull-down menu, select the Diameter item to apply the DIMDIAMETER command to create two diameter dimensions. See Figure 5.55. Note that the DIMDIAMETER command adds the Ø symbol to the dimension text automatically.

<Dimension> **<Diameter>**

Command: **DIMDIAMETER**
Select arc or circle: **[Select A (Figure 5.54).]**
Dimension text = 160
Dimension line location (Mtext/Text/Angle): **[Select B (Figure 5.54).]**

Command: **[Enter]**
DIMDIAMETER
Select arc or circle: **[Select C (Figure 5.54).]**
Dimension text = 24
Dimension line location (Mtext/Text/Angle): **T**
Dimension text <24>: **6 HOLES <>**
Dimension line location (Mtext/Text/Angle): **[Select D (Figure 5.54).]**

In the foregoing dimension text input, 6 HOLES <> means prefixing the text string 6 HOLES to the default dimension text which is, in this case, Ø24. The dimension text becomes 6 HOLES Ø24.

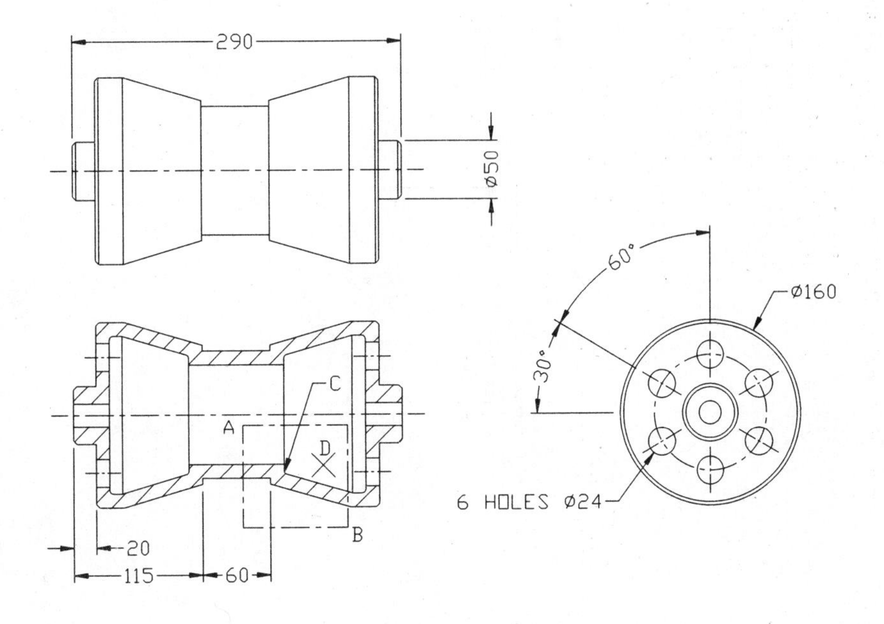

Figure 5.55 Two diameter dimensions created

Radius Dimension

Use the ZOOM command to zoom in to a window described by the rectangle AB of Figure 5.55. Then select the Radius item of the Dimension pull-down menu to use the DIMRADIUS command to create a radial dimension. See Figure 5.56. Note that the DIMRADIUS command adds the R symbol to the dimension text automatically.

<Dimension> **<Radius>**

Command: **DIMRADIUS**
Select arc or circle: **[Select C (Figure 5.55).]**
Dimension text = 3
Dimension line location (Mtext/Text/Angle): **[Select D (Figure 5.55).]**

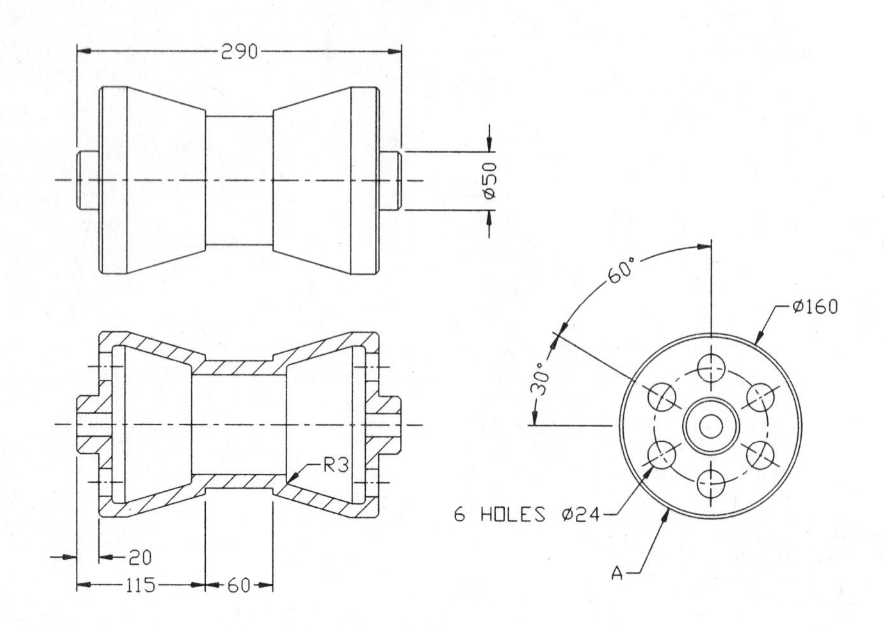

Figure 5.56 Radial dimension created

Center Mark

There are two types of center marks — a cross or a pair of center lines at the center of a circle or arc. A positive value of DIMCEN gives a cross; a negative value of DIMCEN gives a pair of center lines.

Set the DIMCEN variable to -6 and select the Center Mark item of the Dimension pull-down menu to execute the DIMCENTER command to create a center mark. See Figure 5.57.

Command: **DIMCEN**
New value for DIMCEN: **-2.5**

<Dimension> **<Center Mark>**

Command: **DIMCENTER**
Select arc or circle: **[Select A (Figure 5.56).]**

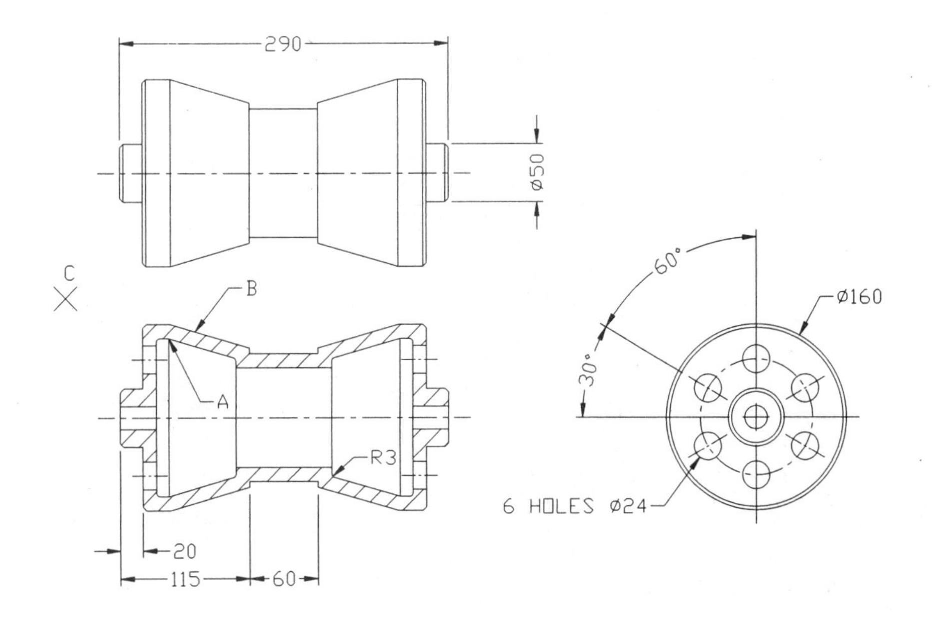

Figure 5.57 Center mark created

In Chapter 3 and other chapters, you learned how to make use of the FROM option in conjunction with the LINE command to produce a pair of center lines that have the linetype "Center." Now you can see that the DIMCENTER command seems to do the same thing. Arguably, the DIMCENTER command is faster. However, there are reasons why you may still want to use the LINE command instead. The LINE command produces two unique line segments whose linetype scale you may adjust to any value you like. Therefore, you can have more dots and dashes for a long center line and fewer for a short center line. On the other hand, the DIMCENTER command gives six line segments no matter how large the circle is. Linetype scale has no effect on the them. As a result, the DIMCENTER command gives very ugly center lines for large circles.

As a guide, you might use the DIMCENTER for small circles and the LINE command for larger circles.

Complete the Dimensions

Use the DIMALIGNED command to add a dimension. See Figure 5.58.

<Dimension> **<Aligned>**

Command: **DIMALIGNED**
First extension line origin or press ENTER to select: **END** of **[Select A (Figure 5.57).]**
Second extension line origin: **PERP** to **[Select B (Figure 5.57).]**
Dimension line location (Mtext/Text/Angle): **[Select C (Figure 5.57).]**
Dimension text = 12

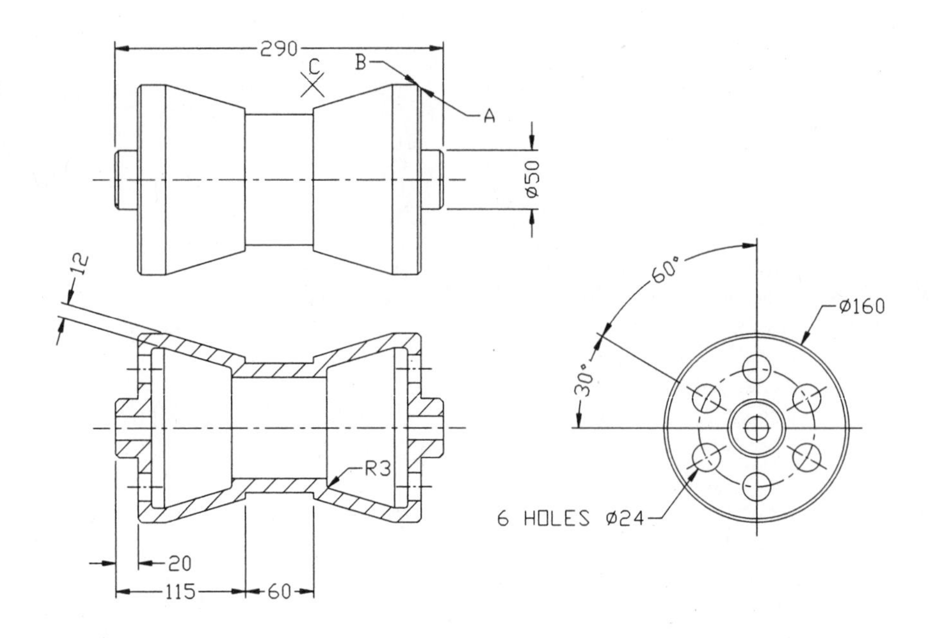

Figure 5.58 Aligned dimension created

Use the DIMLINEAR command to create a linear dimension. See Figure 5.59.

<Dimension> **<Linear>**

Command: **DIMLINEAR**
First extension line origin or press ENTER to select: **END** of [**Select A (Figure 5.58).**]
Second extension line origin: **END** of [**Select B (Figure 5.58).**]
Dimension line location (Mtext/Text/Angle/Horizontal/Vertical/Rotated): **T**
Dimension text <3>: **45%%D X <>**
Dimension line location (Mtext/Text/Angle/Horizontal/Vertical/Rotated): [**Select C (Figure 5.58).**]
Dimension text = 3

In the foregoing text input, 45%%D X <> means adding the prefix 45%%D X to the default dimension text, 3. The dimension text becomes 45° X 3.

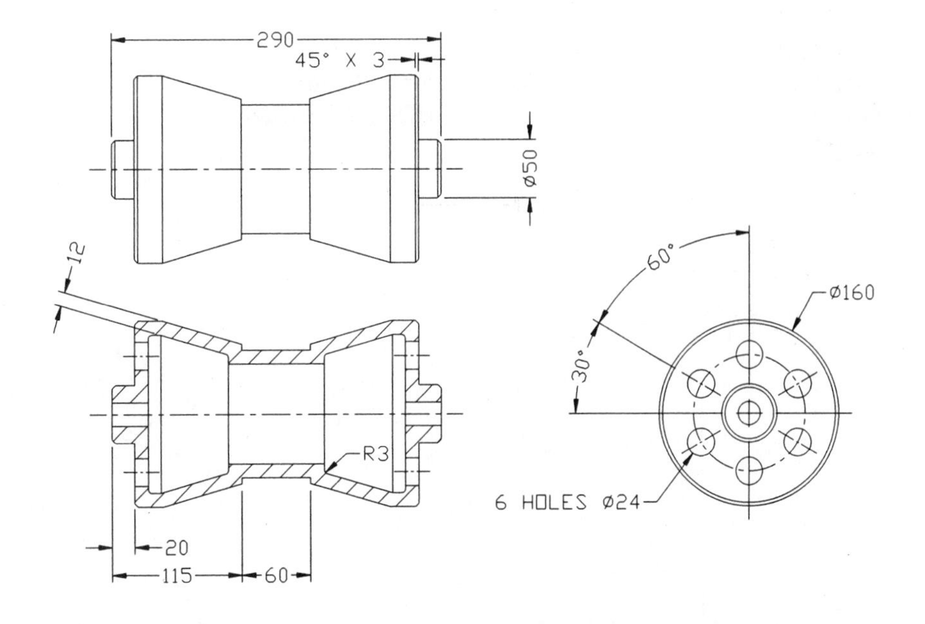

Figure 5.59 Linear dimension created

Add the appropriate dimensions according to Figure 5.60.

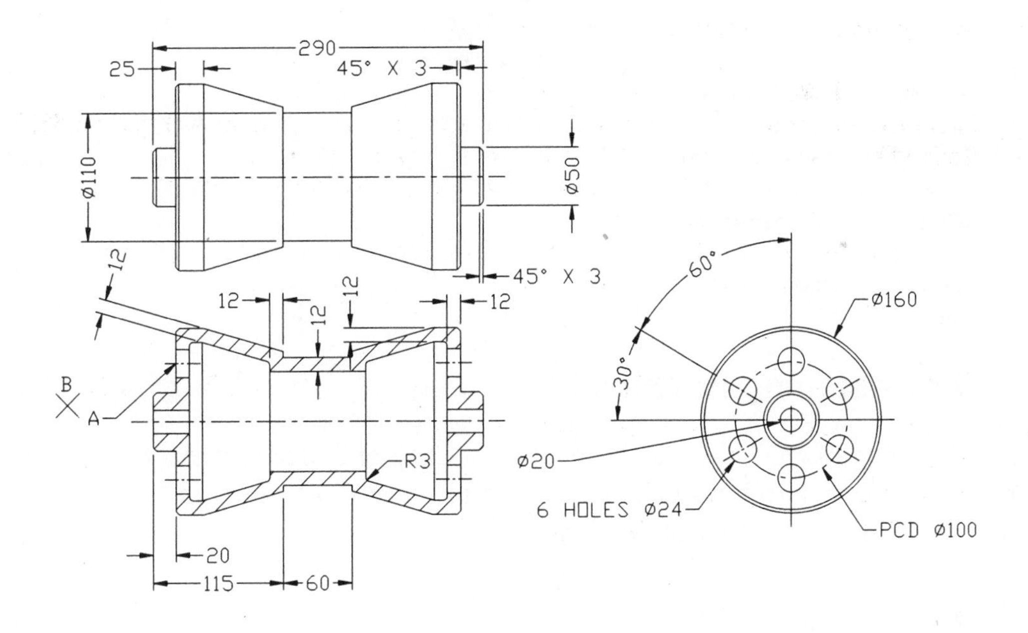

Figure 5.60 Drawing dimensioned

Leader

To complete the drawing, add a leader. See Figure 5.61.

<Dimension> **<Leader>**

Command: **LEADER**

From point: **INT** of **[Select A (Figure 5.60).]**
To point: **[Select B (Figure 5.60).]**
To point (Format/Annotation/Undo)<Annotation>: **[Enter]**
Annotation (or press ENTER for options): **6 HOLES %%C24**
MText: **[Enter]**

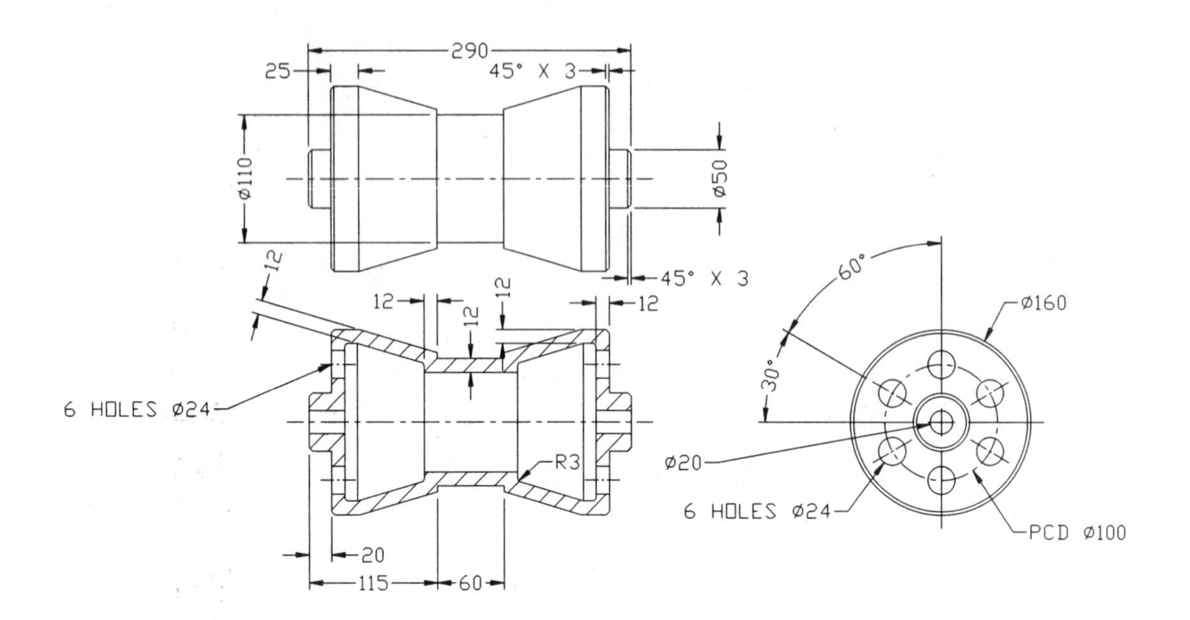

Figure 5.61 Leader added

Save the drawing.

<File> **<Save As...>**

File name: **DIM1.DWG**

5.7 Dimensioning Practice 2

Open the drawing FLOOR that you have saved in Chapter 3. See Figure 5.62.

You will use architectural tick arrowheads to place dimensions in this drawing. In addition, you will add ordinate dimensions.

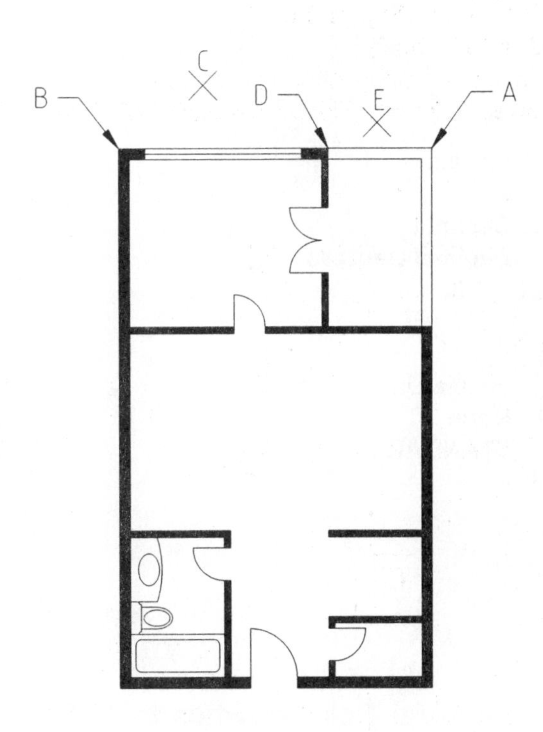

Figure 5.62 Floor plan

Select the Style... item of the Dimension pull-down menu to run the DDIM command to create a dimension style.

<Dimension> **<Style...>**

Command: **DDIM**

[Family **Parent**
Geometry...]

[Dimension Line Extension: **3**
Spacing: **12**
Arrowheads 1st **Architectural Tick**
2nd **Architectural Tick**
Size: **4**
Extension Line Extension: **3**
Origin Offset: **3**
Center **None**
Overall Scale: **50**
OK]

[Format...]

[User Defined **No**
Force Line Inside **No**
Fit: **Best Fit**
Text Inside Horizontal **No**
Outside Horizontal **No**
Horizontal Justification **Centered**
Vertical Justification **Centered**

OK]

[**Annotation...**]

[Primary Units **Units...**]

[Units **Decimal**
Angles **Decimal Degrees**
Dimension Precision **0**
OK]

[Alternate Units Enable Units **No**
Tolerance Method: **None**
Text Style: **STANDARD**
Height: **4**
Gap: **1**
OK]

[**Save**
OK]

Dimensions with Architectural Tick Arrowheads

The type of arrowheads used in a dimension is determined by the DDIM command. After setting a dimension style, run the DIMLINEAR command to add two linear dimensions. See Figure 5.63.

<Dimension> **<Linear>**

Command: **DIMLINEAR**
First extension line origin or press ENTER to select: **END** of [**Select A (Figure 5.62).**]
Second extension line origin: **END** of [**Select B (Figure 5.62).**]
Dimension line location (Mtext/Text/Angle/Horizontal/Vertical/Rotated): [**Select C (Figure 5.62).**]
Dimension text = 6000

Command: **[Enter]**
DIMLINEAR
First extension line origin or press ENTER to select: **END** of [**Select A (Figure 5.62).**]
Second extension line origin: **END** of [**Select D (Figure 5.62).**]
Dimension line location (Mtext/Text/Angle/Horizontal/Vertical/Rotated): **A**
Enter text angle: **45**
Dimension line location (Mtext/Text/Angle/Horizontal/Vertical/Rotated): [**Select E (Figure 5.62).**]
Dimension text = 2000

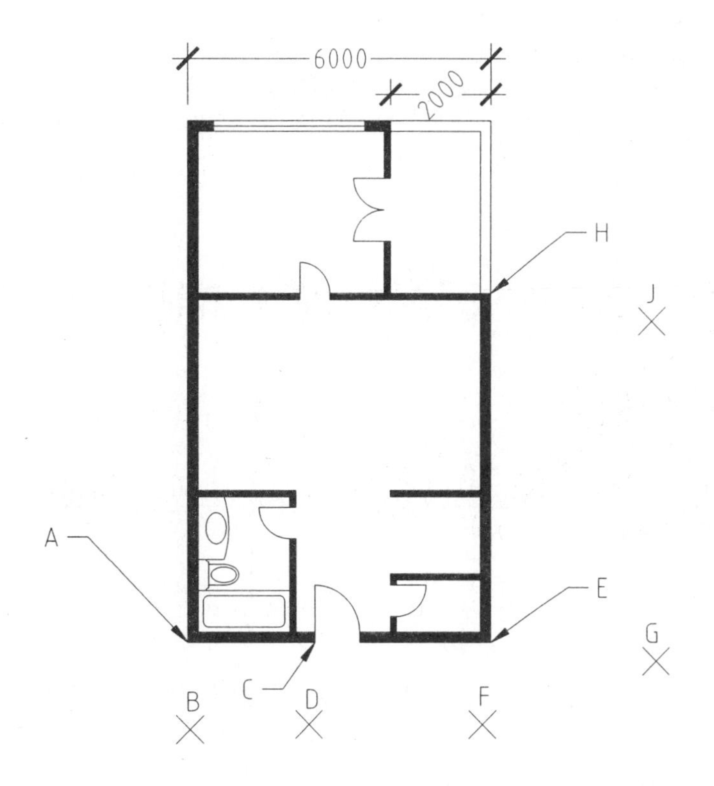

Figure 5.63 Linear dimensions with architectural tick arrowheads

Figure 5.63 shows an example of how to use architectural tick arrowheads in dimensioning a drawing. You can finish the drawing by adding the remaining dimension. In the next paragraph, you will practice another kind of dimension, ordinate dimension.

Ordinate Dimensions

Ordinate dimensions are datum dimensions. They display the X or Y ordinate of a selected point. Because the X or Y ordinate is measured from the current UCS origin, use the UCS command to set the origin position. The ordinate dimension draws a leader line from a selected point. To keep the leader line in an orthogonal direction, turn Ortho mode on. Use the DIMORDINATE command to add ordinate dimensions. See Figure 5.64.

<Tools> **<UCS>** **<Origin>**

Command: **UCS**
Origin/ZAxis/3point/OBject/View/X/Y/Z/Prev/Restore/Save/Del/?/<World>: **O**
Origin point <0,0,0>: **END** of [**Select A (Figure 5.63).**]

[F8]

<Dimension> **<Ordinate>**

Command: **DIMORDINATE**
Select feature: **END** of [**Select A (Figure 5.63).**]
Leader endpoint (Xdatum/Ydatum/Mtext/Text): [**Select B (Figure 5.63).**]
Dimension text = 0

Command: **[Enter]**
DIMORDINATE
Select feature: **END** of **[Select C (Figure 5.63).]**
Leader endpoint (Xdatum/Ydatum/Mtext/Text): **[Select D (Figure 5.63).]**
Dimension text = 2500

Command: **[Enter]**
DIMORDINATE
Select feature: **END** of **[Select E (Figure 5.63).]**
Leader endpoint (Xdatum/Ydatum/Mtext/Text): **[Select F (Figure 5.63).]**
Dimension text = 6000

Command: **[Enter]**
DIMORDINATE
Select feature: **END** of **[Select E (Figure 5.63).]**
Leader endpoint (Xdatum/Ydatum/Mtext/Text): **[Select G (Figure 5.63).]**
Dimension text = 0

Command: **[Enter]**
DIMORDINATE
Select feature: **END** of **[Select H (Figure 5.63).]**
Leader endpoint (Xdatum/Ydatum/Mtext/Text): **[Select J (Figure 5.63).]**
Dimension text = 6700

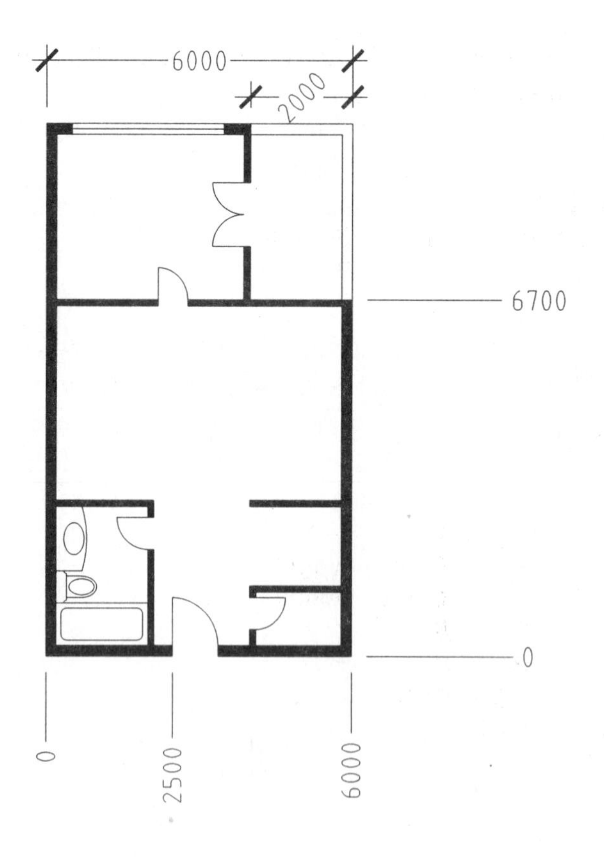

Figure 5.64 Ordinate dimensions created

Refer to Figure 5.64. This drawing serves to illustrate two kinds of dimensions,

dimensions with architectural tick arrowheads and ordinate dimensions. Save the drawing.

<File> **<Save As...>**

File name: **DIM2.DWG**

5.8 Edit Dimension

Open the drawing — SHAFT.DWG. See Figure 5.65.

<File> **<Open...>**

Figure 5.65 Third component of the pipe support assembly

You will add dimensions and edit the third component of the pipe support assembly that you created in Chapter 4.

Add Dimension

Use the DDIM command to set a dimension style.

<Dimension> **<Style...>**

Command: **DDIM**

[Family **Parent**
Geometry...]

[Dimension Line Spacing: **12**
Arrowheads 1st **Closed Filled**
2nd **Closed Filled**
Size: **6**
Extension Line Extension: **3**
Origin Offset: **3**
Center **None**
OK]

[Format...]

[User Defined **No**
Force Line Inside **No**
Fit: **Best Fit**
Text Inside Horizontal **No**
Outside Horizontal **No**

Horizontal Justification **Centered**
Vertical Justification **Centered**
OK]

[Annotation...]

[Primary Units **Units...**]

[Units **Decimal**
Angles **Decimal Degrees**
Dimension Precision **0**
OK]

[Alternate Units Enable Units **No**
Tolerance Method: **None**
Text Style: **STANDARD**
Height: **6**
Gap: **2**
OK]

[Save OK]

Use the DIMLINEAR command to create a linear dimension and the DIMBASELINE command to create a baseline dimension. See Figure 5.66.

<Dimension> <Linear>

Command: **DIMLINEAR**
First extension line origin or press ENTER to select: **END** of **[Select A (Figure 5.65).]**
Second extension line origin: **END** of **[Select B (Figure 5.65).]**
Dimension line location (Mtext/Text/Angle/Horizontal/Vertical/Rotated): **[Select C (Figure 5.65).]**
Dimension text = 3

<Dimension> <Baseline>

Command: **DIMBASELINE**
Specify a second extension line origin or (Undo/<Select>): **END** of **[Select D (Figure 5.65).]**
Dimension text = 346
Specify a second extension line origin or (Undo/<Select>): **[Enter]**
Select base dimension: **[Enter]**

Figure 5.66 Dimensions created

Edit Feature

Definition points are included in each dimension. When a dimension is associative, moving the definition point causes an automatic update to the dimension. You need to include the definition points in your selection set for editing. Edit the drawing with the STRETCH command. See Figure 5.67.

<Modify> **<Stretch>**

Command: **STRETCH**
Select objects to stretch by crossing-window or crossing-polygon...
Select objects: **C**
First corner: [**Select A (Figure 5.66).**]
Other corner: [**Select B (Figure 5.66).**]
Select objects: [**Enter**]
Base point or displacement: **150<0**
Second point of displacement: [**Enter**]

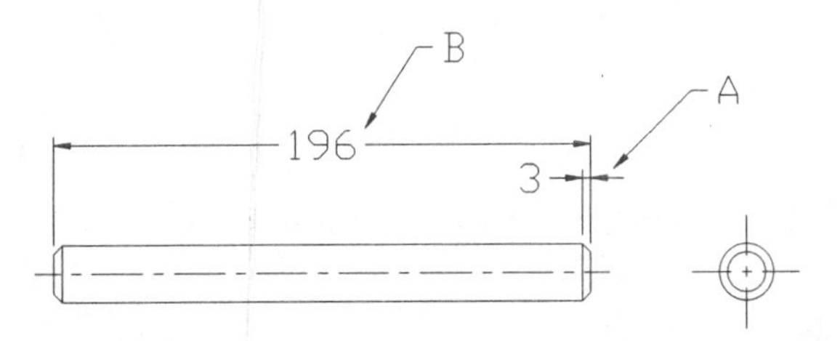

Figure 5.67 Feature stretched and dimension updated automatically

In the foregoing STRETCH command, the definition points for the corresponding dimensions are selected as well. After stretching, the affected dimensions are updated automatically because the dimensions are associative.

Edit Dimension Components

You can use the DIMEDIT command to alter the dimension text string, to rotate the text string, and to rotate the extension lines to an oblique angle. Select the Dimension Edit icon of the Dimension toolbar to apply the DIMEDIT command to edit a dimension text. The <> symbol instructs AutoCAD to supply the measured length between the definition points. The new text string becomes 3 X 45°. See Figure 5.68.

[Dimension] **[Dimension Edit]**

Command: **DIMEDIT**
Dimension Edit (Home/New/Rotate/Oblique): **NEW**

When the text editor appears, type the following:

[Multiline Text Editor
<> X 45%%D
OK]

Select objects: **[Select A (Figure 5.67).]**
Select objects: **[Enter]**

Repeat the DIMEDIT command to rotate the dimension text of a dimension. See Figure 5.68.

Command: **[Enter]**
DIMEDIT
Dimension Edit (Home/New/Rotate/Oblique): **ROTATE**
Enter text angle: **45**
Select objects: **[Select B (Figure 5.67).]**
Select objects: **[Enter]**

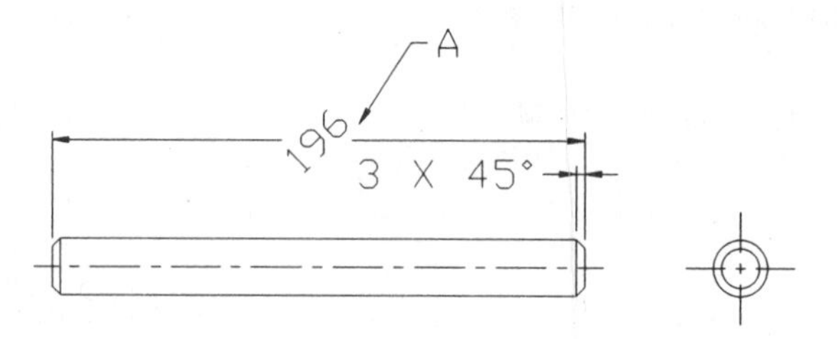

Figure 5.68 Text changed and text rotated

Repeat the DIMEDIT command to place the rotated dimension text at its home position. See Figure 5.69.

Command: **[Enter]**
DIMEDIT
Dimension Edit (Home/New/Rotate/Oblique): **HOME**
Select objects: **[Select A (Figure 5.68).]**
Select objects: **[Enter]**

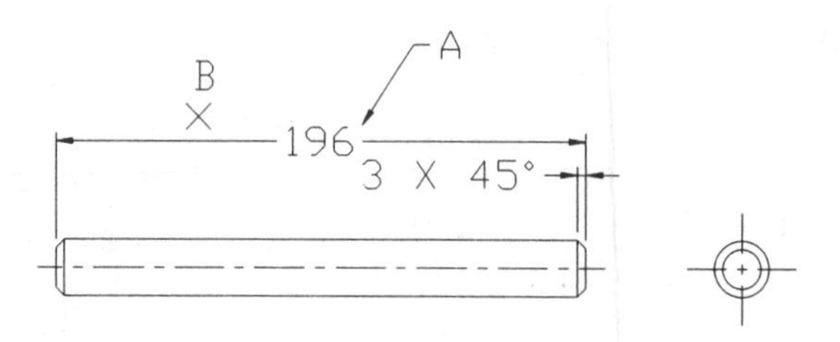

Figure 5.69 Text placed at home position

Repeat the DIMEDIT command to set a horizontal dimension to an oblique angle. See Figure 5.70.

Command: **[Enter]**
DIMEDIT
Dimension Edit (Home/New/Rotate/Oblique): **OBLIQUE**
Select objects: **[Select A (Figure 5.69).]**

Select objects: **[Enter]**
Enter obliquing angle (press ENTER for none): **45**

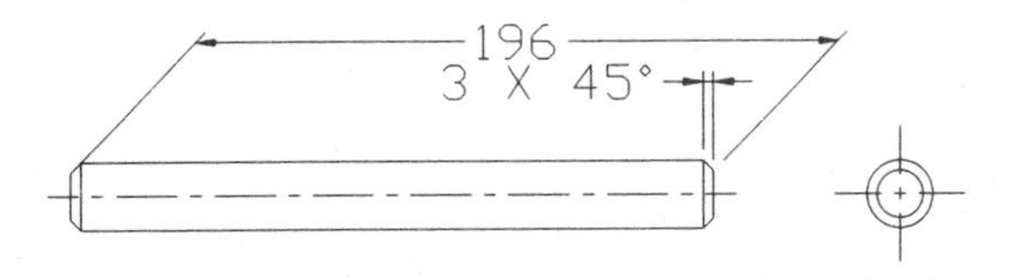

Figure 5.70 Dimension in oblique position

Undo the last command. See Figure 5.69 again.

Command: **U**

Edit Dimension Text Position

Apply the DIMTEDIT command to edit the position of the dimension text. You can select a new location for the dimension text. See Figure 5.71.

[Dimension] **[Dimension Text Edit]**

Command: **DIMTEDIT**
Select dimension: **[Select A (Figure 6.69).]**
Enter text location (Left/Right/Home/Angle): **[Select B (Figure 6.69).]**

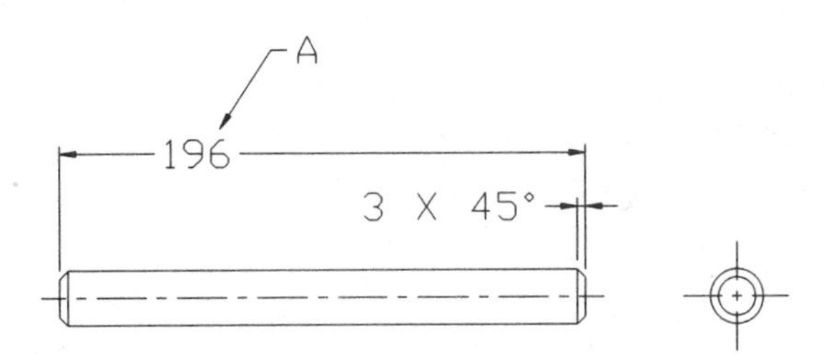

Figure 5.71 Text position of a dimension changed

Override Existing Variable Settings

The way a dimension appears depends on the current dimension style. However, you can change the style of a single dimension without modifying the dimension style by using the DIMOVERRIDE command. Select the Override item of the Dimension pull-down menu to change three dimension variables — DIMTP, DIMTM, and DIMTOL. See Figure 5.72.

<Dimension> **<Override>**

Command: **DIMOVERRIDE**
Dimension variable to override (or Clear to remove overrides): **DIMTP**
Current value <0.0000> New value: **0.2**

Dimension variable to override: **DIMTM**
Current value <0.0000> New value: **0.5**
Dimension variable to override: **DIMTOL**
Current value <Off> New value: **ON**
Dimension variable to override: **[Enter]**
Select objects: **[Select A (Figure 5.71).]**
Select objects: **[Enter]**

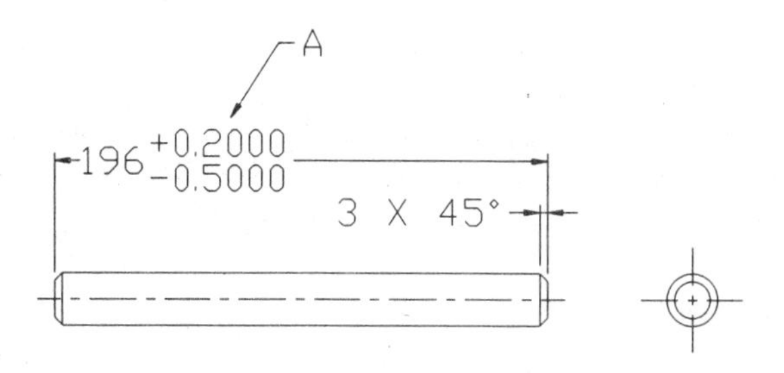

Figure 5.72 Current dimension style overridden

The style of an existing dimension is changed. The change applies only to the selected dimensions. Any subsequently created dimensions will not be affected and will continue to use the current dimension style.

Change Settings and Apply to Existing Dimensions

Change the color of the dimension text by manipulating the DIMCLRT variable.

Command: **DIMCLRT**
New value for DIMCLRT: **1**

The variable controlling the color of the dimension text has changed. It will apply to the dimensions created thereafter. To apply the new variable settings to an existing dimension, run the DIMSTYLE command with the Apply option.

<Dimension> **<Update>**

Command: **DIMSTYLE**
dimension style: STANDARD
dimension style overrides:
DIMCLRT 1 (red)
Dimension Style Edit (Save/Restore/STatus/Variables/Apply/?): **APPLY**
Select objects: **[Select A (Figure 5.72).]**
Select objects: **[Enter]**

Refer to Figure 5.71 and Figure 5.72. The dimension style in Figure 5.72 is overridden. When you apply the current style to the dimension, the latest dimension style is applied to the overridden dimension. Other than a change in dimension text color, the dimension style is the same as in Figure 5.71.

Custom Arrowheads

AutoCAD provides a number of arrowheads for you to choose from. You may choose one of these arrowhead types or create a custom arrowhead type.

Issue the DDIM command. Select the [Geometry...] button and then the Arrowhead 1st item to see a list of arrowheads. See Figure 5.73.

<Dimension> **<Style...>**

Command: **DDIM**

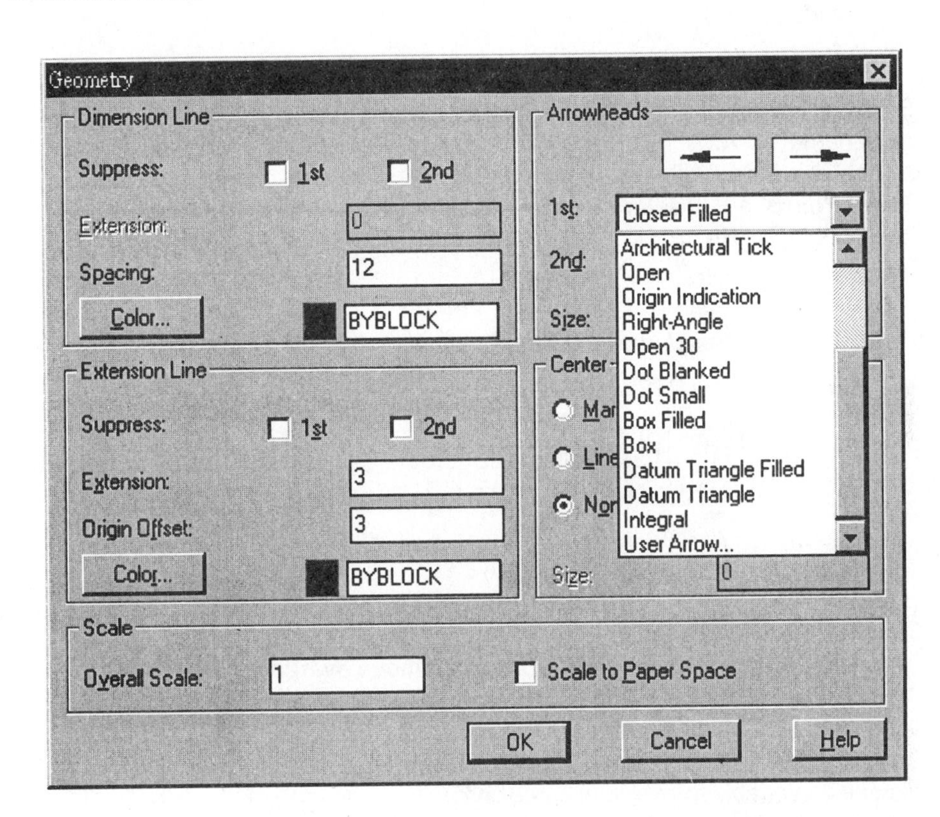

Figure 5.73 Types of arrowheads

To create a custom arrowhead, you have to design your own arrowhead and use the BLOCK command to create one or two blocks, depending on whether you want to use separate blocks or the same block for the two arrowheads.

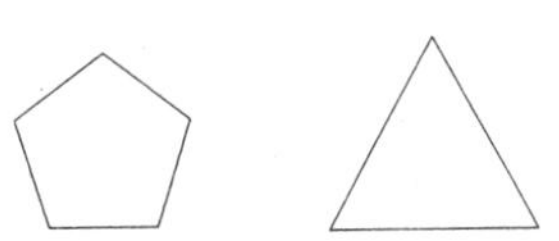

Figure 5.74 Two custom arrowheads

Referring to Figure 5.74, draw the entities. Then use the BLOCK command to create two blocks called ARROW1 and ARROW2, respectively.

<Draw> <Block> <Make...>

Command: **BMAKE**

[Block Definition
Block name: **A1**
Select Objects< [Select the pentagon.]
Select Point< [Select a point near the center of the pentagon.]
Retain Objects **No**
OK]

Command: **[Enter]**
BMAKE

[Block Definition
Block name: **A2**
Select Objects< [Select the triangle.]
Select Point< [Select a point near the center of the triangle.]
Retain Objects **No**
OK]

Now you have two user-defined blocks. Select the Override item of the Dimension pull-down menu to override four dimension variables — DIMSAH (use separate arrowhead blocks for the first and second arrowheads), DIMBLK1 (use the block A1 as the first arrowhead), DIMBLK2 (use the block A2 as the second arrowhead), and DIMASZ (set the scale of the arrowhead blocks). See Figure 5.75.

<Dimension> <Override>

Command: **DIMOVERRIDE**
Dimension variable to override (or Clear to remove overrides): **DIMSAH**
Current value <Off> New value: **ON**
Dimension variable to override: **DIMBLK1**
Current value <> New value: **A1**
Dimension variable to override: **DIMBLK2**
Current value <> New value: **A2**
Dimension variable to override: **DIMASZ**
Current value <6.0000> New value: **1**
Dimension variable to override: **[Enter]**
Select objects: **[Select A (Figure 5.71).]**
Select objects: **[Enter]**

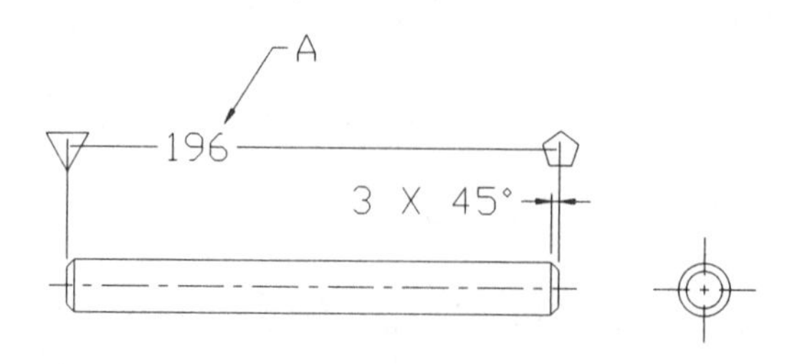

Figure 5.75 Custom arrowheads applied

Restore Dimension Style

The current settings of the DIMSAH, DIMBLK1, DIMBLK2, and DIMASZ variables are changed. If you want to restore to a saved dimension style, run the DIMSTYLE command.

```
Command: DIMSTYLE
dimension style: STANDARD
dimension style overrides:
    DIMCLRT  1 (red)
Dimension Style Edit (Save/Restore/STatus/Variables/Apply/?) <Restore>: R
?/Enter dimension style name or press ENTER to select dimension: STANDARD
```

Apply the current dimension style to a dimension. See Figure 5.71.

```
Command: [Enter]
DIMSTYLE dimension style: STANDARD
Dimension Style Edit (Save/Restore/STatus/Variables/Apply/?): APPLY
Select objects: [Select A (Figure 5.75).]
Select objects: [Enter]
```

Now, your screen should resemble Figure 5.71.

5.9 Tolerance

There are two kinds of tolerances, lateral tolerance and geometric tolerance. Lateral tolerance specifies the allowable dimensional deviation from the basic size. Figure 5.72 shows a lateral tolerance applied to a linear dimension. Geometric tolerance specifies the allowable deviation of form from the basic geometric form. A full description of how to apply tolerances to a drawing is beyond the scope of this book. The following delineation will concentrate on how to apply AutoCAD commands to add tolerances to a drawing.

Lateral Tolerance

You can state the lateral tolerance in two different ways. You can state the basic size together with the allowable deviation from basic size, the tolerance; or you can state the upper and lower limits of size, the limits.

To include tolerances in your drawing, you must either manipulate the variables concerning tolerances directly or make use of the DDIM command.

<Dimension> **<Style...>**

```
Command: DDIM
```

When the Dimension Styles dialog box appears, select the [Annotation...] button. On the Annotation dialog box, select the drop-down box next to the item Method.

See Figure 5.76. There are five methods of tolerancing. If you select the item [None], there will be no tolerance included in the dimension. Symmetrical, Deviation, and Limits concern lateral tolerances. If you pick the Symmetrical item, the tolerance plus value and the tolerance minus value are identical. Both the Deviation item and the Limits item allow you to input different plus and minus tolerance values. The Basic item concerns geometric tolerance. It will be discussed in the next section.

Choose the Deviation method, and enter 0.005 in the box next to the item Upper Value and 0.007 in the box next to the item Lower Value. See Figure 5.77.

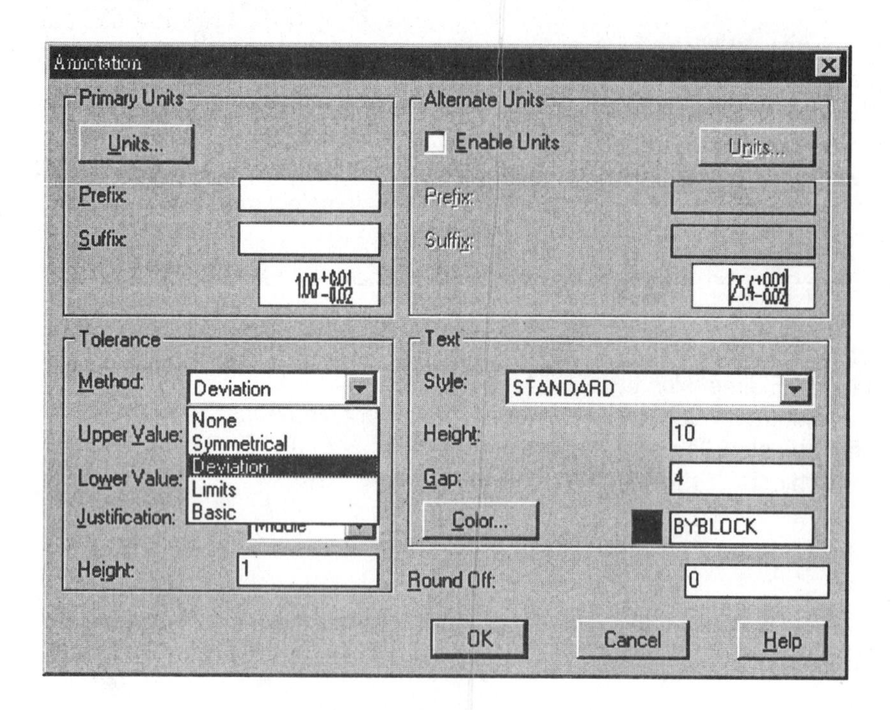

Figure 5.76 Five methods of tolerancing

Annotation
Primary Units
Units...
Prefix:
Suffix:
Alternate Units
Enable Units
Units...
Prefix:
Suffix:
Tolerance
Method: Deviation
Upper Value: 0.005
Lower Value: 0.007
Justification: Middle
Height: 1
Text
Style: STANDARD
Height: 10
Gap: 4
Color... BYBLOCK
Round Off: 0
OK
Cancel
Help

Figure 5.77 Tolerance values set

There are two ways to exit the command. You can select the [Save] button to save the current variable setting to a dimension style and then select the [OK] button to exit, or you can select the [OK] button to exit.

If you save the setting before exiting the command, all existing dimensions that use the saved dimension style will be changed according to the latest dimension setting. If you exit the command without saving to a dimension style, existing dimensions are not affected and new dimensions will use the new setting.

Now select the [OK] button to exit. Refer to your screen. The two dimensions are not affected by the current dimension variable setting.

To apply the tolerance settings to a linear dimension, execute the DIMSTYLE command. See Figure 5.78.

<Dimension> **<Update>**

Command: **DIMSTYLE**
dimension style: STANDARD
dimension style overrides:
    DIMTM 0.0070
    DIMTOL On
    DIMTP 0.0050
Dimension Style Edit (Save/Restore/STatus/Variables/Apply/?) <Restore>: **A**
Select objects: 1 found
Select objects:

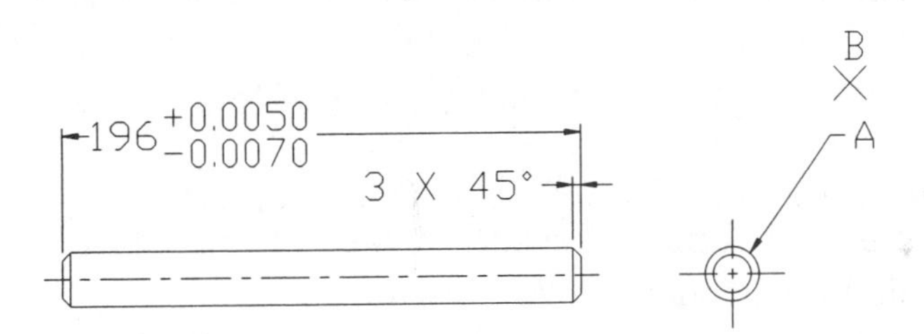

Figure 5.78 Tolerance settings applied

Now a dimension is updated using the latest setting. The other dimension remains unchanged and uses the setting of the saved dimension style.

Use the DIMDIAMETER command to create a diameter dimension. Because you have used deviation type tolerance, this dimension will include tolerance. See Figure 5.79.

<Dimension> **<Diameter>**

Command: **DIMDIAMETER**
Select arc or circle: **[Select A (Figure 5.78).]**
Dimension text = 20
Dimension line location (Mtext/Text/Angle): **[Select B (Figure 5.78).]**

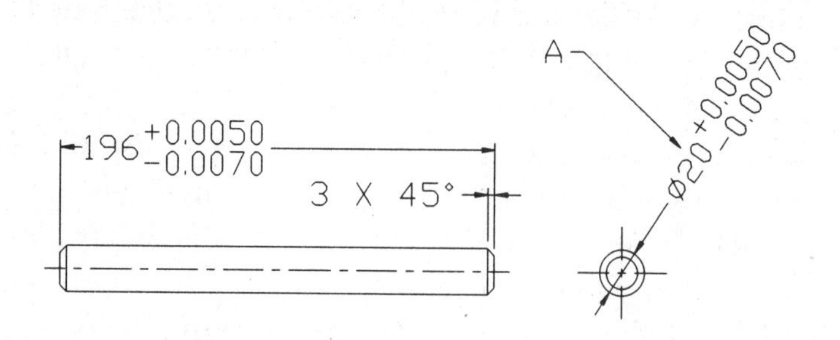

Figure 5.79 Diameter dimension with tolerance created

Now you have three dimensions. One dimension remains unchanged. The second dimension updates in accordance with the current variable setting. The third dimension uses the latest setting.

Geometric Tolerance

Lateral tolerances control size variation. If you specify only lateral tolerances to a drawing, the geometric aspects of the objects are not controlled. Refer to Figure 5.79. You have controlled the length of the shaft to be between 196.005 and 195.993 and the diameter of the shaft to be between 20.005 and 19.993. Apart from size control, there is no control on the roundness and straightness of the shaft. To impose a comprehensive control over the form and shape, you can use geometric tolerances.

Geometric Reference Frame and Basic Dimension

Geometric tolerances control the form and shape. They specify the allowable deviation from theoretically perfect form and shape. In order to specify the theoretically perfect form, you have to define a geometric reference frame.

A geometric reference frame is a set of dimensions to define the theoretically perfect shape of a component. These dimensions are called basic dimensions. They are distinguished from other dimensions by their enclosure within a rectangular box. Basic dimensions are not toleranced because they specify the theoretically perfect form.

Select the Style... item of the Dimension item to use the DDIM command to set the tolerance method to Basic. See Figure 5.80.

<Dimension> **<Style...>**

Command: **DDIM**

When the Dimension Styles dialog box appears, select the [Annotation] button to set the annotation style. In the Annotation dialog box, choose the Basic tolerance method. See Figure 5.80.

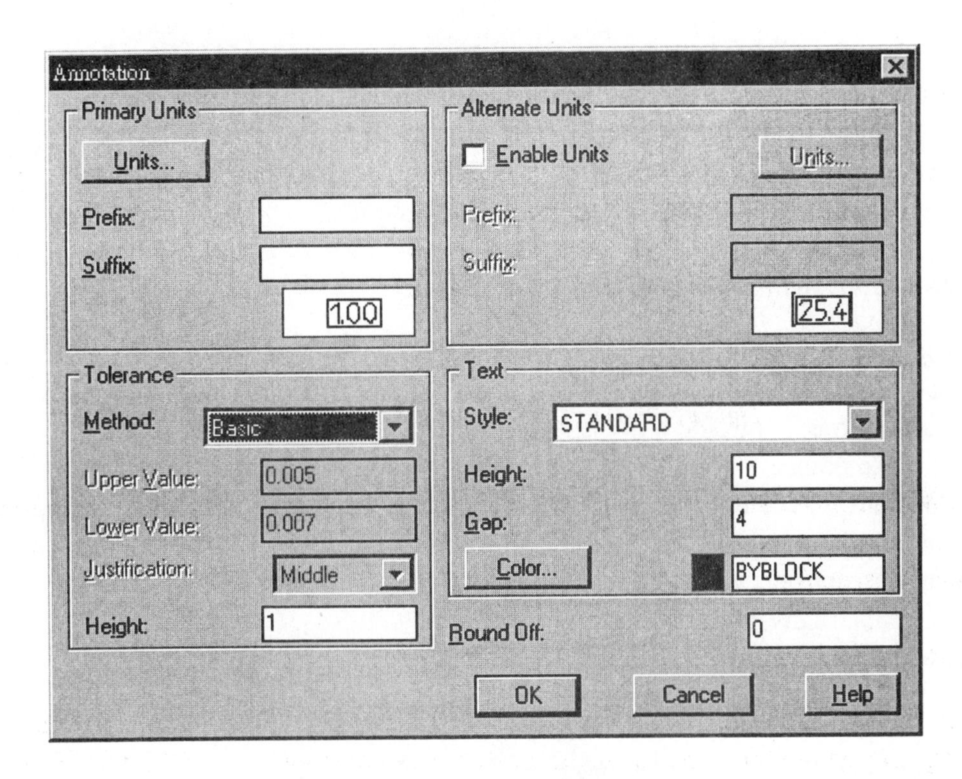

Figure 5.80 Basic method selected

Select the [OK] button to exit the command. Then use the DIMSTYLE command to apply the current dimension variable settings to the diameter dimension. See Figure 5.81.

<Dimension> **<Update>**

Command: **DIMSTYLE**
dimension style: STANDARD
dimension style overrides:
DIMGAP -4.0000
DIMTM 0.0070
DIMTP 0.0050
Dimension Style Edit (Save/Restore/STatus/Variables/Apply/?) <Restore>: **APPLY**
Select objects: [**Select A (Figure 5.79).**]
Select objects: [**Enter**]

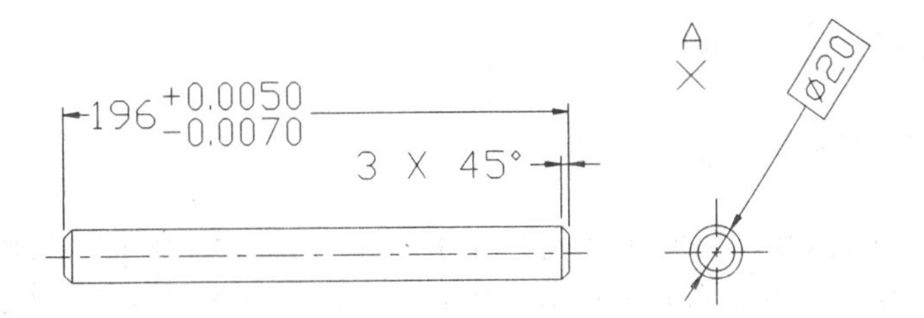

Figure 5.81 Diameter dimension updated to basic dimension

Now you have changed the diameter dimension to a basic dimension. This basic dimension specifies a theoretically perfect cylinder 20 units in diameter. For this component, the theoretically perfect form is defined. You need to add appropriate geometric tolerance to control the roundness and straightness. Collectively, you can specify a cylindricity tolerance. See the next paragraph.

Geometric Tolerance Symbols

As we have said, a basic dimension does not carry any tolerance because its function is to define a theoretically perfect form in the Geometric Reference Frame. In conjunction with basic dimensions, you have to specify allowable geometric tolerance by using a Geometric Tolerance Frame.

A Geometric Tolerance Frame consists of two or more rectangular compartments. The first compartment contains a symbol that depicts the type of tolerance. The second compartment consists of a value delineating the size of the tolerance zone. Further compartments, if any, state the datum reference.

AutoCAD allows you to specify 14 different types of geometric tolerances. Run the TOLERANCE command. See Figure 5.82.

<Dimension> **<Tolerance>**

Command: **TOLERANCE**

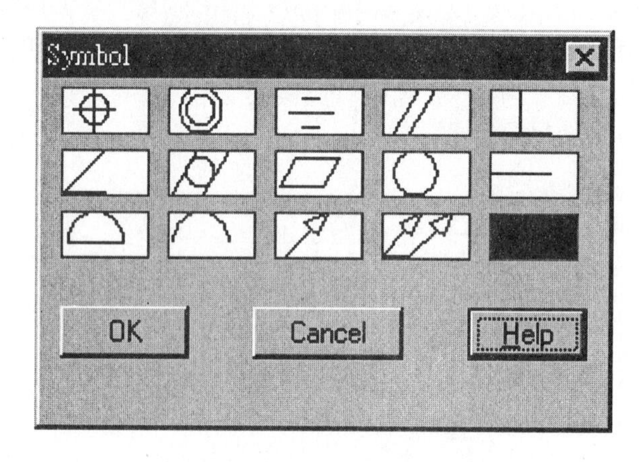

Figure 5.82 Fourteen types of geometric tolerances

Refer to Figure 5.82. The tolerance symbols are (from top left):

Position	specifies the allowable deviation from perfect position.
Concentricity	specifies the allowable deviation from perfectly concentric.
Symmetry	specifies the allowable deviation from perfectly symmetrical.
Parallelism	specifies the allowable deviation from perfectly parallel.
Perpendicularity	specifies the allowable deviation from perfectly perpendicular.

Angularity	specifies the allowable deviation from the stated perfect angle.
Cylindricity	specifies the allowable deviation from perfectly cylindrical.
Flatness	specifies the allowable deviation from perfectly flat.
Roundness	specifies the allowable deviation from perfectly round.
Straightness	specifies the allowable deviation from perfectly straight.
Surface profile	specifies the allowable deviation from the stated perfect surface.
Line profile	specifies the allowable deviation from the stated perfect line.
Circular runout	specifies the allowable runout.
Total runout	specifies the allowable total runout.

Select the symbol that your requirements call for. In this case, you need to specify a cylindricity tolerance. Select the second icon of the second row from the left. Select the [OK] button, and another dialog box appears. See Figure 5.83.

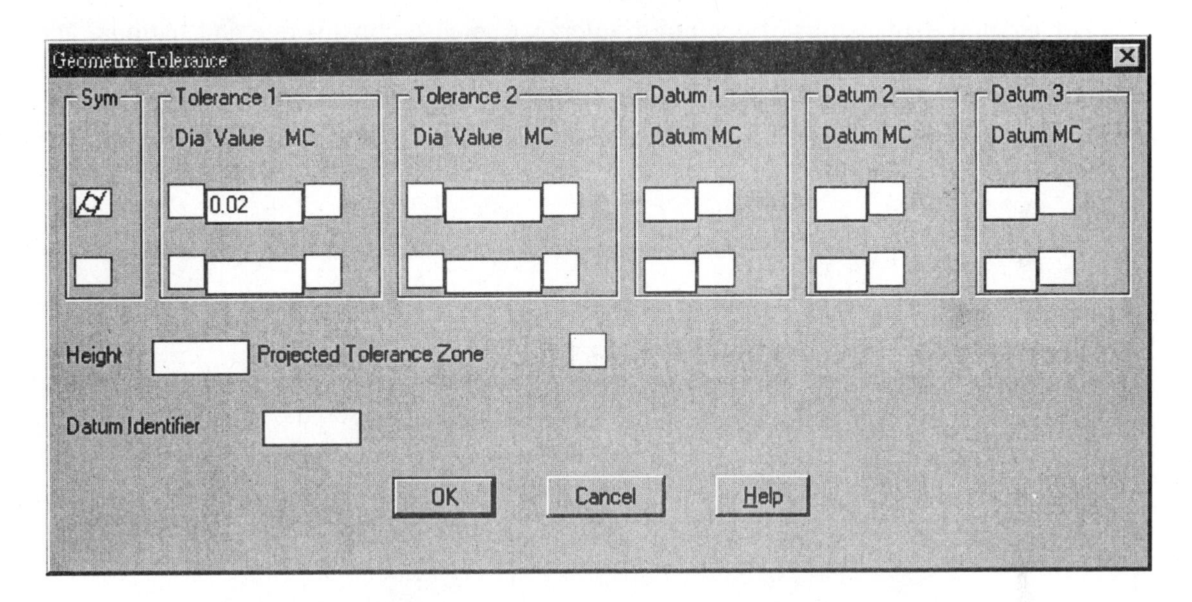

Figure 5.83 Dialog box for specifying tolerance values and datum

From the second dialog box, you can specify the size of the tolerance zone in the box under the item Value. Enter a value of 0.02 here. This specifies a tolerance zone of 0.02 width on the theoretically perfect cylindrical form.

Refer to the dialog box. You can select the box below the item Dia if the tolerance zone is cylindrical. In this case, leave this box unselected.

Then you can specify the type of condition for the tolerance zone in the box under the item MC. There are three conditions: Maximum Material Condition, Least Material Condition, and Regardless of Feature Size. Leave this box unselected.

After you select the [OK] button, the following prompt appears. Then select point A (Figure 5.81) on the screen. See Figure 5.84.

Enter tolerance location: **[Select A (Figure 5.81).]**

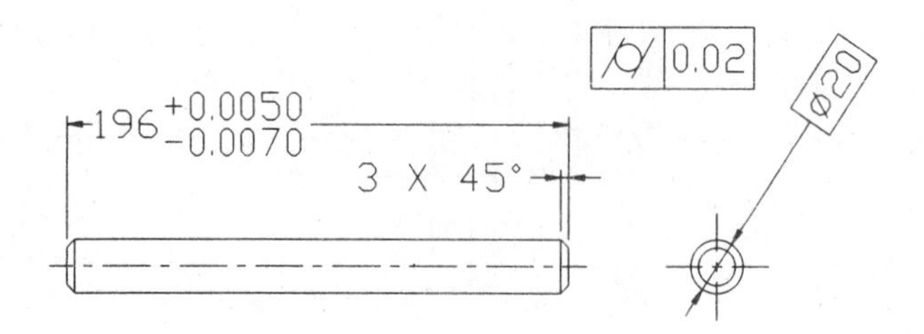

Figure 5.84 Cylindricity position tolerance added

The drawing is complete. Save your drawing.

<File> **<Save...>**

For a detail explanation of geometric tolerance, please consult relevant national and international standards. Figure 5.85 shows another example of using basic dimensions and geometric tolerance.

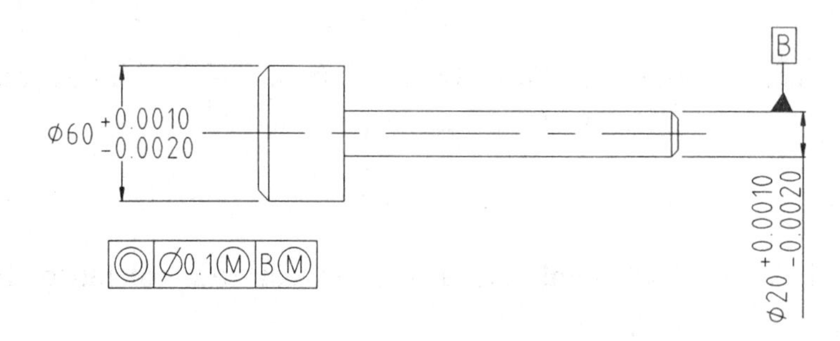

Figure 5.85 An example of geometric tolerance

5.10 Key Points and Exercises

Annotations are textual information that you add to a drawing. Before adding annotations, you have to set a text style. Text can be placed in a single line or in multiline. You can edit and check the spelling of annotations. To speed up regeneration, you can control text fill and text quality and can use quick text mode.

Objects in a drawing can be inquired. You can retrieve information about entities in a drawing and find out the areas of selected regions, the distances of objects, and the location of points.

Dimensional information is an integral part of the drawing database. You can generate dimensions by using dimensioning commands. By default, each dimension is an anonymous block and is associative to the selected objects that are dimensioned. To control the appearance of a dimension, you must set a dimension style or manipulate the dimension variables.

In addition to annotations and dimensions, you can add tolerances to a drawing. There are two kinds of tolerances, lateral tolerances and geometric tolerances. Lateral tolerances control the allowable dimensional deviation from basic size. Geometric tolerances control the allowable devation of form from the specified perfect shape.

In this chapter, you learned how to inquire drawing objects and practiced including annotations and adding dimensions to a drawing. In so doing, you learned various methods and techniques. Completing the following exercises will enhance your learning. Some of these exercises are continuations of the exercises in previous chapters.

Exercise 5.1

What text height value should you set to a text style to obtain a variable text height? How can you increase the regeneration speed of a drawing that contains a lot of annotations?

Exercise 5.2

What commands will you use to discover the information about the entities in a drawing? How can you find out the areas of selected regions, the distance between selected points, and the location of selected points?

Exercise 5.3

State the dimensioning principles. Describe the components of a dimension. Explain how you will set a dimension style by using the DDIM command.

Exercise 5.4

Explain what is meant by lateral tolerance and geometric tolerance. How are these tolerances added to a drawing?

Exercise 5.5

Open the drawing EX3_1.DWG again. Create a new layer called DIM, and add dimensions in accordance with Figure 3.162. Save the drawings as EX5_1.DWG.

Exercise 5.6

Open the drawing EX3_3.DWG. Add a layer, and add dimensions in accordance with Figure 3.165. Save the drawing as EX5_2.DWG.

Exercise 5.7

Open the drawing EX3_4.DWG. Add dimensions in accordance with Figure 3.166. Save the drawing as EX5_3.DWG.

Exercise 5.8

Open the drawing EX3_5.DWG. Add dimensions in accordance with Figure 3.167. Save the drawing as EX5_4.DWG.

Exercise 5.9

Open the drawing EX3_6.DWG. Add dimensions in accordance with Figure 3.168. Save the drawing as EX5_5.DWG.

Exercise 5.10

Open the drawing EX3_7.DWG. Add dimensions in accordance with Figure 3.169. Save the drawing as EX5_6.DWG.

Chapter 6

3D Solid Modeling

6.1 Solid Modeling Applications
6.2 AutoCAD Native Solids
6.3 Unite and Subtract of Primitives
6.4 Extrude, Revolve, and Intersect
6.5 Slicing
6.6 Cutting by Intersection
6.7 3D Coordinate Systems
6.8 Key Points and Exercises

Aims and Objectives

The aim of this chapter is to introduce to you the principles of various solid model applications, to explain how AutoCAD solids are created, and to give you practice in constructing 3D solid models. After studying this chapter, you should be able to:

- describe various solid modeling principles,
- explain how AutoCAD native solids are created,
- construct 3D primitive solids, extruded solids, and revolved solids,
- use Boolean operations to construct complex 3D solid models,
- fillet and chamfer 3D solids,
- slice and cut 3D solid models, and
- use 3D coordinate systems.

Overview

There are three kinds of 3D models in a computer — the wireframe model, surface model, and solid model. A 3D wireframe model is the most primitive type of 3D object. It is a set of unassociated line and arc segments that are put together in a 3D space. The line and arc segments serve only to give the pattern of a 3D object. There is no relationship between them. Thus, the model does not have any surface information or volume information. It has only data that describe the edge of the 3D object. Because of the limited information provided by the model, the use of wireframe models is very confined.

The second type of 3D model, the surface model, is a set of surfaces that are put together in a 3D space to give the figure of a 3D object. When compared to a 3D wireframe model, a surface model has, in addition to edge data, information on the contour and silhouette of the surfaces. You can use the surface models in a computerized manufacturing system or to generate photo-realistic rendering or animation. To build 3D surface models, you can use the Mechanical Desktop R2 application. Using AutoCAD R14, you can create 3D polygonal meshed objects.

In regard to information, a 3D solid model is superior to the other two models because a solid model in a computer is integrated mathematical data that contains information not only on the surfaces and edges but also on the volume of the object that the model describes. In addition to visualization and manufacturing, you can use the data of a solid model for design calculation. Basically, you can use AutoCAD R14 to create native solids and use Mechanical Desktop R2 to create parametric solid models. In brief, Mechanical Desktop R2 is an application that runs on top of AutoCAD R14 to provide professional tools for product and engineering design. This application consists of four major components — NURBS surface modeling, parametric solid modeling, virtual assembly, and associative engineering drafting. If you want to learn more about Mechanical Desktop, you can refer to *"Mastering Mechanical Desktop R3"*.

6.1 Solid Modeling Applications

There are many ways to represent a solid in a computer. The following paragraphs briefly describe some of these methods.

Pure Primitive Instancing Method

This method predefines a limited range of solid objects (see Figure 6.1). To make a solid, the user supplies the values of the parameters to the system.

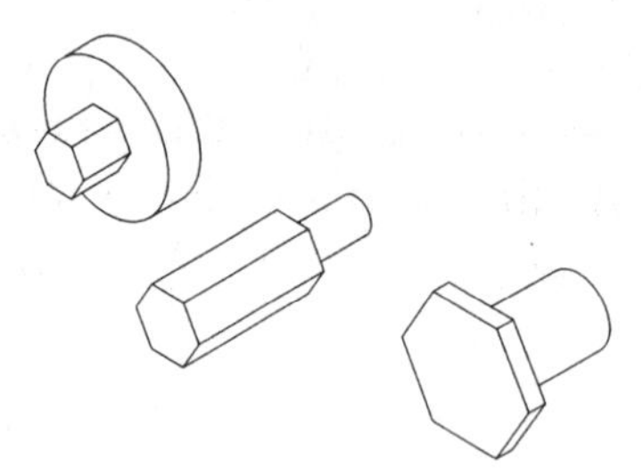

Figure 6.1 Range of solids that share common features — hexagonal prism and cylinder

Generalized Sweep Method

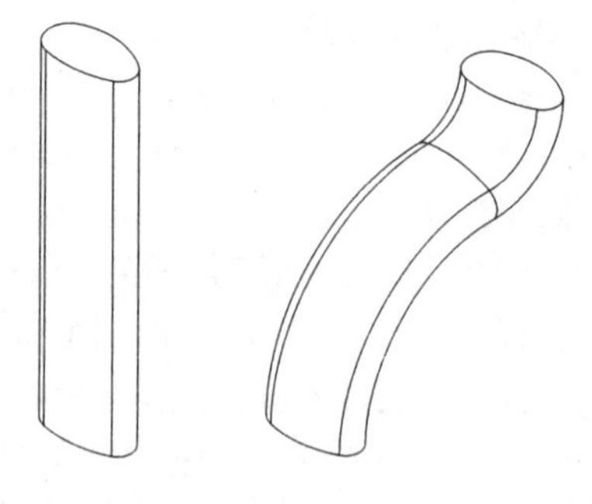

Figure 6.2 Lamina swept along a 3D curve to form a solid

This method creates solid objects by sweeping a 2D or 3D lamina along a 3D curve (see Figure 6.2).

Spatial Occupancy Enumeration Method

This method divides the entire 3D space into a number of cubical cells (see Figure 6.3). Solids are represented by listing the cells that the solids occupy in 3D space. Accuracy of the solid that is being represented is a function of the size of the cubical cells. However, a smaller cubical size increases the file size tremendously.

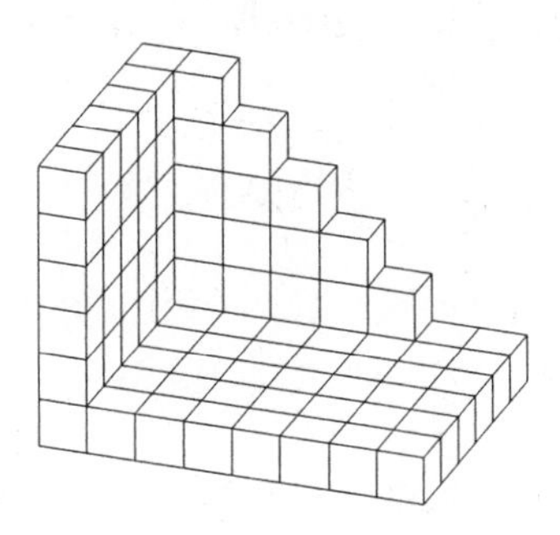

Figure 6.3 Solid object defined by occupancy of specific cubical cells in 3D space

Cellular Decomposition Method

This method is similar to the spatial occupancy enumeration method in that the solids are represented by listing the cells in 3D space. However, the cells are not necessarily identical or cubical in shape (see Figure 6.4). As a result, this method requires less memory.

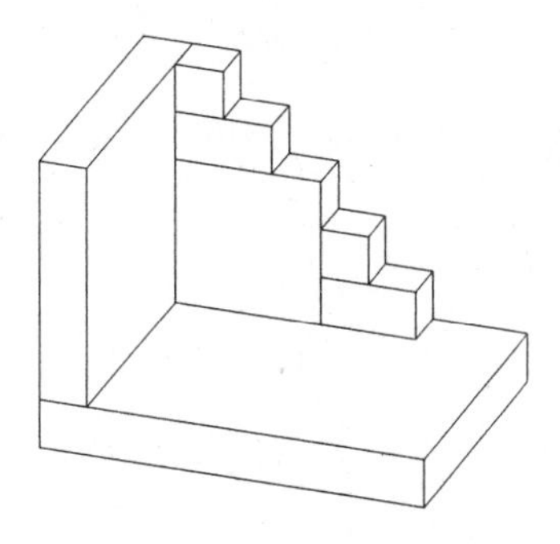

Figure 6.4 Solid defined by decomposing into cells

Constructive Solid Geometry (CSG) Method

This method provides a range of primitive solids in a way that is similar to the pure primitive instancing method mentioned above. In addition, it provides an extra facility for the user to perform Boolean operations on the primitive solids to produce a complex solid. Basically, AutoCAD uses CSG method in defining solids. Explanations are given below.

6.2 AutoCAD Native Solids

Constructive solid geometry is a building-block approach to create 3D solid models. The system provides tools to create solids of simple geometric shape as building blocks and provides Boolean operations to combine the solids. Any complex solid model is constructed by combining simpler solids together. The simplest solids are called primitives. The six primitive types available are box, sphere, cylinder, cone, wedge, and torus. See Figure 6.5.

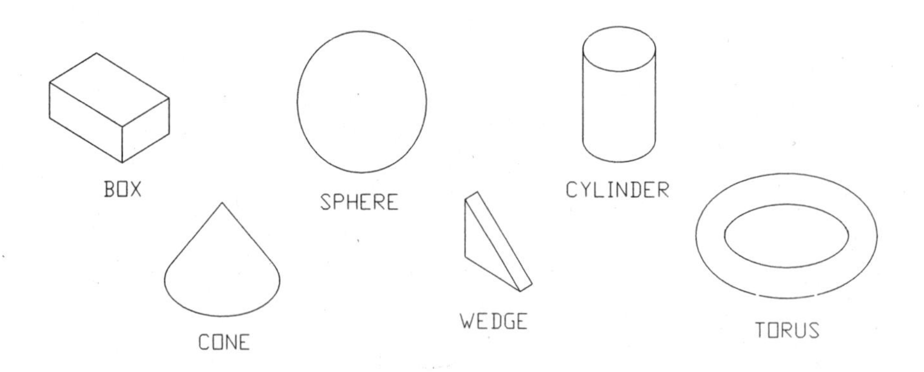

Figure 6.5 Primitive solids

The repertoire of complex solids that can be composed from these six types of primitive solids is very limited. To extend the scope of model creation, AutoCAD allows you to make solid objects by extruding a 2D object along a line or a path and revolving a 2D object about an axis. The 2D object can be either a region or a closed polyline. Figure 6.6 shows a revolved solid, a solid extruded along a line, and a solid extruded along a path.

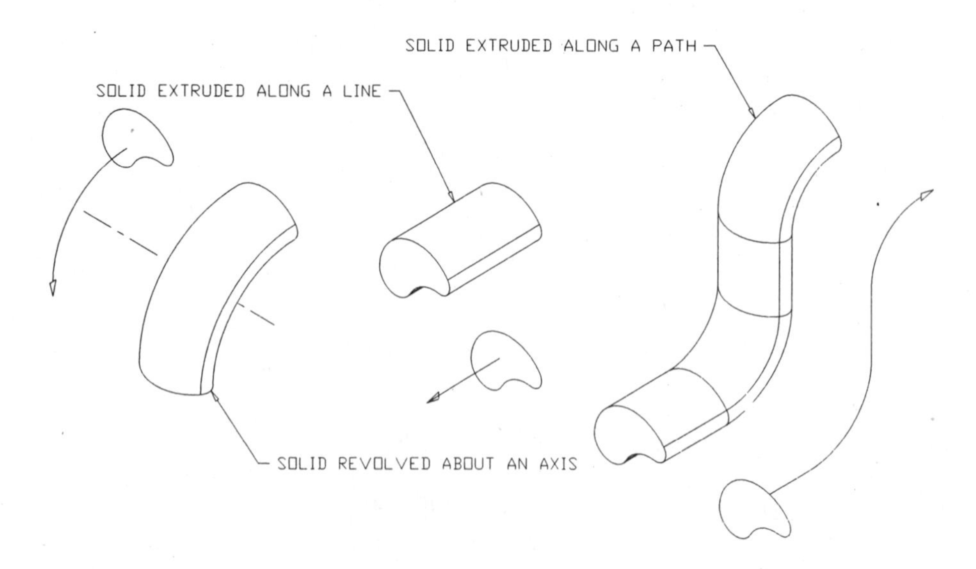

Figure 6.6 A revolved solid and extruded solids

To compose two solid objects together to form a complex solid, you can use one of the three types of Boolean operations — union, subtraction, and intersection. A union of two solids creates a solid that encloses all the volume enclosed by the first and the second solid. A subtraction of two solids creates a solid that encloses the volume of the first solid but not that of the second solid. An intersection of two solids creates a solid that encloses the volume of the first solid that is also contained in the second solid. Figure 6.7 illustrates the effect of these Boolean operations on a solid box and a solid cone.

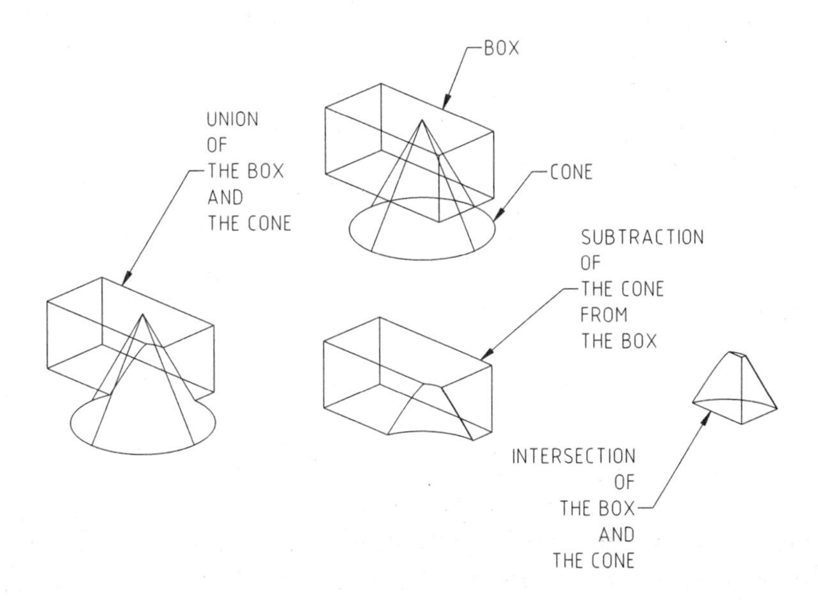

Figure 6.7 Boolean operations

As a further refinement to a solid, you can chamfer or fillet the edge. Figure 6.8 shows a solid box, a chamfered box, and a filleted box.

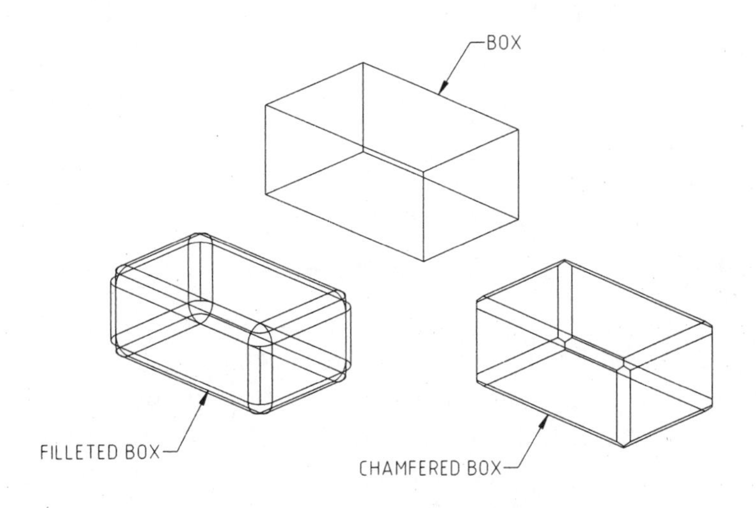

Figure 6.8 Solid box, filleted box, and chamfered box

Isolines and Silhouette Display

The appearance of 3D solid objects in a computer is affected by the setting of two system variables, ISOLINES and DISPSILH. Isolines are series of lines on the curved surfaces. The ISOLINES variable controls the density of the isolines. If you set it to zero, there will be no isolines. If you set it to a higher value, there will be more isolines. The DISPSILH variable controls the display of silhouette curves of solid objects. It has two settings. If it is 1, the silhouette edges appear. If it is 0, the silhouette edges do not show. Silhouettes provide a better picture of the 3D object. However, regeneration of the display will take a longer time. Figure 6.9 shows a box, cone, and cylinder displayed with ISOLINES = 5 and DISPSILH = 0.

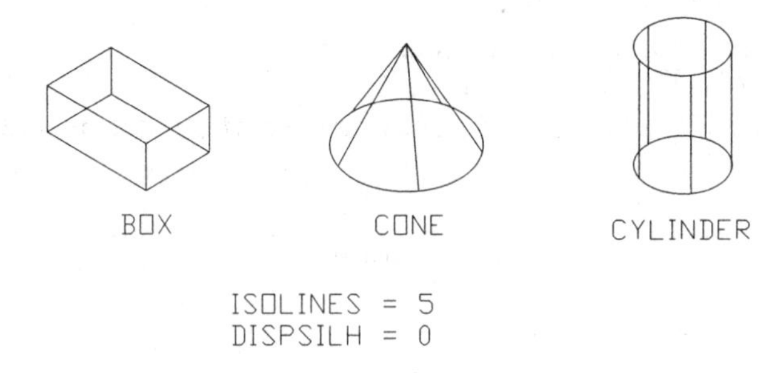

Figure 6.9 ISOLINES variable

Figure 6.10 shows the same box, cone, and cylinder displayed with ISOLINES = 0 and DISPSILH = 1.

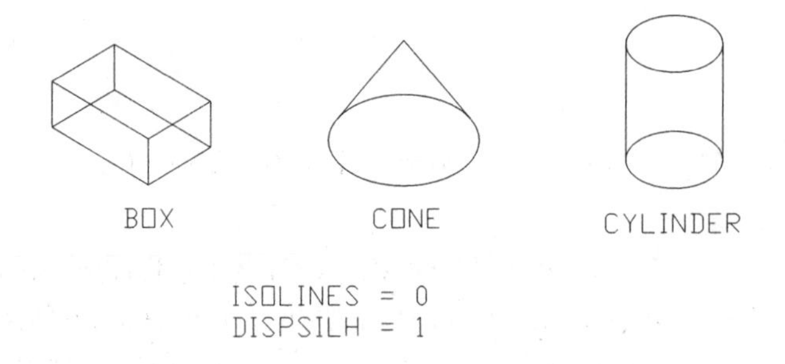

Figure 6.10 DISPSILH variable

Refer to Figure 6.9 and Figure 6.10. The box, which consists of only planar surfaces, is not affected by the setting of the variables. The cone and cylinder, which consist of curved surfaces, are affected. In Figure 6.9, the cone and cylinder have isolines but not silhouette edges. In Figure 6.10, the cone and cylinder have silhouette edges but not isolines.

Compatibility

An AutoCAD native solid is 100% compatible with objects created by Mechanical Desktop. Mechanical Desktop creates 3D NURBS surfaces and 3D parametric solids. You can use a 3D NURBS surface to cut a native solid, as well as to convert a native solid into a set of 3D NURBS surfaces. The commands that are available in Mechanical Desktop are AMSOLCUT and AM2SF.

If you have installed Mechanical Desktop properly, you can use the Mechanical Desktop menu and select the Edit Solid item from the Surface pull-down menu to run the AMSOLCUT command. See Figure 6.11.

<Surface> **<Edit Solid>**

Command: **AMSOLCUT**
Select solid to cut: **[Select a native solid.]**
Select surface: **[Select a NURBS surface.]**
Portion to remove: Flip/<Accept>: **[Enter, if the direction is correct.]**

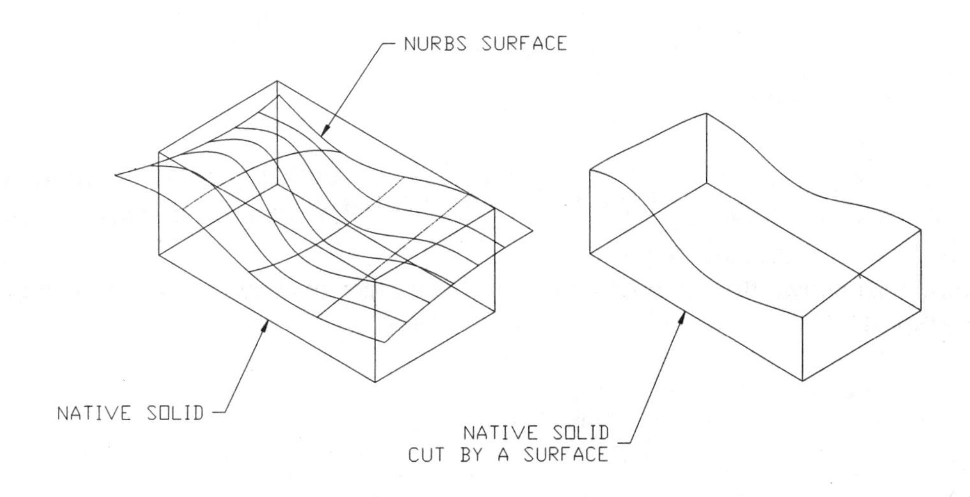

Figure 6.11 Native solid cut by a NURBS surface

To change a native solid to a set of NURBS surfaces for further surface creation, such as making variable fillets, free-form surfaces, and other derived surfaces, you can use the AM2SF command by selecting the Form ACAD item of the Create Surface cascading menu of the Surface pull-down menu. See Figure 6.12.

<Surface> **<Create Surface>** **<From ACAD>**

Command: **AM2SF**
Face/<objects>: **[Enter]**
Select objects: **[Select a native solid.]**
Select objects: **[Enter]**

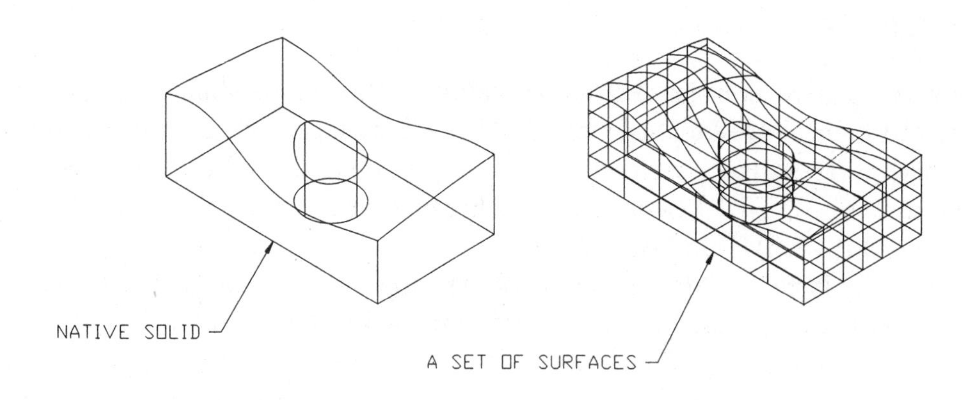

Figure 6.12 Native solid converted to a set of NURBS surfaces

You can convert a native solid to a base solid of Mechanical Desktop and convert a parametric solid to a static native solid. To convert a Mechanical Desktop solid part to a native solid, you can use the EXPLODE command.

<Modify> **<Explode>**

Command: **EXPLODE**

To use a native solid as a base solid feature for further parametric solid feature creation in Mechanical Desktop, you can use the AMNEW command. This command is part of the Mechanical Desktop package.

<Parts> **<Part>** **<Convert Solid>**

Command: **AMNEW**
Create a new Part/Scene/subAssembly/<Instance>: **P**
Select (or) <PART1>: [**Select the native solid to be converted.**]
Name <PART1>: [**Enter**]

Summary of AutoCAD Native Solids

In a computer, 3D solid models provide unique data that contain information on the edges, surfaces, and volume that the model describes. AutoCAD native solids use constructive solid geometry techniques. You can compose a complex solid from the primitive solids and from the solids of extrusion and revolving. The method of composition can be to unite two solids, to subtract one solid from another, or to form an intersection of two solids. You can also fillet or chamfer the edges of a solid.

AutoCAD solids are compatible with NURBS surfaces and parametric solids created by Mechanical Desktop. A NURBS surface can be used to cut a native solid. A native solid can be converted into a set of NURBS surfaces. In addition, a native solid can be converted into a new part in AutoCAD Designer, and a parametric solid can be converted into a static native solid model.

6.3 Unite and Subtract of Primitives

Using AutoCAD Native solids to construct a 3D solid model in a computer is very much like making a model with building blocks physically. You can build up a complex object by putting together simpler objects. However, if you compare the two processes, you may find that the former is far more flexible because you can perform subtraction and intersection in addition to union.

To build a complex solid using constructive solid geometry, you need to go through three basic steps. First, you should analyze the complex solid to determine what primitive solids are required and how these primitives are combined. Next, you have to build the primitive solids. Finally, you need to perform Boolean operations on the primitive solids to combine them into one complex solid. For example, to make a rectangular block with a hole, you need to make a box and a cylinder and then subtract the cylinder from the box.

Model of an Angle Block

Figure 6.13 shows the rendered solid model of an angle block that you are going to construct. You will carry out union and subtraction on a number of primitive solids to obtain this model. To learn how to obtain a rendered image, see Chapter 10.

Figure 6.13 Angle block

Analysis

Given a solid model to construct, the first thing to do is to analyze the model. Think critically about the composition of the model and break it down into simple objects that can be represented by the primitive solids provided by AutoCAD: boxes, spheres, cylinders, cones, wedges, and tori. Together with thinking about what primitives are to be created, you have to think about what Boolean operations you will apply to combine the primitives: union, subtraction, and intersection.

Refer to Figure 6.13. This model consists of rectangular boxes, a wedge, and circular holes. First, you make two boxes and unite them together. Then you make a wedge and unite it to the two boxes. Finally, you make five cylinders and subtract them from the union of the boxes and wedge. Figure 6.14 shows the primitives tweaked apart and assembled together to form the model.

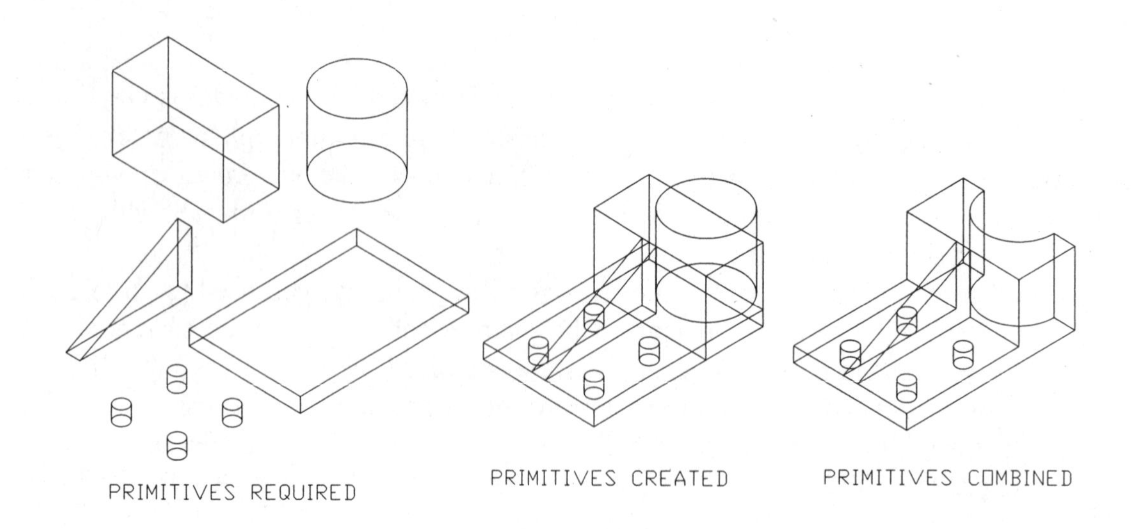

Figure 6.14 Primitive solids

Drawing Preparation

Use the New command to start a new drawing. Start from scratch and use metric units.

<File> **<New...>**

Select the Layer... item from the Format pull-down menu to make a new layer called Solid. Change its color to yellow and set it as the current layer.

<Format> **<Layer...>**

Command: **LAYER**

[Layer

Name	Color	Linetype
0	White	Continuous
Solid	Yellow	Continuous

Current Layer: **Solid**

OK]

Model Construction

From the Solids cascading menu of the Draw pull-down menu, select the Box item to use the BOX command to make a box of 80 x 120 x 10. See Figure 6.14.

<Draw> **<Solids>** **<Box>**

Command: **BOX**
Center/<Corner of box>: [**Select a point near the lower left corner of your screen.**]
Cube/Length/<other corner>: **@80,120**
Height: **10**

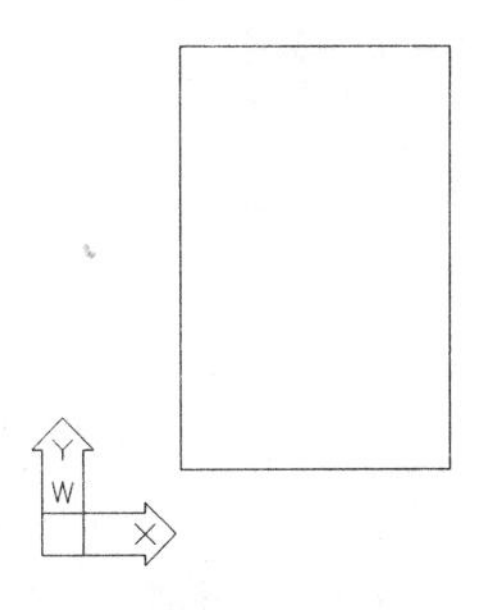

Figure 6.15 Box created but shown as a 2D rectangle

Refer to Figure 6.15. Your screen shows a 2D rectangular object instead of the pictorial view of a box that you may have expected. Do not worry. It is a 3D box and not a 2D rectangle. The reason why you see a 2D rectangle instead of a box is that you are looking from the top of a box, which appears as a rectangle.

To obtain a pictorial view, you can select the SE Isometric item from the 3D Viewpoint cascading menu of the View pull-down menu to use the VPOINT command to set the display to an isometric view looking from the southeast direction. See Figure 6.16.

<View> **<3D Viewpoint>** **<SE Isometric>**

Command: **VPOINT**
Rotate/<View point>: **1,-1,1**

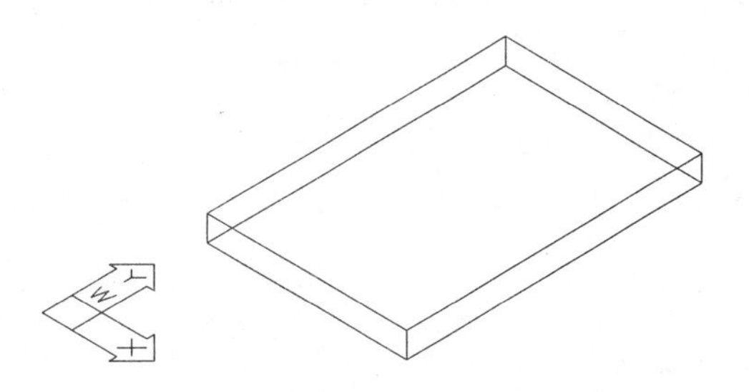

Figure 6.16 SE isometric view

Refer to Figure 6.16. There is a UCS icon at the lower left corner of the screen. This icon depicts the orientation and position of the imaginary construction plane on which you are currently working. If you cannot find this icon, you should use the UCSICON command to turn it on by checking a tick against the On item of the Display and UCS Icon cascading menu of the View pull-down menu.

<View> **<Display>** **<UCS Icon>** **<v On>**

Command: **UCSICON**
ON/OFF/All/Noorigin/ORigin: **ON**

Although a pictorial view gives a better description of the 3D model, there is one drawback — 3D elusion. To elaborate, there are chances that two spatially wide apart objects may be seen as placing together. To overcome this problem, you can divide the display viewport into tiles of viewports and display the object from a different viewing direction in each tiled viewport. Select the Layout... item of the Tiled Viewports cascading menu of the View pull-down menu to use the VPORTS command to produce four tiled viewports. See Figure 6.17.

<View> <Tiled Viewports> <Layout...>

Tiled Viewport Layout

Single
Three: Vertical
Three: Horizontal
Four: Equal
Two: Vertical
Three: Right
Three: Left
Four: Left
Two: Horizontal
Three: Above
Three: Below
Four: Right

Previous Next OK Cancel

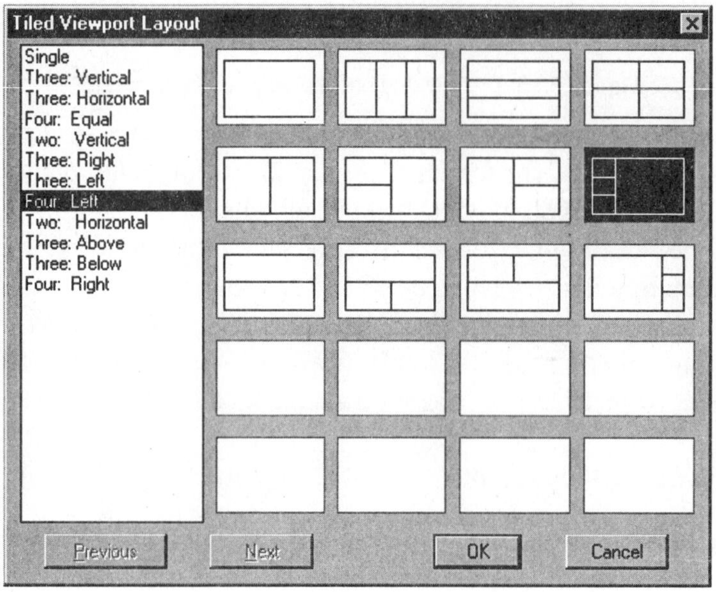

Figure 6.17 Tiled viewports layout

Refer to Figure 6.17. There are twelve preset viewport arrangements. From the dialog box, select the Four: Left icon and then the [OK] button to divide the viewport into four tiles. See Figure 6.18.

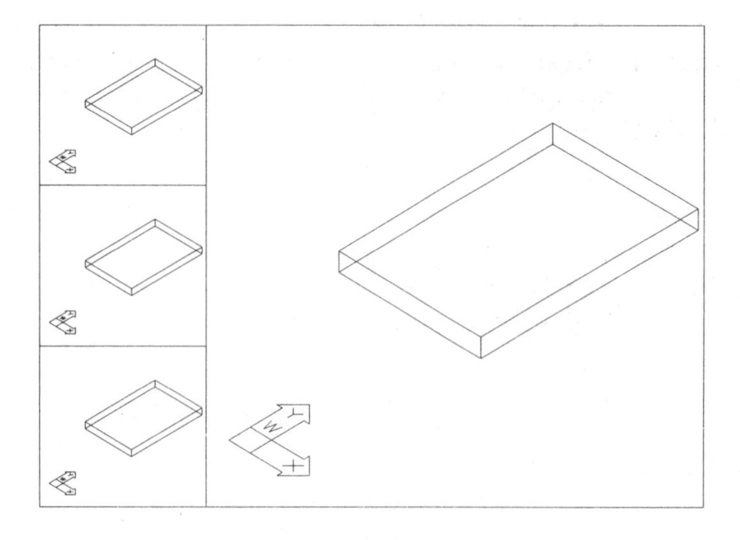

Figure 6.18 Four tiled viewports

Initially, the displays in the four tiled viewports are the same. To view the 3D model from different directions in each viewport, you have to select each tiled viewport one by one and change the direction of viewing by using the VPOINT command. See Figure 6.19.

Select the upper left tiled viewport.

<View> **<3D Viewpoint>** **<Top>**

Command: **VPOINT**
Rotate/<View point>: **0,0,1**

Select the middle left tiled viewport.

<View> **<3D Viewpoint>** **<Front>**

Command: **VPOINT**
Rotate/<View point>: **0,-1,0**

Select the lower left tiled viewport.

<View> **<3D Viewpoint>** **<Right>**

Command: **VPOINT**
Rotate/<View point>: **1,0,0**

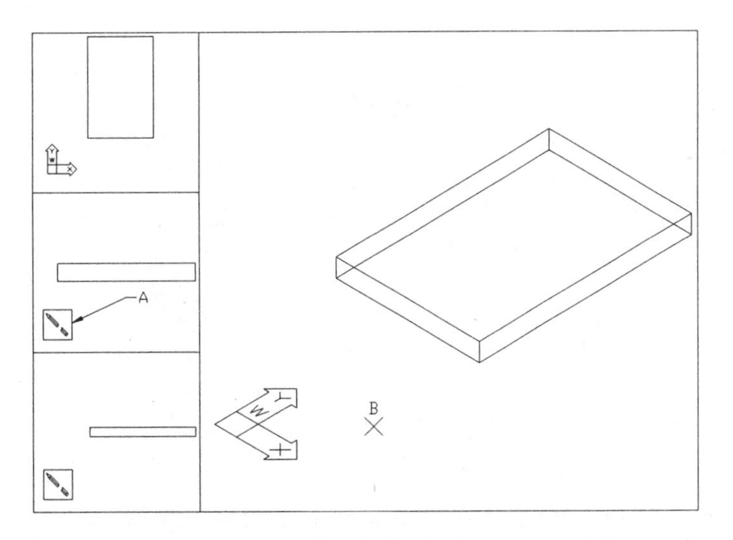

Figure 6.19 Tiled viewports showing top, front, right, and isometric views

Refer to Figure 6.19. You can see in the middle left viewport and lower left viewport that the UCS icons (see A in Figure 6.19) change to the shape of a broken pencil. This is a special icon to remind you that the imaginary working planes in these views are perpendicular to the viewport and that you may not readily see what you draw. For example, you will see a line when you actually draw a circle because the circle will appear as a line if your viewing direction is parallel to the plane on which the circle resides. Therefore, you should avoid working on a viewport with this kind of UCS icon displayed.

Although they enable you to view the model in four different directions, the drawback of tiled viewports is that the effective working area of the viewports are decreased. Therefore, you may want to change the display back to a single viewport. To do so, select B of Figure 6.19 and then the 1 Viewport item of the Tiled Viewports cascading menu of the View pull-down menu to use the VPORTS command to set a single viewport. See Figure 6.17 again.

<View> <Tiled Viewports> <1 Viewport>

Command: **VPORTS**
Save/Restore/Delete/Join/SIngle/?/2/<3>/4: **SI**

Rather than setting the display to a preset viewing direction, you can view your 3D model from anywhere you like. Select the Select... item of the 3D Viewpoint cascading menu of the View pull-down menu to use the DDVPOINT command to set the viewing direction 300° from the X axis and 25° from the XY plane. See Figure 6.20 and Figure 6.21.

<View> <3D Viewpoint> <Select...>

Command: **DDVPOINT**

[Viewpoint Presets
From X Axis: **300** XY Plane: **25**
OK]

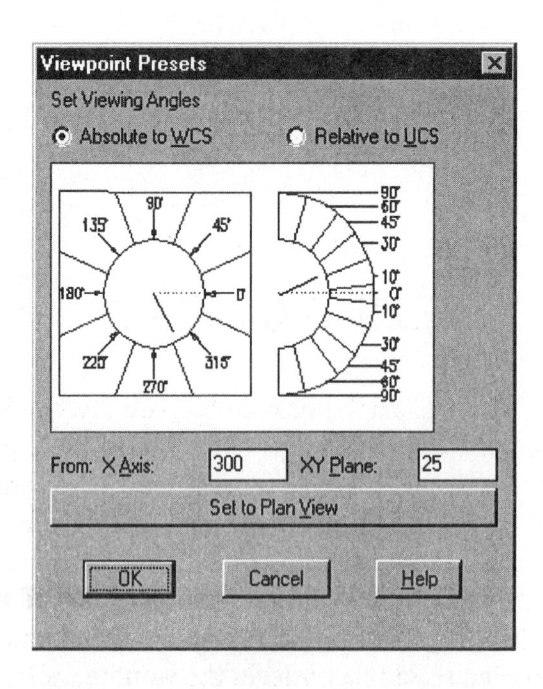

Figure 6.20 DDVPOINT command dialog box

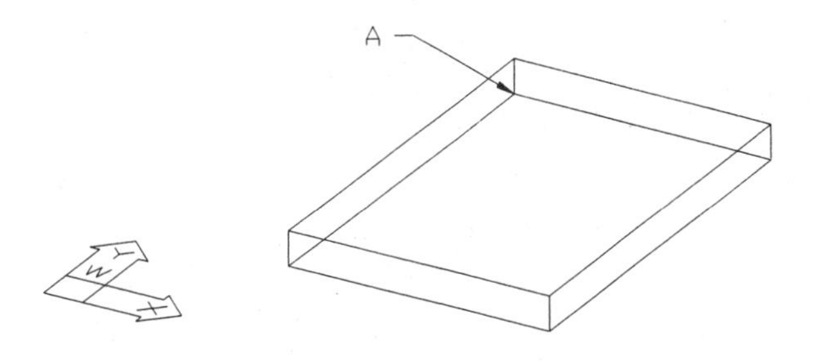

Figure 6.21 Viewing from 300° from the X axis and 25° from the XY plane

To sum up, you can divide the viewport into a number of tiles by using the VPORTS command, and you can set the viewing direction by using the VPOINT command or the DDVPOINT command.

To continue with model making, select the Box item of the Solids cascading menu of the Draw pull-down menu to use the BOX command to make another solid box. See Figure 6.22.

<Draw> **<Solids>** **<Box>**

Command: **BOX**
Center/<Corner of box>: **END** of **[Select A (Figure 6.21).]**
Cube/Length/<other corner>: **@80,-40,50**

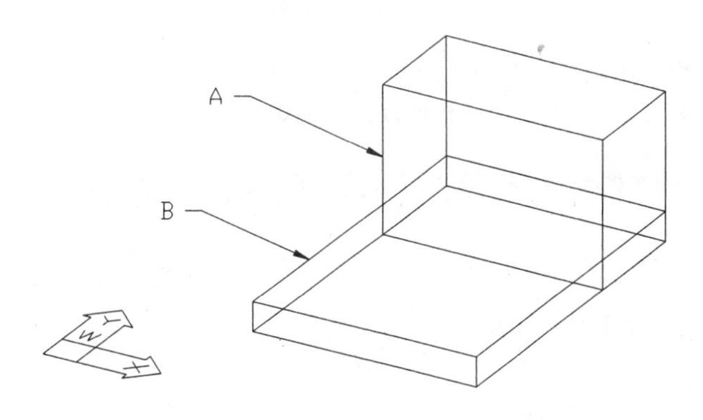

Figure 6.22 Second box created

Refer to Figure 6.22. It shows a solid box sitting partially within another solid box. In reality, this is not possible. You just cannot put a solid object partly or wholly inside another solid object without damaging any of them. But in the computer, they are simply two unrelated data that represent two solid objects. Therefore, you may place them anywhere you like. You can manipulate each of them independently. To combine the two solid boxes to form a complex solid, select the Union item of the Boolean cascading menu of the Modify pull-down menu to apply the UNION command. A union of two solid boxes creates a complex solid that encloses the volume enclosed by the first box and that enclosed by the second. Compare Figure 6.22 with Figure 6.23 to see the difference.

<Modify> **<Boolean>** **<Union>**

Command: **UNION**
Select objects: **[Select A and B (Figure 6.22).]**
Select objects: **[Enter]**

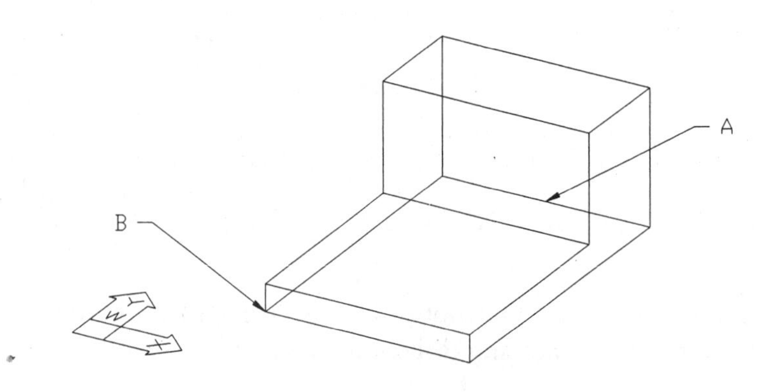

Figure 6.23 Union of two boxes

After uniting, the two boxes become a single complex solid. From now on, you will not be able to separate the two boxes.

Now you will construct the holes and circular groove. Select the Cylinder item of the Solids cascading menu of the Draw pull-down menu to use the CYLINDER command to create two solid cylinders.

<Draw> **<Solids>** **<Cylinder>**

Command: **CYLINDER**
Elliptical/<center point>: **MID** of **[Select A (Figure 6.23).]**
Diameter/<Radius>: **25**
Center of other end/<Height>: **50**

Command: **[Enter]**
CYLINDER
Elliptical/<center point>: **FROM**
Base point: **END** of **[Select B (Figure 6.23).]**
<Offset>: **@20,20**
Diameter/<Radius>: **5**
Center of other end/<Height>: **10**

Select the Array item from the Modify pull-down menu to create a rectangular array of the last cylinder. See Figure 6.24.

<Modify> **<Array>**

Command: **ARRAY**
Select objects: **LAST**
Select objects: **[Enter]**
Rectangular or Polar array (<R>/P): **R**
Number of rows (---): **2**
Number of columns (|||): **2**
Unit cell or distance between rows (---): **30**
Distance between columns (|||): **40**

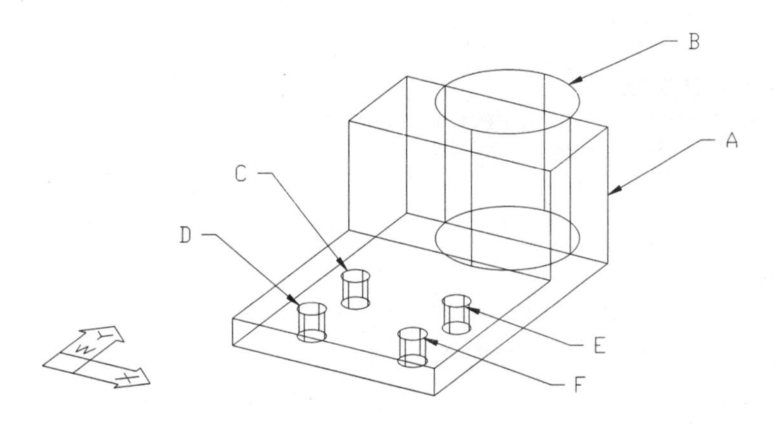

Figure 6.24 Cylinders created

Like the boxes in Figure 6.22, the cylinders in Figure 6.24 are separate objects from the union of the two boxes. They sit partly or wholly inside the united boxes. To construct the holes and the circular groove, select the Subtract item of the Boolean cascading menu of the Modify pull-down menu to use the SUBTRACT command. See Figure 6.25.

<Modify> **<Boolean>** **<Subtract>**

Command: **SUBTRACT**
Select solids and regions to subtract from...
Select objects: **[Select A (Figure 6.24).]**
Select objects: **[Enter]**
Select solids and regions to subtract...
Select objects: **[Select B, C, D, E, and F (Figure 6.24).]**
Select objects: **[Enter]**

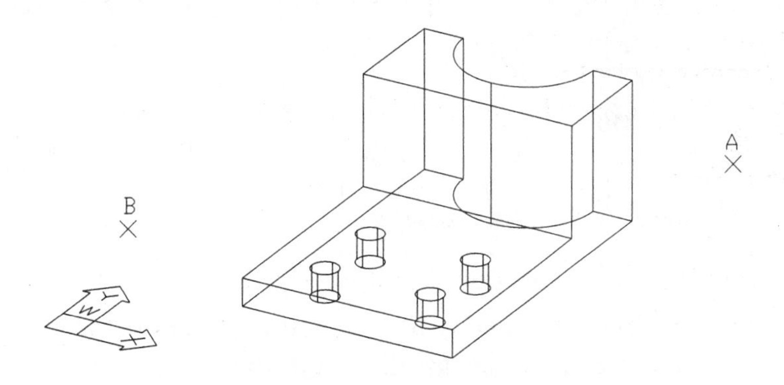

Figure 6.25 Holes cut

Compare Figure 6.24 with Figure 6.25. The cylinders are combined with the union of the two boxes. The large cylinder becomes a circular groove, and the four smaller cylinders become four holes.

The last feature is a wedge. A solid wedge is a triangular block. Select the Wedge item of the Solids cascading menu of the Draw pull-down menu to create two wedges. See Figure 6.26.

<Draw> <Solids> <Wedge>

Command: **WEDGE**
Center/<Corner of wedge>: **[Select A (Figure 6.25).]**
Cube/Length/<other corner>: **@80,10**
Height: **40**

Command: **[Enter]**
WEDGE
Center/<Corner of wedge>: **[Select B (Figure 6.25).]**
Cube/Length/<other corner>: **@10,80**
Height: **40**

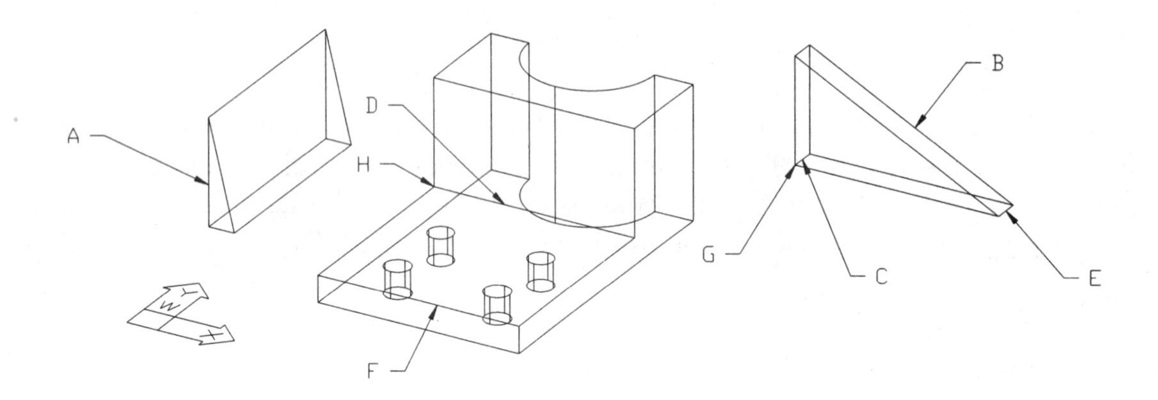

Figure 6.26 Wedges created

Refer to Figure 6.26. The two wedges are not the same, although they have the same base size. Wedge A is 10 units times 80 units. Wedge B is 80 units times 10 units. As can be seen, the taller edge is always on the ZY plane.

Comparing Figure 6.26 with Figure 6.13 reveals that wedge A is not required. Use the ERASE command to delete this wedge.

<Modify> <Erase>

Command: **ERASE**
Select objects: **[Select A (Figure 6.26).]**
Select objects: **[Enter]**

To align wedge B with the main body of the solid model, select the Align item of the 3D Operation cascading menu of the Modify pull-down menu to apply the ALIGN command. See Figure 6.27.

<Modify> <3D Operation> <Align>

Command: **ALIGN**
Select objects: **[Select B (Figure 6.26).]**
Select objects: **[Enter]**
Specify 1st source point: **MID** of **[Select C (Figure 6.26).]**
Specify 1st destination point: **MID** of **[Select D (Figure 6.26).]**
Specify 2nd source point: **MID** of **[Select E (Figure 6.26).]**

Specify 2nd destination point: **MID** of [**Select F (Figure 6.26).**]
Specify 3rd source point or <continue>: **END** of [**Select G (Figure 6.26).**]
Specify 3rd destination point: **END** of [**Select H (Figure 6.26).**]

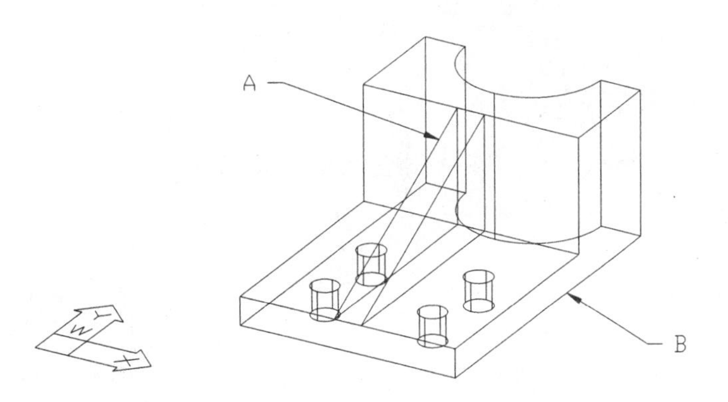

Figure 6.27 Wedge aligned

To complete the model, use the UNION command to unite the wedge with the main body. There will be no noticeable change after uniting the solids.

<Modify> **<Boolean>** **<Union>**

Command: **UNION**
Select objects: [**Select A and B (Figure 6.27).**]
Select objects: [**Enter**]

The model is complete. See also Figure 6.13. Save your drawing.

<File> **<Save>**

File name: **ABLK.DWG**

In constructing this model, you learned how to create boxes, wedges, and cylinders and learned how to combine them to form a complex solid by using the UNION and SUBTRACT commands. In addition, you learned how to divide the viewport into tiles and display 3D views.

Model of a Bearing Block

After working on the model of an angle block, you will construct the model of a bearing block. Figure 6.28 shows the rendered image of the completed model. In making this model, you will further enhance your knowledge of model building by using all the kinds of primitives available — boxes, spheres, cylinders, cones, wedges, and tori.

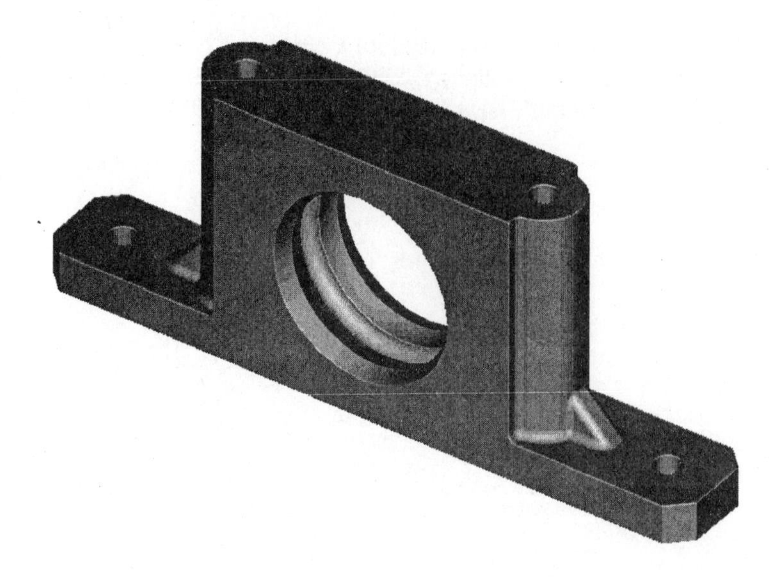

Figure 6.28 Bearing block

Analysis

Refer to Figure 6.28. Take some time to think critically about what primitives you will need to create and what Boolean operations you will need to perform on the primitives in order to construct the model.

The model is symmetrical about its central axis. If you remove the fillet edges and the chamfer edges, you may find that its outer skin consists of three boxes, two cylinders, and two wedges, along with cylindrical, conical, toroidal, and spherical openings. See Figure 6.29.

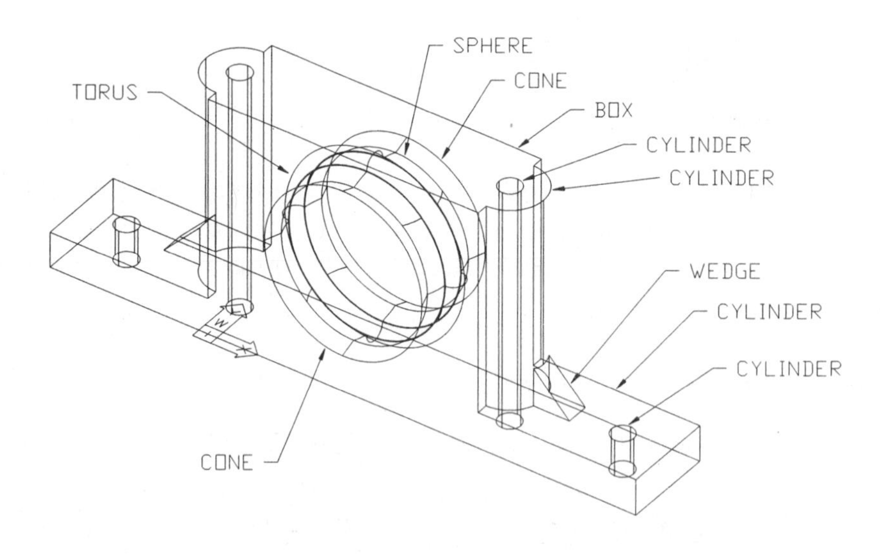

Figure 6.29 Solid model with fillet and chamfer edges removed

To make it clear what primitives you need to create, they are exploded and shown in Figure 6.30.

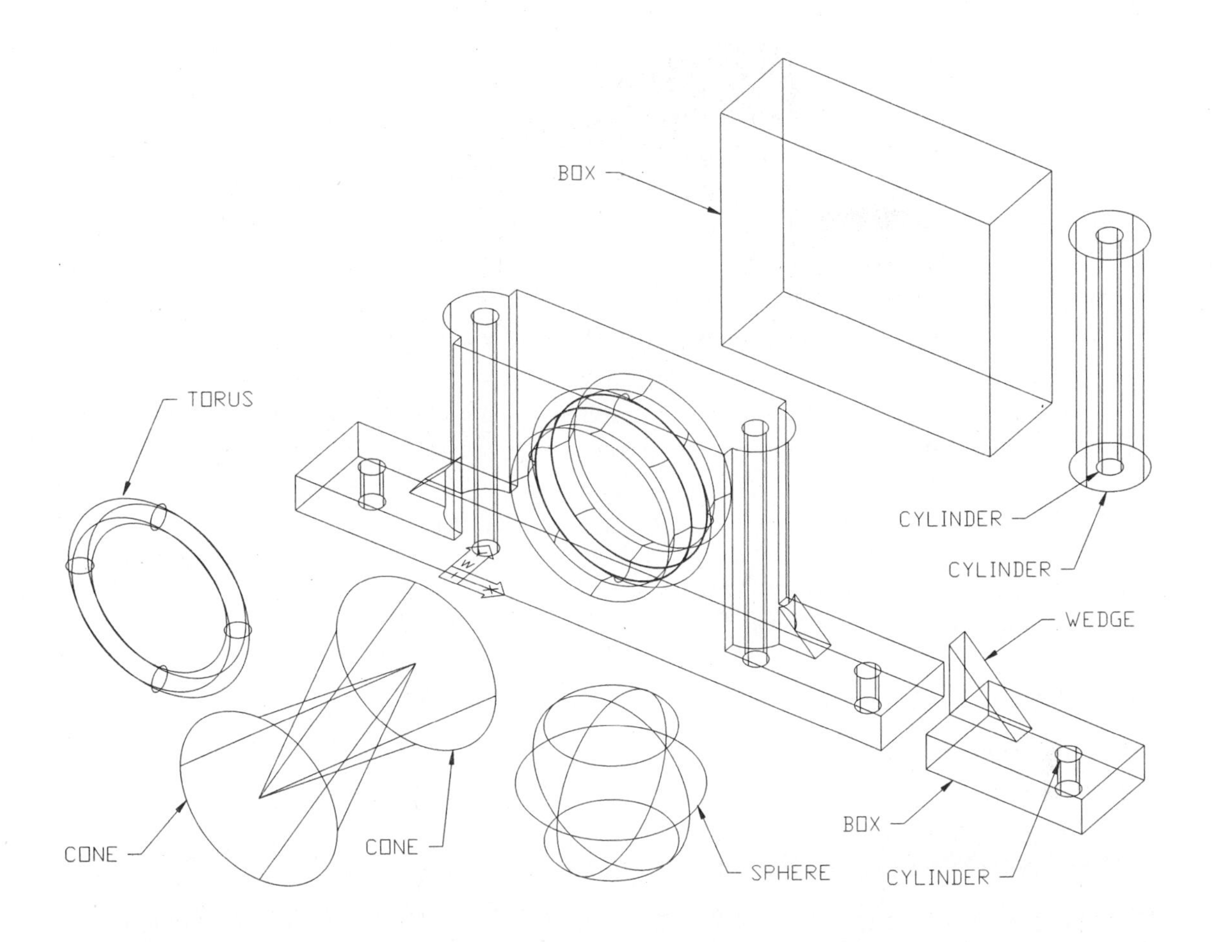

Figure 6.30 Exploded view of primitives required

To build the model, you will first construct a box, a cone, a sphere, and a torus. Then you will subtract the cone, sphere, and torus from the box. Next, you will create and mirror another box, three cylinders, and a wedge. Then you will unite the larger cylinders, the boxes, and the wedges to the main body. After that, you will subtract the remaining cylinders from the complex solid. Finally, you will refine the complex solid model by filleting and chamfering some edges.

Drawing Preparation

Use the NEW command to start a new drawing from scratch, and use metric units.

<File> **<New...>**

Apply the LAYER command to make a new layer called Solid and set its color to yellow. You will create the solid model on this layer.

<Format> **<Layer...>**

Command: **LAYER**

[Layer

Name	Color	Linetype
0	White	Continuous
Solid	Yellow	Continuous

OK]

UCS Icon Display

When working in 3D space, you will get a better picture of the location of the construction plane by setting the UCS icon to display at the origin of the current user coordinate system (UCS).

Select and check the On and Origin items of the UCS Icon item of the Display cascading menu of the View pull-down menu.

<View> **<Display>** **<UCS Icon>** <v **On>**

Command: **UCSICON**
ON/OFF/All/Noorigin/ORigin: **ON**

<View> **<Display>** **<UCS Icon>** <v **Origin>**

Command: **UCSICON**
ON/OFF/All/Noorigin/ORigin: **OR**

Isometric Viewing Position

Because the objects that you are going to create are 3D, set the display to a 3D isometric view looking from the southeast direction by selecting the SE Isometric item of the 3D Viewpoint cascading menu of the View pull-down menu to run the VPOINT command.

<View> **<3D Viewpoint>** **<SE Isometric>**

Command: **VPOINT**
Rotate/<View point>: **1,-1,1**

Primitives Creation and Subtraction

Now you will use solid creation commands to make the primitives and use Boolean operation commands to combine the primitives into a single complex solid.

Use the BOX command to build a solid box. This is the main body of the complex solid model. See Figure 6.31.

<Draw> **<Solids>** **<Box>**

Command: **BOX**
Center/<Corner of box>: **0,0**
Cube/Length/<other corner>: **@120,40,100**

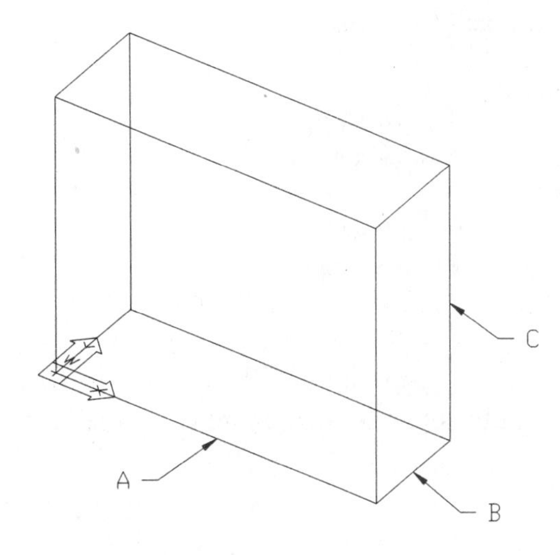

Figure 6.31 Box created

Select the Sphere item of the Solids cascading menu of the Draw pull-down menu to use the SPHERE command to create a solid sphere. This solid constitutes the spherical hole in the main body of the model. See Figure 6.32.

<Draw> **<Solids>** **<Sphere>**

Command: **SPHERE**
Center of sphere: **.X** of **MID** of [**Select A (Figure 6.31)**.]
(need YZ): **.Y** of **MID** of [**Select B (Figure 6.31)**.]
(need Z): **MID** of [**Select C (Figure 6.31)**.]
Diameter/<Radius> of sphere: **35**

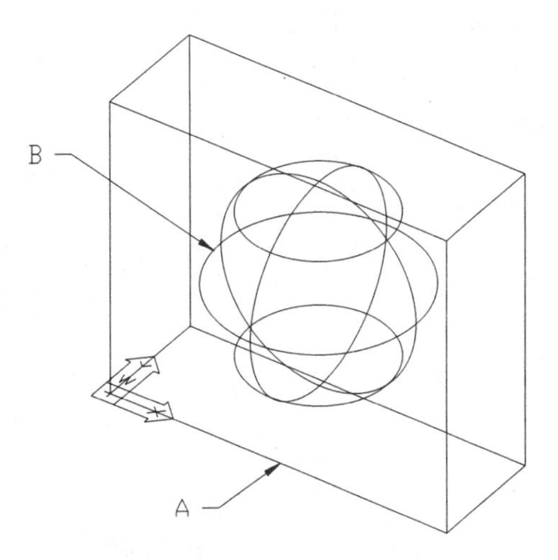

Figure 6.32 Solid sphere created

To use the solid sphere to construct a spherical hole in the solid box, apply the SUBTRACT command by selecting the Subtract item of the Boolean cascading menu of the Modify pull-down menu. See Figure 6.33.

<Modify> <Boolean> <Subtract>

Command: **SUBTRACT**
Select solids and regions to subtract from...
Select objects: **[Select A (Figure 6.32).]**
Select objects: **[Enter]**
Select solids and regions to subtract...
Select objects: **[Select B (Figure 6.32).]**
Select objects: **[Enter]**

After subtraction, the sphere combines with the box to form a complex solid. From now on, you will not be able to separate the sphere from the box.

Changing the UCS

So far, you have been creating solid objects on the default XY plane. In some models like this one, you need to produce solid objects on planes not parallel to the XY plane.

The next solid, a solid torus, resides on a plane perpendicular to the current UCS. To create the torus, you need to change the UCS by selecting the X Axis Rotate item of the UCS cascading menu of the Tools pull-down menu to rotate the UCS about the X axis for 90°. See Figure 6.33.

<Tools> <UCS> <X Axis Rotate>

Command: **UCS**
Origin/ZAxis/3point/OBject/View/X/Y/Z/Prev/Restore/Save/Del/?/<World>: **X**
Rotation angle about X axis: **90**

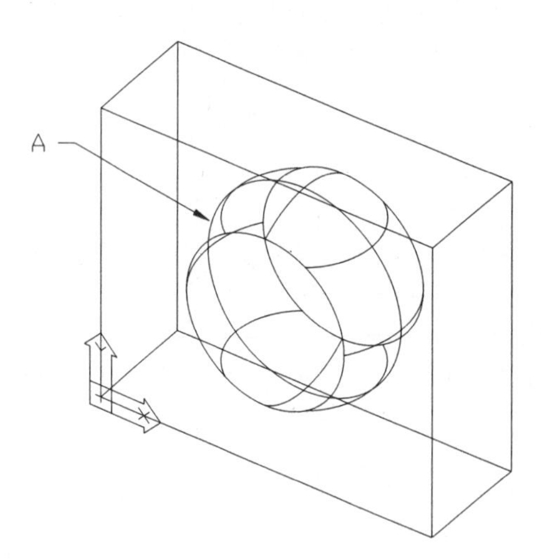

Figure 6.33 Sphere subtracted from box and UCS rotated about X axis

To prepare a solid for making a circular groove, use the TORUS command to make a solid torus. See Figure 6.34.

<Draw> <Solids> <Torus>

Command: **TORUS**

Center of torus: **CEN** of [**Select A (Figure 6.33), center of the sphere.**]
Diameter/<Radius> of torus: **35**
Diameter/<Radius> of tube: **5**

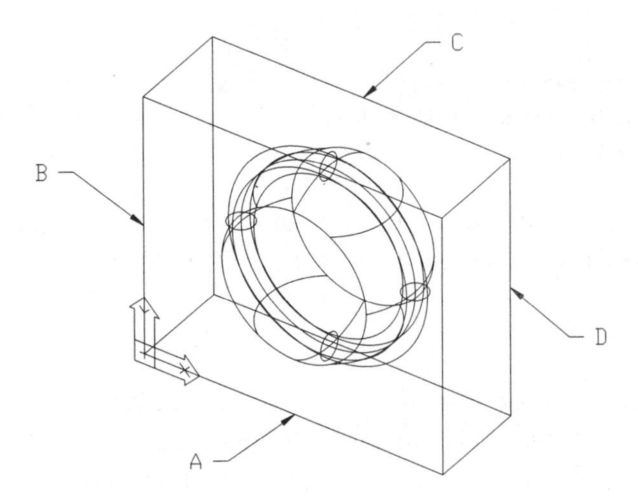

Figure 6.34 Solid torus created

Use the CONE command to make two solid cones. They will be used to construct two conical features. See Figure 6.35.

<Draw> **<Solids>** **<Cone>**

Command: **CONE**
Elliptical/<center point>: **.X** of **MID** of [**Select A (Figure 6.34).**]
(need YZ): **MID** of [**Select B (Figure 6.34).**]
Diameter/<Radius>: **35**
Apex/<Height>: **-100**

Command: [**Enter**]
CONE
Elliptical/<center point>: **.X** of **MID** of [**Select C (Figure 6.34).**]
(need YZ): **MID** of [**Select D (Figure 6.34).**]
Diameter/<Radius>: **35**
Apex/<Height>: **100**

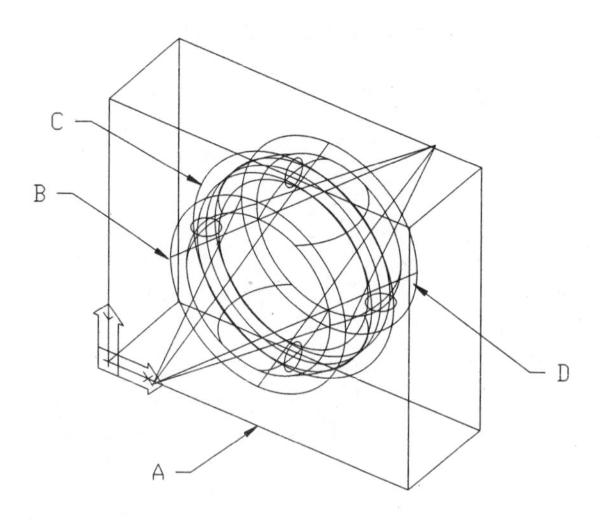

Figure 6.35 Two solid cones created

Now you have a complex solid of a box having a spherical hole, a solid torus, and two solid cones. The solid torus and the solid cones are separate solids from the complex solid. To cut conical recesses and a circular groove in the complex solid, use the SUBTRACT command. After that, set the UCS back to the system default — World. See Figure 6.36.

<Modify> **<Boolean>** **<Subtract>**

Command: **SUBTRACT**
Select solids and regions to subtract from...
Select objects: [**Select A (Figure 6.35).**]
Select objects: [**Enter**]
Select solids and regions to subtract...
Select objects: [**Select B, C, and D (Figure 6.35).**]
Select objects: [**Enter**]

<Tools> **<UCS>** **<World>**

Command: **UCS**
Origin/ZAxis/3point/OBject/View/X/Y/Z/Prev/Restore/Save/Del/?/<World>: **W**

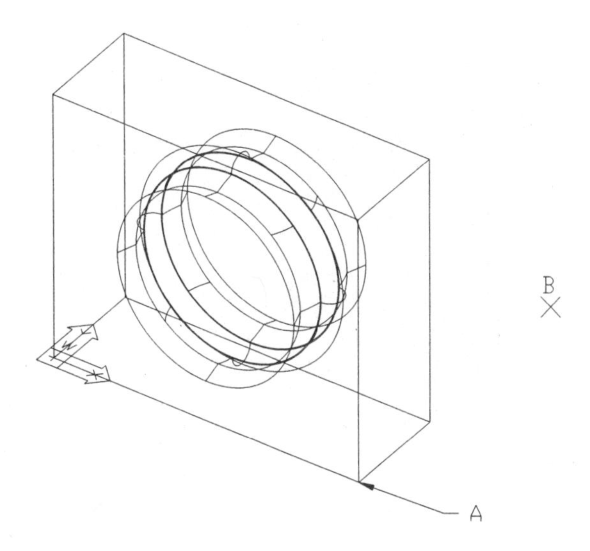

Figure 6.36 Cones and torus cut from main body

Now you have two solid cones, a solid torus, and a solid sphere subtracted from a solid box. To continue, use the BOX command to make a solid box and the WEDGE command to make a solid wedge. See Figure 6.37.

<Draw> **<Solids>** **<Box>**

Command: **BOX**
Center/<Corner of box>: **END** of **[Select A (Figure 6.36).]**
Cube/Length/<other corner>: **@70,40,15**

<Draw> **<Solids>** **<Wedge>**

Command: **WEDGE**
Center/<Corner of wedge>: **[Select B (Figure 6.36).]**
Cube/Length/<other corner>: **@30,10,30**

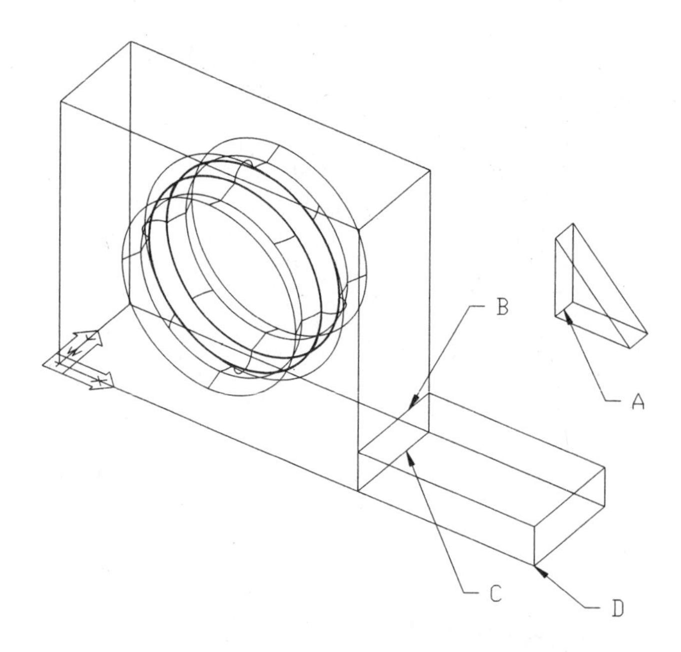

Figure 6.37 Box and wedge created

Use the MOVE command to move the solid wedge to its proper position, and use the CYLINDER command to create three cylinders. See Figure 6.38.

<Modify> **<Move>**

Command: **MOVE**
Select objects: [**Select A (Figure 6.37).**]
Select objects: [**Enter**]
Base point or displacement: **MID** of [**Select A (Figure 6.37).**]
Second point of displacement: **MID** of [**Select B (Figure 6.37).**]

<Draw> **<Solids>** **<Cylinder>**

Command: **CYLINDER**
Elliptical/<center point>: **MID** of [**Select C (Figure 6.37).**]
Diameter/<Radius>: **15**
Center of other end/<Height>: **100**

Command: [**Enter**]
CYLINDER
Elliptical/<center point>: **MID** of [**Select C (Figure 6.37).**]
Diameter/<Radius>: **5**
Center of other end/<Height>: **100**

Command: [**Enter**]
CYLINDER
Elliptical/<center point>: **FROM**
Base point: **END** of [**Select D (Figure 6.37).**]
<Offset>: **@-20,20**
Diameter/<Radius>: **5**
Center of other end/<Height>: **15**

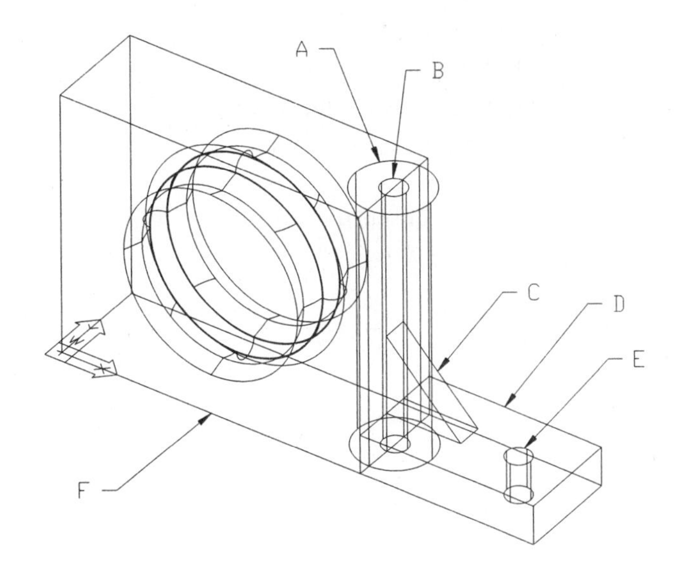

Figure 6.38 Wedge moved and cylinders created

The bearing block is symmetrical about its central axis. Use the MIRROR3D command to mirror the wedge, the box, and the cylinders. See Figure 6.39.

<Modify> **<3D Operation>** **<Mirror 3D>**

Command: **MIRROR3D**
Select objects: **[Select A, B, C, D, and E (Figure 6.38).]**
Select objects: **[Enter]**
Plane by Object/Last/Zaxis/View/XY/YZ/ZX/<3points>: **YZ**
Point on YZ plane: **MID** of **[Select F (Figure 6.38).]**
Delete old objects? <N> **N**

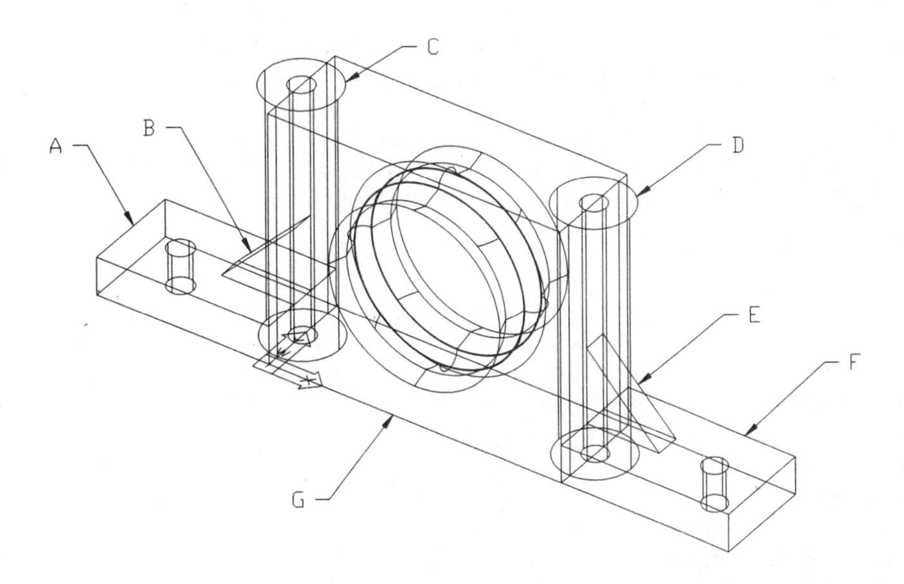

Figure 6.39 Wedge, box, and cylinders mirrored

Rather than using the MIRROR3D command, you can use the 3DARRAY command to duplicate objects in 3D. Like the ARRAY command, the 3DARRAY command copies objects rectangularly and polarly. A rectangular 3D array copies objects in a matrix of rows in X axis, columns in Y axis, and levels in Z axis. A polar 3D array copies objects about an axis of rotation in 3D.

Use the UNION command to combine the two large cylinders, two wedges, and two boxes with the main body. See Figure 6.40.

<Modify> **<Boolean>** **<Union>**

Command: **UNION**
Select objects: **[Select A, B, C, D, E, F, and G (Figure 6.39).]**
Select objects: **[Enter]**

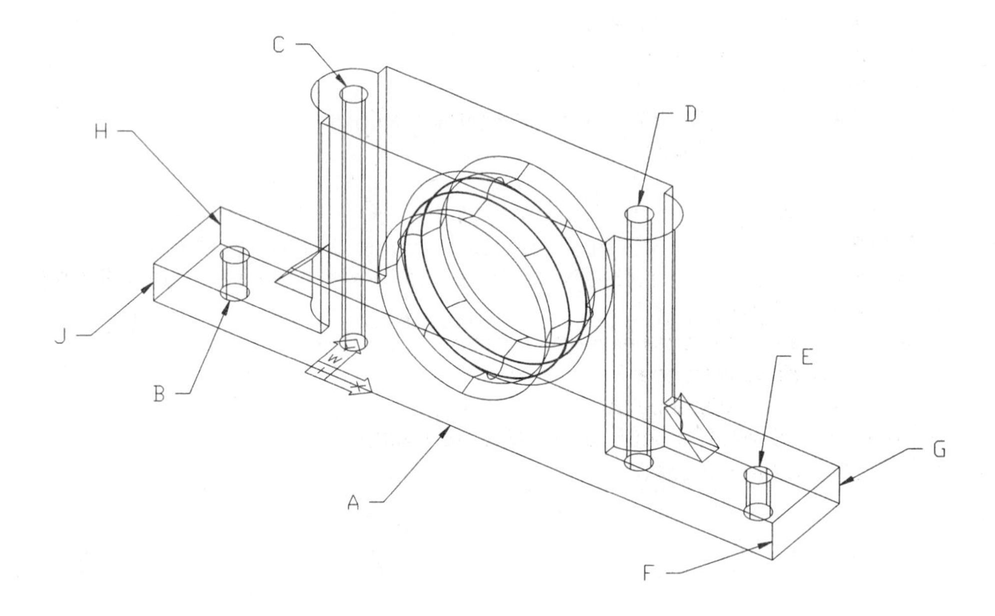

Figure 6.40 Wedges, cylinders, and boxes united with the main body

To remind you, the four cylinders that appear as holes have not been subtracted yet. Use the SUBTRACT command to finish the model. You will not find any noticeable change on your screen after the subtraction.

<Modify> **<Boolean>** **<Subtract>**

Command: **SUBTRACT**
Select solids and regions to subtract from...
Select objects: **[Select A (Figure 6.40).]**
Select objects: **[Enter]**
Select solids and regions to subtract...
Select objects: **[Select B, C, D, and E (Figure 6.40).]**
Select objects: **[Enter]**

Chamfer and Fillet

Basically, the model is complete. To further refine the model, you will chamfer and fillet the edges. Select the Chamfer item from the Modify pull-down menu to apply the CHAMFER command. This command operates on 3D solids as well as on 2D wireframes. When operated on 2D wireframes, it produces a bevel at the intersection of two non-parallel lines. When operated on 3D solids, it produces a bevel edge. See Figure 6.41.

<Modify> **<Chamfer>**

Command: **CHAMFER**
Polyline/Distance/Angle/Trim/Method/<Select first line>: **[Select F (Figure 6.40).]**
Select base surface:
Next/<OK>: **[OK, if face FG is highlighted.]**
Enter base surface distance: **8**
Enter other surface distance: **8**
Loop/<Select edge>: **[Select F and G (Figure 6.40).]**
Loop/<Select edge>: **[Enter]**

Command: **[Enter]**
CHAMFER
Polyline/Distance/Angle/Trim/Method/<Select first line>: **[Select H (Figure 6.40).]**
Select base surface:
Next/<OK>: **[OK, if face HJ is highlighted.]**
Enter base surface distance: **8**
Enter other surface distance: **8**
Loop/<Select edge>: **[Select H and J (Figure 6.40).]**
Loop/<Select edge>: **[Enter]**

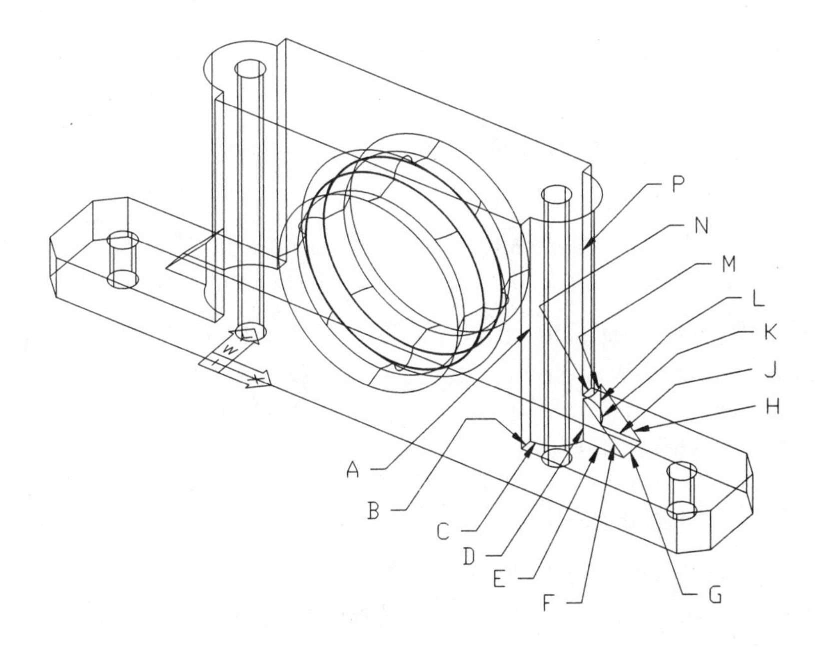

Figure 6.41 Four edges chamfered

After chamfering, select the Fillet item of the Modify pull-down menu to use the FILLET command to fillet 14 edges. Like the CHAMFER command, the FILLET command operates on 3D solids as well as 2D wireframes. When operated on 2D wireframes, it produces an arc at the intersection of two non-parallel lines. When operated on 3D solids, it produces a rounded edge. See Figure 6.42.

The sequence of filleting affects the fillet outcome (see next section). Therefore, do the fillets all at one time. While filleting, you may have to use the ZOOM command to zoom in closer to select the edges.

<Modify> **<Fillet>**

Command: **FILLET**
Polyline/Radius/Trim/<Select first object>: **[Select A (Figure 6.41).]**
Enter radius: **3**
Chain/Radius/<Select edge>: **[Select B, C, D, E, F, G, H, J, K, L, M, N, and P (Figure 6.41).]**
Chain/Radius/<Select edge>: **[Enter]**

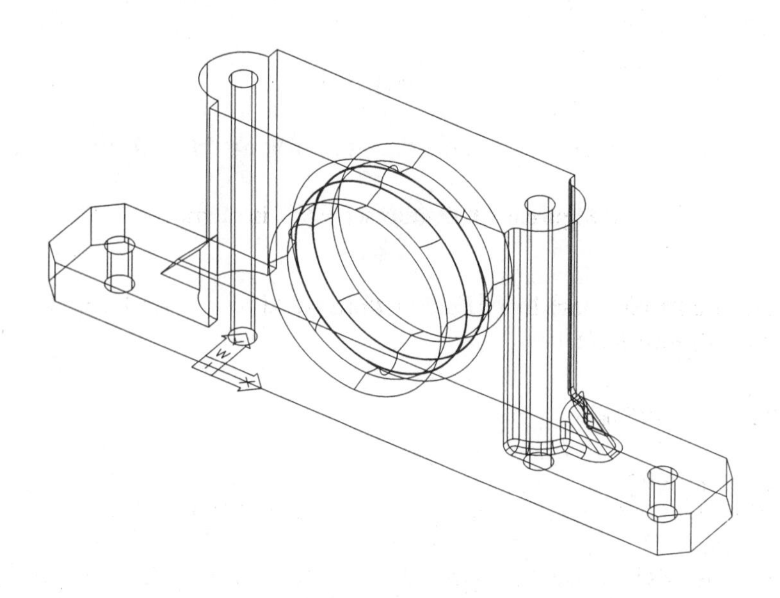

Figure 6.42 Fourteen edges filleted

In order to fillet the other side of the model, set the display to view the model from the other side. Select the SW Isometric item of the 3D Viewpoint cascading menu of the View pull-down menu to use the VPOINT command to set the display to an isometric view looking from the southwest direction. See Figure 6.43.

<View> **<3D Viewpoint>** **<SW Isometric>**

Command: **VPOINT**
Rotate/<View point>: **-1,-1,1**

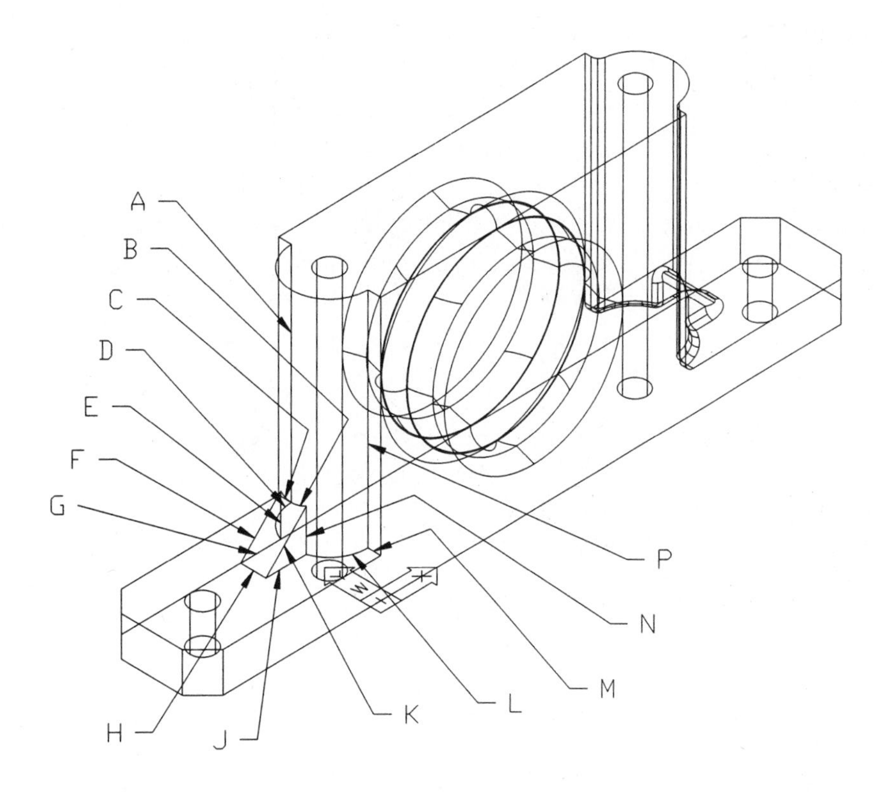

Figure 6.43 Display set to SW isometric

To complete the model, use the FILLET command to fillet 14 edges of the other side of the model. See Figure 6.44.

<Modify> **<Fillet>**

Command: **FILLET**
Polyline/Radius/Trim/<Select first object>: **[Select A (Figure 6.43).]**
Enter radius: **3**
Chain/Radius/<Select edge>: **[Select B, C, D, E, F, G, H, J, K, L, M, N, and P (Figure 6.43).]**
Chain/Radius/<Select edge>: **[Enter]**

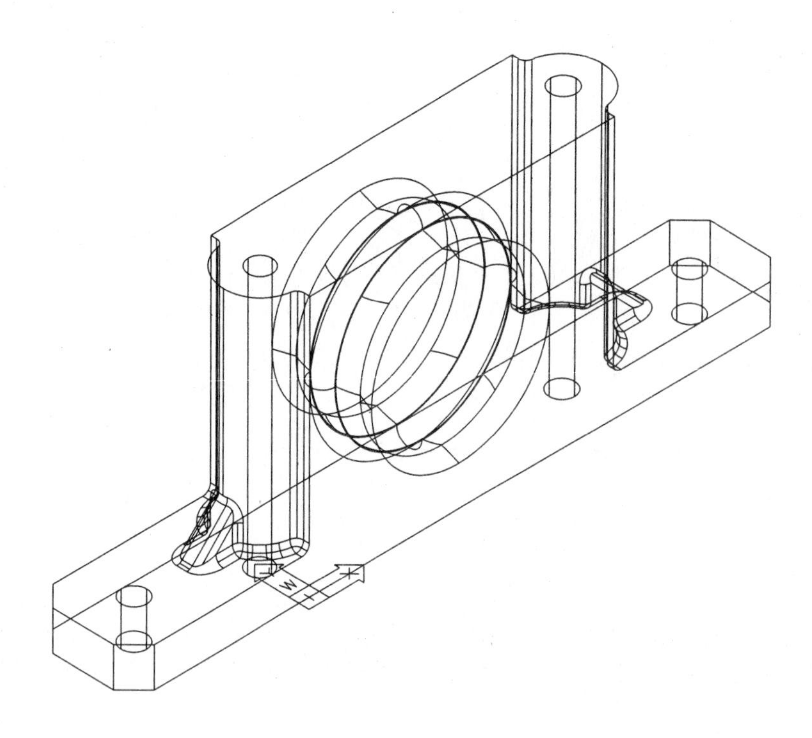

Figure 6.44 14 edges filleted

The solid model of the bearing block is complete. See also Figure 6.28. Save your drawing.

<File> **<Save...>**

File name: **BBLK.DWG**

In constructing this model, you built six kinds of primitive solids and combined them to form a complex solid. You also filleted and chamfered the edges.

Sequence of Filleting and Chamfering

Fillets and chamfers are commonly added to the edges of a solid model. When you need to fillet or chamfer a number of edges, you have to plan the sequence of work because the order in which the edges are treated affects the outcome. To appreciate this sequential effect, create three identical simple solid models and then round off the edges.

Start a new drawing and create three solid boxes. The corners of the first box are (0,0,0) and (50,50,50). The corners of the second box are (50,10,0) and (100,50,10). The corners of the third box are (50,40,0) and (100,50,20). Unite the three boxes. Then array them for a distance of 150 units in the X-direction. See Figure 6.45.

You will fillet the edges of these three models in different orders to see the effect of filleting sequence on the filleting outcome.

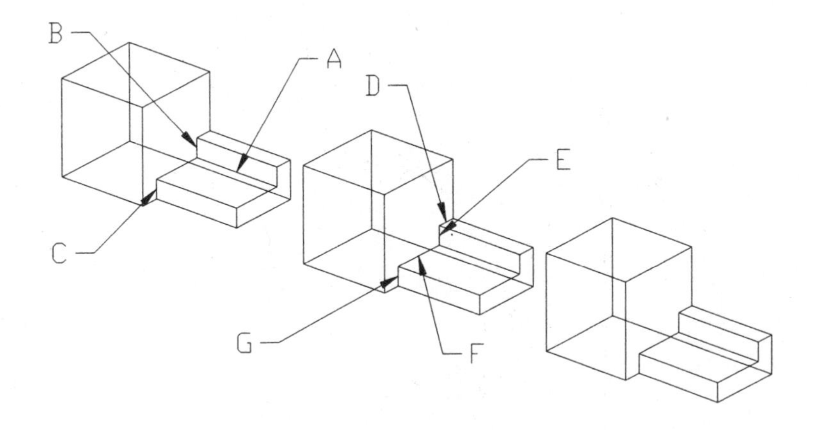

Figure 6.45 Three solid boxes

Use the FILLET command to fillet the edges A, B, and C of the left model and to fillet the edges D, E, F, and G of the middle model. The fillet radius is 4 units. Compare your models with Figure 6.46.

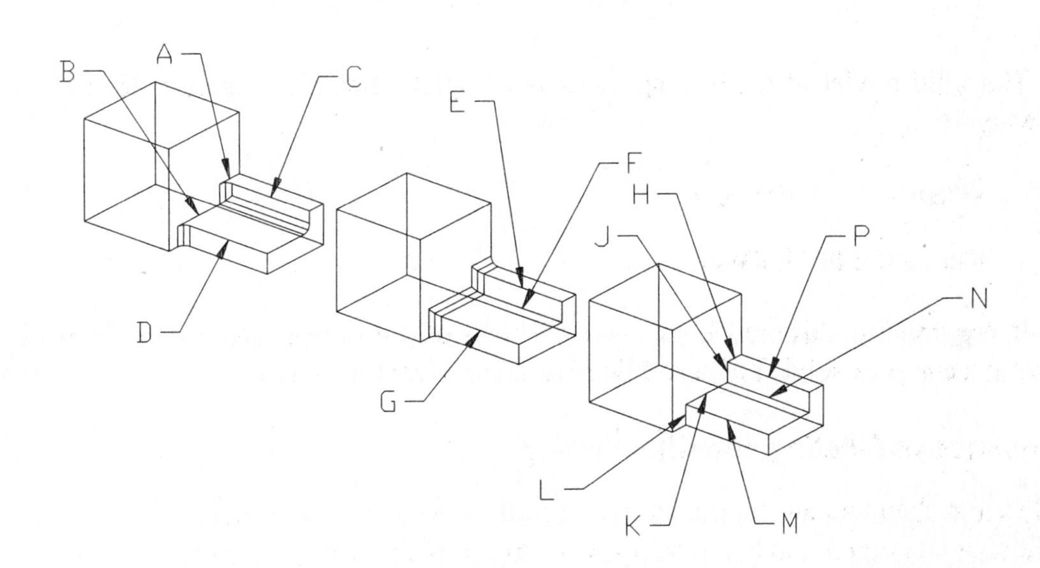

Figure 6.46 Edges filleted

To continue, fillet the edges H, J, K, L, M, N, and P of the right model in one operation. Then, in one operation again, fillet the edges A, B, C, and D of the left model and the edges E, F, and G of the middle mode. When you fillet the right and middle models, you have to choose the CHAIN option to select the edges.

After running the command, you will find that the edges of the left model cannot be filleted in one operation. If you leave out the edges C and D, the outcome should look like Figure 6.47.

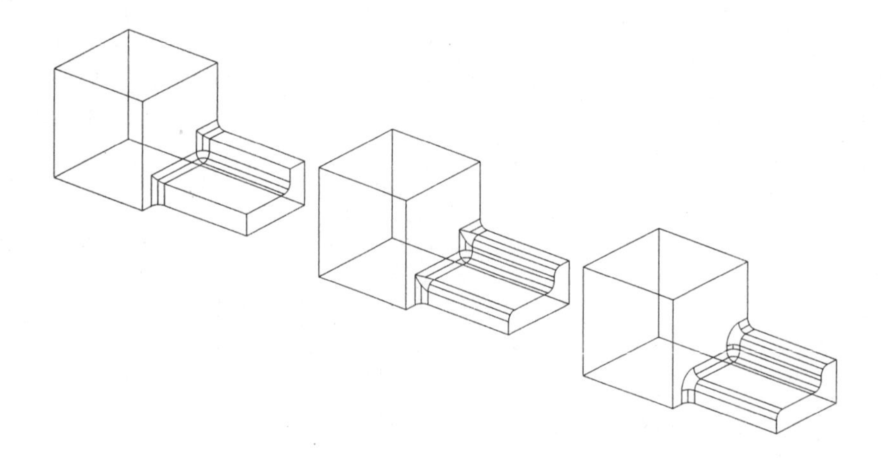

Figure 6.47 More edges filleted

As you can see in Figure 6.47, the outcomes of the middle and right models are different. Now, you may fillet the edges C and D (Figuer 6.46) again by using the CHAIN option. The left model should now be the same as the middle model.

From these results, you should realize that if you have to fillet a number of edges, you must be cautious about the sequence of fillet application.

6.4 Extrude, Revolve, and Intersect

The primitive solids available for constructing a complex solid model are boxes, cylinders, spheres, tori, and wedges. In the last section, you created complex solid models from them by using the Boolean operations union and subtraction.

With these simple primitive solids, the types of complex solids you can build are quite limited. To produce a wider repertoire of solid models, you can perform extrusion and revolving. Both types of solids require a closed, 2D planar area that can be either a closed polyline or a region.

An extruded solid is formed by extruding a closed, 2D planar area in a direction perpendicular to the 2D plane or along a path. A revolved solid is formed by revolving a closed, 2D planar area about an axis.

In this section, you will carry out a number of exercises. First, you will make the model of an engine mount. In making the model, you will practice region modeling, extrude a 2D region in a direction perpendicular to the regions to form extruded solids, and create the solid model of an engine mount by combining the extruded solids. Next, you will make the model of a gear box from two kinds of extruded solids: solids extruded along a line and solids extruded along a path. Then you will create the solid model of a pulley by revolving a 2D region about an axis. After that, you will create the model of a watch case by intersecting an extruded solid and a revolved solid. Finally, you will create the solid model of a fluid power valve by combining extruded solids and a revolved solid.

Model of an Engine Mount

Figure 6.48 shows the model of an engine mount that you are going to construct. You will use a different approach from primitive solids used in the foregoing model.

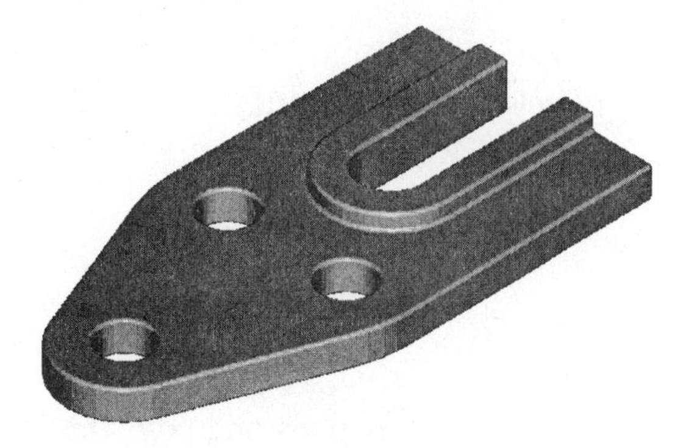

Figure 6.48 Engine Mount

Analysis

As depicted earlier, you need to analyze the model to break it down into simpler objects. In thinking about how to construct this model, think about using extruded solids instead of using primitive solids like boxes, spheres, cones, cylinders, wedges, and tori.

Refer to Figure 6.49. You will construct three complex regions, extrude the regions to form extruded solids, and combine the extruded solids together.

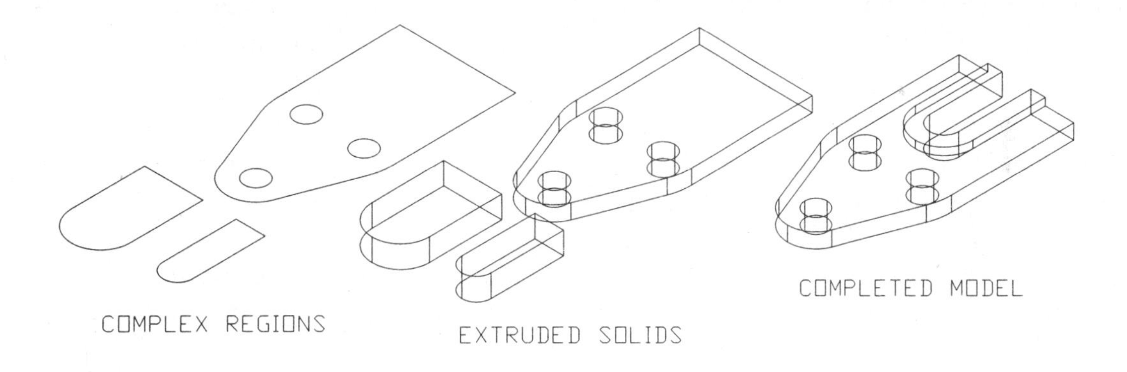

Figure 6.49 Complex regions and extruded solids

Drawing Preparation

As in the foregoing models, start a new drawing from scratch with metric units by using the NEW command.

<File> **<New...>**

Use the LAYER command to create two additional layers, Wire and Solid, with color cyan and yellow, respectively. Set layer Wire as the current layer.

<Format> **<Layer...>**

Command: **LAYER**

[Layer

Name	Color	Linetype
0	White	Continuous
Wire	Cyan	Continuous
Solid	Yellow	Continuous

OK]

Region Modeling

In Chapter 3, you learned how to create regions and combine regions to form complex regions by using the Boolean operations: union, subtract, and intersect. A region is a closed, 2D planar area. You can construct a region from a combination of lines, polylines, circles, ellipses, elliptical arcs, and arcs. Much as with solid objects, you can perform Boolean operations on regions to create complex regions.

Use the RECTANG command to create a rectangle 85 units times 80 units. See Figure 6.50.

<Draw> **<Rectangle>**

Command: **RECTANG**
Chamfer/Elevation/Fillet/Thickness/Width/<First corner>: **[Select a point near the lower left corner of your screen.]**
Other corner: **@85,80**

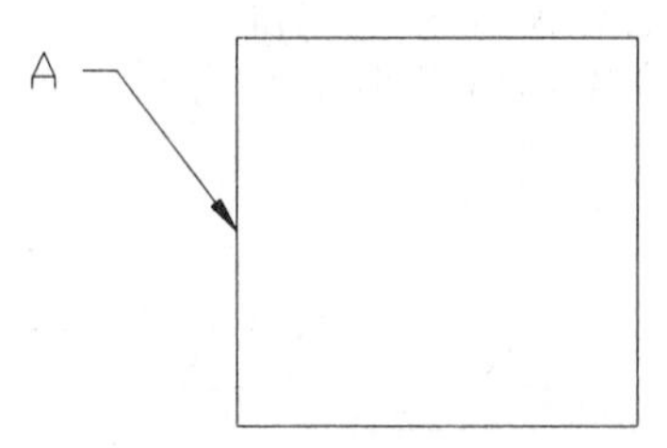

Figure 6.50 Rectangle created

Apply the CIRCLE command to create two circles. See Figure 6.51.

<Draw> **<Circle>** **<Center, Radius>**

Command: **CIRCLE**
3P/2P/TTR/<Center point>: **MID** of **[Select A (Figure 6.50).]**
Diameter/<Radius>: **40**

Command: **[Enter]**
CIRCLE
3P/2P/TTR/<Center point>: **FROM**

Base point: **MID** of [**Select A (Figure 6.50).**]
<Offset>: **@55<180**
Diameter/<Radius>: **20**

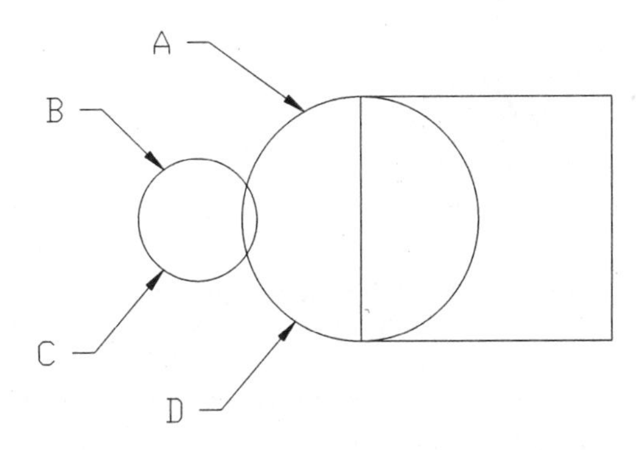

Figure 6.51 Circles created

Use the LINE command to draw two lines that are tangential to the two circles drawn. See Figure 6.52.

<Draw> **<Line>**

Command: **LINE**
From point: **TAN** to [**Select A (Figure 6.51).**]
To point: **TAN** to [**Select B (Figure 6.51).**]
To point: [**Enter**]

Command: [**Enter**]
LINE
From point: **TAN** to [**Select C (Figure 6.51).**]
To point: **TAN** to [**Select D (Figure 6.51).**]
To point: [**Enter**]

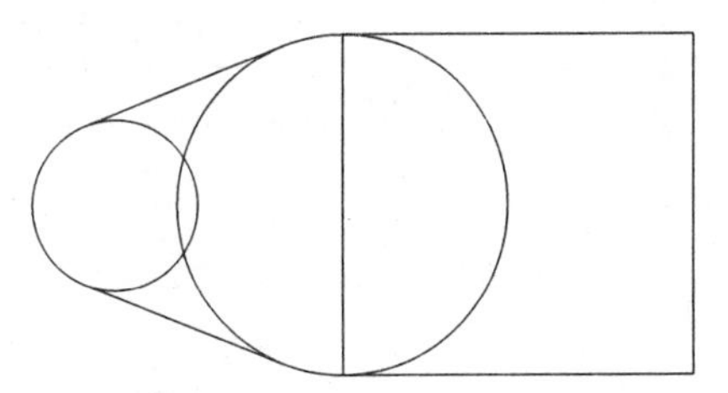

Figure 6.52 Tangential lines created

In accordance with Figure 6.53, use the TRIM command to trim the circles and the rectangle. Now you have a closed, non-intersecting loop.

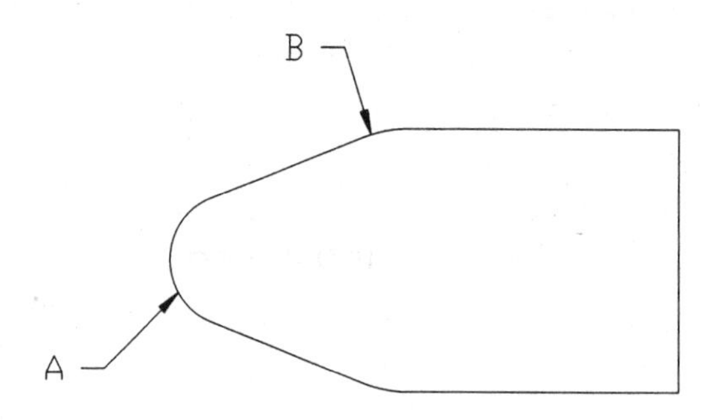

Figure 6.53 Circles and rectangle trimmed

Use the CIRCLE command to create three circles. See Figure 6.54.

<Draw> **<Circle>** **<Center, Radius>**

Command: **CIRCLE**
3P/2P/TTR/<Center point>: **CEN** of **[Select A (Figure 6.53).]**
Diameter/<Radius>: **8**

Command: **[Enter]**
CIRCLE
3P/2P/TTR/<Center point>: **FROM**
Base point: **CEN** of **[Select B (Figure 6.53).]**
<Offset>: **@20<90**
Diameter/<Radius>: **8**

Command: **[Enter]**
CIRCLE
3P/2P/TTR/<Center point>: **FROM**
Base point: **CEN** of **[Select B (Figure 6.53).]**
<Offset>: **@20<270**
Diameter/<Radius>: **8**

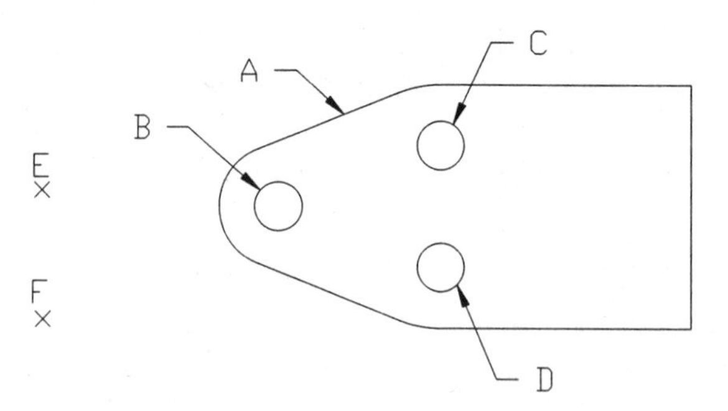

Figure 6.54 Circles created

Now you have three closed areas. To form regions from them, you can apply the REGION command. To reiterate, you can use any combination of lines, arcs, and polylines to define a closed area for subsequent conversion into a region. The main criteria for each single region is that the entities must form a single closed loop and must not cross each other. After conversion of the closed areas into regions, there should be no noticeable change on the screen. If you want to break a region down into lines and arcs, you can run the EXPLODE command.

Before making the regions, you have to decide whether to retain or to delete the original object after it has been used. To delete the original object, you have to set the system variable DELOBJ to 1. If DELOBJ is 0, the original objects are retained after an operation. Because the entities are not required after they are used to make the regions, set the variable to 1.

<Tools> **<Inquiry>** **<Set Variables>**

Command: **SETVAR**
Variable name or ?: **DELOBJ**
New value for DELOBJ: **1**

<Draw> **<Region>**

Command: **REGION**
Select objects: [**Select all the objects.**]
Select objects: [**Enter**]
4 loops extracted.
4 regions created.

To combine the four regions to form a complex region, use the SUBTRACT command to subtract the three circular regions from the larger region.

<Modify> **<Boolean>** **<Subtract>**

Command: **SUBTRACT**
Select solids and regions to subtract from...
Select objects: [**Select A (Figure 6.24).**]
Select objects: [**Enter**]
Select solids and regions to subtract...
Select objects: [**Select B, C, and D (Figure 6.54).**]
Select objects: [**Enter**]

There should be no noticeable change after subtraction. Now you have a complex region instead of four regions.

Use the RECTANG command to create two rectangles. See Figure 6.55.

<Draw> **<Rectangle>**

Command: **RECTANG**
Chamfer/Elevation/Fillet/Thickness/Width/<First corner>: [**Select E (Figure 6.54).**]
Other corner: **@50,40**

Command: [**Enter**]
RECTANG
Chamfer/Elevation/Fillet/Thickness/Width/<First corner>: [**Select F (Figure 6.54).**]
Other corner: **@50,20**

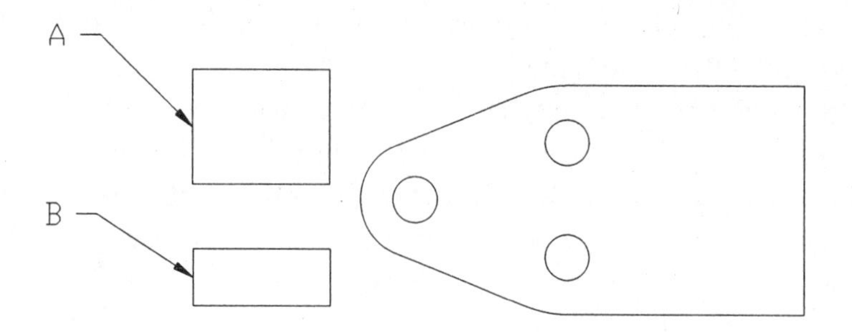

Figure 6.55 Rectangles created

Create two circles. See Figure 6.56.

<Draw> **<Circle>** **<Center, Radius>**

Command: **CIRCLE**
3P/2P/TTR/<Center point>: **MID** of **[Select A (Figure 6.55).]**
Diameter/<Radius>: **20**

Command: **[Enter]**
CIRCLE
3P/2P/TTR/<Center point>: **MID** of **[Select B (Figure 6.55).]**
of Diameter/<Radius>: **10**

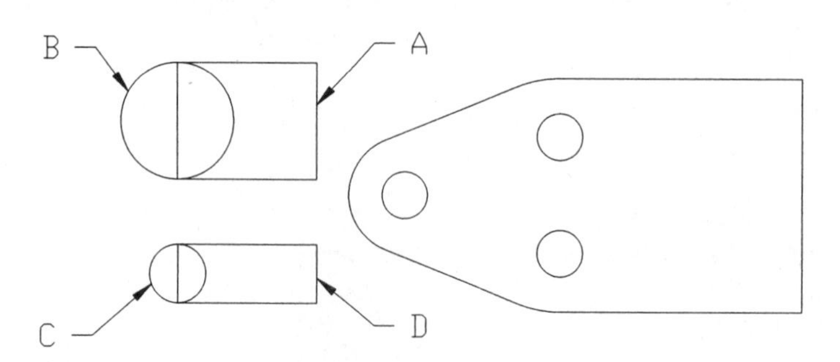

Figure 6.56 Circles created

Use the REGION command to form four regions from the two circles and two rectangles. Then unite them to form two complex regions by using the UNION command. See Figure 6.57.

<Draw> **<Region>**

Command: **REGION**
Select objects: **[Select A, B, C, and D (Figure 6.56).]**
Select objects: **[Enter]**
4 loops extracted.
4 Regions created.

<Modify> **<Boolean>** **<Union>**

Command: **UNION**
Select objects: [**Select A and B (Figure 6.56).**]
Select objects: [**Enter**]

Command: [**Enter**]
UNION
Select objects: [**Select C and D (Figure 6.56).**]
Select objects: [**Enter**]

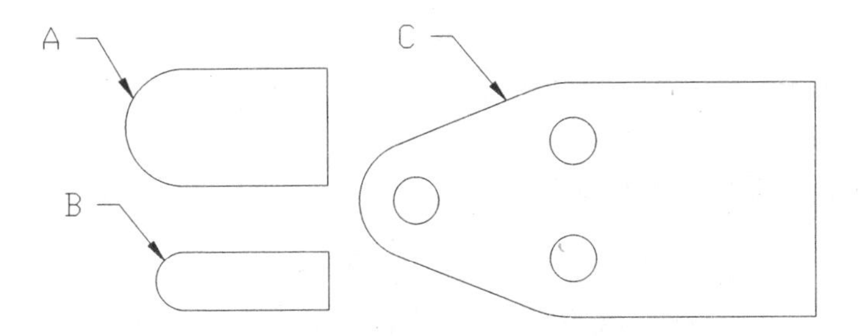

Figure 6.57 Regions formed and united

Now you have three 2D complex regions.

Extruded Solids

You will extrude these three complex regions into extruded solids. Before extruding, you have to decide whether you want to retain the regions or delete the regions after they are used for extruding. Keeping the regions has the advantage that you may use them to reconstruct the model if something goes wrong. To retain the regions, you should set the DELOBJ variable to 0.

<Tools> **<Inquiry>** **<Set Variables>**

Command: **SETVAR**
Variable name or ?: **DELOBJ**
New value for DELOBJ: **0**

From the Layer Control box of the Object Properties toolbar, select the layer Solid to set it as the current layer. You will create the solids on this layer.

[Object Properties] **[Layer Control]**

Current layer: **Solid**

Select the Extrude item of the Solids cascading menu of the Draw pull-down menu to use the EXTRUDE command to extrude the three complex regions.

<Draw> **<Solids>** **<Extrude>**

Command: **EXTRUDE**
Select objects: [**Select A and B (Figure 6.57).**]

Select objects: **[Enter]**
Path/<Height of Extrusion>: **15**
Extrusion taper angle: **0**

Command: **[Enter]**
EXTRUDE
Select objects: **[Select C (Figure 6.57).]**
Select objects: **[Enter]**
Path/<Height of Extrusion>: **10**
Extrusion taper angle: **0**

In order to see clearly the 3D solid, turn off the layer Wire and set the display to an isometric view. See Figure 6.58.

[Object Properties] **[Layer Control]**

Off Wire

<View> **<3D Viewpoint>** **<SW Isometric>**

Command: **VPOINT**
Rotate/<View point>: **-1,-1,1**

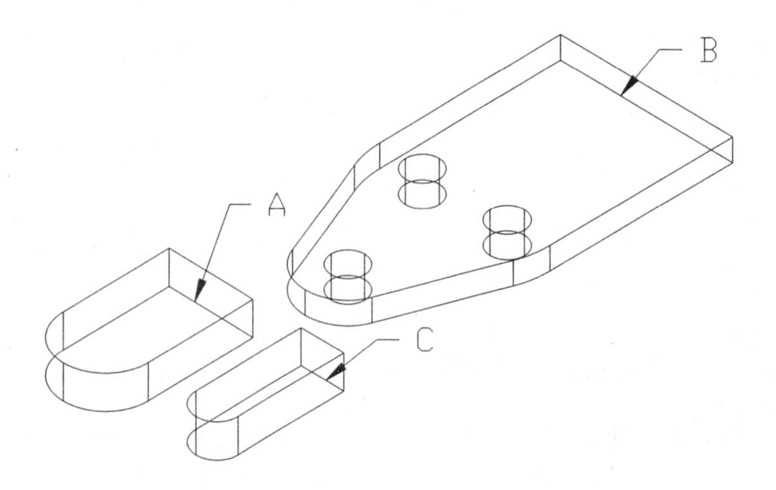

Figure 6.58 Complex regions extruded and display set to SW isometric

The extruded solids for making the complex solid model are complete. Use the MOVE command to put them together in the proper position. See Figure 6.59.

<Modify> **<Move>**

Command: **MOVE**
Select objects: **[Select A (Figure 6.58).]**
Select objects: **[Enter]**
Base point or displacement: **MID** of **[Select A (Figure 6.58).]**
Second point of displacement: **MID** of **[Select B (Figure 6.58).]**

Command: **[Enter]**
MOVE

Select objects: **[Select C (Figure 6.58).]**
Select objects: **[Enter]**
Base point or displacement: **MID** of **[Select C (Figure 6.58).]**
Second point of displacement: **MID** of **[Select B (Figure 6.58).]**

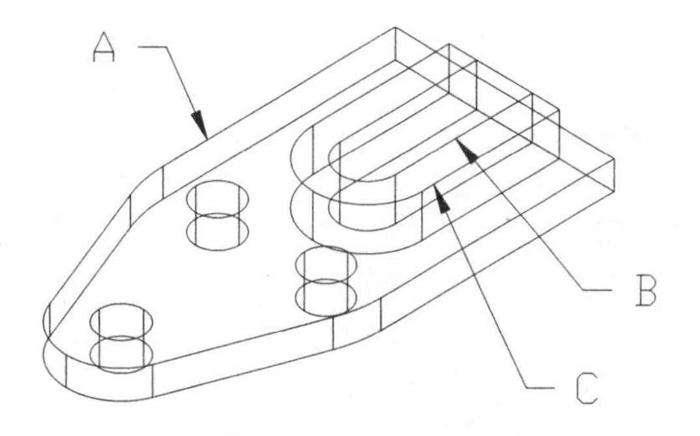

Figure 6.59 Extruded solids moved

After moving two extruded solids to their proper positions, use the UNION command and the SUBTRACT command to combine the three solids into a complex solid. See Figure 6.60.

<Modify> **<Boolean>** **<Union>**

Command: **UNION**
Select objects: **[Select A and C (Figure 6.59).]**
Select objects: **[Enter]**

<Modify> **<Boolean>** **<Subtract>**

Command: **SUBTRACT**
Select solids and regions to subtract from...
Select objects: **[Select A (Figure 6.59).]**
Select objects: **[Enter]**
Select solids and regions to subtract...
Select objects: **[Select B (Figure 6.59).]**
Select objects: **[Enter]**

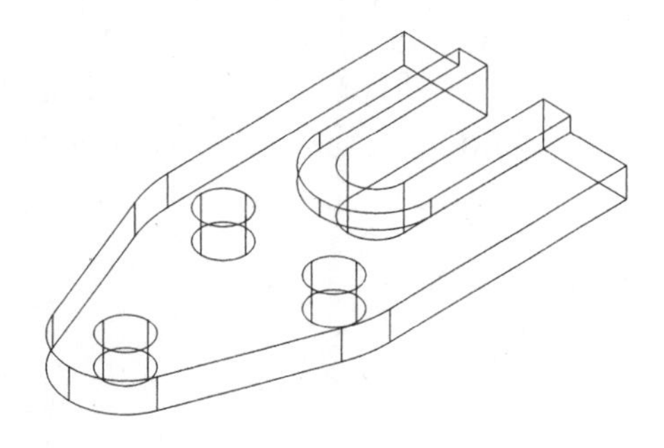

Figure 6.60 Completed model

The model is complete. See also Figure 6.48. Save the drawing.

<File> **<Save>**

File name: **BRCK.DWG**

In making this model, you made 2D regions, extruded the regions to become extruded solids, and combined the extruded solids.

Model of a Gear Box Cover

Figure 6.61 shows the model of a gear box cover. In making this model, you will construct 2D regions and extrude 2D regions in a direction perpendicular to the 2D regions as well as extruding 2D regions along a path.

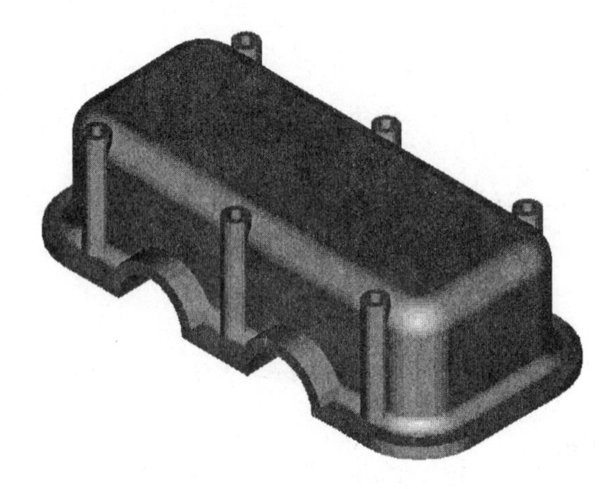

Figure 6.61 Gear box cover

Analysis

Besides adding a lot of flexibility in model making, the provision of extruded solids speeds up modeling. As we have said, you can create a solid by extruding a 2D region or closed, non-intersecting polyline in a direction perpendicular to the 2D region or along a path residing on a plane perpendicular to the 2D region. Refer to Figure 6.61, and think critically about how you can construct this model by using the two kinds of extruded solids — solids extruded along a line and solids extruded along a path.

Figure 6.62 shows a suggested way of composing the model. You will construct three 2D regions and a polyline. From these 2D objects, you will make four extruded solids. In addition, you will make solid cylinders. Finally, you will compose the required model from these objects.

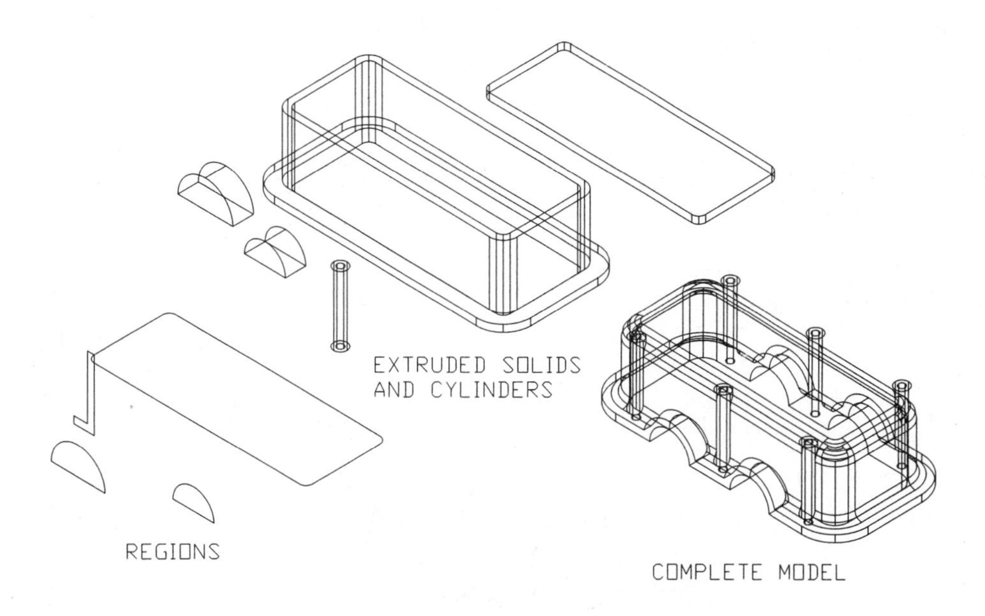

Figure 6.62 2D regions, extruded solids, and model

Model Construction

Use the NEW command to start a new drawing from scratch with metric units.

<File> **<New...>**

Create an additional layer Solid with color yellow and set layer Solid as the current layer.

<Format> **<Layer...>**

Command: **LAYER**

On layer Solid, use the RECTANG command to create a rectangle with four round corners. A rectangle created by the RECTANG command is a polyline.

<Draw> **<Rectangle>**

Command: **RECTANG**
Chamfer/Elevation/Fillet/Thickness/Width/<First corner>: **F**
Fillet radius for rectangles: **20**
Chamfer/Elevation/Fillet/Thickness/Width/<First corner>: **[Select a point near the lower left corner of the screen.]**
Other corner: **@320,120**

Set the display to an isometric view. See Figure 6.63.

<View> **<3d Viewpoint>** **<SE Isometric>**

Command: **VPOINT**
Rotate/<View point>: **1,-1,1**

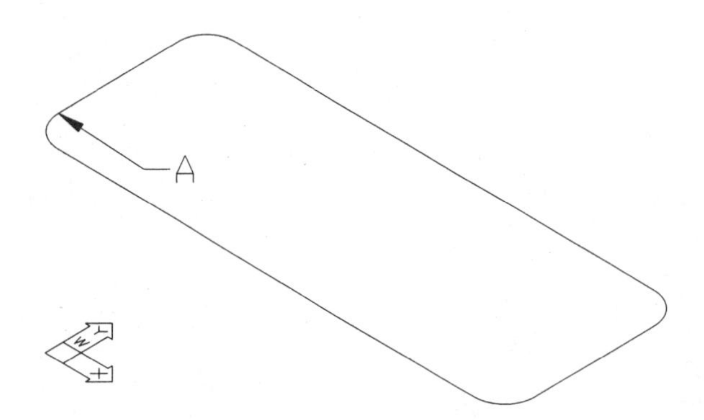

Figure 6.63 Rectangle created, and display set to an isometric view

The rectangle that you create will serve two purposes: for use as a path for extruding a 2D object and for extrusion to become a solid.

The next object will be constructed on a plane perpendicular to the rectangle. Set the UCS to a new position and orientation by selecting the Z Axis Vector item of the UCS cascading menu of the Tools pull-down menu to use the UCS command. Select A of Figure 6.63 as the new UCS origin, and locate the Z axis of the UCS by setting a vector of 1 unit in the 270° direction.

<Tools> **<UCS>** **<Z Axis Vector>**

Command: **UCS**
Origin/ZAxis/3point/OBject/View/X/Y/Z/Prev/Restore/Save/Del/?/<World>: **ZAXIS**
Origin point: **END** of [**Select A (Figure 6.63).**]
Point on positive portion of Z-axis: **@1<270**

To see the effect clearly, turn on the UCS icon and display it at the current origin position. See Figure 6.64.

<View> **<Display>** **<UCS Icon>** **<v On>**

Command: **UCSICON**
ON/OFF/All/Noorigin/ORigin: **ON**

<View> **<Display>** **<UCS Icon>** **<v Origin>**

Command: **UCSICON**
ON/OFF/All/Noorigin/ORigin: **OR**

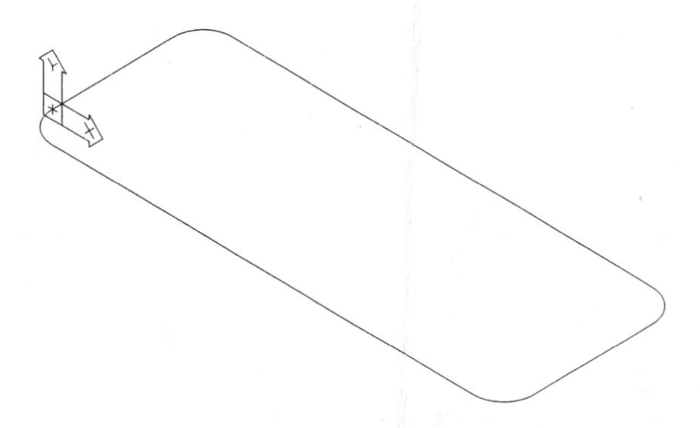

Figure 6.64 UCS set to new position

Use the RECTANG command to construct two rectangles. Do you still remember that you have set the fillet radius to 10 when you make the last rectangle? In order to make rectangles with sharp corners, set the fillet radius to 0. See Figure 6.65.

<Draw> **<Rectangle>**

Command: **RECTANG**
Chamfer/Elevation/Fillet/Thickness/Width/<First corner>: **F**
Fillet radius for rectangles: **0**
Chamfer/Elevation/Fillet/Thickness/Width/<First corner>: **0,0**
Other corner: **-10,-100**

Command: **[Enter]**
RECTANG
Chamfer/Elevation/Fillet/Thickness/Width/<First corner>: **0,-100**
Other corner: **@-30,10**

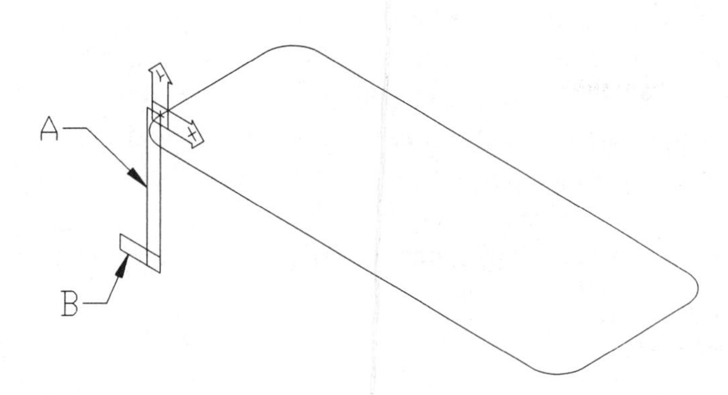

Figure 6.65 Rectangles created

Rectangles are closed polylines. You can use them to make extruded solids without converting them to regions. However, if you want to make a complex region from two or more rectangles, you have to change the rectangular polylines to regions.

Use the REGION command to convert the two rectangles to regions, and use the UNION command to unite them to form a complex region. See Figure 6.66.

<Draw> **<Region>**

Command: **REGION**
Select objects: **[Select A and B (Figure 6.65).]**
Select objects: **[Enter]**
2 loops extracted.
2 Regions created.

<Modify> **<Boolean>** **<Union>**

Command: **UNION**
Select objects: **[Select A and B (Figure 6.65).]**
Select objects: **[Enter]**

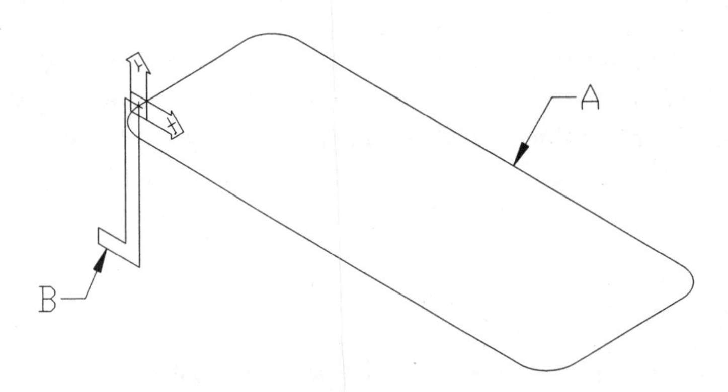

Figure 6.66 Rectangles converted to regions and united into a complex region

Now you have a rectangle with round corners and a complex region. As we have said before, the rectangle will serve two purposes. You will first use it as a path for the complex region and then extrude it to become an extruded solid. In so doing, you will face a problem. After you have used the rectangle as a path to make an extruded solid, the resulting solid coincides with the path and you may find it difficult to select this rectangle for extrusion in a linear direction. To solve this problem, you can create an additional layer, turn off the new layer, and put the extruded solid on that layer so that the rectangle is readily selectable. To solve this problem quickly, you can use the GROUP command to put the rectangle in an entity group. Later on, you can select it by specifying the entity group name.

Select the Object Group... item from the Tools pull-down menu to use the GROUP command.

<Tools> **<Object Group...>**

Command: **GROUP**

[Object Grouping
Group Name: **TOP**
New<]

Select objects for grouping:
Select objects: **[Select A (Figure 6.66).]**

Select objects: **[Enter]**

[OK]

Now you have put the rectangle in an entity group called TOP. You can extrude the complex region along it and then extrude it by specifying its entity group name.

Select the Extrude item from the Solids cascading menu of the Draw pull-down menu to create an extruded solid from the complex region and the rectangular path. See Figure 6.67.

<Draw> **<Solids>** **<Extrude>**

Command: **EXTRUDE**
Select objects: **[Select B (Figure 6.66).]**
Select objects: **[Enter]**
Path/<Height of Extrusion>: **P**
Select path: **[Select A (Figure 6.66).]**

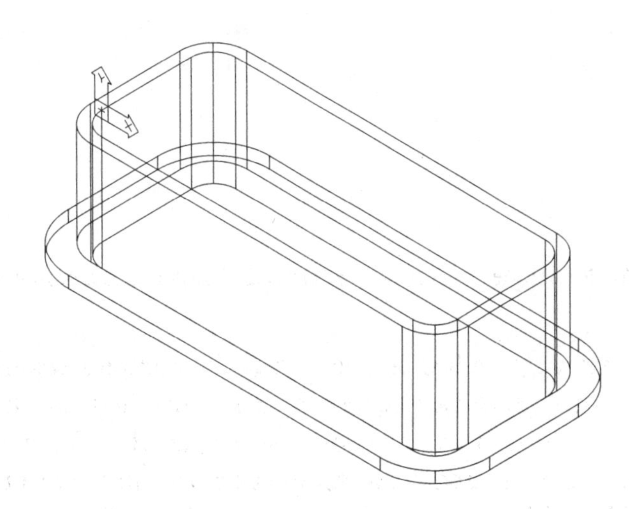

Figure 6.67 Complex region extruded along a path

After extrusion of complex region along the rectangular path, an extruded solid forms and overlaps with the rectangle.

In order to use this rectangle as a cross section for extrusion, specify the name of the group that contains this rectangle while you use the EXTRUDE command. See Figure 6.68.

<Draw> **<Solids>** **<Extrude>**

Command: **EXTRUDE**
Select objects: **G**
Enter group name: **TOP**
Select objects: **[Enter]**
Path/<Height of Extrusion>: **-10**
Extrusion taper angle: **0**

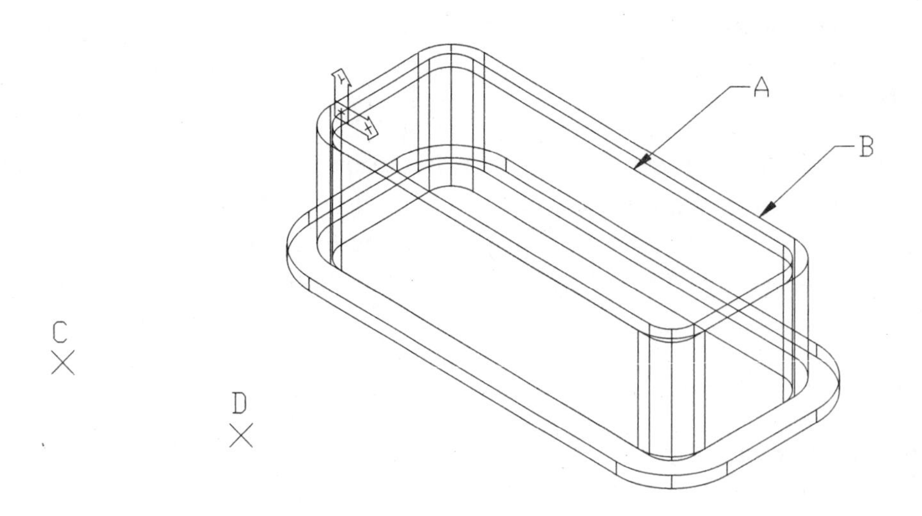

Figure 6.68 Rectangle extruded

Two kinds of extruded solids are created. To unite them, select the Union item of the Boolean cascading menu of the Modify pull-down menu to use the UNION command.

<Modify> **<Boolean>** **<Union>**

Command: **UNION**
Select objects: [**Select A and B (Figure 6.68).**]
Select objects: [**Enter**]

The main body of the gear box cover is complete. You will make two regions from two circular arcs and two line segments. Select the Center, Start, End item of the Arc cascading menu of the Draw pull-down menu to create two arcs. See Figure 6.69.

<Draw> **<Arc>** **<Center, Start, End>**

Command: **ARC**
Center/<Start point>: **C**
Center: [**Select C (Figure 6.68).**]
Start point: **@40<0**
Angle/Length of chord/<End point>: **@40<180**

<Draw> **<Arc>** **<Center, Start, End>**

Command: **ARC**
Center/<Start point>: **C**
Center: [**Select D (Figure 6.68).**]
Start point: **@30<0**
Angle/Length of chord/<End point>: **@30<180**

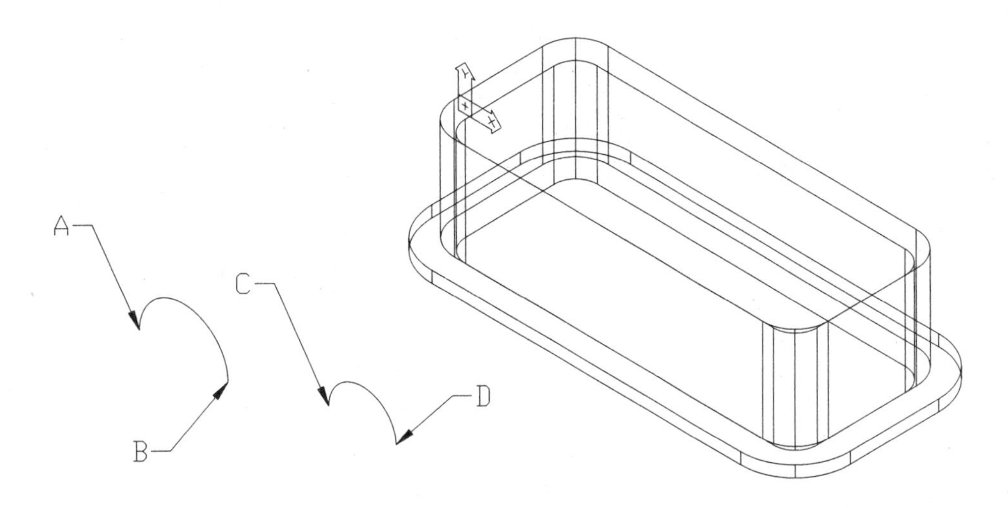

Figure 6.69 Arcs created

Select the Line item of the Draw pull-down menu to create two line segments to form two closed loops with the two arcs drawn. Then select the Region item of the Draw pull-down menu to create two regions from the two lines and arcs. See Figure 6.70.

<Draw> **<Line>**

Command: **LINE**
From point: **END** of [**Select A (Figure 6.69).**]
To point: **END** of [**Select B (Figure 6.69).**]
To point: [**Enter**]

<Draw> **<Line>**

Command: **LINE**
From point: **END** of [**Select C (Figure 6.69).**]
To point: **END** of [**Select D (Figure 6.69).**]
To point: [**Enter**]

<Draw> **<Region>**

Command: **REGION**
Select objects: [**Select the two arcs and two lines.**]
Select objects: [**Enter**]
2 loops extracted.
2 Regions created.

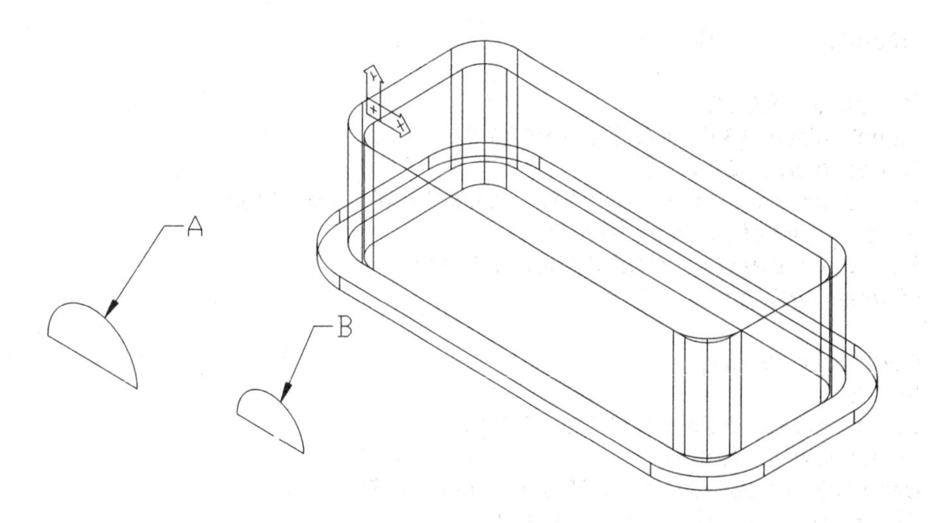

Figure 6.70 Regions created

Using the two regions as cross sections, apply the EXTRUDE command to create two extruded solids. See Figure 6.71.

<Draw> **<Solids>** **<Extrude>**

Command: **EXTRUDE**
Select objects: [**Select A and B (Figure 6.70).**]
Select objects: [**Enter**]
Path/<Height of Extrusion>: **30**
Extrusion taper angle: **0**

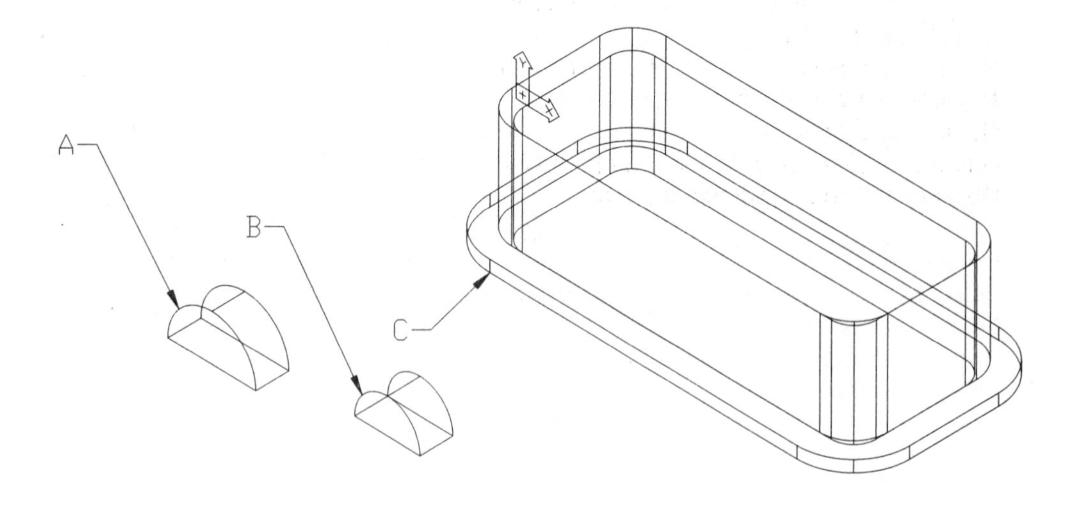

Figure 6.71 Regions extruded

To translate the two extruded solids to their proper position, select the Move item of the Modify pull-down menu to use the MOVE command.

<Modify> <Move>

Command: **MOVE**
Select objects: [**Select A (Figure 6.71).**]
Select objects: [**Enter**]
Base point or displacement: **CEN** of [**Select A (Figure 6.71).**]
Second point of displacement: **FROM**
Base point: **END** of [**Select C (Figure 6.71).**]
<Offset>: **@80<0**

Command: [**Enter**]
MOVE
Select objects: [**Select B (Figure 6.71).**]
Select objects: [**Enter**]
Base point or displacement: **CEN** of [**Select B (Figure 6.71).**]
Second point of displacement: **FROM**
Base point: **END** of [**Select C (Figure 6.71).**]
<Offset>: **@80<0**

There should be three more copies of these two extruded solids. All together, there are two rows and two columns. To achieve this, set the UCS to World and use the ARRAY command to make a rectangular array. See Figure 6.72.

<Tools> <UCS> <World>

Command: **UCS**
Origin/ZAxis/3point/OBject/View/X/Y/Z/Prev/Restore/Save/Del/?/<World>: **W**

<Modify> <Array>

Command: **ARRAY**
Select objects: [**Select the two extruded solids.**]
Select objects: [**Enter**]
Rectangular or Polar array (<R>/P): **R**
Number of rows (---): **2**
Number of columns (|||): **2**
Unit cell or distance between rows (---): **150**
Distance between columns (|||): **120**

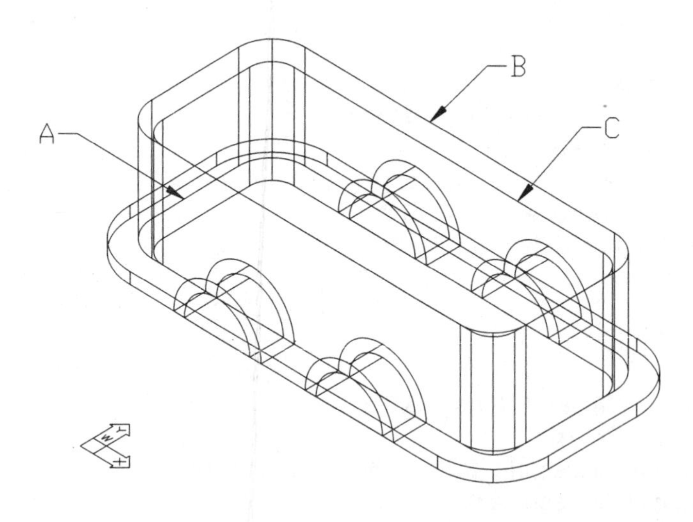

Figure 6.72 Extruded solids moved and arrayed

Use the FILLET command to create three chains of fillet edges. See Figure 6.73. When you fillet, you will use the Chain option to select a chain of edges collectively by selecting an edge of the chain. The fillet radius of the chain of edges at A and C is 10 units. These are internal fillets. The fillet radius of the chain of edges at B is 20. These are external fillets.

<Modify> **<Fillet>**

Command: **FILLET**
Polyline/Radius/Trim/<Select first object>: **[Select A (Figure 6.72).]**
Enter radius: **10**
Chain/Radius/<Select edge>: **C**
Edge/Radius/<Select edge chain>: **[Select A and C (Figure 6.72).]**
Edge/Radius/<Select edge chain>: **[Enter]**

Command: **[Enter]**
FILLET
Polyline/Radius/Trim/<Select first object>: **[Select B (Figure 6.72).]**
Enter radius: **20**
Chain/Radius/<Select edge>: **C**
Edge/Radius/<Select edge chain>: **[Select B (Figure 6.72).]**
Edge/Radius/<Select edge chain>: **[Enter]**

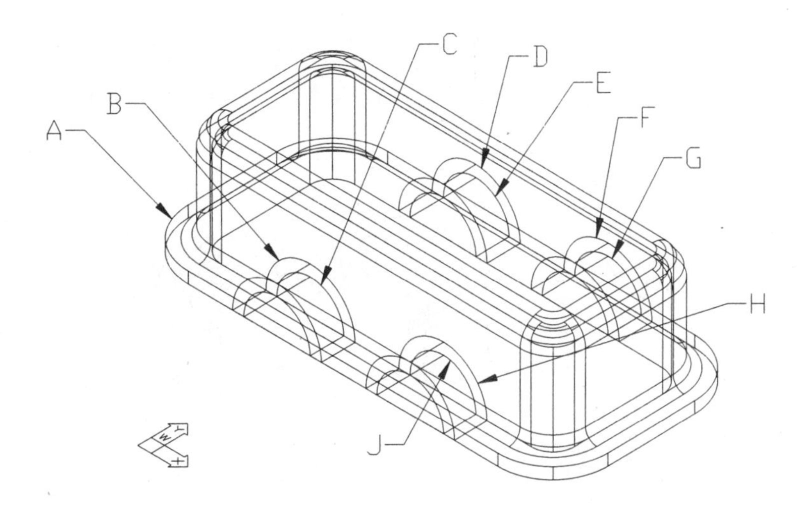

Figure 6.73 Edges filleted

After filleting, use the UNION command to unite four extruded solids to make four semi-circular bosses, and use the SUBTRACT command to subtract four extruded solids to make four semi-circular holes. See Figure 6.74.

In using the UNION command, select the main body and the four larger extruded solids.

<Modify> **<Boolean>** **<Union>**

Command: **UNION**
Select objects: [**Select A, B, D, F, and H (Figure 6.73).**]
Select objects: [**Enter**]

Check your drawing to ensure that the four larger extruded solids are united. Then use the SUBTRACT command to subtract the four smaller extruded solids from the main body.

<Modify> **<Boolean>** **<Subtract>**

Command: **SUBTRACT**
Select solids and regions to subtract from...
Select objects: [**Select A (Figure 6.73).**]
Select objects: [**Enter**]
Select solids and regions to subtract...
Select objects: [**Select C, E, G, and J (Figure 6.73).**]
Select objects: [**Enter**]

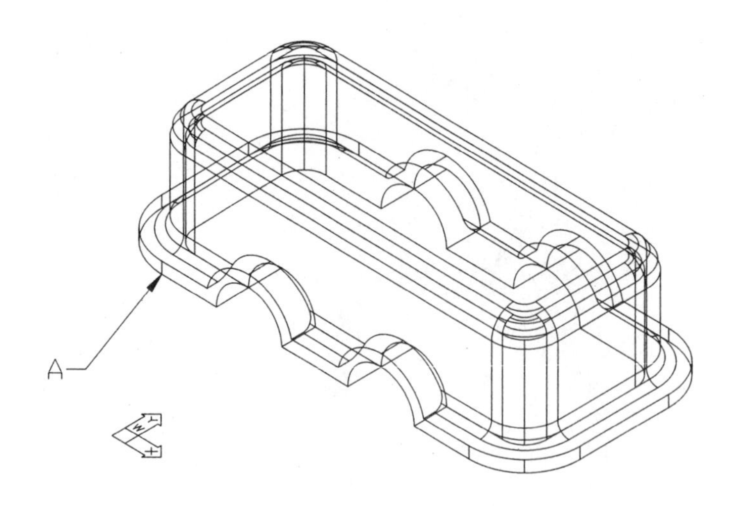

Figure 6.74 Extruded and arrayed solids united and subtracted

There should be six vertical circular bosses with concentric holes on the completed model. Use the CYLINDER command to make two solid cylinders. See Figure 6.75. You will use the large cylinder to form a boss and the smaller cylinder to form a hole.

<Draw> **<Solids>** **<Cylinder>**

Command: **CYLINDER**
Elliptical/<center point>: **FROM**
Base point: **END** of **[Select A (Figure 6.74).]**
<Offset>: **@10,20**
Diameter/<Radius>: **5**
Center of other end/<Height>: **100**

Command: **[Enter]**
CYLINDER
Elliptical/<center point>: **FROM**
Base point: **END** of **[Select A (Figure 6.74).]**
<Offset>: **@10,20**
Diameter/<Radius>: **10**
Center of other end/<Height>: **100**

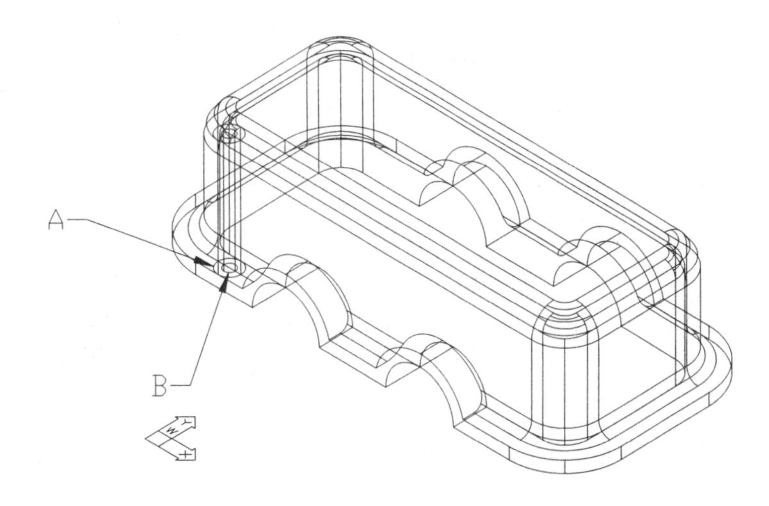

Figure 6.75 Cylinders created

Now you have one set of boss and hole. To create the remaining five bosses and holes, use the ARRAY command to make a rectangular array of two rows and three columns. See Figure 6.76.

<Modify> **<Array>**

Command: **ARRAY**
Select objects: [**Select A and B (Figure 6.75).**]
Select objects: [**Enter**]
Rectangular or Polar array (<R>/P): **R**
Number of rows (---): **2**
Number of columns (|||): **3**
Unit cell or distance between rows (---): **140**
Distance between columns (|||): **130**

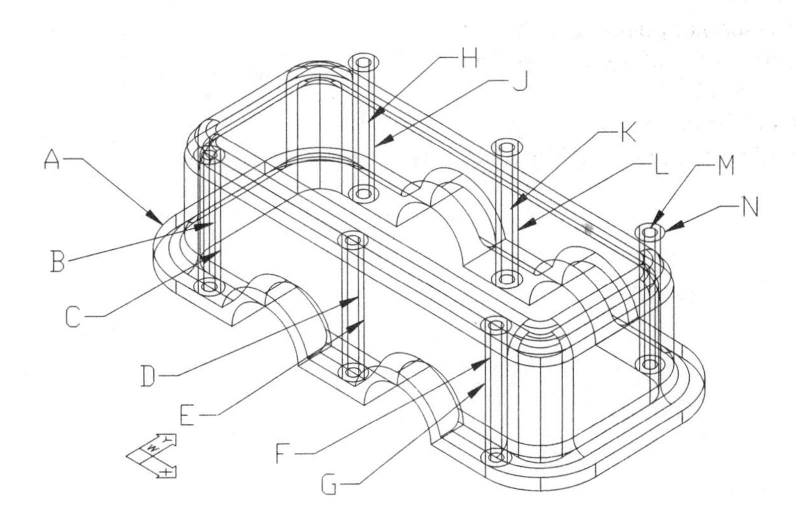

Figure 6.76 Cylinders arrayed

To complete the solid model, use the UNION command to unite the six larger cylinders to form six bosses and the SUBTRACT command to subtract the six smaller cylinders to form six holes. See Figure 6.77.

<Modify> **<Boolean>** **<Union>**

Command: **UNION**
Select objects: **[Select A, C, E, G, J, L, and N (Figure 6.76).]**
Select objects: **[Enter]**

<Modify> **<Boolean>** **<Subtract>**

Command: **SUBTRACT**
Select solids and regions to subtract from...
Select objects: **[Select A (Figure 6.76).]**
Select objects: **[Enter]**
Select solids and regions to subtract...
Select objects: **[Select B, D, F, H, K, and M (Figure 6.76).]**
Select objects: **[Enter]**

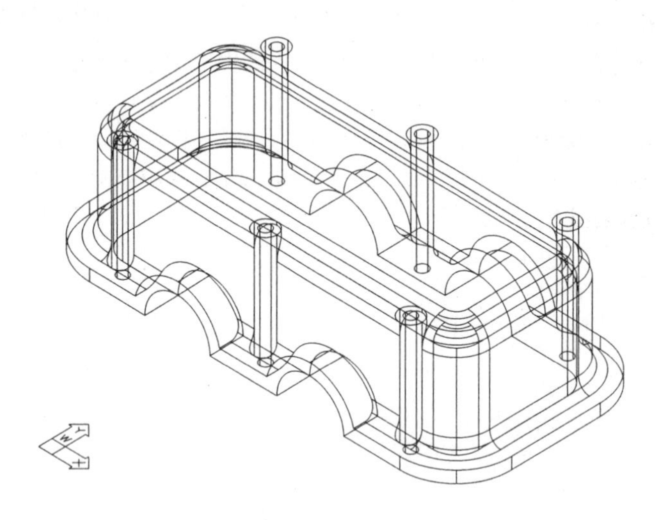

Figure 6.77 Cylinders united and subtracted

The model of a gear box cover is complete. See also Figure 6.61. Save your drawing.

<File> **<Save>**

File name: **GBOX.DWG**

In making this model, you constructed two kinds of extruded solids: solids extruded linearly and solids extruded along a path.

Model of a Pulley

With a 2D region or closed, non-intersecting polyline as cross section, you can revolve as well as extrude. Figure 6.78 shows a model of a pulley that can be readily created by revolving a region or a closed polyline.

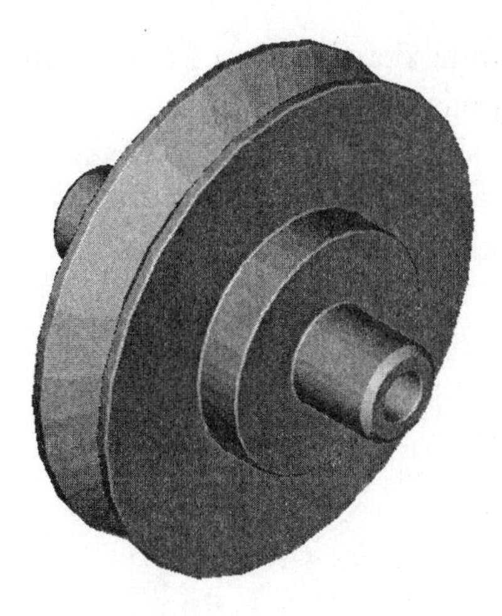

Figure 6.78 Pulley

Analysis

Refer to Figure 6.78. You can make the model in many ways. One way is to combine cylinder and cone primitives together and then chamfer the edges. In so doing, you have to create a number of primitives. To do it in a much quicker way, you can construct a 2D region or polyline that represents a section cut along the axis of the model and revolve the region or polyline about an axis that passes through the center of the final model. Figure 6.79 shows the region and the model.

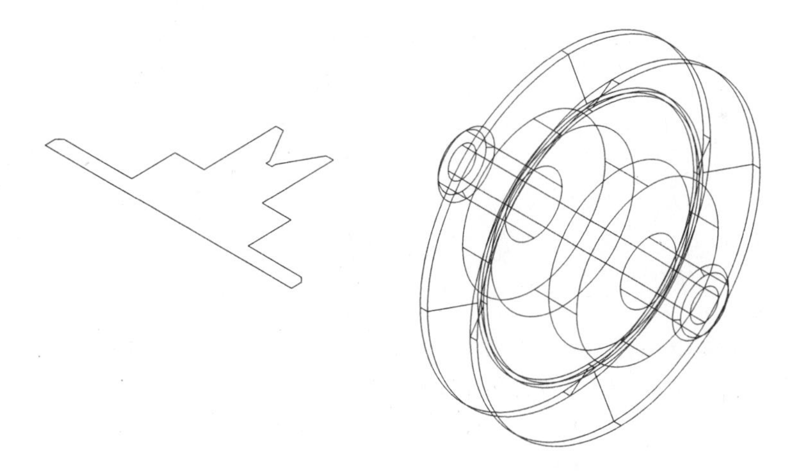

Figure 6.79 Cross section along axis and model

Model Construction

Start a new drawing from scratch and use metric units.

<File> **<New...>**

According to the dimensions delineated in Figure 6.80, create a set of line segments. Do not include the dimensions in your drawing. After making the drawing, use the REGION command to form a region from these line segments.

<Draw> **<Line>**

In making the drawing, you may need to use the TRIM command, the EXTEND command, and other modify commands.

<Draw> **<Region>**

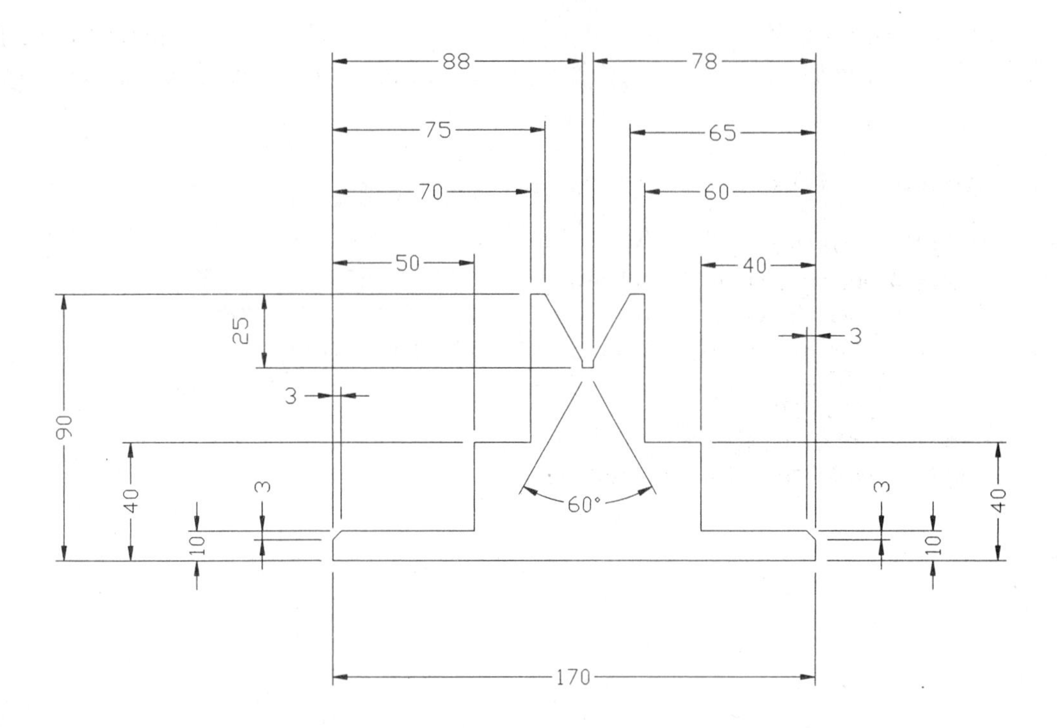

Figure 6.80 Region

After completing the 2D region for the model of a pulley, select the SE Isometric item of the 3D Viewpoint to set the display to an isometric view. See Figure 6.81.

<View> **<3D Viewpoint>** **<SE Isometric>**

```
Command: VPOINT
Rotate/<View point>: 1,-1,1
```

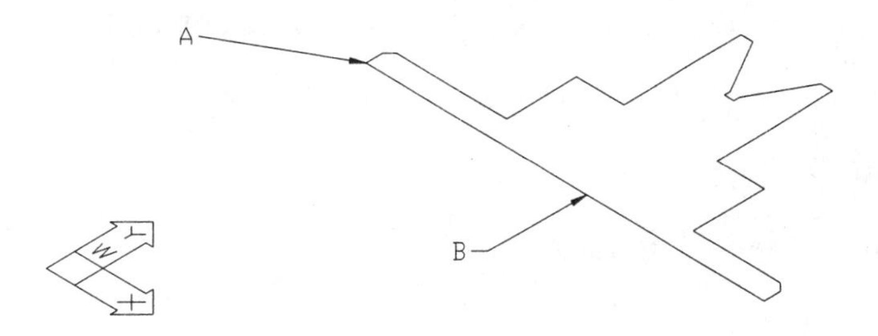

Figure 6.81 Display set to an isometric view

Select the Revolve item of the Solids cascading menu of the Draw menu to use the REVOLVE command to revolve the 2D region to become a revolved solid. See Figure 6.82.

<Draw> **<Solids>** **<Revolve>**

Command: **REVOLVE**
Select objects: [**Select B (Figure 6.81).**]
Select objects: [**Enter**]
Axis of revolution - Object/X/Y/<Start point of axis>: **FROM**
Base point: **END** of [**Select A (Figure 6.81).**]
<Offset>: **@10<270**
<End point of axis>: **@1<0**
Angle of revolution <full circle>: [**Enter**]

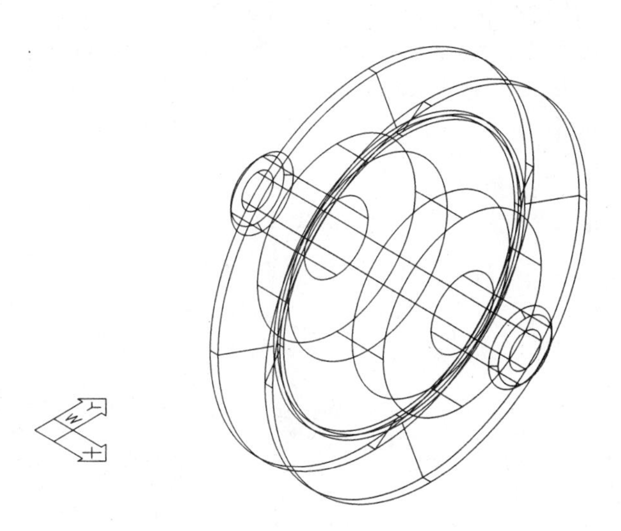

Figure 6.82 Revolved solid created

The model is complete. See also Figure 6.78. Save your drawing.

<File> **<Save>**

File name: **PULLEY.DWG**

In constructing this model, you made a closed, 2D region and revolved it to become a solid of revolution.

Model of a Watch Case

Figure 6.83 shows the model of a watch case.

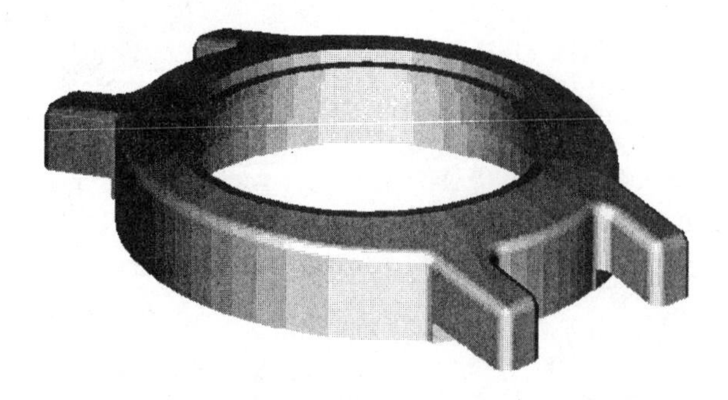

Figure 6.83 Watch case

Analysis

In considering how to construct this model, think about the use of an extruded solid and a revolved solid. Refer to Figure 6.84. You can make this model by intersecting an extruded solid and a revolved solid. You will prepare two profiles that represent the top view and the cross section, respectively. From the profiles, you will produce two regions. The region that is made from the top view will be used for making an extruded solid. The region that is made from the cross section will be used for making a revolved solid. The final complex solid model of the watch case is the intersection of these two solids.

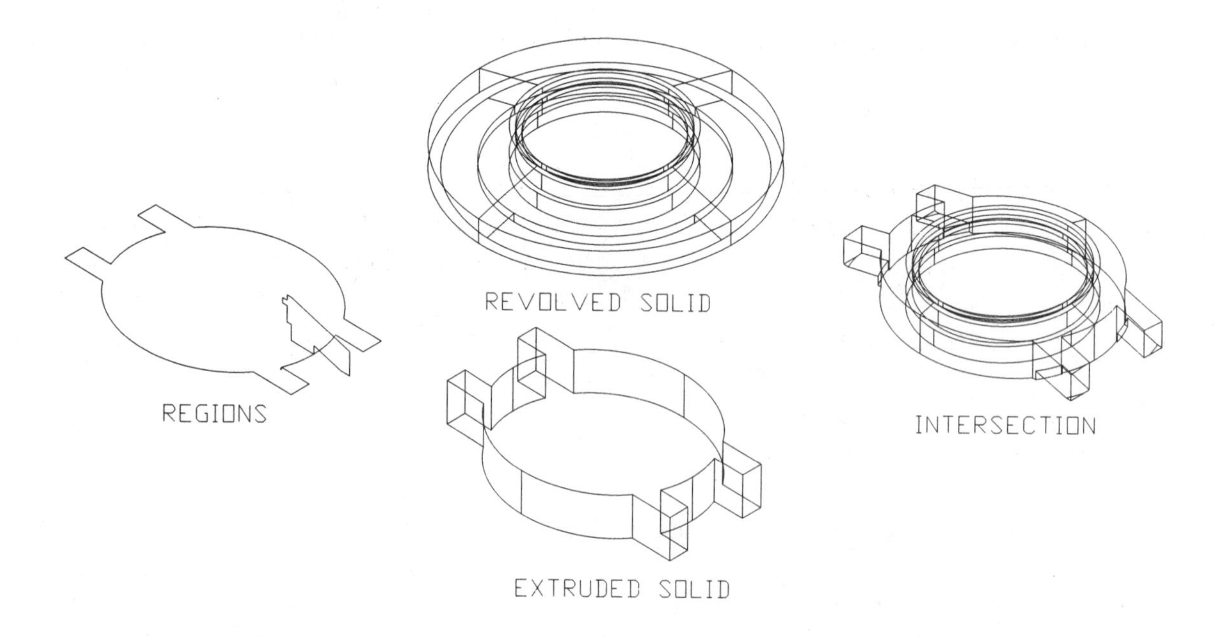

Figure 6.84 Regions, extruded solid, revolved solid, and solid of intersection

Figure 6.85 shows the dimension layout of profiles of the watch case.

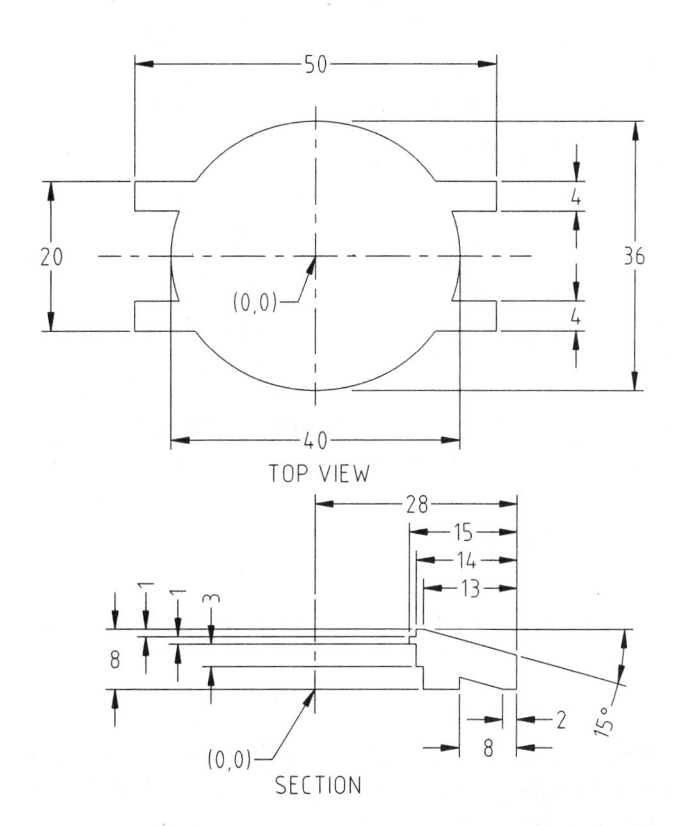

Figure 6.85 Profiles of watch case

Model Construction

Use the NEW command to start a new drawing from scratch with metric units.

<File> **<New...>**

Prepare your drawing. First, create a new layer called Solid, and set it to current with the LAYER command. Then place the UCS icon at the origin position by using the UCSICON command. Refer to Figure 6.85 and Figure 6.86. Create the required entities by using the ELLIPSE, LINE, and TRIM commands. Next, convert them either to a closed polyline using the PEDIT command or to a region using the REGION command.

<Draw> **<Region>**

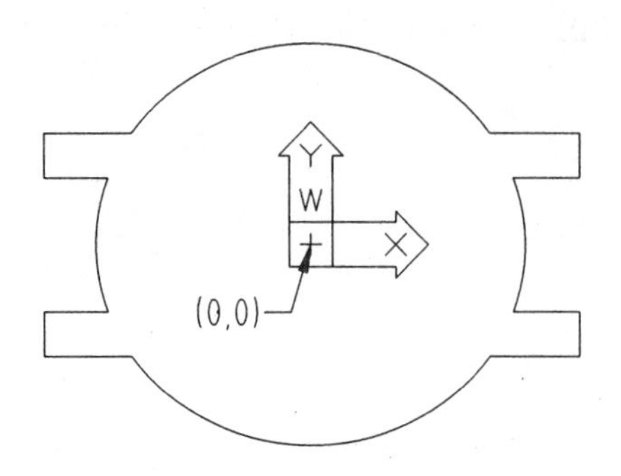

Figure 6.86 Profile of top view created

Set the display to an isometric viewing position by using the VPOINT command. Then set UCS to rotate 90° about the X axis. See Figure 6.87.

<View> **<3D Viewpoint>** **<SE Isometric>**

Command: **VPOINT**
Rotate/<View point>: **1,-1,1**

<Tools> **<UCS>** **<X Axis Rotate>**

Command: **UCS**
Origin/ZAxis/3point/OBject/View/X/Y/Z/Prev/Restore/Save/Del/?/<World>: **X**
Rotation angle about X axis: **90**

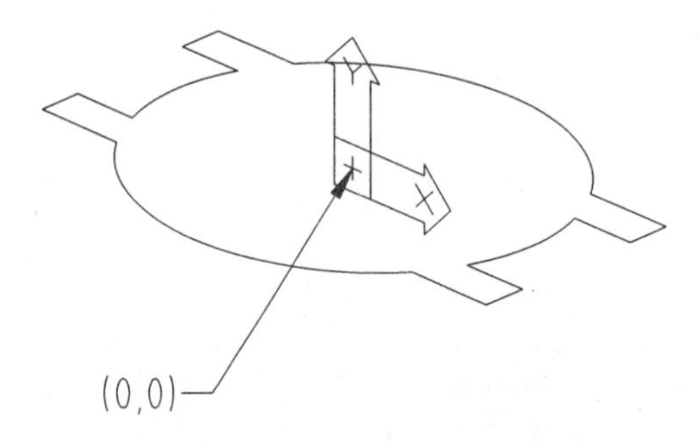

Figure 6.87 Display set to isometric, and UCS rotated

Referring to Figure 6.85 and Figure 6.88, create another set of entities. Then convert them to either a polyline or a region.

<Draw> **<Region>**

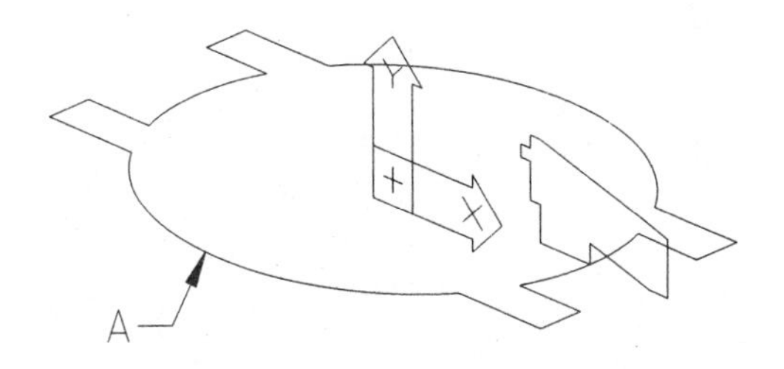

Figure 6.88 Profile of cross section created

The wireframes for making the solid model of the watch case are complete. The solid model is an intersection of two solids: extruded solid and revolved solid. To produce the extruded solid, run the EXTRUDE command. See Figure 6.89.

<Draw> **<Solids>** **<Extrude>**

Command: **EXTRUDE**
Select objects: **[Select A (Figure 6.88).]**
Select objects: **[Enter]**
Path/<Height of Extrusion>: **8**
Extrusion taper angle: **0**

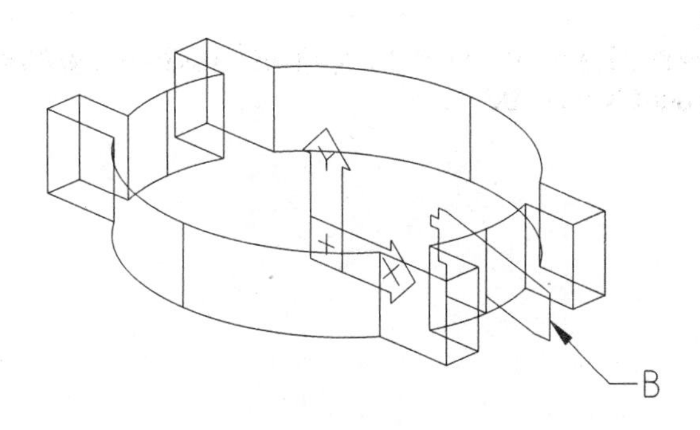

Figure 6.89 Extruded solid created

To produce the revolved solid, issue the REVOLVE command. See Figure 6.90.

<Draw> **<Solids>** **<Revolve>**

Command: **REVOLVE**
Select objects: **[Select B (Figure 6.89).]**
Select objects: **[Enter]**
Axis of revolution - Object/X/Y/<Start point of axis>: **Y**
Angle of revolution <full circle>: **[Enter]**

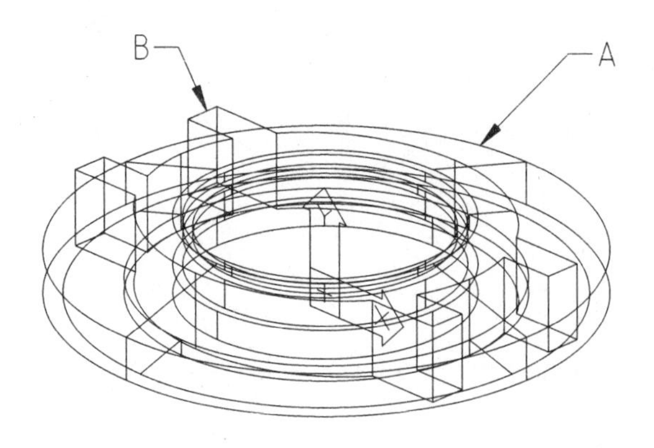

Figure 6.90 Revolved solid created

With the extruded solid and the revolved solid ready, use the INTERSECT command to form a solid of intersection. See Figure 6.91.

<Modify> **<Boolean>** **<Intersect>**

Command: **INTERSECT**
Select objects: **[Select A and B (Figure 6.90).]**
Select objects: **[Enter]**

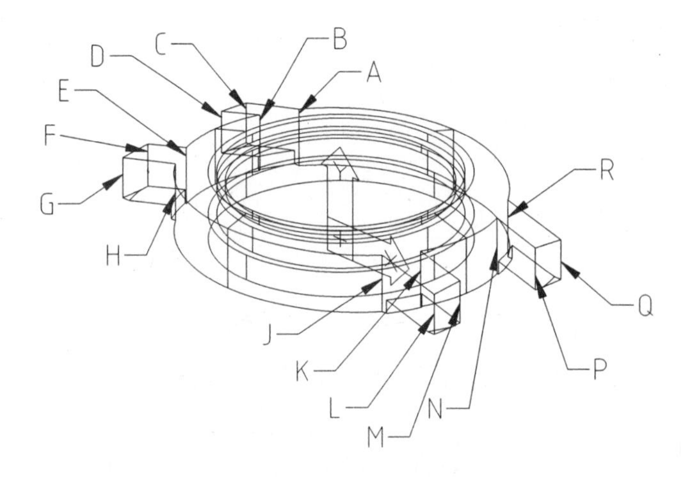

Figure 6.91 Solid of intersection created

The main body of the watch case is complete. To round off the edges, run the FILLET command. See Figure 6.92.

<Modify> **<Fillet>**

Command: **FILLET**
Polyline/Radius/Trim/<Select first object>: **[Select A (Figure 6.91).]**
Enter radius: **2**
Chain/Radius/<Select edge>: **[Select A, B, C, D, E, F, G, H, J, K, L, M, N, P, Q, and R (Figure 6.91).]**
Chain/Radius/<Select edge>: **[Enter]**

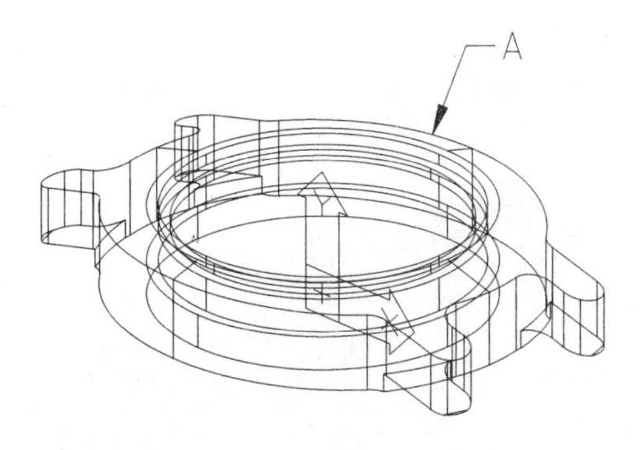

Figure 6.92 Eighteen edges filleted

Repeat the FILLET command to create a chain of fillets on the top face of the model. See Figure 6.93.

<Modify> **<Fillet>**

Command: **FILLET**
Polyline/Radius/Trim/<Select first object>: **[Select A (Figure 6.92).]**
Enter radius: **2**
Chain/Radius/<Select edge>: **C**
Edge/Radius/<Select edge chain>: **[Select A (Figure 6.92).]**
Edge/Radius/<Select edge chain>: **[Enter]**

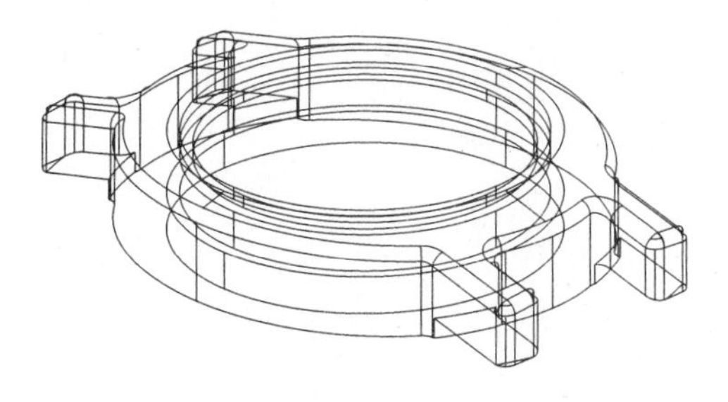

Figure 6.93 Edges filleted

The model of the watch case is complete. See also Figure 6.83. Save your drawing.

<File> **<Save>**

File name: **WATCH.DWG**

In making this model, you constructed 2D regions that represent the top view and the cross section of the watch case. Then you produced an extruded solid and a revolved solid. With the extruded solid and the revolved solid, you formed the main body of the watch case by intersection.

Model of a Fluid Power Valve

Figure 6.94 shows the model of a fluid power valve casing.

Figure 6.94 Fluid power valve casing

Analysis

There are many ways to construct this model. Take some time to think how you can make the main body of this model by intersecting three extruded solids. Figure 6.95 shows the three extruded solids required.

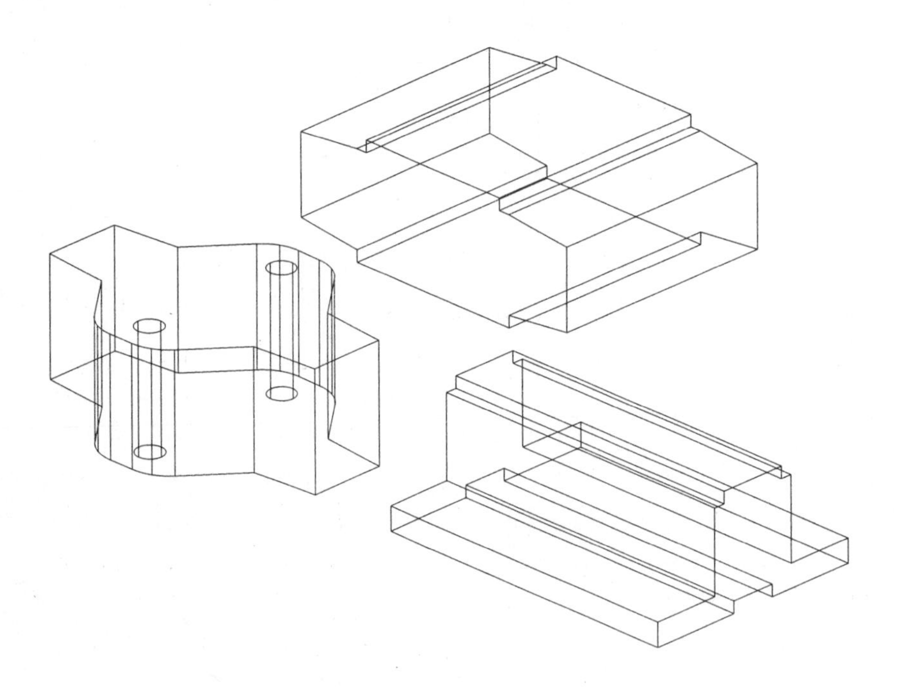

Figure 6.95 Extruded solids

The core of the hydraulic valve is made from a solid of revolution. Figure 6.96 shows the cross section and the revolved solid created from the cross section.

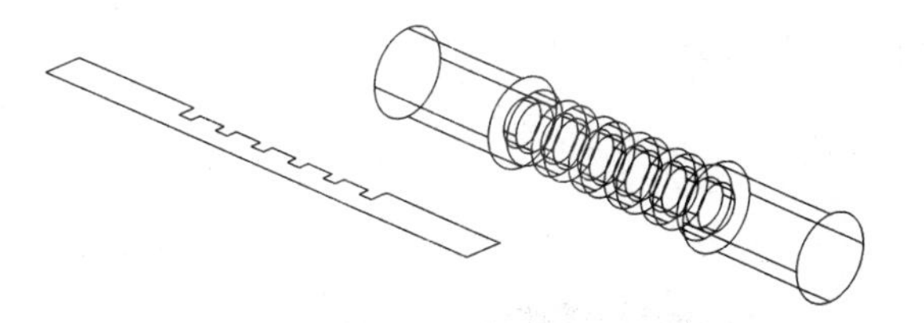

Figure 6.96 Core cross section and revolved solid

To produce the three extruded solids for the main body, you need three profiles that can be either regions or closed polylines. Figure 6.97 shows the shapes and dimensions of the profiles. From these profiles, you will create three solids of extrusion. Intersection of the three extruded solids forms the main body of the valve.

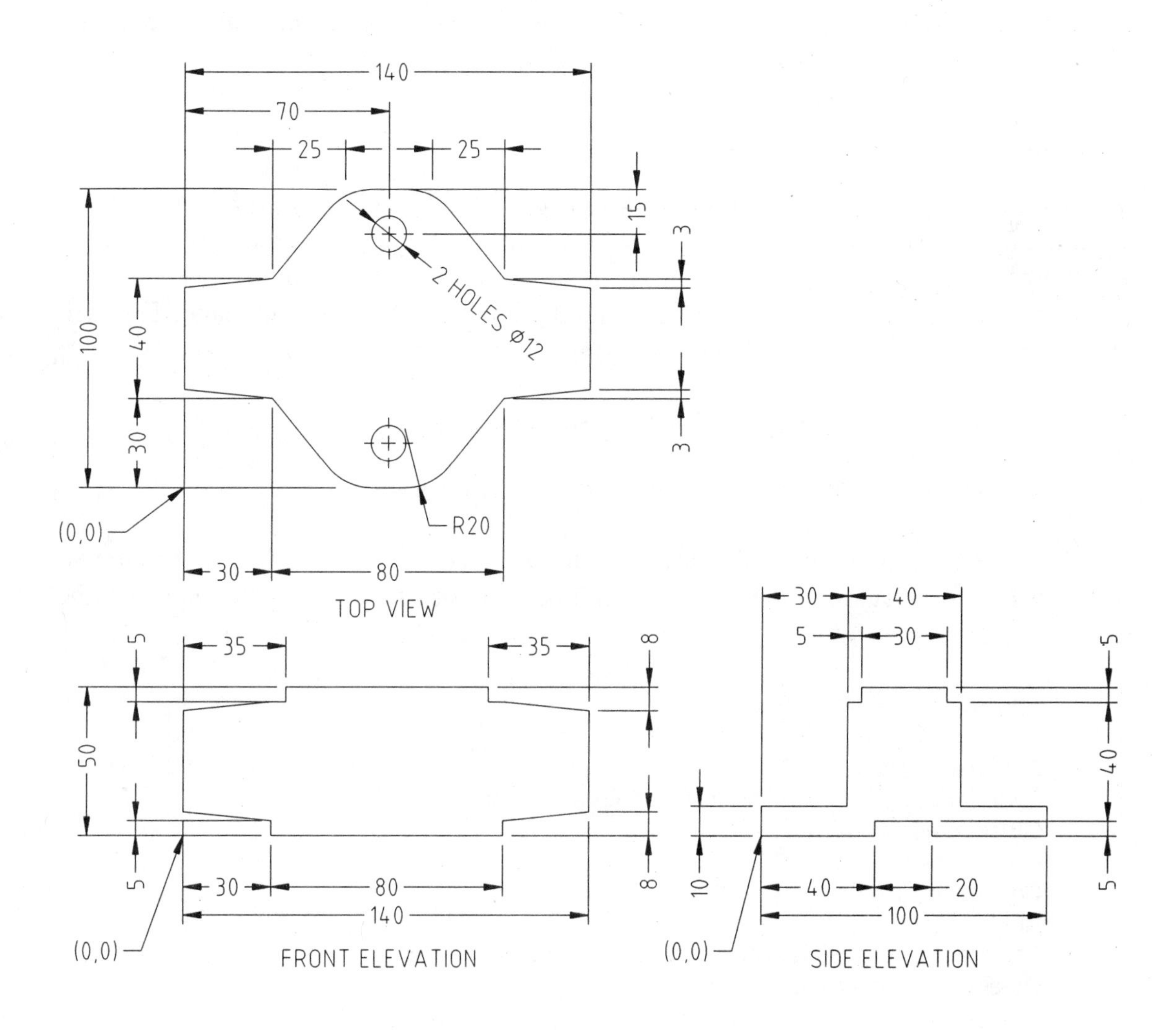

Figure 6.97 Profiles required

Model Construction

Figure 6.98 shows the profile for the top view that you will make on the XY plane.

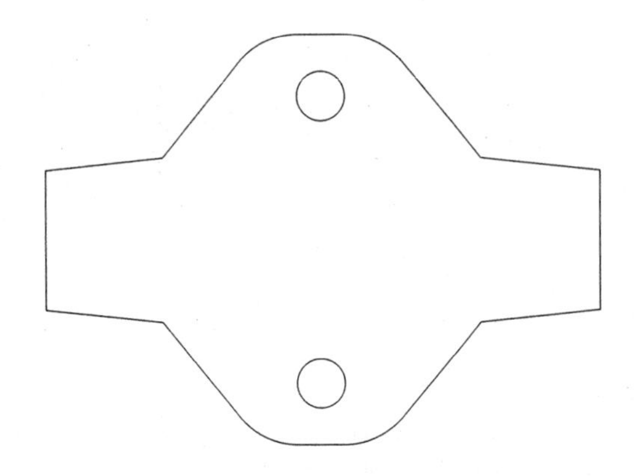

Figure 6.98 Profile of top view

Use the NEW command to begin a new drawing. Start from scratch and use metric units.

<File> **<New...>**

Make three layers called Spool, Wireframe, and Solid. The colors of layers Spool, Wireframe, and Solid are red, cyan, and yellow, respectively. The current layer should be Wireframe. You will create the profiles for the three extruded solids on this layer.

Set the UCS icon to lie on the origin position with the UCSICON command. This will give you a better idea of where you are working.

<View> **<Display>** **<UCS Icon>** **<v On>**

<View> **<Display>** **<UCS Icon>** **<v Origin>**

To create the profile, you will start by using the RECTANG command to create two rectangles. After that, use the ZOOM command to obtain a complete view of the rectangles. See Figure 6.99.

<Draw> **<Rectangle>**

Command: **RECTANG**
Chamfer/Elevation/Fillet/Thickness/Width/<First corner>: **0,30**
Other corner: **@140,40**

Command: **[Enter]**
RECTANG
Chamfer/Elevation/Fillet/Thickness/Width/<First corner>: **30,0**
Other corner: **@80,100**

<View> **<Zoom>** **<Extents>**

Command: **ZOOM**
All/Center/Dynamic/Extents/Previous/Scale(X/XP)/Window/<Realtime>: **E**

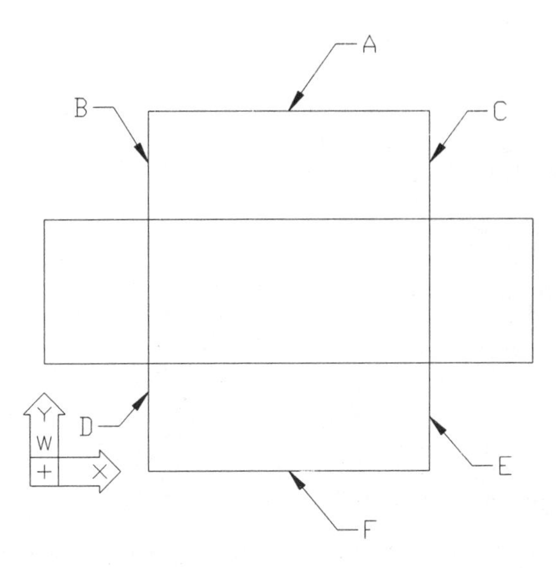

Figure 6.99 Rectangles created

To obtain the required profile, you need to chamfer and fillet the rectangles. Use the CHAMFER command to set the chamfer distances to 25 and 30 units, respectively, and to chamfer the corners of a rectangle. See Figure 6.100.

<Modify> **<Chamfer>**

Command: **CHAMFER**
Polyline/Distance/Angle/Trim/Method/<Select first line>: **D**
Enter first chamfer distance: **25**
Enter second chamfer distance: **30**

Command: **[Enter]**
CHAMFER
(TRIM mode) Current chamfer Dist1 = 25.0000, Dist2 = 30.0000
Polyline/Distance/Angle/Trim/Method/<Select first line>: [**Select A (Figure 6.99).**]
Select second line: [**Select B (Figure 6.99).**]

Command: **[Enter]**
CHAMFER
(TRIM mode) Current chamfer Dist1 = 25.0000, Dist2 = 30.0000
Polyline/Distance/Angle/Trim/Method/<Select first line>: [**Select A (Figure 6.99).**]
Select second line: [**Select C (Figure 6.99).**]

Command: **[Enter]**
CHAMFER
(TRIM mode) Current chamfer Dist1 = 25.0000, Dist2 = 30.0000
Polyline/Distance/Angle/Trim/Method/<Select first line>: [**Select F (Figure 6.99).**]
Select second line: [**Select D (Figure 6.99).**]

Command: **[Enter]**
CHAMFER
(TRIM mode) Current chamfer Dist1 = 25.0000, Dist2 = 30.0000
Polyline/Distance/Angle/Trim/Method/<Select first line>: [**Select F (Figure 6.99).**]
Select second line: [**Select E (Figure 6.99).**]

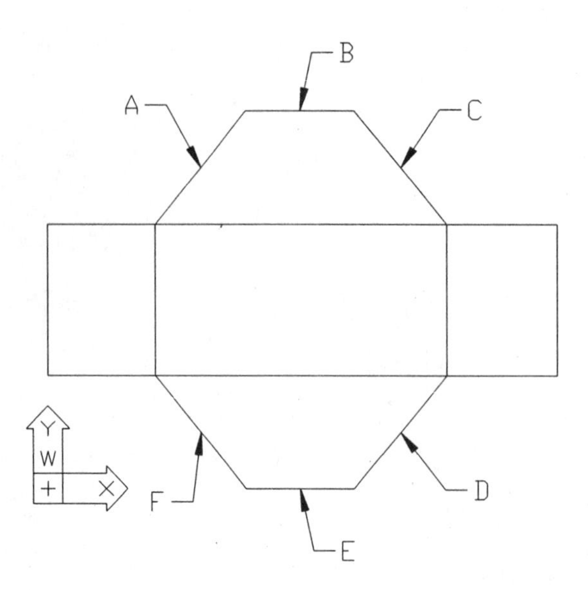

Figure 6.100 Rectangle chamfered

After chamfering, run the FILLET command to set the fillet radius to 20 units and to round off four corners. See Figure 6.101.

<Modify> **<Fillet>**

Command: **FILLET**
Polyline/Radius/Trim/<Select first object>: **R**
Enter fillet radius: **20**

Command: **[Enter]**
FILLET
(TRIM mode) Current fillet radius = 20.0000
Polyline/Radius/Trim/<Select first object>: **[Select A (Figure 6.100).]**
Select second object: **[Select B (Figure 6.100).]**

Command: **[Enter]**
FILLET
(TRIM mode) Current fillet radius = 20.0000
Polyline/Radius/Trim/<Select first object>: **[Select B (Figure 6.100).]**
Select second object: **[Select C (Figure 6.100).]**

Command: **[Enter]**
FILLET
(TRIM mode) Current fillet radius = 20.0000
Polyline/Radius/Trim/<Select first object>: **[Select D (Figure 6.100).]**
Select second object: **[Select E (Figure 6.100).]**

Command: **[Enter]**
FILLET
(TRIM mode) Current fillet radius = 20.0000
Polyline/Radius/Trim/<Select first object>: **[Select E (Figure 6.100).]**
Select second object: **[Select F (Figure 6.100).]**

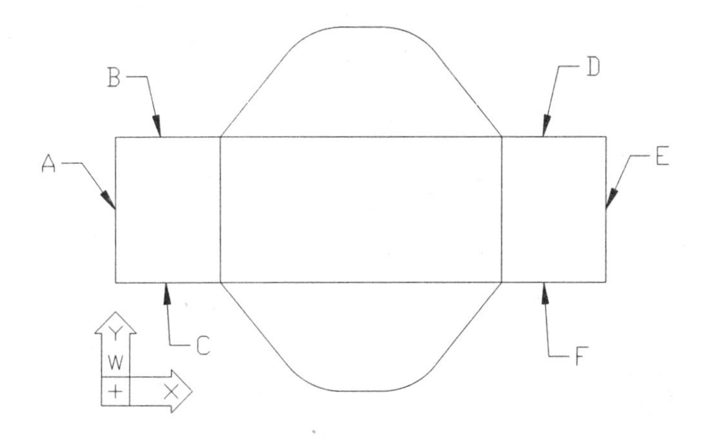

Figure 6.101 Corners filleted

One of the rectangles is complete. To continue, run the CHAMFER command to set the chamfer distances to 3 and 30, respectively, and to chamfer the four corners of the other rectangle. See Figure 6.102.

<Modify> **<Chamfer>**

Command: **CHAMFER**
Polyline/Distance/Angle/Trim/Method/<Select first line>: **D**
Enter first chamfer distance: **3**
Enter second chamfer distance: **30**

Command: **[Enter]**
CHAMFER
(TRIM mode) Current chamfer Dist1 = 3.0000, Dist2 = 30.0000
Polyline/Distance/Angle/Trim/Method/<Select first line>: **[Select A (Figure 6.101).]**
Select second line: **[Select B (Figure 6.101).]**

Command: **[Enter]**
CHAMFER
(TRIM mode) Current chamfer Dist1 = 3.0000, Dist2 = 30.0000
Polyline/Distance/Angle/Trim/Method/<Select first line>: **[Select A (Figure 6.101).]**
Select second line: **[Select C (Figure 6.101).]**

Command: **[Enter]**
CHAMFER
(TRIM mode) Current chamfer Dist1 = 3.0000, Dist2 = 30.0000
Polyline/Distance/Angle/Trim/Method/<Select first line>: **[Select E (Figure 6.101).]**
Select second line: **[Select D (Figure 6.101).]**

Command: **[Enter]**
CHAMFER
(TRIM mode) Current chamfer Dist1 = 3.0000, Dist2 = 30.0000
Polyline/Distance/Angle/Trim/Method/<Select first line>: **[Select E (Figure 6.101).]**
Select second line: **[Select F (Figure 6.101).]**

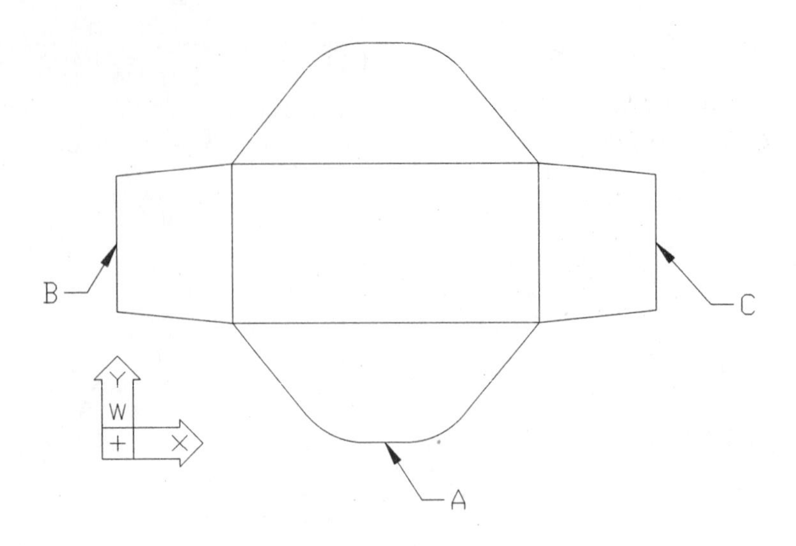

Figure 6.102 Rectangle chamfered

The other rectangle is also complete. To proceed, issue the CIRCLE command to create a circle of radius 6 units, and use the MIRROR command to add another circle. See Figure 6.103.

<Draw> **<Circle>** **<Center, Radius>**

Command: **CIRCLE**
3P/2P/TTR/<Center point>: **FROM**
Base point: **MID** of [**Select A (Figure 6.102).**]
<Offset>: **@15<90**
Diameter/<Radius>: **6**

<Modify> **<Mirror>**

Command: **MIRROR**
Select objects: **LAST**
Select objects: **[Enter]**
First point of mirror line: **MID** of [**Select B (Figure 6.102).**]
Second point: **MID** of [**Select C (Figure 6.102).**]
Delete old objects? **N**

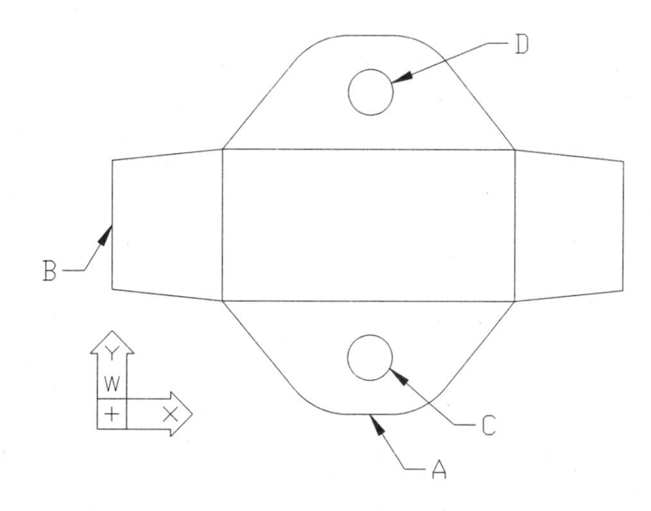

Figure 6.103 Circles created

Set the DELOBJ variable to 1 because the entities are not required after they are used to make the regions.

Command: **DELOBJ**
New value for DELOBJ: **1**

Run the REGION command to form four regions from the four closed areas: two circles, a chamfered rectangle, and a chamfered and filleted rectangle.

<Draw> **<Region>**

Command: **REGION**
Select objects: [**Select A, B, C, and D (Figure 6.103).**]
Select objects: [**Enter**]
4 loops extracted.
4 regions created.

As we have said, you will not find any noticeable change on the screen display. To know what kinds of entities are on the screen, you can use the LIST command. If you have used the REGION command correctly, the LIST command will tell you that the entities are regions.

Now you have four regions. To combine them to form a complex region that resembles Figure 6.98, you have to apply the UNION command to unite the regions A and B and then use the SUBTRACT command to subtract the regions C and D from the united region. Because DELOBJ is 1, the original regions are deleted. See Figure 6.104.

<Modify> **<Boolean>** **<Union>**

Command: **UNION**
Select objects: [**Select A and B (Figure 6.103).**]
Select objects: [**Enter**]

<Modify> **<Boolean>** **<Subtract>**

Command: **SUBTRACT**
Select solids and regions to subtract from...
Select objects: **[Select A (Figure 6.103).]**
Select objects: **[Enter]**
Select solids and regions to subtract...
Select objects: **[Select C and D (Figure 6.103).]**
Select objects: **[Enter]**

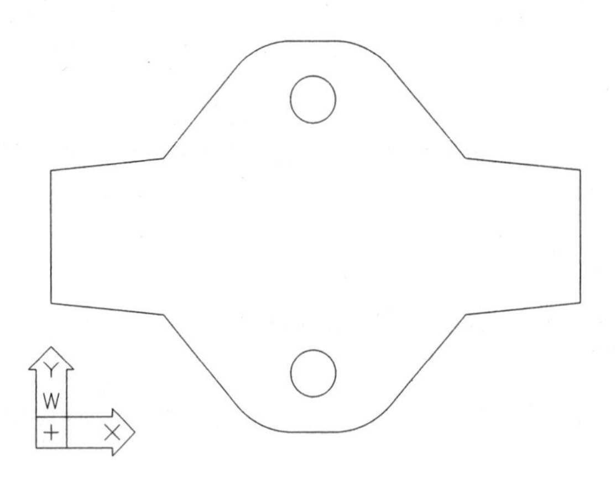

Figure 6.104 Region representing top view created

The profile for the top view is complete. The second profile that you will be working with resides on the ZX plane of the WCS. To get a better view, set the display to an isometric view by using the VPOINT command. To set the new entities on the ZX plane, set the UCS to rotate 90° about the X axis by applying the UCS command. See Figure 6.105.

<View> **<3D Viewpoint>** **<SE Isometric>**

Command: **VPOINT**
Rotate/<View point>: **1,-1,1**

<Tools> **<UCS>** **<X Axis Rotate>**

Command: **UCS**
Origin/ZAxis/3point/OBject/View/X/Y/Z/Prev/Restore/Save/Del/?/<World>: **X**
Rotation angle about X axis: **90**

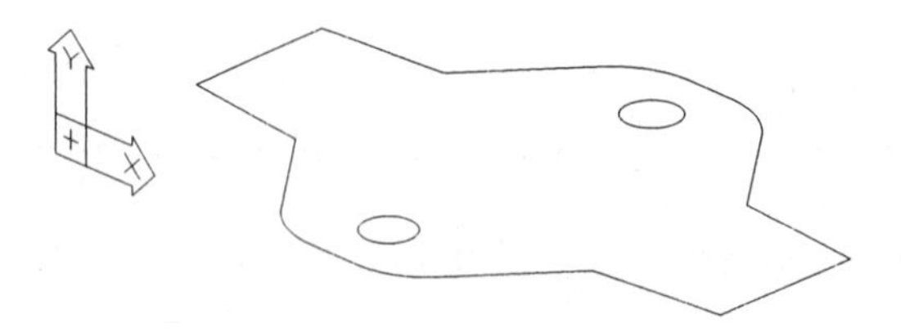

Figure 6.105 New viewing position and new UCS

Check your UCS icon against Figure 6.105. Then use the RECTANG command to create three rectangles. See Figure 6.106.

<Draw> **<Rectangle>**

Command: **RECTANG**
Chamfer/Elevation/Fillet/Thickness/Width/<First corner>: **0,5**
Other corner: **@140,40**

Command: **[Enter]**
RECTANG
Chamfer/Elevation/Fillet/Thickness/Width/<First corner>: **30,0**
Other corner: **@80,5**

Command: **[Enter]**
RECTANG
Chamfer/Elevation/Fillet/Thickness/Width/<First corner>: **35,45**
Other corner: **@70,5**

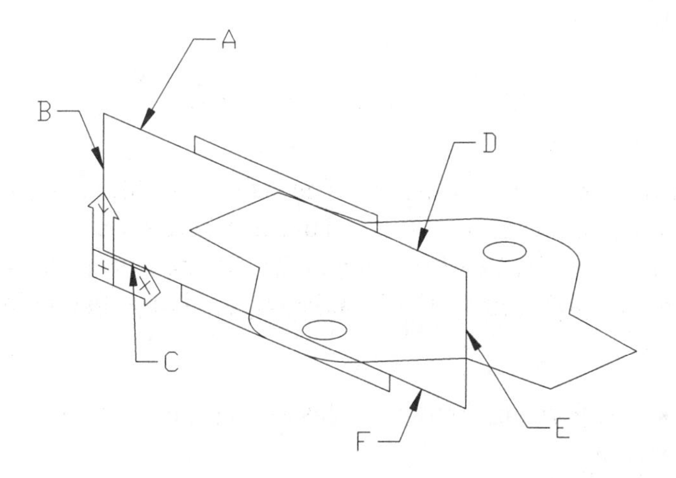

Figure 6.106 Rectangles created

The large rectangle needs to be chamfered. Run the CHAMFER command to set the chamfer distances, and then repeat it four times to chamfer four corners. See Figure 6.107.

<Modify> **<Chamfer>**

Command: **CHAMFER**
Polyline/Distance/Angle/Trim/Method/<Select first line>: **D**
Enter first chamfer distance: **3**
Enter second chamfer distance: **30**

Command: **[Enter]**
CHAMFER
(TRIM mode) Current chamfer Dist1 = 3.0000, Dist2 = 30.0000
Polyline/Distance/Angle/Trim/Method/<Select first line>: **[Select B (Figure 6.106).]**
Select second line: **[Select A (Figure 6.106).]**

Command: **[Enter]**
CHAMFER
(TRIM mode) Current chamfer Dist1 = 3.0000, Dist2 = 30.0000
Polyline/Distance/Angle/Trim/Method/<Select first line>: **[Select B (Figure 6.106).]**
Select second line: **[Select C (Figure 6.106).]**

Command: **[Enter]**
CHAMFER
(TRIM mode) Current chamfer Dist1 = 3.0000, Dist2 = 30.0000
Polyline/Distance/Angle/Trim/Method/<Select first line>: **[Select E (Figure 6.106).]**
Select second line: **[Select D (Figure 6.106).]**

Command: **[Enter]**
CHAMFER
(TRIM mode) Current chamfer Dist1 = 3.0000, Dist2 = 30.0000
Polyline/Distance/Angle/Trim/Method/<Select first line>: **[Select E (Figure 6.106).]**
Select second line: **[Select F (Figure 6.106).]**

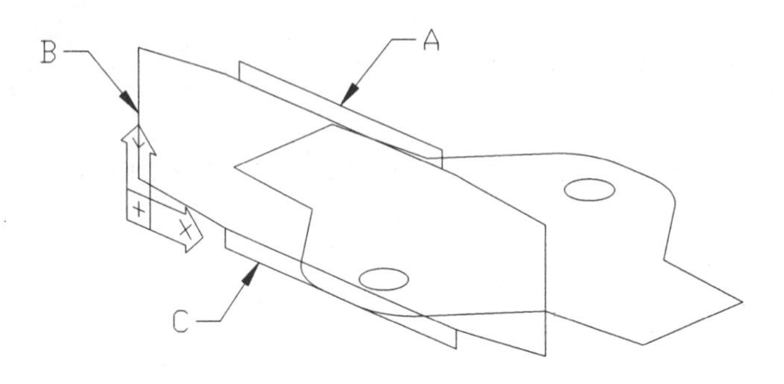

Figure 6.107 Rectangle chamfered

After chamfering, make three regions from the three closed areas by using the REGION command. Then apply the UNION command on the regions to combine them into a single region. See Figure 6.108.

<Draw> **<Region>**

Command: **REGION**
Select objects: **[Select A, B, and C (Figure 6.107).]**
Select objects: **[Enter]**
3 loops extracted.
3 regions created.

<Modify> **<Boolean>** **<Union>**

Command: **UNION**
Select objects: **[Select A, B, and C (Figure 6.107).]**
Select objects: **[Enter]**

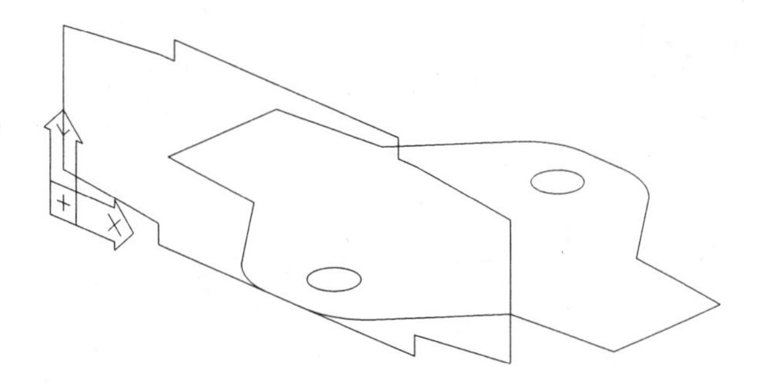

Figure 6.108 Region representing front view created

The profile for the front view is complete. Before making the third region, rotate the UCS about the Y axis for 90°. The third profile resides on a new UCS plane. See Figure 6.109.

<Tools> **<UCS>** **<Y Axis Rotate>**

Command: **UCS**
Origin/ZAxis/3point/OBject/View/X/Y/Z/Prev/Restore/Save/Del/?/<World>: **Y**
Rotation angle about Y axis: **90**

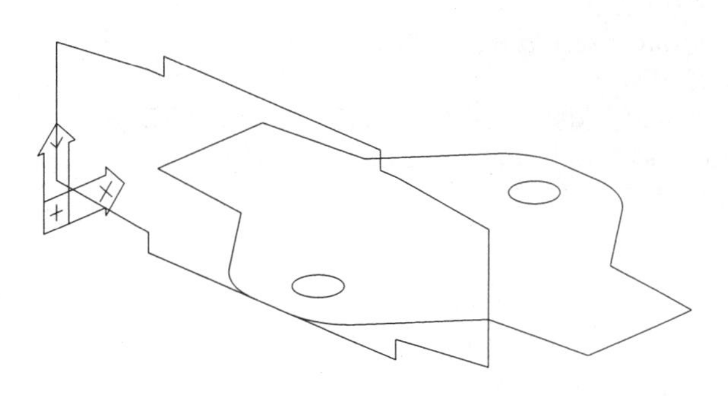

Figure 6.109 New viewing position

After setting the UCS, create three rectangles with the RECTANG command. See Figure 6.110.

<Draw> **<Rectangle>**

Command: **RECTANG**
Chamfer/Elevation/Fillet/Thickness/Width/<First corner>: **0,0**
Other corner: **@40,10**

Command: **[Enter]**
RECTANG
Chamfer/Elevation/Fillet/Thickness/Width/<First corner>: **30,5**
Other corner: **@40,40**

Command: **[Enter]**
RECTANG

Chamfer/Elevation/Fillet/Thickness/Width/<First corner>: **35,45**
Other corner: **@30,5**

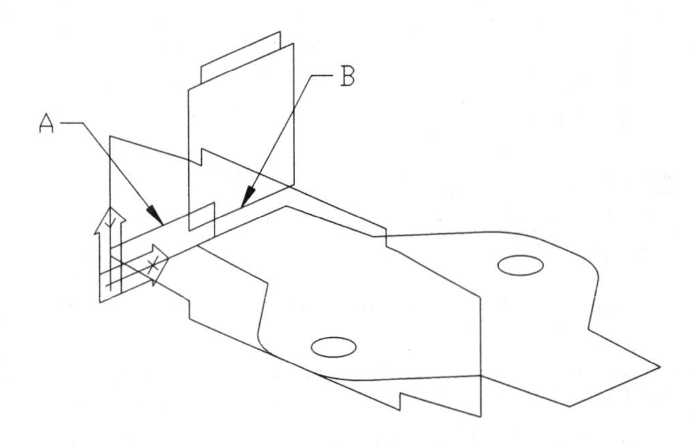

Figure 6.110 Rectangles created

Use the MIRROR3D command to mirror a rectangle. See Figure 6.111.

<Modify> **<3D Operation>** **<Mirror 3D>**

Command: **MIRROR3D**
Select objects: **[Select A (Figure 6.110).]**
Select objects: **[Enter]**
Plane by Object/Last/Zaxis/View/XY/YZ/ZX/<3points>: **YZ**
Point on YZ plane: **MID** of **[Select B (Figure 6.110).]**
Delete old objects? **N**

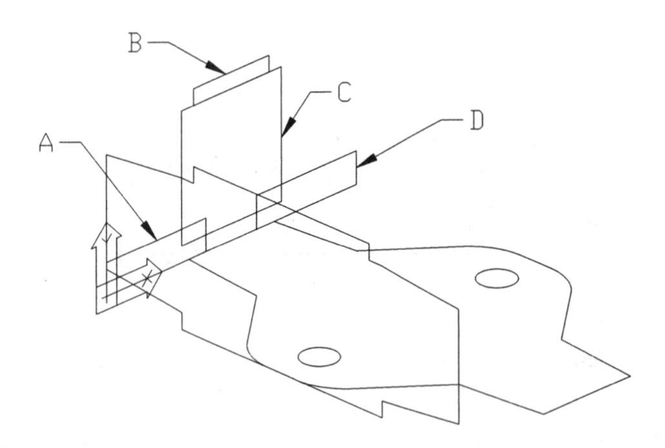

Figure 6.111 Rectangle mirrored

After mirroring, convert the four rectangles into four regions by using the REGION command. Then run the UNION command to unite the regions to form a single region. See Figure 6.112.

<Draw> **<Region>**

Command: **REGION**
Select objects: [**Select A, B, C and D (Figure 6.111).**]
Select objects: [**Enter**]
4 loops extracted.
4 regions created.

<Modify> **<Boolean>** **<Union>**

Command: **UNION**
Select objects: [**Select A, B, C and D (Figure 6.111).**]
Select objects: [**Enter**]

The profiles for making three extruded solids are complete. To make a core, which is a revolved solid, you have to create the fourth region. Set a new UCS. See Figure 6.112.

<Tools> **<UCS>** **<World>**

Command: **UCS**
Origin/ZAxis/3point/OBject/View/X/Y/Z/Prev/Restore/Save/Del/?/<World>: **W**

<Tools> **<UCS>** **<Origin>**

Command: **UCS**
Origin/ZAxis/3point/OBject/View/X/Y/Z/Prev/Restore/Save/Del/?/<World>: **O**
Origin point: **0,50,25**

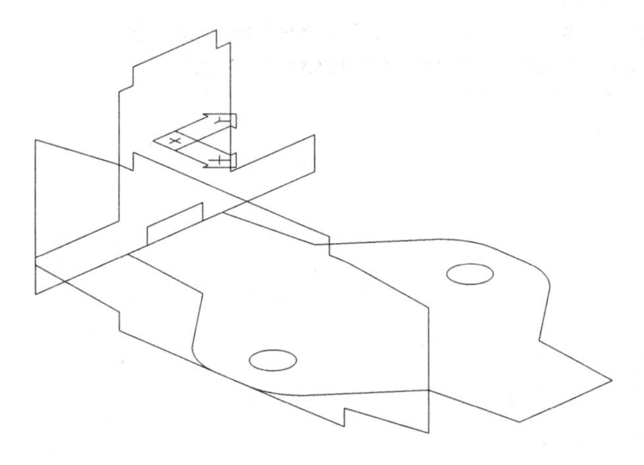

Figure 6.112 Region representing side view created

From the Object Properties toolbar, select the Layer Control box and the layer Spool to set the current layer to Spool. You will put the wireframe for the spool on this layer.

[Object Properties] **[Layer Control]**

Current layer: **Spool**

Run the PLAN command to set the viewing position to the plan view of the current UCS. Then create a region by referring to Figure 6.113.

<View> **<3D Viewpoint>** **<Plan View>** **<Current UCS>**

Command: **PLAN**
<Current UCS>/Ucs/World: **C**

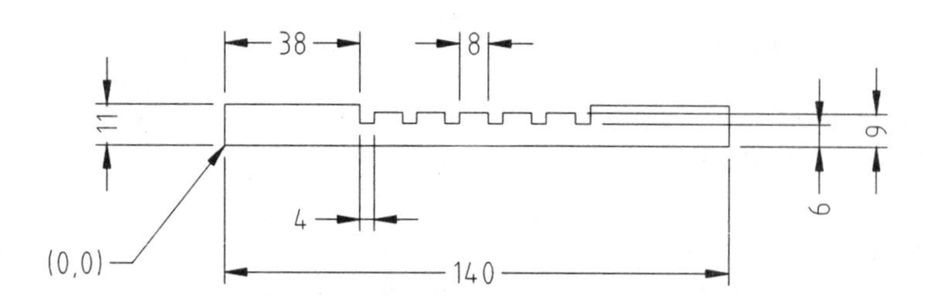

Figure 6.113 Dimensions for fourth region

Use the RECTANG command and the ARRAY command to create a series of rectangles. See Figure 6.114.

<Draw> **<Rectangle>**

Command: **RECTANG**
Chamfer/Elevation/Fillet/Thickness/Width/<First corner>: **0,0**
Other corner: **@140,11**

Command: **[Enter]**
RECTANG
Chamfer/Elevation/Fillet/Thickness/Width/<First corner>: **38,9**
Other corner: **@64,10**

Command: **[Enter]**
RECTANG
Chamfer/Elevation/Fillet/Thickness/Width/<First corner>: **38,6**
Other corner: **@4,15**

<Modify> **<Array>**

Command: **ARRAY**
Select objects: **LAST**
Select objects: **[Enter]**
Rectangular or Polar array (<R>/P): **R**
Number of rows (---): **1**
Number of columns (|||): **6**
Distance between columns (|||): **12**

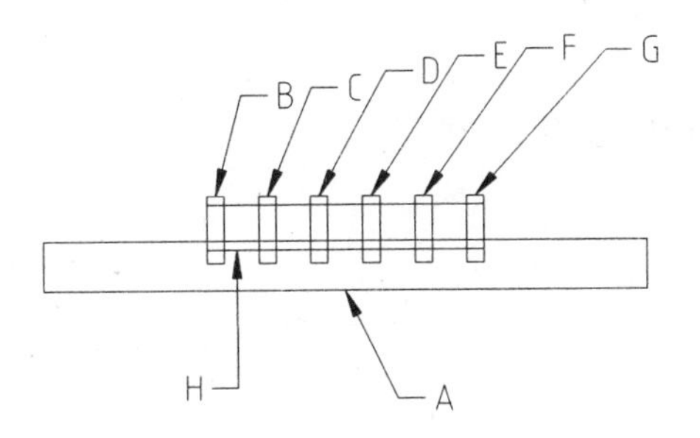

Figure 6.114 Eight rectangles created

Convert the rectangles into regions by applying the REGION command. Then use the SUBTRACT command to subtract the smaller regions from the large region.

<Draw> **<Region>**

Command: **REGION**
Select objects: [**Select A, B, C, D, E, F, G, and H (Figure 6.114).**]
Select objects: [**Enter**]
8 loops extracted.
8 regions created.

<Modify> **<Boolean>** **<Subtract>**

Command: **SUBTRACT**
Select solids and regions to subtract from...
Select objects: [**Select A (Figure 6.114).**]
Select objects: [**Enter**]
Select solids and regions to subtract...
Select objects: [**Select B, C, D, E, F, G, and H (Figure 6.114).**]
Select objects: [**Enter**]

The four regions for making the complex solid are complete. Before starting to produce the solid model, set the display to an isometric view by using the VPOINT command. See Figure 6.115.

<View> **<3D Viewpoint>** **<SE Isometric>**

Command: **VPOINT**
Rotate/<View point>: **1,-1,1**

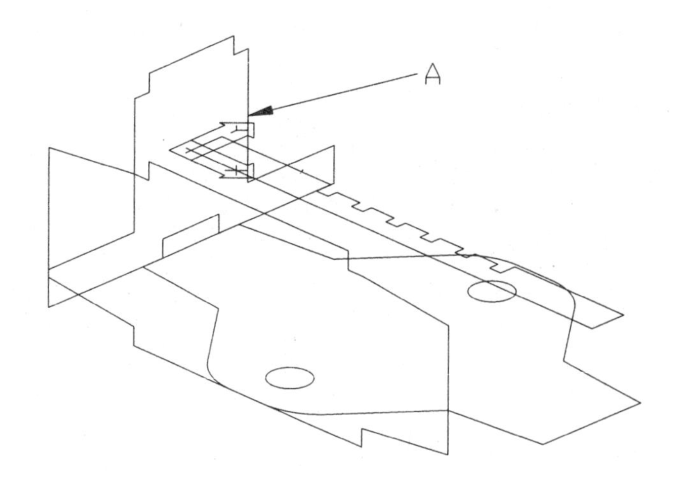

Figure 6.115 Region for revolved solid created

Extrusion and Intersection

After completing the wireframes, you will start to build the solid model. To make the main body of the model, you will extrude three regions and then form an intersection of the extruded solids.

Select the layer Solid from the Layer Control box of the Object Properties toolbar to set it as the current layer. You will create the solids on this layer.

[Object Properties] **[Layer Control]**

Current layer: **Solid**

Set the system variable DELOBJ to 0 so that the region or polyline that is used for extrusion or revolving is retained. Retaining the original wireframes enables you to construct the solid again quickly if anything goes wrong during solid creation.

Command: **DELOBJ**
New value for DELOBJ: **0**

Using the region on the ZY plane, run the EXTRUDE command to produce a solid of extrusion. See Figure 6.116.

<Draw> **<Solids>** **<Extrude>**

Command: **EXTRUDE**
Select objects: **[Select A (Figure 6.115).]**
Select objects: **[Enter]**
Path/<Height of Extrusion>: **140**
Extrusion taper angle: **0**

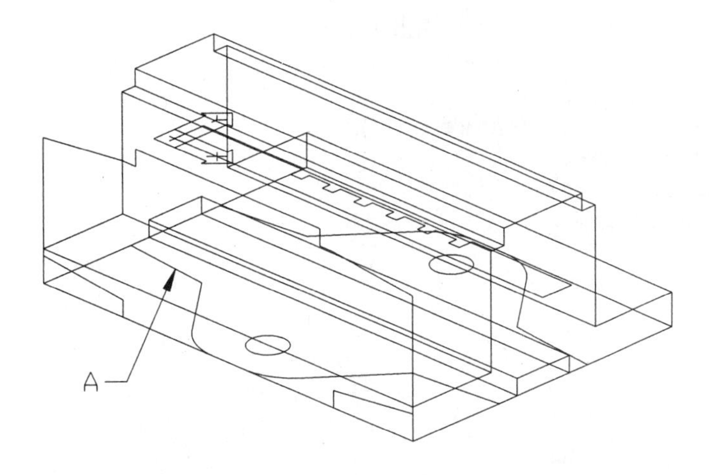

Figure 6.116 Region representing side view extruded

In specifying the height of extrusion, you need to be aware that the direction, whether positive or negative, depends on the Z-direction of the UCS on which the region or polyline is made, not the current UCS. You may refer to Figure 6.111 to check the UCS orientation of the last extruded solid.

To make the second extruded solid, run the EXTRUDE command. See Figure 6.117.

<Draw> **<Solids>** **<Extrude>**

Command: **EXTRUDE**
Select objects: [**Select A (Figure 6.116).**]
Select objects: [**Enter**]
Path/<Height of Extrusion>: **50**
Extrusion taper angle: **0**

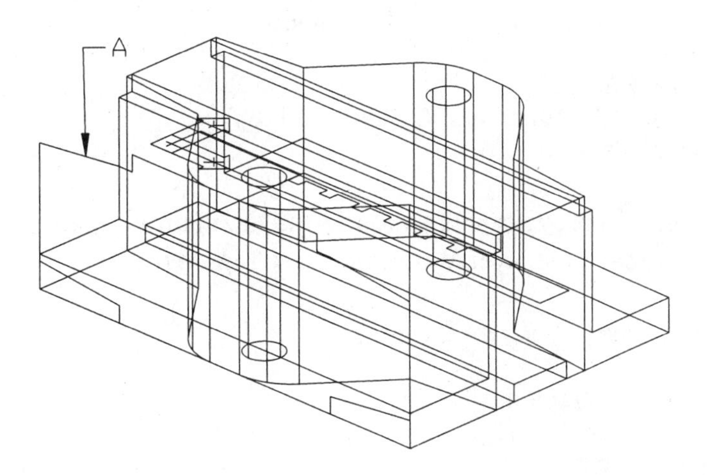

Figure 6.117 Region representing top view extruded

Apply the EXTRUDE command once more to create the third solid of extrusion. See Figure 6.118. To find out the UCS orientation of this region, refer to Figure 6.108. The extrusion should be in the negative direction.

<Draw> **<Solids>** **<Extrude>**

Command: **EXTRUDE**
Select objects: **[Select A (Figure 6.117).]**
Select objects: **[Enter]**
Path/<Height of Extrusion>: **-100**
Extrusion taper angle: **0**

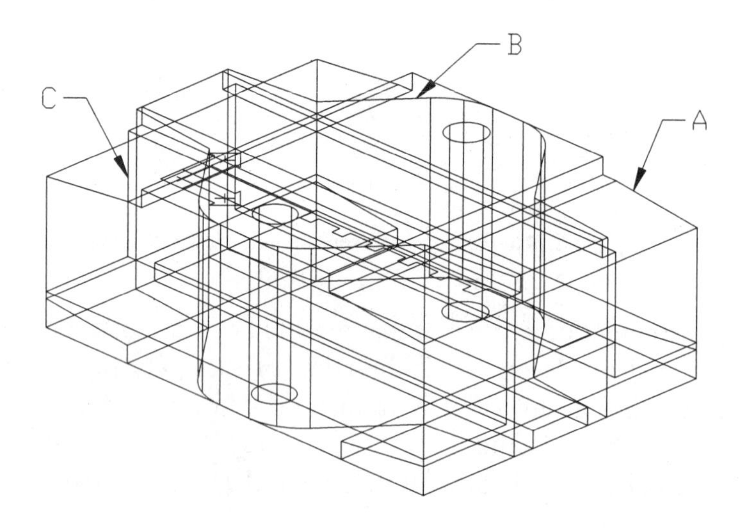

Figure 6.118 Region representing front view extruded

After making three extruded solids from the regions, run the INTERSECT command to produce a solid of intersection from them. A solid of intersection is a solid that has the common volume of a set of solids.

<Modify> **<Boolean>** **<Intersect>**

Command: **INTERSECT**
Select objects: **[Select A, B, and C (Figure 6.118).]**
Select objects: **[Enter]**

The intersection of three extruded solids forms the main body of the hydraulic valve. After intersection, turn off the layer Wireframe. See Figure 6.119.

[Object Properties] **[Layer Control]**

Off Wireframe

Current layer: **Solid**

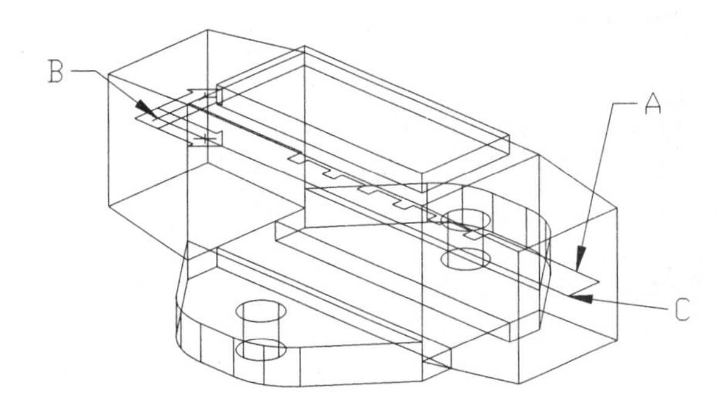

Figure 6.119 Intersection of three extruded solids

Revolving and Subtraction

To make the core of the fluid power valve, you will revolve a region to form a solid of revolution and subtract the revolved solid from the main body. Run the REVOLVE command. See Figure 6.120.

<Draw> **<Solids>** **<Revolve>**

Command: **REVOLVE**
Select objects: [**Select A (Figure 6.119).**]
Select objects: [**Enter**]
Axis of revolution - Object/X/Y/<Start point of axis>: **END** of [**Select B (Figure 6.119).**]
<End point of axis>: **END** of [**Select C (Figure 6.119).**]
Angle of revolution <full circle>: [**Enter**]

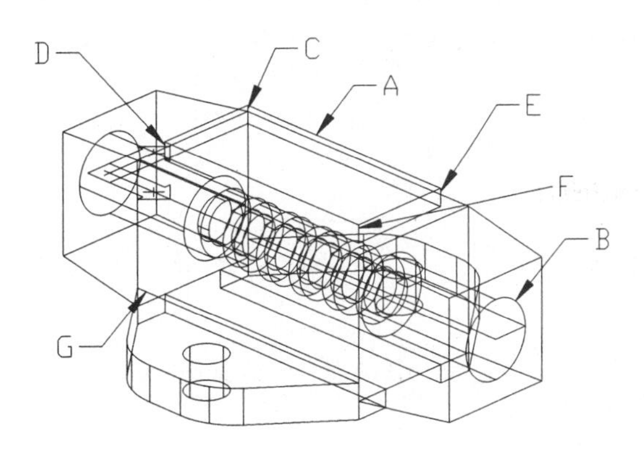

Figure 6.120 Revolved solid created

After revolving the wireframe, you do not need it any more. Turn off the layer SPOOL. To cut a core on the main body using the revolved solid, run the SUBTRACT command.

[Object Properties] **[Layer Control]**

Off Spool

<Modify> **<Boolean>** **<Subtract>**

Command: **SUBTRACT**
Select solids and regions to subtract from...
Select objects: **[Select A (Figure 6.120).]**
Select objects: **[Enter]**
Select solids and regions to subtract...
Select objects: **[Select B (Figure 6.120).]**
Select objects: **[Enter]**

Filleting and Adding Holes

To complete the fluid power valve, you have to round off some corners and cut five holes.

Create a series of fillets on the solid model by applying the FILLET command. See Figure 6.121.

<Modify> **<Fillet>**

Command: **FILLET**
Polyline/Radius/Trim/<Select first object>: **[Select A (Figure 6.120).]**
Enter radius: **2**
Chain/Radius/<Select edge>: **[Select A and B (Figure 6.120).]**
Chain/Radius/<Select edge>: **[Enter]**

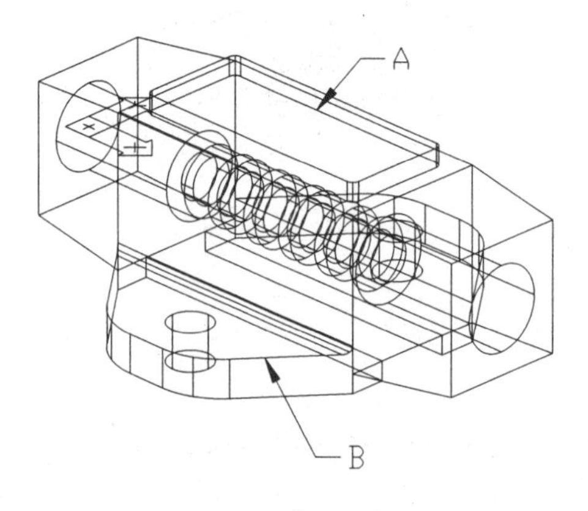

Figure 6.121 Five edges filleted

Run the FILLET command to make two chains of fillets. See Figure 6.122.

<Modify> **<Fillet>**

Command: **FILLET**
Polyline/Radius/Trim/<Select first object>: **[Select A (Figure 6.121).]**
Enter radius: **2**
Chain/Radius/<Select edge>: **C**
Edge/Radius/<Select edge chain>: **[Select A and B (Figure 6.121).]**
Edge/Radius/<Select edge chain>: **[Enter]**

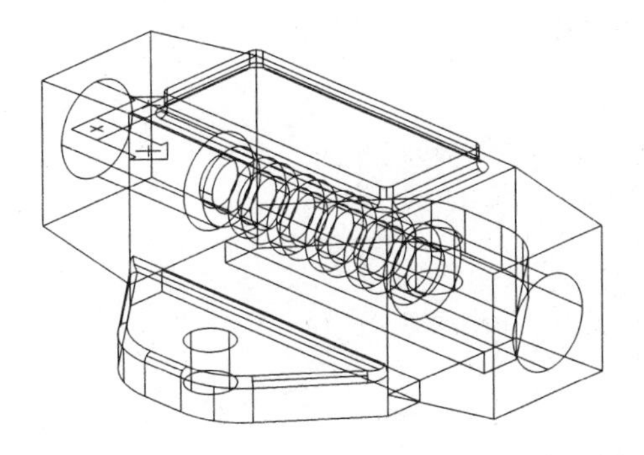

Figure 6.122 Fillet chains created

Before filleting the other side of the valve, select the NW Isometric item of the 3D Viewpoint cascading menu of the View pull-down menu to set the display to a northwest isometric view. See Figure 6.123.

<View> **<3D Viewpoint>** **<NW Isometric>**

Command: **VPOINT**
Rotate/<View point>: **-1,1,1**

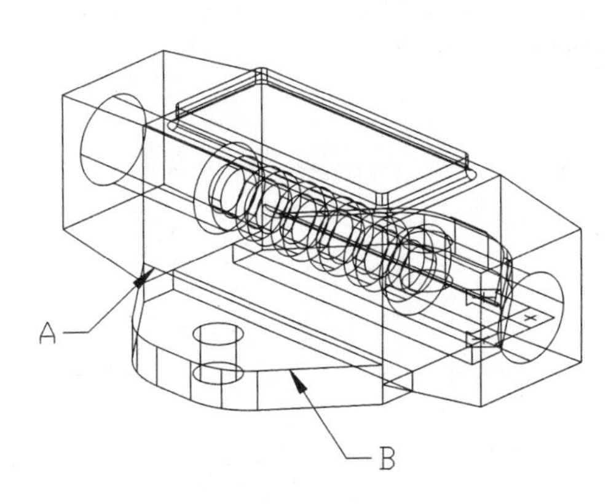

Figure 6.123 New viewing position

Run the FILLET command again. Then set the UCS back to WORLD by using the UCS command. See Figure 6.124.

<Modify> **<Fillet>**

Command: **FILLET**
Polyline/Radius/Trim/<Select first object>: **[Select A (Figure 6.123).]**
Enter radius: **2**
Chain/Radius/<Select edge>: **C**
Edge/Radius/<Select edge chain>: **[Select B (Figure 6.123).]**
Edge/Radius/<Select edge chain>: **[Enter]**

<Tools> **<UCS>** **<World>**

Command: **UCS**
Origin/ZAxis/3point/OBject/View/X/Y/Z/Prev/Restore/Save/Del/?/<World>: **W**

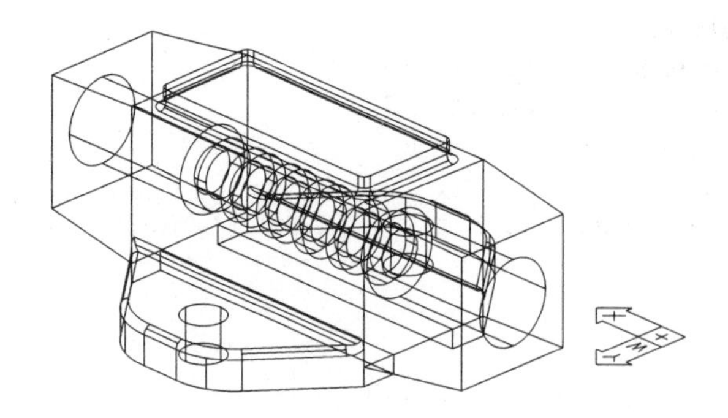

Figure 6.124 Edges filleted

Set the display to the previous viewing direction with the VPOINT command. Then create five cylinders by using the CYLINDER command. See Figure 6.125.

<View> **<3D Viewpoint>** **<SE Isometric>**

Command: **VPOINT**
Rotate/<View point>: **1,-1,1**

<Draw> **<Solids>** **<Cylinder>**

Command: **CYLINDER**

Center location	Radius	Height
46,56,25	**3**	**40**
70,56,25	**3**	**40**
94,56,25	**3**	**40**
58,44,25	**3**	**40**
82,44,25	**3**	**40**

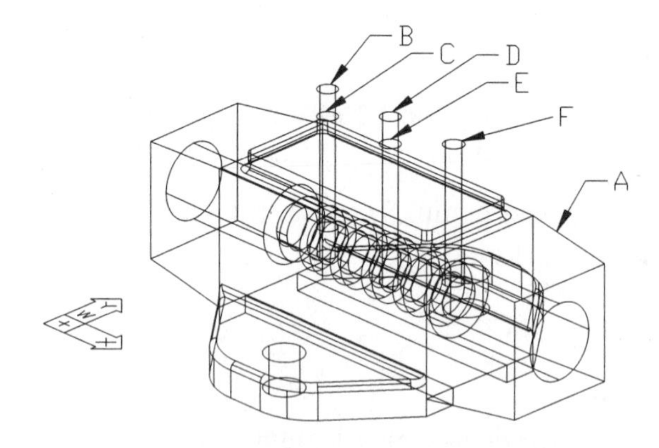

Figure 6.125 Cylinders created

To cut the holes on the valve body, execute the SUBTRACT command to subtract the five cylinders. See Figure 6.126.

<Modify> **<Boolean>** **<Subtract>**

Command: **SUBTRACT**
Select solids and regions to subtract from...
Select objects: [**Select A (Figure 6.125).**]
Select objects: [**Enter**]
Select solids and regions to subtract...
Select objects: [**Select B, C, D, E, and F(Figure 6.125).**]
Select objects: [**Enter**]

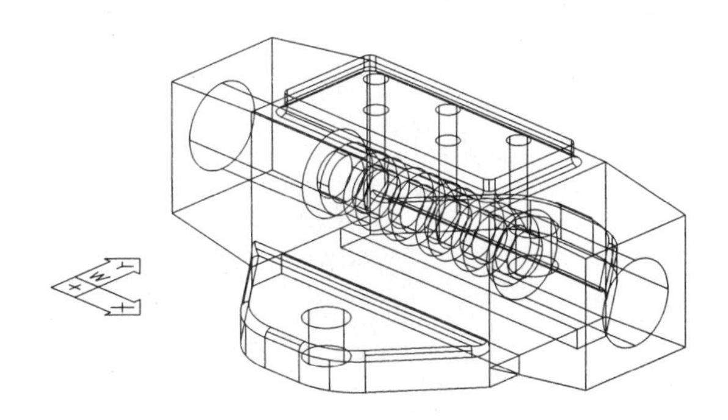

Figure 6.126 Cylinders subtracted

The solid model of the fluid power valve is complete. See also Figure 6.94. Save your drawing.

<File> **<Save>**

File name: **VALVE.DWG**

In making this model, you constructed 2D regions, extruded solids, and revolved solids. You formed the model of the fluid power valve by intersection and subtraction.

6.5 Slicing

A solid can be sliced into two parts. Open the file of the bearing block that you saved earlier in this chapter: BBLK.DWG. See Figure 6.127. You will cut it into two solid models.

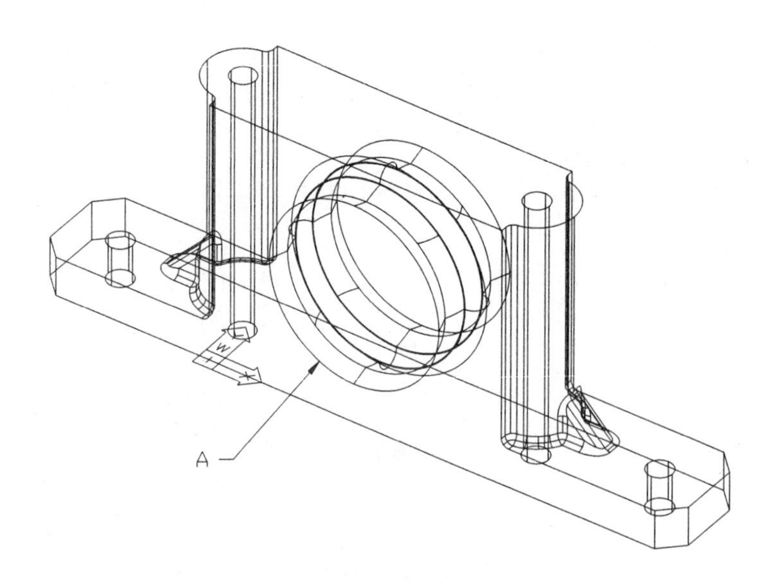

Figure 6.127 Bearing block

Before proceeding, you will save one more copy of this model for use in the next section. Apply the SAVEAS command.

<File> **<Save As...>**

File name: **BBLK1.DWG**

Your current working file now becomes BBLK1.DWG. Select the Slice item of the Solids cascading menu of the Draw pull-down menu to run the SLICE command, and take the Both option. See Figure 6.128.

<Draw> **<Solids>** **<Slice>**

Command: **SLICE**
Select objects: [**Select A (Figure 6.127).**]
Select objects: [**Enter**]
Slicing plane by Object/Zaxis/View/XY/YZ/ZX/<3points>: **XY**
Point on XY plane: **CEN** of [**Select A (Figure 6.127).**]
Both sides/<Point on desired side of the plane>: **B**

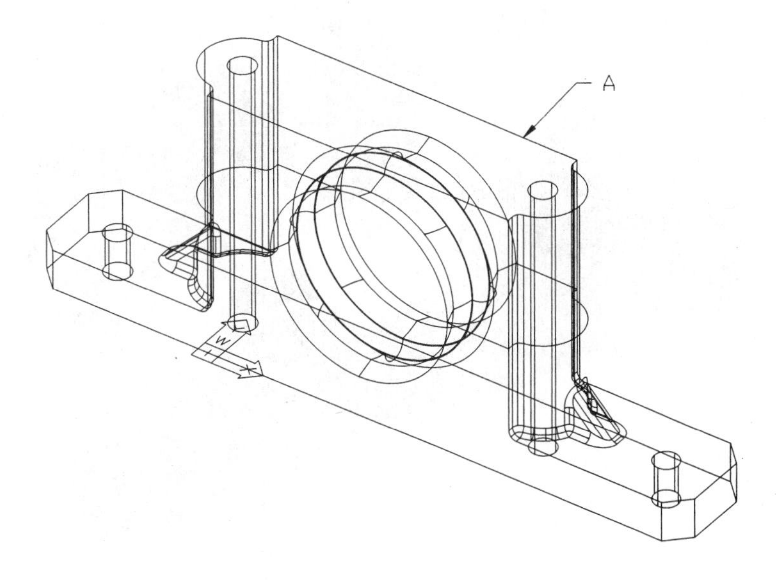

Figure 6.128 Solid model sliced

The model is now sliced into two pieces. To see the effect of slicing, move the pieces apart. See Figure 6.129 and Figure 6.130.

<Modify> **<Move>**

Command: **MOVE**
Select objects: [**Select A (Figure 6.128).**]
Select objects: [**Enter**]
Base point or displacement: **45,45**
Second point of displacement: [**Enter**]

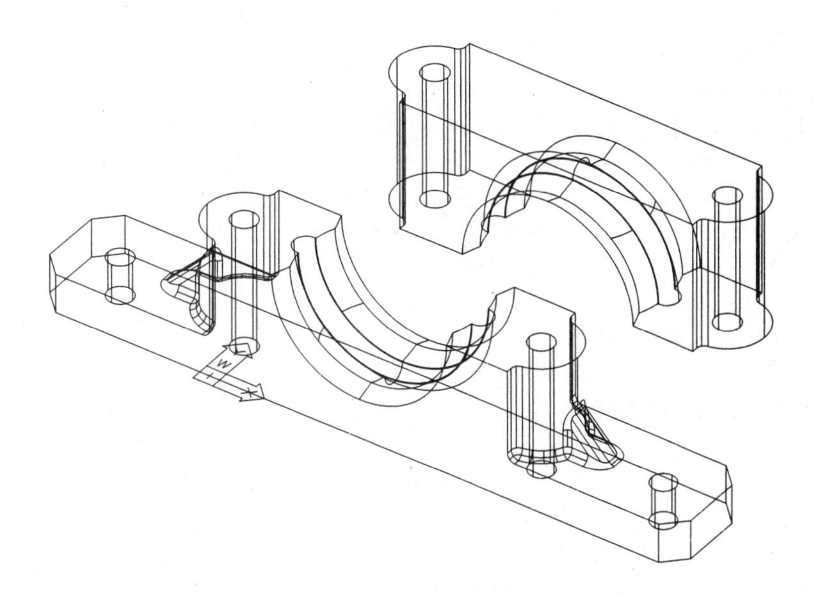

Figure 6.129 Upper part moved apart

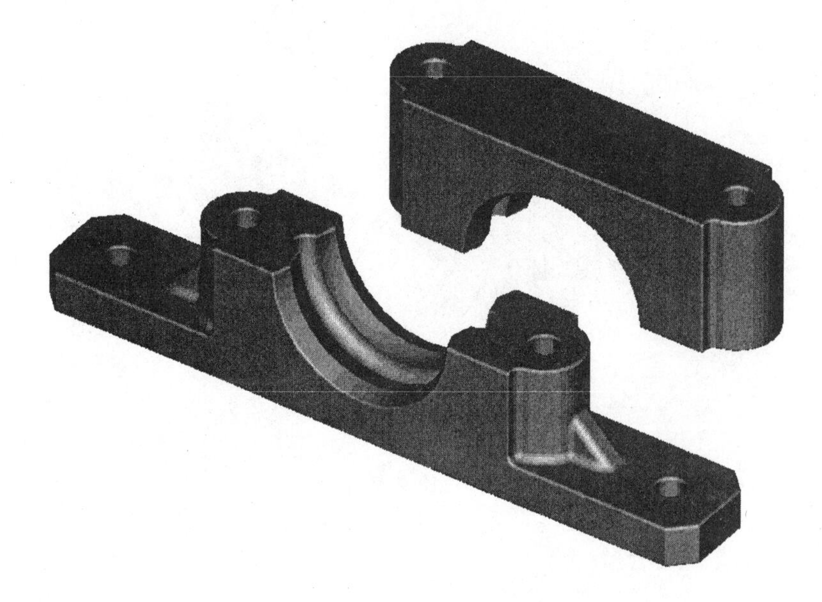

Figure 6.130 Rendered drawing of sliced solid

The sliced model is complete. Save your drawing.

<File> **<Save...>**

Command: **QSAVE**

Here, you learned how to cut a solid model into two pieces.

6.6 Cutting by Intersection

In the last section, you saved the solid model of the bearing block in two files. In one drawing file, you sliced the solid model into two parts. Because the SLICE command gives a straight cutting edge, the cutting edge is straight and flat.

Now you will retrieve the saved file of the uncut solid model and cut it into two pieces with an irregular cutting edge by intersection. Use the OPEN command to open the saved file.

<File> **<Open...>**

Command: **BBLK.DWG**

Set the UCS to WORLD, and then rotate it for 90° about the X axis. See Figure 6.131.

<Tools> **<UCS>** **<World>**

<Tools> **<UCS>** **<X Axis Rotate>**

Command: **UCS**
Origin/ZAxis/3point/OBject/View/X/Y/Z/Prev/Restore/Save/Del/?/<World>: **X**
Rotation angle about X axis: **90**

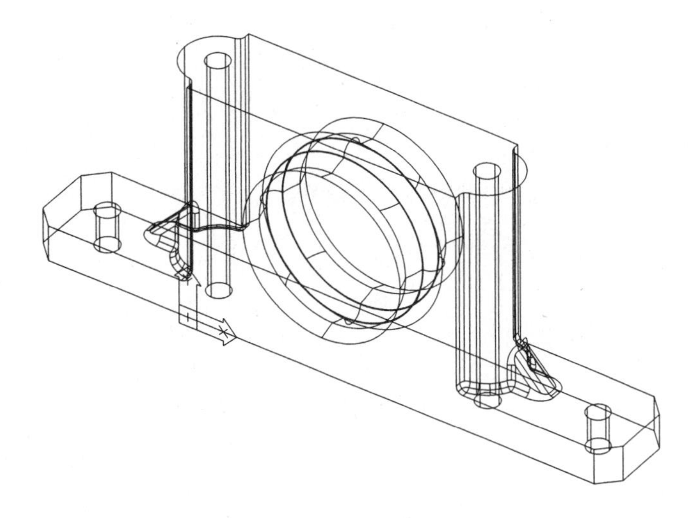

Figure 6.131 UCS rotated

In order to make a cut solid with an irregular cutting edge, you need to create an extruded solid with an irregular edge and intersect the extruded solid with the existing solid. After intersection, a portion of the original solid will be removed. To cut a solid and to keep both sides of the solid, you have to make a copy of the original solid, make two extruded solids, and intersect the two extruded solids with the two copies of the original solid. To prepare two extruded solids, create two polylines by using the PLINE command. See Figure 6.132.

<Draw> **<Polyline>**

```
Command: PLINE
From point: -80,0
Current line-width is 0.0000
Arc/Close/Halfwidth/Length/Undo/Width/<Endpoint of line>: -80,55
Arc/Close/Halfwidth/Length/Undo/Width/<Endpoint of line>: 15,55
Arc/Close/Halfwidth/Length/Undo/Width/<Endpoint of line>: 15,50
Arc/Close/Halfwidth/Length/Undo/Width/<Endpoint of line>: 105,50
Arc/Close/Halfwidth/Length/Undo/Width/<Endpoint of line>: 105,55
Arc/Close/Halfwidth/Length/Undo/Width/<Endpoint of line>: 200,55
Arc/Close/Halfwidth/Length/Undo/Width/<Endpoint of line>: 200,0
Arc/Close/Halfwidth/Length/Undo/Width/<Endpoint of line>: C

Command: [Enter]
PLINE
From point: -80,55
Current line-width is 0.0000
Arc/Close/Halfwidth/Length/Undo/Width/<Endpoint of line>: 15,55
Arc/Close/Halfwidth/Length/Undo/Width/<Endpoint of line>: 15,50
Arc/Close/Halfwidth/Length/Undo/Width/<Endpoint of line>: 105,50
Arc/Close/Halfwidth/Length/Undo/Width/<Endpoint of line>: 105,55
Arc/Close/Halfwidth/Length/Undo/Width/<Endpoint of line>: 200,55
Arc/Close/Halfwidth/Length/Undo/Width/<Endpoint of line>: 200,100
Arc/Close/Halfwidth/Length/Undo/Width/<Endpoint of line>: -80,100
Arc/Close/Halfwidth/Length/Undo/Width/<Endpoint of line>: C
```

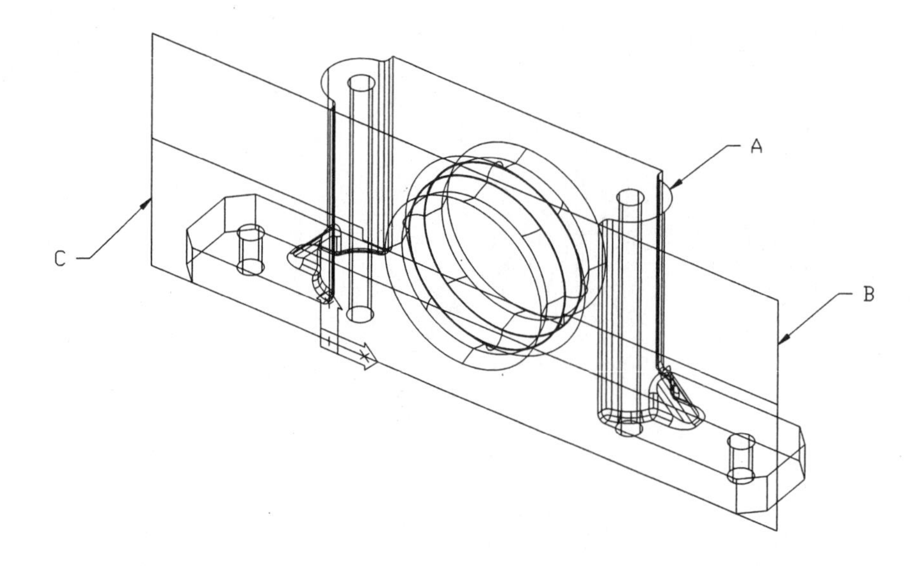

Figure 6.132 Polylines created

Make a copy of the original solid by using the COPY command. Because the copied solid and the original solid are at the same position, there should no noticeable change on the screen.

<Modify> **<Copy>**

Command: **COPY**
Select objects: **[Select A (Figure 6.132).]**
Select objects: **[Enter]**
<Base point or displacement>/Multiple: **0,0**
Second point of displacement: **[Enter]**

To produce two extruded solids from the two polylines, run the EXTRUDE command. See Figure 6.133.

<Draw> **<Solids>** **<Extrude>**

Command: **EXTRUDE**
Select objects: **[Select B and C (Figure 6.132).]**
Select objects: **[Enter]**
Path/<Height of Extrusion>: **-40**
Extrusion taper angle: **0**

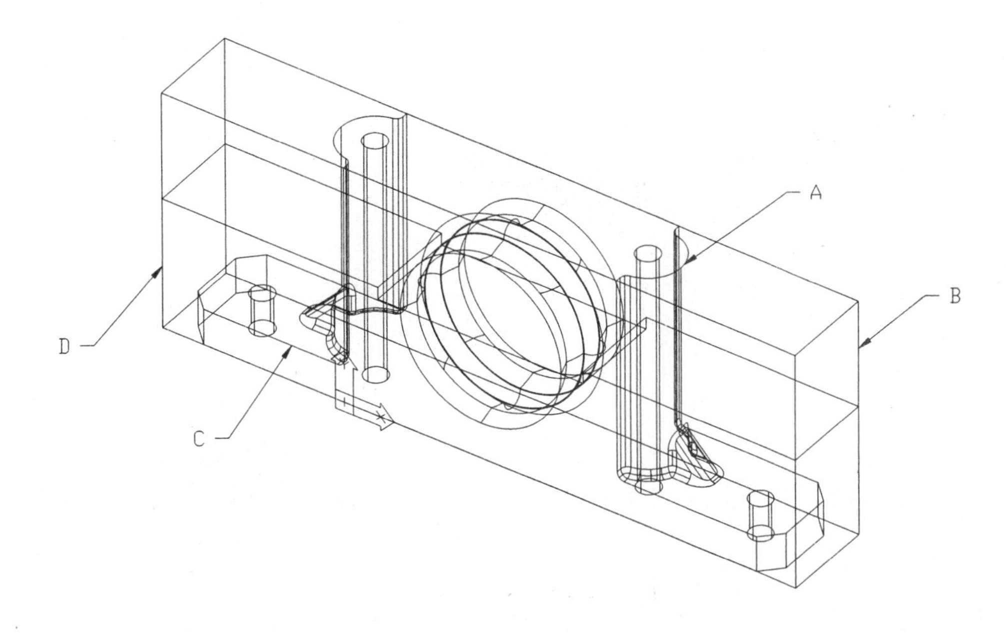

Figure 6.133 Solid copied, and two extruded solids created

Now you have two copies of the original solid and two extruded solids. To cut one copy of the original solid, apply the INTERSECT command. Select the upper extruded solid and the upper part of the complex solid. You will obtain the upper, cut half of the model.

<Modify> **<Boolean>** **<Intersect>**

Command: **INTERSECT**
Select objects: **[Select A and B (Figure 6.133).]**
Select objects: **[Enter]**

After intersecting, the lower half of a copy of the original complex solid is removed. Because you have two copies of the original solid, you will use the second copy to intersect with the lower extruded solid. The result is that you have two parts created, each one from a different copy of the solid.

Apply the INTERSECT command once again. This time, select the lower extruded solid and the lower part of the complex solid. See Figure 6.134.

<Modify> **<Boolean>** **<Intersect>**

Command: **INTERSECT**
Select objects: **[Select C and D (Figure 6.133).]**
Select objects: **[Enter]**

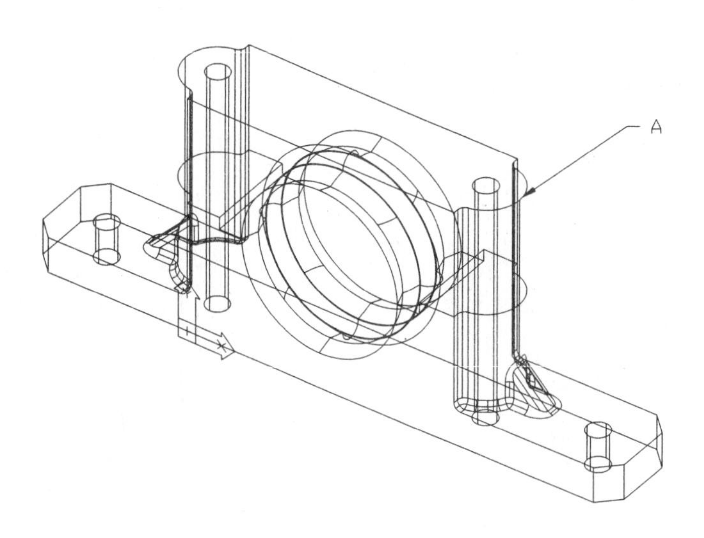

Figure 6.134 Solid cut into two pieces by intersection

In order to see clearly the effect of cutting, run the MOVE command to separate the upper piece. See Figure 6.135.

<Modify> **<Move>**

Command: **MOVE**
Select objects: [**Select A (Figure 6.134).**]
Select objects: [**Enter**]
Base point or displacement: **45,45**
Second point of displacement: [**Enter**]

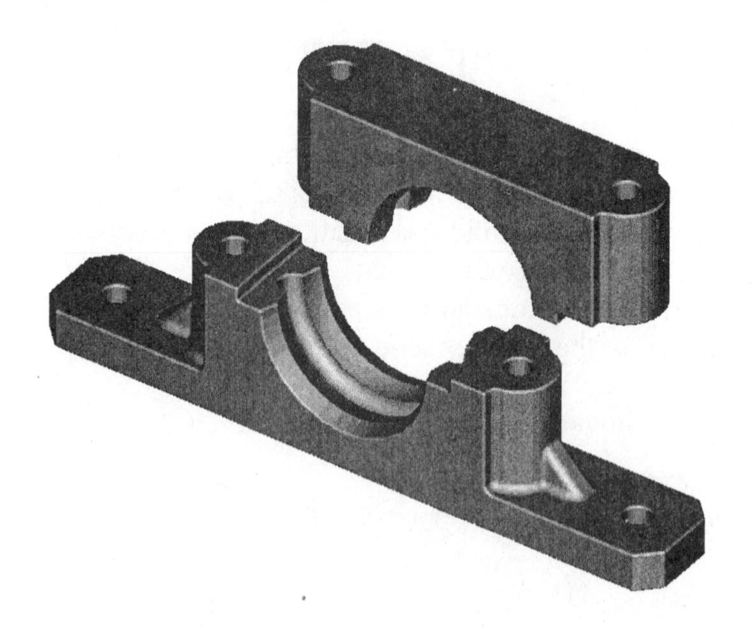

Figure 6.135 Cut model moved apart

The cutting is complete. Save your drawing.

<File> <Save...>

Here, you learned how to cut a solid model into two pieces with an irregular edge.

6.7 3D Coordinate Systems

To specify 3D points, you can use the 3D Cartesian coordinate system, the 3D cylindrical coordinate system, or the 3D spherical coordinate system.

3D Cartesian Coordinates

A 3D Cartesian coordinate specifies the X, Y, and Z values of a point.

A point (10,10,10) is specified in absolute Cartesian coordinates. It is 10 units in the X direction, 10 units in the Y direction, and 10 units in the Z direction. Measurement is from the current origin point.

A point (@10,10,10) is specified in relative Cartesian coordinates. It is 10 units in the X direction, 10 units in the Y direction, and 10 units in the Z direction. Measurement is from the last selected point.

3D Cylindrical Coordinates

A 3D cylindrical coordinate is very similar to a 2D polar coordinate. It has an additional entry to state the distance measured perpendicularly from the XY plane.

A point (10<45,10) is specified in absolute cylindrical coordinates. It is 10 units in the 45° direction and 10 units in the Z direction. Measurement is from current origin point.

A point (@20<60,5) is specified in relative cylindrical coordinates. It is 20 units in the 60° direction and 5 units in the Z direction. Measurement is from the last selected point.

3D Spherical Coordinates

A 3D spherical coordinate is also very similar to a 2D polar coordinate. It has an additional entry to specify the angle from the XY plane.

A point (10<45<30) is specified in absolute spherical coordinates. It is 10 units in the 45° direction on the XY plane in the 30° direction from the XY plane. Measurement is from current origin point.

A point (@10<45<30) is specified in relative spherical coordinates. It is 10 units in the 45° direction on the XY plane in the 30° direction from the XY plane. Measurement is from the last selected point.

6.8 Key Points and Exercises

There are three kinds of 3D models: 3D wireframe models, 3D surface models, and 3D solid models. Among them, solid models are superior to the other two kinds because 3D solid models contain information on the surfaces, edges, and volume of the object that the model describes. There are many ways to represent a solid in a computer. AutoCAD native solids use a constructive solid geometry method. You can construct simple solid primitives by specifying the parameters of the primitives, solids by extruding a 2D closed region along a path or in a direction perpendicular to the 2D plane, and solids by revolving a 2D closed region. To construct a complex solid model, you can combine primitive solids, extruded solids, and revolved solids by using Boolean operations (union, subtract, and intersection), filleting, and chamfering. To cut a solid into two pieces, you can slice it along a plane or use intersection.

In this chapter, you learned how to create 3D solid primitives and how to construct a complex 3D solid by adding and subtracting simpler 3D solids. You gained a thorough understanding of how to construct regions and 2D profiles, how to produce extruded solids and revolved solids from 2D profiles, and how to use the intersection process in model creation and cutting.

The solid model created by using AutoCAD R14 is explicit. The two major limitations of explicit solid models are that we cannot separate two solids once they are combined by a Boolean operation and that we cannot change the parameters of a solid once it is created. To overcome these two limitations, we can use Mechanical Desktop. It runs on top of AutoCAD and is an advanced engineering design tool that enables us to create parametric solid models and virtual assemblies, and automates the production of engineering documents.

Enhance your experience in 3D solid modeling by doing the following exercises.

Exercise 6.1

What are the three kinds of 3D models in a computer?

Exercise 6.2

What kinds of solid primitives are available? Apart from such primitives, what other shapes of solids can you construct?

Exercise 6.3

How can you combine simple solid objects to form complex objects? What are the two kinds of methods to treat the edge of a model?

Exercise 6.4

How can you cut a solid along a straight edge? Along an irregular cutting edge?

Exercise 6.5

Explain the three kinds of 3D coordinate systems: 3D Cartesian, 3D cylindrical, and 3D spherical.

Exercise 6.6

Figure 6.136 shows the engineering drawing views of an angle bracket. Before looking at Figure 6.137, which is the suggested breakdown, take some time to analyze the component to find out what primitive solids you need to make this object.

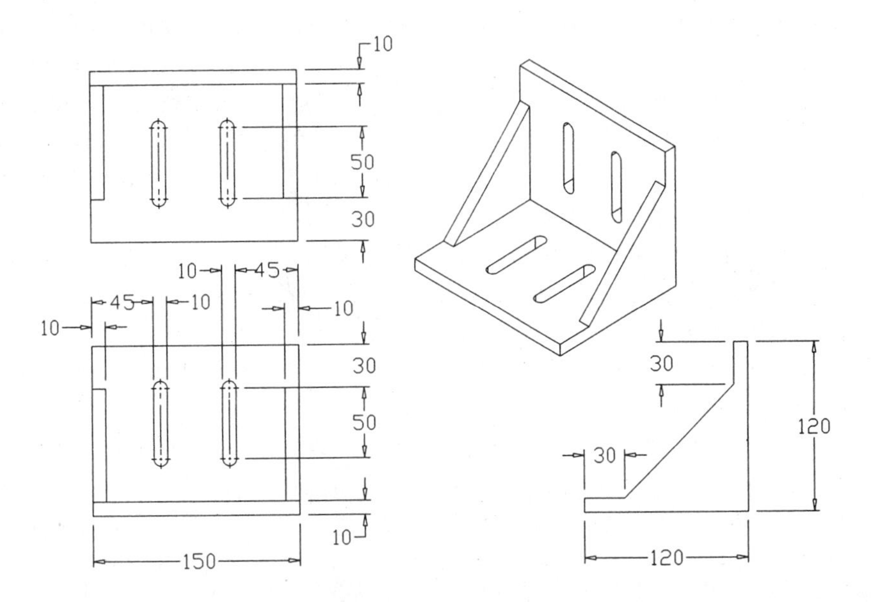

Figure 6.136 Angle bracket

Figure 6.137 is the suggested breakdown of the angle bracket. Refer to the dimensions shown in Figure 6.136, start a new drawing from scratch with metric default, and create the primitives one by one. Place them in a correct position relative to each other.

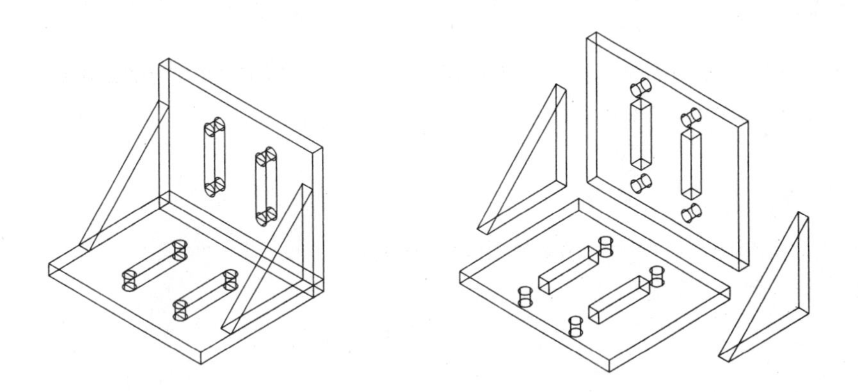

Figure 6.137 Suggested break down of angle bracket

To form the complex solid, unite the two larger boxes and wedges. Then subtract the remaining primitives. See Figure 6.138. Finally, fillet the edges A, B, C, D, E, F, G, H, J, K, and L.

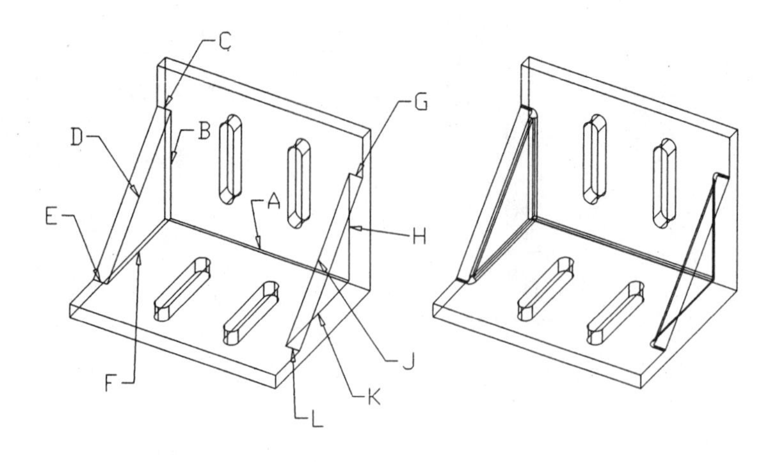

Figure 6.138 Edges filleted

The angle bracket is complete. Save the drawing as EX6_1.DWG.

Exercise 6.7

Start a new drawing from scratch. Use metric default. Create the solid model of an adjusting pin, which is shown in Figure 6.139. Take some time to analyze the model to break it down into simple primitive solids. Then compare your breakdown with Figure 6.140. Figure 6.140 is the suggested breakdown of the pin. There are five solid cylinders, one solid cone, and one solid box.

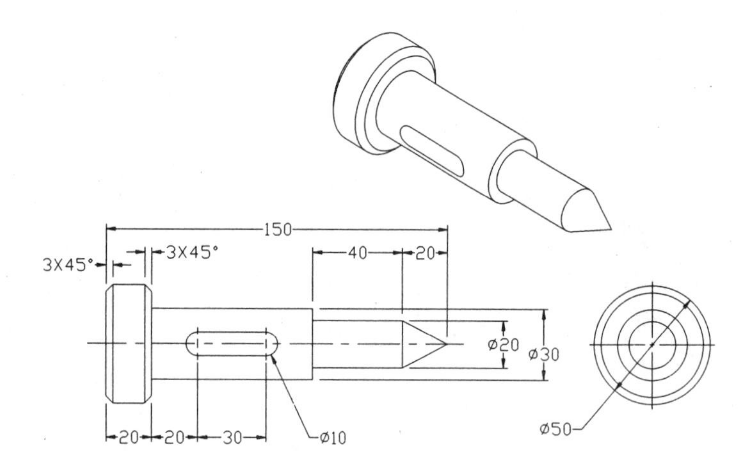

Figure 6.139 Adjusting pin

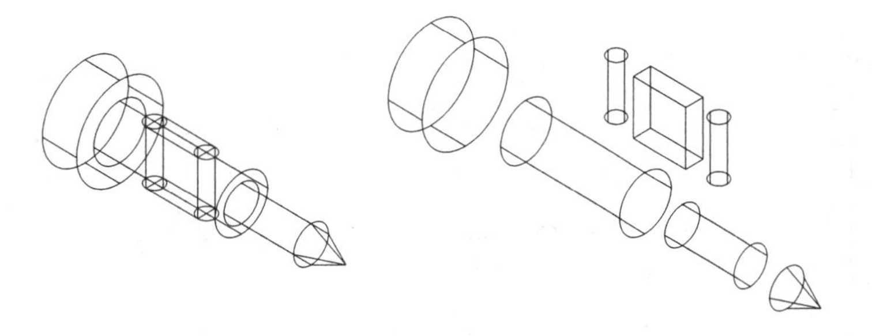

Figure 6.140 Suggested breakdown of adjusting pin

Refer to the dimensions shown in Figure 6.139 to create the primitives. Then unite the three larger cylinders and the solid cone. Next, subtract the remaining solids. Finally, chamfer two edges. See Figure 6.141. Save your drawing as EX6_2.DWG.

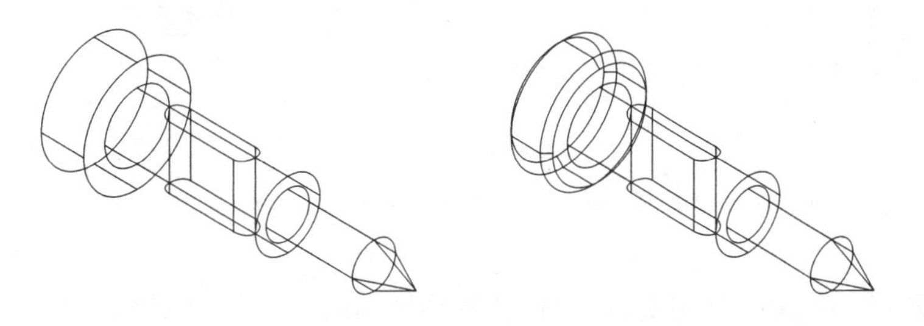

Figure 6.141 Edges chamfered

Exercise 6.8

Figure 6.142 shows the engineering drawing of the solid model of the lower suspension arm of a scale model car. Take some time to analyze the solid model to find out what primitives and Boolean operations you would use.

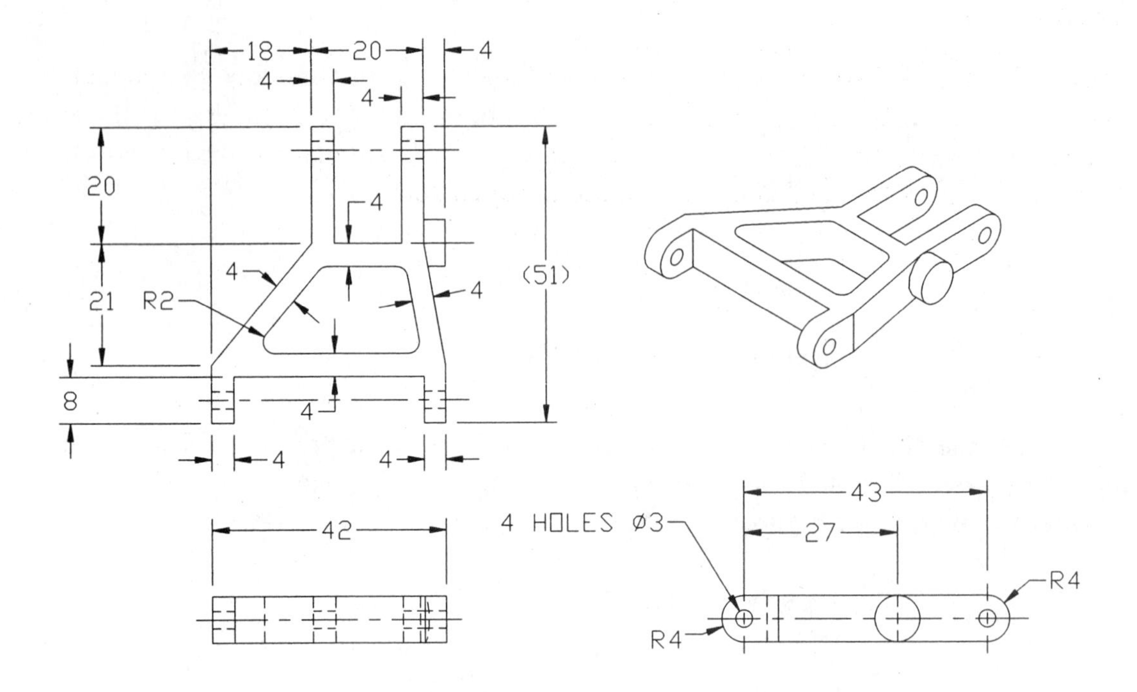

Figure 6.142 Lower suspension arm of a scale model car

Figure 6.143 is a suggested way of making the model. Create two regions and extrude them to become solids of extrusion. Then make a number of solid cylinders. After making all the necessary primitive solids, use Boolean operations to compose the complex solid model. Save your drawing as EX6_3.DWG.

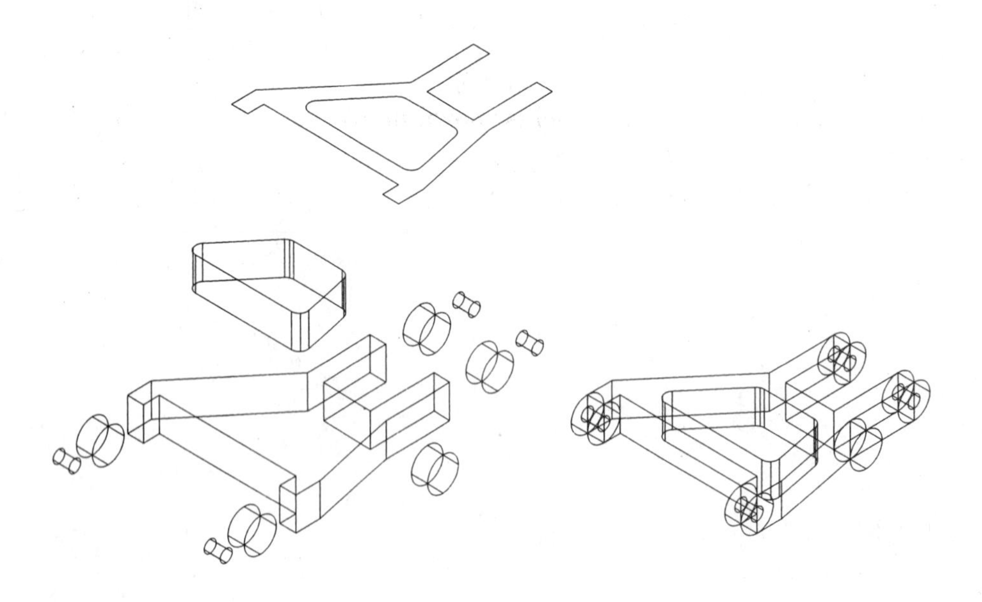

Figure 6.143 Suggested breakdown of lower suspension arm

Exercise 6.9

Start a new drawing. Create a solid model of the upper suspension arm of a model car. See Figure 6.144. Instead of using the union and subtraction operations, you will use the intersection operation. Before looking at Figure 6.145, the suggested breakdown, think for a moment how you can apply the intersection operation.

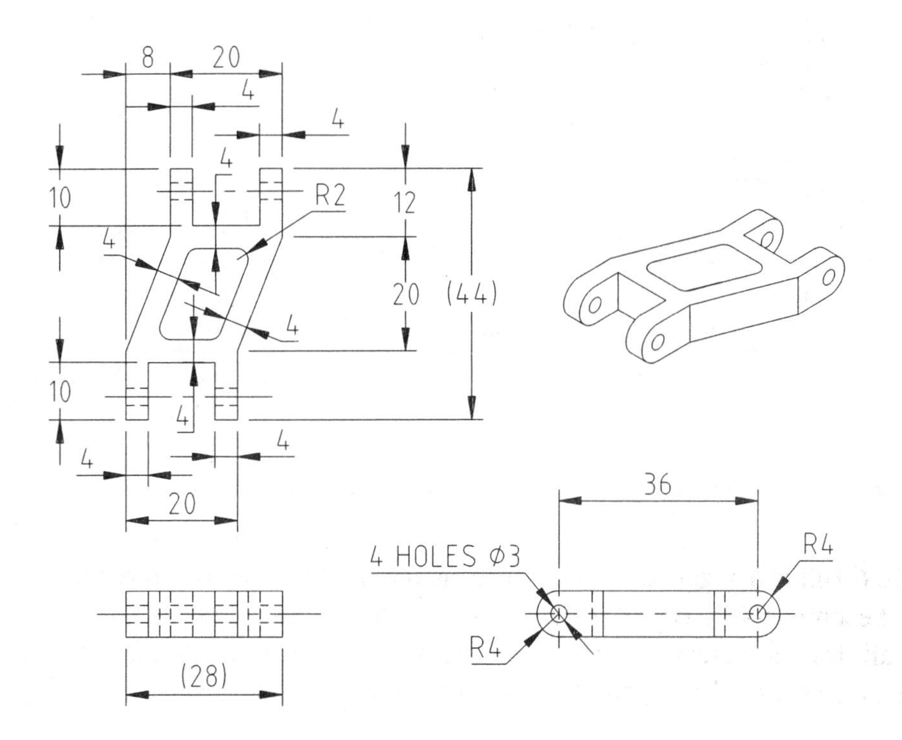

Figure 6.144 Upper suspension arm of scale model car

A suggested breakdown of the model is shown in Figure 6.145. You can produce two sets of wireframe, and create the main body of the model by extrusion and then intersection. After intersection, add four holes. Save you work as EX6_4.DWG.

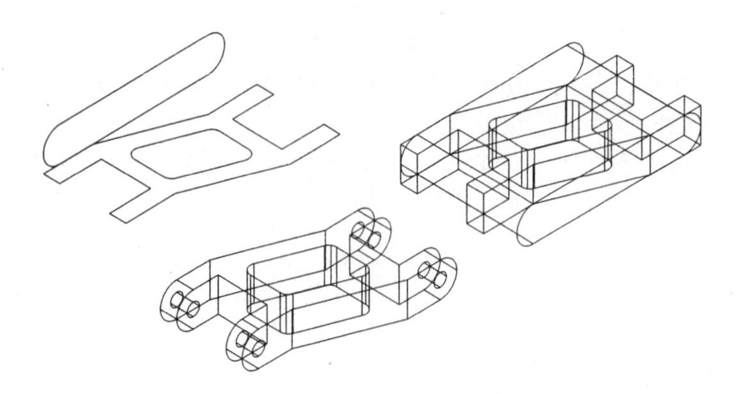

Figure 6.145 Suggested breakdown of upper suspension arm

Exercise 6.10

Start a new drawing to produce the solid model of the U-bracket of a model car. See Figure 6.146. Analyze the model to find out what primitive solids are required. Then compare your analysis with the suggested breakdown shown in Figure 6.147.

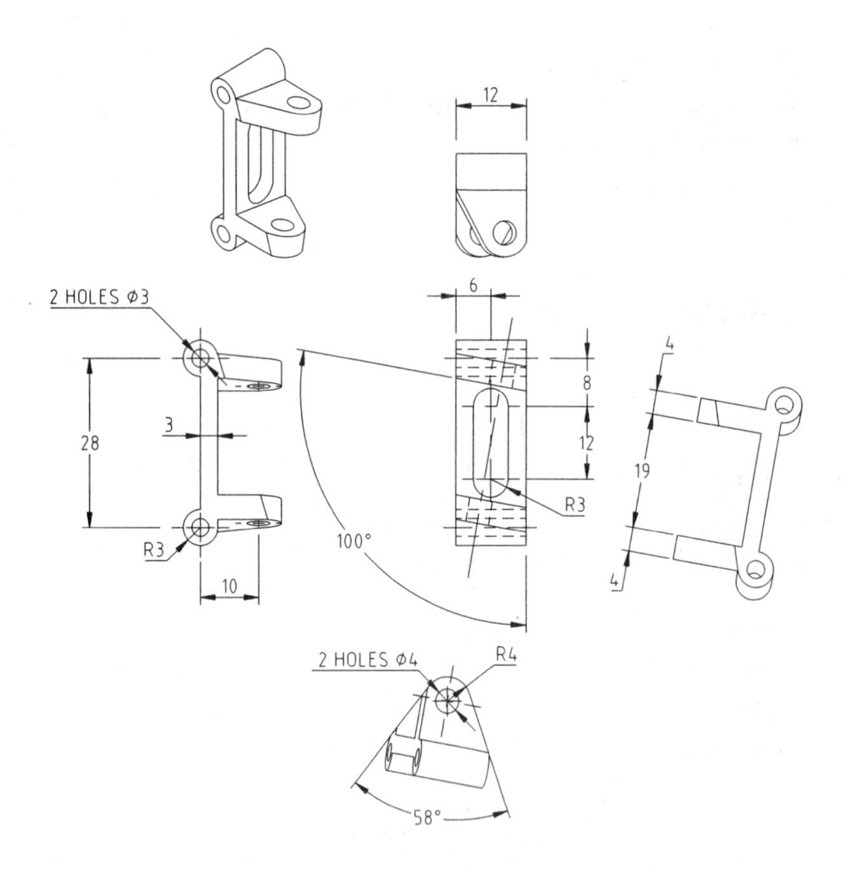

Figure 6.146 U-bracket of scale model car

Figure 6.147 shows the suggested breakdown of the U-bracket. To make the model, you can start by making the primitives shown on the left of the figure. Then you have to rotate the two triangular pieces and compose the solid model by using union and subtraction. To complete the model, you may make a large solid box and perform intersection. Save your drawing as EX6_5.DWG.

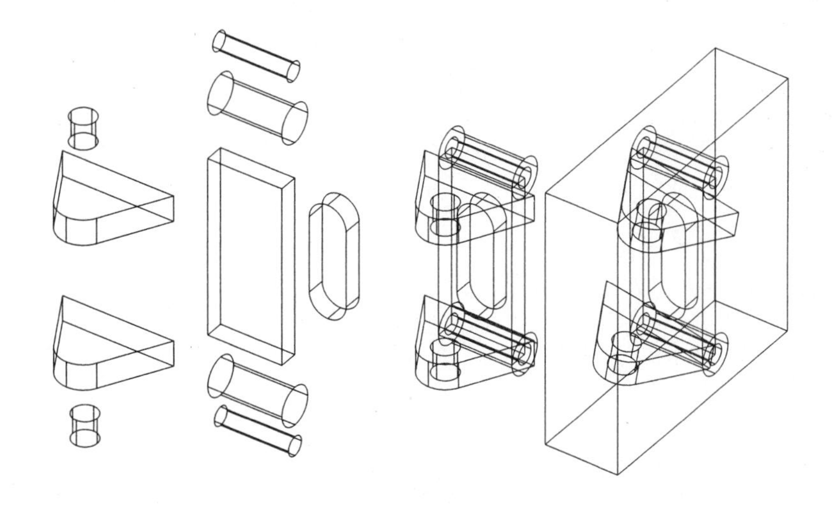

Figure 6.147 Suggested breakdown of U-bracket

Exercise 6.11

Start a new drawing to create a solid model for the transmission box of a model car. Read Figure 6.148 carefully, and prepare a breakdown of the model.

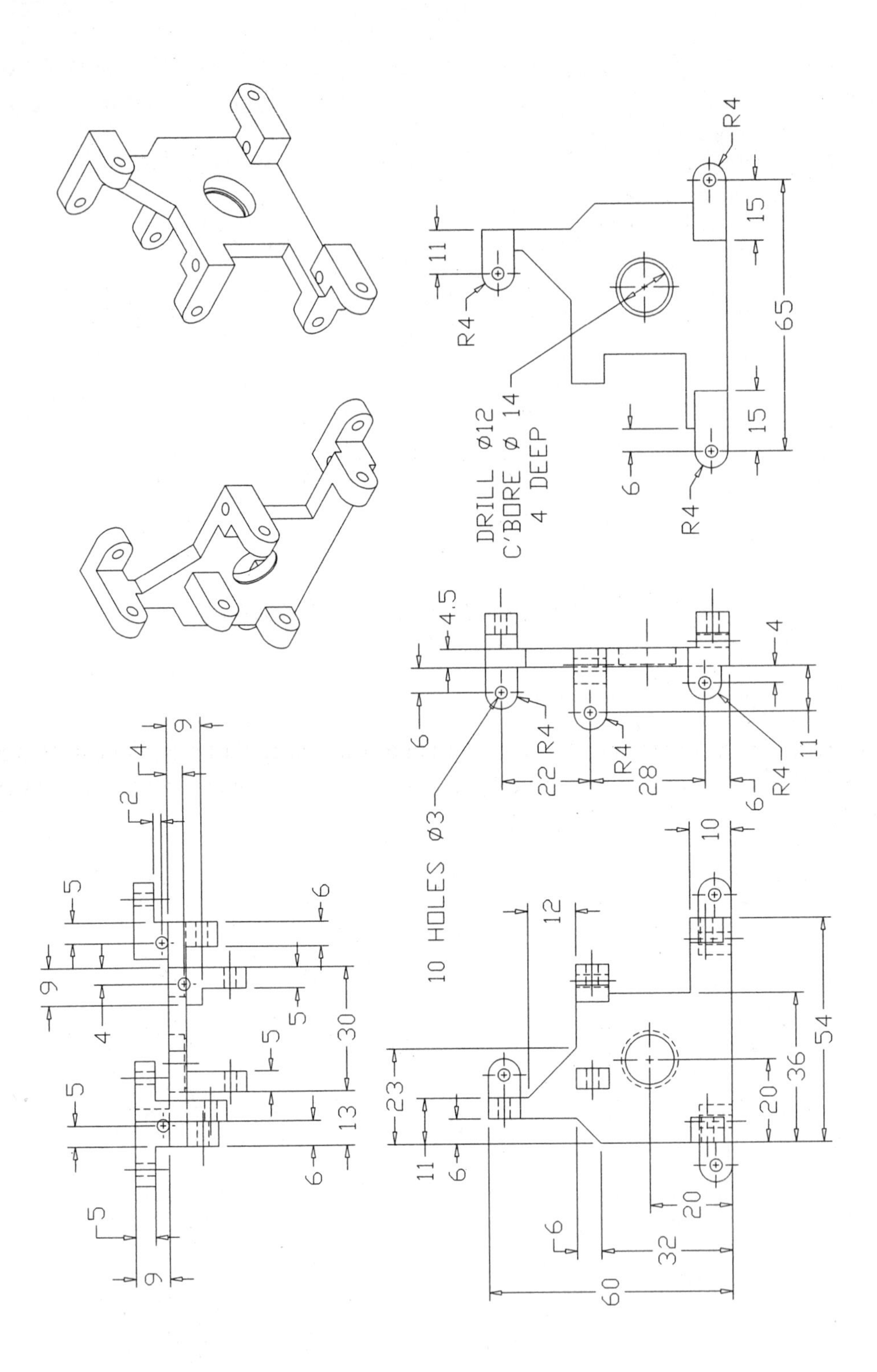

Figure 6.148 Half transmission box of scale model car

Figure 6.149 shows the suggested breakdown of the model. The main body is a solid of extrusion. Other features are solid boxes and solid cylinders that are either united to or subtracted from the main body. Create the primitives accordingly, and compose them together. Save your work as EX6_6.DWG.

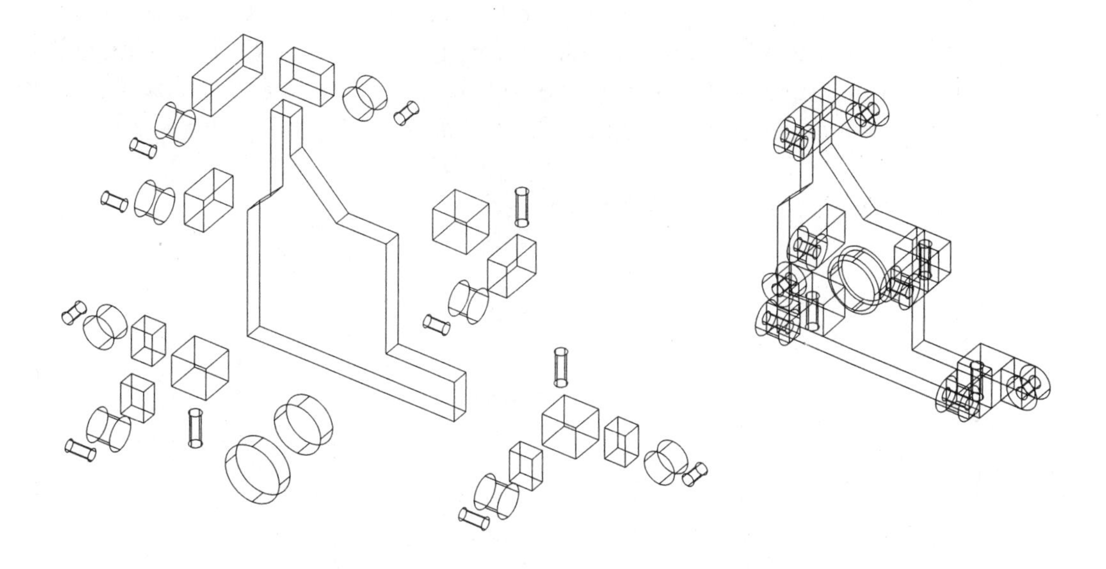

Figure 6.149 Suggested breakdown of transmission box

Exercise 6.12

Start a new drawing to produce the solid model of the rear hub of a model car. See Figure 6.150. Take a moment to analyze its constituents. Then compare your breakdown with Figure 6.151.

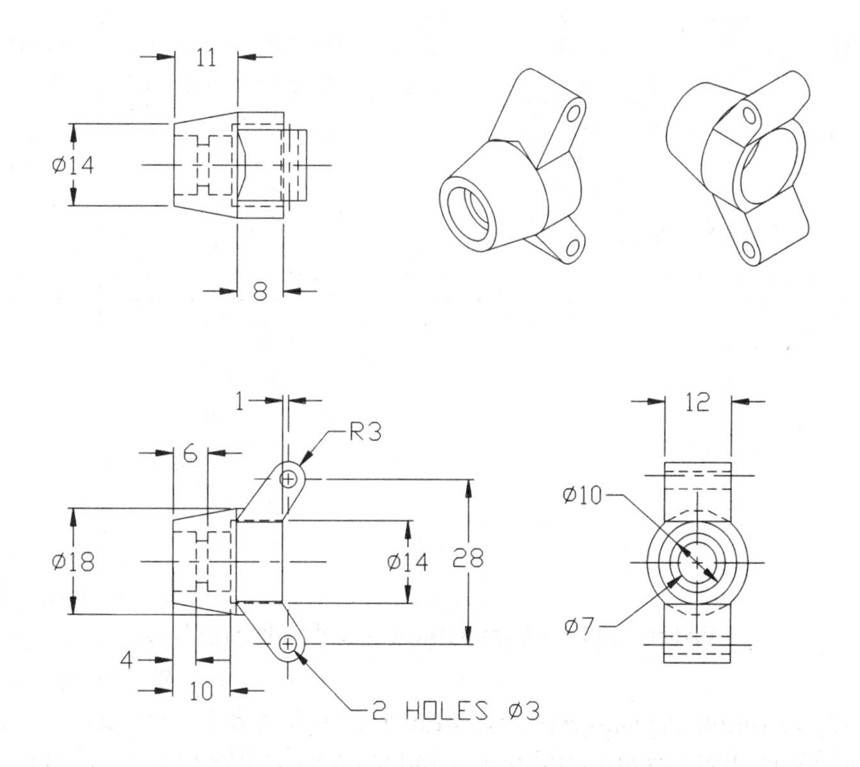

Figure 6.150 Rear hub of scale model car

The model consists mainly of two revolved solids and one extruded solid. Figure 6.151 shows the dimensions of the wireframes for making the revolved solids and the extruded solid. The figure also shows how the wireframes relate to each other. Create the solids accordingly, compose the complex model, and save your work as EX6_7.DWG.

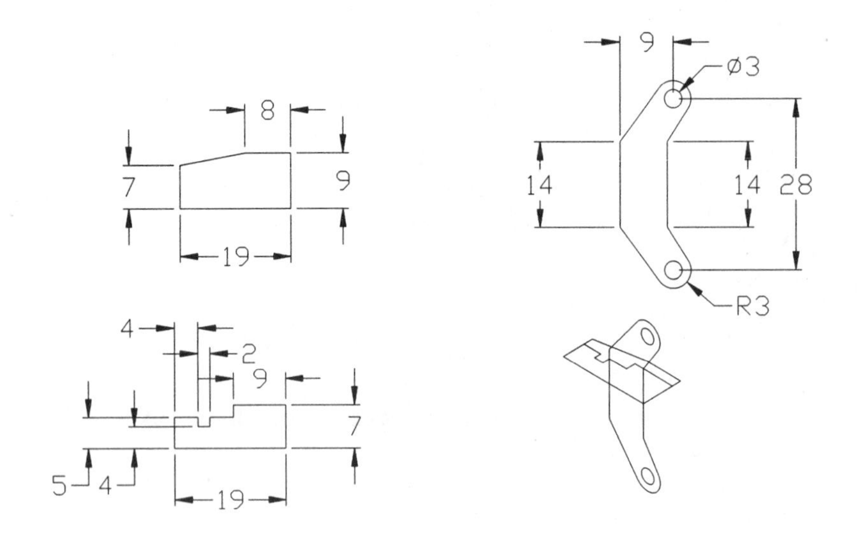

Figure 6.151 Wireframes for making rear hub

Exercise 6.13

Figure 6.152 shows the solid model of the pipe support base. Analyze the model to find out what 3D objects you will create and what Boolean operations you will use.

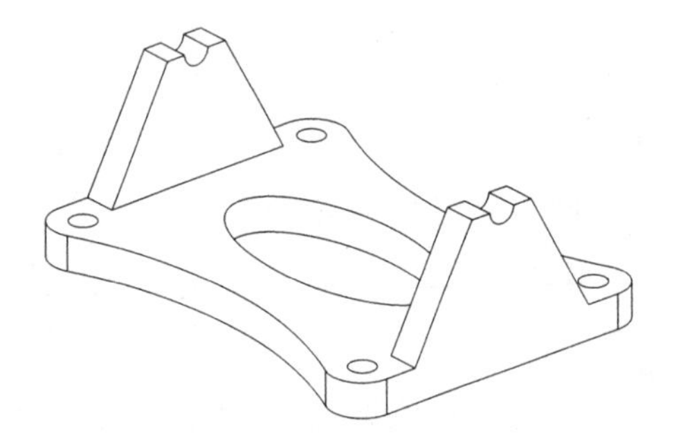

Figure 6.152 Pipe support base

Figure 6.153 shows a suggested breakdown of the model. Use the dimensions given in Figure 6.154 to make a region according to the top view and a region according to the side view. From the regions, make extruded solids. To compose the model, combine the extruded solids by uniting. Save the drawing as EX6_8.DWG.

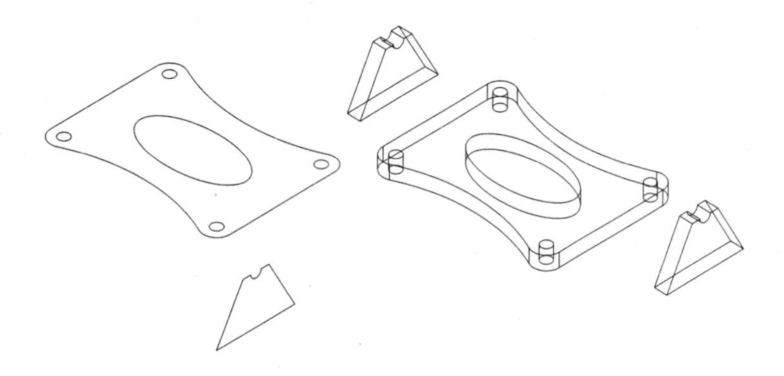

Figure 6.153 Regions and extruded solids

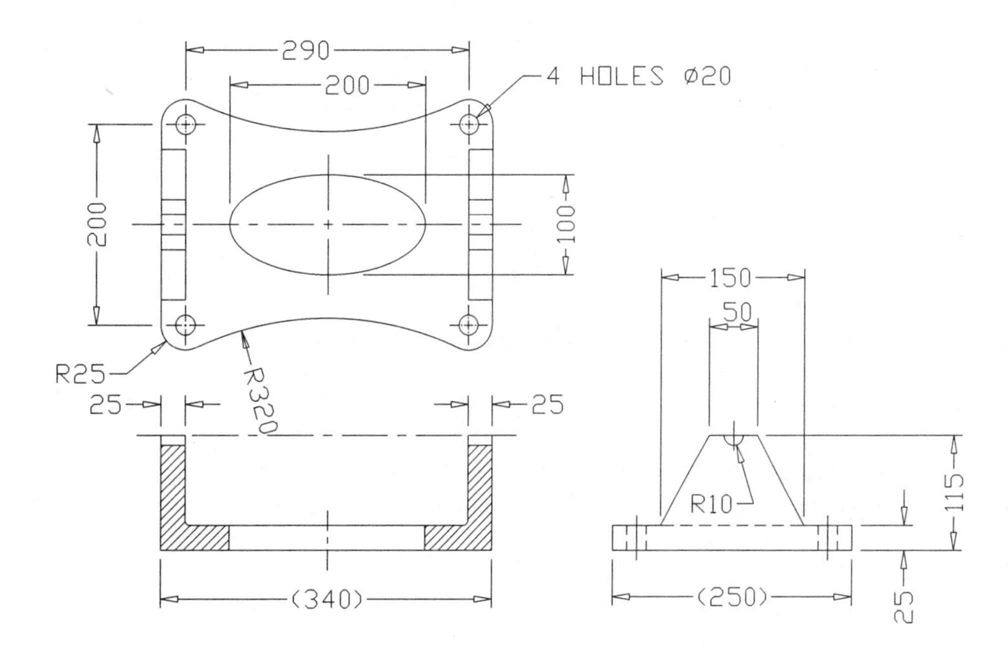

Figure 6.154 Dimensions of model

Exercise 6.14

Figure 6.155 shows the model of the pipe support roller. Think for a moment how you can construct this model. Then refer to Figure 6.156 for a breakdown and to Figure 6.157 for the dimensions. Save your drawing as EX6_9.DWG.

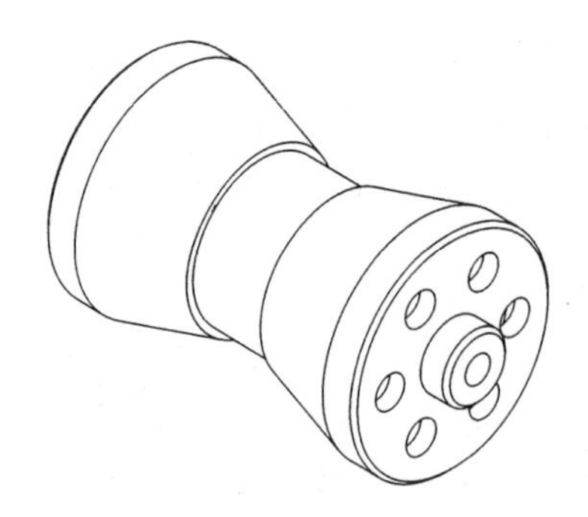

Figure 6.155 Pipe support roller

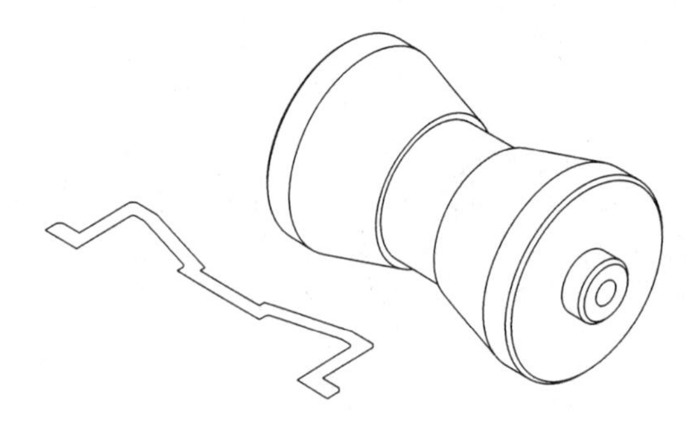

Figure 6.156 Region and revolved solid

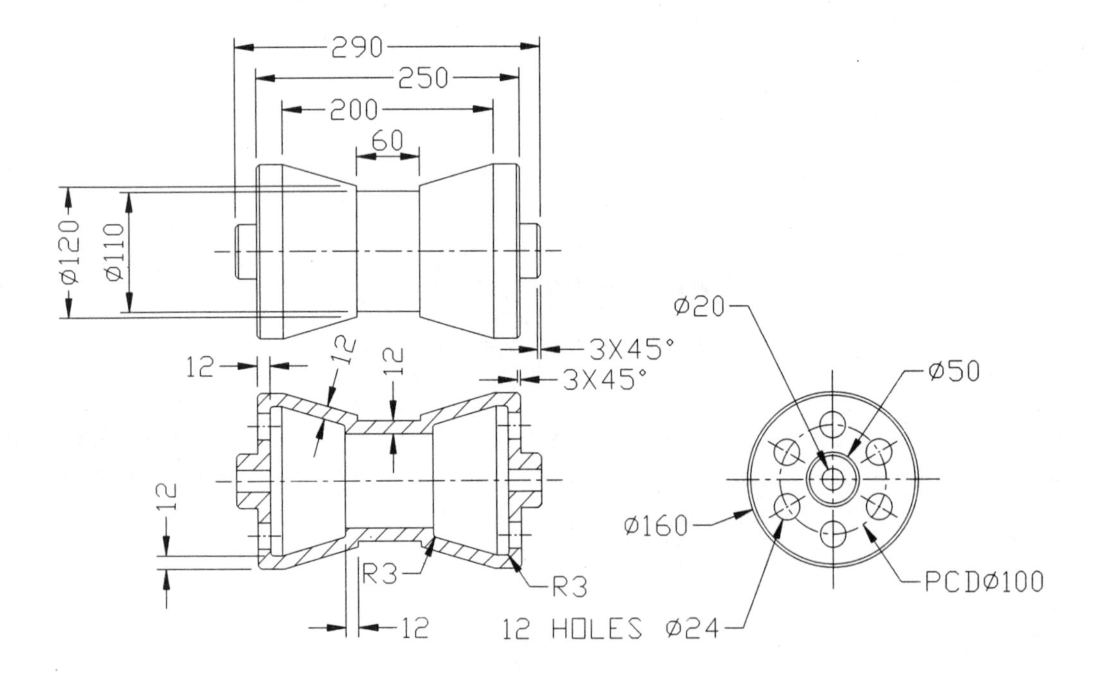

Figure 6.157 Dimensions of pipe support roller

Exercise 6.15

Figure 6.158 shows the orthographic projection views of the pipe support shaft. Construct a solid model and save the file as EX6_10.DWG.

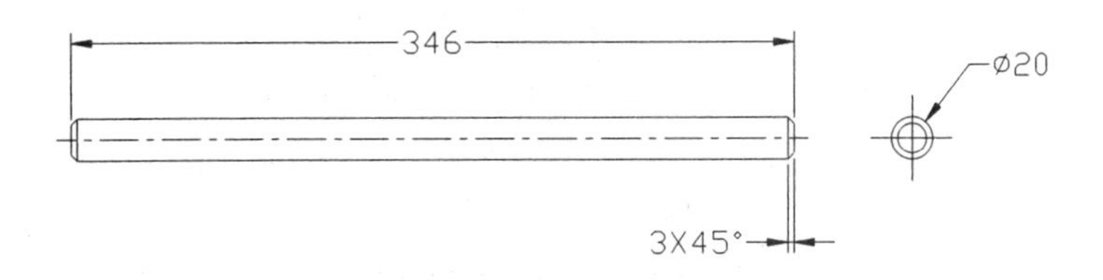

Figure 6.158 Pipe support shaft

Exercise 6.16

Start a new drawing. Use the XREF command to reference the drawing files EX6_8.DWG, EX6_9.DWG, and EX6_10.DWG externally. Align them properly according to Figure 6.159. Save your drawing as EX6_11.DWG.

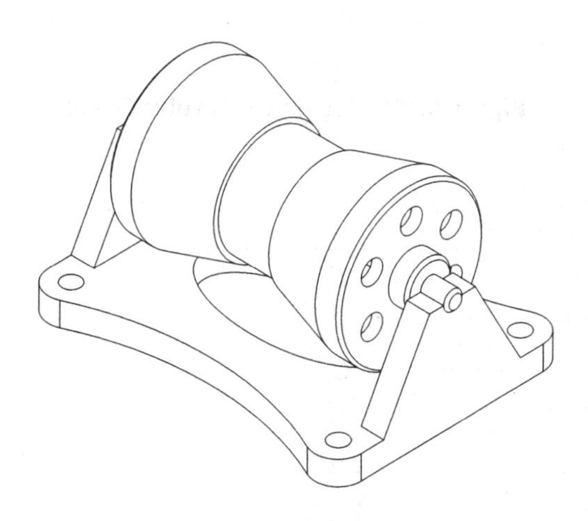

Figure 6.159 Pipe support assembly

Chapter 7

Solid Modeling Projects: Mechanical

7.1 Drawing Preparation
7.2 Upper Casing
7.3 Lower Casing
7.4 Interference, Mass Properties, and Sections
7.5 Buttons
7.6 Ball Cover
7.7 Assembly of Solid Models
7.8 Key Points and Exercises

Aims and Objectives

The aim of this chapter is to introduce the methods and techniques for constructing thin-shell and complex mechanical engineering solid parts. After studying this chapter, you should be able to:

- construct thin-shell solids with bosses and webs, and
- construct complex mechanical engineering parts.

Overview

In the last chapter, you constructed a number of 3D solid models by combining primitive solids, revolved solids, and extruded solids. You also sliced, filleted, and chamfered 3D solids. In this chapter, you will work on an electro-mechanical product. The main aim of constructing the solid models of the electro-mechanical product is, in addition to enhancing your knowledge, to gain an understanding of making a 3D thin-shell solid model with internal webs and bosses.

This product is a set of mouse casing. It consists of electronic and mechanical components that are enclosed within a polymeric casing. The casing consists of four parts: upper casing, lower casing, buttons, and ball cover. Figure 7.1 shows an exploded view of the product.

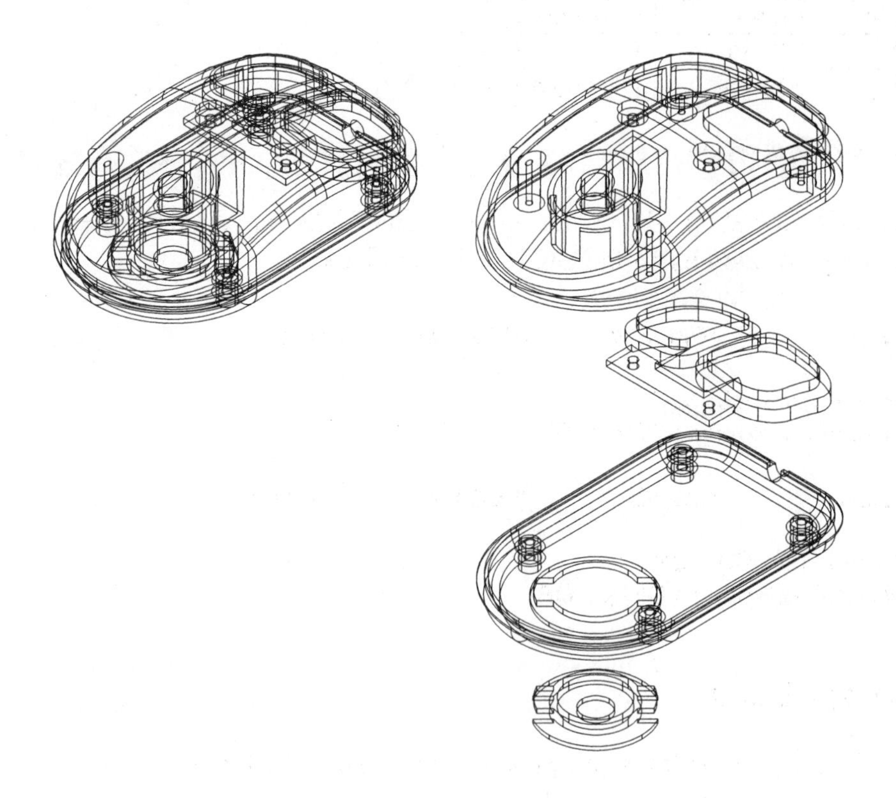

Figure 7.1 Mouse casing

In the following delineation, you will first work on the upper and lower casings. After that, you will put them together to check if there is any interference between them, and you will create sections to see how the two parts mate. Then you will find out the mass properties of the models. Finally, you will work on the buttons and the ball cover to complete the assembly.

7.1 Drawing Preparation

Start a new drawing by using the NEW command. Use metric default settings.

<File> **<New...>**

Use the LAYER command to create four additional layers: S_l, S_u, W_l, and W_u. Select layer W_u as the current layer. You will use layer W_u for the wireframes of the upper casing, layer S_u for the solids of the upper casing, layer W_l for the wireframes of the lower casing, and layer S_l for the solids of the lower casing.

<Format> **<Layer...>**

Command: **LAYER**

Name	Color	Linetype
0	White	Continuous
S_l	Green	Continuous

S_u	Blue	Continuous
W_l	Cyan	Continuous
W_u	Yellow	Continuous

Current layer: **W_u**

As is routine for 3D model creation, turn on the UCS icon display and set it to display at the origin position by using the UCSICON command.

<View> **<Display>** **<UCS Icon>** <v **On>**

Command: **UCSICON**
ON/OFF/All/Noorigin/ORigin: **ON**

<View> **<Display>** **<UCS Icon>** <v **Origin>**

Command: **UCSICON**
ON/OFF/All/Noorigin/ORigin: **OR**

7.2 Upper Casing

The upper face of the model is spherical in shape. To make the main solid body, you can intersect a solid sphere and an extruded solid. See Figure 7.2.

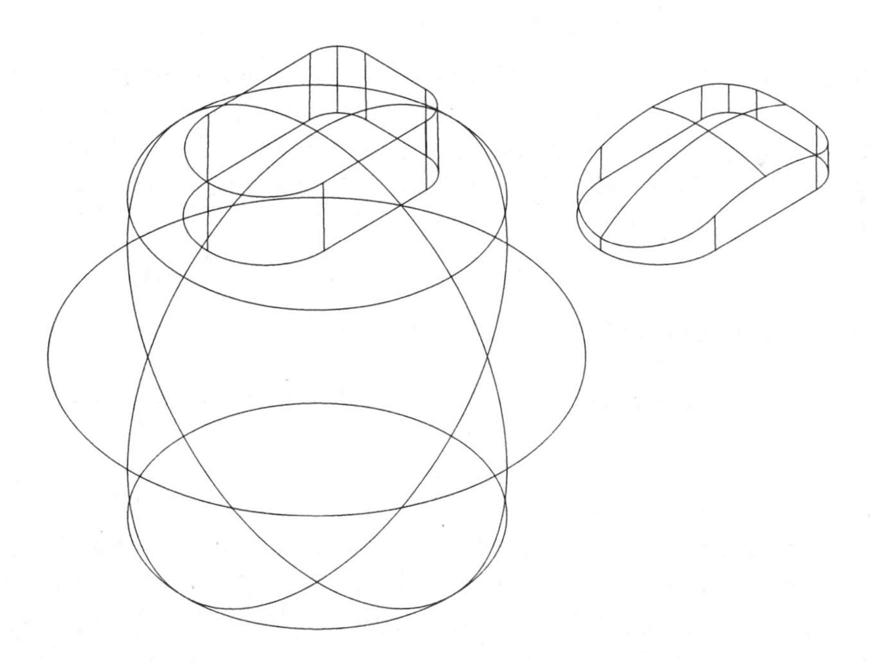

Figure 7.2 Solid sphere, extruded solid, and solid of intersection

The casing is a thin shell. It houses electronic and mechanical components. To locate and fix the components, there are bosses and webs. If you use Mechanical Desktop R2, you can simply make a thin shell from a solid body and then add the bosses or webs to it.

With AutoCAD R14, you will use an approach that is different from that in the previous chapter because the upper face of the model is a curved surface. A general direction for modeling a thin-shell object with internal bosses and webs is to treat the model as two solid parts: outer skin and inner core. That is, you should make two solid models instead of one. First, you will model the inner core. To make the inner core, you should regard the inner core as a solid object, not a void. In this sense, the webs and bosses on the final model should appear as recesses in the core solid. See Figure 7.3.

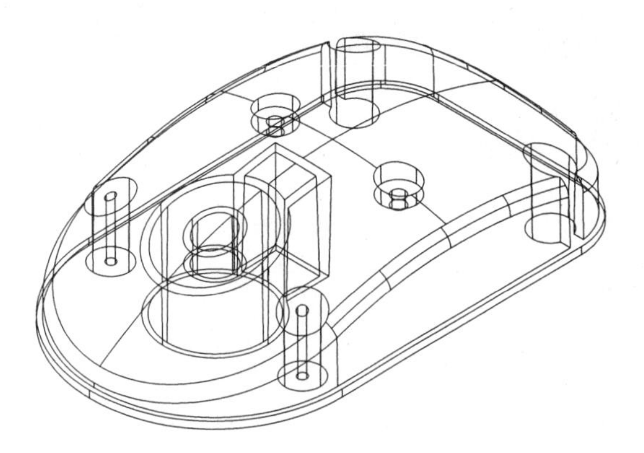

Figure 7.3 Solid that represents the core of the upper casing

Then you should make the model as if it did not have any core. That is, you will make the solid model according to the outer dimensions. See Figure 7.4.

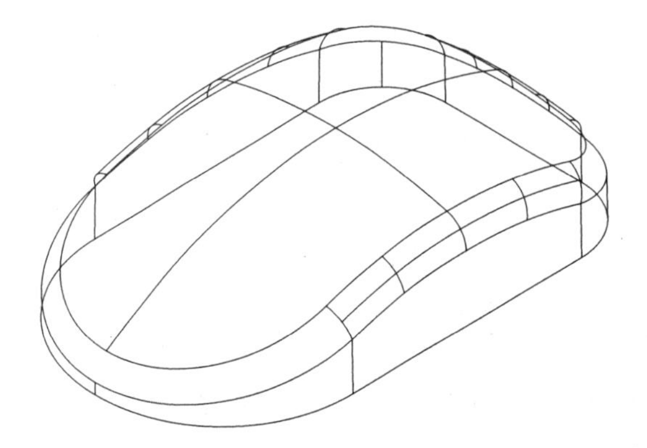

Figure 7.4 Solid that represents the outer skin of the upper casing

After you have made both the inner core and the outer skin as solid objects, you can then subtract the inner-core solid from the outer-skin solid to yield the final model.

Wireframes for the Upper Casing

Now you will start by working on the wireframes for the upper casing. You can divide the wireframes for the upper casing into four groups. Figure 7.5 through Figure 7.8 delineate their X and Y coordinates.

In accordance with Figure 7.5, create three wireframes for making the outer skin, the recess, and the inner core.

Set the system variable DELOBJ to 1 so that the entities that are used to make the regions are deleted.

Command: **DELOBJ**
New value for DELOBJ: **1**

Use the REGION command to convert the wireframes to form three regions.

<Draw> **<Region>**

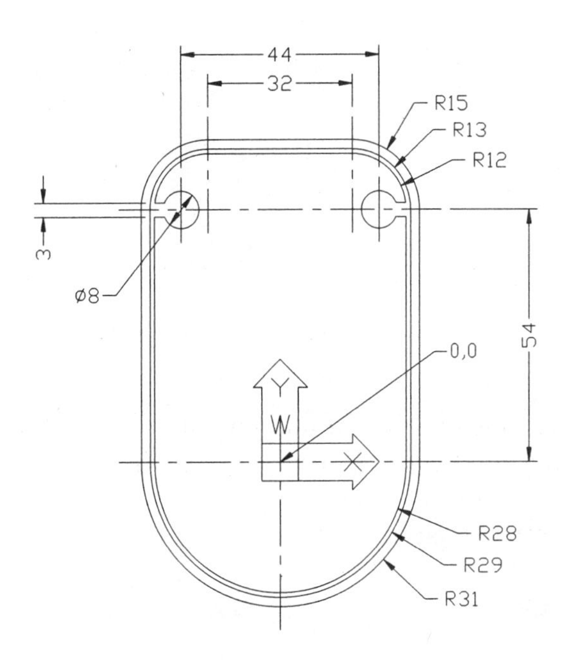

Figure 7.5 Regions for the outer skin, recess, and inner core created

To see the position of the entities more easily, set the display to two-viewport.

<View> **<Tiled Viewports>** **<2 Viewports>**

Select the right viewport and set it to show a southeast isometric view.

<View> **<3D Viewpoint>** **<SE Isometric>**

Create two more regions as shown in Figure 7.6. After that, move the regions a distance of 6 units in the Z-direction. The regions will be used for making the button openings.

<Modify> **<Move>**

Command: **MOVE**
Select objects: **[Select A and B (Figure 7.6).]**

Select objects: **[Enter]**
Base point or displacement: **0,0,6**
Second point of displacement: **[Enter]**

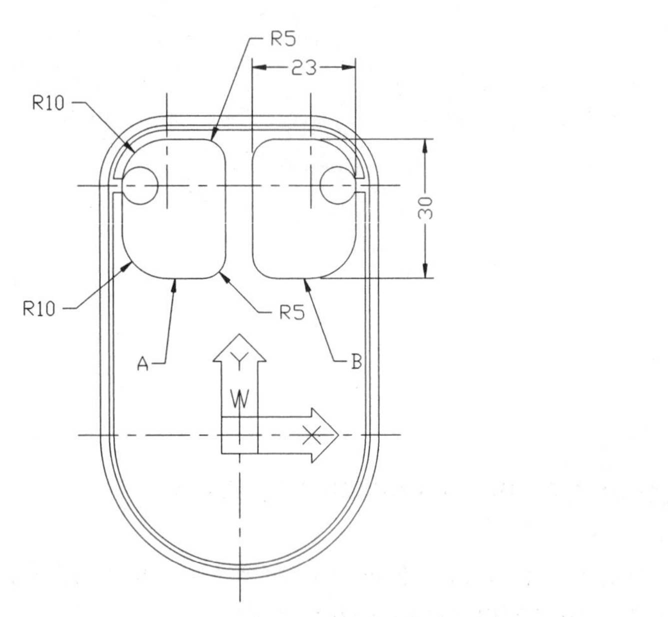

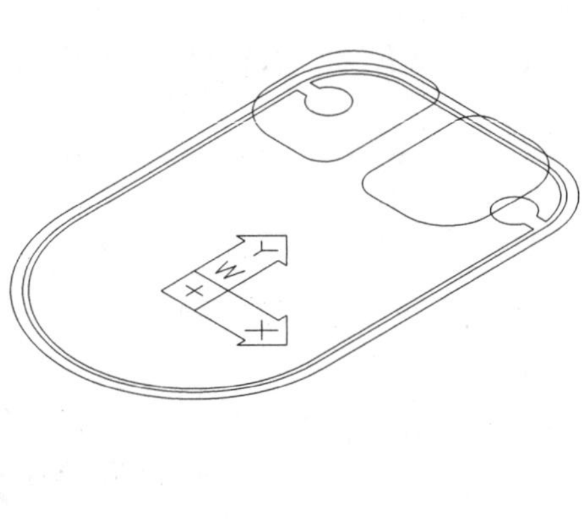

Figure 7.6 Wireframes for button openings created

In accordance with Figure 7.7, create three regions that will be used for making the mouse ball housing. Subtract regions B and C from region A to form a complex region.

<Modify> **<Boolean>** **<Subtract>**

Command: **SUBTRACT**
Select solids and regions to subtract from...
Select objects: **[Select A (Figure 7.7).]**
Select objects: **[Enter]**
Select solids and regions to subtract...
Select objects: **[Select B and C (Figure 7.7).]**
Select objects: **[Enter]**

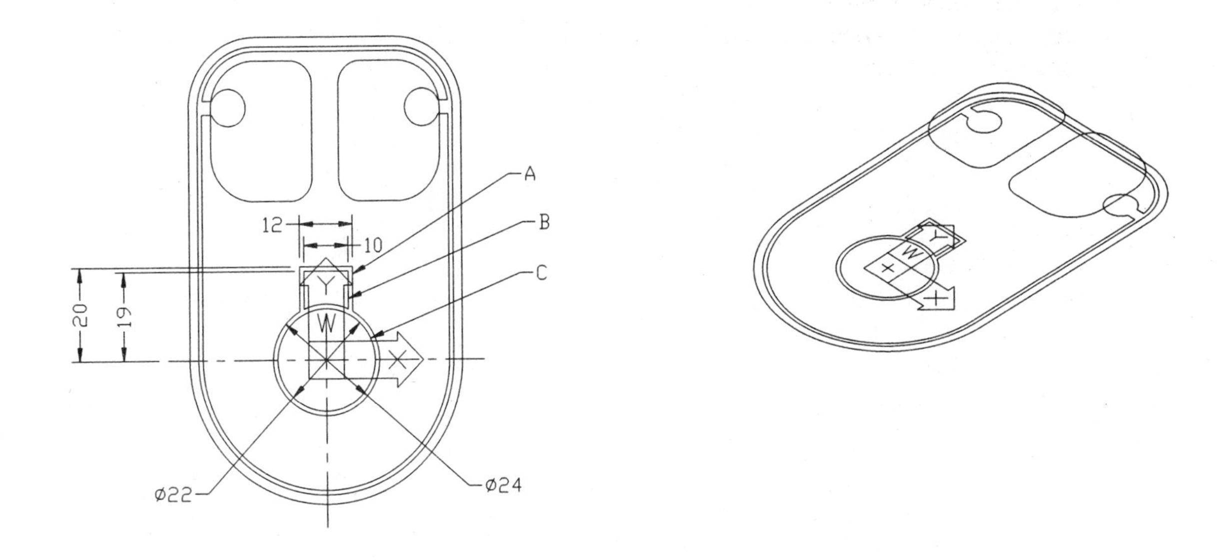

Figure 7.7 Wireframes for mouse ball housing created

In accordance with Figure 7.8, create six circles. Four of them will be used for making the bosses. Two will be used for making the mouse ball stopper.

<Draw> **<Circle>** **<Center, Radius>**

Convert the circles to regions. Then subtract region B from region A. The subtracted complex region will be used for making the mouse ball stopper.

<Draw> **<Region>**

<Modify> **<Boolean>** **<Subtract>**

Command: **SUBTRACT**
Select solids and regions to subtract from...
Select objects: **[Select A (Figure 7.8).]**
Select objects: **[Enter]**
Select solids and regions to subtract...
Select objects: **[Select B (Figure 7.8).]**
Select objects: **[Enter]**

Move the subtracted region a distance of 11 units in the Z-direction, and move two circular regions a distance of 17 units in the Z-direction.

<Modify> **<Move>**

Command: **MOVE**
Select objects: **[Select A (Figure 7.8).]**
Select objects: **[Enter]**
Base point or displacement: **0,0,11**
Second point of displacement: **[Enter]**

Command: **[Enter]**

MOVE
Select objects: **[Select C and D (Figure 7.8).]**
Select objects: **[Enter]**
Base point or displacement: **0,0,17**
Second point of displacement: **[Enter]**

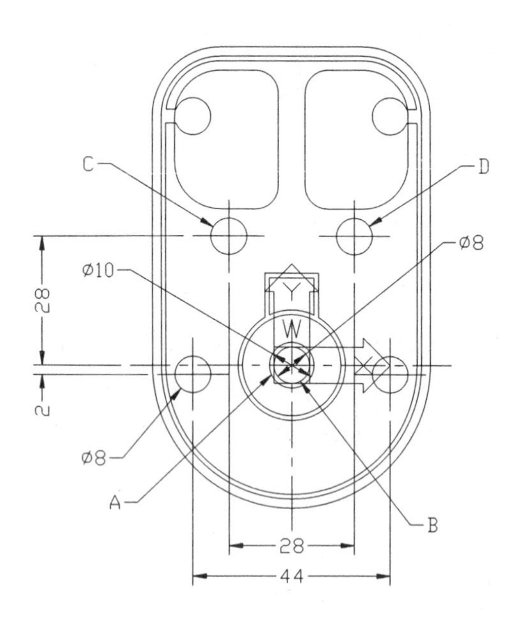

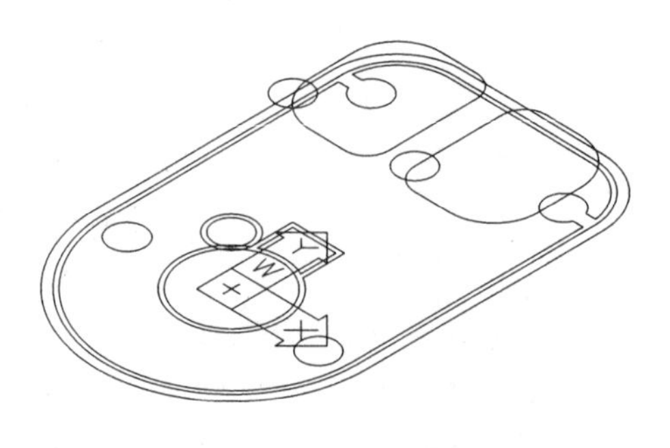

Figure 7.8 Two complex regions created and one moved

The wireframes for the upper casing are complete.

Solid Model for the Upper Casing

After creating the wireframes for the upper casing, you will use the wireframes to make the solid model. As mentioned earlier, this is a thin-shell model with internal webs and bosses. You will build two solids — one solid for the outer skin (as if there were no internal void) and one solid to represent the void.

Set the current layer to S_u. You will use the wireframes residing on layer W_u to build a solid model on layer S_u.

[Object Properties] **[Layer Control]**

Current layer: **Solid**

Set to a single viewport. Then set the display to SE isometric view. See Figure 7.9.

<View> **<Tiled Viewports>** **<1 Viewport>**

<View> **<3D Viewpoint>** **<SE Isometric>**

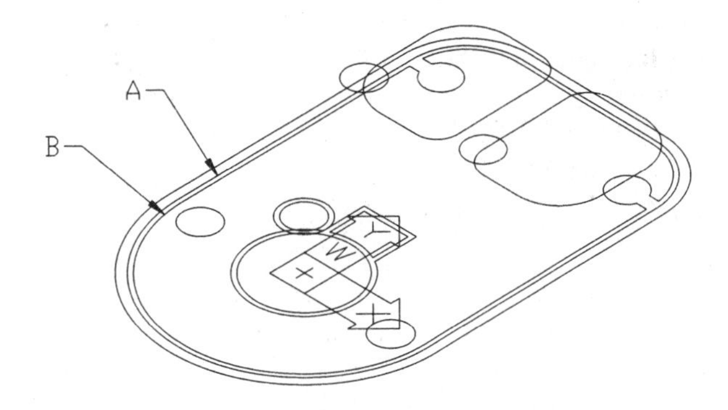

Figure 7.9 Wireframes for upper casing

When you make the wireframes, you have to set the variable DELOBJ to 1 so the wireframes used to make the regions are deleted. This way, you will not duplicate the wireframes. Now you will use the wireframes for extruding or revolving. You should set the variable to 0 in order to keep the wireframes. Keeping the wireframes enables you to rebuild the model quickly if there is any mistake.

Command: **DELOBJ**
New value for DELOBJ: **0**

To begin, you will build the solid model that represents the inner core of the casing. The inner core consists mainly of a recess, an extruded side wall, and a spherical top face.

Use the EXTRUDE command to extrude a region a distance of 1.5 in Z-direction to form the recess.

<Draw> **<Solids>** **<Extrude>**

Command: **EXTRUDE**
Select objects: [**Select A (Figure 7.9).**]
Select objects: [**Enter**]
Path/<Height of Extrusion>: **1.5**
Extrusion taper angle: **1**

Repeat the EXTRUDE command to form the main body of the inner core.

Command: [**Enter**]
EXTRUDE
Select objects: [**Select A (Figure 7.9).**]
Select objects: [**Enter**]
Path/<Height of Extrusion>: **30**
Extrusion taper angle: **1**

Use the SPHERE command to create a solid sphere to form the upper spherical face. See Figure 7.10.

<Draw> **<Solids>** **<Sphere>**

Command: **SPHERE**

Center of sphere: **0,27,-76**
Diameter/<Radius> of sphere: **97**

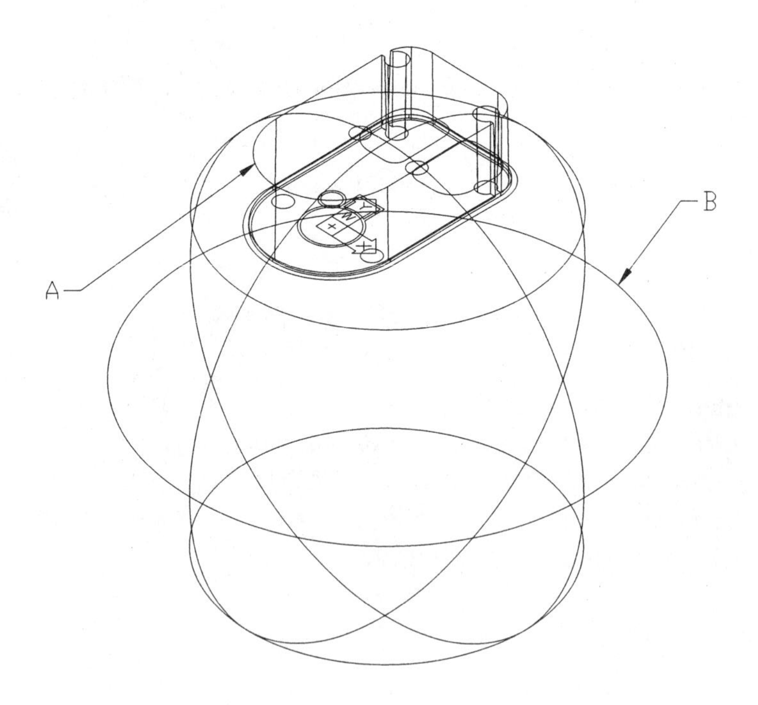

Figure 7.10 Extruded solids and solid sphere created

To form the main body of the core, use the INTERSECT command to create a solid of intersection from an extruded solid and a solid sphere. See Figure 7.11.

<Modify> **<Boolean>** **<Intersect>**

Command: **INTERSECT**
Select objects: [**Select A and B (Figure 7.10).**]
Select objects: [**Enter**]

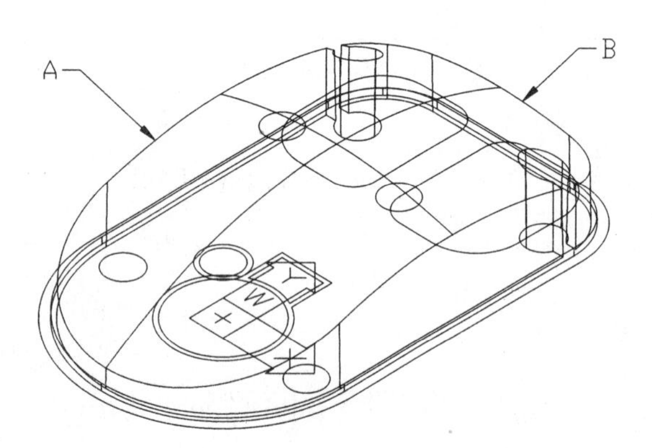

Figure 7.11 Extruded solid intersected with solid sphere

Run the FILLET command to round off the upper edge of the solid of intersection.

See Figure 7.12.

<Modify> **<Fillet>**

Command: **FILLET**
Polyline/Radius/Trim/<Select first object>: **[Select A (Figure 7.11).]**
Enter radius: **3**
Chain/Radius/<Select edge>: **C**
Edge/Radius/<Select edge chain>: **[Select A and B (Figure 7.11).]**
Edge/Radius/<Select edge chain>: **[Enter]**

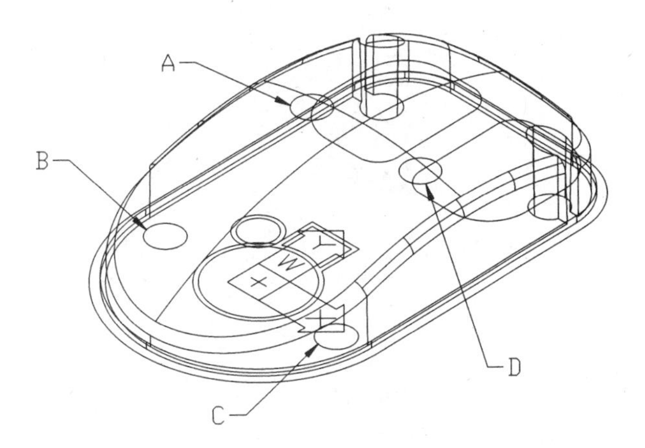

Figure 7.12 Upper edges of solid of intersection filleted

Use the CYLINDER command to create four solid cylinders with centers at A, B, C, and D. After that, use the EXTRUDE command to extrude four circles. Together, they will form two bosses with holes and two stepped bosses. See Figure 7.13.

<Draw> **<Solids>** **<Cylinder>**

Command: **CYLINDER**
Elliptical/<center point>: **CEN** of **[Select A (Figure 7.12).]**
Diameter/<Radius>: **1.5**
Center of other end/<Height>: **-2**

Command: **[Enter]**
CYLINDER
Elliptical/<center point>: **CEN** of **[Select B (Figure 7.12).]**
Diameter/<Radius>: **1**
Center of other end/<Height>: **30**

Command: **[Enter]**
CYLINDER
Elliptical/<center point>: **CEN** of **[Select C (Figure 7.12).]**
Diameter/<Radius>: **1**
Center of other end/<Height>: **30**

Command: **[Enter]**
CYLINDER
Elliptical/<center point>: **CEN** of **[Select D (Figure 7.12).]**

Diameter/<Radius>: **1.5**
Center of other end/<Height>: **-2**

<Draw> **<Solids>** **<Extrude>**

Command: **EXTRUDE**
Select objects: [**Select A, B, C, and D (Figure 7.12).**]
Select objects: [**Enter**]
Path/<Height of Extrusion>: **30**
Extrusion taper angle: **-1**

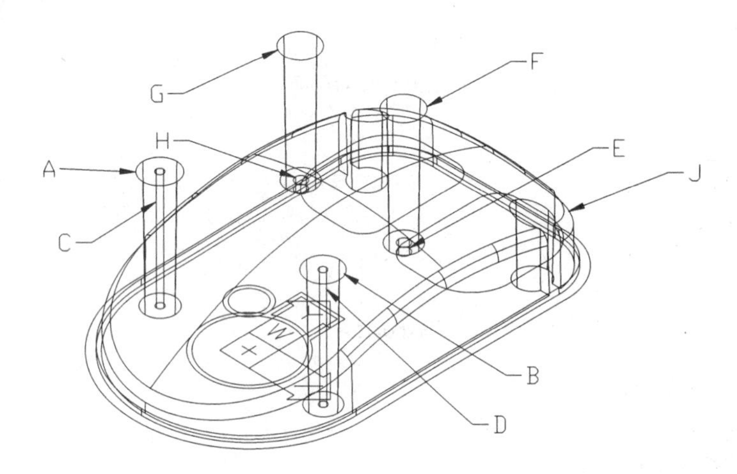

Figure 7.13 Solid cylinders and extruded solids created

Use the SUBTRACT command to subtract two cylinders from two extruded circles. Then repeat the SUBTRACT command to subtract the resulting hollow extruded solids, cylinders, and extruded solids from the core solid. See Figure 7.14.

<Modify> **<Boolean>** **<Subtract>**

Command: **SUBTRACT**
Select solids and regions to subtract from...
Select objects: [**Select A and B (Figure 7.13).**]
Select objects: [**Enter**]
Select solids and regions to subtract...
Select objects: [**Select C and D (Figure 7.13).**]
Select objects: [**Enter**]

Command: [**Enter**]
SUBTRACT
Select solids and regions to subtract from...
Select objects: [**Select J (Figure 7.13).**]
Select objects: [**Enter**]
Select solids and regions to subtract...
Select objects: [**Select A, E, F, G, and H (Figure 7.13).**]
Select objects: [**Enter**]

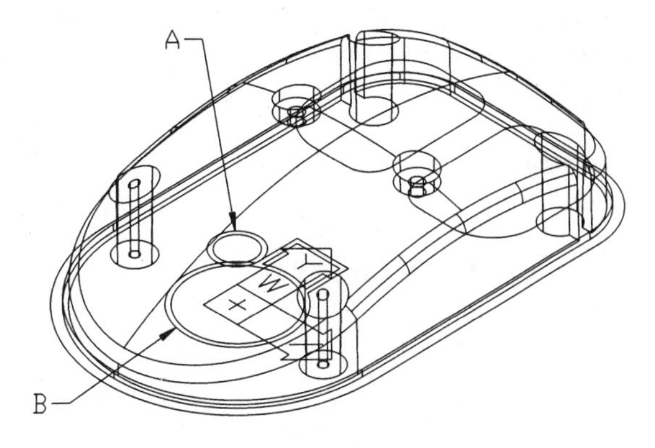

Figure 7.14 Solids subtracted and united

Now you have two bosses with holes and two stepped bosses. To prepare the ball housing and the ball stopper, use the EXTRUDE command to extrude two complex regions. See Figure 7.15.

<Draw> **<Solids>** **<Extrude>**

Command: **EXTRUDE**
Select objects: [**Select A and B (Figure 7.14).**]
Select objects: [**Enter**]
Path/<Height of Extrusion>: **30**
Extrusion taper angle: **-1**

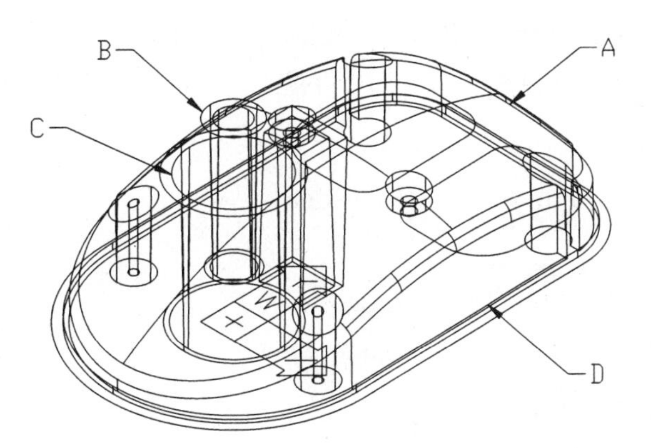

Figure 7.15 Ball housing and ball stopper prepared

Subtract the last extruded solid from the core solid and unite the recess solid to the core solid. See Figure 7.16.

<Modify> **<Boolean>** **<Subtract>**

Command: **SUBTRACT**
Select solids and regions to subtract from...
Select objects: [**Select A (Figure 7.15).**]
Select objects: [**Enter**]

Select solids and regions to subtract...
Select objects: [**Select B and C (Figure 7.15).**]
Select objects: [**Enter**]

<Modify> **<Boolean>** **<Union>**

Command: **UNION**
Select objects: [**Select A and D (Figure 7.15).**]
Select objects: [**Enter**]

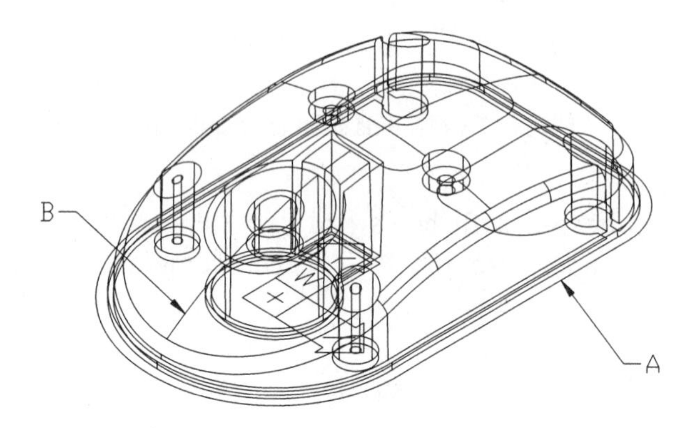

Figure 7.16 Core solid completed

The solid that represents the core of the upper casing is complete. To continue, you will work on the outer shell.

Run the EXTRUDE command to extrude the outer profile, and use the SPHERE command to create a solid sphere. Their intersection will form the main body of the outer skin. See Figure 7.17.

<Draw> **<Solids>** **<Extrude>**

Command: **EXTRUDE**
Select objects: [**Select A (Figure 7.16).**]
Select objects: [**Enter**]
Path/<Height of Extrusion>: **30**
Extrusion taper angle: **1**

<Draw> **<Solids>** **<Sphere>**

Command: **SPHERE**
Center of sphere: **CEN** of [**Select B (Figure 7.16).**]
Diameter/<Radius> of sphere: **100**

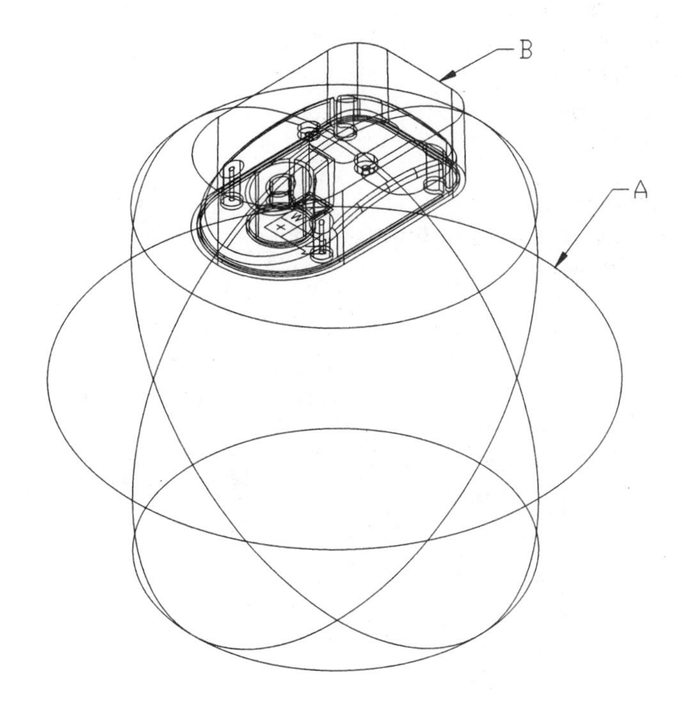

Figure 7.17 Outer profile extruded, and solid sphere created

To make the main body of the outer skin, run the INTERSECT command to form a solid of intersection between the extruded solid and the solid sphere. See Figure 7.18.

<Modify> **<Boolean>** **<Intersect>**

Command: **INTERSECT**
Select objects: [**Select A and B (Figure 7.17).**]
Select objects: [**Enter**]

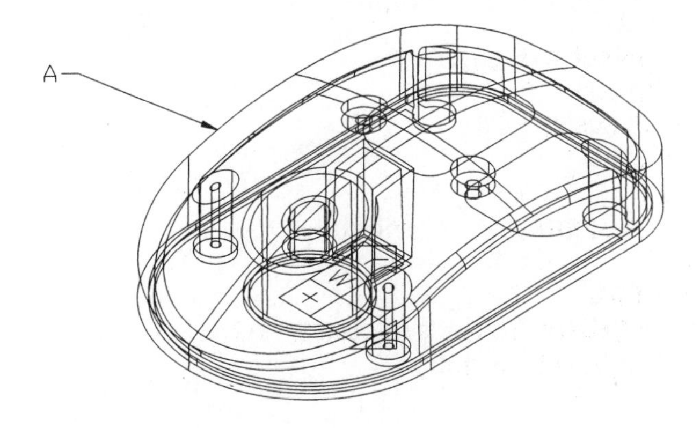

Figure 7.18 Solids intersected and subtracted

Run the FILLET command to fillet the upper edge of the outer skin. The fillet radius is the sum of the internal fillet radius and the thickness of the shell. See Figure 7.19.

<Modify> **<Fillet>**

Command: **FILLET**
Polyline/Radius/Trim/<Select first object>: **[Select A (Figure 7.18).]**
Enter radius: **6**
Chain/Radius/<Select edge>: **C**
Edge/Radius/<Select edge chain>: **[Select A (Figure 7.18).]**
Edge/Radius/<Select edge chain>: **[Enter]**

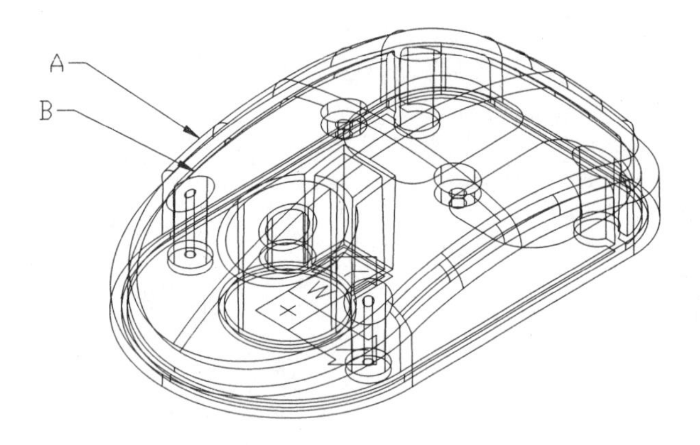

Figure 7.19 Outer skin solid completed

The solid that represents the outer skin is complete. Now you have two solids: a solid that represents the outer skin and a solid that represents the inner core. Use the SUBTRACT command to subtract the inner-core solid from the outer-skin solid. See Figure 7.20.

<Modify> **<Boolean>** **<Union>**

Command: **UNION**
Select objects: **[Select A and D (Figure 7.15).]**
Select objects: **[Enter]**

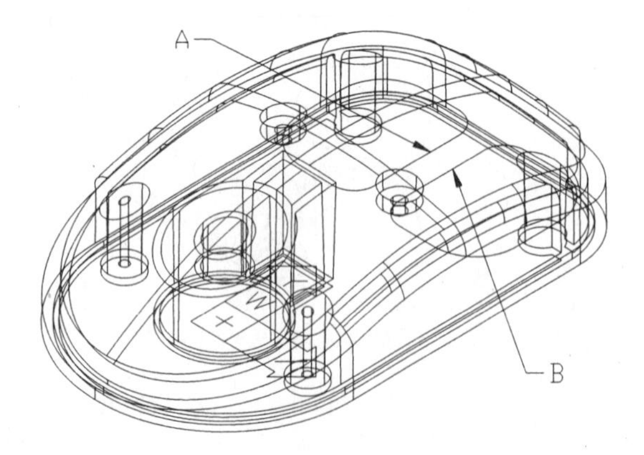

Figure 7.20 Inner-core solid subtracted from outer solid

Now, you have a thin solid shell. Compare Figure 7.20 with Figure 7.19. You will not find any significant difference visually. To continue, set to two-viewport display again, with the left viewport showing the top view and the right viewport showing a southeast isometric view.

<View> **<Tiled Viewports>** **<2 Viewports>**

Select the left viewport.

<View> **<3D Viewpoint>** **<Plan View>** **<World UCS>**

Extrude two regions for making button openings.

<Draw> **<Solids>** **<Extrude>**

```
Command: EXTRUDE
Select objects: [Select A (Figure 7.20).]
Select objects: [Enter]
Path/<Height of Extrusion>: 20
Extrusion taper angle: 0
```

The wireframes on layer W_u are not required. Turn layer W_u off. See Figure 7.21.

[Object Properties] **[Layer Control]**

Off layer: **W_u**

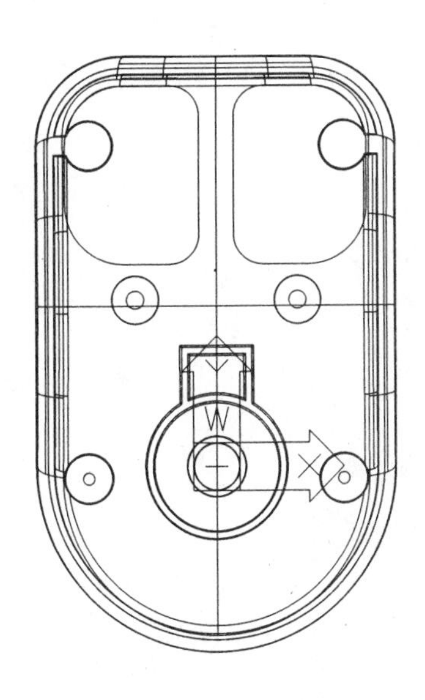

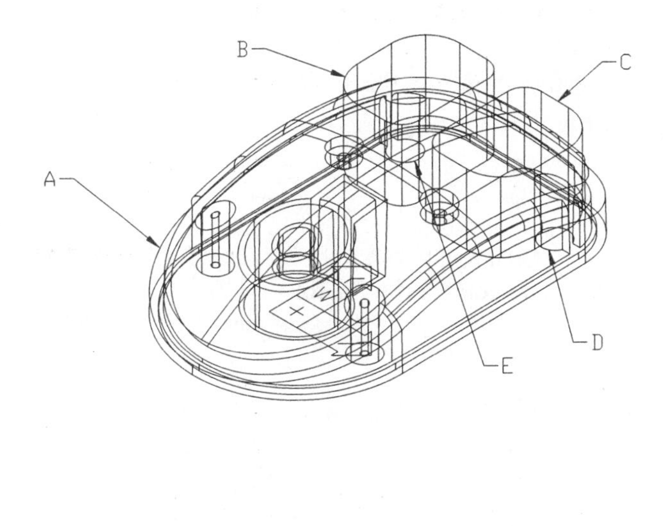

Figure 7.21 Solids for button openings created, and layer W_u turned off

To create two button openings, use the SUBTRACT command to subtract the extruded solids from the main solid body.

<Modify> **<Boolean>** **<Subtract>**

Command: **SUBTRACT**
Select solids and regions to subtract from...
Select objects: [**Select A (Figure 7.21).**]
Select objects: [**Enter**]
Select solids and regions to subtract...
Select objects: [**Select B and C (Figure 7.21).**]
Select objects: [**Enter**]

Create three cylinders for making two holes and a semi-circular slot.

<Draw> **<Solids>** **<Cylinder>**

Command: **CYLINDER**
Elliptical/<center point>: **CEN** of [**Select B (Figure 7.21).**]
Diameter/<Radius>: **1**
Center of other end/<Height>: **30**

Command: [**Enter**]
CYLINDER
Elliptical/<center point>: **CEN** of [**Select C (Figure 7.21).**]
Diameter/<Radius>: **1**
Center of other end/<Height>: **30**

Command: [**Enter**]
CYLINDER
Elliptical/<center point>: **0,40**
Diameter/<Radius>: **2.5**
Center of other end/<Height>: **C**
Center of other end: **@50<90**

After creating the cylinders, construct three boxes for making internal slots in the casing. See Figure 7.22.

<Draw> **<Solids>** **<Box>**

Command: **BOX**
Center/<Corner of box>: **-4,7**
Cube/Length/<other corner>: **4,18**
Height: **8**

Command: [**Enter**]
BOX
Center/<Corner of box>: **-4,-7**
Cube/Length/<other corner>: **4,-17**
Height: **11**

Command: [**Enter**]
BOX
Center/<Corner of box>: **-7,4**
Cube/Length/<other corner>: **-17,-4**
Height: **11**

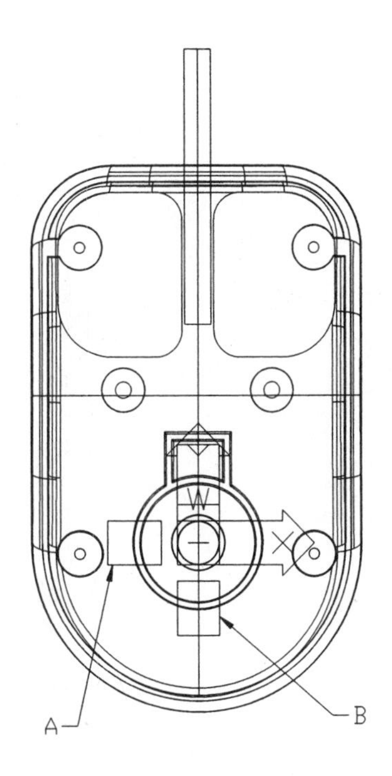

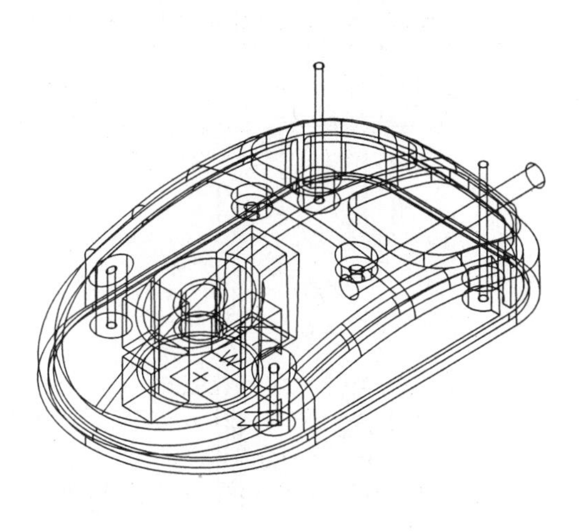

Figure 7.22 Solid boxes and cylinders created

Rotate two solid boxes to their final position. See Figure 7.23.

<Modify> **<Rotate>**

Command: **ROTATE**
Select objects: [**Select A and B (Figure 7.22).**]
Select objects: [**Enter**]
Base point: **0,0**
<Rotation angle>/Reference: **45**

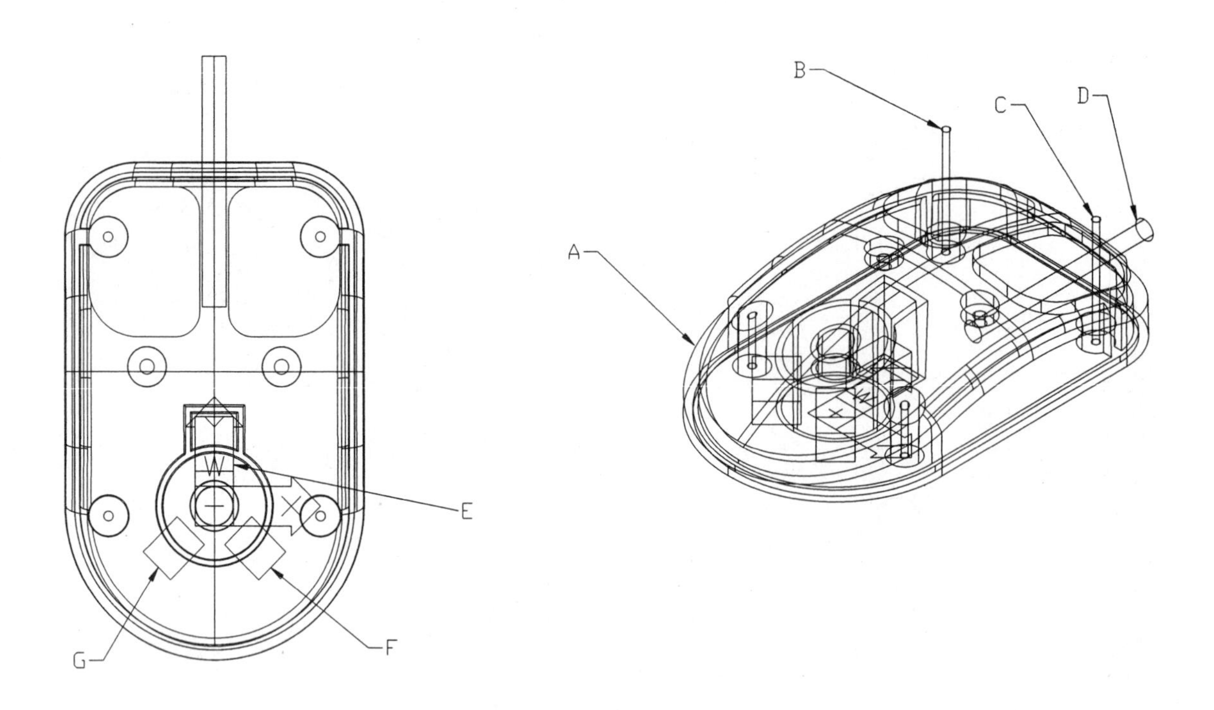

Figure 7.23 Solid boxes rotated

Use the SUBTRACT command to subtract the boxes and cylinders from the main solid body. See Figure 7.24. Figure 7.25 shows the rendered image viewing from below.

<Modify> **<Boolean>** **<Subtract>**

Command: **SUBTRACT**
Select solids and regions to subtract from...
Select objects: **[Select A (Figure 7.23).]**
Select objects: **[Enter]**
Select solids and regions to subtract...
Select objects: **[Select B, C, D, E, F, and G (Figure 7.23).]**
Select objects: **[Enter]**

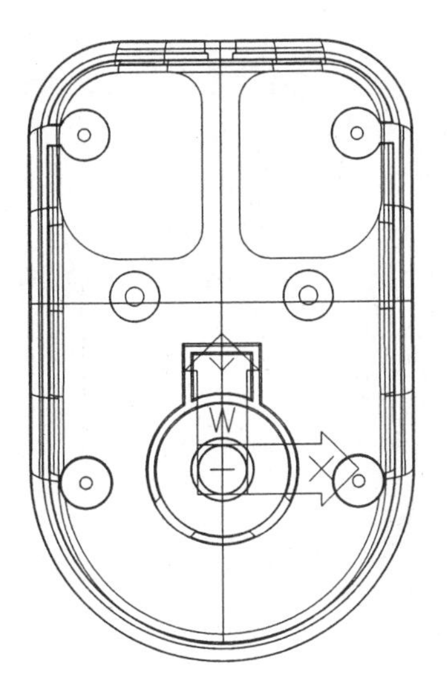

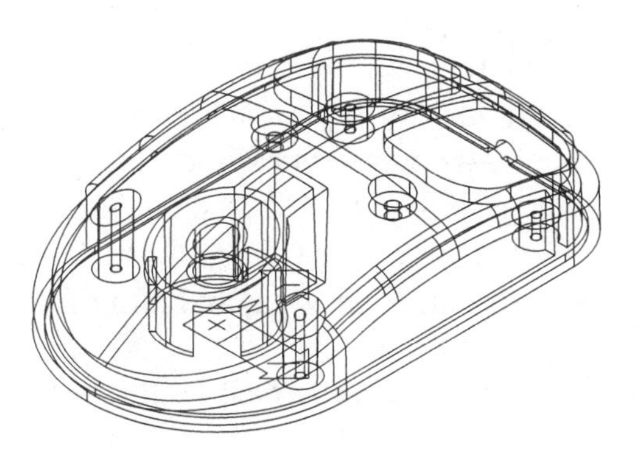

Figure 7.24 Completed upper casing

Figure 7.25 Rendered drawing

The solid model for the upper casing of the mouse is complete.

7.3 Lower Casing

Basically, the method used to create the lower casing is similar to that for the upper casing. You will carry out model creation in three major steps. You will make the outer skin first and then the inner core. Finally, you will subtract the inner core from the outer skin.

You will construct the model mainly by extrusion. To form extruded solids requires wireframes. Therefore, you need to create the wireframes before making the solids. Because the lower casing and the upper casing are two complementary parts that assemble together, some of the profiles, and of course the wireframes, will be the same for both the upper casing and the lower casing. In order to reduce the effort needed to make the wireframes, you can copy some of the wireframes from the upper casing.

Wireframes of the Lower Casing

To reiterate, layers S_u and W_u are used for the solids and wireframes of the upper casing, respectively. You should turn off layer S_u to clear the solid of the upper casing. To copy the wireframes from the upper casing, you need to turn on W_u. To work on the lower casing, you need to set the current layer to W_l.

From the Layer Control icon of the Object Properties toolbar, set the current layer to W_l, turn on layer W_u, and turn off layer S_u. See Figure 7.26.

[Object Properties] **[Layer Control]**

On	S_l
Off	S_u
On	W_l
On	W_u

Current layer: **W_l**

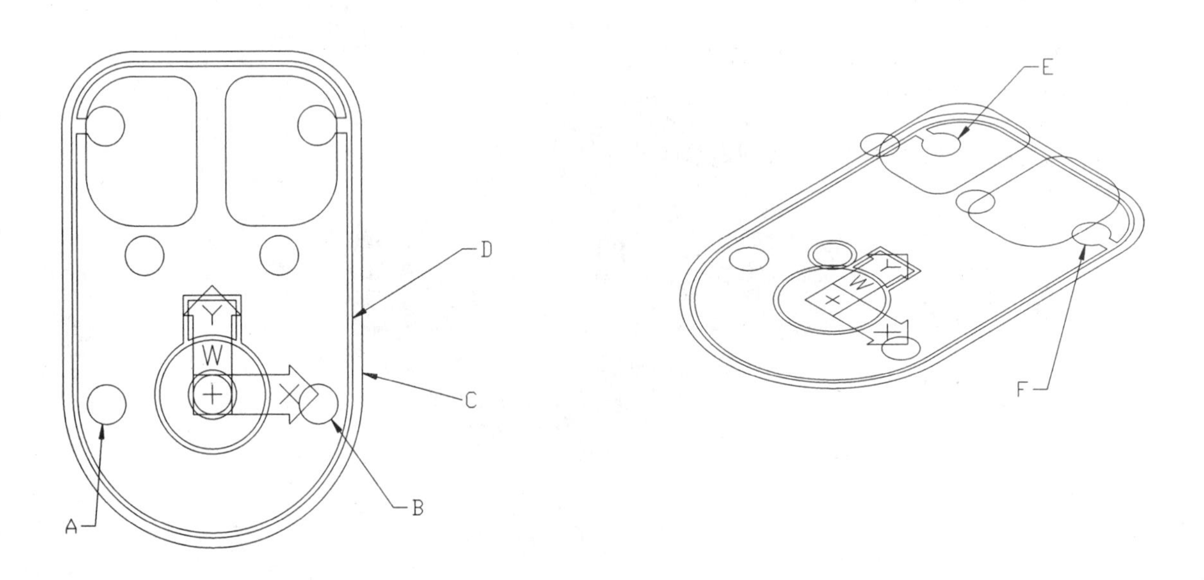

Figure 7.26 Wireframes for upper casing

In order not to duplicate the entities while making regions, set the DELOBJ variable to 1 for not keeping the originals.

Command: **DELOBJ**
New value for DELOBJ: **1**

Use the COPY command to copy two circles and two regions. They are common to the upper and lower casings. After copying, change the copied entities to layer W_l. Then create two more circles on layer W_l. After that, the wireframes for the upper casing are not required, so turn off layer W_u. See Figure 7.27.

<Modify> **<Copy Object>**

Command: **COPY**
Select objects: **[Select A, B, C, and D (Figure 7.26).]**
Select objects: **[Enter]**
<Base point or displacement>/Multiple: **0,0**
Second point of displacement: **[Enter]**

<Modify> **<Properties...>**

Select objects: **P**
Select objects: **[Enter]**
[Properties **Layer...** **W_l**
OK]

<Draw> **<Circle>** **<Center, Radius>**

Command: **CIRCLE**
3P/2P/TTR/<Center point>: **CEN** of **[Select E (Figure 7.26).]**
Diameter/<Radius>: **4**

Command: **[Enter]**
CIRCLE
3P/2P/TTR/<Center point>: **CEN** of **[Select F (Figure 7.26).]**
Diameter/<Radius>: **4**

[Object Properties] **[Layer Control]**

On	S_l
Off	S_u
On	W_l
Off	W_u

Current layer: **W_l**

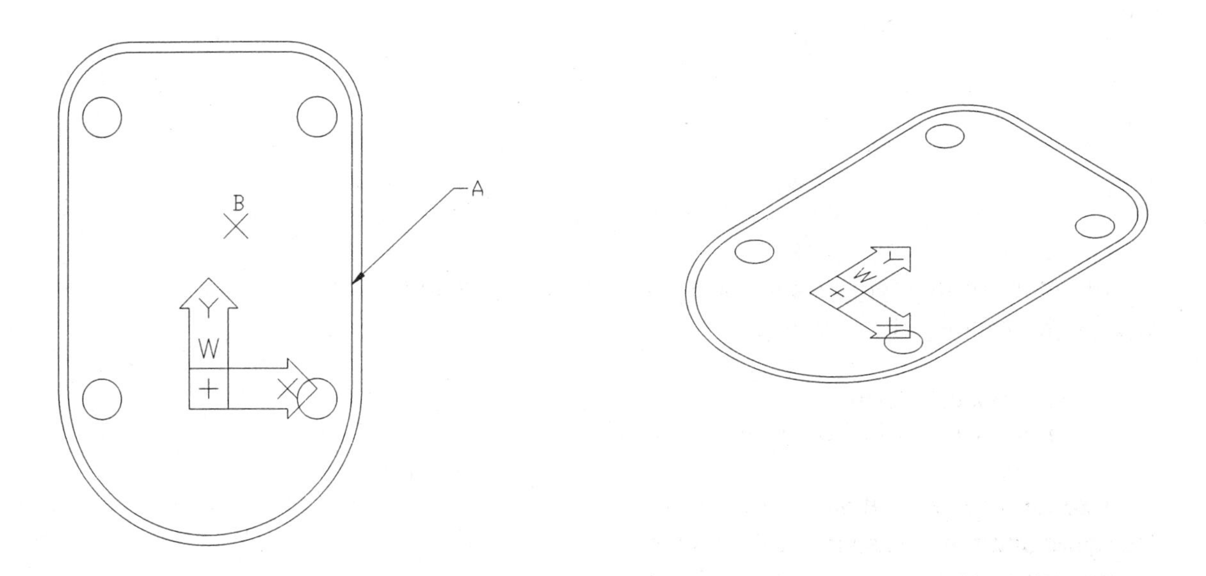

Figure 7.27 Wireframes for outer profile, recess, and bosses created

Now you have the regions for the outer profile, and you have the recess and four circles for making the bosses. You need a region for the inner core of the lower casing. Use the OFFSET command to offset a region. See Figure 7.28.

<Modify> **<Offset>**

Command: **OFFSET**
Offset distance or Through: **1**
Select object to offset: [**Select A (Figure 7.27).**]
Side to offset? [**Select B (Figure 7.27).**]
Select object to offset: [**Enter**]

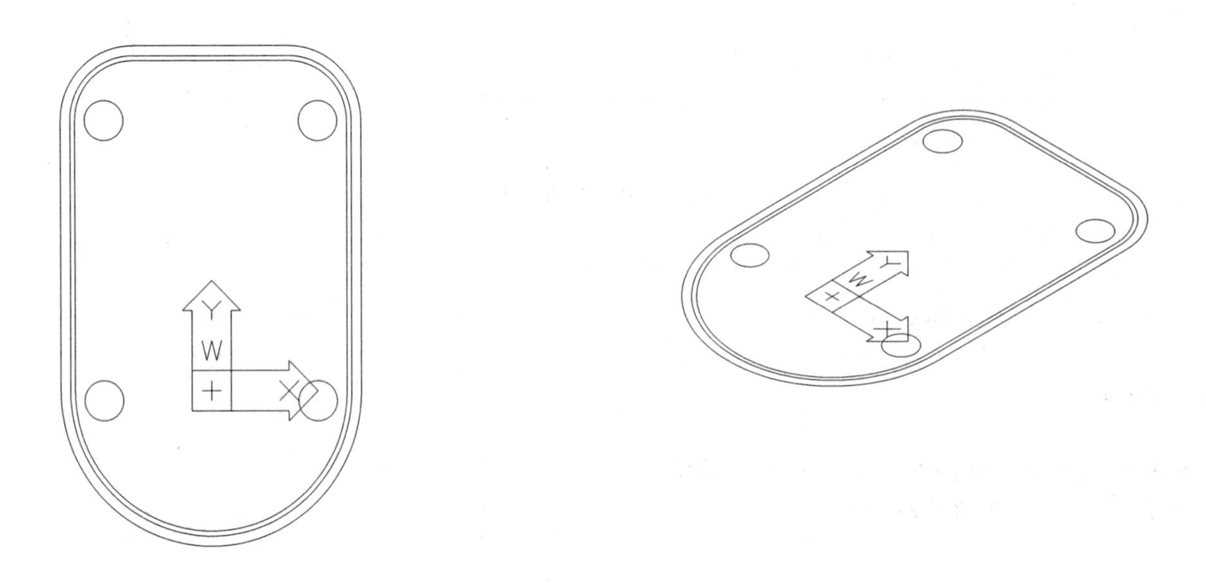

Figure 7.28 Wireframes for inner core created

In accordance with Figure 7.29, create two regions for making the mouse ball opening.

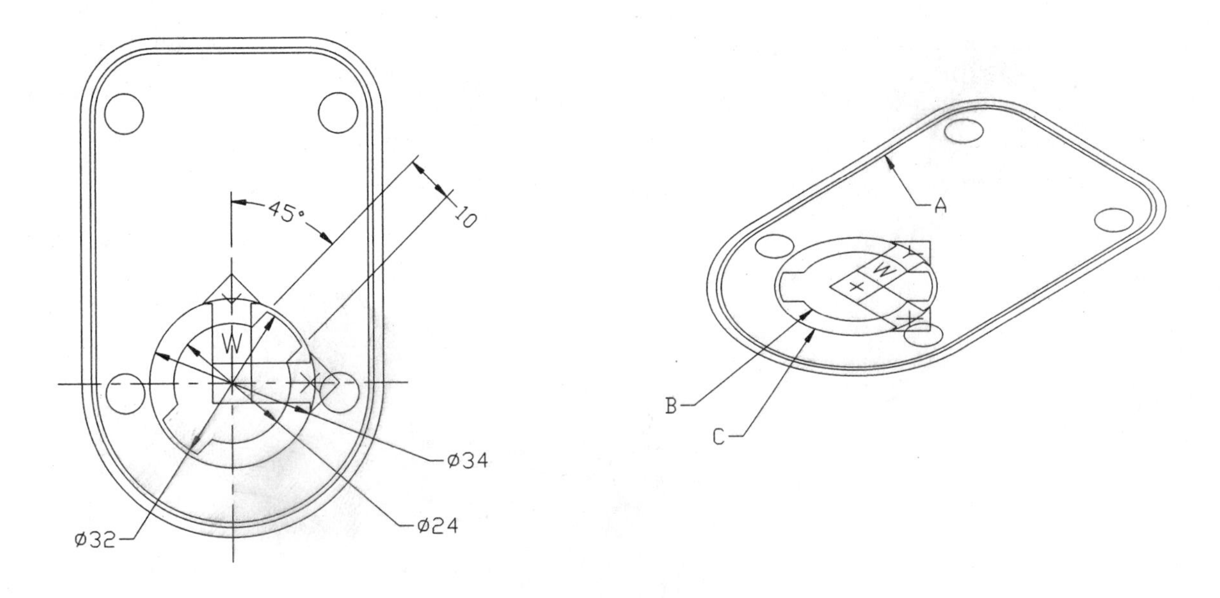

Figure 7.29 Wireframes for mouse opening created

Because the mouse ball locates at the lower face, which is 6.5 units in the minus Z-direction, and the inner profile locates at the upper face, which is 1.5 units in the Z-direction, use the MOVE command to move three regions to the required position. See Figure 7.30.

<Modify> **<Move>**

Command: **MOVE**
Select objects: **[Select A (Figure 7.29).]**
Select objects: **[Enter]**
Base point or displacement: **0,0,1.5**
Second point of displacement: **[Enter]**

Command: **[Enter]**
MOVE
Select objects: **[Select B and C (Figure 7.29).]**
Select objects: **[Enter]**
Base point or displacement: **0,0,-6.5**
Second point of displacement: **[Enter]**

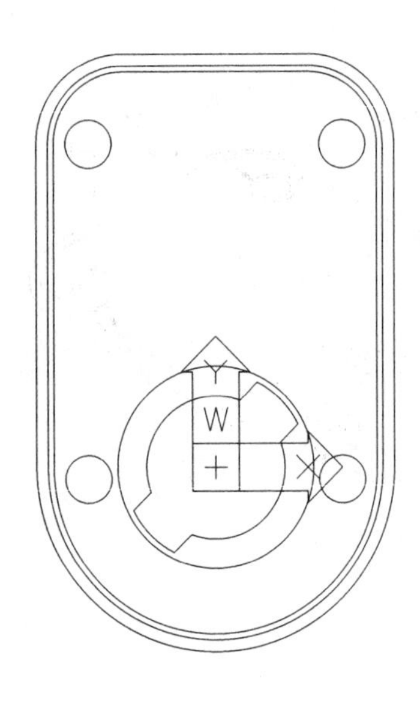

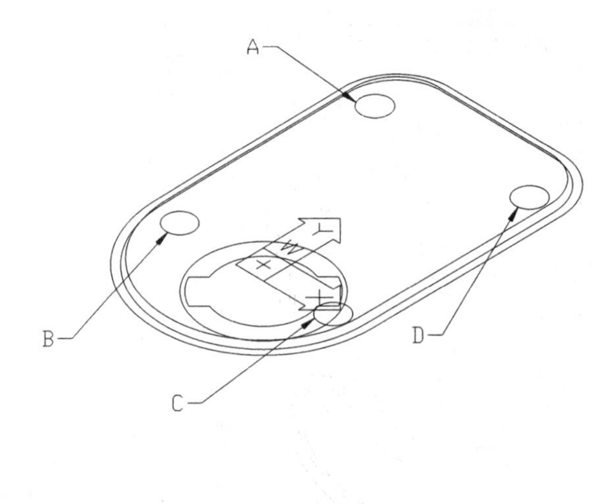

Figure 7.30 Wireframes for inner core and ball opening moved

To prepare for making a step in each of the bosses, create four more circles. See Figure 7.31.

<Draw> **<Circle>** **<Center, Radius>**

Command: **CIRCLE**
3P/2P/TTR/<Center point>: **CEN** of **[Select A (Figure 7.30).]**
Diameter/<Radius>: **3**

Command: **[Enter]**
CIRCLE
3P/2P/TTR/<Center point>: **CEN** of **[Select B (Figure 7.30).]**
Diameter/<Radius>: **3**

Command: **[Enter]**
CIRCLE
3P/2P/TTR/<Center point>: **CEN** of **[Select C (Figure 7.30).]**
Diameter/<Radius>: **3**

Command: **[Enter]**
CIRCLE
3P/2P/TTR/<Center point>: **CEN** of **[Select D (Figure 7.30).]**
Diameter/<Radius>: **3**

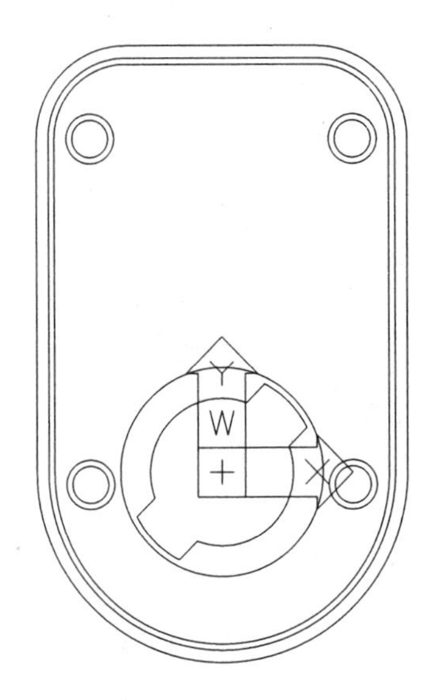

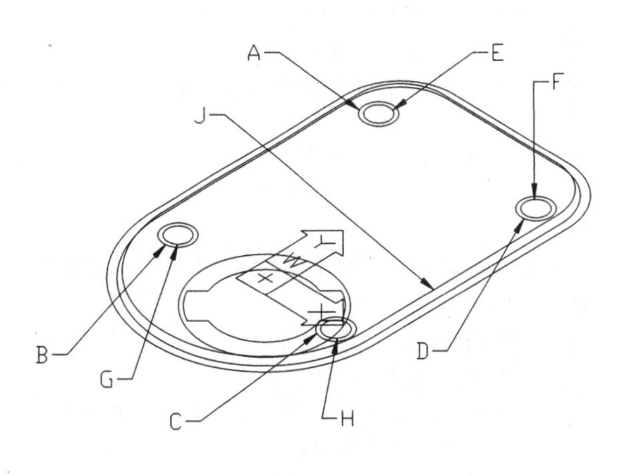

Figure 7.31 Wireframes for making steps in bosses created

The wireframes for the lower casing are complete. Return to a single-viewport display that shows the south-east isometric view.

<View>	**<Tiled Viewports>**	**<1 Viewport>**
<View>	**<3D Viewpoint>**	**<SE Isometric>**

Solid Model for the Lower Casing

To reiterate, the general approach for making the lower casing is similar to that for the upper casing. You will create the void as core solid and then the skin without the void. Finally, you will subtract the solid core from the solid skin. Run the LAYER command to set the current layer to S_l. You will create the solids for the lower casing on this layer.

[Object Properties] **[Layer Control]**

On	S_l
Off	S_u
On	W_l
Off	W_u

Current layer: **S_l**

The main body of the core is an extruded solid with filleted edges. In order to keep the wireframes for backup purpose, set DELOBJ to zero for not deleting them.

Command: **DELOBJ**
New value for DELOBJ: **0**

Create four extruded solids for making the bosses.

<Draw> **<Solids>** **<Extrude>**

Command: **EXTRUDE**
Select objects: [**Select A, B, C, and D (Figure 7.31).**]
Select objects: [**Enter**]
Path/<Height of Extrusion>: **-5**
Extrusion taper angle: **-1**

Repeat the command to make four extruded solids for making the steps in the bosses.

Command: [**Enter**]
EXTRUDE
Select objects: [**Select E, F, G, and H (Figure 7.31).**]
Select objects: [**Enter**]
Path/<Height of Extrusion>: **2**
Extrusion taper angle: **1**

Extrude the inner profile to form the main body of the inner core. See Figure 7.32.

Command: [**Enter**]
EXTRUDE
Select objects: [**Select J (Figure 7.31).**]
Select objects: [**Enter**]
Path/<Height of Extrusion>: **-5**
Extrusion taper angle: **1**

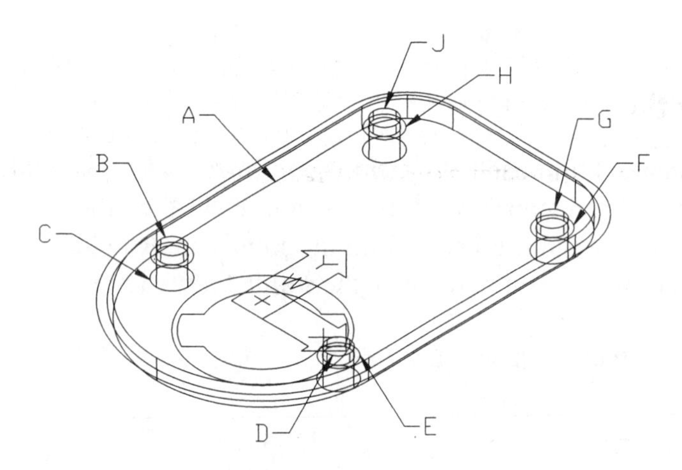

Figure 7.32 Extruded solids created

The solids for the core are constructed. Use the FILLET command to create a chain of fillets at the lower edges of the main body of the core. Then use the SUBTRACT command to subtract eight solids to form four stepped bosses. See Figure 7.33.

<Modify> **<Fillet>**

Command: **FILLET**
Polyline/Radius/Trim/<Select first object>: [**Select A (Figure 7.32).**]
Enter radius: **3**
Chain/Radius/<Select edge>: **C**

Edge/Radius/<Select edge chain>: **[Select A (Figure 7.32).]**
Edge/Radius/<Select edge chain>: **[Enter]**

<Modify> <Boolean> <Subtract>

Command: **SUBTRACT**
Select solids and regions to subtract from...
Select objects: **[Select A (Figure 7.32).]**
Select objects: **[Enter]**
Select solids and regions to subtract...
Select objects: **[Select B, C, D, E, F, G, H, and J (Figure 7.32).]**
Select objects: **[Enter]**

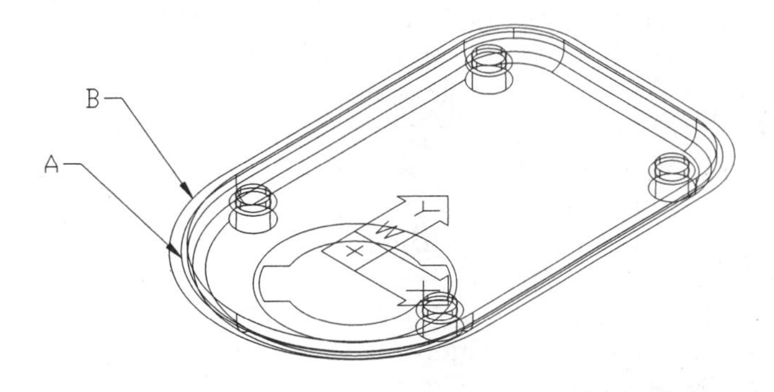

Figure 7.33 Edges filleted and solids subtracted

The core solid that represents the void of the model of the lower casing is complete. To continue, you will work on the outer skin solid. Extrude two regions: one for 1.5 units in the Z-direction for making the recess and one for 6.5 units in the minus Z-direction for making the main body of the outer skin. See Figure 7.34.

<Draw> <Solids> <Extrude>

Command: **EXTRUDE**
Select objects: **[Select A (Figure 7.33).]**
Select objects: **[Enter]**
Path/<Height of Extrusion>: **1.5**
Extrusion taper angle: **0**

Command: **[Enter]**
EXTRUDE
Select objects: **[Select B (Figure 7.33).]**
Select objects: **[Enter]**
Path/<Height of Extrusion>: **-6.5**
Extrusion taper angle: **1**

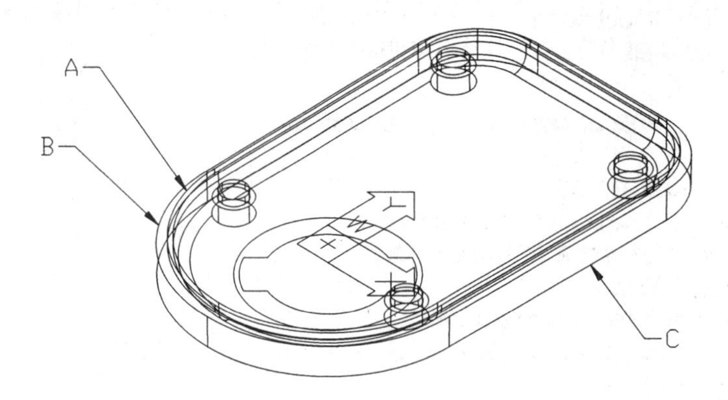

Figure 7.34 Extruded solids created

The main body and the recess are created. Use the UNION command to unite them to form the main body of the outer skin.

<Modify> **<Boolean>** **<Union>**

Command: **UNION**
Select objects: [**Select A and B (Figure 7.34).**]
Select objects: [**Enter**]

After uniting, use the FILLET command to round off the lower edges. Because the fillet radius for the inner core is 3 units and the thickness of the model is 3 units, the fillet radius is 6 units. See Figure 7.35.

<Modify> **<Fillet>**

Command: **FILLET**
Polyline/Radius/Trim/<Select first object>: [**Select C (Figure 7.34).**]
Enter radius: **6**
Chain/Radius/<Select edge>: **C**
Edge/Radius/<Select edge chain>: [**Select C (Figure 7.34).**]
Edge/Radius/<Select edge chain>: [**Enter**]

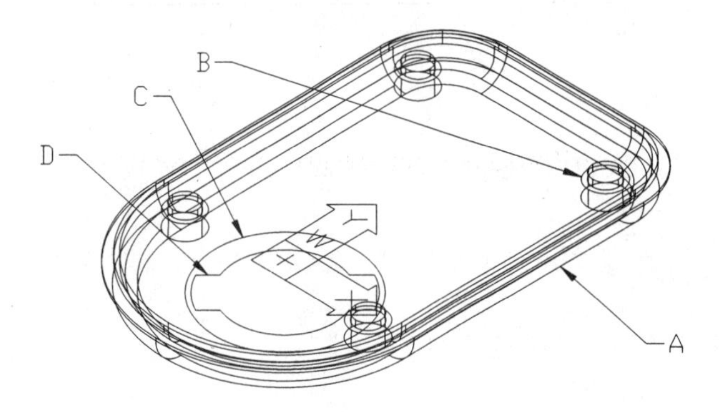

Figure 7.35 Extruded solids united, and edges filleted

The outer solid skin and the main core solid are complete. Use the SUBTRACT command to subtract the inner core from the outer skin to construct a solid shell. After subtraction, you will not find much difference visually.

<Modify> **<Boolean>** **<Subtract>**

Command: **SUBTRACT**
Select solids and regions to subtract from...
Select objects: [**Select A (Figure 7.35).**]
Select objects: [**Enter**]
Select solids and regions to subtract...
Select objects: [**Select B (Figure 7.35).**]
Select objects: [**Enter**]

Extrude two regions to form the opening for the mouse ball. See Figure 7.36.

<Draw> **<Solids>** **<Extrude>**

Command: **EXTRUDE**
Select objects: [**Select C (Figure 7.35).**]
Select objects: [**Enter**]
Path/<Height of Extrusion>: **1.25**
Extrusion taper angle: **0**

Command: [**Enter**]
EXTRUDE
Select objects: [**Select D (Figure 7.35).**]
Select objects: [**Enter**]
Path/<Height of Extrusion>: **3**
Extrusion taper angle: **1**

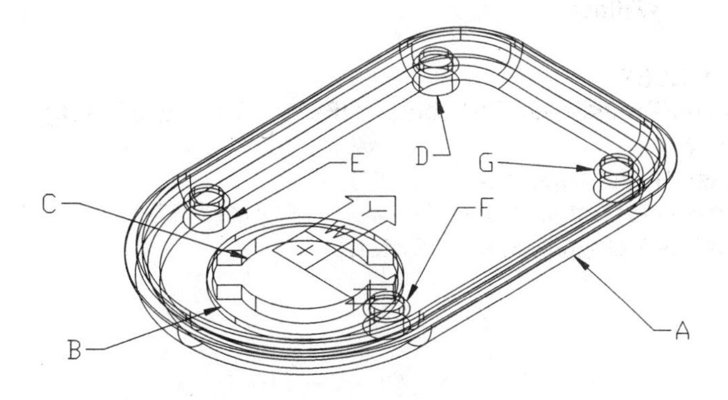

Figure 7.36 Solid shell created, and solids extruded

Now, the wireframes for the lower casing are not required. Turn off layer W_1.

[Object Properties] **[Layer Control]**

On	S_l
Off	S_u
Off	W_l
Off	W_u

Current layer: **S_I**

To cut the mouse ball opening on the thin shell, use the SUBTRACT command.

<Modify> **<Boolean>** **<Subtract>**

Command: **SUBTRACT**
Select solids and regions to subtract from...
Select objects: [**Select A (Figure 7.36).**]
Select objects: [**Enter**]
Select solids and regions to subtract...
Select objects: [**Select B and C (Figure 7.36).**]
Select objects: [**Enter**]

After cutting the ball opening, create eight solid cylinders for making counter-bored holes in the stepped bosses. See Figure 7.37.

<Draw> **<Solids>** **<Cylinder>**

Command: **CYLINDER**
Elliptical/<center point>: **.XY** of **CEN** of [**Select D (Figure 7.36).**]
(need Z): **-6.5**
Diameter/<Radius>: **1.5**
Center of other end/<Height>: **8**

Command: [**Enter**]
CYLINDER
Elliptical/<center point>: **.XY** of **CEN** of [**Select D (Figure 7.36).**]
(need Z): **-6.5**
Diameter/<Radius>: **3**
Center of other end/<Height>: **3**

Command: [**Enter**]
CYLINDER
Elliptical/<center point>: **.XY** of **CEN** of [**Select E (Figure 7.36).**]
(need Z): **-6.5**
Diameter/<Radius>: **1.5**
Center of other end/<Height>: **8**

Command: [**Enter**]
CYLINDER
Elliptical/<center point>: **.XY** of **CEN** of [**Select E (Figure 7.36).**]
(need Z): **-6.5**
Diameter/<Radius>: **3**
Center of other end/<Height>: **3**

Command: [**Enter**]
CYLINDER
Elliptical/<center point>: **.XY** of **CEN** of [**Select F (Figure 7.36).**]
(need Z): **-6.5**
Diameter/<Radius>: **1.5**
Center of other end/<Height>: **8**

Command: [**Enter**]
CYLINDER

Elliptical/<center point>: **.XY** of **CEN** of [**Select F (Figure 7.36).**]
(need Z): **-6.5**
Diameter/<Radius>: **3**
Center of other end/<Height>: **3**

Command: [**Enter**]
CYLINDER
Elliptical/<center point>: **.XY** of **CEN** of [**Select G (Figure 7.36).**]
(need Z): **-6.5**
Diameter/<Radius>: **1.5**
Center of other end/<Height>: **8**

Command: [**Enter**]
CYLINDER
Elliptical/<center point>: **.XY** of **CEN** of [**Select G (Figure 7.36).**]
(need Z): **-6.5**
Diameter/<Radius>: **3**
Center of other end/<Height>: **3**

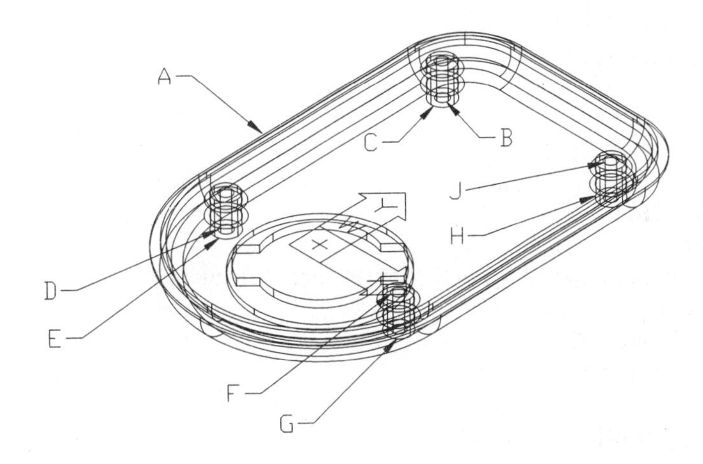

Figure 7.37 Mouse opening cut, and eight solid cylinders created

To make the counter-bored holes, use the SUBTRACT command to subtract the eight solid cylinders from the thin shell. See Figure 7.38.

<Modify> **<Boolean>** **<Subtract>**

Command: **SUBTRACT**
Select solids and regions to subtract from...
Select objects: [**Select A (Figure 7.37).**]
Select objects: [**Enter**]
Select solids and regions to subtract...
Select objects: [**Select B, C, D, E, F, G, H, and J (Figure 7.37).**]
Select objects: [**Enter**]

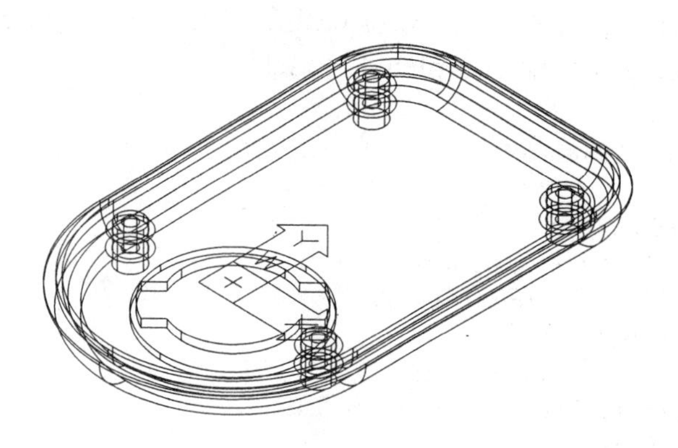

Figure 7.38 Four counter-bored holes cut on the thin shell

Finally, you will add a slot to the rear vertical face. Create a solid box and a solid cylinder. The size and location of the solid cylinder are the same as those of the cylinder used for making a slot in the upper casing. The diagonal points of the solid box are the quadrant points of the cylinders. See Figure 7.39.

<Draw> <Solids> <Cylinder>

Command: **CYLINDER**
Elliptical/<center point>: **0,40**
Diameter/<Radius>: **2.5**
Center of other end/<Height>: **C**
Center of other end: **@50<90**

<Draw> <Solids> <Box>

Command: **BOX**
Center/<Corner of box>: **2.5,40**
Cube/Length/<other corner>: **-2.5,90**
Height: **10**

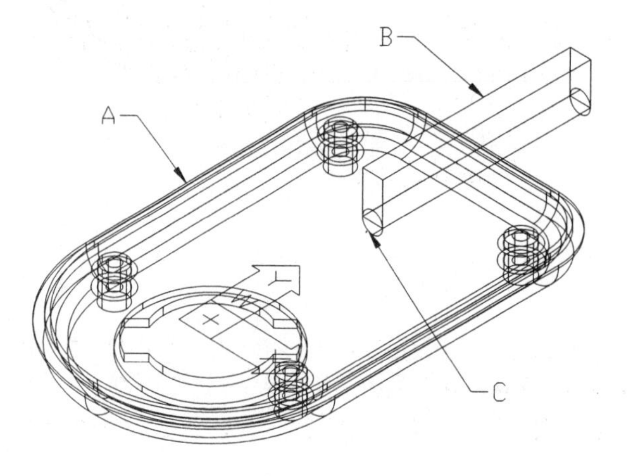

Figure 7.39 Solid box and solid cylinder created

Subtract the solid box and the solid cylinder from the thin shell to form a slot. See

Figure 7.40. The solid model for the lower casing of the mouse is complete. Figure 7.41 shows the rendered image of the completed model.

<Modify> **<Boolean>** **<Subtract>**

Command: **SUBTRACT**
Select solids and regions to subtract from...
Select objects: **[Select A (Figure 7.39).]**
Select objects: **[Enter]**
Select solids and regions to subtract...
Select objects: **[Select B and C (Figure 7.39).]**
Select objects: **[Enter]**

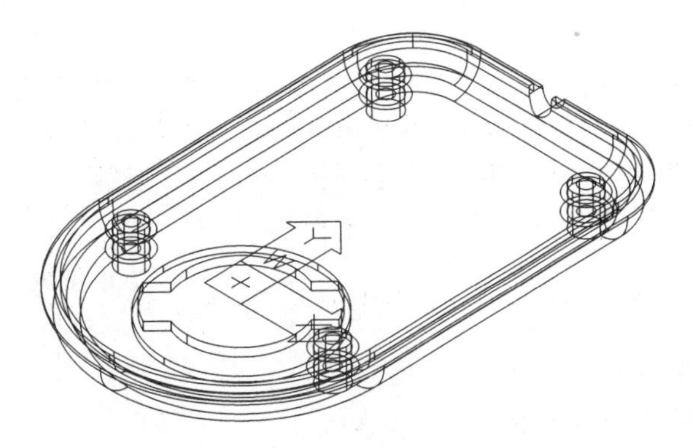

Figure 7.40 Completed model of lower casing

Figure 7.41 Rendered drawing of lower casing

Now that you have made the solid models for the upper and lower casings, you will learn how to check interference between adjacent solid models, find out the mass properties of a solid model, and create a 2D cross section across a solid model.

7.4 Interference, Mass Properties, and Sections

Interference checking is important to ensure that the solid parts mate together properly in an assembly. Evaluating the mass properties of a solid model is valuable to the designer analyzing the final product. Creating sections enables the designer to visualize the internal structure of the solid model more precisely. Before you can check whether there is any interference between the upper casing and the lower casing, you have to turn on the layer S_u to display both the upper casing and the lower casing. See Figure 7.42.

[Object Properties] **[Layer Control]**

On	S_l
On	S_u
Off	W_l
Off	W_u

Current layer: **S_l**

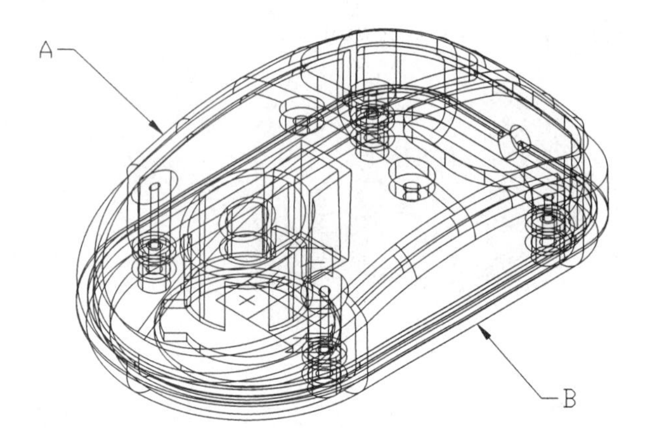

Figure 7.42 Upper casing and lower casing put together

From the drawing shown, it is very hard to tell whether there is any interference between the upper and lower casings. To be certain, you should use the INTERFERE command. Make a layer called Uty. Then run the INTERFERE command on the upper and lower casings. If the command reports any interference, create the solid of interference.

<Format> **<Layer...>**

Command: **LAYER**

Name	Color	Linetype
0	White	Continuous
S_l	Green	Continuous
S_u	Blue	Continuous
Uty	Magenta	Continuous
W_l	Cyan	Continuous
W_u	Yellow	Continuous

Current layer: **Uty**

<Draw> **<Solids>** **<Interfere>**

Command: **INTERFERE**
Select the first set of solids:
Select objects: [**Select A (Figure 7.42).**]
Select objects: [**Enter**]
Select the second set of solids:
Select objects: [**Select B (Figure 7.42).**]
Select objects: [**Enter**]
Comparing 1 solid against 1 solid.
Interfering solids (first set): 1
(second set): 1
Interfering pairs : 1
Create interference solids ? <N>: **Y**

The volume of interference is represented as a solid. To see this solid clearly, you have to turn off the layers S_u and S_l. See Figure 7.43.

[Object Properties] **[Layer Control]**

Off	S_l
Off	S_u
On	Uty
Off	W_l
Off	W_u

Current layer: **Uty**

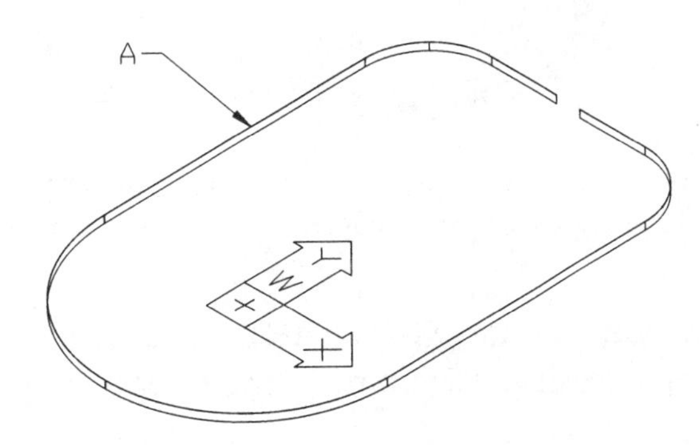

Figure 7.43 Solid of interference between upper casing and lower casing created

As you can see in Figure 7.43, the solid of interference is very small. Therefore, you can hardly notice it. Despite such small interference, it will prevent the upper and lower casings from fitting together properly. To ensure a perfect match between them, some solid from either casing has to be removed. If you study the design carefully, you will find that the solid of interference should be removed from the lower casing.

Use the GROUP command to put the solid of interference in an object group called INTER. Then turn on the layer S_l and set it as the current layer. See Figure 7.44.

<Tool> **<Object Group...>**

Command: **GROUP**

[Group name: **INTER**
New]

Select objects for grouping:
Select objects: **[Select A (Figure 7.43).]**
Select objects: **[Enter]**

[OK]

[Object Properties] **[Layer Control]**

On	S_l
Off	S_u
On	Uty
Off	W_l
Off	W_u

Current layer: **S_l**

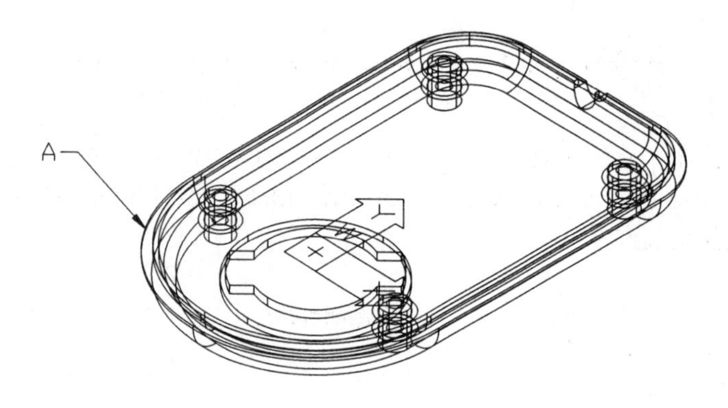

Figure 7.44 Lower casing and solid of interference

To remove interference, use the SUBTRACT command to subtract the solid of interference from the lower casing. Then turn on the layer S_u again and set the current layer to Uty. See Figure 7.45.

<Modify> **<Boolean>** **<Subtract>**

Command: **SUBTRACT**
Select solids and regions to subtract from...
Select objects: **[Select A (Figure 7.44).]**
Select objects: **[Enter]**
Select solids and regions to subtract...
Select objects: **G**
Enter group name: **INTER**
Select objects: **[Enter]**

[Object Properties] **[Layer Control]**

On	S_l
On	S_u
On	Uty

Off	W_l
Off	W_u

Current layer: **Uty**

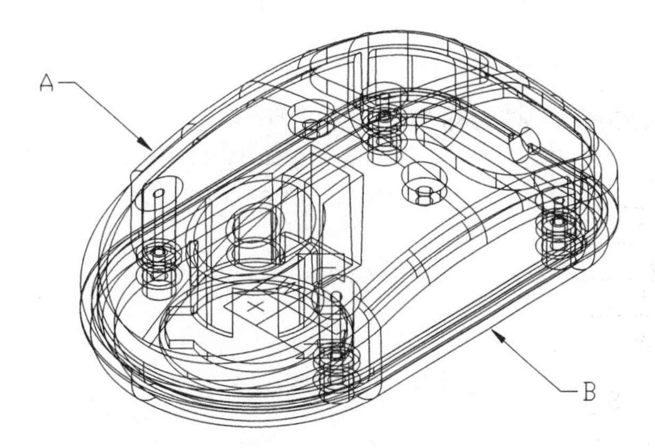

Figure 7.45 Upper casing and modified lower casing

After subtracting the extra solid that causes interference, you can run the INTERFERE command again to ensure that the upper casing and the lower casing do not interfere with each other.

<Draw> **<Solids>** **<Interfere>**

Command: **INTERFERE**
Select the first set of solids:
Select objects: [**Select A (Figure 7.45).**]
Select objects: [**Enter**]
Select the second set of solids:
Select objects: [**Select B (Figure 7.45).**]
Select objects: [**Enter**]
Comparing 1 solid against 1 solid.
Solids do not interfere.

As expected, there is no interference between the upper and lower casings. Now you can safely assemble them together. Because a solid model consists of volume data as well as edge and surface data, you can determine the mass properties of a solid. From the Inquiry toolbar, select the Mass Properties icon to use the MASSPROP command to find out the mass properties of the upper casing. The default density is one.

[Inquiry] **[Mass Properties]**

Command: **MASSPROP**
Select objects: [**Select A (Figure 7.45).**]
Select objects: [**Enter**]

---------------- SOLIDS ----------------
Mass: ...
Volume: ...

Bounding box: ...
Centroid: ...
Moments of inertia: ...
Products of inertia: ...
Radii of gyration: ...
Principal moments and X-Y-Z directions about centroid: ...
Write to a file ? <N>: **N**

In the process of product design, you might want to examine the internal structure of a solid. To aid visualizing, you can create a section by using the SECTION command. Run the command to create a cross section across the two solids.

<Draw> **<Solids>** **<Section>**

Command: **SECTION**
Select objects: **[Select A and B (Figure 7.45).]**
Select objects: **[Enter]**
Section plane by Object/Zaxis/View/XY/YZ/ZX/<3points>: **YZ**
Point on YZ plane: **0,0,0**

To see the cross sections clearly, turn off the layers S_u, and S_1. See Figure 7.46.

[Object Properties] **[Layer Control]**

Off	S_l
Off	S_u
On	Uty
Off	W_l
Off	W_u

Current layer: **Uty**

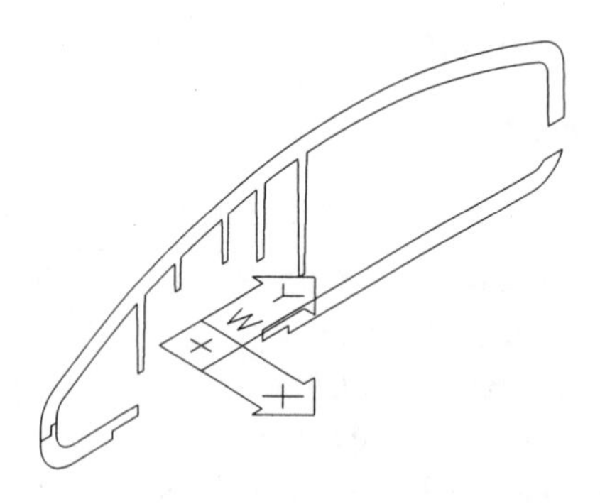

Figure 7.46 Section across two solids

After seeing the cross section, turn off layer Uty, turn on layers S_u and S_1, and set current layer to S_u.

[Object Properties] **[Layer Control]**

On	S_l

On	S_u
Off	Uty
Off	W_l
Off	W_u

Current layer: **S_u**

Checking is complete. Proceed to work on the buttons.

7.5 Buttons

Now you will create the solid model for the buttons. Figure 7.47 shows the model of the buttons.

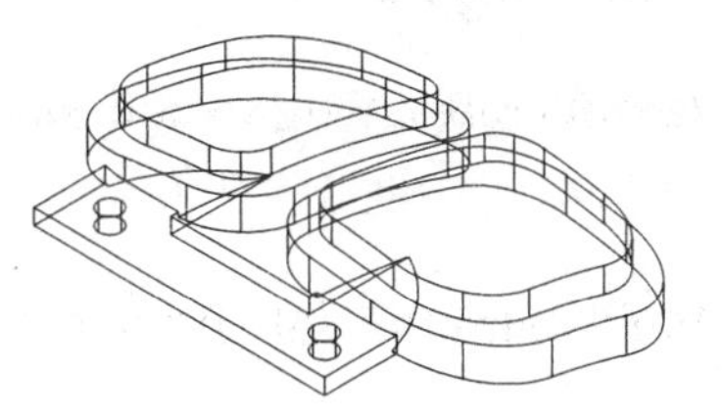

Figure 7.47 Buttons

You can divide this model into three major parts: left button, right button, and hinge lever. Because the left button and the right button are mirror images of each other, you need to make one button and then do mirroring. See Figure 7.48.

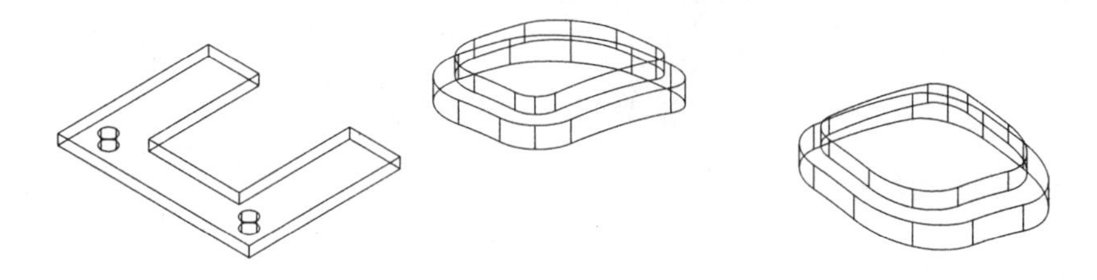

Figure 7.48 Three major parts

Examine the buttons carefully. The buttons fit into the button openings of the upper casing. Therefore, the profile of the upper surface of the buttons should conform to that of the upper casing. In addition, each button has a step.

To make a button, you need two extruded solids and three solid spheres. The extruded solids form the button main body and the step on the button. The three solid spheres represent the outer face, the step, and the inner face. See Figure 7.49.

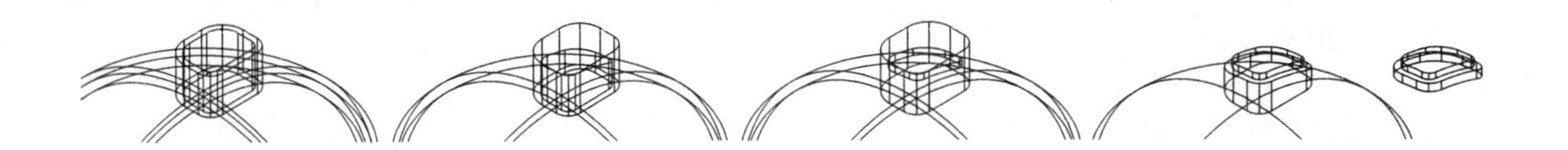

Figure 7.49 Steps in making a button

Refer to 7.49. From left to right, there are five steps in making the button. (1) Make two extruded solids and three solid spheres. (2) Intersect the inner extruded solid with the outer sphere. (3) Unite with the second sphere. (4) Intersect with the second extruded solid. (5) Subtract the third sphere.

Wireframes for the Buttons

Create two additional layers, W_b and S_b. Layer W_b is used for making the wireframes, and layer S_b is used for making the solid model.

Set the current layer to W_b, turn off layers S_1 and S_u, and turn on layer W_u. See Figure 7.50.

<Format> **<Layer...>**

Command: **LAYER**

Name		Color	Linetype
0	On	White	Continuous
S_b	On	Red	Continuous
S_l	Off	Green	Continuous
S_u	Off	Blue	Continuous
Uty	Off	Magenta	Continuous
W_b	On	Green	Continuous
W_l	Off	Cyan	Continuous
W_u	On	Yellow	Continuous

Current layer: **W_b**

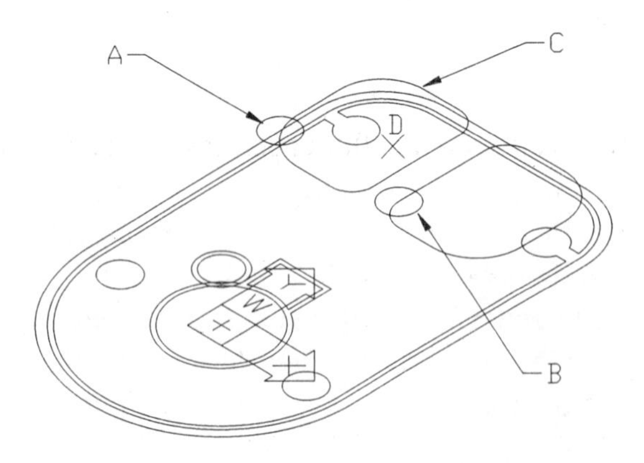

Figure 7.50 Wireframes for upper casing

Set the DELOBJ variable to 1 to delete the originals when making regions from a set of wireframes.

Command: **DELOBJ**
New value for DELOBJ: **1**

Create two circles concentric to the stepped bosses of the upper casing.

<Draw> **<Circle>** **<Center, Radius>**

Command: **CIRCLE**
3P/2P/TTR/<Center point>: **CEN** of **[Select A (Figure 7.50).]**
Diameter/<Radius>: **1.5**

Command: **[Enter]**
CIRCLE
3P/2P/TTR/<Center point>: **CEN** of **[Select B (Figure 7.50).]**
Diameter/<Radius>: **1.5**

There is a small clearance between the button and the button opening. Offset the button opening wireframe and change it to layer W_b. This wireframe will be used to make the button main body.

<Modify> **<Offset>**

Command: **OFFSET**
Offset distance or Through: **1**
Select object to offset: **[Select C (Figure 7.50).]**
Side to offset? **[Select D (Figure 7.50).]**
Select object to offset: **[Enter]**

<Modify> **<Properties...>**

Select objects: **LAST**
Select objects: **[Enter]**

[Modify Line
Properties
Layer... **W_b**
OK]

After offsetting, the wireframes for the upper casing are not needed. Turn off layer W_u. See Figure 7.51.

[Object Properties] **[Layer Control]**

On	S_b
Off	S_l
Off	S_u
Off	Uty
On	W_b
Off	W_l

Off	W_u

Current layer: **W_b**

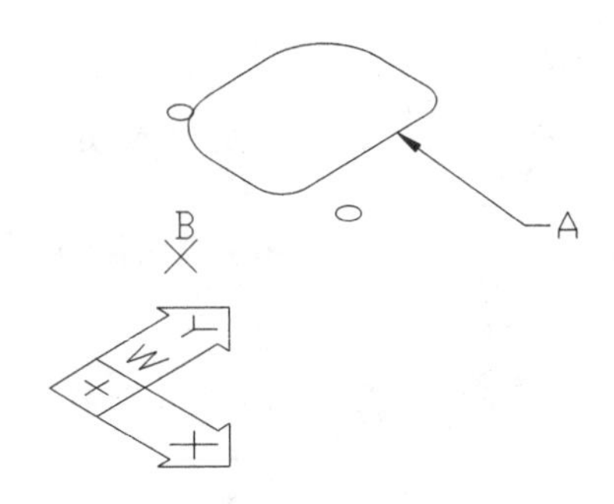

Figure 7.51 Button opening offset and circles copied

Set the display to two viewports with the left viewport showing the top view and the right viewport showing a southeast isometric view. Then offset the button wireframe for making a step in the button.

<Modify> **<Offset>**

Command: **OFFSET**
Offset distance or Through: **3**
Select object to offset: **[Select A (Figure 7.51).]**
Side to offset? **[Select B (Figure 7.51).]**
Select object to offset: **[Enter]**

After offsetting the wireframe, create the U-shape wireframe for the hinge lever according to Figure 7.52. Then convert the wireframe and the two circles to regions.

<Draw> **<Region>**

Subtract regions C (Figure 7.52) and D (Figure 7.52) from region B (Figure 7.52) to form a complex region.

<Modify> **<Boolean>** **<Subtract>**

Command: **SUBTRACT**
Select solids and regions to subtract from...
Select objects: **[Select A (Figure 7.52).]**
Select objects: **[Enter]**
Select solids and regions to subtract...
Select objects: **[Select B and C (Figure 7.52).]**
Select objects: **[Enter]**

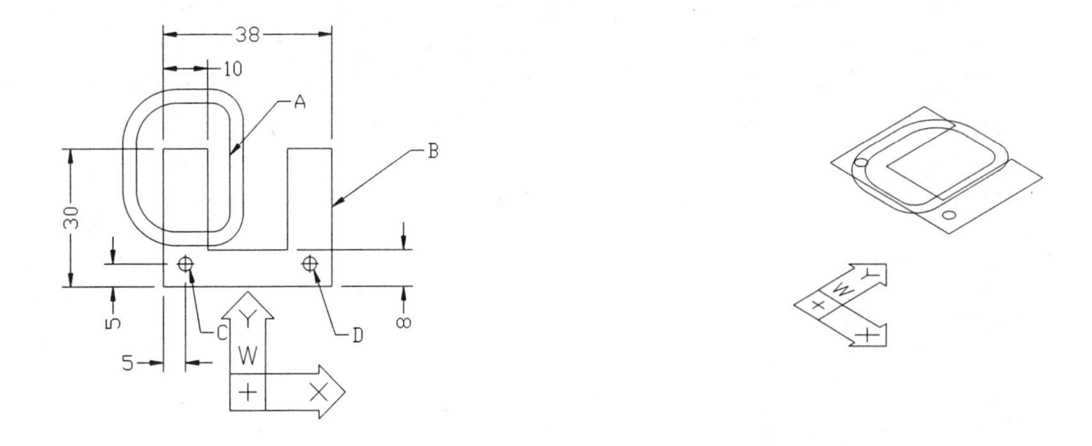

Figure 7.52 Button opening offset again, and hinge lever wireframe created

The wireframes for the button are complete. Proceed to make the solid model.

Solid Model for the Buttons

Using the wireframes that reside on the layer W_b, you will start to build the solid model on layer S_b.

[Object Properties] **[Layer Control]**

On	S_b
Off	S_l
Off	S_u
Off	Uty
On	W_b
Off	W_l
Off	W_u

Current layer: **S_b**

Set the DELOBJ variable to 0 to keep the originals after using them for extrusion.

Command: **DELOBJ**
New value for DELOBJ: **0**

Set to a single-viewport display and extrude the button inner wireframe.

<Draw> **<Solids>** **<Extrude>**

Command: **EXTRUDE**
Select objects: [**Select A (Figure 7.52).**]
Select objects: [**Enter**]
Path/<Height of Extrusion>: **20**
Extrusion taper angle: **0**

The upper face of the button is spherical and should match the profile of the upper casing. Therefore, the solid spheres that you will create to make the button should be concentric with the spherical profile of the upper casing. In order to select the center point of the spherical surface of the upper casing, turn on layer S_u. See Figure 7.53.

[Object Properties] **[Layer Control]**

On	S_b
Off	S_l
On	S_u
Off	Uty
On	W_b
Off	W_l
Off	W_u

Current layer: **S_b**

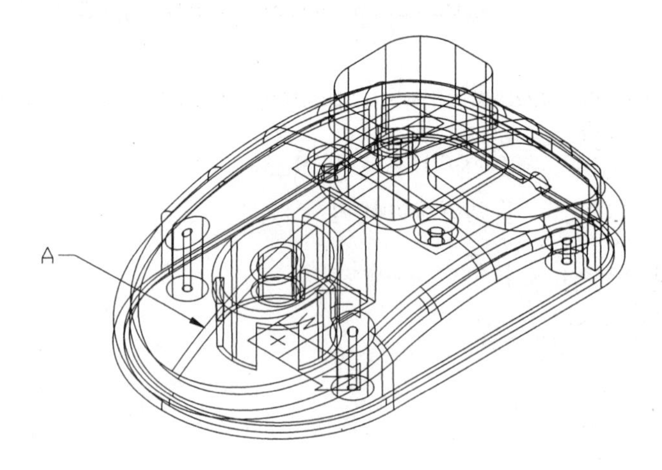

Figure 7.53 Button inner wireframe extruded, and layer S_u turned on

Selecting the center point of the spherical face of the upper casing as center, you will create three solid spheres. These spheres are used for making the outer spherical profile, the step on the button, and the underside of the button. See Figure 7.54.

<Draw> **<Solids>** **<Sphere>**

Command: **SPHERE**
Center of sphere: **CEN** of [**Select A (Figure 7.53).**]
Diameter/<Radius> of sphere: **100**

Command: [**Enter**]
SPHERE
Center of sphere: **CEN** of [**Select A (Figure 7.53).**]
Diameter/<Radius> of sphere: **97**

Command: [**Enter**]
SPHERE
Center of sphere: **CEN** of [**Select A (Figure 7.53).**]
Diameter/<Radius> of sphere: **93**

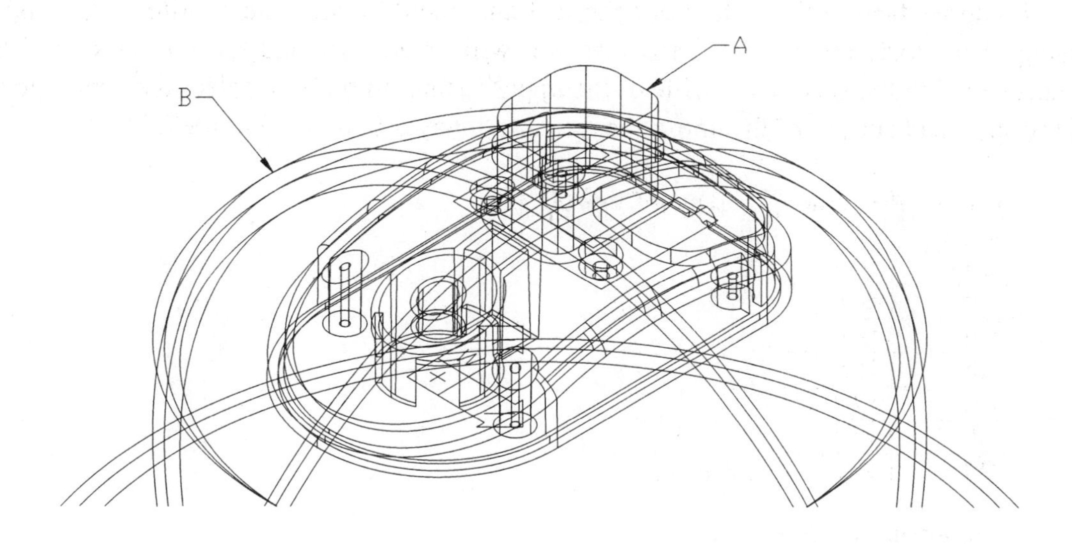

Figure 7.54 Solid spheres created

After you have constructed three solid spheres, the solid model for the upper casing is not needed. Turn off layer S_u.

[Object Properties] **[Layer Control]**

On	S_b
Off	S_l
Off	S_u
Off	Uty
On	W_b
Off	W_l
Off	W_u

Current layer: **S_b**

To form the main body of a button, intersect the extruded solid with the outer solid sphere. See Figure 7.55.

<Modify> **<Boolean>** **<Intersect>**

Command: **INTERSECT**
Select objects: [**Select A and B (Figure 7.54).**]
Select objects: [**Enter**]

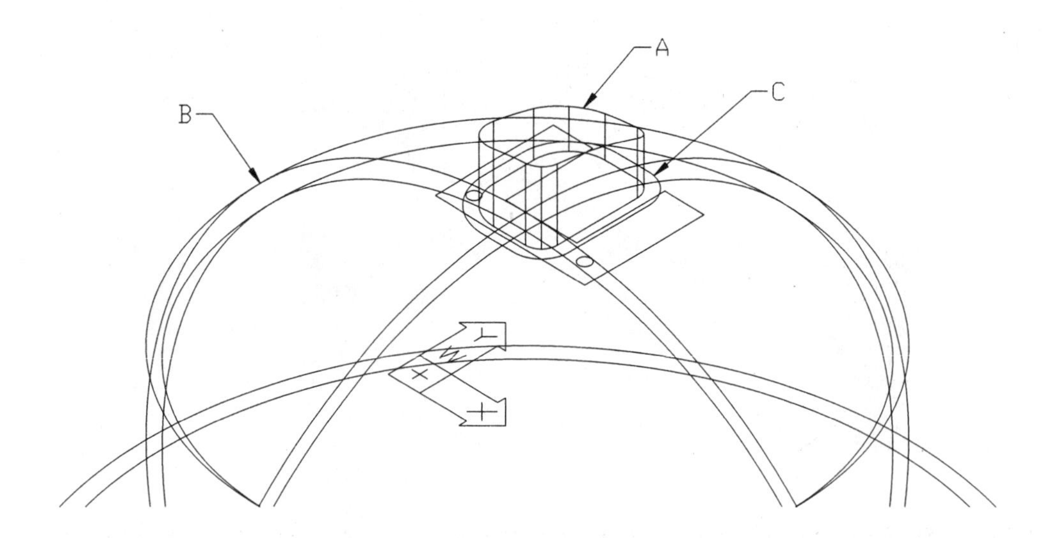

Figure 7.55 Upper casing turned off, and extruded solid intersected with outer sphere

To join the intersected solid to the second sphere to form a step, use the UNION command.

<Modify> **<Boolean>** **<Union>**

Command: **UNION**
Select objects: **[Select A and B (Figure 7.55).]**
Select objects: **[Enter]**

To make the body of the step on the button, extrude the wireframe that represents the outer edge of the button. See Figure 7.56.

<Draw> **<Solids>** **<Extrude>**

Command: **EXTRUDE**
Select objects: **[Select C (Figure 7.55).]**
Select objects: **[Enter]**
Path/<Height of Extrusion>: **20**
Extrusion taper angle: **0**

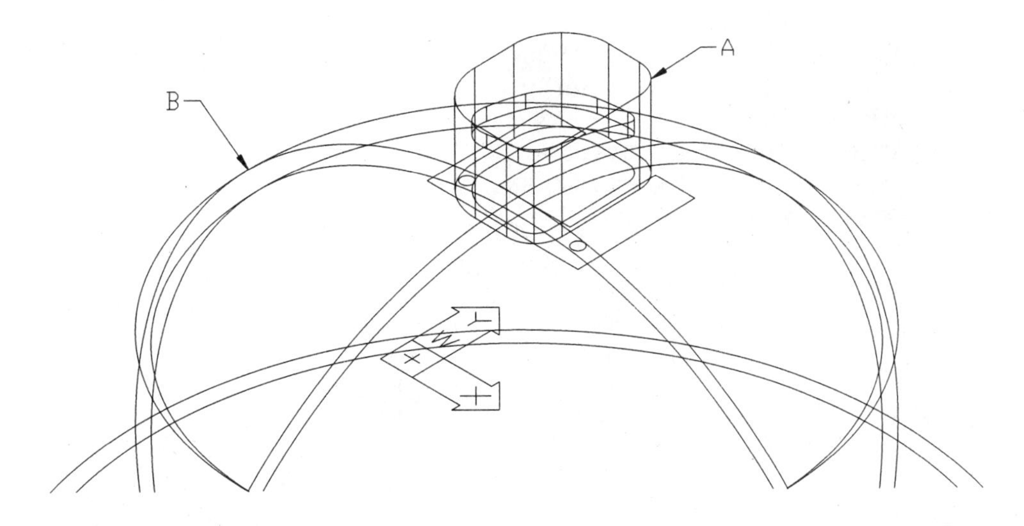

Figure 7.56 Solid of intersection united with second sphere, and button outer wireframe extruded

Intersect the second solid sphere with the united solid to form a step in the button. See Figure 7.57.

<Modify> **<Boolean>** **<Intersect>**

Command: **INTERSECT**
Select objects: [**Select A and B (Figure 7.56).**]
Select objects: [**Enter**]

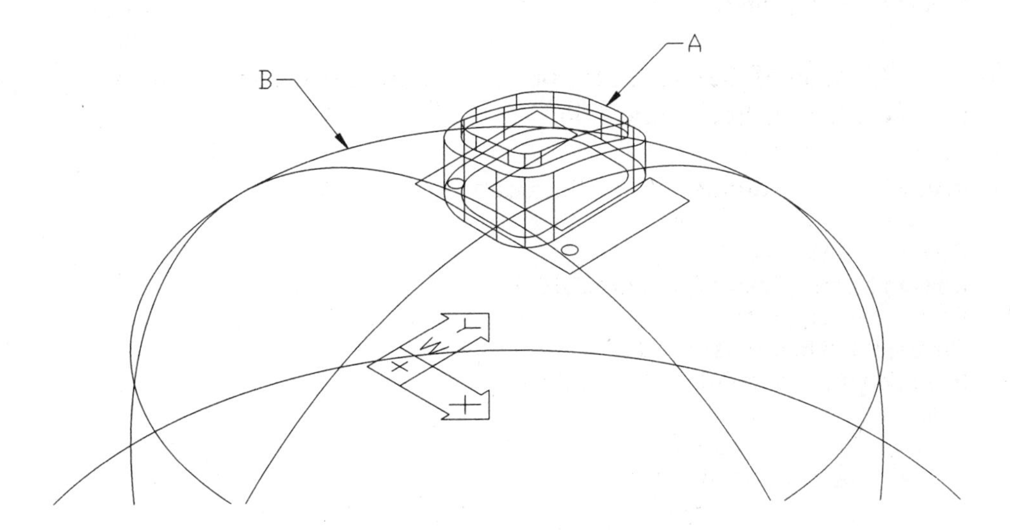

Figure 7.57 Step in button created

To cut away the underside of the button, use the SUBTRACT command to subtract the third sphere from the intersected solid. See Figure 7.58.

<Modify> **<Boolean>** **<Subtract>**

Command: **SUBTRACT**
Select solids and regions to subtract from...
Select objects: [**Select A (Figure 7.57).**]
Select objects: [**Enter**]
Select solids and regions to subtract...
Select objects: [**Select B (Figure 7.57).**]
Select objects: [**Enter**]

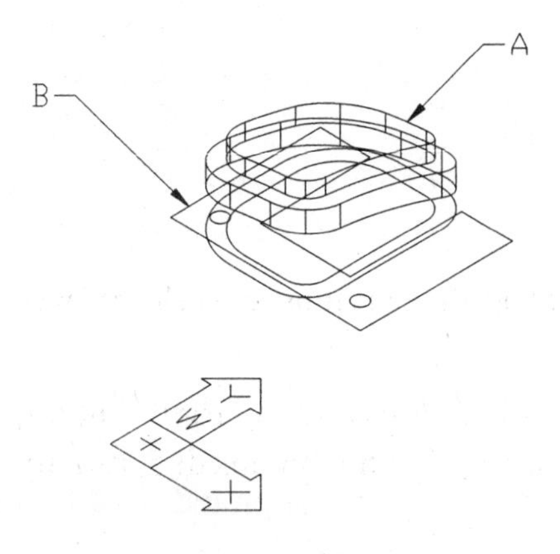

Figure 7.58 Left button completed

The left button is complete. Mirror it to form the right button, and extrude the U-shape wireframe to form the hinge lever. See Figure 7.59.

<Modify> **<Mirror>**

Command: **MIRROR**
Select objects: [**Select A (Figure 7.58).**]
Select objects: [**Enter**]
First point of mirror line: **0,0**
Second point: **0,1**
Delete old objects? <N> **N**

<Draw> **<Solids>** **<Extrude>**

Command: **EXTRUDE**
Select objects: [**Select B (Figure 7.58).**]
Select objects: [**Enter**]
Path/<Height of Extrusion>: **-2**
Extrusion taper angle: **0**

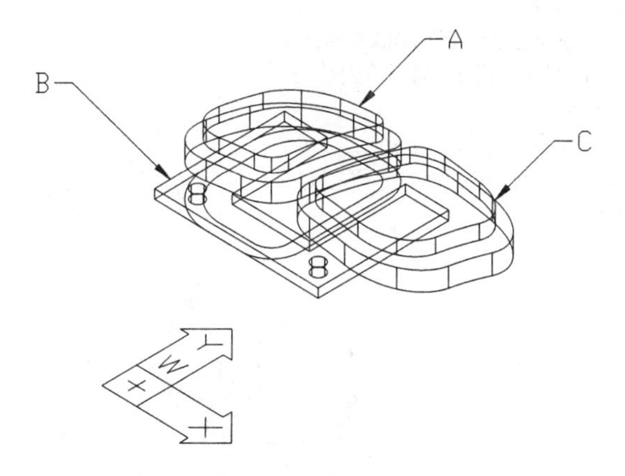

Figure 7.59 Button mirrored, and hinge lever created

Now you have three solids: left button, right button, and U-shape hinge lever. Because the wireframes on layer W_b are not needed, turn off layer W_b.

[Object Properties] **[Layer Control]**

On	S_b
Off	S_l
Off	S_u
Off	Uty
Off	W_b
Off	W_l
Off	W_u

Current layer: **S_b**

To complete the model of the buttons, unite the two buttons with the hinge lever. See Figure 7.60.

<Modify> **<Boolean>** **<Union>**

Command: **UNION**
Select objects: **[Select A, B, and C (Figure 7.59).]**
Select objects: **[Enter]**

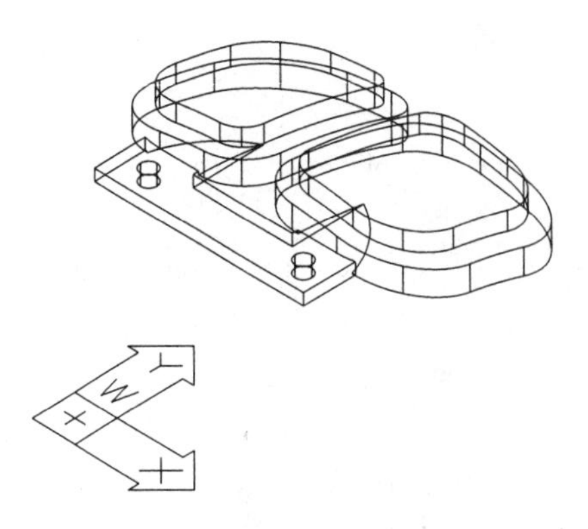

Figure 7.60 Completed model

The model of the buttons is complete. Figure 7.61 shows a rendered view of the completed model.

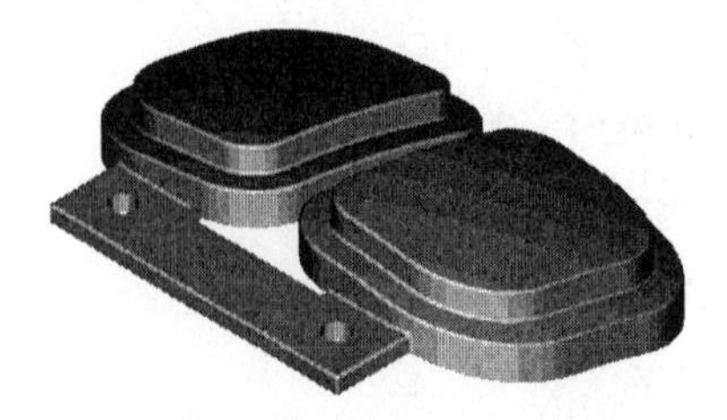

Figure 7.61 Rendered view of buttons

Interference Checking

After making the buttons, you should check for any interference. Turn on layers S_l and S_u. Then use the INTERFERE command twice to check for any interference with the upper casing or the lower casing. See Figure 7.62.

[Object Properties] **[Layer Control]**

On	S_b
On	S_l
On	S_u
Off	Uty
Off	W_b
Off	W_l
Off	W_u

Current layer: **S_b**

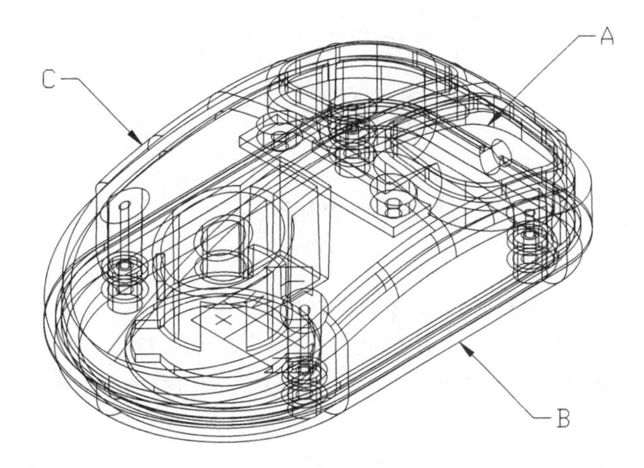

Figure 7.62 Upper casing, lower casing, and buttons

Use the INTERFERE command to check the interference between the button and the upper casing and the button and the lower casing.

<Draw> **<Solids>** **<Interfere>**

Command: **INTERFERE**
Select the first set of solids:
Select objects: [**Select A (Figure 7.62).**]
Select objects: [**Enter**]
Select the second set of solids:
Select objects: [**Select B (Figure 7.62).**]
Select objects: [**Enter**]
Comparing 1 solid against 1 solid.
Solids do not interfere.

Command: [**Enter**]
INTERFERE
Select the first set of solids:
Select objects: [**Select A (Figure 7.62).**]
Select objects: [**Enter**]
Select the second set of solids:
Select objects: [**Select C (Figure 7.62).**]
Select objects: [**Enter**]
Comparing 1 solid against 1 solid.
Solids do not interfere.

There is no interference. Proceed to make the ball cover.

7.6 Ball Cover

Figure 7.63 shows the model of the ball cover. Take some time to analyze this model to determine how it can be constructed.

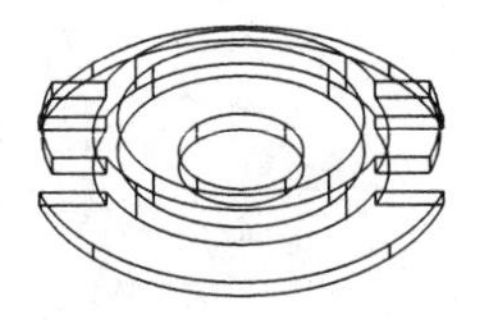

Figure 7.63 Ball cover

To make this model you can create three regions or closed polylines. See Figure 7.64. Using these regions, you will make a revolved solid and two extruded solids. Then you will subtract the extruded solids from the revolved solid to form the final model.

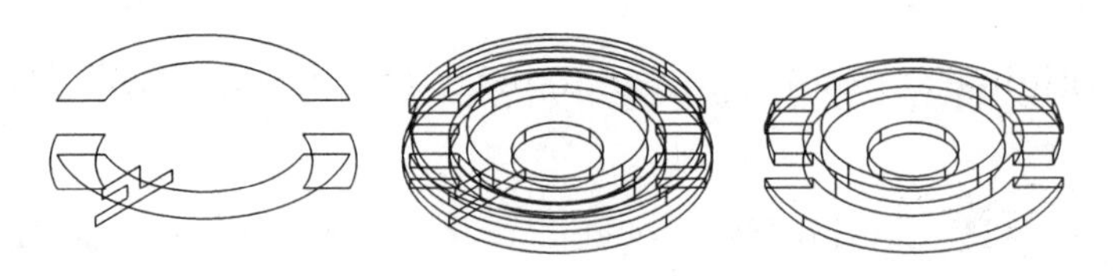

Figure 7.64 Wireframes, revolve solid, extruded solids, and final model

Wireframes for the Ball Cover

To make the revolved solid and extruded solids requires three sets of wireframes. Make two layers, S_c and W_c, for making the solids and wireframes, respectively. Set layer W_c as the current layer.

Because the ball cover has to match the ball cover opening of the lower casing, turn on layer W_l. To clear the screen, turn off layers S_b, S_l, and S_u. See Figure 7.65.

<Format> **<Layer...>**

Command: **LAYER**

Name		Color	Linetype
0	On	White	Continuous
S_b	Off	Red	Continuous
S_c	On	Yellow	Continuous
S_l	Off	Green	Continuous
S_u	Off	Blue	Continuous
Uty	Off	Magenta	Continuous
W_b	Off	Green	Continuous
W_c	On	Blue	Continuous
W_l	On	Cyan	Continuous
W_u	Off	Yellow	Continuous

Current layer: **W_c**

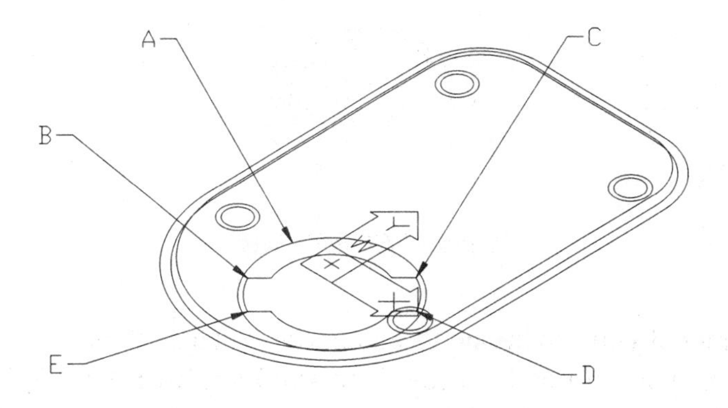

Figure 7.65 Wireframe of lower casing

To delete the originals after using them to make regions, set DELOBJ to 1.

Command: **DELOBJ**
New value for DELOBJ: **1**

Using the wireframes of the lower casing as references, create two circles and four lines.

<Draw> **<Circle>** **<Center, Radius>**

Command: **CIRCLE**
3P/2P/TTR/<Center point>: **CEN** of **[Select A (Figure 7.65).]**
Diameter/<Radius>: **12**

Command: **[Enter]**
CIRCLE
3P/2P/TTR/<Center point>: **CEN** of **[Select A (Figure 7.65).]**
Diameter/<Radius>: **17**

<Draw> **<Line>**

Command: **LINE**
From point: **END** of **[Select B (Figure 7.65).]**
To point: **END** of **[Select C (Figure 7.65).]**
To point: **[Enter]**

Command: **[Enter]**
LINE
From point: **END** of **[Select D (Figure 7.65).]**
To point: **END** of **[Select E (Figure 7.65).]**
To point: **[Enter]**

Turn off layer W_1 because the wireframes on this layer are not needed anymore. See Figure 7.66.

[Object Properties] **[Layer Control]**

Off	S_b

On	S_c
Off	S_l
Off	S_u
Off	Uty
Off	W_b
On	W_c
Off	W_l
Off	W_u

Current layer: **W_c**

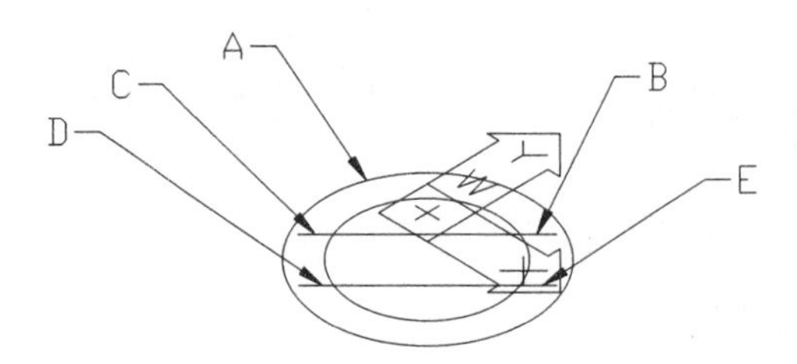

Figure 7.66 Circles and lines created

Extend four lines to meet the outer circle. See Figure 7.67.

<Modify> <Trim>

Command: **EXTEND**
Select boundary edges: (Projmode = UCS, Edgemode = No extend)
Select objects: [**Select A (Figure 7.66).**]
Select objects: [**Enter**]
<Select object to extend>/Project/Edge/Undo: [**Select B, C, D, and E (Figure 7.66).**]
<Select object to extend>/Project/Edge/Undo: [**Enter**]

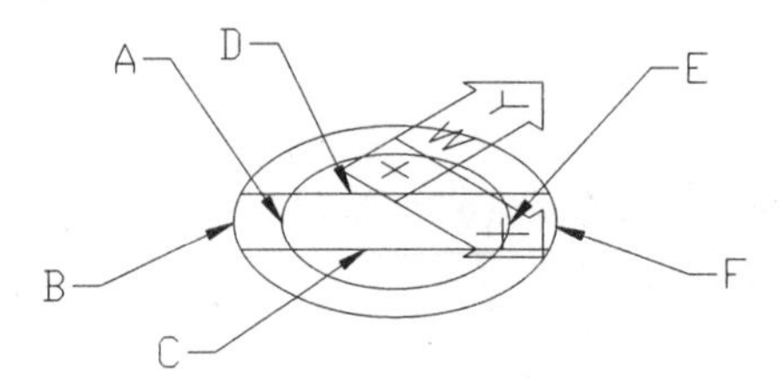

Figure 7.67 Lines extended

Trim the lines and circles to form two closed loops. See Figure 7.68.

<Modify> <Trim>

Command: **TRIM**
Select cutting edges: (Projmode = UCS, Edgemode = No extend)
Select objects: [**Select A, B, C, and D (Figure 7.67).**]
Select objects: [**Enter**]
<Select object to trim>/Project/Edge/Undo: [**Select A, B, C, D, E, and F (Figure 7.67).**]

<Select object to trim>/Project/Edge/Undo: **[Enter]**

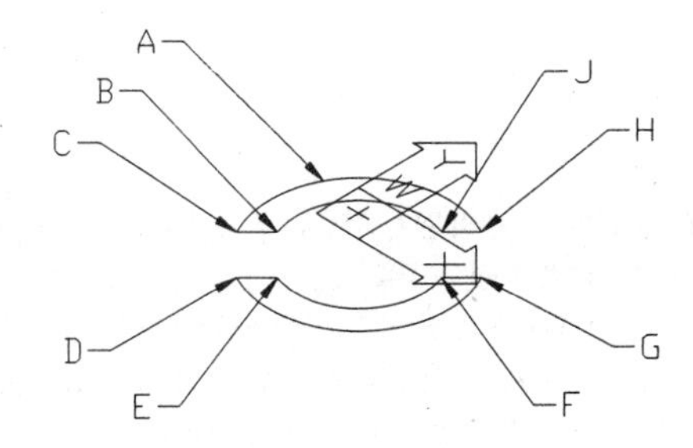

Figure 7.68 Loops formed

Add four arcs and copy four lines. See Figure 7.69.

<Draw> **<Arc>** **<Center, Start, End>**

Command: **ARC**
Center/<Start point>: **C**
Center: **CEN** of **[Select A (Figure 7.68).]**
Start point: **END** of **[Select B (Figure 7.68).]**
Angle/Length of chord/<End point>: **END** of **[Select E (Figure 7.68).]**

Command: **[Enter]**
ARC
Center/<Start point>: **C**
Center: **CEN** of **[Select A (Figure 7.68).]**
Start point: **END** of **[Select C (Figure 7.68).]**
Angle/Length of chord/<End point>: **END** of **[Select D (Figure 7.68).]**

Command: **[Enter]**
ARC
Center/<Start point>: **C**
Center: **CEN** of **[Select A (Figure 7.68).]**
Start point: **END** of **[Select F (Figure 7.68).]**
Angle/Length of chord/<End point>: **END** of **[Select J (Figure 7.68).]**

Command: **[Enter]**
ARC
Center/<Start point>: **C**
Center: **CEN** of **[Select A (Figure 7.68).]**
Start point: **END** of **[Select G (Figure 7.68).]**
Angle/Length of chord/<End point>: **END** of **[Select H (Figure 7.68).]**

<Draw> **<Line>**

Command: **LINE**
From point: **END** of **[Select B (Figure 7.68).]**
To point: **END** of **[Select C (Figure 7.68).]**
To point: **[Enter]**

Command: **[Enter]**

LINE
From point: **END** of [**Select D (Figure 7.68).**]
To point: **END** of [**Select E (Figure 7.68).**]
To point: [**Enter**]

Command: [**Enter**]
LINE
From point: **END** of [**Select F (Figure 7.68).**]
To point: **END** of [**Select G (Figure 7.68).**]
To point: [**Enter**]

Command: [**Enter**]
LINE
From point: **END** of [**Select H (Figure 7.68).**]
To point: **END** of [**Select J (Figure 7.68).**]
To point: [**Enter**]

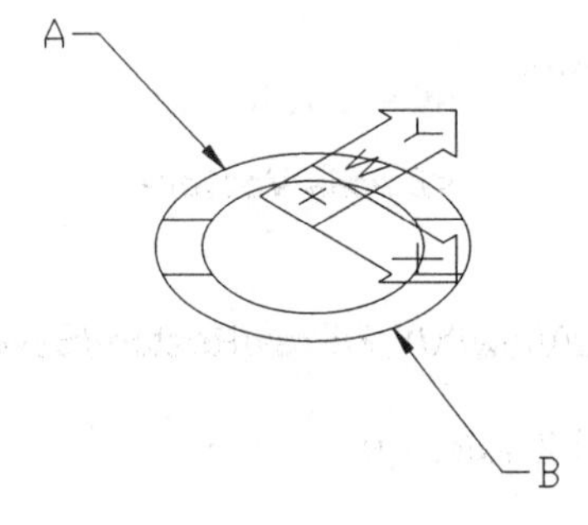

Figure 7.69 Arcs and lines created

Make sure you get the four lines correct. They are not readily visible because they overlap the existing lines. Now you have four loops. The loops are formed from eight arcs and eight lines. Select all the entities to form four regions.

<Draw> **<Region>**

Command: **REGION**
Select objects: [**Select 8 lines and 8 arcs.**]
16 found
Select objects: [**Enter**]
4 Regions created.

There should be no noticeable change on your screen after you make the regions. Check to be sure that you have four regions.

After making the regions, move two regions a distance of 4.5 units in the Z-direction.

<Modify> **<Move>**

Command: **MOVE**
Select objects: [**Select A and B (Figure 7.69).**]
Select objects: [**Enter**]
Base point or displacement: **0,0,4.5**
Second point of displacement: [**Enter**]

Turn on layer Uty. On this layer, you have created a section across the upper and lower casings. Set the UCS to this plane. See Figure 7.70.

[Object Properties] **[Layer Control]**

Off	S_b
On	S_c
Off	S_l
Off	S_u
On	Uty
Off	W_b
On	W_c
Off	W_l
Off	W_u

Current layer: **W_c**

Set the UCS to a new position.

<Tools> **<UCS>** **<Z Axis Vector>**

Command: **UCS**
Origin/ZAxis/3point/OBject/View/X/Y/Z/Prev/Restore/Save/Del/?/<World>: **ZA**
Origin point: **0,0,0**
Point on positive portion of Z-axis: **1,0**

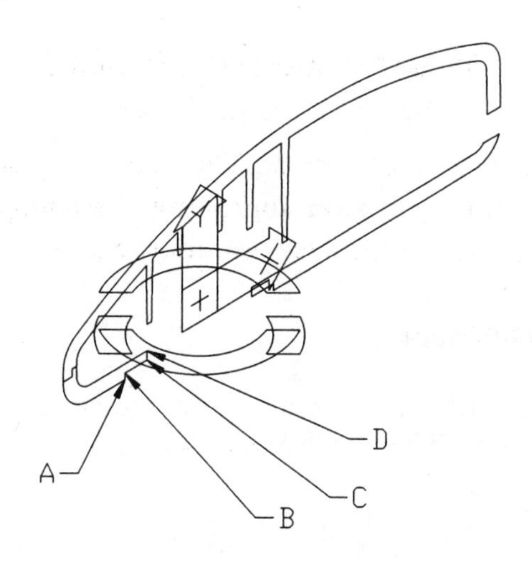

Figure 7.70 Regions moved, layer Uty turned on, and UCS changed

On the new UCS XY plane, construct a polyline. See Figure 7.71.

<Draw> **<Polyline>**

Command: **PLINE**
From point: **END** of **[Select A (Figure 7.70).]**
Arc/Close/Halfwidth/Length/Undo/Width/<Endpoint of line>: **END** of **[Select B (Figure 7.70).]**

Arc/Close/Halfwidth/Length/Undo/Width/<Endpoint of line>: **END** of **[Select C (Figure 7.70).]**
Arc/Close/Halfwidth/Length/Undo/Width/<Endpoint of line>: **END** of **[Select D (Figure 7.70).]**
Arc/Close/Halfwidth/Length/Undo/Width/<Endpoint of line>: **@4<180**
Arc/Close/Halfwidth/Length/Undo/Width/<Endpoint of line>: **@1.5<90**
Arc/Close/Halfwidth/Length/Undo/Width/<Endpoint of line>: **@6<0**
Arc/Close/Halfwidth/Length/Undo/Width/<Endpoint of line>: **@3<270**
Arc/Close/Halfwidth/Length/Undo/Width/<Endpoint of line>: **@5<0**
Arc/Close/Halfwidth/Length/Undo/Width/<Endpoint of line>: **@1.5<270**
Arc/Close/Halfwidth/Length/Undo/Width/<Endpoint of line>: **C**

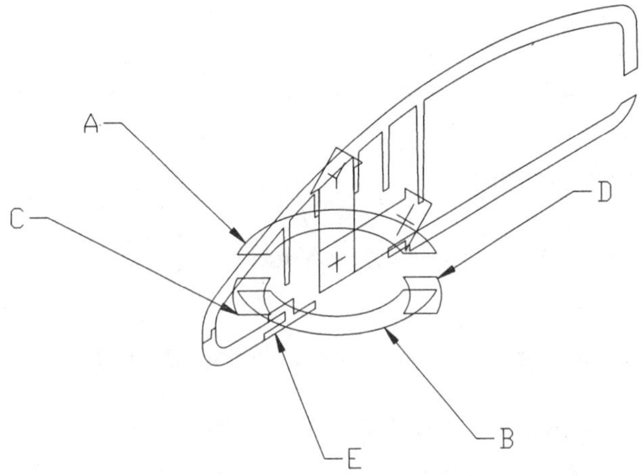

Figure 7.71 Polyline created

Now you have two regions and a polyline. Regions and polylines can be used for making revolved and extruded solids.

Solid Model for the Ball Cover

The wireframes for the ball cover are complete. Set the current layer to S_c and turn off layer Uty.

[Object Properties] **[Layer Control]**

Off	S_b
On	S_c
Off	S_l
Off	S_u
Off	Uty
Off	W_b
On	W_c
Off	W_l
Off	W_u

Current layer: **S_c**

Set DELOBJ to zero to keep the originals after using them for extrusion or revolving.

Command: **DELOBJ**
New value for DELOBJ: **0**

Extrude the two regions and revolve the polyline. See Figure 7.72.

<Draw> **<Solids>** **<Extrude>**

Command: **EXTRUDE**
Select objects: [**Select A and B (Figure 7.71).**]
Select objects: [**Enter**]
Path/<Height of Extrusion>: **-3.25**
Extrusion taper angle: **0**

<Draw> **<Solids>** **<Extrude>**

Command: **EXTRUDE**
Select objects: [**Select C and D (Figure 7.71).**]
Select objects: [**Enter**]
Path/<Height of Extrusion>: **1.25**
Extrusion taper angle: **0**

<Draw> **<Solids>** **<Extrude>**

Command: **REVOLVE**
Select objects: [**Select E (Figure 7.71).**]
Select objects: [**Enter**]
Axis of revolution - Object/X/Y/<Start point of axis>: **0,0**
<End point of axis>: **0,1**
Angle of revolution <full circle>: [**Enter**]

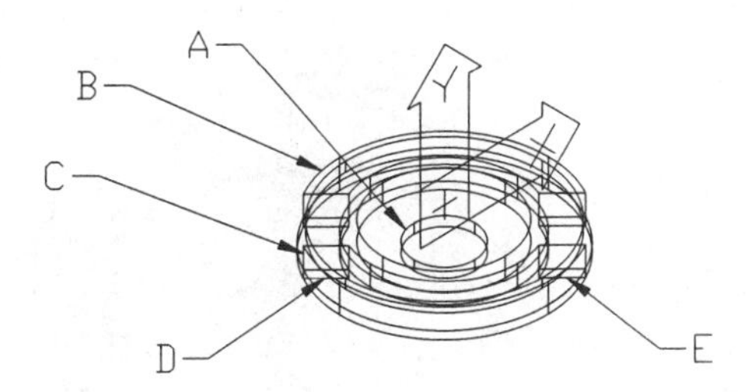

Figure 7.72 Layer Uty turned off and extruded, and revolved solids created

To form the model, subtract the extruded solids from the revolved solid. See Figure 7.73.

<Modify> **<Boolean>** **<Subtract>**

Command: **SUBTRACT**
Select solids and regions to subtract from...
Select objects: [**Select A (Figure 7.72).**]
Select objects: [**Enter**]
Select solids and regions to subtract...

Select objects: [**Select B, C, D, and E (Figure 7.72).**]
Select objects: [**Enter**]

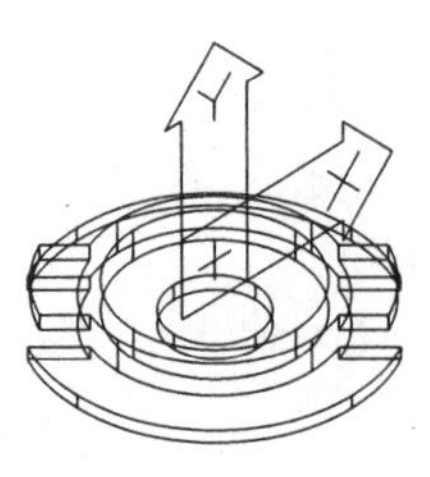

Figure 7.73 Extruded solids subtracted from revolved solid

The ball cover model is complete. Figure 7.74 is a rendered view of the ball cover.

Figure 7.74 Rendered view of ball cover

Interference Checking

Turn off layer W_c and turn on layers S_b, S_l, and S_u. See Figure 7.75. Use the INTERFERE command to check the interference between the ball cover and the other three models. If there is any interference, form the solid of interference and subtract the solid from one of the mating models. If there is no interference, the models are complete.

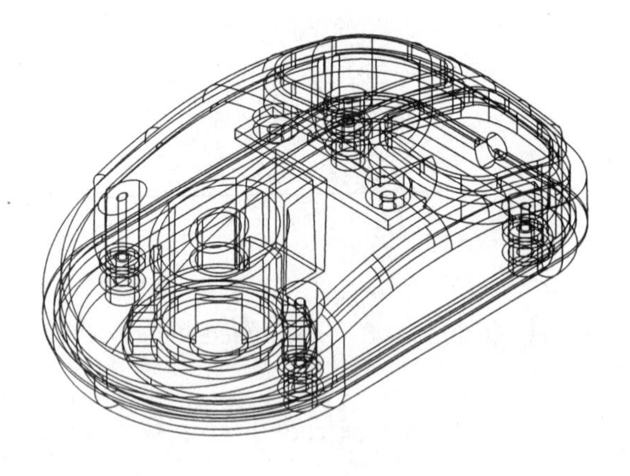

Figure 7.75 Four solid models

Save your drawing.

<File> **<Save>**

File name: **MOUSE.DWG**

7.7 Assembly of Solid Models

The solid models for the components of a mouse casing are complete. Now you have four solid models in a single drawing file. If you wish to have separate drawing files for each model, you can use the WBLOCK command to export the solids. To put them back together in an assembly, you can use the XREF command to attach them.

In combining solids from a number of drawing files, you can use the XREF command to attach the solid or use the INSERT command to create an internal block. For both cases, you cannot edit the referenced or inserted objects because they are instances of external drawings or internal blocks. To edit an instance of the external drawing, you can use the Bind option of the XREF command. After binding, the external drawing becomes an internal block. However, you still cannot edit an instance of an internal block. To edit it, you need to run the EXPLODE command. When you use this command on an instance of an inserted solid, you must be very careful not to over-explode it.

The process of exploding an inserted solid works as follows: When you explode the inserted solid once, you create a copy of the inserted solid at the insertion point. If you explode it twice or more, you will break it down into separate line and arc entities. Of course, you can UNDO the EXPLODE command.

If you are not sure whether you have exploded an instance or not, you can use the LIST command to check the data type. It should be a solid. To reduce the memory size after exploding the inserted solid, you have to use the PURGE command to remove the unreferenced inserted block. After inserting, exploding, and purging, you can use the INTERFERE command to check if there is any interference between the inserted solids. For more details about block and external references, please refer to Chapter 4.

7.8 Key Points and Exercises

An electro-mechanical product usually has electronic and mechanical parts that are enclosed in a polymeric casing. Basically, the casing is a solid thin shell that has bosses and webs. To make the solid model by using AutoCAD R14, you work in three stages. First, you construct the solid model without the core. In other words, you disregard the core and internal details. Second, you construct the core as if it were a solid. Finally, you subtract the solid that represents the core from the solid that represents the outer skin.

With more than two solid models put together to form an assembly, you need to check whether they exhibit interference or not. Besides interference checking, you can evaluate mass properties and generate 2D cross sections.

In this chapter, you worked on an electro-mechanical product. Through this project, you gained an appreciation of how to create a thin-shell solid model with internal webs and bosses. In addition, you learned how to check two sets of solids for interference, how to evaluate the mass properties of a solid model, and how to check the minute details of a model by creating a section. To make a thin shell automatically from a solid object, you can use Mechanical Desktop R2.

Exercise 7.1

Delineate the steps you will take to construct a thin-shell solid model that has bosses and webs.

Exercise 7.2

Why is it necessary to check interference, evaluate mass properties, and generate cross sections?

Exercise 7.3

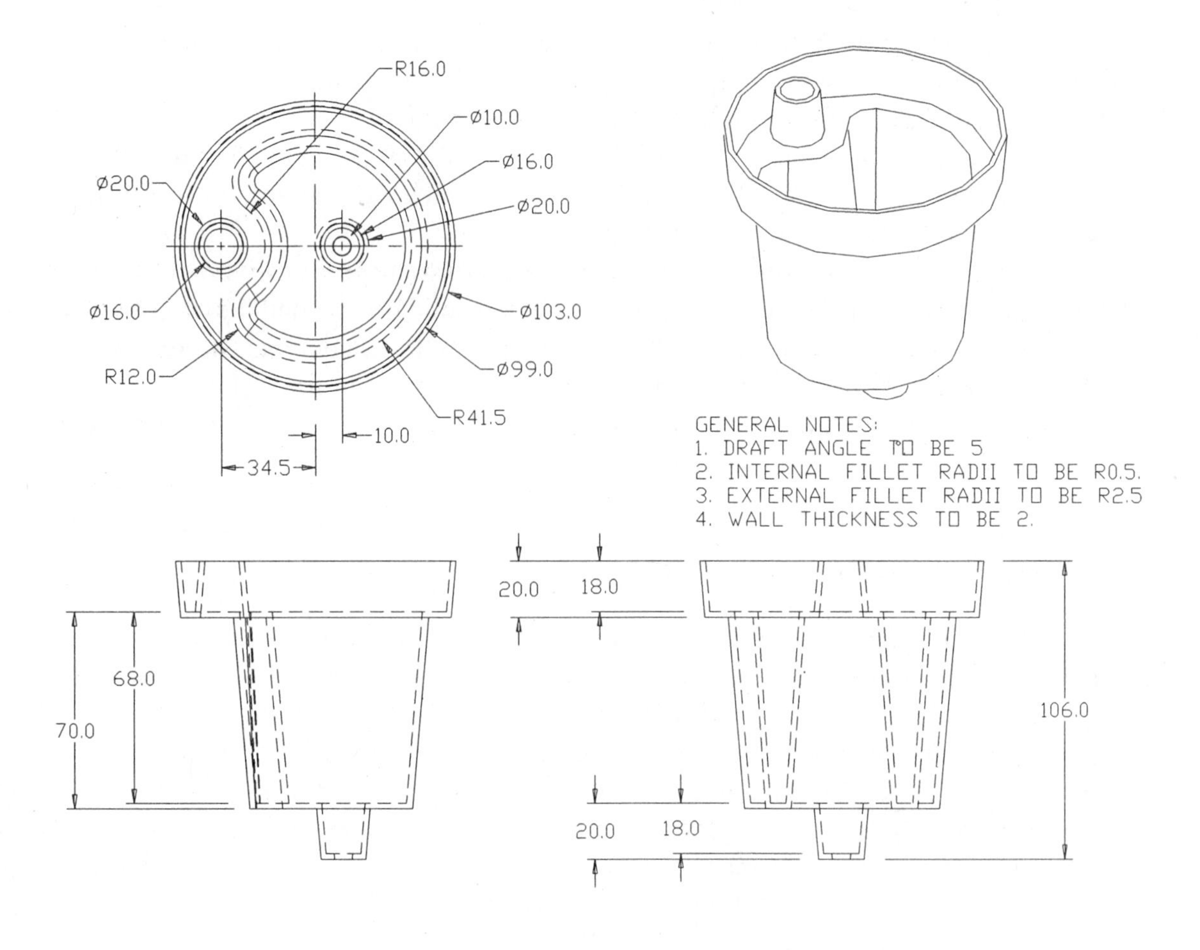

Figure 7.76 Casing for water filtration unit

Figure 7.76 shows the engineering drawing of the casing for a water filtration unit. It is a thin-shell solid model.

Start a new drawing. Make two layers called Wire and Solid, and set the layer Wire as the current layer. You will build the wireframes on this layer.

On the layer Wire, create a number of circles and arcs. Change them to nine regions or polylines. Then move the wires A and B a distance of (0,0,-20), the wires C, D, and E a distance of (0,0,-86), and the wires F and G a distance of (0,0,-18). See Figure 7.77.

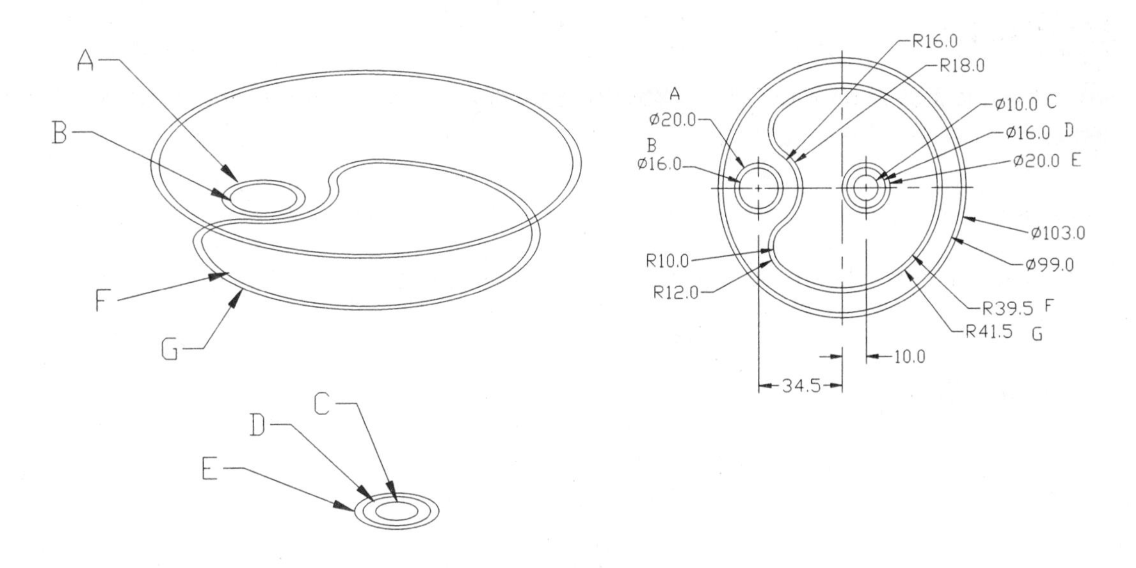

Figure 7.77 Set of circles and arcs drawn

Set the current layer to Solid. Then extrude the ∅103 circle a distance of -20 units, the wire G a distance of -70 units, and the wire E a distance of -20 units. The draft angle of all extrusion is 5°. Unite the three extruded solids. This is the outer skin of the solid model. See Figure 7.78.

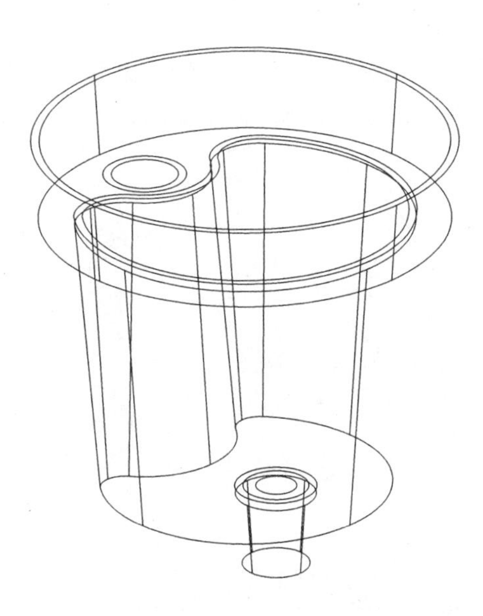

Figure 7.78 Outer skin created

To create the inner core of the solid, extrude the Ø99 circle a distance of -18 units, the wire F a distance of -68 units, and the wire D a distance of -18 units. The draft angle of all extrusion is 5°. Unite the three extruded solids. This is the core solid. Subtract it from the outer skin. See Figure 7.79.

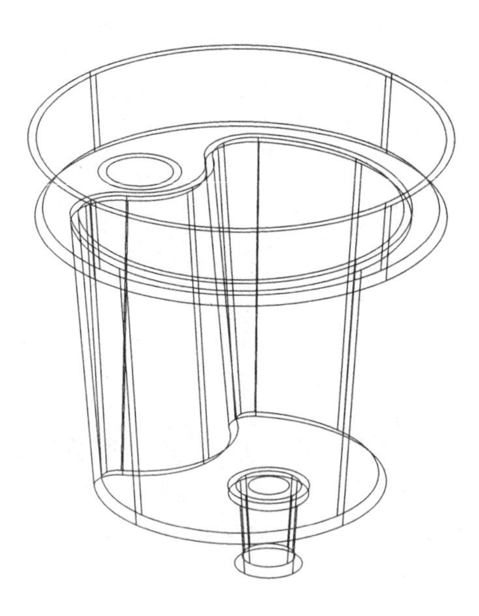

Figure 7.79 Inner solid core created

To continue, extrude the wires A and B a distance of 20 units and the wire C a distance of -20 units. The draft angle of all extrusion is 5°. See Figure 7.80.

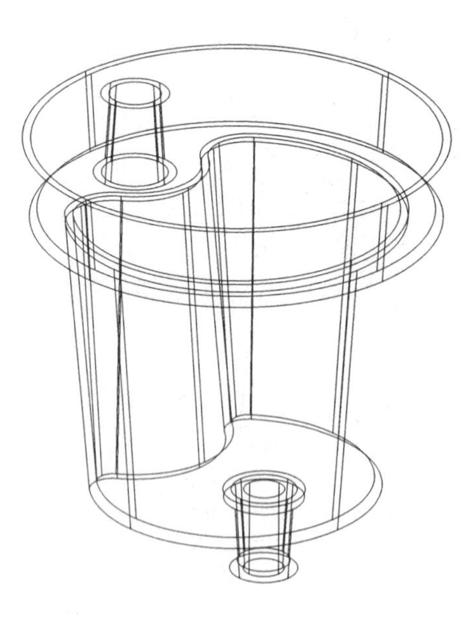

Figure 7.80 Three more wires extruded

To complete the model, unite the extruded solid A to the main body, and subtract the extruded solids B and C. To see the model more clearly, set the ISOLINES variable to 0 and the variable DISPSILH to 1. See Figure 7.81.

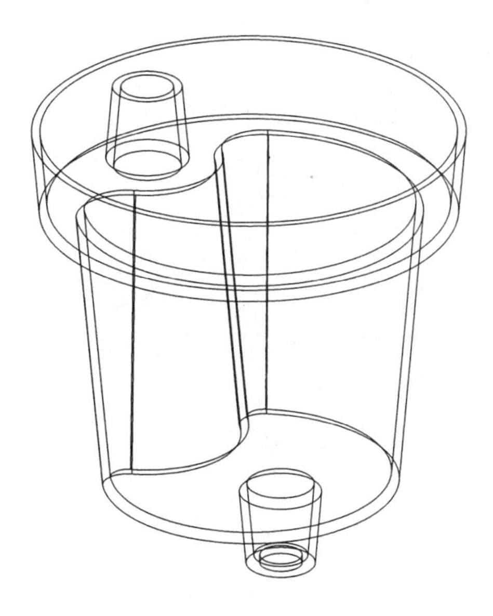

Figure 7.81 Completed water filtration model

To modify the model further, fillet the internal edges with a radius of R0.5 and the external edges with a radius of R2.5. See Figure 7.82. Save the drawing as FILTER.DWG.

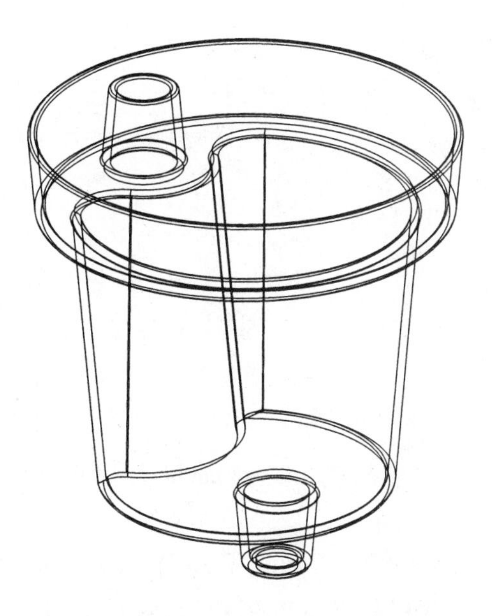

Figure 7.82 Water filtration model filleted

Exercise 7.2

Figure 7.83 shows the model of a personal computer front panel. It is a thin-shell solid model. To construct this model, you will make a skin solid and a core solid and then subtract the core solid from the skin solid.

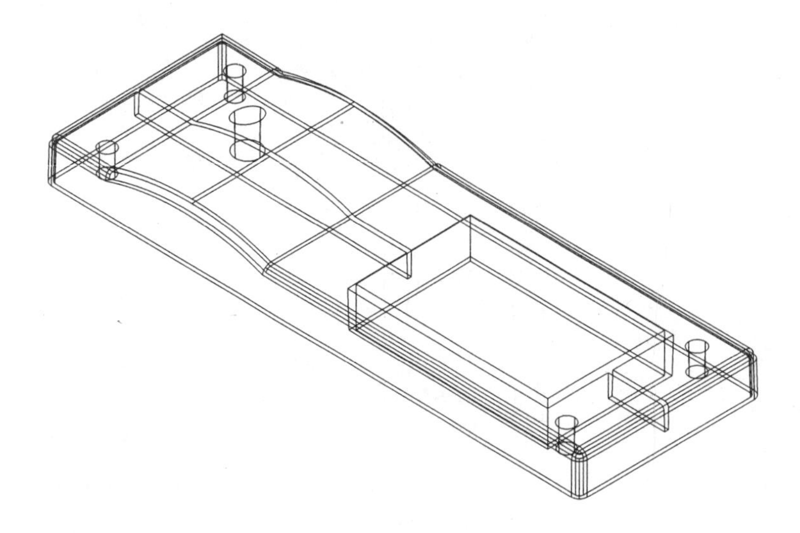

Figure 7.83 Front panel of a personal computer

Start a new drawing and make the four layers Core, Skin, Wire_c, and Wire_s for holding the core solids, skin solids, core wireframes, and skin wireframes, respectively.

Set layer Wire_s as the current layer and create a set of wireframes according to Figure 7.84. Convert them to two regions.

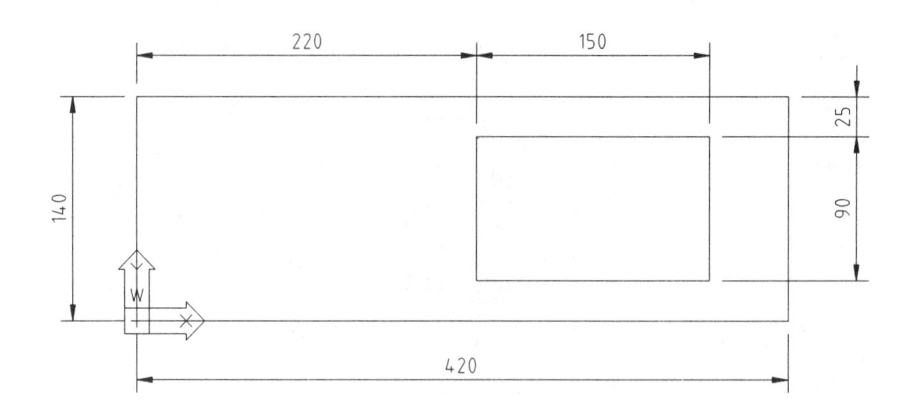

Figure 7.84 Two regions created

Rotate the UCS for 90° about the X axis. Then create a set of wireframes according to Figure 7.85. Convert the wireframes to a region.

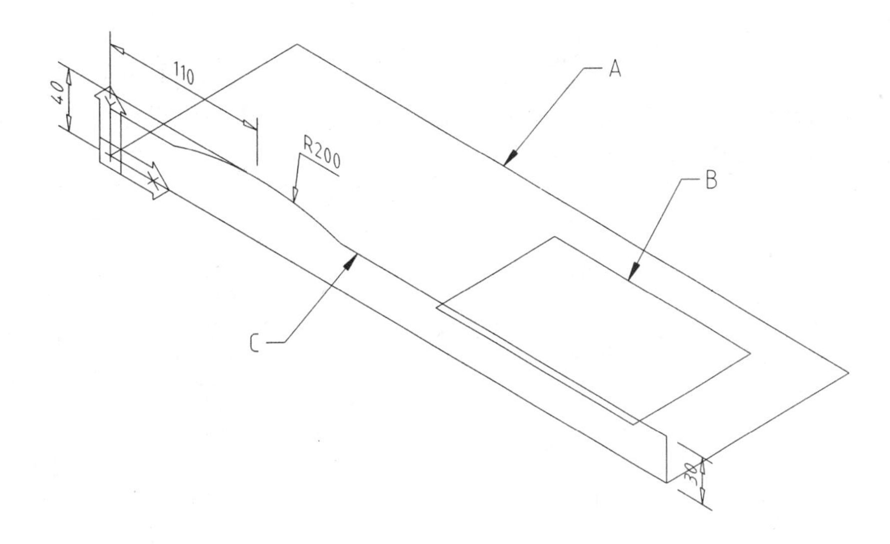

Figure 7.85 Third region created on the new UCS

Set the current layer to Solid_s. Extrude region A (Figure 7.85) a distance of 40 units with an extrusion taper angle of 1°, region B (Figure 7.85) a distance of 40 units with zero taper angle, and region C (Figure 7.85) a distance of 140 units with zero taper angle. See Figure 7.86.

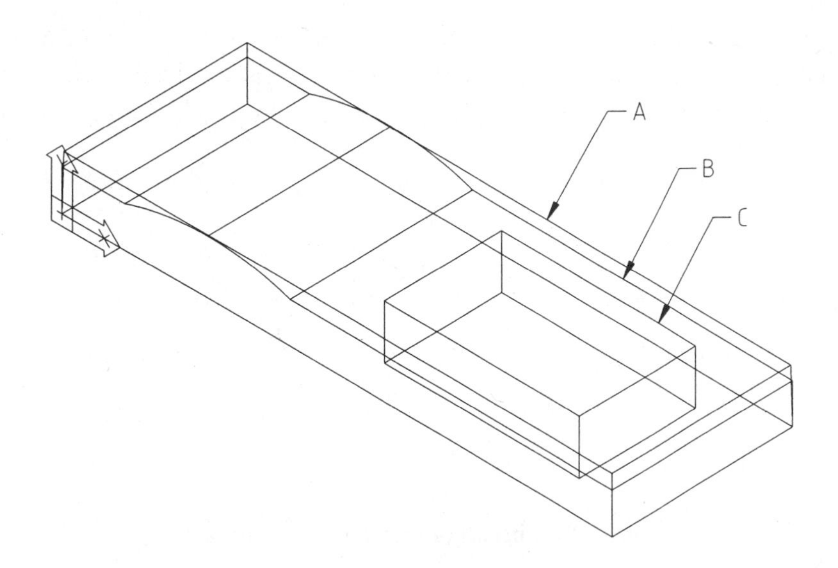

Figure 7.86 Three solids of extrusion

After extruding the wireframes, turn off layer Wire_s and create a solid of intersection between solid A (Figure 7.86) and solid B (Figure 7.86). Then subtract solid C (Figure 7.86) from the solid of intersection. After that, fillet the edges with a fillet radius of 6 units. The skin solid is complete. See Figure 7.87.

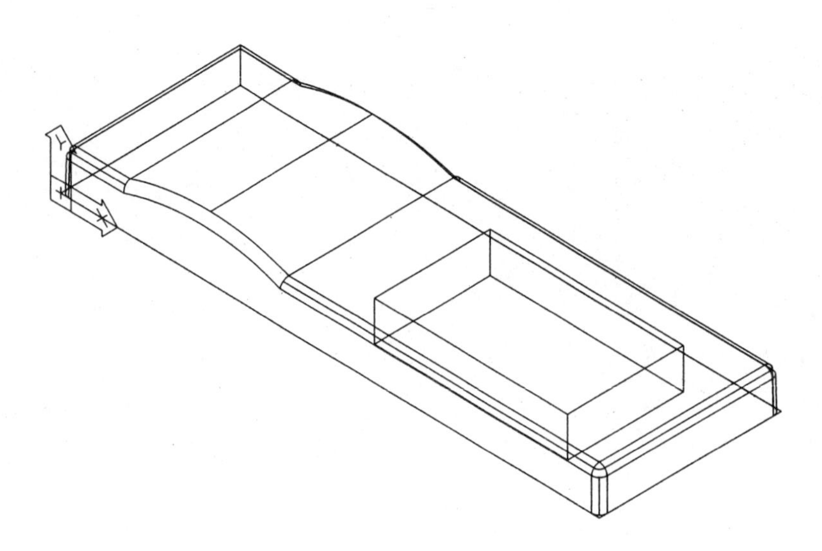

Figure 7.87 Skin solid created

Set the current layer to Wire_c and turn off the layer Skin. Then create a set of wireframes according to Figure 7.88. Convert the wireframes to a region.

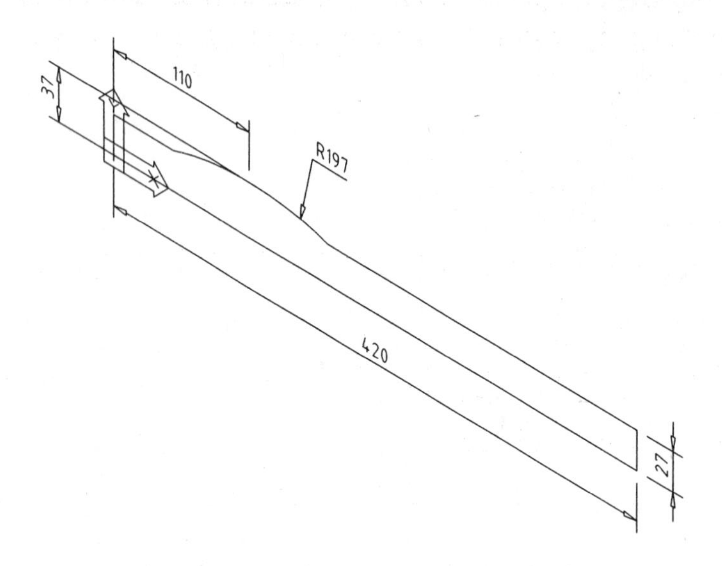

Figure 7.88 First set of wireframes for core solid

Set the UCS to World and create the wireframes according to Figure 7.89. Convert them to two regions.

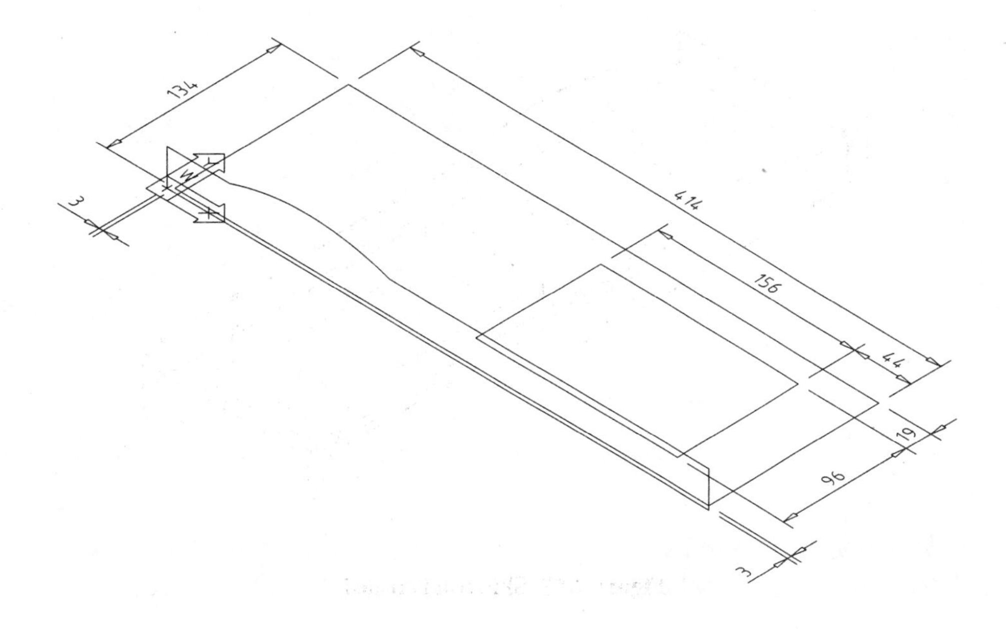

Figure 7.89 Second set of wireframes for core solid

Set the UCS origin to a point (*0,0,10). In accordance with Figure 7.90, create the five circles for making the bosses. Then create a rectangle for making a web according to Figure 7.91.

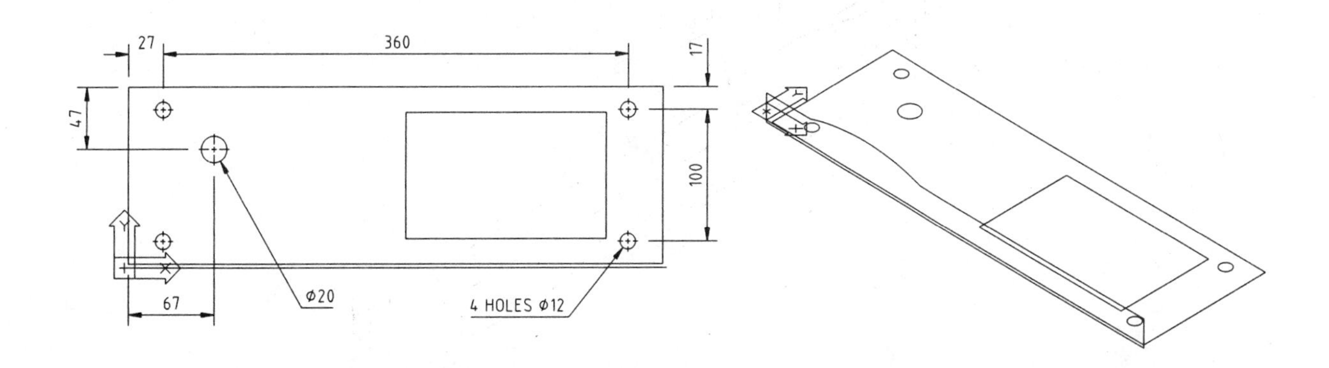

Figure 7.90 Wireframes for core solid created

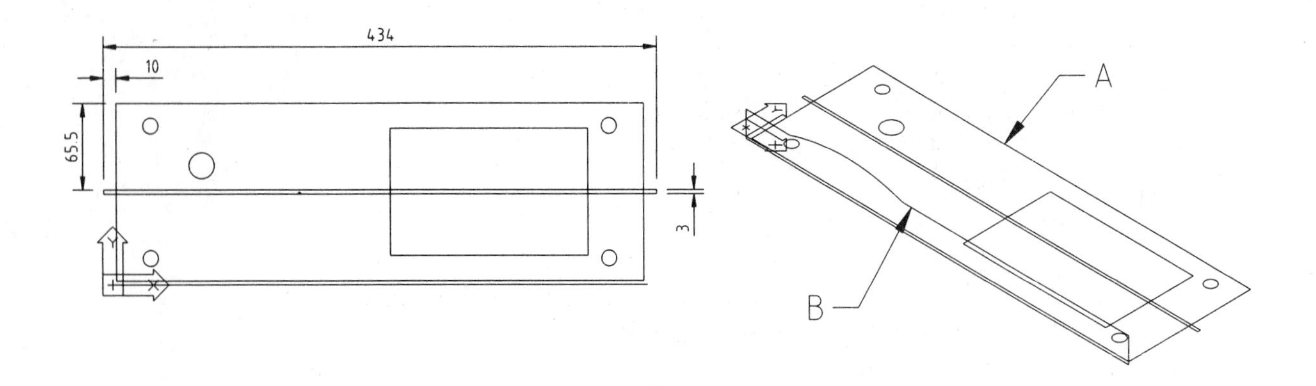

Figure 7.91 Rectangle created

Extrude region A (Figure 7.91) a distance of 40 units with 1° taper angle and region B (Figure 7.91) a distance of 140 units with zero taper angle. See Figure 7.92.

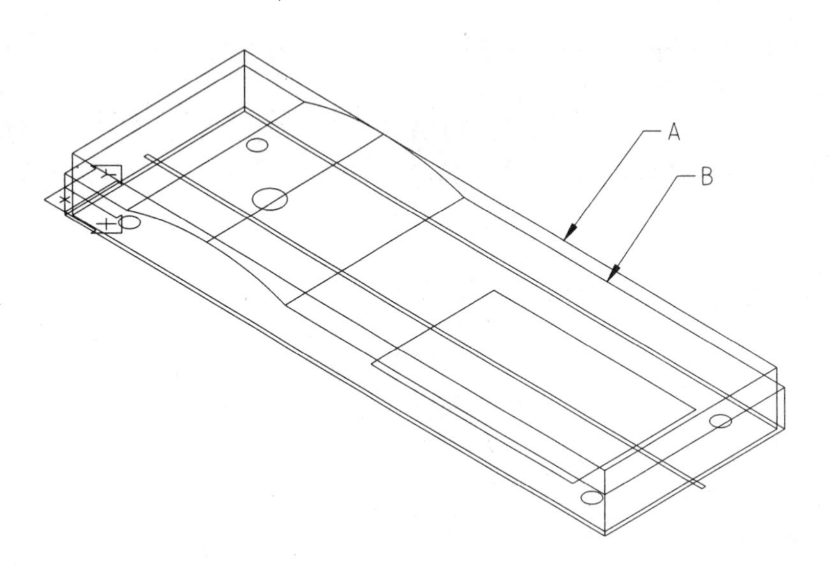

Figure 7.92 Two solids of extrusion created

Form a solid of intersection between solid A and B (Figure 7.92) and fillet the edges. The fillet radius is 3 units. The main body of the core solid is complete. See Figure 7.93.

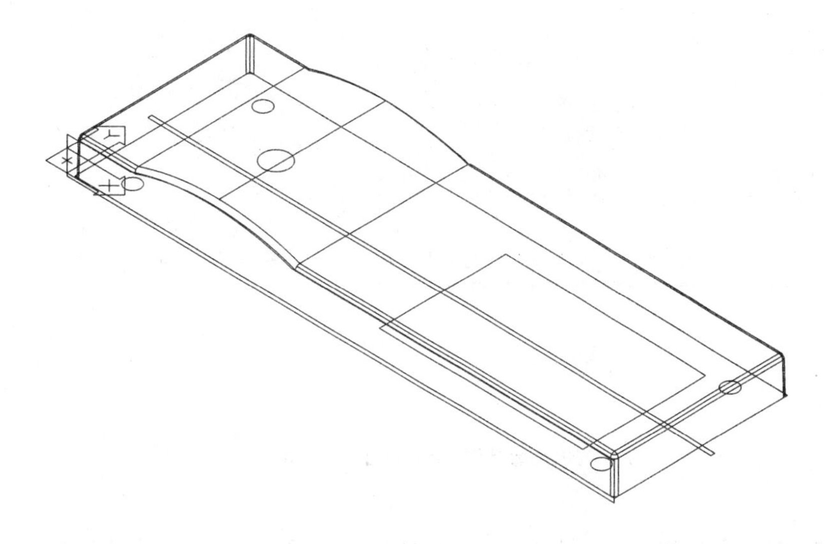

Figure 7.93 Main body of core solid created

Extrude the bosses and webs as solids. Height of extrusion is 40 units, and taper angle is -1°. Turn off layer Wire_c. See Figure 7.94.

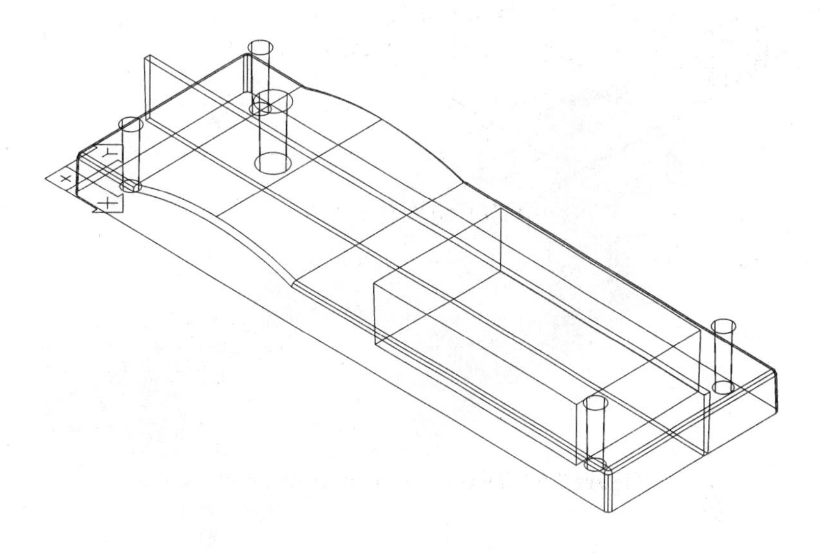

Figure 7.94 Bosses and webs extruded as solids

Subtract the boss and web solids from the core solid. See Figure 7.95.

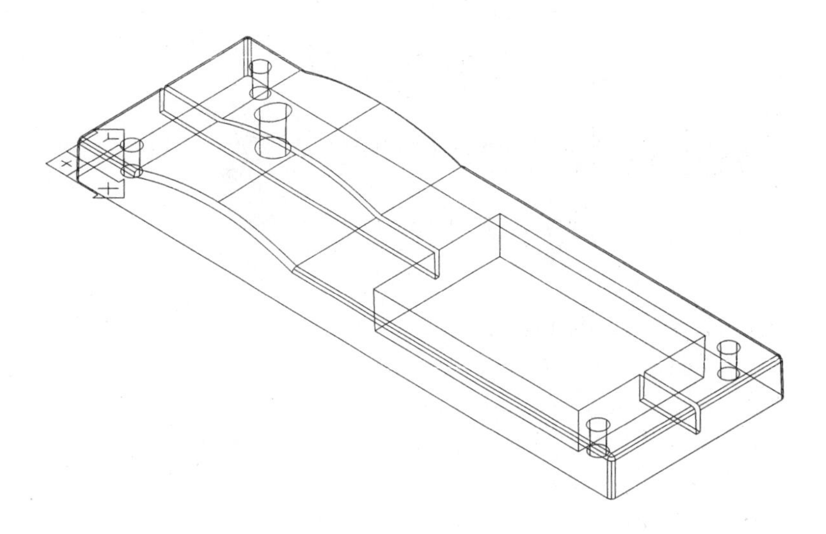

Figure 7.95 Boss and web solids subtracted from core solid

The core solid is complete. To construct the final model, turn on layer Solid_s. Then, subtract the core solid from the skin solid.

Figure 7.96 shows the rendered drawing of the completed model. Save the drawing as PANEL.DWG.

Figure 7.96 Completed model for PC front panel

Chapter 8

Solid Modeling Projects: Architectural

8.1 Drawing Preparation and Analysis
8.2 Wireframes for the First Level
8.3 Wireframes for the Staircase
8.4 Wireframes for the Second Level
8.5 Wireframes for the Roofs
8.6 3D Model for the First Level and the Staircase
8.7 3D Model for the Second Level
8.8 3D Model for the Roofs
8.9 Complete Model
8.10 Key Points and Exercises

Aims and Objectives

The aim of this chapter is to introduce the methods and techniques for constructing the solid model of a house. After studying this chapter, you should be able to:

- convert multilines to regions for subsequent extrusion to 3D solids, and
- use AutoCAD solid modeling as a tool to construct architectural models.

Overview

A solid model in a computer is integrated mathematical data that contains information about the edges, the surfaces, and the volume of the object that the model describes. Basically, the set of constructive solid geometry commands of AutoCAD is an engineering design tool for mechanical and manufacturing engineering. Given such a useful and powerful tool, we should not limit its use to the disciplines mentioned. Here, you will use constructive solid geometry modeling tools to create the 3D model of a house. The completed model is shown in Figure 8.1.

The aims of constructing the model of a house are to explore and appreciate the use of solid modeling techniques on architectural projects, to widen the scope of engineering application, and to broaden your outlook on model creation. Perhaps you will think of other applications too.

In order to make this project easy to follow, the model has been simplified, and many minute details have been omitted. This is not intended to imply that architectural projects should be simplified in such a way. After completing the model, you may, by all means, increase the detail of the model by adding more features to it.

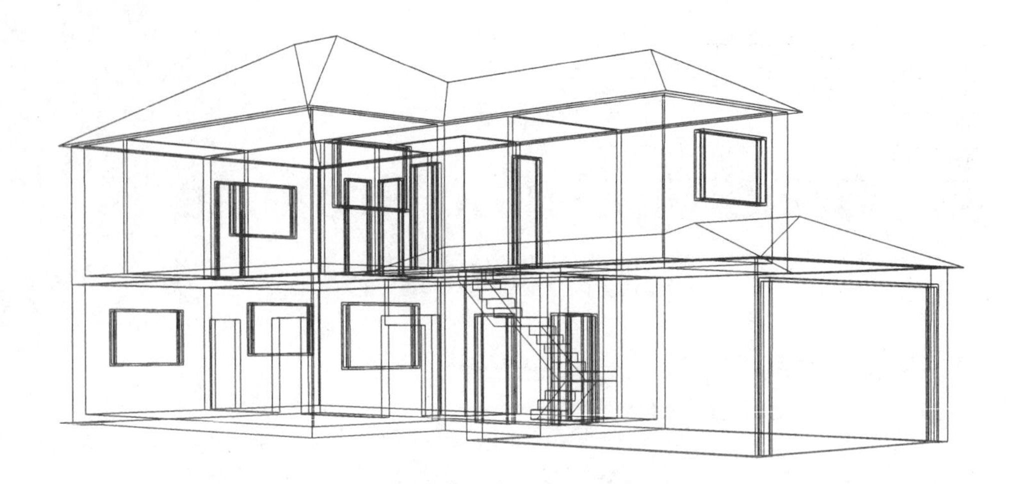

Figure 8.1 Completed model

8.1 Drawing Preparation and Analysis

Start a new drawing by using the quick start up wizard.

<File> **<New...>**

Command: **NEW**

[Create New Drawing
Use a Wizard **Quick Setup** **OK**]

[Step 1: Units
Decimal **Next>>**]

[Step 2: Area
Width: **20000** Length: **10000**
Done]

Analysis

Although the model of the house has been simplified and many details have been omitted, you must still build the internal floors, internal partitions, and staircase, in addition to the exterior of the house. You can divide the model into four major parts — Walls and floor of the first level, staircase, walls and floor of the second level, and roofs.

Create six layers: F1, F2, STAIR, ROOF, W1, and W2. Layer F1 is used for the floor of the first level; layer F2 is used for the floor of the second level; layer STAIR is used for the staircase; layer ROOF is used for the roof; layer W1 is used for the wall of the first level; and layer W2 is used for the wall of the second level.

<Format> **<Layer...>**

Command: **LAYER**

Name	Color	Linetype
0	White	Continuous
F1	Yellow	Continuous

F2	Magenta	Continuous
Roof	Blue	Continuous
Stair	Red	Continuous
W1	Cyan	Continuous
W2	Green	Continuous

Current layer: **W1**

8.2 Wireframes for the First Level

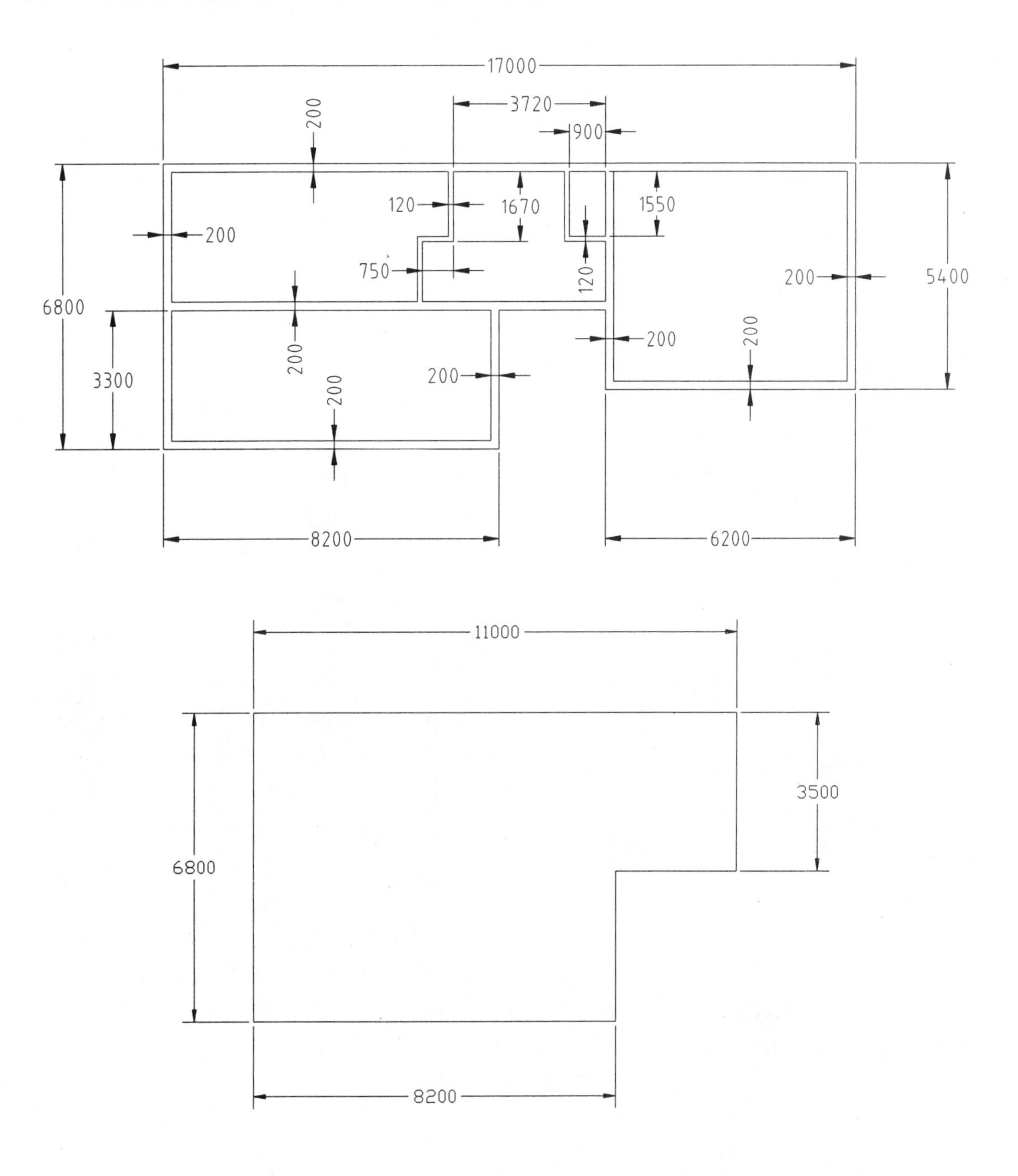

Figure 8.2 Dimensioned layout of the walls and floor of the first level

Figure 8.2 shows the dimensioned layout of the walls and floor of the first level of the house. A very quick way to create the 2D wireframes for the walls is to make use of the MLINE command. The MLINE command lets you produce a series of multiple parallel lines simultaneously. It is particularly useful for creating walls. Before you may use the MLINE command, you should take the trouble to set up the styles of the multilines and save them by using the MLSTYLE command.

<Format> **<Multiline Style...>**

Command: **MLSTYLE**

[Multiline Style
Name: **120WALL**
Add]

[Element Properties...]

[Offset Color Ltype
60 BYLAYER BYLAYER
-60 BYLAYER BYLAYER
OK]

[Multiline Properties...]

[Line **Start End**
Angle **90° 90°**
Fill: **Yes**
Color: **Cyan**
OK]

[Save...]

[Multiline Style
Name: **200WALL**
Add]

[Element Properties...]

[Offset Color Ltype
100 BYLAYER BYLAYER
-100 BYLAYER BYLAYER
OK]

[Multiline Properties...]

[Line **Start End**
Angle **90° 90°**
Fill: **Yes**
Color: **Cyan**
OK]

[Save...]

[OK]

After setting up two multiline styles, run the MLINE command to create a multiline. Select the 200WALL style that you have created, and set the justification to TOP. See Figure 8.3.

<Draw> **<Multiline>**

Command: **MLINE**
Justification/Scale/STyle/<From point>: **ST**
Mstyle name (or ?): **200WALL**
Justification/Scale/STyle/<From point>: **J**
Top/Zero/Bottom: **T**
Justification/Scale/STyle/<From point>: **0,0**
<To point>: **@6800<90**
Undo/<To point>: **@17000<0**
Close/Undo/<To point>: **@5400<270**
Close/Undo/<To point>: **@6200<180**
Close/Undo/<To point>: **@5400<90**
Close/Undo/<To point>: **[Enter]**

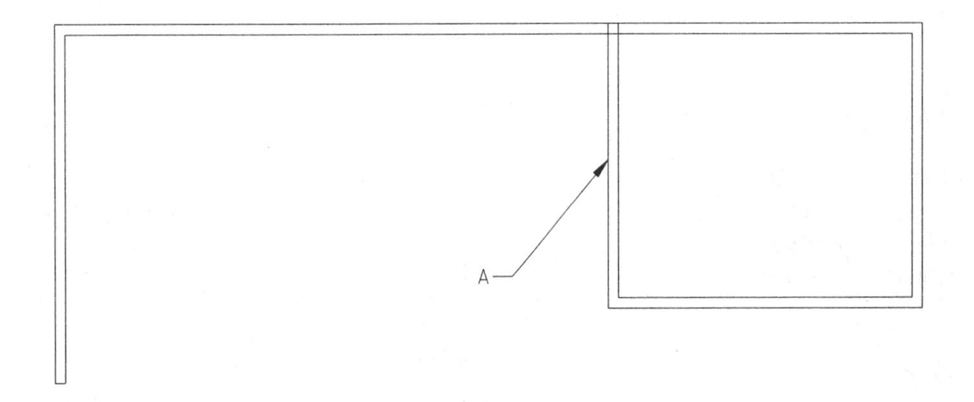

Figure 8.3 Multilines created

A multiline, as its name implies, has a number of parallel lines. Before you create a multiline, you need to decide which element of the multiline is to coincide with the cursor pick point. The setting is governed by the method of justification. If you choose top justification, the uppermost multiline element will align with the selected position of the cursor. If you choose zero justification, the zero offset multiline element will align with the cursor. Finally, if you choose bottom justification, the lowermost multiline element will align with the cursor.

Repeat the MLINE command to create two more multilines. In making these lines, use bottom justification. See Figure 8.4.

Command: **[Enter]**
MLINE
Justification/Scale/STyle/<From point>: **J**
Top/Zero/Bottom: **B**
Justification/Scale/STyle/<From point>: **0,3300**
Close/Undo/<To point>: **PERP** of **[Select at A (Figure 8.3).]**
Close/Undo/<To point>: **[Enter]**

Command: **[Enter]**
MLINE
Justification/Scale/STyle/<From point>: **0,0**
<To point>: **@8200<0**
Undo/<To point>: **@3300<90**
Close/Undo/<To point>: **[Enter]**

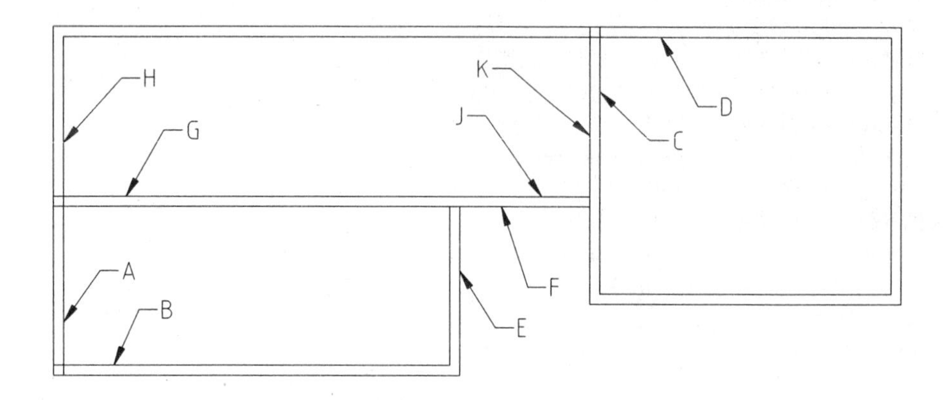

Figure 8.4 Two more multilines created

Compare bottom justification with top justification. Note the difference in multiline element location.

You have created three multilines. Now, you need to edit the joints between them. Run the MLEDIT command (see Figure 8.5) to edit a corner joint and four open tee joints. In making the open tee joints, the sequence of selecting is important. See Figure 8.6.

<Modify> <Object> <Multiline...>

Command: **MLEDIT**

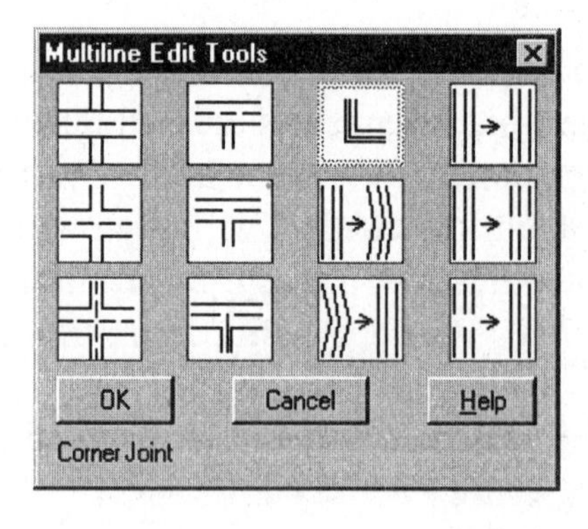

Figure 8.5 Multiline Edit Tools dialog box

[Corner Joint]

Select first mline: [**Select A (Figure 8.4).**]
Select second mline: [**Select B (Figure 8.4).**]
Select first mline(or Undo): **[Enter]**

Command: [**Enter**]
MLEDIT

[**Open Tee**]

Select first mline: [**Select C (Figure 8.4).**]
Select second mline: [**Select D (Figure 8.4).**]
Select first mline: [**Select E (Figure 8.4).**]
Select second mline: [**Select F (Figure 8.4).**]
Select first mline: [**Select G (Figure 8.4).**]
Select second mline: [**Select H (Figure 8.4).**]
Select first mline: [**Select J (Figure 8.4).**]
Select second mline: [**Select K (Figure 8.4).**]
Select first mline(or Undo): [**Enter**]

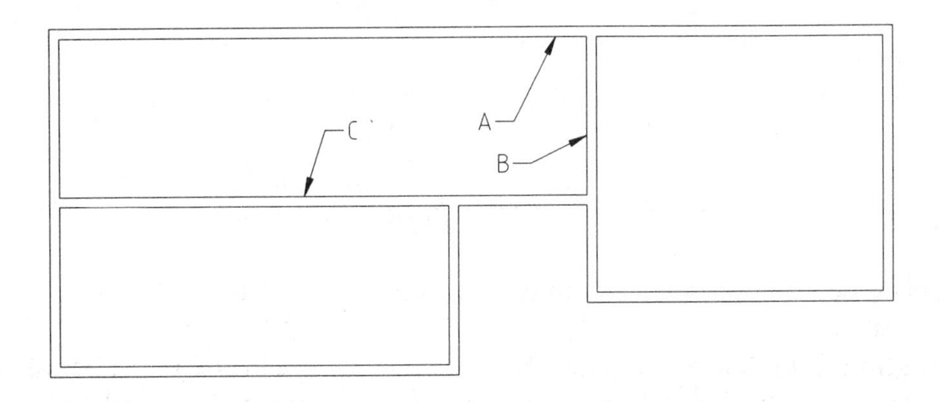

Figure 8.6 Five joints edited

You have created the main walls of the first level. Run the MLINE command again to create the thinner walls. This time, use the 120WALL style. See Figure 8.7.

<Draw> **<Multiline>**

Command: **MLINE**
Justification/Scale/STyle/<From point>: **ST**
Mstyle name (or ?): **120WALL**
Justification/Scale/STyle/<From point>: **J**
Top/Zero/Bottom: **T**
Justification/Scale/STyle/<From point>: **FROM**
Base point: **END** of [**Select at A (Figure 8.6).**]
<Offset>: **@900<180**
<To point>: **@1550<270**
Undo/<To point>: **PERP** to [**Select at B (Figure 8.6).**]
Close/Undo/<To point>: [**Enter**]

Command: [**Enter**]
MLINE
Justification/Scale/STyle/<From point>: **FROM**
Base point: **END** of [**Select at A (Figure 8.6).**]
<Offset>: **@3720<180**
<To point>: **@1670<270**
Undo/<To point>: **@750<180**

Close/Undo/<To point>: **PERP** to **[Select at C (Figure 8.6).]**
Close/Undo/<To point>: **[Enter]**

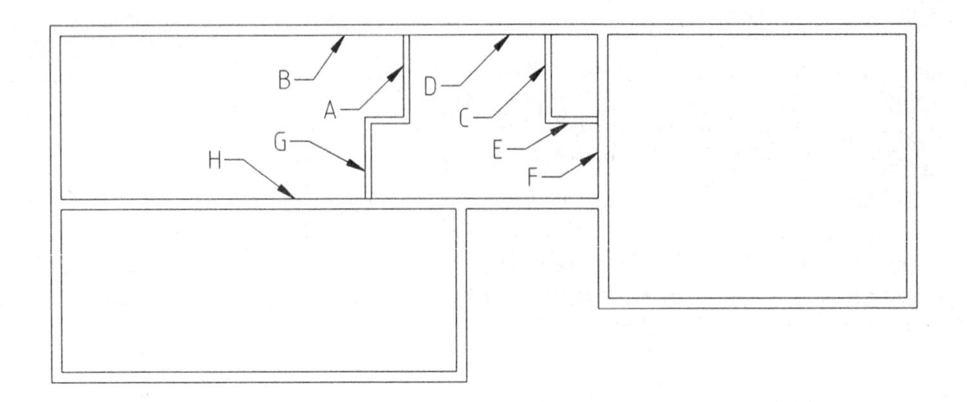

Figure 8.7 Two more multilines created

Again, you need to edit the joints between the multilines. Run the MLEDIT command. The 2D walls of the first level are complete. See Figure 8.8.

<Modify> **<Object>** **<Multiline...>**

Command: **MLEDIT**

	First mline	Second mline
Open Tee	A	B
Open Tee	C	D
Open Tee	E	F
Open Tee	G	H

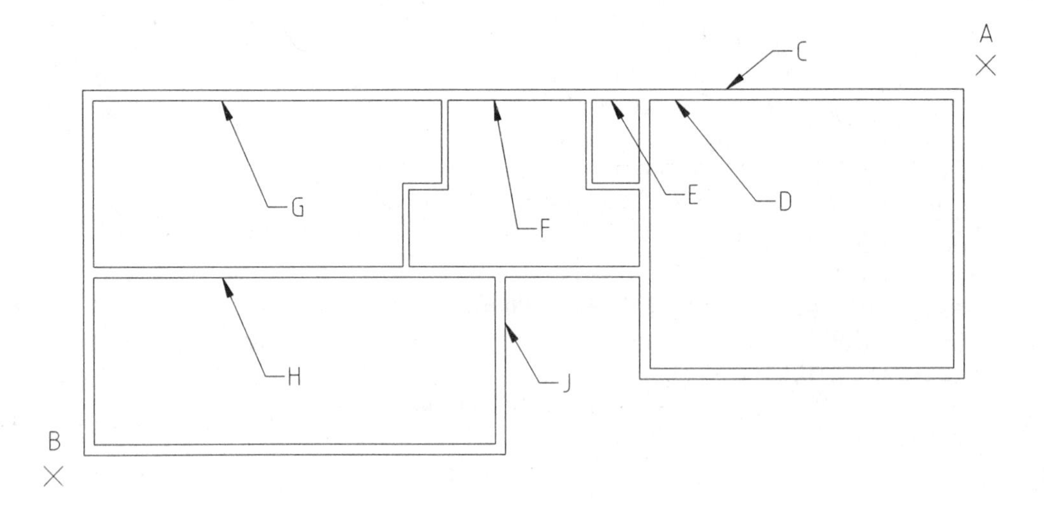

Figure 8.8 Multilines edited

If you want a 2D floor plan, then the first level is complete. Because you want to extrude the wireframes to 3D model, you have to explode the multilines to become separate line segments (using the EXPLODE command) and have to put them together to form regions.

<Modify> **<Explode**

Command: **EXPLODE**
Select objects: [**Select at A (Figure 8.8).**]
Other corner: [**Select at B (Figure 8.8).**]
Select objects: [**Enter**]

The system variable DELOBJ determines whether or not the original objects are deleted after they are used in an operation. Set its value to 1, so that the originals are deleted.

Command: **DELOBJ**
New value for DELOBJ: **1**

Form regions from the line segments with the REGION command. Then use the SUBTRACT command to create a complex region for extrusion to become a three-dimensional solid model.

<Draw> **<Region>**

Command: **REGION**
Select objects: [**Select at A (Figure 8.8).**]
Other corner: [**Select at B (Figure 8.8).**]
Select objects: [**Enter**]
6 loops extracted.
6 Regions created.

<Modify> **<Boolean>** **<Subtract>**

Command: **SUBTRACT**
Select solids and regions to subtract from...
Select objects: [**Select C (Figure 8.8).**]
Select objects: [**Enter**]
Select solids and regions to subtract...
Select objects: [**Select D, E, F, G, and H (Figure 8.8).**]
Select objects: [**Enter**]

After exploding, region making, and Boolean operation, there should be no visual changes on the screen. The drawing on the screen should be the same as before.

Now the wireframes for the walls of the first level are complete. To continue with the first level, set the current layer to F1 by using the LAYER command. Then create a 2D region for the floor.

[Object Properties] **[Layer Control]**

On	F1
On	F2
On	Roof

On	Stair
On	W1
On	W2

Current layer: **F1**

Run the RECTANG command to create two rectangles. See Figure 8.9.

<Draw> **<Rectangle>**

Command: **RECTANG**
Chamfer/Elevation/Fillet/Thickness/Width/<First corner>: **0,0**
Other corner: **8200,6800**

Command: **[Enter]**
RECTANG
Chamfer/Elevation/Fillet/Thickness/Width/<First corner>: **END** of **[Select J (Figure 8.8).]**
Other corner: **.X** of **END** of of **[Select D (Figure 8.8).]**
(need Y): **END** of of **[Select C (Figure 8.8).]**

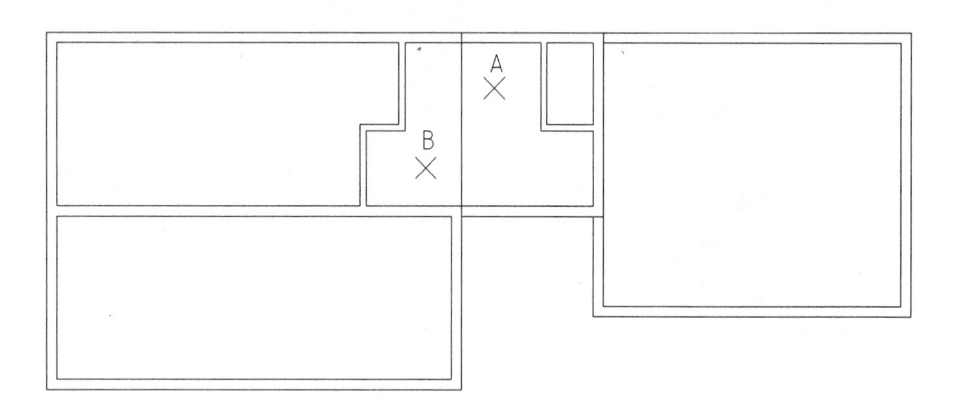

Figure 8.9 Two rectangles created

Run the REGION command to convert the rectangles into regions. Then run the UNION command to unite them to form a complex region. See Figure 8.10.

<Draw> **<Region>**

Command: **REGION**
Select objects: **[Select at A (Figure 8.9).]**
Other corner: **[Select at B (Figure 8.9).]**
Select objects: **[Enter]**
2 loops extracted.
2 Regions created.

<Modify> **<Boolean>** **<Union>**

Command: **UNION**
Select objects: **[Select at A (Figure 8.9).]**
Other corner: **[Select at B (Figure 8.9).]**
Select objects: **[Enter]**

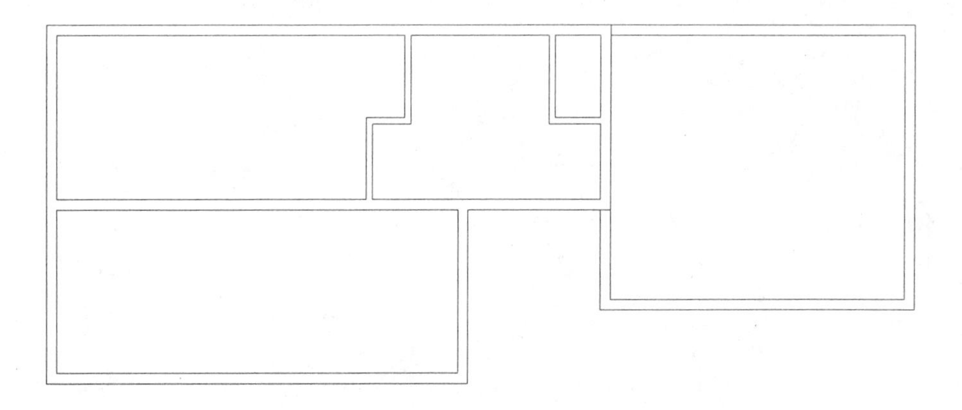

Figure 8.10 Floor wireframes created

The 2D wireframes for the walls and floor of the first floor are complete. You will run the EXTRUDE command on them later.

8.3 Wireframes for the Staircase

The staircase runs in three directions. You can divide it into three sections and create three sets of wireframes. With the wireframes, you will form three extruded solids and unite them to form a model of the staircase. Set the current layer to STAIR, which will hold the wireframes for the staircase.

[Object Properties] **[Layer Control]**

On	F1
On	F2
On	Roof
On	Stair
On	W1
On	W2

Current layer: **Stair**

Unlike the wireframes for the walls and floor that lie on the XY plane of the WCS, the wireframes for the staircase lie on the ZY plane. Therefore, you should set the viewing position to a 3D view. Set the display to southeast isometric view, and place the UCS icon to display at the origin position by using the UCSICON command. See Figure 8.11.

<View> **<3D Viewpoint>** **<SE Isometric>**

<View> **<Display>** **<UCS Icon>** **<v On>**

<View> **<Display>** **<UCS Icon>** **<v Origin>**

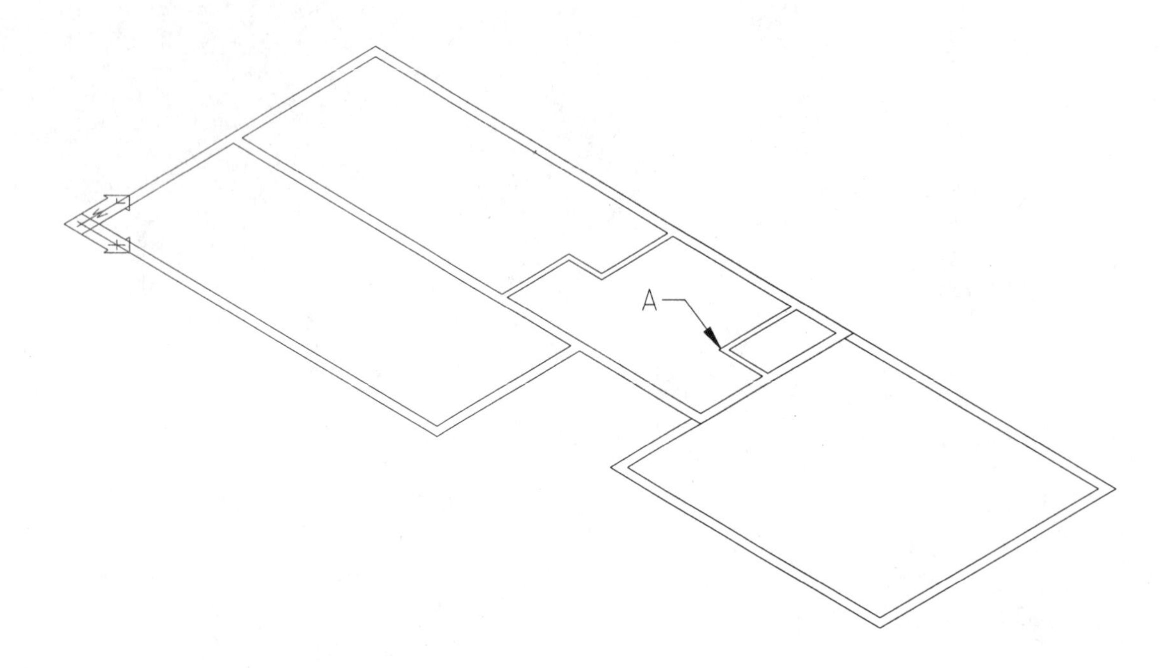

Figure 8.11 New viewing position

Set the UCS to a new position using the UCS command. Place the origin at a point 180 units in the Z-direction from the point A (Figure 8.11), and set the orientation of the Z axis to point to the current X-direction. See Figure 8.12.

<Tools> **<UCS>** **<Z Axis Vector>**

Command: **UCS**
Origin/ZAxis/3point/OBject/View/X/Y/Z/Prev/Restore/Save/Del/?/<World>: **ZAXIS**
Origin point: **FROM**
Base point: **END** of [**Select A (Figure 8.11).**]
<Offset>: **@0,0,180**
Point on positive portion of Z-axis: **@1,0**

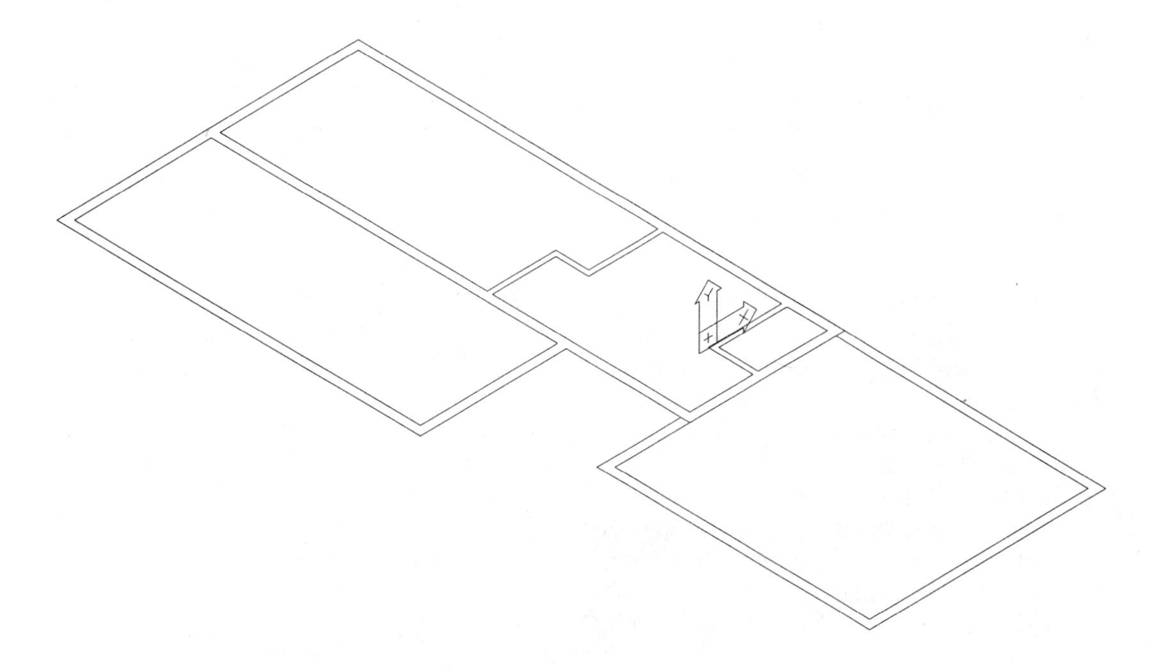

Figure 8.12 New UCS position

For your reference, the dimensions of the three sections of the staircase are shown in Figure 8.13.

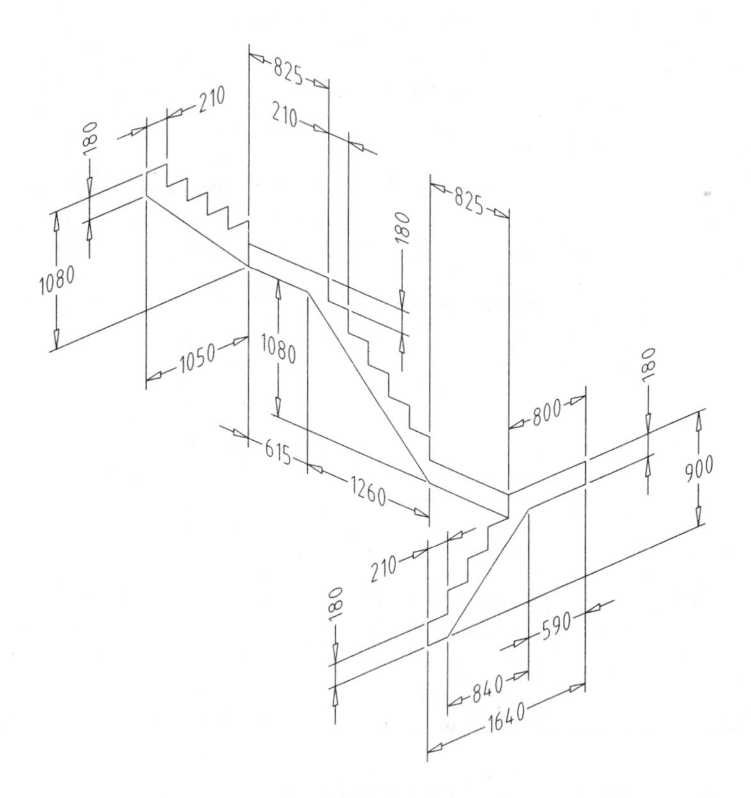

Figure 8.13 Staircase in three sections

Using the dimensions shown in Figure 8.13, run the PLINE command to create a polyline for the first section of the staircase. See Figure 8.14.

<Draw> **<Polyline>**

Command: **PLINE**
From point: **30,0**
Arc/Close/Halfwidth/Length/Undo/Width/<Endpoint of line>: **@180<90**
Arc/Close/Halfwidth/Length/Undo/Width/<Endpoint of line>: **@210<0**
Arc/Close/Halfwidth/Length/Undo/Width/<Endpoint of line>: **@180<90**
Arc/Close/Halfwidth/Length/Undo/Width/<Endpoint of line>: **@210<0**
Arc/Close/Halfwidth/Length/Undo/Width/<Endpoint of line>: **@180<90**
Arc/Close/Halfwidth/Length/Undo/Width/<Endpoint of line>: **@210<0**
Arc/Close/Halfwidth/Length/Undo/Width/<Endpoint of line>: **@180<90**
Arc/Close/Halfwidth/Length/Undo/Width/<Endpoint of line>: **@210<0**
Arc/Close/Halfwidth/Length/Undo/Width/<Endpoint of line>: **@180<90**
Arc/Close/Halfwidth/Length/Undo/Width/<Endpoint of line>: **@800<0**
Arc/Close/Halfwidth/Length/Undo/Width/<Endpoint of line>: **@180<270**
Arc/Close/Halfwidth/Length/Undo/Width/<Endpoint of line>: **@590<180**
Arc/Close/Halfwidth/Length/Undo/Width/<Endpoint of line>: **210,0**
Arc/Close/Halfwidth/Length/Undo/Width/<Endpoint of line>: **C**

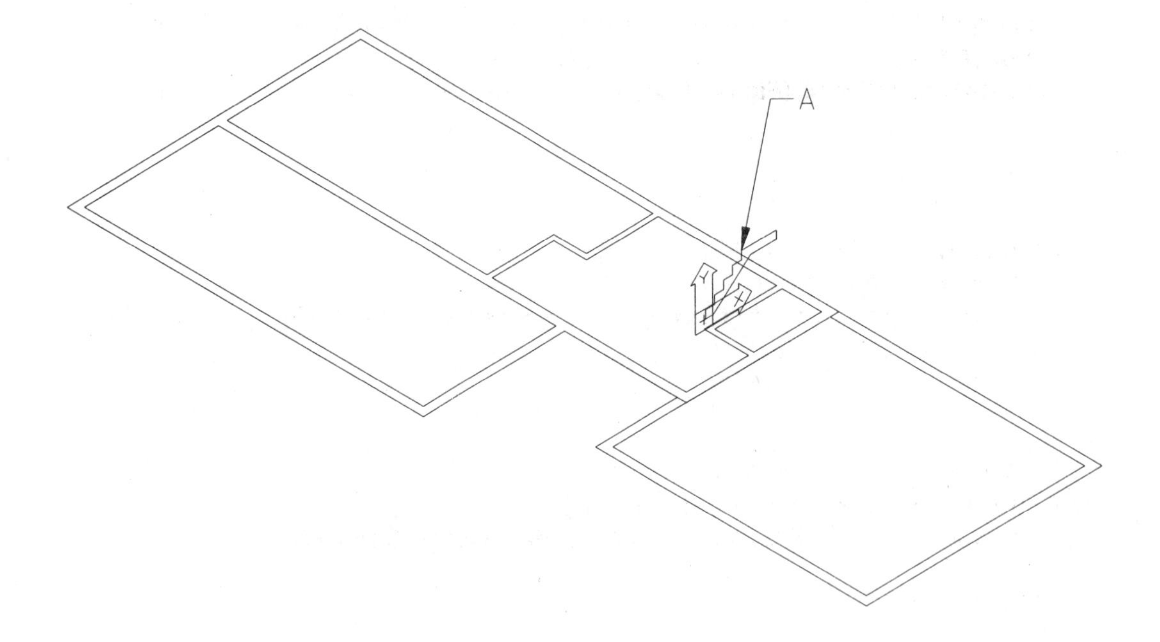

Figure 8.14 Wireframe for the first section of the staircase drawn

Set the UCS to a new position by using the UCS command. Then run the PLINE command to create the wireframe for the second section of the staircase. See Figure 8.15.

<Tools> **<UCS>** **<Z Axis Vector>**

Command: **UCS**
Origin/ZAxis/3point/OBject/View/X/Y/Z/Prev/Restore/Save/Del/?/<World>: **ZAXIS**
Origin point: **END** of [**Select A (Figure 8.14)**.]
Point on positive portion of Z-axis: **@-1.0**

<Draw> **<Polyline>**

Command: **PLINE**
From point: **0,0**
Arc/Close/Halfwidth/Length/Undo/Width/<Endpoint of line>: **@825<180**
Arc/Close/Halfwidth/Length/Undo/Width/<Endpoint of line>: **@180<90**
Arc/Close/Halfwidth/Length/Undo/Width/<Endpoint of line>: **@210<180**
Arc/Close/Halfwidth/Length/Undo/Width/<Endpoint of line>: **@180<90**
Arc/Close/Halfwidth/Length/Undo/Width/<Endpoint of line>: **@210<180**
Arc/Close/Halfwidth/Length/Undo/Width/<Endpoint of line>: **@180<90**
Arc/Close/Halfwidth/Length/Undo/Width/<Endpoint of line>: **@210<180**
Arc/Close/Halfwidth/Length/Undo/Width/<Endpoint of line>: **@180<90**
Arc/Close/Halfwidth/Length/Undo/Width/<Endpoint of line>: **@210<180**
Arc/Close/Halfwidth/Length/Undo/Width/<Endpoint of line>: **@180<90**
Arc/Close/Halfwidth/Length/Undo/Width/<Endpoint of line>: **@210<180**
Arc/Close/Halfwidth/Length/Undo/Width/<Endpoint of line>: **@180<90**
Arc/Close/Halfwidth/Length/Undo/Width/<Endpoint of line>: **@825<180**
Arc/Close/Halfwidth/Length/Undo/Width/<Endpoint of line>: **@180<270**
Arc/Close/Halfwidth/Length/Undo/Width/<Endpoint of line>: **@615<0**
Arc/Close/Halfwidth/Length/Undo/Width/<Endpoint of line>: **-825,-180**
Arc/Close/Halfwidth/Length/Undo/Width/<Endpoint of line>: **0,-180**
Arc/Close/Halfwidth/Length/Undo/Width/<Endpoint of line>: **C**

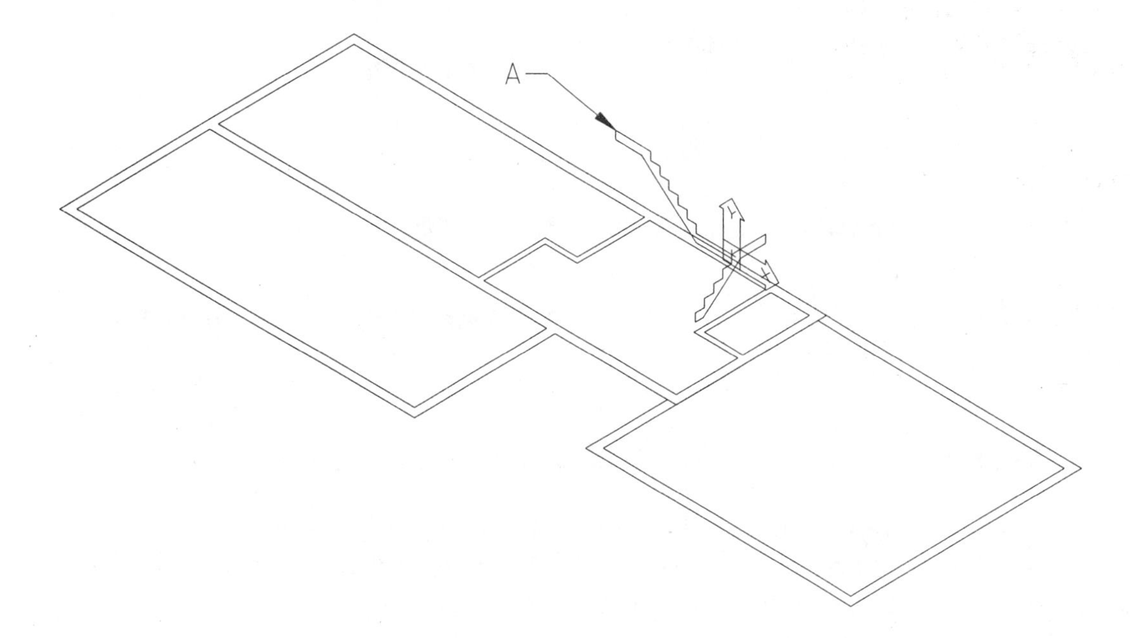

Figure 8.15 Wireframe for the second section of the staircase drawn

Before constructing the third section of the staircase, set the UCS to a new position by using the UCS command.

<Tools> **<UCS>** **<Z Axis Vector>**

Command: **UCS**
Origin/ZAxis/3point/OBject/View/X/Y/Z/Prev/Restore/Save/Del/?/<World>: **ZAXIS**

Origin point: **END** of **[Select A (Figure 8.15).]**
Point on positive portion of Z-axis: **@1,0**

By referring to the dimensions shown in Figure 8.13, use the PLINE command to create the wireframe for the third section of the staircase. The outcome should resemble Figure 8.16.

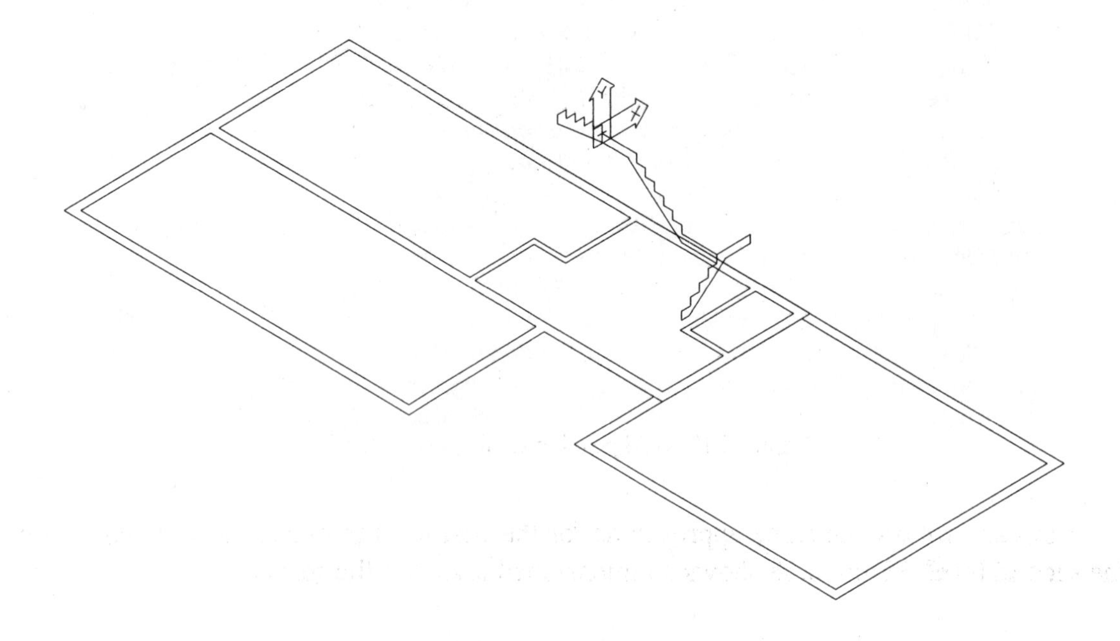

Figure 8.16 Wireframe for the third section of the staircase drawn

The 3D wireframes for the staircase of the house are complete. You will extrude them to become solids later.

8.4 Wireframes for the Second Level

On the second level of the house, there are also two object types: walls and floor. This level is similar to the first level, except that the XY plane lies on the positive direction of the Z axis of the WCS. Run the UCS command. See Figure 8.17.

<Tools> **<UCS>** **<World>**

Command: **UCS**
Origin/ZAxis/3point/OBject/View/X/Y/Z/Prev/Restore/Save/Del/?/<World>: **W**

<Tools> **<UCS>** **<Origin>**

Command: **UCS**
Origin/ZAxis/3point/OBject/View/X/Y/Z/Prev/Restore/Save/Del/?/<World>:**O**
Origin point: **0,0,2880**

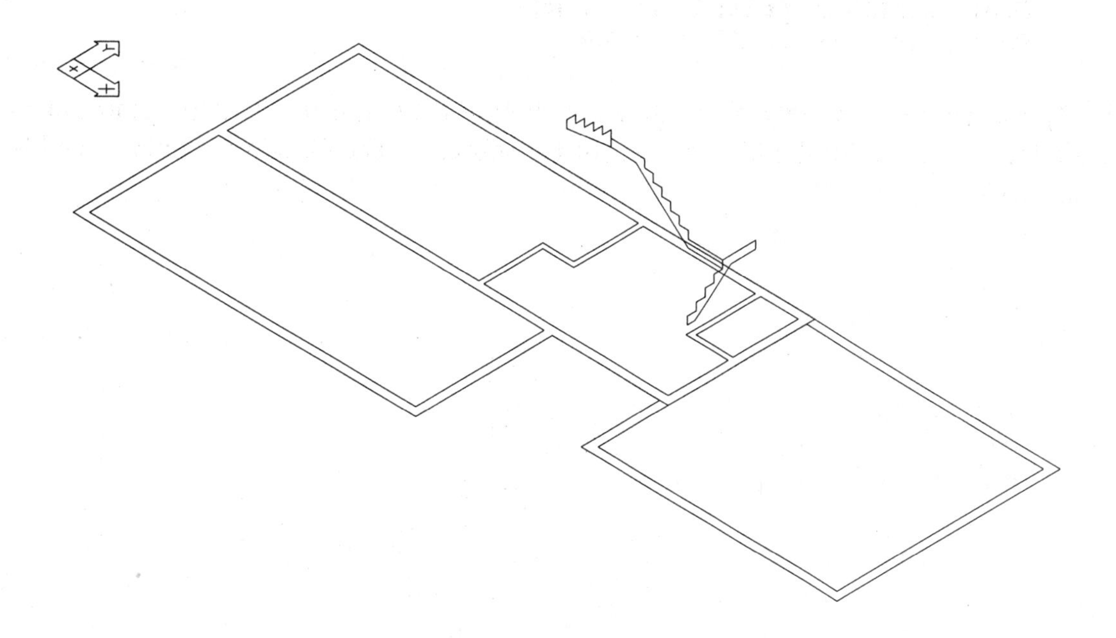

Figure 8.17 UCS position for the second level

You may follow the same approach as for the first level to create the wireframes for the second level. Figure 8.18 shows a dimensioned layout of the second level.

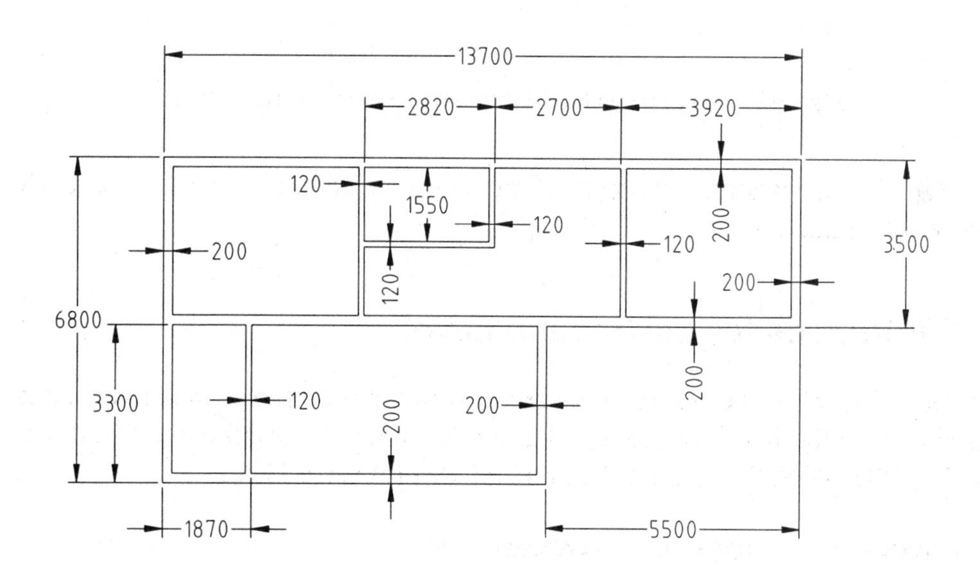

Figure 8.18 Dimensioned layout for the second level

Use the PLAN command to set the viewing direction to the plan view of the current UCS.

<View> **<3D Viewpoint>** **<Plan View>** **<Current UCS>**

Command: **PLAN**

<Current UCS>/Ucs/World: **[Enter]**

Turn off layers W1, F1, and Stair. Then set the current layer to W2. You will place the walls of the second floor on this layer.

[Object Properties] **[Layer Control]**

Off	F1
On	F2
On	Roof
Off	Stair
Off	W1
On	W2

Current layer: **W2**

Refer to the dimensioned drawing to create the layout drawings for the second level. First, use the MLINE command to draw the necessary multilines. Then edit the joints of the multilines by using the MLEDIT command. Next, explode the multilines into separate line segments by using the EXPLODE command. Finally, convert the line segments into a series of regions by using the REGION command, and run the SUBTRACT command on the regions to form a complex region. You will extrude the complex region later. See Figure 8.19.

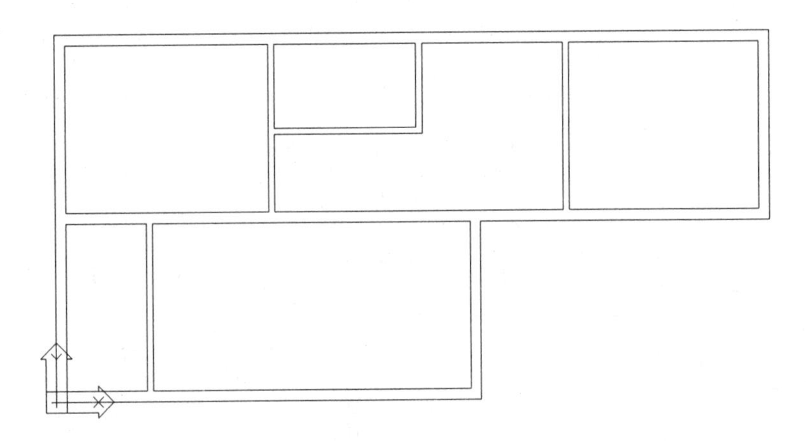

Figure 8.19 Multilines edited, exploded, and converted into regions and then into a complex region

The wireframes for the walls of the second level are complete. Now, you will work on the floor of the second level. Turn on layer Stair and W1, turn off layer W2, and set the current layer to F2.

[Object Properties] **[Layer Control]**

Off	F1
On	F2
On	Roof
On	Stair
On	W1

Off	W2

Current layer: **F2**

To create the wireframes for the floor of the second level, use the RECTANG command to create three rectangles.

<Draw> **<Rectangle>**

Command: **RECTANG**

Rectangle	First corner	Other corner
A	0,0	8200,6800
B	8200,3300	13700,6800
C	7080,4960	9780,6800

Set the display to a southeast isometric view. Check your drawing against Figure 8.20.

<View> **<3D Viewpoint>** **<SE Isometric>**

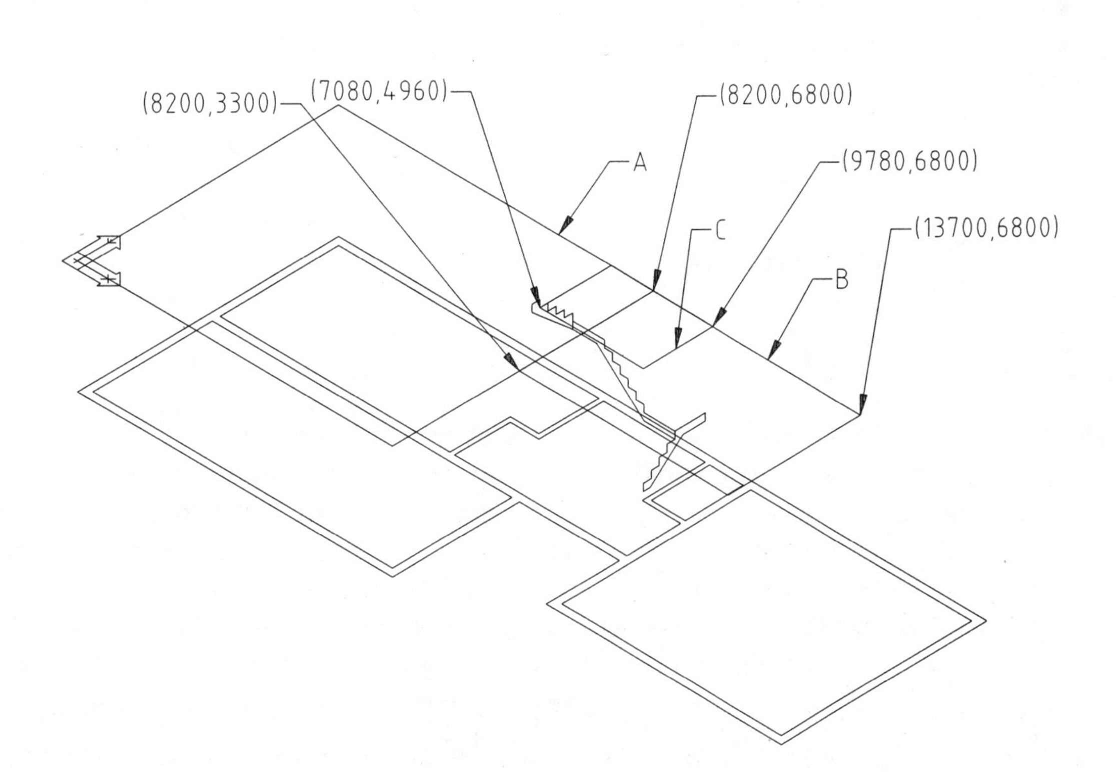

Figure 8.20 Three rectangles created

Use the REGION command to convert the rectangles to three regions. Then run the UNION command to unite the rectangles A and B (Figure 8.20) into a single region. Next, use the SUBTRACT command to subtract the rectangle C (Figure 8.20) from the united region. The wireframe of the floor for the second level is complete. See Figure 8.21.

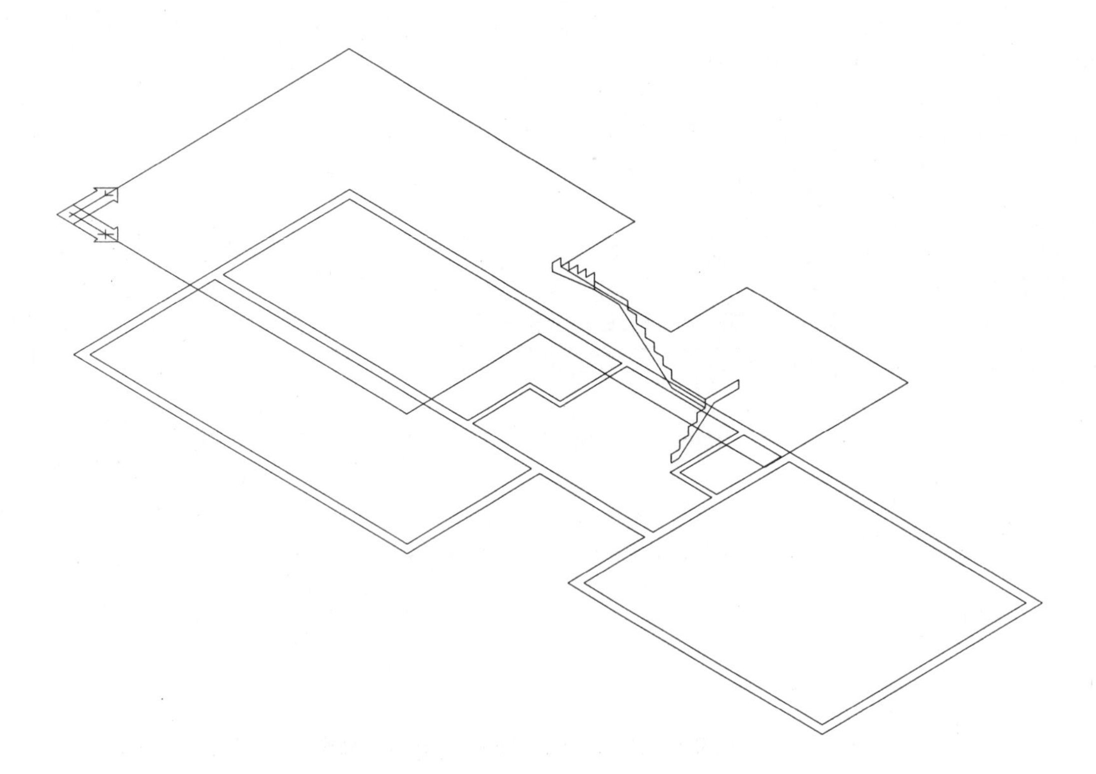

Figure 8.21 Wireframe for the floor of the second level created

8.5 Wireframes for the Roofs

There are two roofs: one for the first level and another for the second level. The roof for the second level consists of two parts. In total, you will need to create seven wireframes in three groups. From the wireframes, you will create seven extruded solids. The roof of the first level is formed by the intersection of three extruded solids. The roof of the second level consists of two parts. Each part is the intersection of two extruded solids. The completed roof of the second level is the union of two solids of intersection.

Figure 8.22 shows the dimensions of the wireframes for the roofs. In the figure, the origin positions of the UCS for the three sets of wireframes are delineated in terms of WCS coordinates. For your reference, the lower left corner of the floor of the first level is at (0,0,0) of the WCS.

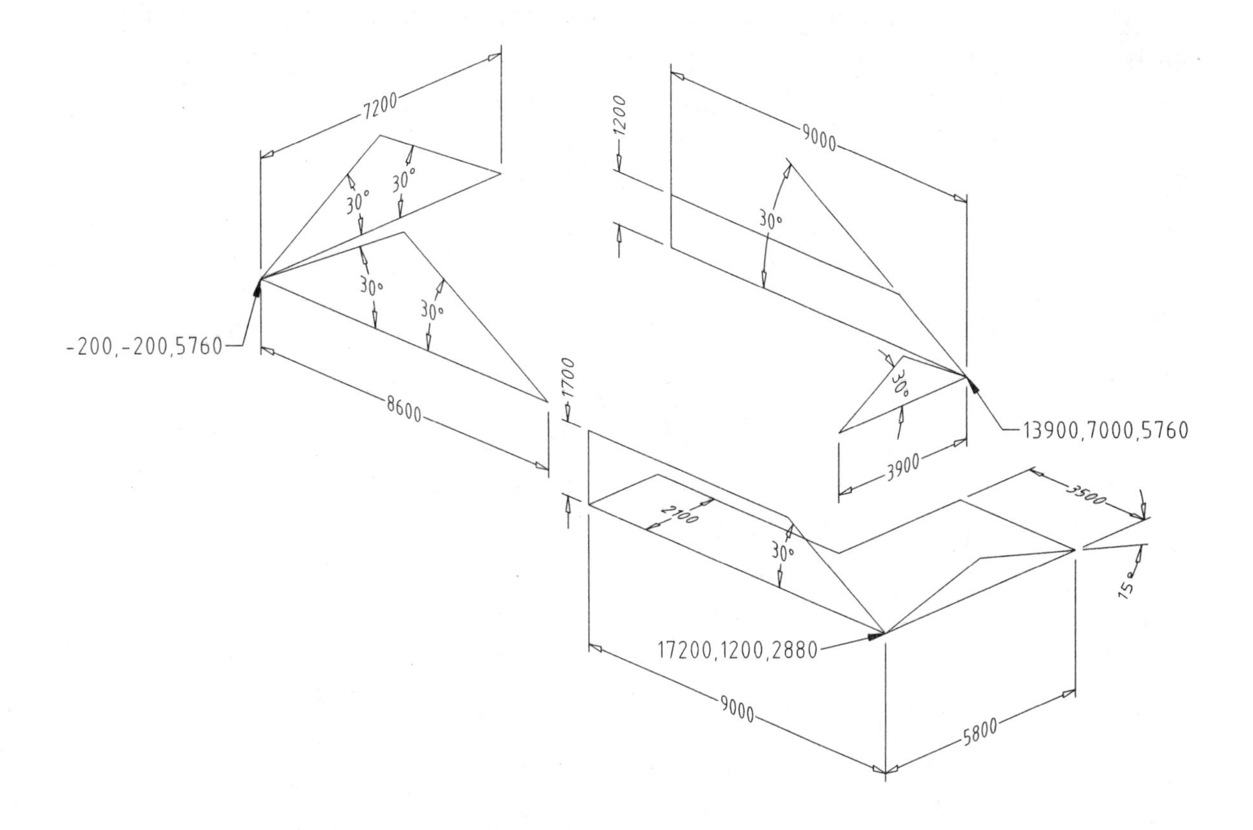

Figure 8.22 Dimensioned layout of the roofs

Set the current layer to ROOF, turn on layer F1, and turn off the layers STAIR, W1, W2, and F2. On your screen, you will have the entities on layer F1 left behind.

[Object Properties] **[Layer Control]**

On	F1
Off	F2
On	Roof
Off	Stair
Off	W1
Off	W2

Current layer: **Roof**

Set the UCS to a new position by using the UCS command. The symbol (*) refers to the absolute WCS coordinates regardless of current UCS position.

<Tools> **<UCS>** **<Z Axis Vector>**

Command: **UCS**
Origin/ZAxis/3point/OBject/View/X/Y/Z/Prev/Restore/Save/Del/?/<World>: **ZA**
Origin point: ***-200,-200,5760**
Point on positive portion of Z-axis: **@1,0**

According to the dimensions shown in Figure 8.22, create a 2D wireframe on the XY plane of the new UCS. See Figure 8.23.

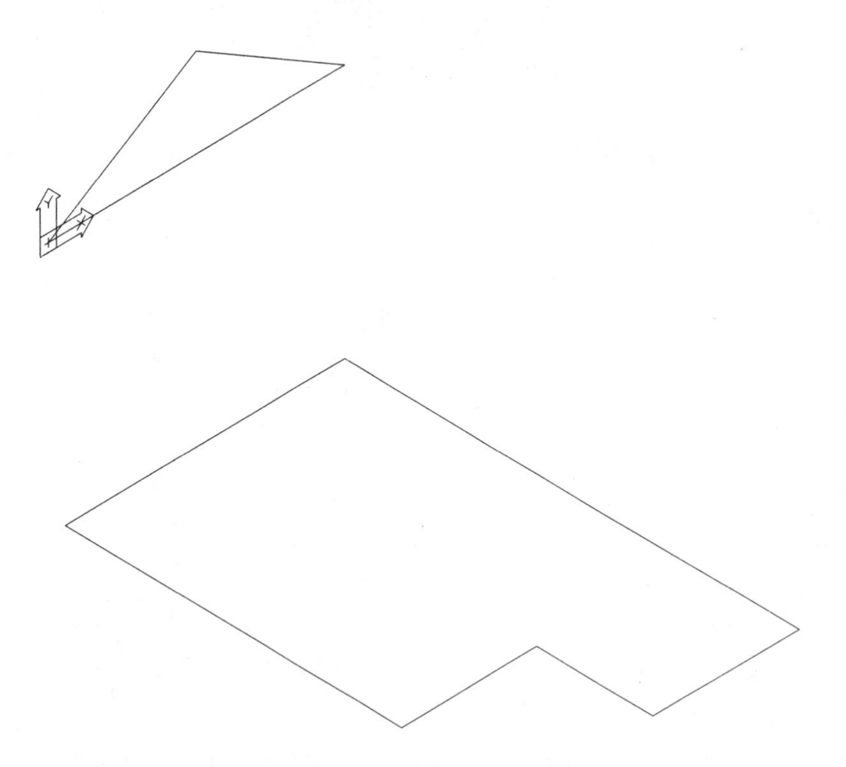

Figure 8.23 First wireframe for the roofs created

Set the UCS to rotate -90° about the current Y axis with the UCS command.

<Tools> **<UCS>** **<Y Axis Rotate>**

Command: **UCS**
Origin/ZAxis/3point/OBject/View/X/Y/Z/Prev/Restore/Save/Del/?/<World>: **Y**
Rotation angle about Y axis: **-90**

Draw a wireframe on this new UCS location. See Figure 8.24.

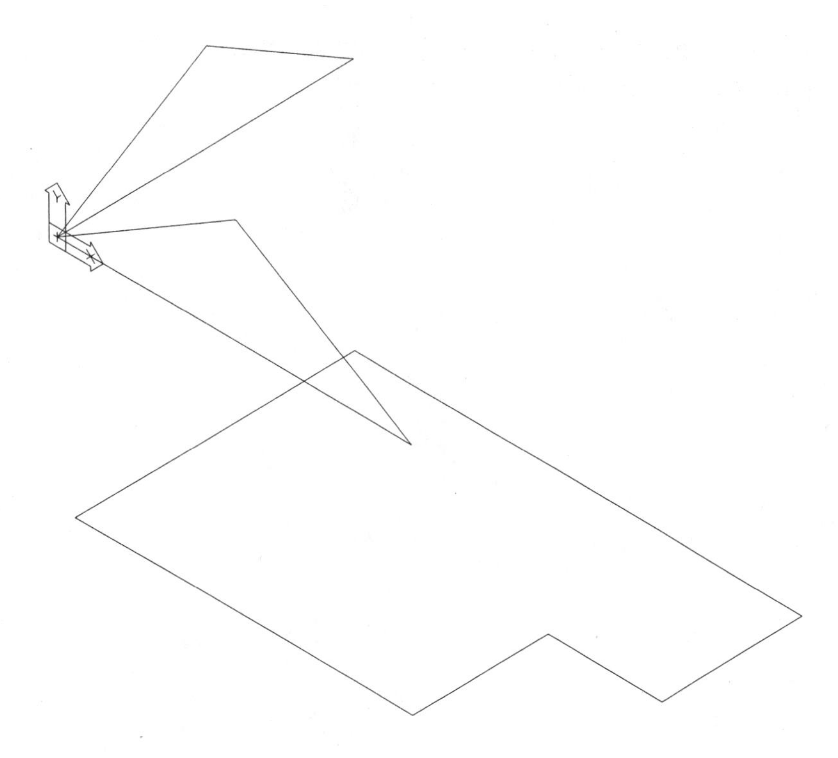

Figure 8.24 Second wireframe for the roofs created

Set the UCS location again for the second part of the roof for the second level.

<Tools> **<UCS>** **<Z Axis Vector>**

Command: **UCS**
Origin/ZAxis/3point/OBject/View/X/Y/Z/Prev/Restore/Save/Del/?/<World>: **ZA**
Origin point: ***13900,7000,5760**
Point on positive portion of Z-axis: **@1,0**

Refer to the dimensioned drawing shown in Figure 8.22 to create the third wireframe. See Figure 8.25.

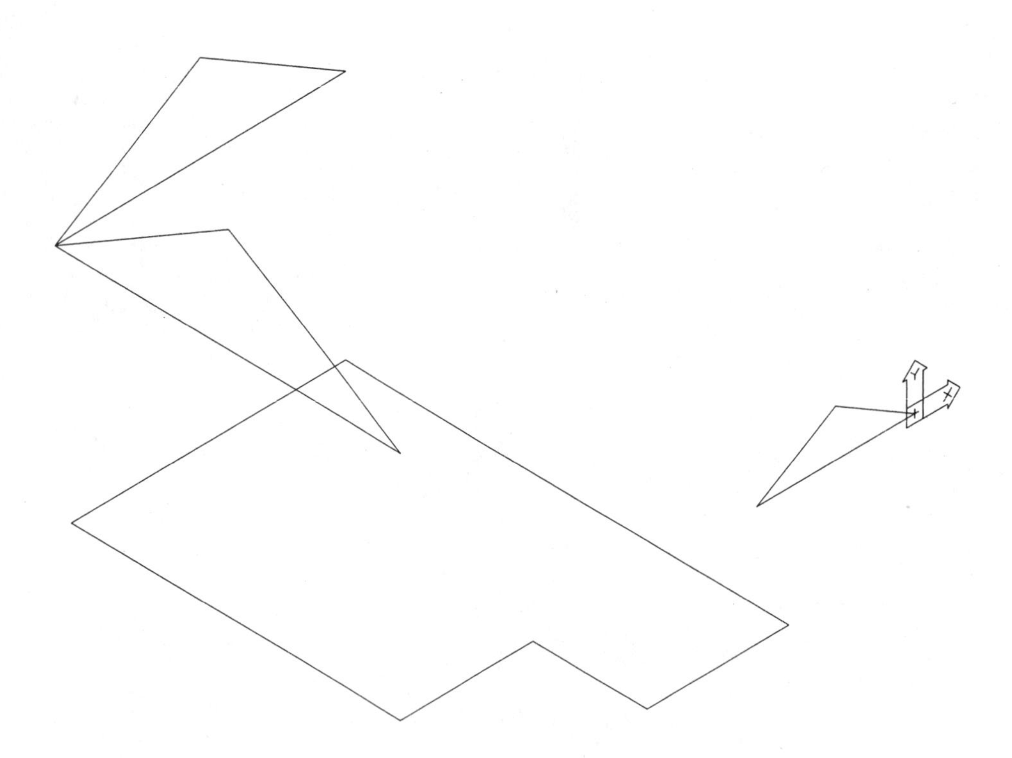

Figure 8.25 Third wireframe for the roofs created

Set the UCS again with the UCS command. Rotate it -90° about the current Y axis.

<Tools> **<UCS>** **<Y Axis Rotate>**

Command: **UCS**
Origin/ZAxis/3point/OBject/View/X/Y/Z/Prev/Restore/Save/Del/?/<World>: **Y**
Rotation angle about Y axis: **-90**

Create the fourth wireframe. See Figure 8.26.

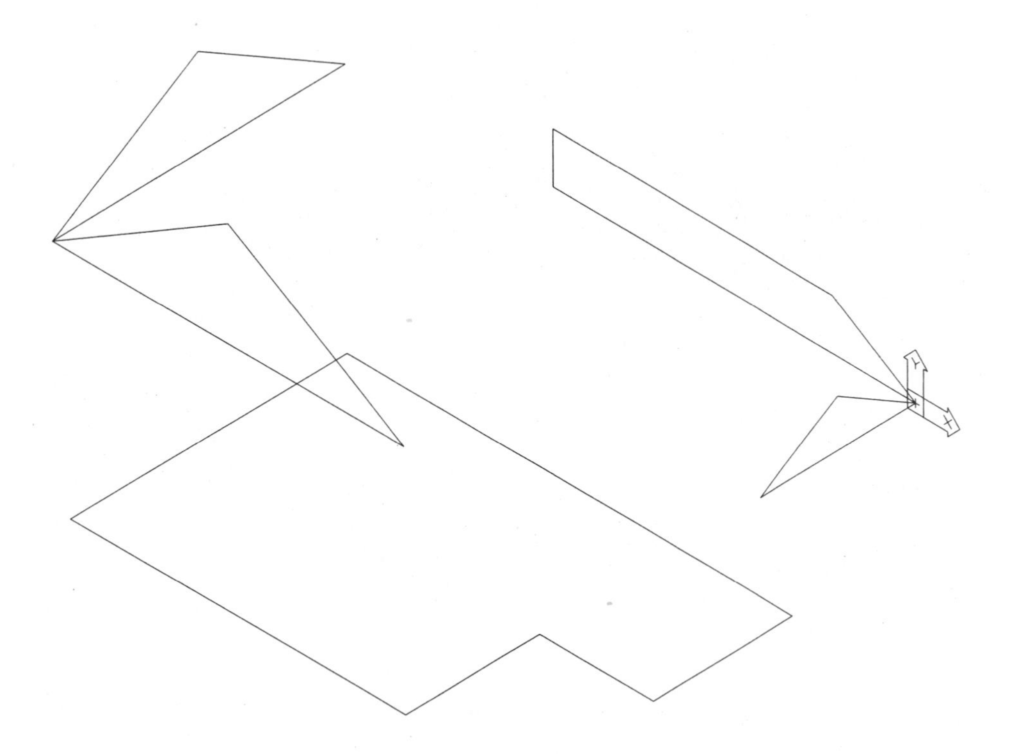

Figure 8.26 Fourth wireframe for the roofs created

Before proceeding to making the wireframes for the roof of the first level, turn off layer F1 so that only those entities related to the roof are left on the screen.

[Object Properties] **[Layer Control]**

Off	F1
Off	F2
On	Roof
Off	Stair
Off	W1
Off	W2

Current layer: **Roof**

Switch to the roof for the first level. Just as for the other two sets of wireframes, set the UCS to a new position.

<Tools> **<UCS>** **<Z Axis Vector>**

Command: **UCS**
Origin/ZAxis/3point/OBject/View/X/Y/Z/Prev/Restore/Save/Del/?/<World>: **ZA**
Origin point: ***17200,1200,2880**
Point on positive portion of Z-axis: **@1,0**

Refer to Figure 8.22 to create the fifth wireframe. See Figure 8.27.

Figure 8.27 Fifth wireframe for the roofs created

Set the UCS to a new position again by using the UCS command.

<Tools> **<UCS>** **<Y Axis Rotate>**

Command: **UCS**
Origin/ZAxis/3point/OBject/View/X/Y/Z/Prev/Restore/Save/Del/?/<World>: **Y**
Rotation angle about Y axis: **-90**

Create the sixth wireframe. See Figure 8.28.

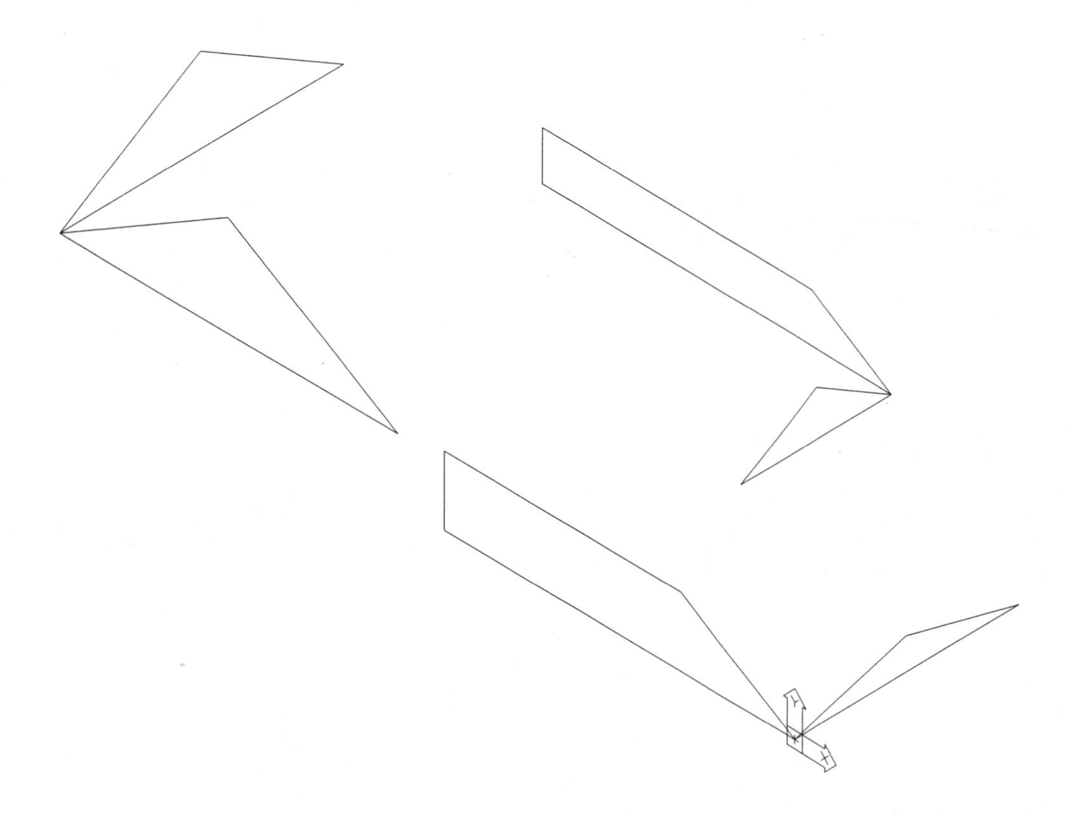

Figure 8.28 Sixth wireframe for the roofs created

Now switch to the last wireframe. Set the UCS again by using the UCS command.

<Tools> **<UCS>** **<X Axis Rotate>**

Command: **UCS**
Origin/ZAxis/3point/OBject/View/X/Y/Z/Prev/Restore/Save/Del/?/<World>: **X**
Rotation angle about X axis <0>: **-90**

Create the seventh wireframe. See Figure 8.29.

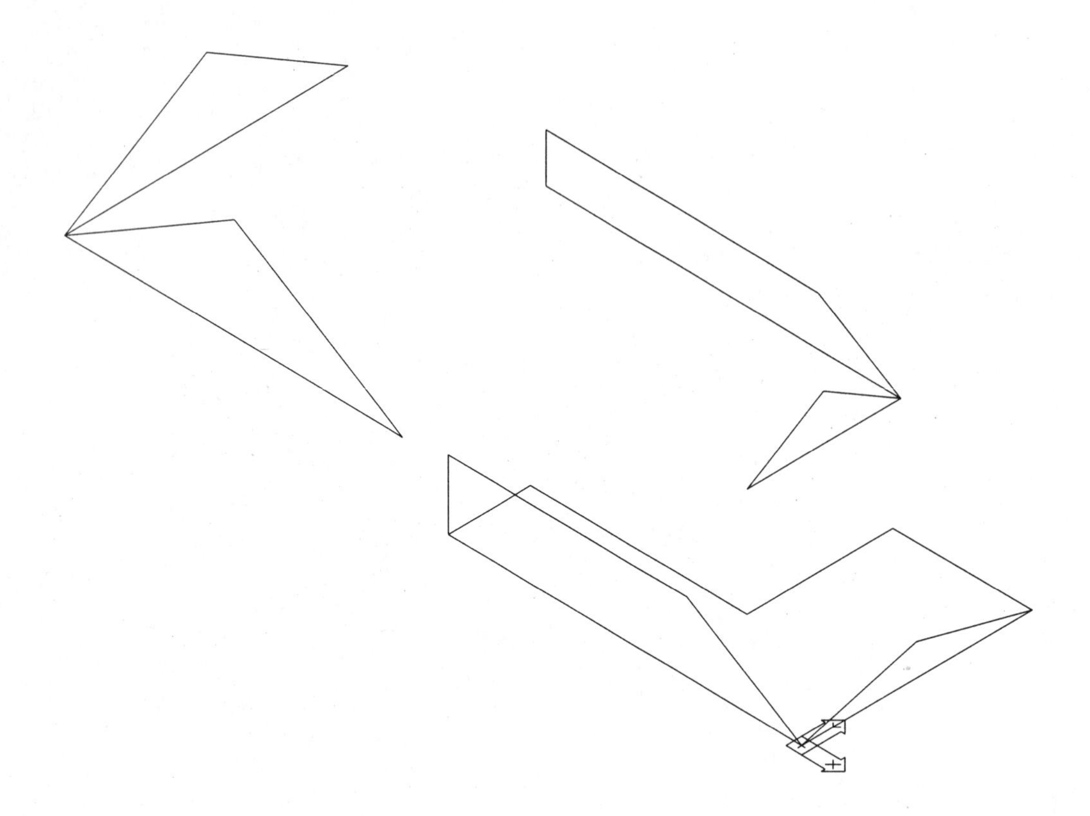

Figure 8.29 Wireframes for the roofs completed

All the wireframes for the model of the house are complete. Reset the UCS to WORLD by applying the UCS command. Then turn on all the layers. See Figure 8.30.

<Tools> **<UCS>** **<World>**

Command: **UCS**
Origin/ZAxis/3point/OBject/View/X/Y/Z/Prev/Restore/Save/Del/?/<World>: **[Enter]**

[Object Properties] **[Layer Control]**

On	F1
On	F2
On	Roof
On	Stair
On	W1
On	W2

Current layer: **Roof**

All the wireframes are complete.

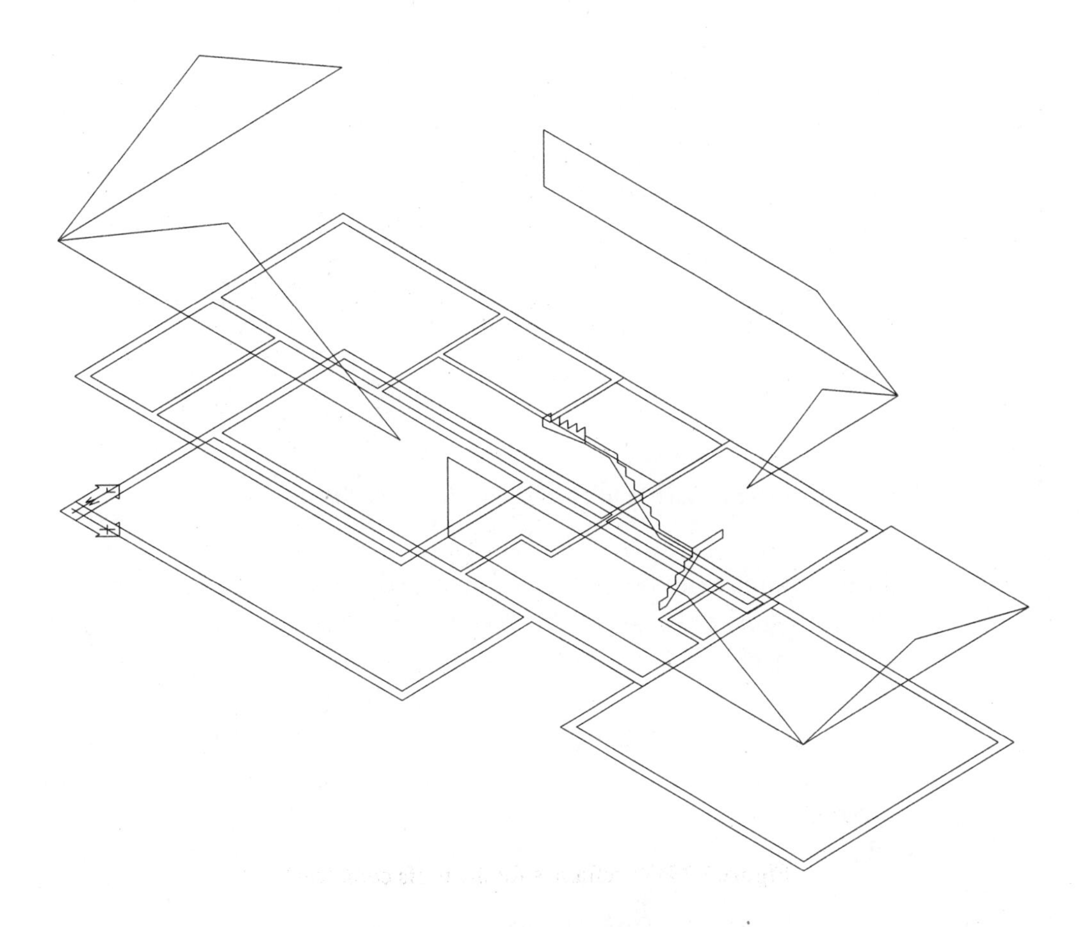

Figure 8.30 Wireframes required for the house

8.6 3D Model for the First Level and the Staircase

After finishing all the wireframes, you can create the model for the first level of the house. You will work first on the staircase, then on the floor, and then on the walls. Before advancing to the second level, you will cut openings for the windows and doors on the walls.

Set the current layer to layer STAIR, and turn off layers ROOF, W1, W2, and F2. See Figure 8.31.

[Object Properties] **[Layer Control]**

On	F1
Off	F2
Off	Roof
On	Stair
Off	W1
Off	W2

Current layer: **Stair**

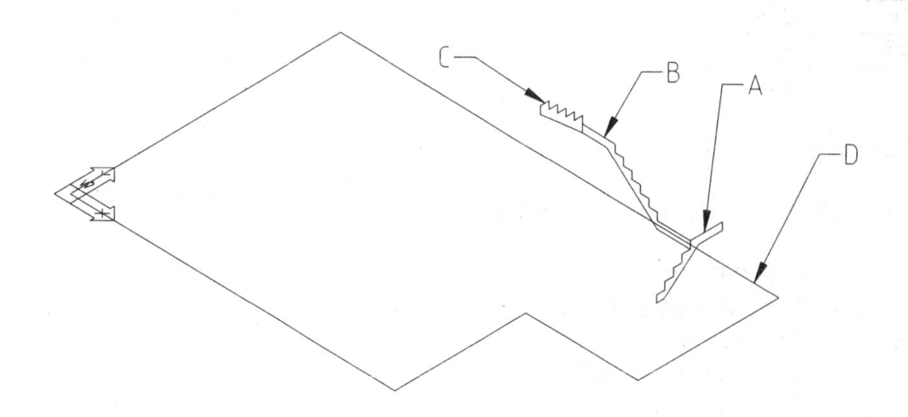

Figure 8.31 Wireframes for the first level and the staircase

The model for the staircase is divided into three sections, and you have created three wireframes for them. To make the staircase, you will extrude them to 3D solids. When you extrude a region or a polyline, the direction of extrusion is in the Z-direction of the UCS in which the region or polyline is created, not the current UCS. Therefore, you may refer to Figures 8.14, 8.15, and 8.16 to check the UCS of the object. For the first and the second sections, the extrusion should be in the negative direction. For the third section, the extrusion should be in the positive direction.

Use the EXTRUDE command to extrude the wireframes for the staircase sections.

<Draw> **<Solids>** **<Extrude>**

Command: **EXTRUDE**
Select objects: **[Select A (Figure 8.31).]**
Select objects: **[Enter]**
Path/<Height of Extrusion>: **-825**
Extrusion taper angle: **0**

Command: **[Enter]**
EXTRUDE
Select objects: **[Select B (Figure 8.31).]**
Select objects: **[Enter]**
Path/<Height of Extrusion>: **-800**
Extrusion taper angle: **0**

Command: **[Enter]**
EXTRUDE
Select objects: **[Select C (Figure 8.31).]**
Select objects: **[Enter]**
Path/<Height of Extrusion>: **825**

Set the current layer to F1 and extrude the wireframes for the floor of the first level.

[Object Properties] **[Layer Control]**

On	F1
Off	F2
Off	Roof

On	Stair
Off	W1
Off	W2

Current layer: **F1**

<Draw> **<Solids>** **<Extrude>**

Command: **EXTRUDE**
Select objects: **[Select D (Figure 8.31).]**
Select objects: **[Enter]**
Path/<Height of Extrusion>: **180**
Extrusion taper angle: **0**

After extruding four wireframes, turn on layer W1 and set it as the current layer. See Figure 8.32.

[Object Properties] **[Layer Control]**

On	F1
Off	F2
Off	Roof
On	Stair
On	W1
Off	W2

Current layer: **W1**

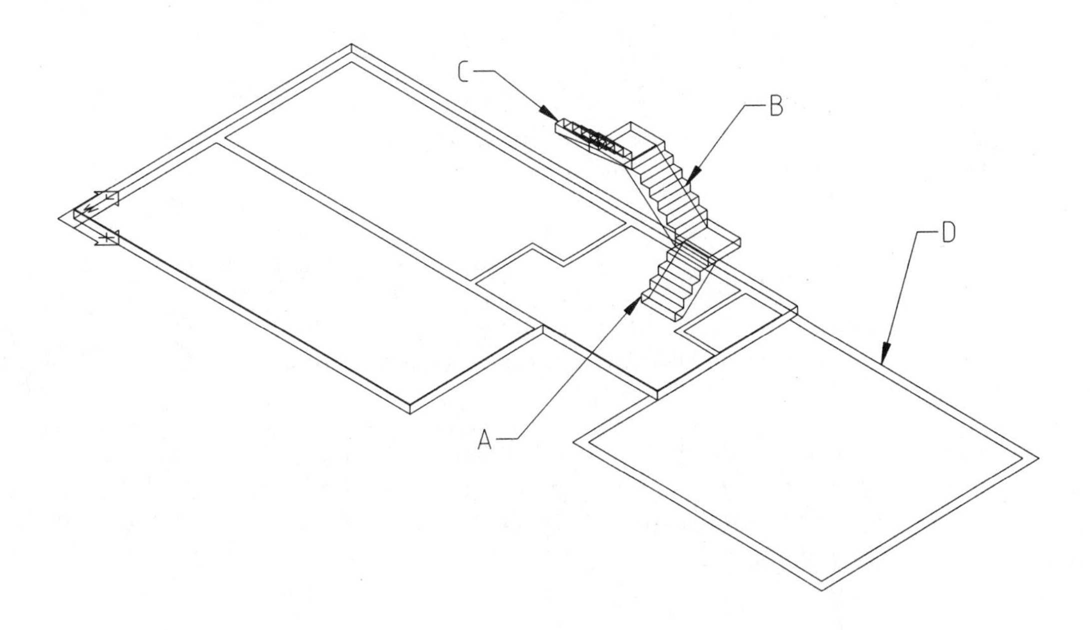

Figure 8.32 Three staircase sections and the floor of the first level created

Unite the three sections of staircase and extrude the wireframes for the walls of the first floor. See Figure 8.33.

<Modify> **<Boolean>** **<Union>**

Command: **UNION**
Select objects: **[Select A, B, and C (Figure 8.32).]**
Select objects: **[Enter]**

<Draw> **<Solids>** **<Extrude>**

Command: **EXTRUDE**
Select objects: **[Select D (Figure 8.32).]**
Select objects: **[Enter]**
Path/<Height of Extrusion>: **2880**
Extrusion taper angle: **0**

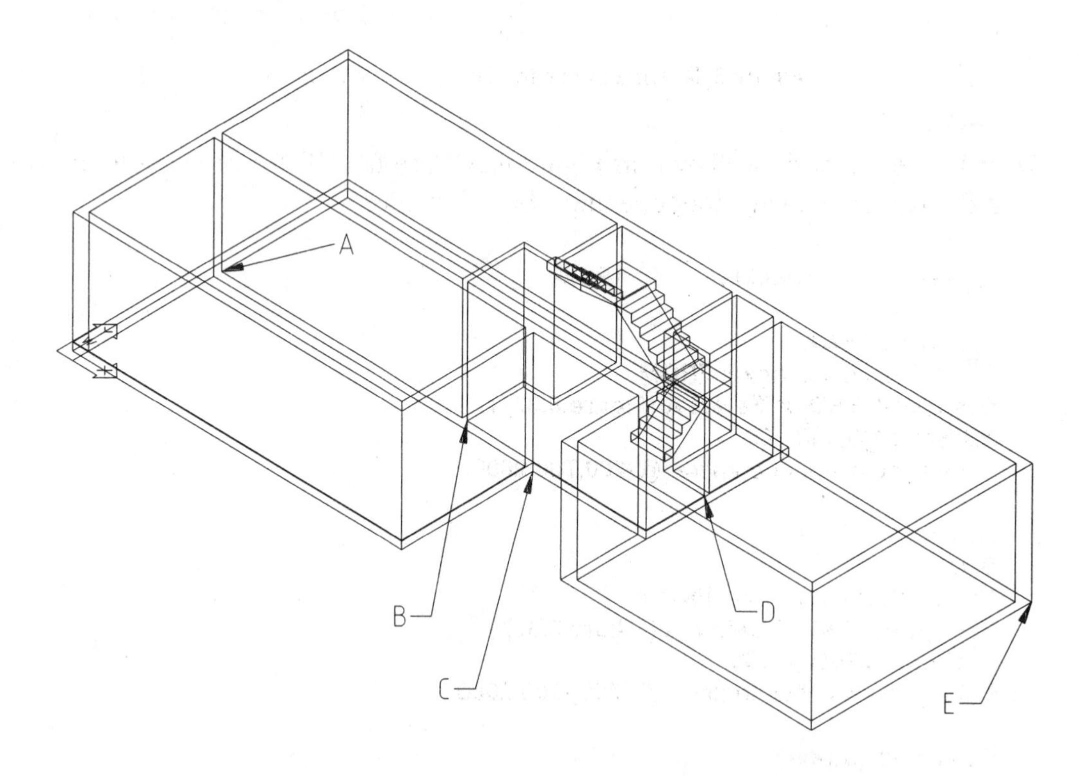

Figure 8.33 Staircase sections united and walls extruded

The walls should have openings for the doors and windows. Figure 8.34 shows a dimensioned layout of these openings. The height of the doors and that of the gate are 2000 units and 2600 units, respectively. The window opening is 1000 units from the top of the floor and is 1200 units high.

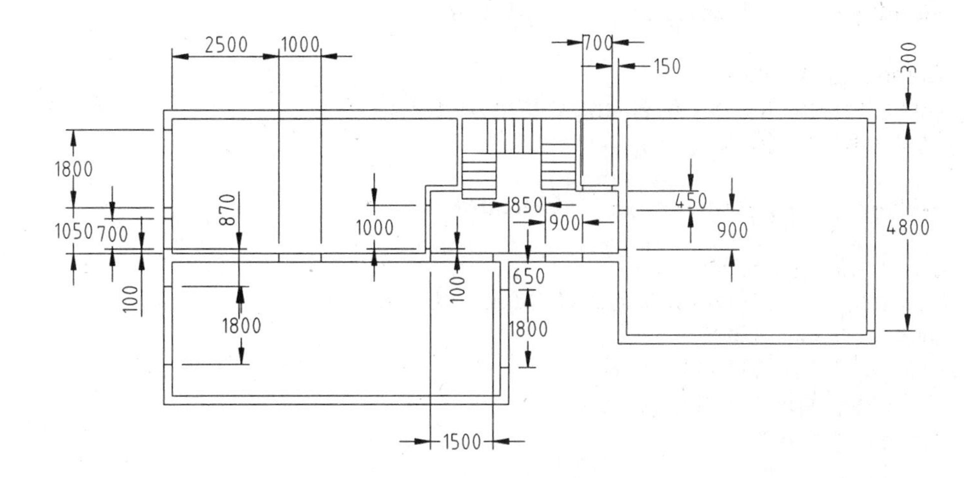

Figure 8.34 Dimensions for the windows and doors

Refer to the dimensions shown in Figure 8.34. Use the BOX command to construct eight solid boxes for making door openings. See Figure 8.35.

<Draw> **<Solids>** **<Box>**

Command: **BOX**
Center/<Corner of box>: **FROM**
Base point: **END** of **[Select A (Figure 8.33).]**
<Offset>: **@0,100,180**
Cube/Length/<other corner>: **@-200,700,2000**

Command: **[Enter]**
BOX
Center/<Corner of box>: **FROM**
Base point: **END** of **[Select A (Figure 8.33).]**
<Offset>: **@2500,0,180**
Cube/Length/<other corner>: **@1000,-200,2000**

Command: **[Enter]**
BOX
Center/<Corner of box>: **FROM**
Base point: **END** of **[Select B (Figure 8.33).]**
<Offset>: **@0,100,180**
Cube/Length/<other corner>: **@-120,1000,2000**

Command: **[Enter]**
BOX
Center/<Corner of box>: **FROM**
Base point: **END** of **[Select B (Figure 8.33).]**
<Offset>: **@0,0,180**
Cube/Length/<other corner>: **@1500,-200,2000**

Command: **[Enter]**
BOX
Center/<Corner of box>: **FROM**
Base point: **END** of **[Select C (Figure 8.33).]**
<Offset>: **@850,0,180**

Cube/Length/<other corner>: **@900,200,2000**

Command: **[Enter]**
BOX
Center/<Corner of box>: **FROM**
Base point: **END** of [**Select D (Figure 8.33).**]
<Offset>: **@-150,0,180**
Cube/Length/<other corner>: **@-700,120,2000**

Command: **[Enter]**
BOX
Center/<Corner of box>: **FROM**
Base point: **END** of [**Select D (Figure 8.33).**]
<Offset>: **@0,-450,180**
Cube/Length/<other corner>: **@200,-900,2000**

Command: **[Enter]**
BOX
Center/<Corner of box>: **FROM**
Base point: **END** of [**Select E (Figure 8.33).**]
<Offset>: **@0,-300,0**
Cube/Length/<other corner>: **@-200,-4800,2600**

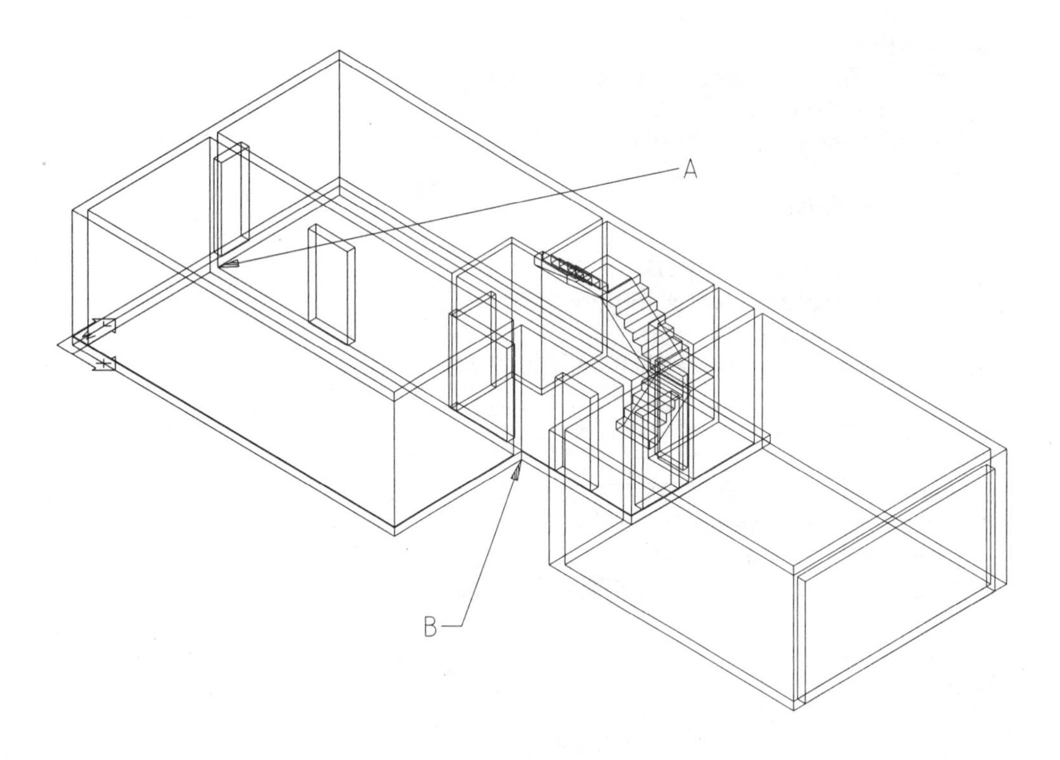

Figure 8.35 Door openings prepared

Using the dimensions shown in Figure 8.34, run the BOX command to construct three solid boxes for making window openings. See Figure 8.36.

<Draw> **<Solids>** **<Box>**

Command: **BOX**
Center/<Corner of box>: **FROM**
Base point: **END** of [**Select A (Figure 8.35).**]
<Offset>: **@0,1050,1180**
Cube/Length/<other corner>: **@-200,1800,1200**

Command: **[Enter]**
BOX
Center/<Corner of box>: **FROM**
Base point: **END** of [**Select A (Figure 8.35).**]
<Offset>: **@0,-770,1180**
Cube/Length/<other corner>: **@-200,-1800,1200**

Command: **[Enter]**
BOX
Center/<Corner of box>: **FROM**
Base point: **END** of [**Select B (Figure 8.35).**]
<Offset>: **@0,-650,1180**
Cube/Length/<other corner>: **@-200,-1800,1200**

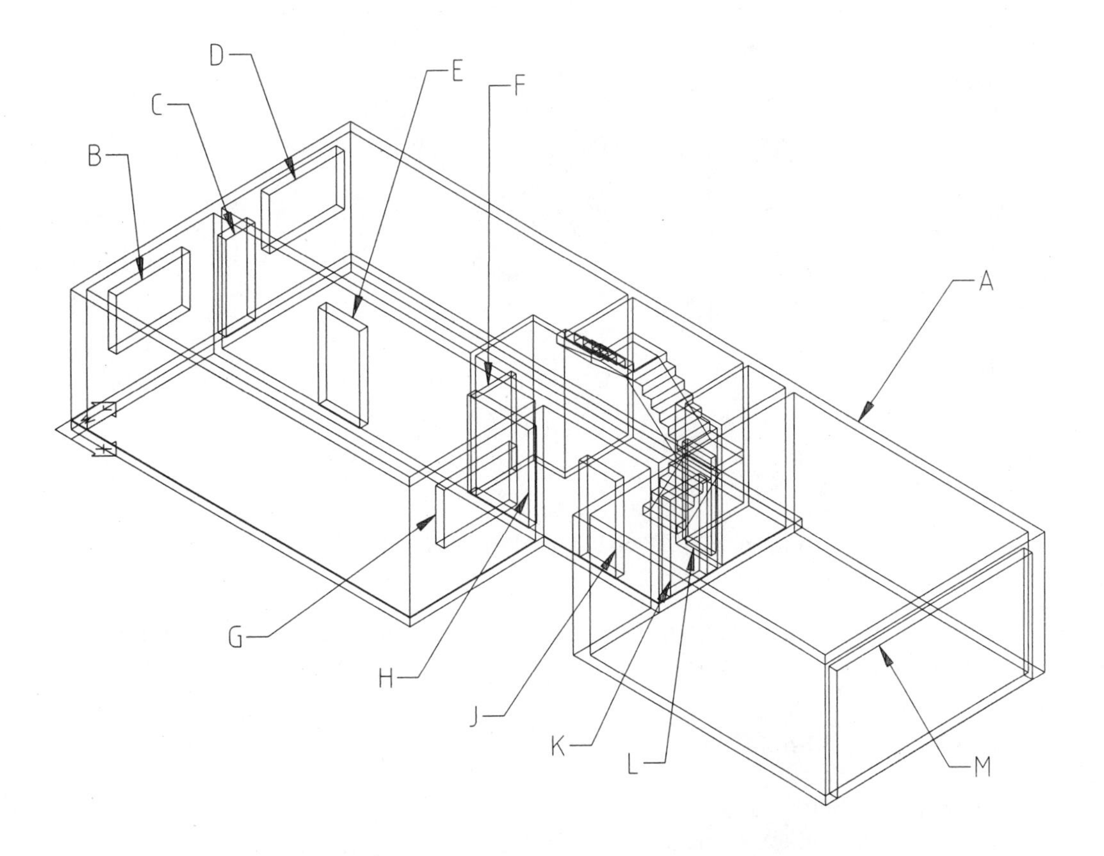

Figure 8.36 Window openings for the first level prepared

You have created 11 solid boxes to use as openings in the walls. At this stage, they are separate entities. To cut the openings, run the SUBTRACT command to subtract the boxes from the wall. You may not find any significant visual change to the model after subtraction. See Figure 8.37.

<Modify> **<Boolean>** **<Subtract>**

Command: **SUBTRACT**
Select solids and regions to subtract from...
Select objects: [**Select A (Figure 8.36).**]
Select objects: [**Enter**]
Select solids and regions to subtract...
Select objects: [**Select B, C, D, E, F, G, H, J, K, L, and M (Figure 8.36).**]
Select objects: [**Enter**]

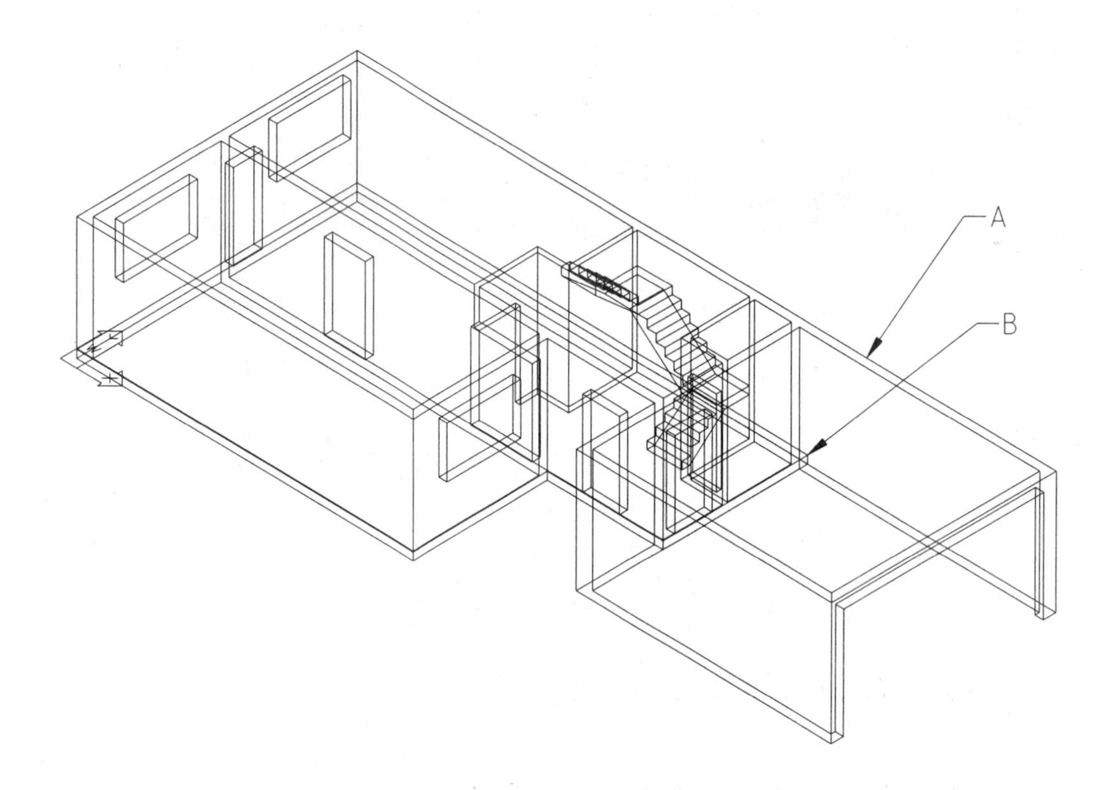

Figure 8.37 Door and window openings created

Run the UNION command to unite the wall and the floor. After uniting the floor and the walls of the first level, turn off layer F1. See Figure 8.38.

<Modify> **<Boolean>** **<Union>**

Command: **UNION**
Select objects: [**Select A and B (Figure 8.37).**]
Select objects: [**Enter**]

[**Object Properties**] [**Layer Control**]

Off	F1
Off	F2
Off	Roof

On	Stair
On	W1
Off	W2

Current layer: **W1**

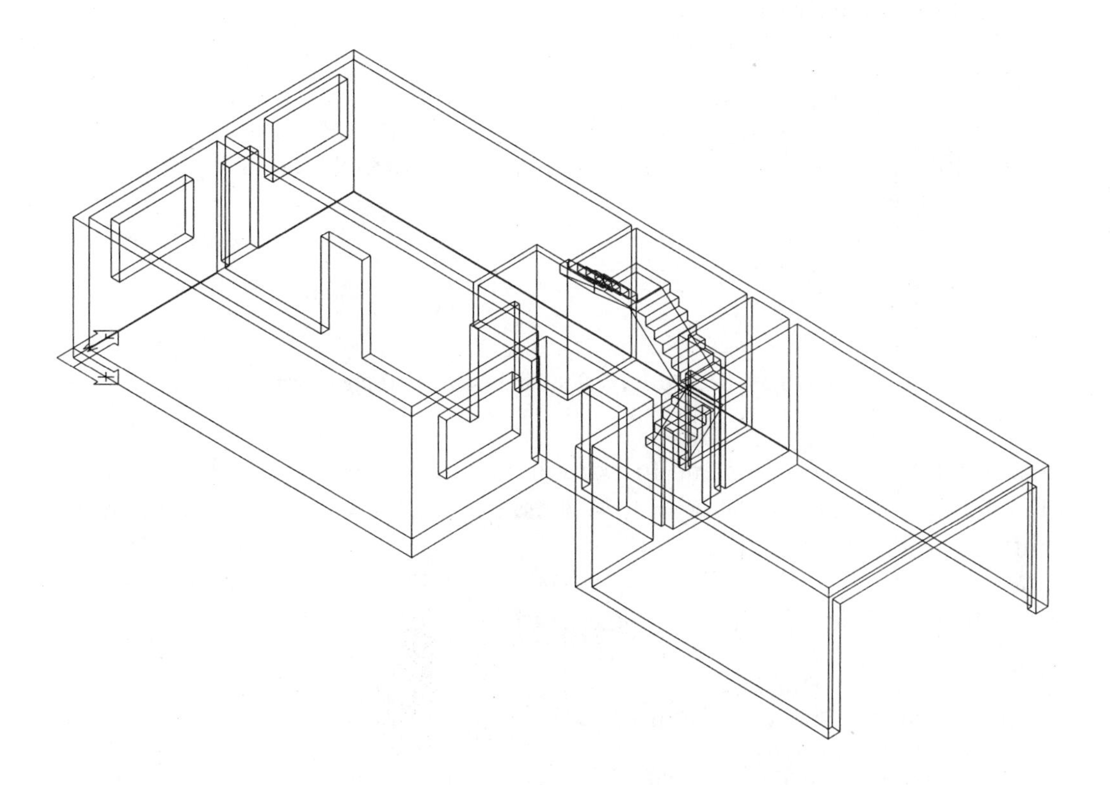

Figure 8.38 Completed first level and staircase

The models of the first level and the staircase are complete.

8.7 3D Model for the Second Level

First you will create the model for the floor, and then the model for the walls, of the second level. After that, you will cut openings for the windows and doors.

Turn on layers F2 and W2 and turn off layers F1, Stair, and W1. Then set the current layer to F2. See Figure 8.39.

[Object Properties] **[Layer Control]**

Off	F1
On	F2
Off	Roof
Off	Stair
Off	W1
On	W2

Current layer: **F2**

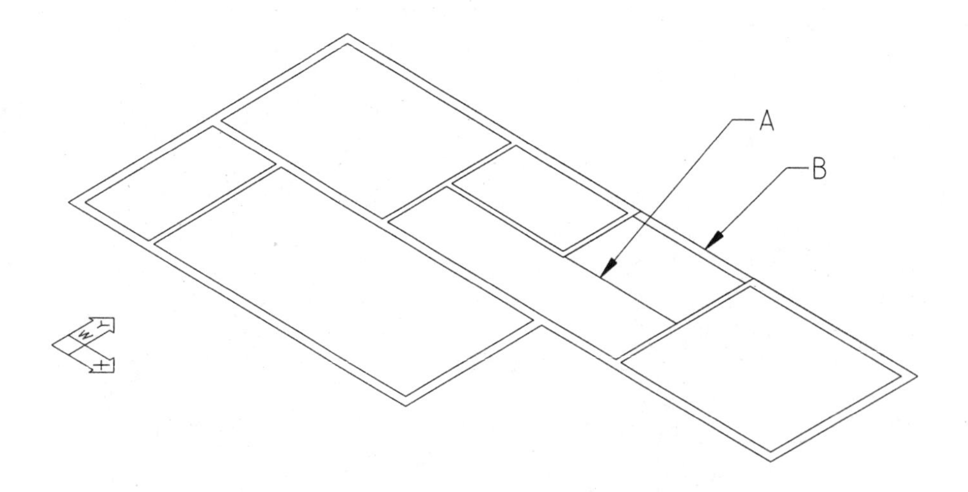

Figure 8.39 Second-level floor and wall wireframes turned on

Run the EXTRUDE command to extrude the wireframes for the floor of the second level. After that, set the current layer to W2 and extrude the wireframes for the wall of the second floor. See Figure 8.40.

<Draw> **<Solids>** **<Extrude>**

Command: **EXTRUDE**
Select objects: **[Select A (Figure 8.39).]**
Select objects: **[Enter]**
Path/<Height of Extrusion>: **180**
Extrusion taper angle: **0**

[Object Properties] **[Layer Control]**

Off	F1
On	F2
Off	Roof
On	Stair
Off	W1
On	W2

Current layer: **W2**

<Draw> **<Solids>** **<Extrude>**

Command: **EXTRUDE**
Select objects: **[Select B (Figure 8.39).]**
Select objects: **[Enter]**
Path/<Height of Extrusion>: **2880**
Extrusion taper angle: **0**

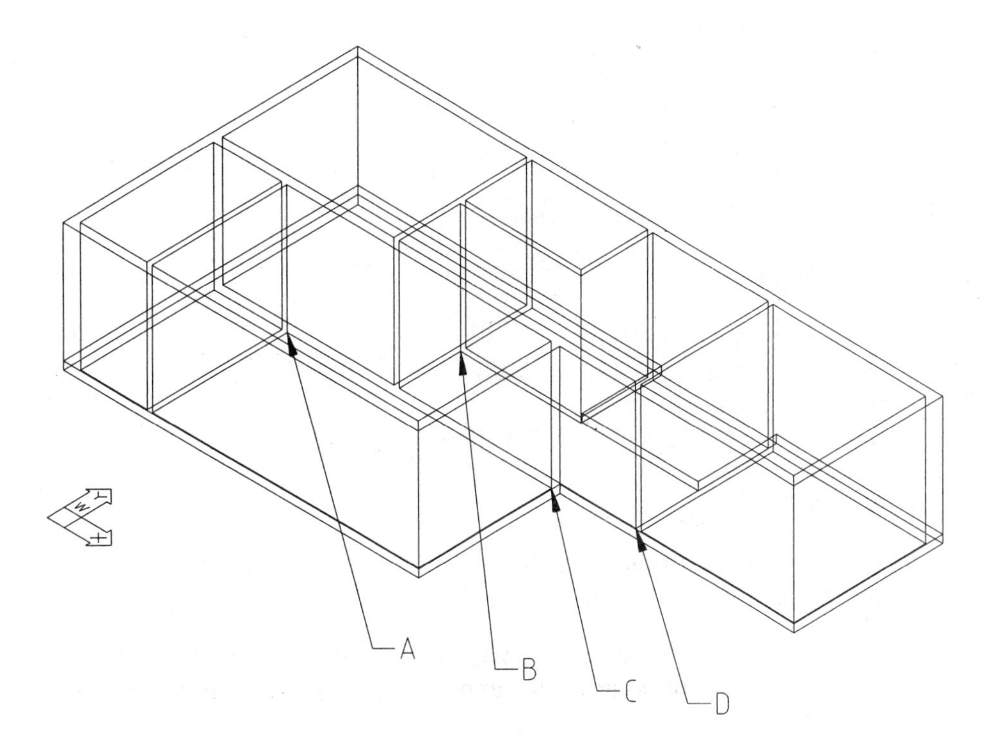

Figure 8.40 Floor and walls of the second level created

Next, you will cut window and door openings on the walls. Figure 8.41 is a dimensioned layout of the windows and doors. The lower edges of the windows are 1000 units from the floor surface, and the windows are 1200 units high. The height of the doors is 2000 units.

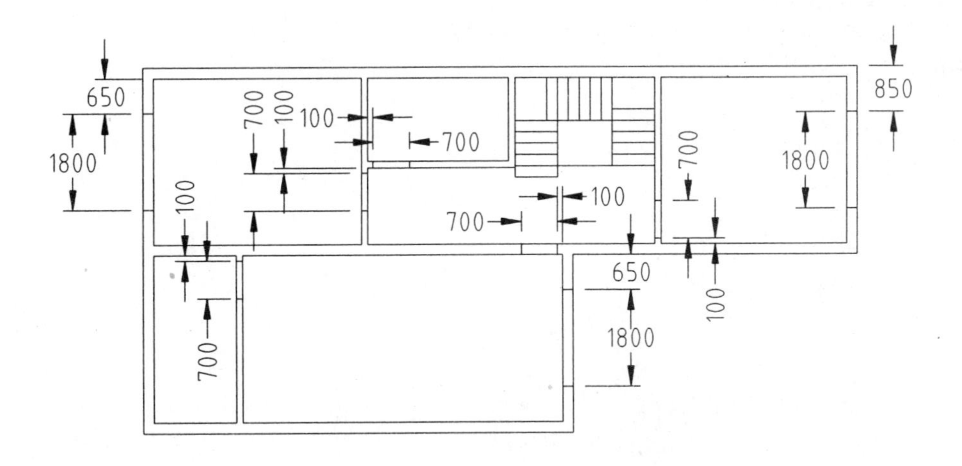

Figure 8.41 Dimensioned layout of the window and door openings

Using the dimensions shown in Figure 8.41, run the BOX command to prepare five solid boxes for use as door openings. See Figure 8.42.

<Draw> **<Solids>** **<Box>**

Command: **BOX**
Center/<Corner of box>: **FROM**
Base point: **END** of [**Select A (Figure 8.40).**]
<Offset>: **@0,-100,180**
Cube/Length/<other corner>: **@-120,-700,2000**

Command: [**Enter**]
BOX
Center/<Corner of box>: **FROM**
Base point: **END** of [**Select B (Figure 8.40).**]
<Offset>: **@0,-100,180**
Cube/Length/<other corner>: **@-120,-700,2000**

Command: [**Enter**]
BOX
Center/<Corner of box>: **FROM**
Base point: **END** of [**Select B (Figure 8.40).**]
<Offset>: **@100,0,180**
Cube/Length/<other corner>: **@700,120,2000**

Command: [**Enter**]
BOX
Center/<Corner of box>: **FROM**
Base point: **END** of [**Select C (Figure 8.40).**]
<Offset>: **@-100,0,180**
Cube/Length/<other corner>: **@-700,200,2000**

Command: [**Enter**]
BOX
Center/<Corner of box>: **FROM**
Base point: **END** of [**Select D (Figure 8.40).**]
<Offset>: **@0,100,180**
Cube/Length/<other corner>: **@120,700,2000**

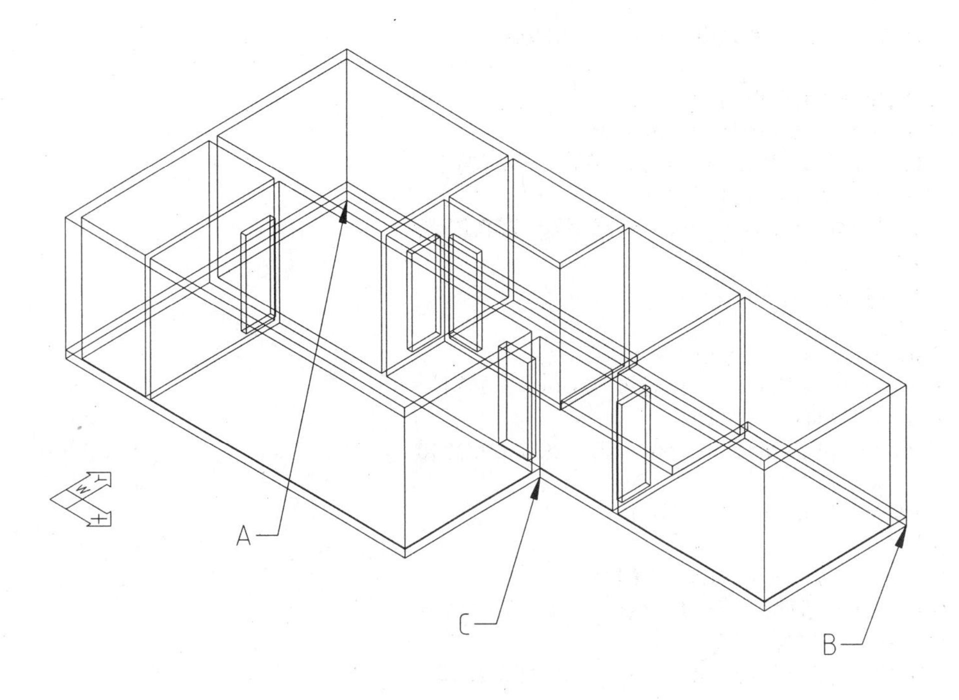

Figure 8.42 Solid boxes for making the door openings prepared

Using the dimensions shown in Figure 8.41, create three solid boxes to use as window openings by using the BOX command. See Figure 8.43.

<Draw> **<Solids>** **<Box>**

Command: **BOX**
Center/<Corner of box>: **FROM**
Base point: **END** of **[Select A (Figure 8.42).]**
<Offset>: **@0,-650,1180**
Cube/Length/<other corner>: **@-200,-1800,1200**

Command: **[Enter]**
BOX
Center/<Corner of box>: **FROM**
Base point: **END** of **[Select B (Figure 8.42).]**
Cube/Length/<other corner>: **@-200,-1800,1200**

Command: **[Enter]**
BOX
Center/<Corner of box>: **FROM**
Base point: **END** of **[Select C (Figure 8.42).]**
<Offset>: **@0,-650,1180**
Cube/Length/<other corner>: **@-200,-1800,1200**

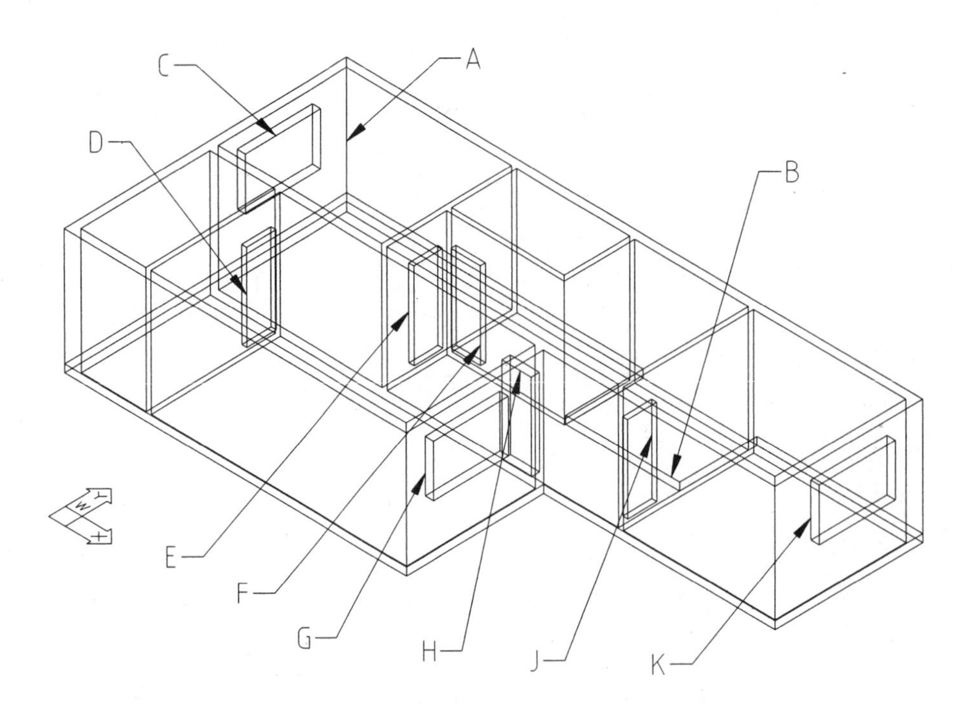

Figure 8.43 Solid boxes for making the window openings prepared

Run the UNION command to unite the walls and the floor of the second level. Then cut the window and door openings from the wall by using the SUBTRACT command. See Figure 8.44.

<Modify> **<Boolean>** **<Union>**

Command: **UNION**
Select objects: [**Select A and B (Figure 8.43).**]
Select objects: [**Enter**]

<Modify> **<Boolean>** **<Subtract>**

Command: **SUBTRACT**
Select solids and regions to subtract from...
Select objects: [**Select A (Figure 8.43).**]
Select objects: [**Enter**]
Select solids and regions to subtract...
Select objects: [**Select C, D, E, F, G, H, J, and K (Figure 8.43).**]
Select objects: [**Enter**]

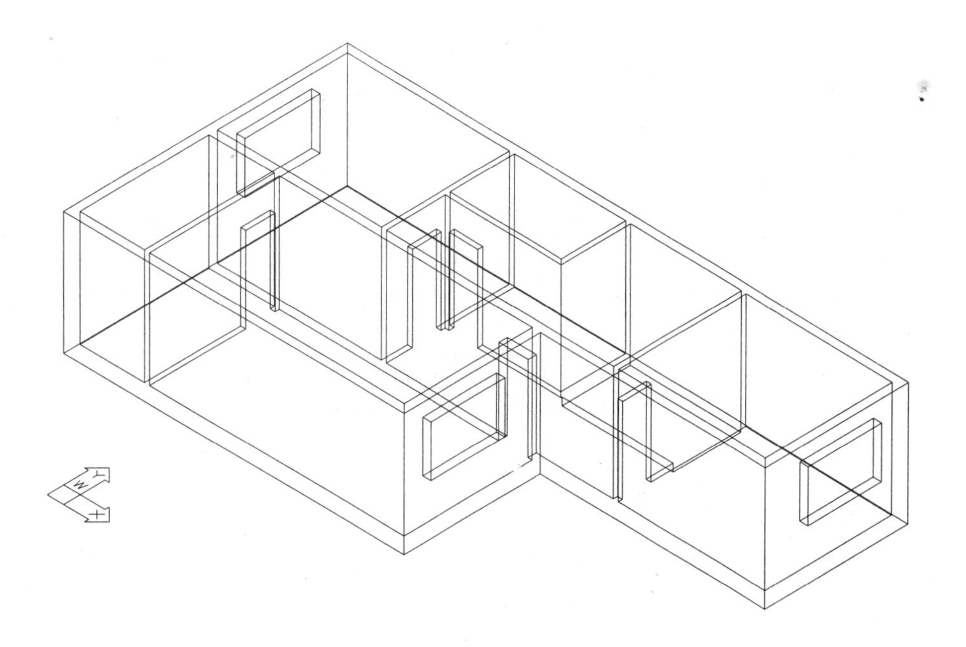

Figure 8.44 Completed second level

The second level of the house is complete.

8.8 3D Model for the Roofs

The final stage in creating the model is to work on the roofs, which have seven wireframes in three groups. The roof for the second level needs two groups of wireframes, and the roof for the first level needs one group of wireframes.

You will extrude all the seven wireframes to solids of extrusion. After extrusion, you will create three solids of intersection. Finally, you will unite two solids of intersection to become the roof of the second level and will leave the third solid of intersection as the roof of the first level.

Turn on layer Roof, set it as the current layer, and turn off all other layers. See Figure 8.45.

[Object Properties] **[Layer Control]**

Off	F1
Off	F2
Off	Roof
On	Stair
Off	W1
Off	W2

Current layer: **Roof**

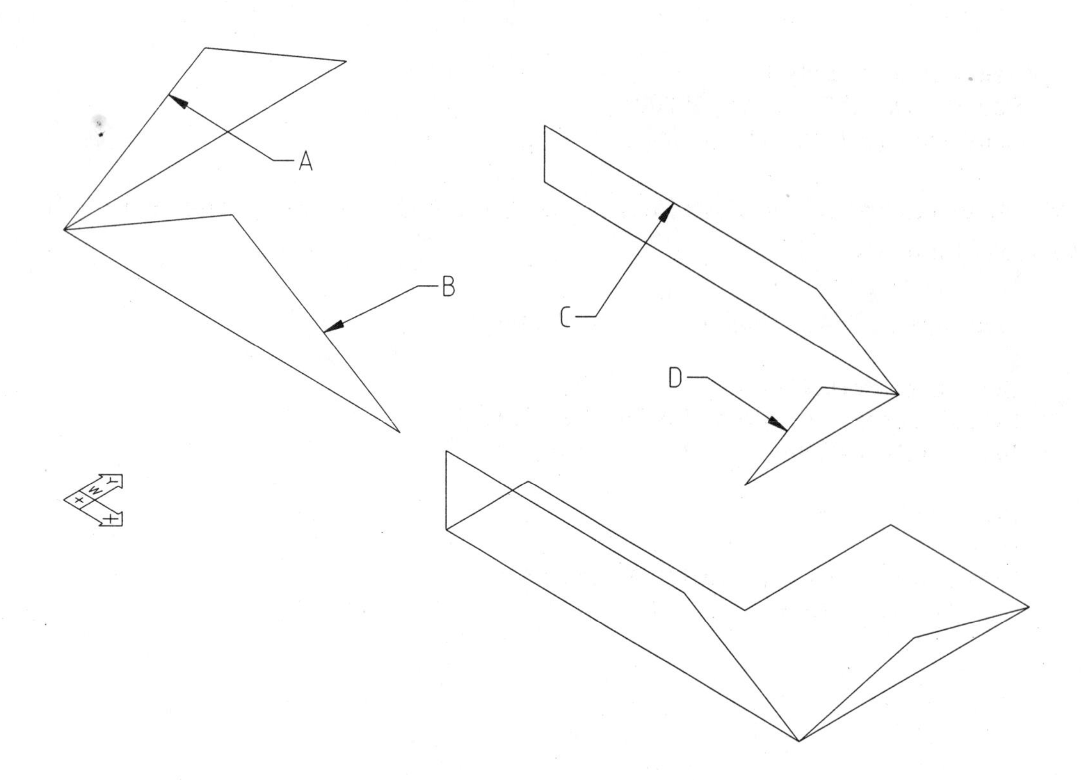

Figure 8.45 Wireframes for the roofs

Run the EXTRUDE command to extrude the wireframes for the roof of the second level. As mentioned before, the direction of extrusion depends on the Z-direction of the UCS of the wireframes themselves, not the current UCS. See Figure 8.46.

<Draw> **<Solids>** **<Extrude>**

Command: **EXTRUDE**
Select objects: **[Select A (Figure 8.45).]**
Select objects: **[Enter]**
Path/<Height of Extrusion>: **8600**
Extrusion taper angle: **0**

Command: **[Enter]**
EXTRUDE
Select objects: **[Select B (Figure 8.45).]**
Select objects: **[Enter]**
Path/<Height of Extrusion>: **-7200**
Extrusion taper angle: **0**

Command: **[Enter]**
EXTRUDE
Select objects: **[Select C (Figure 8.45).]**
Select objects: **[Enter]**
Path/<Height of Extrusion>: **3900**
Extrusion taper angle: **0**

Command: **[Enter]**
EXTRUDE
Select objects: **[Select D (Figure 8.45).]**

Select objects: **[Enter]**
Path/<Height of Extrusion>: **-9000**
Extrusion taper angle: **0**

You have created four solids of extrusion. Run the INTERSECT command to obtain two solids of intersection.

<Modify> **<Boolean>** **<Intersect>**

Command: **INTERSECT**
Select objects: **[Select A and B (Figure 8.45).]**
Select objects: **[Enter]**

Command: **[Enter]**
INTERSECT
Select objects: **[Select C and D (Figure 8.45).]**
Select objects: **[Enter]**

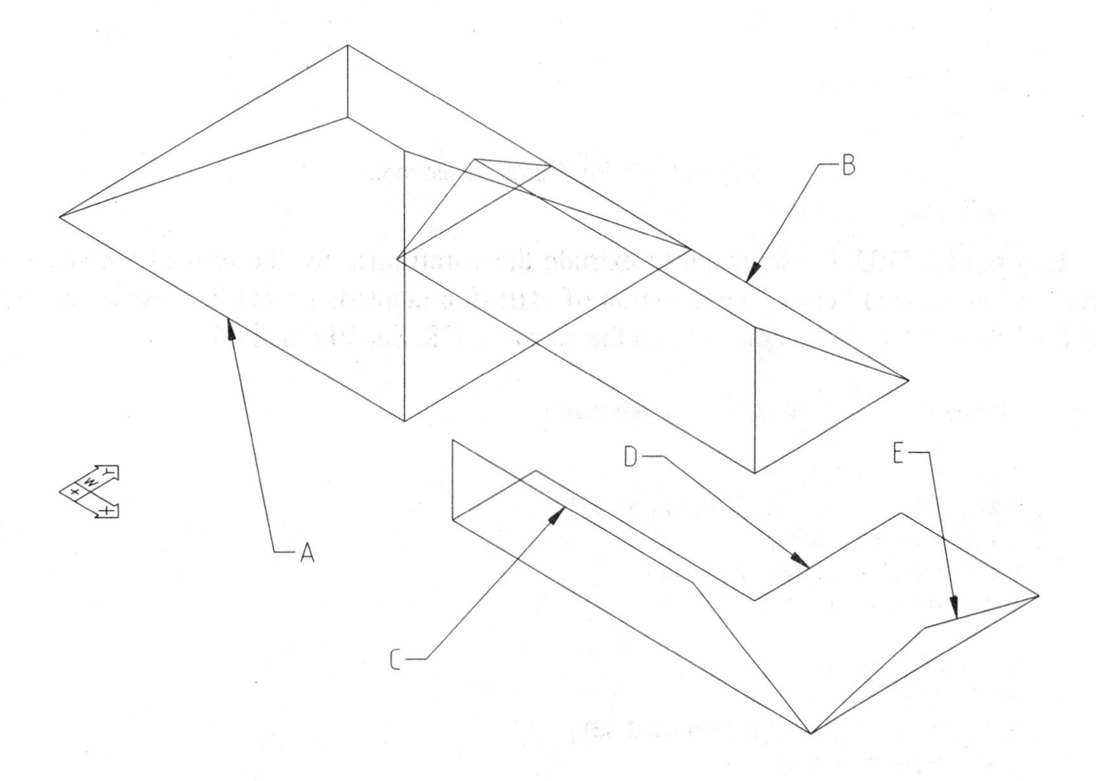

Figure 8.46 Two solids of intersection formed from four extruded solids

The two solids of intersection are components of the roof for the second level. Run the UNION command to unite them.

<Modify> **<Boolean>** **<Union>**

Command: **UNION**
Select objects: **[Select A and B (Figure 8.46).]**
Select objects: **[Enter]**

After uniting, the roof for the second level is complete. There are three wireframes left. Run the EXTRUDE command on them to form three solids of extrusion. See Figure 8.47.

<Draw> **<Solids>** **<Extrude>**

Command: **EXTRUDE**
Select objects: **[Select C (Figure 8.46).]**
Select objects: **[Enter]**
Path/<Height of Extrusion>: **-5800**
Extrusion taper angle: **0**

Command: **[Enter]**
EXTRUDE
Select objects: **[Select D (Figure 8.46).]**
Select objects: **[Enter]**
Path/<Height of Extrusion>: **1700**
Extrusion taper angle: **0**

Command: **[Enter]**
EXTRUDE
Select objects: **[Select E (Figure 8.46).]**
Select objects: **[Enter]**
Path/<Height of Extrusion>: **-9000**
Extrusion taper angle: **0**

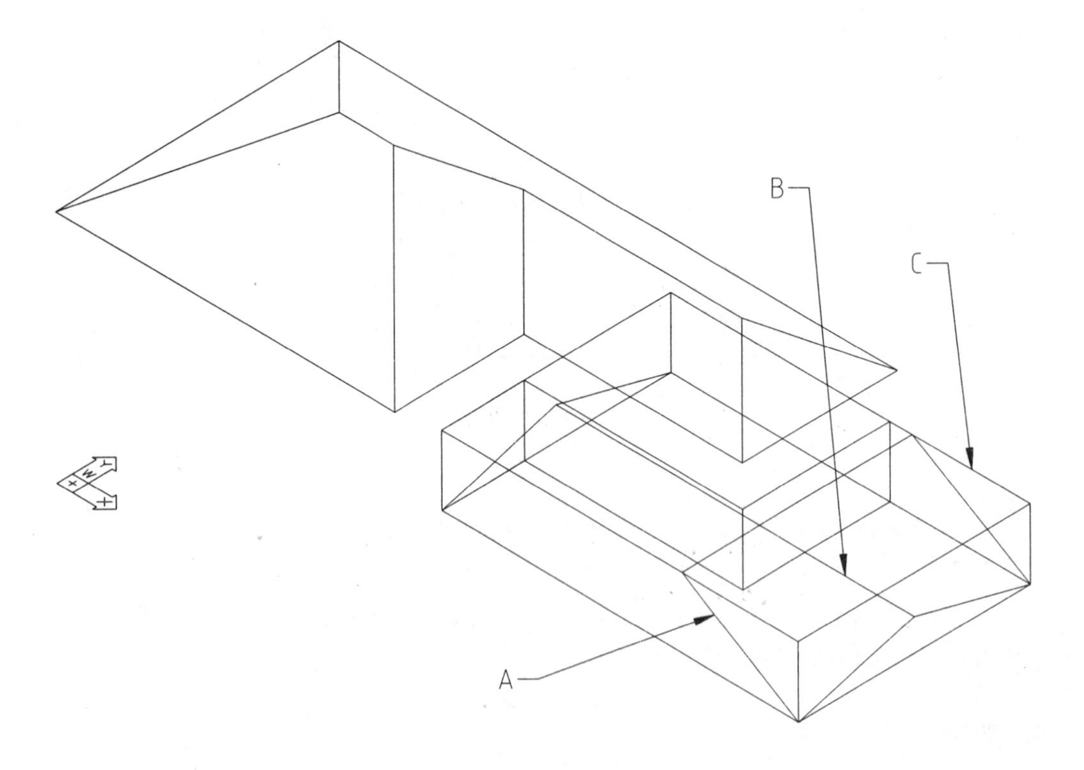

Figure 8.47 Roof of the second level completed, and wireframes for the roof of the first level extruded

Complete the roof for the first level by using the INTERSECT command on the three solids of extrusion. See Figure 8.48.

<Modify> **<Boolean>** **<Intersect>**

Command: **INTERSECT**
Select objects: [**Select A, B, and C (Figure 8.47).**]
Select objects: [**Enter**]

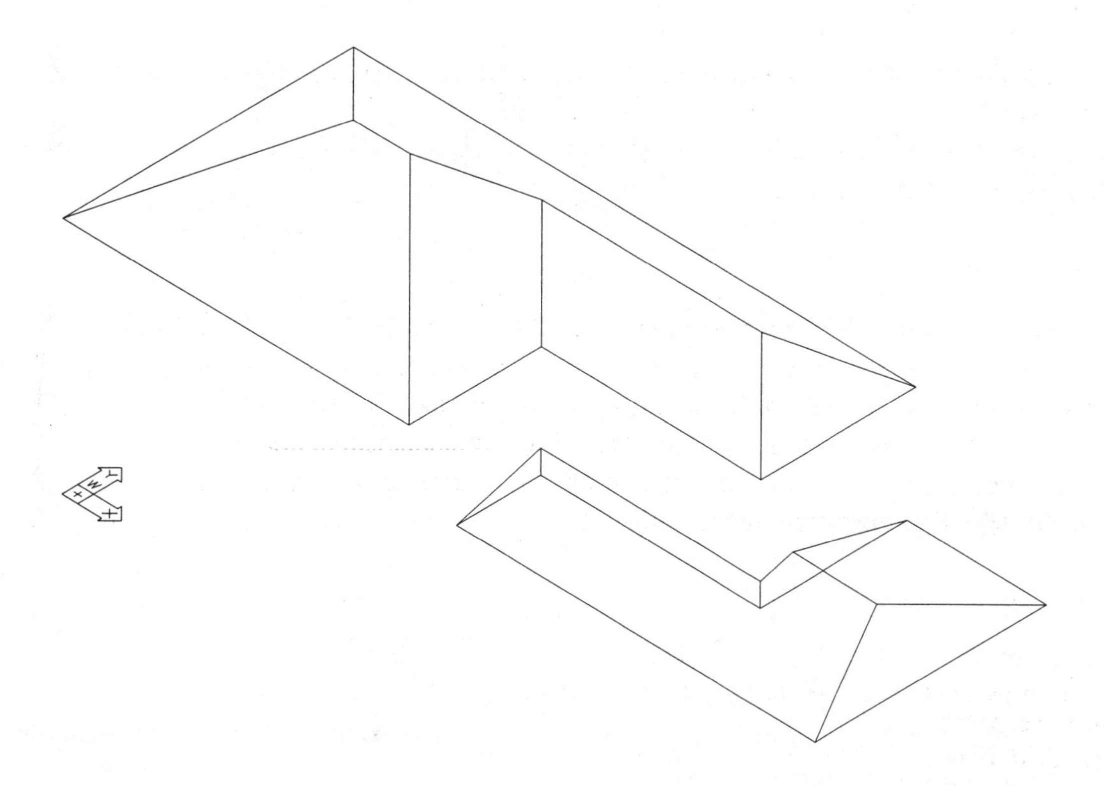

Figure 8.48 Completed roofs

The roofs for both levels are complete.

8.9 Complete Model

To view the entire model of the house, turn on all the layers and set current layer to 0. See Figure 8.49. The basic structure of the house is complete. Save your drawing.

<File> **<Save>**

File name: **HOUSE.DWG**

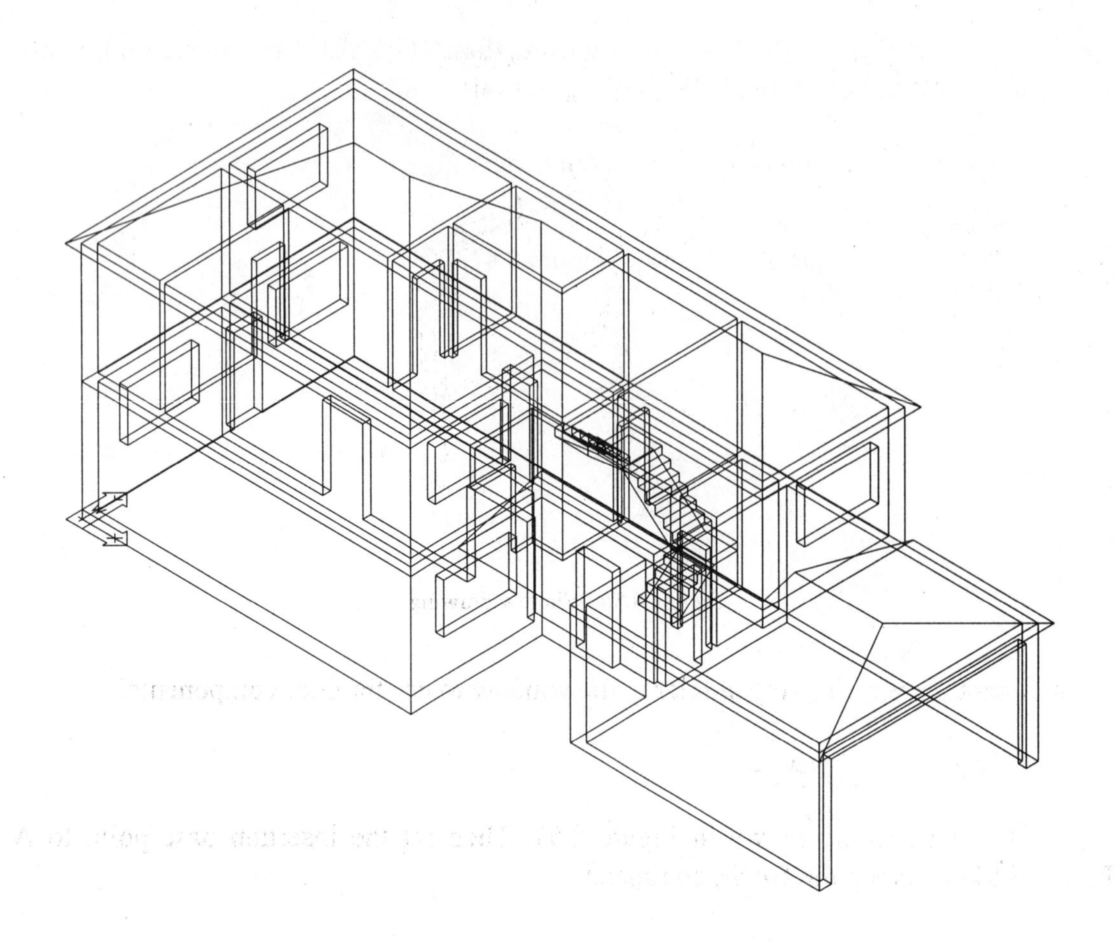

Figure 8.49 All layers turned on

To add details to the model, you will create four new drawings for the windows, doors, and gate to insert into the main drawing.

Start a new drawing with the NEW command.

<File> **<New...>**

Using the dimensions shown in Figure 8.50, run the BOX command and the SUBTRACT command to create the window frame and the window panel. After you have completed the drawing, use the BASE command to set the insertion base point.

<Draw> **<Block>** **<Base>**

Command: **BASE**
Base point: **[Pick at A (Figure 8.50).]**

Execute the SAVE command to save the drawing under the name WINDOW. This drawing will be inserted into the house as window components.

<File> **<Save>**

File name: **WINDOW.DWG**

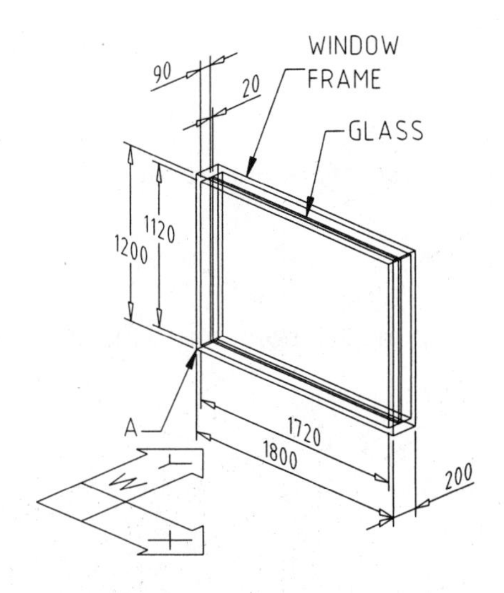

Figure 8.50 Window drawing

Start another new drawing similar to the window to use for door components.

<File> **<New...>**

Use the dimensions shown in Figure 8.51. Then set the insertion base point to A (Figure 8.51) by using the BASE command.

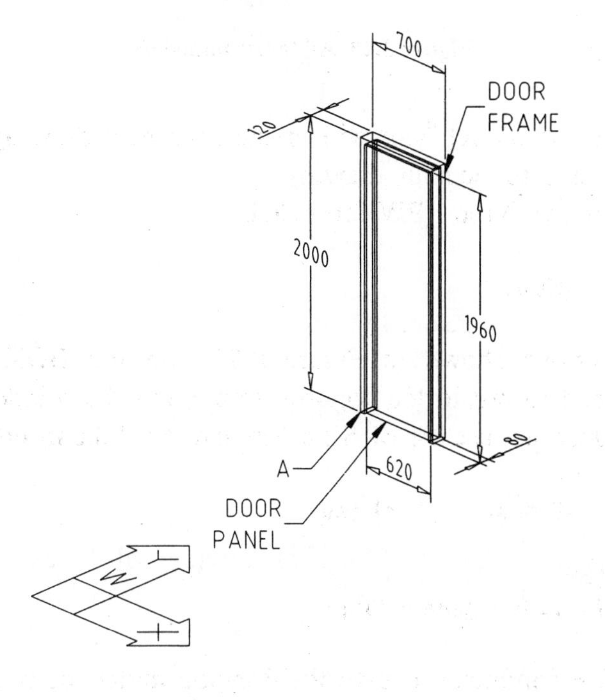

Figure 8.51 Door drawing

Save your drawing.

<File> **<Save>**

File name: **DOOR700.DWG**

As shown in Figure 8.52, start another new drawing to use for the main door.

<File> **<New...>**

Do not forget to set the insertion base point to A (Figure 8.52).

<File> **<Save>**

File name: **DOOR900.DWG**

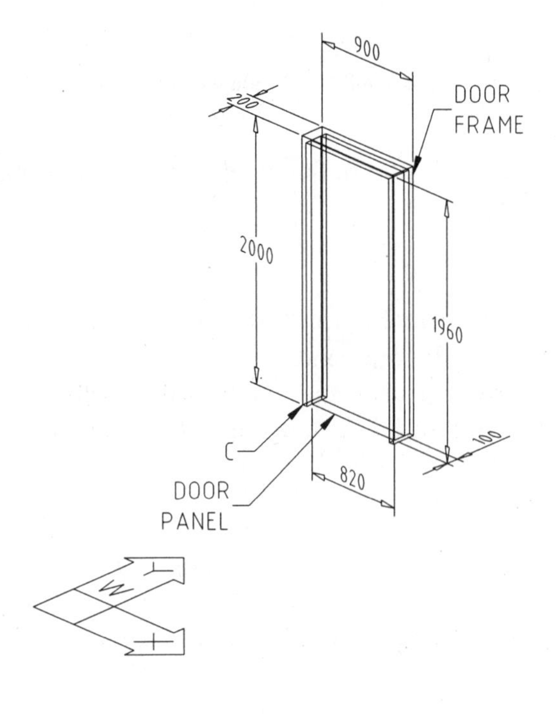

Figure 8.52 Second door drawing

Using Figure 8.53, make a drawing called GATE.

<File> **<New...>**

The base point is at A (Figure 8.53). This drawing will be used for the main gate of the house.

<File> **<Save>**

File name: **GATE.DWG**

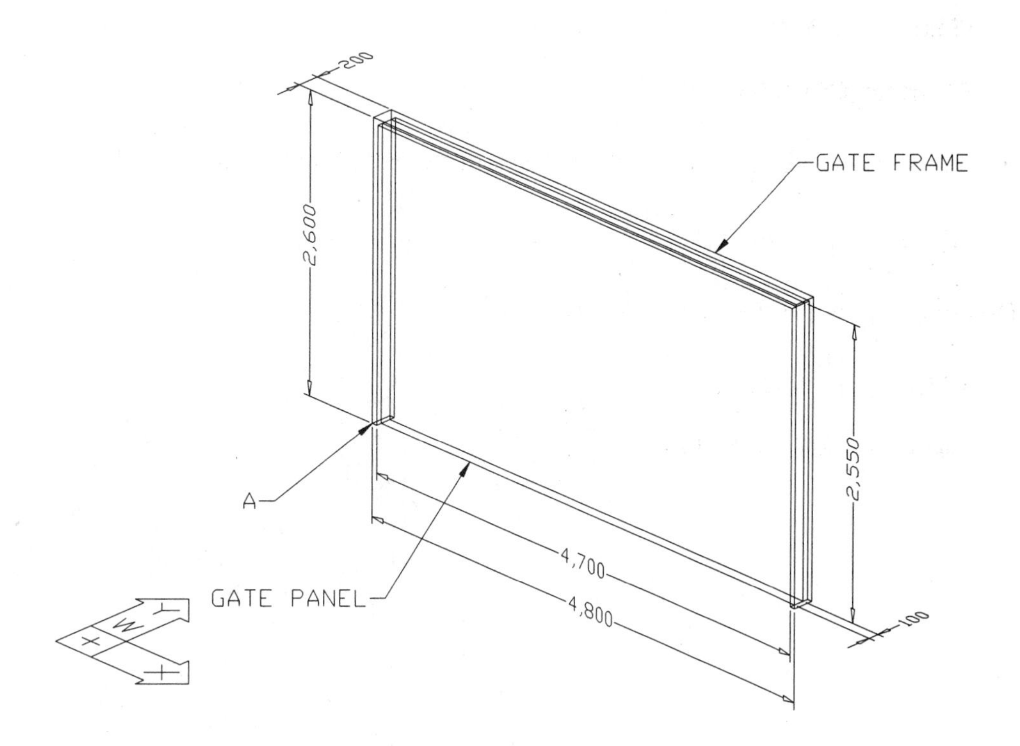

Figure 8.53 Gate drawing

Now you have prepared four drawings for insertion. Open the drawing of the house again. You will insert the windows, doors, and gate to the house.

<File> **<Open...>**

File name: **HOUSE.DWG**

Set the current layer to W1, and turn off all other layers. See Figure 8.54.

[Object Properties] **[Layer Control]**

Off	F1
Off	F2
Off	Roof
Off	Stair
On	W1
Off	W2

Current layer: **W1**

Set the display with the DDVPOINT command.

<View> **<3D Viewpoint>** **<Select...>**

Command: **DDVPOINT**

[Viewpoint Presets

From X Axis: **315** XY Plane: **25**
OK]

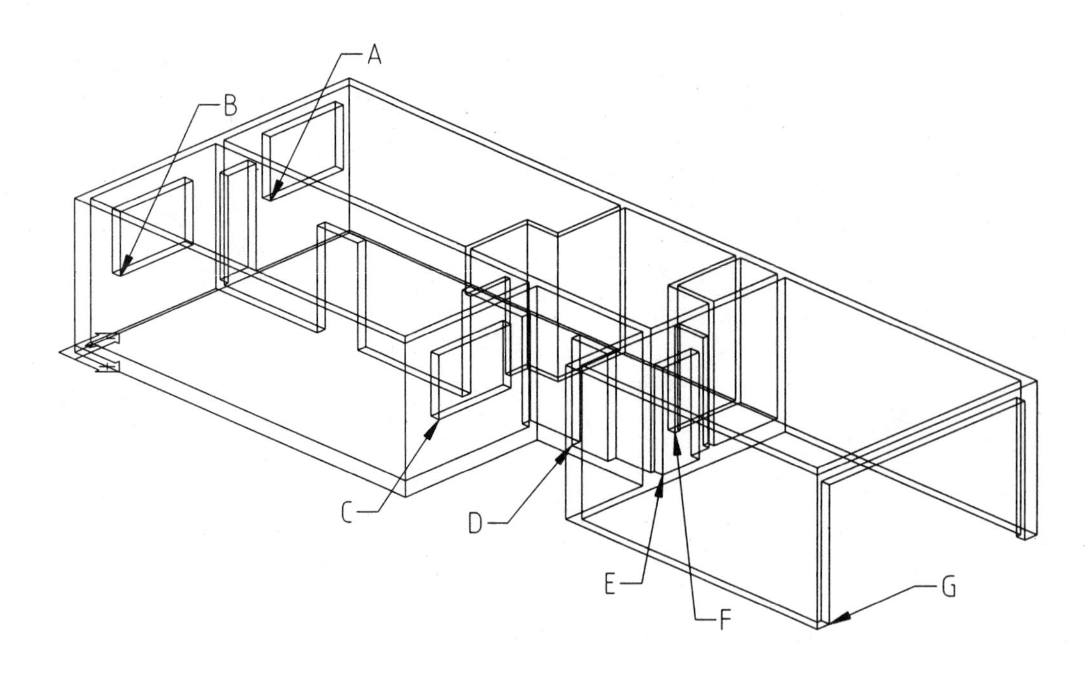

Figure 8.54 First level of the house

From the Insert pull-down menu, select the Block... item to run the DDINSERT command to insert the drawing WINDOW into the house.

<Insert> **<Block...>**

Command: **DDINSERT**

[File: **WINDOW**
Specify Parameters on Screen
OK]

Insertion point: **END** of [**Select A (Figure 8.54).**]
X scale factor <1> / Corner / XYZ: **1**
Y scale factor (default=X): **1**
Rotation angle: **90**

Repeat the DDINSERT command to insert the drawings DOOR700, DOOR900, and GATE into this drawing. See Figure 8.55.

Command: **[Enter]**
DDINSERT

File name	Insertion point	X-scale	Y-scale	Rotation angle
WINDOW	**B**	**1**	**1**	**90**
WINDOW	**C**	**1**	**1**	**90**
DOOR900	**D**	**1**	**1**	**0**
DOOR900	**E**	**1**	**1**	**90**

DOOR700	**F**	**1**	**1**	**0**
GATE	**G**	**1**	**1**	**90**

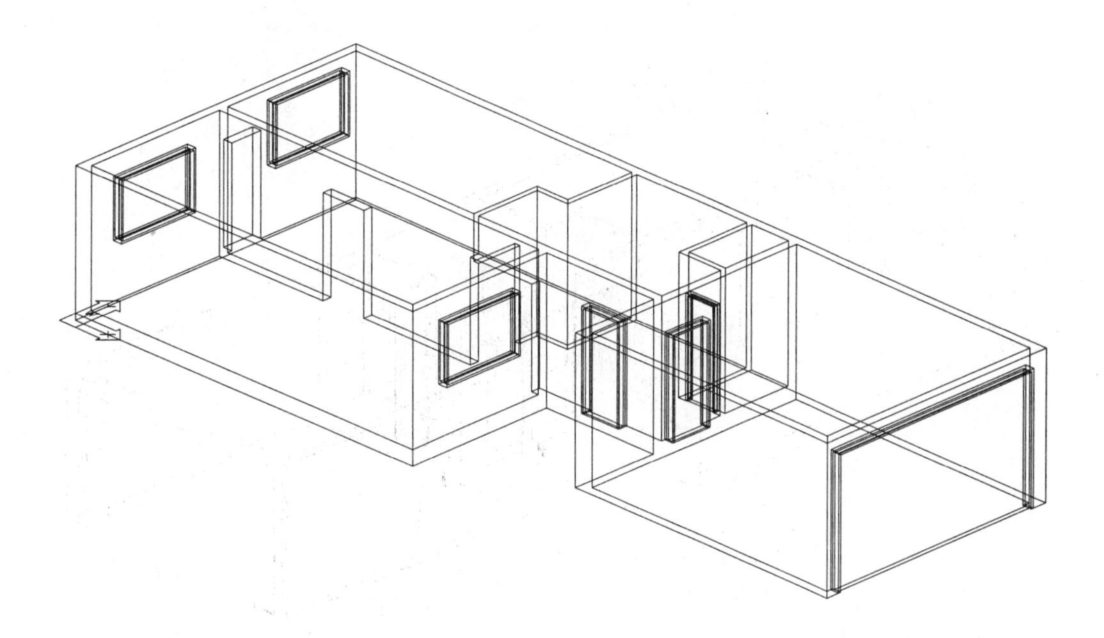

Figure 8.55 Windows, doors, and gate inserted into the first level

After completing the first level, turn on layer W2, set the current layer to W2, and turn off all other layers. See Figure 8.56.

[Object Properties] **[Layer Control]**

Off	F1
Off	F2
Off	Roof
Off	Stair
Off	W1
On	W2

Current layer: **W2**

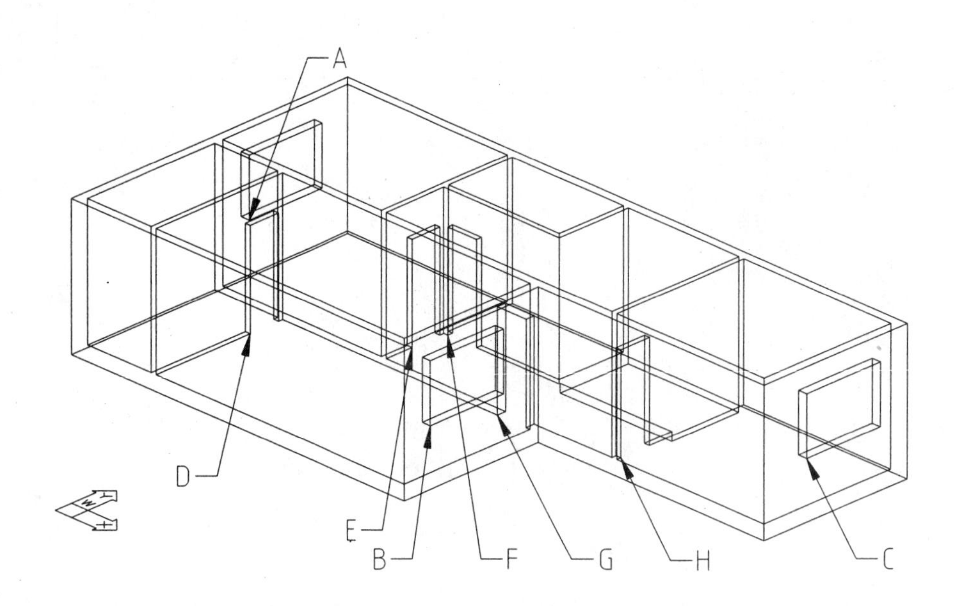

Figure 8.56 Second level of the house

The treatment of the second level is similar to that of the first level. Use the DDINSERT command to insert the drawings WINDOW and DOOR700. See Figure 8.57.

<Insert> **<Block...>**

Command: **DDINSERT**

Block name	Insertion point	X-scale	Y-scale	Rotation angle
WINDOW	**A**	**1**	**1**	**90**
WINDOW	**B**	**1**	**1**	**90**
WINDOW	**C**	**1**	**1**	**90**
DOOR700	**D**	**1**	**1**	**90**
DOOR700	**E**	**1**	**1**	**90**
DOOR700	**F**	**1**	**1**	**0**
DOOR700	**G**	**1**	**1**	**0**
DOOR700	**H**	**1**	**1**	**90**

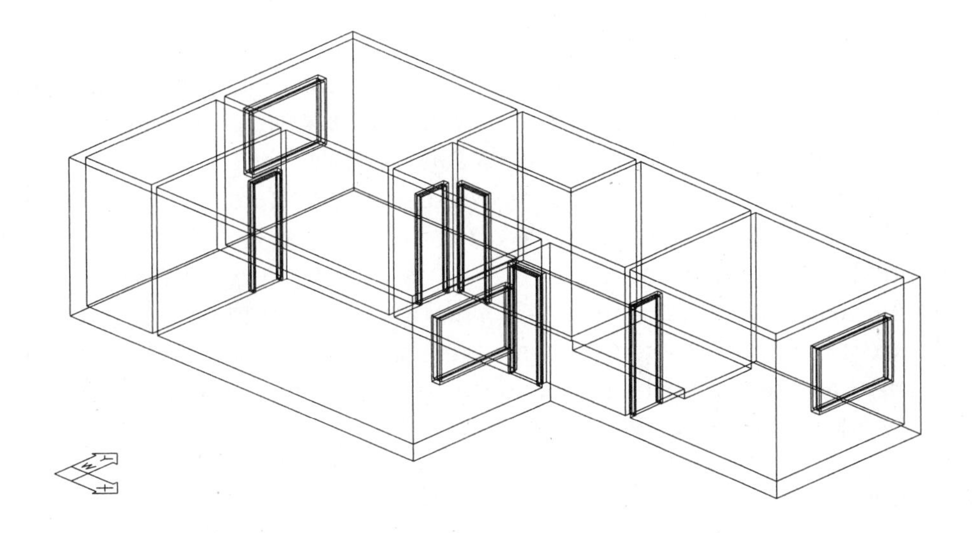

Figure 8.57 Windows and doors inserted into the second level

The model of the house is complete. Turn on all the layers. See Figure 8.58.

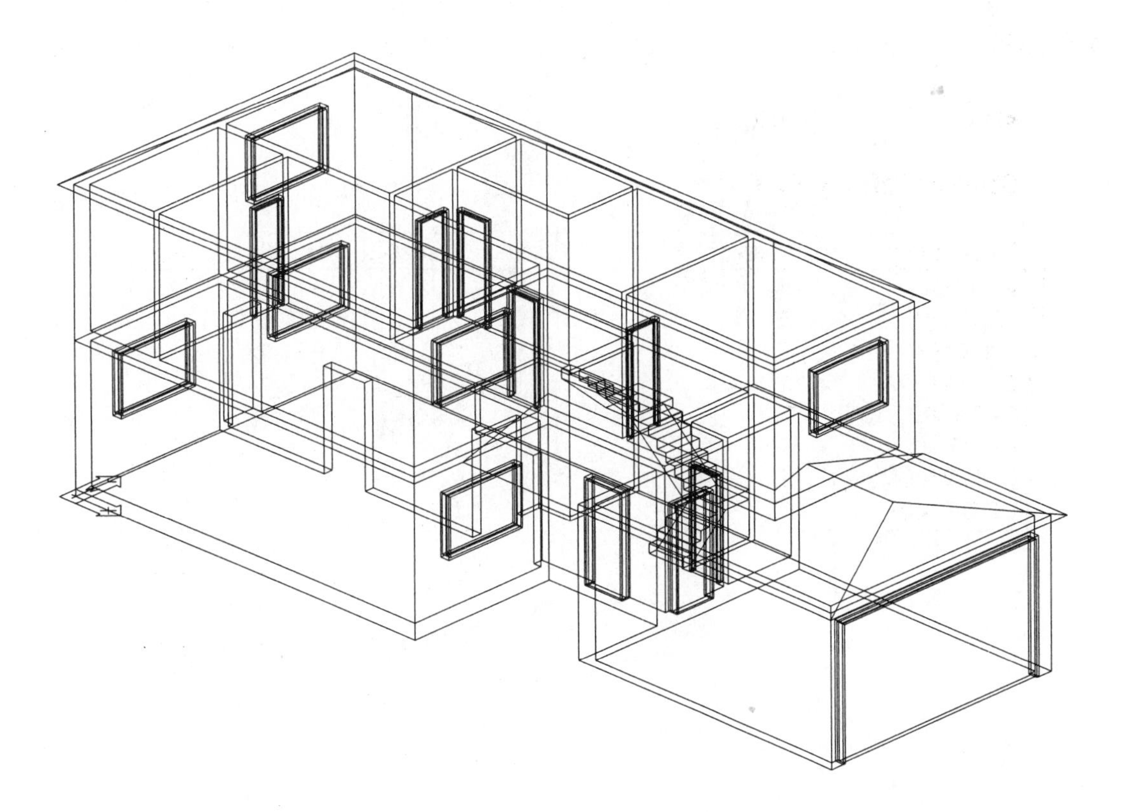

Figure 8.58 Completed house

8.10 Key Points and Exercises

To build the solid model of a house, you can start by constructing multilines. Then you explode them and convert them into regions. From the regions, you can create complex regions and extrude the regions to become 3D solid models.

In this chapter, you worked on the project of a house. Here, you applied the set of solid modeling commands, together with 2D entity creation and editing commands, to create the model of a house. In Chapter 3, you learned how to create 2D layouts with multilines. In this project, you gained an appreciation of how to convert multilines into regions for extrusion, how to cut openings on the extruded walls of the model house, and how to apply the solid modeling techniques to architectural projects. Enhance your learning by working on the following exercises.

Exercise 8.1

Describe how multilines are used to construct the solid model of the walls of a house.

Exercise 8.2

Figure 8.59 shows the model of a house.

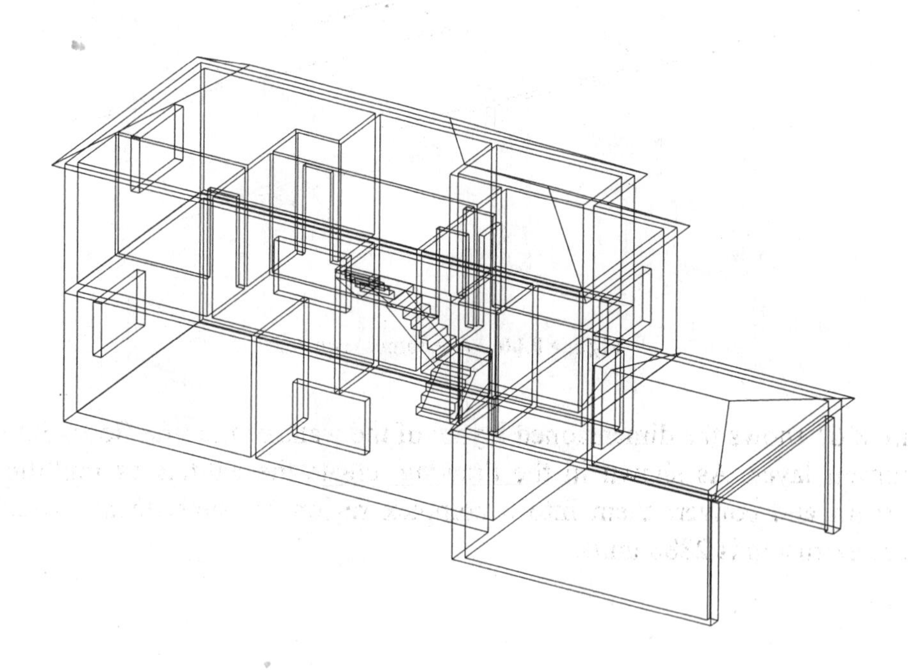

Figure 8.59 Completed house

Start a new drawing. Set up six additional layers:

Layer	Purpose
F1	Floor of the first level
F2	Floor of the second level

Roof	Roof
Stair	Staircase
W1	Wall of the first level
W2	Wall of the second level

All together, there are six groups of wireframes. Figure 8.60 shows all the wireframes required.

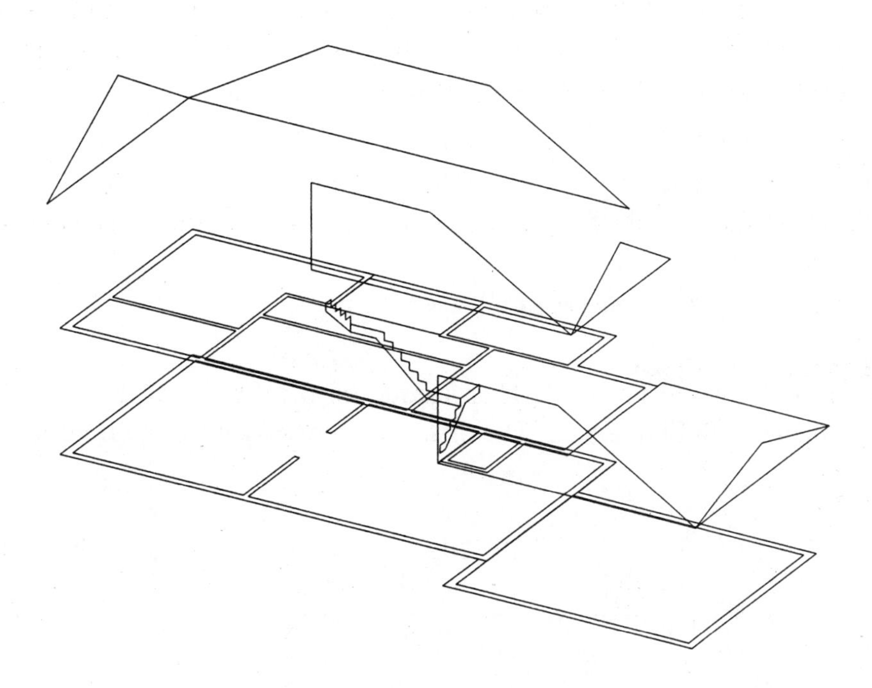

Figure 8.60 Wireframes required

Figure 8.61 shows the dimensioned layout of the walls of the first floor. Set layer W1 as the current layer. As shown in the drawing, create the entities as multilines. Then explode them and convert them into a complex region for subsequent extrusion. The distance of extrusion is 2880 units.

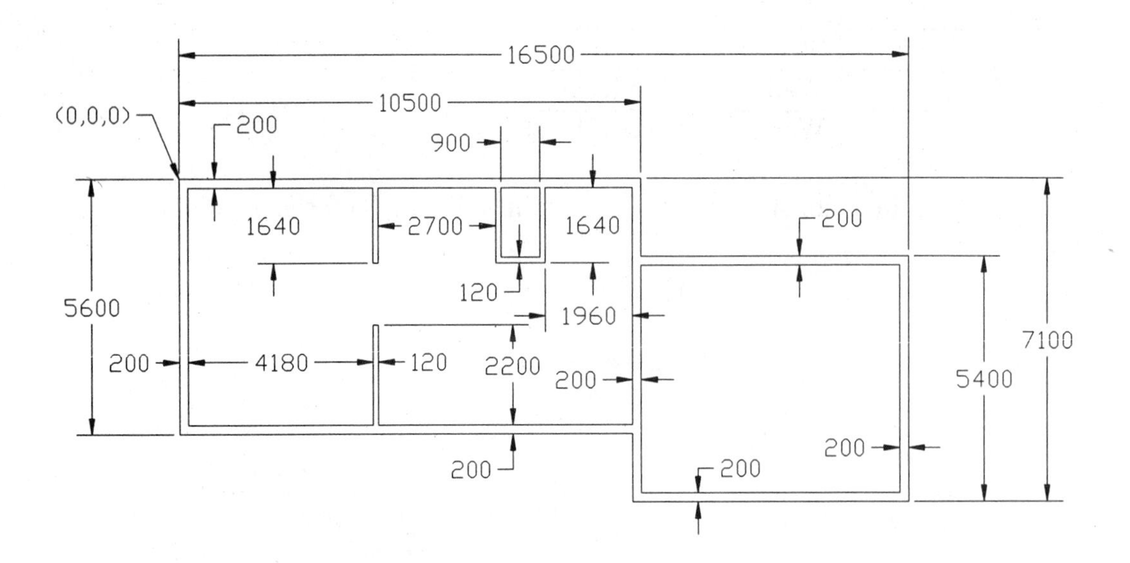

Figure 8.61 Walls of the first level

Set the current layer to layer F1. Then create the wireframe for the floor of the first level in accordance with Figure 8.62. The distance of extrusion for the floor is 180 units.

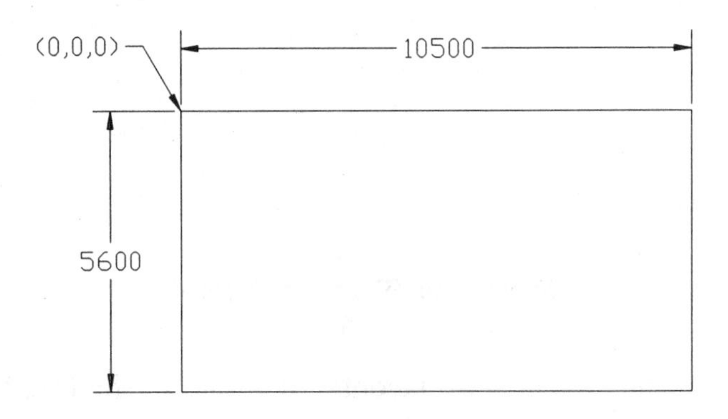

Figure 8.62 Floor of the first level

Place the UCS origin at (*0,0,2880). Set the current layer to layer W2. Then create a complex region in accordance with Figure 8.63. This represents the walls for the second level. The distance of extrusion is 2880 units.

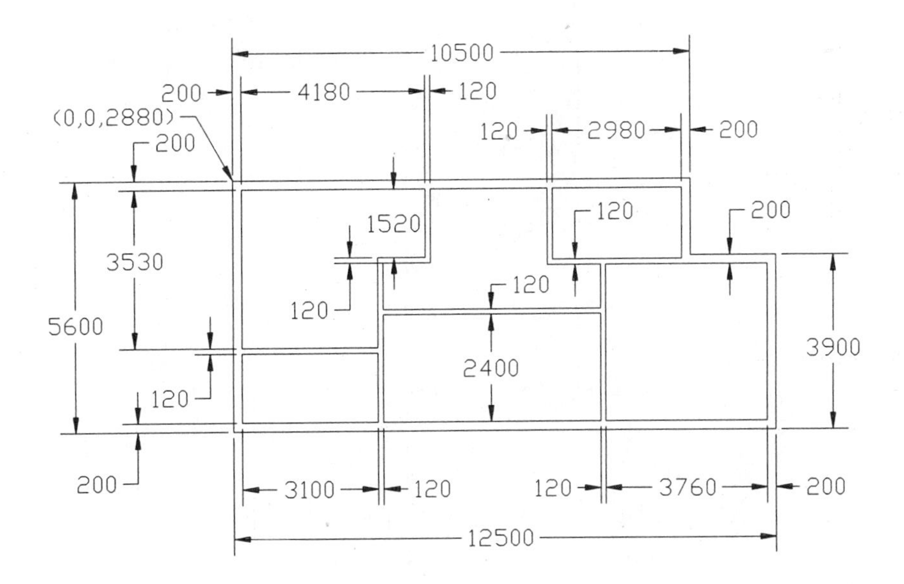

Figure 8.63 Walls of the second level

Keep the UCS origin at (*0,0,2880), and set the current layer to layer F2. As shown in Figure 8.64, create a region for the floor of the second level. The distance of extrusion is 180 units.

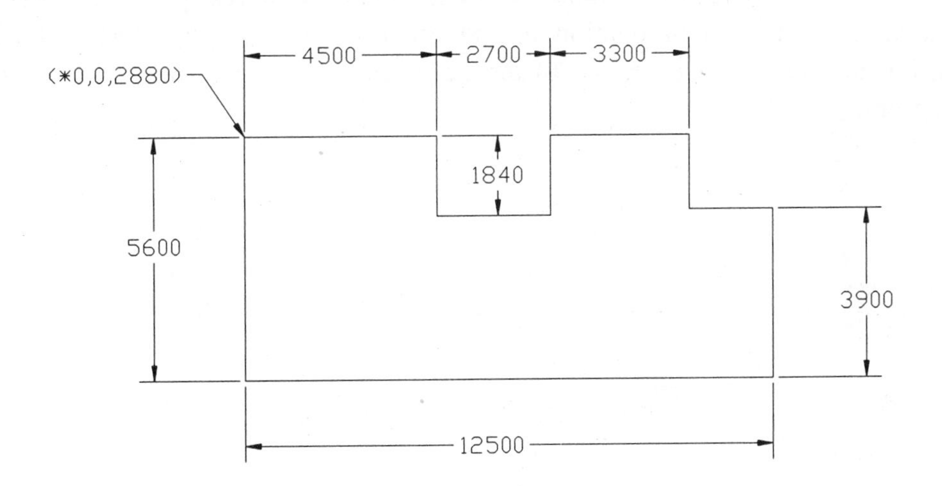

Figure 8.64 Floor of the second level

Set the current layer to STAIR. Then create three sets of wires for making the solid model of the staircase. The dimensions of the staircase are identical to those for the house you created in this chapter. Therefore, you may use the WBLOCK command to export the wireframe of that house to a file and then use the INSERT command to import the wireframe to this house. For your reference, Figure 8.65 shows the dimensions of the staircase. The distance of extrusion for each section of the staircase is 825 units.

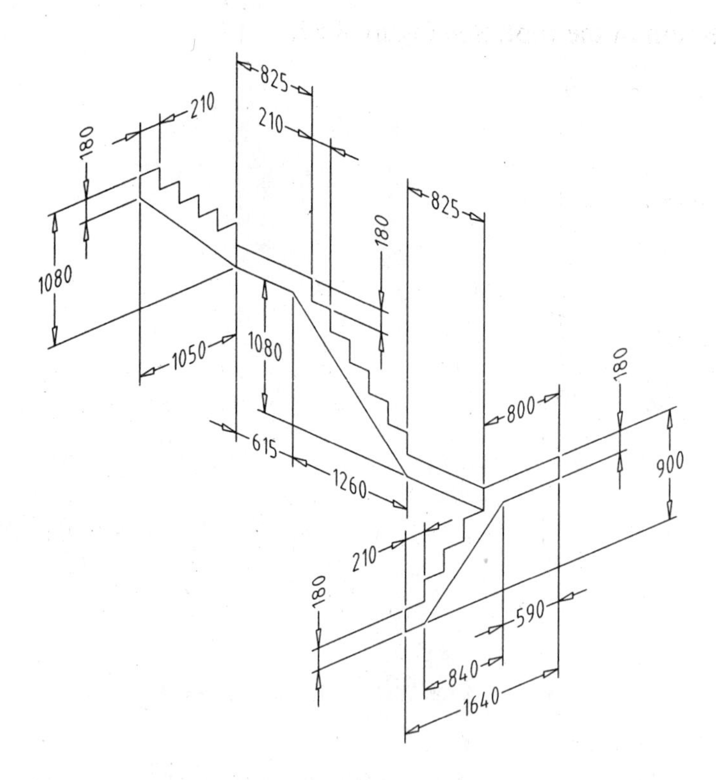

Figure 8.65 Staircase

Whether you prefer to insert the wires or to create the wires, you need to know the position of the staircase in relation to the other objects. Figure 8.66 indicates the relationship between the wireframes of the staircase and the wireframes of the walls of the first floor.

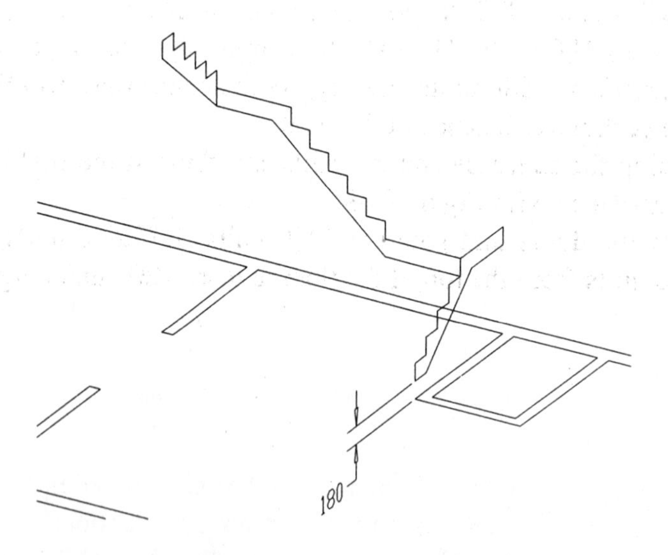

Figure 8.66 Staircase position

After making the wires for the staircase, set the current layer to ROOF. Then create seven sets of wires for the roofs. Just as in making the staircase, you have to locate the

UCS for each section of the roof. See Figure 8.67.

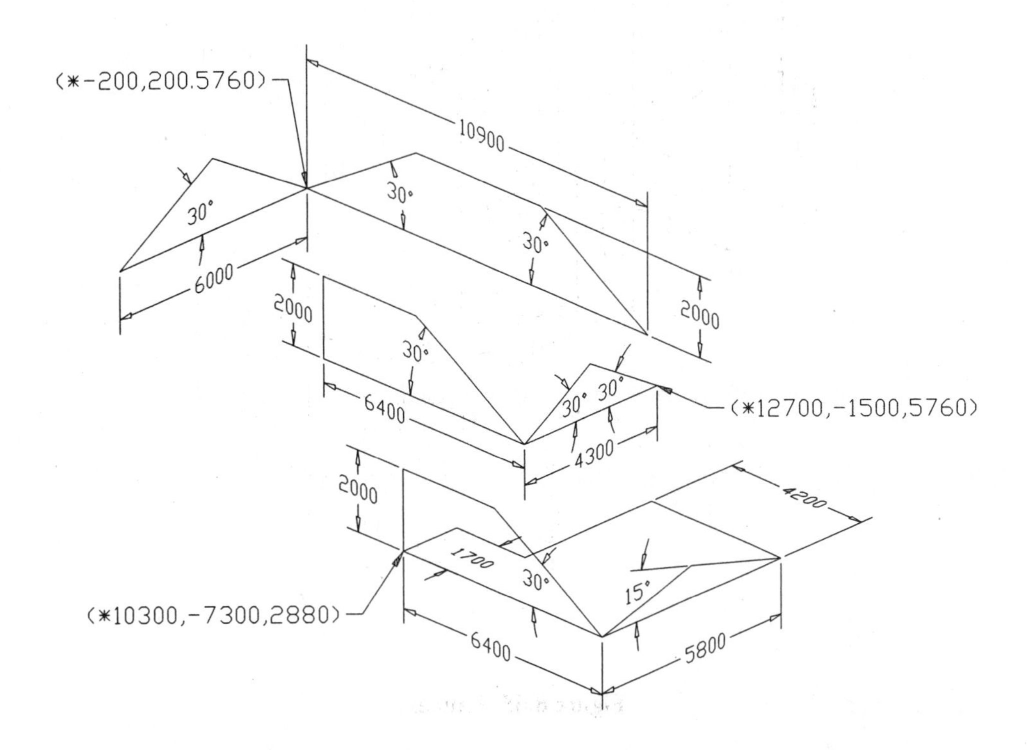

Figure 8.67 Roofs

The wireframes are complete. You can start to build the solid model for the house. Turn off layers W2, F2, and ROOF. Set the current layer to F1 and extrude the wires on layer F1 a distance of 180 units. Then set the current layer to W1 and extrude the wires on layer W1 a distance of 2880 units. Finally, set the current layer to STAIR and extrude the three staircase sections a distance of 825 units.

Upon completing the staircase and the walls and floor of the first level, cut a number of openings in accordance with Figure 8.68.

The heights of the doors and gate are 2000 units and 2600 units, respectively. The windows are 1000 units from the top of the floor and are 1200 units high.

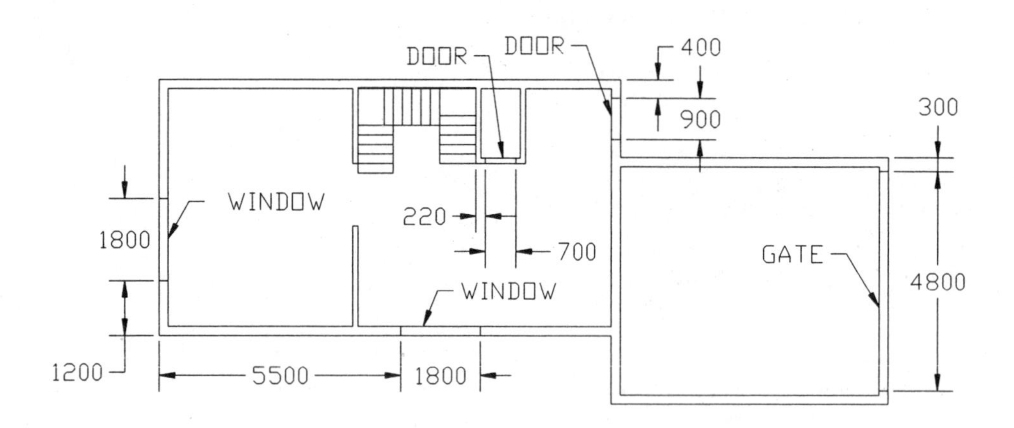

Figure 8.68 Window, door, and gate openings for the first level

Figure 8.69 shows the completed first level and the staircase.

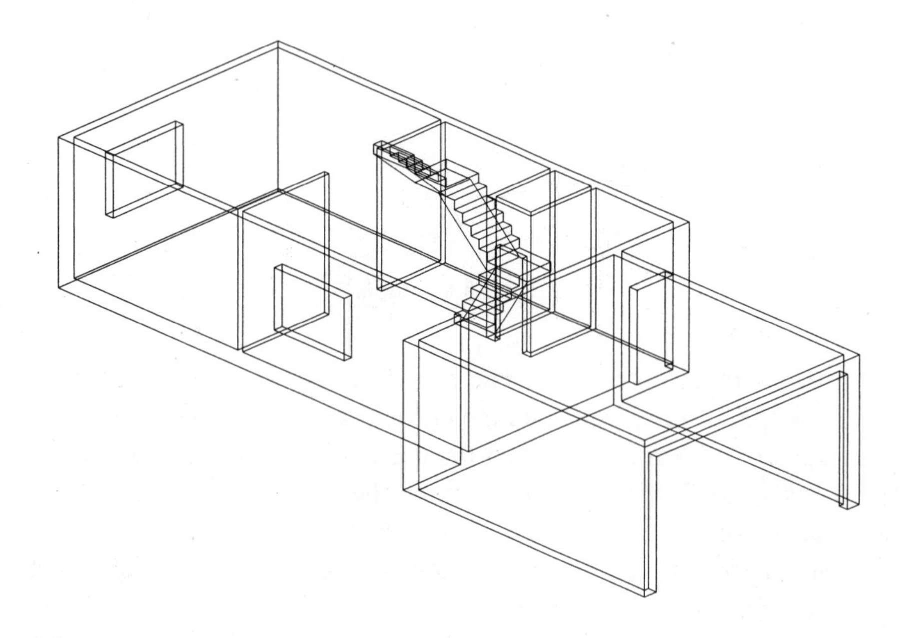

Figure 8.69 Completed first level and staircase

The first level and the staircase are complete. Turn off the layers W1, F1, and STAIR. Set the current layer to layer W2 and extrude the entities on this layer a distance of 2880 units. Set the current layer to F2 and extrude the floor wireframes a distance of 180 units.

After making the floor and walls of the second floor, cut a number of door and window openings in accordance with Figure 8.70.

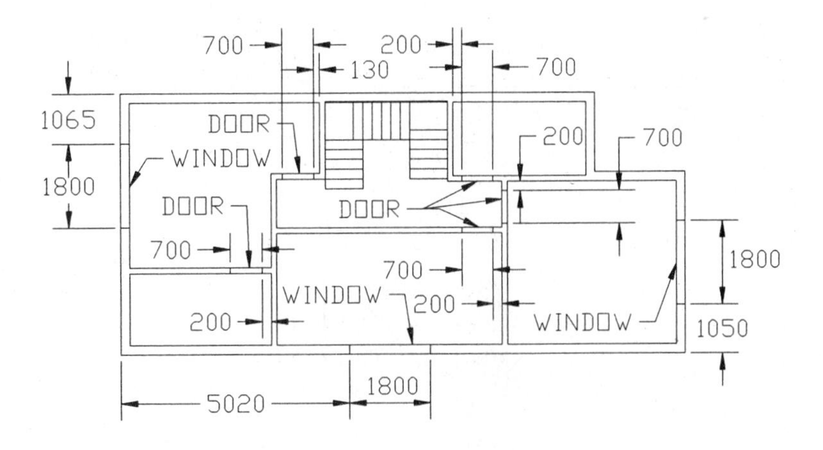

Figure 8.70 Window and door openings for the second level

Figure 8.71 shows the completed second level of the house.

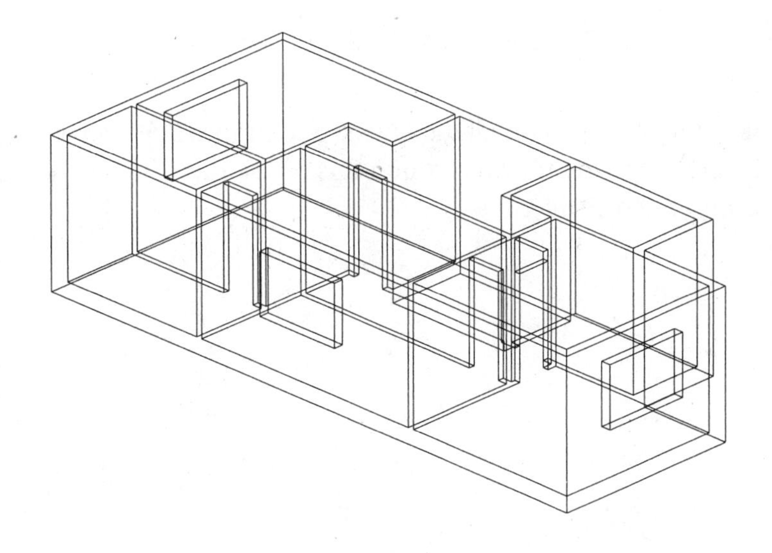

Figure 8.71 Completed second level

After completing the second level of the house, set the current layer to ROOF and turn off the layers W2 and F2. Extrude the seven wires. Then form three solids of intersection and unite two to form the roof of the second level. See Figure 8.72.

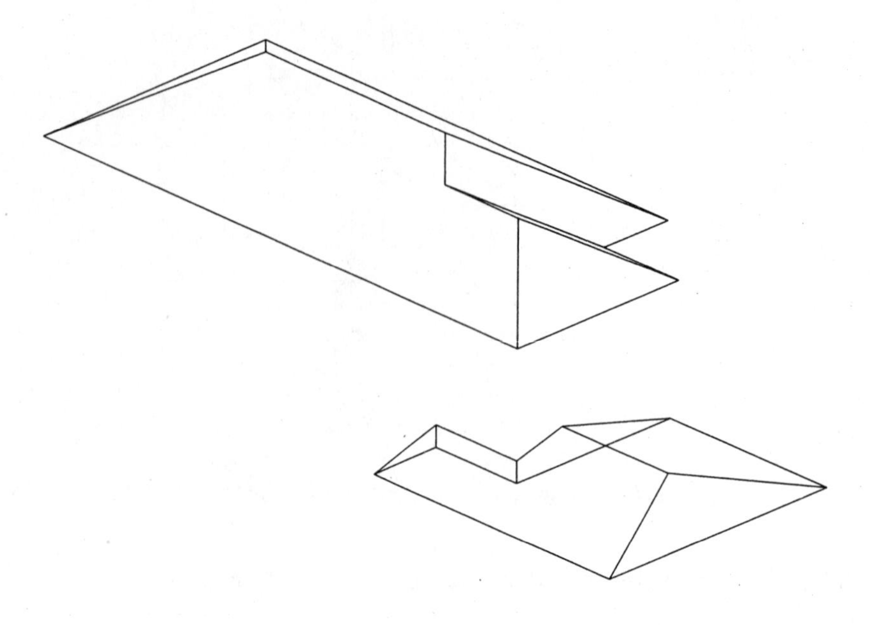

Figure 8.72 Completed roofs

The house is complete. Turn on all the layers. See Figure 8.59. Save your drawing. To further refine the drawing, you may insert into this house the window, doors and gate that you prepared earlier in the chapter.

Chapter 9

Documentation and Plotting

9.1 Engineering Drawing Title Block
9.2 Documentation for a 2D drawing
9.3 Documentation for a 3D Solid Model
9.4 Advanced Drawing Setup
9.5 Plot a Document
9.6 Batch Plotting
9.7 Key Points and Exercises

Aims and Objectives

The aims of this chapter are to illustrate the use of attributes in preparing an engineering drawing title block, to delineate the steps required to produce engineering documents for 2D drawing and 3D solid models, to explain the use of Advanced Setup Wizard, to depict plotting procedure, and to outline the steps in batch plotting. After studying this chapter, you should be able to:

- construct an engineering drawing title block with attributes defined,
- prepare an engineering document from a 2D drawing,
- prepare an engineering document from a 3D solid model,
- plot a document, and
- perform batch plotting.

Overview

AutoCAD has two working environments: model space environment and paper space environment. The model space environment is for you to construct the main constituent of a drawing. The paper space environment is for you to prepare a document for plotting. You can switch from one environment to another by setting the system variable TILEMODE. Document preparation involves two major tasks: adding a title block and creating floating viewports in paper space environment.

In the previous chapters, you used only the model space environment and created all the 2D objects and 3D models there. In this chapter, you will use the paper space environment to create engineering documents for both 2D drawings and 3D models. In addition, you will learn how to plot a drawing. Theoretically, you may plot a drawing in either model space or paper space. Practically, it is better to plot in paper space with a document properly prepared.

9.1 Engineering Drawing Title Block

To document a drawing is to construct a title block and some floating viewports in the paper space environment. Because it is time-consuming repeatedly to construct a title block for every document, you should prepare a drawing containing the title block and insert the title block to the document in paper space.

Typically, a title block should include the following information:

1. Title — Title of the project
2. Part Name — Name of the drawing part
3. Drawing Number — Number of the drawing
4. Drawn By — Name of the draftsperson
5. Date — Date of the drawing
6. File — Name of the drawing file
7. Revision Number — Number of the revision
8. Revised By — Name of the person revising the drawing
9. Revision Date — Date of revising the drawing
10. Finish — Surface finish to be applied
11. Material — Material of the drawing part
12. Scale — Plot scale of the drawing document

To make a title block for insertion, use the NEW command to start a new drawing.

<File> **<New...>**

Figure 9.1 shows a typical title block. In making the title block, you might ask what width of the margin between the four borders and the edges of the paper is. The answer is that it depends on how wide your plotting device needs to be to clamp the paper while it prints. You can determine this by drawing a rectangle. Then plot the rectangle with your plotting device, using the "Scaled to Fit" option. After plotting, measure the lengths of the lines printed. These lengths are the width and height of the rectangle that you should use to create a formal title block.

Construct the title block according to Figure 9.1. When creating the border lines of the title block, keep the size of the rectangle slightly less than the plotting size of the plotter.

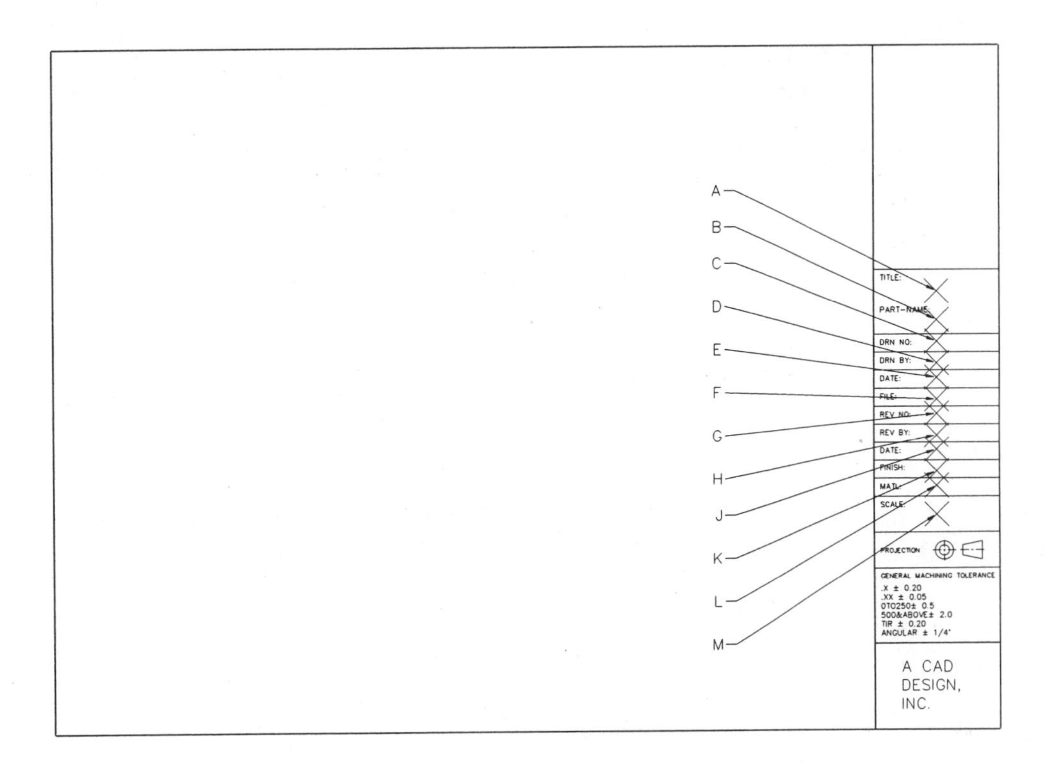

Figure 9.1 Typical drawing title block

Define Attribute Templates

Refer to Figure 9.1. The title block carries miscellaneous information. To save your time to input the data, you can define attribute templates in the title block so that you can add the information interactively during insertion. As explained in Chapter 4, you can use the ATTDEF command or the DDATTDEF command to define attribute templates.

To use the ATTDEF command to define an attribute template, type the command name at the command line interface.

Command: **ATTDEF**
Attribute modes -- Invisible:N Constant:N Verify:N Preset:N
Enter (ICVP) to change, or press ENTER when done: **[Enter]**
Attribute tag: **TITLE**
Attribute prompt: **PROJECT TITLE**
Default attribute value: **[Enter]**
Justify/Style/<Start point>: **[Select A (Figure 9.1).]**
Height: **3**
Rotation angle: **0**

To use the DDATTDEF command to prepare a template, select the Define Attributes... item of the Block cascading menu of the Draw pull-down menu.

<Draw> **<Block>** **<Define Attributes...>**

Command: **DDATTDEF**

[Attribute Definition

Attribute
Tag: **PART-NAME**
Prompt: **Part name...**
Text Options
Justification: **Left**
Text Style: **Standard**
Height< **3**
Rotation< **0**
Insertion Point
Pick Point<]

Start point: [**Select B (Figure 9.1).**]

[**OK**]

Now you have two attribute templates. Referring to the following table, repeat the DDATTDEF command to add ten more attributes. See Figure 9.2.

Command: [**Enter**]
DDATTDEF

Tag	Prompt	Pick Point<
NUMBER	Drawing number...	Select C (Figure 9.1).
NAME	Your name...	Select D (Figure 9.1).
DATE	Drawing date...	Select E (Figure 9.1).
FILE	Drawing file number...	Select F (Figure 9.1).
NUMBER	Revision number...	Select G (Figure 9.1).
NAME	Revised by...	Select H (Figure 9.1).
DATE	Date of revision...	Select J (Figure 9.1).
FINISH	Surface finish...	Select K (Figure 9.1).
MATERIAL	Material...	Select L (Figure 9.1).
SCALE	Plot scale...	Select M (Figure 9.1).

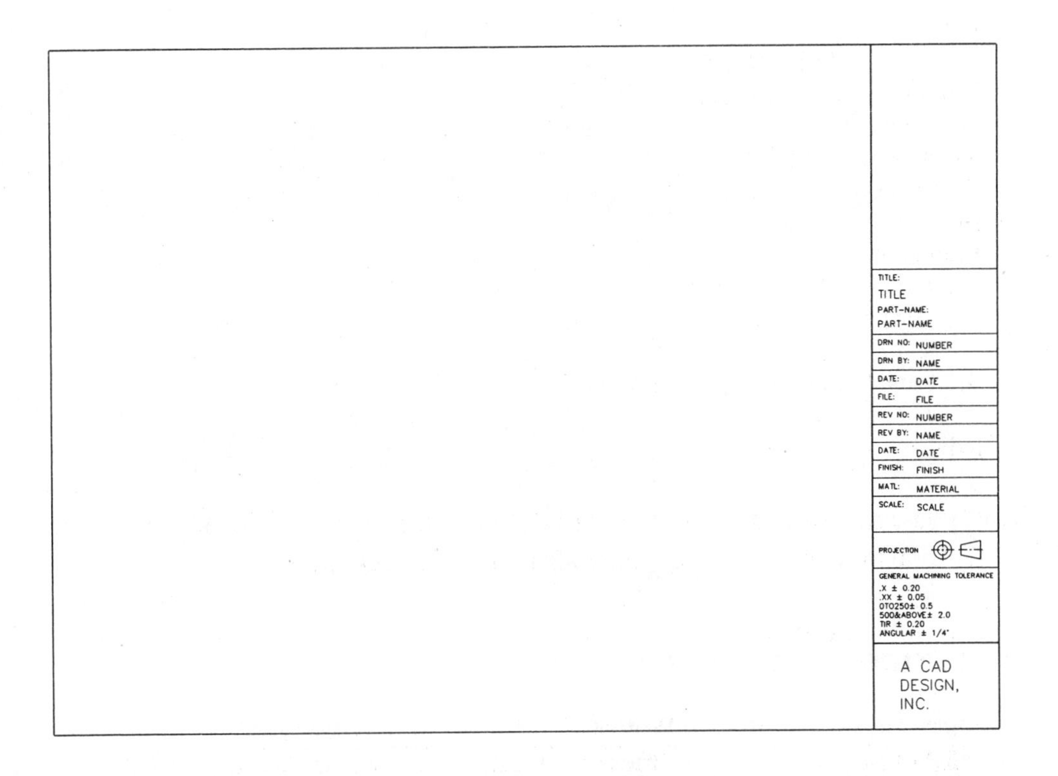

Figure 9.2 Attribute templates created

After defining all the templates, save the drawing. The file name is TITLE.

<File> **<Save...>**

File name: **TITLE**

The title block is complete. It is now ready for insertion to paper space environment for making an engineering document.

9.2 Documentation for a 2D Drawing

Use the OPEN command to open the assembly drawing that you created in Chapter 4. See Figure 9.3. You will prepare an engineering document for this 2D drawing.

<File> **<Open...>**

Filename: **PIPESP.DWG**

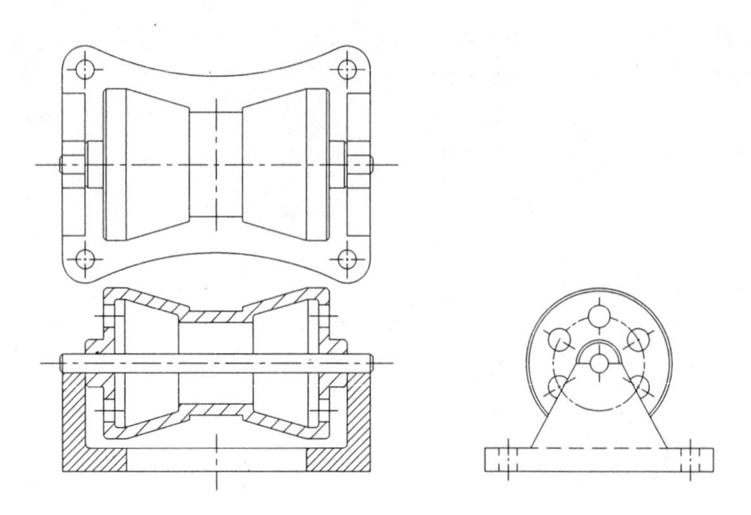

Figure 9.3 Pipe support assembly

Switch to Paper Space Environment

The entities created so far in this drawing are all in the model space environment. To switch from model space environment to paper space environment, you can manipulate the TILEMODE variable or select the [TILE] button of the status bar.

The default tilemode value is 1 (model space environment). Change the variable setting to 0 (paper space environment).

[Status Bar] **[TILE]**

Command: **TILEMODE**
New value for TILEMODE: **0**

After you switch to paper space environment, the original entities that you created in the drawing seemingly disappear because they are in another environment.

To remind you that you are working in paper space environment, AutoCAD replaces the normal UCS icon by a special icon that looks like a set square. See Figure 9.4.

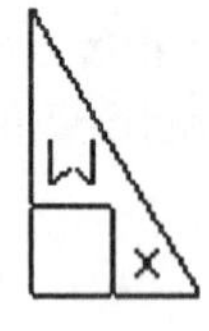

Figure 9.4 Special icon to depict paper space environment

Add a Title Block in Paper Space

The default screen display size is the size defined by the LIMITS command. This size depends on the drawing limits of the template drawing that you used to create the drawing. This display size might not be suitable for preparing a drawing. Before you can create floating viewports in the paper space, you must either zoom to a known size or place a title block of known size on the drawing.

Now you will add a title block. Use the LAYER command to create a layer called TITLE for holding the title block. Set this layer current.

<Format> **<Layer...>**

Command: **LAYER**

[Layer

Name	Color	Linetype
Title	White	Continuous

Current layer: **Title**

[OK]

In order to bring out a dialog box for attribute value input when inserting the title block drawing, set the ATTDIA variable to 1.

Command: **ATTDIA**
New value for ATTDIA: **1**

Apply the DDINSERT command to insert a title block into your drawing. Set the insertion point at (0,0), the insertion scale for both the X and the Y axis to be 1, and the rotation angle to be 0.

<Insert> **<Block...>**

Command: **DDINSERT**

[Insert **File...**]

File name: **TITLE.DWG**

[Options
Specify Parameters on Screen **NO**
Insertion Point:
X: **1** Y: **1** Z: **1**
Scale:
X: **1** Y: **1** Z: **1**
Rotation:
Angle: **0**
OK]

When the attribute input dialog box appears, supply attribute values according to Figure 9.5.

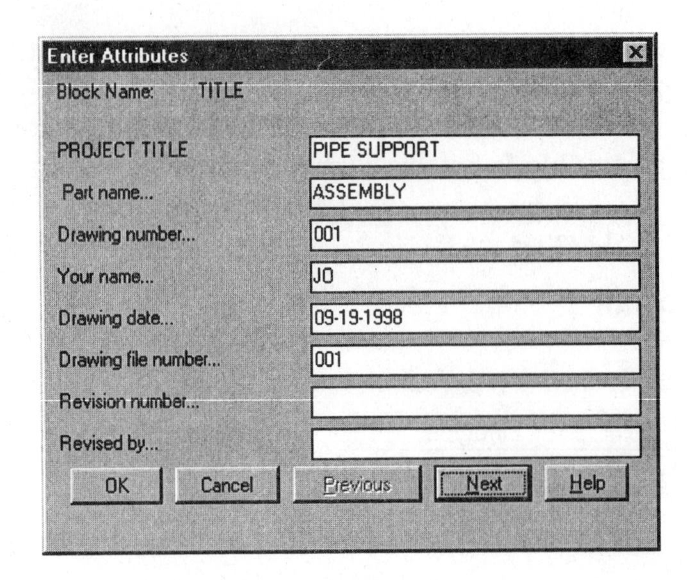

Figure 9.5 Add attribute values

Select the [Next] button and enter attribute values according to Figure 9.6.

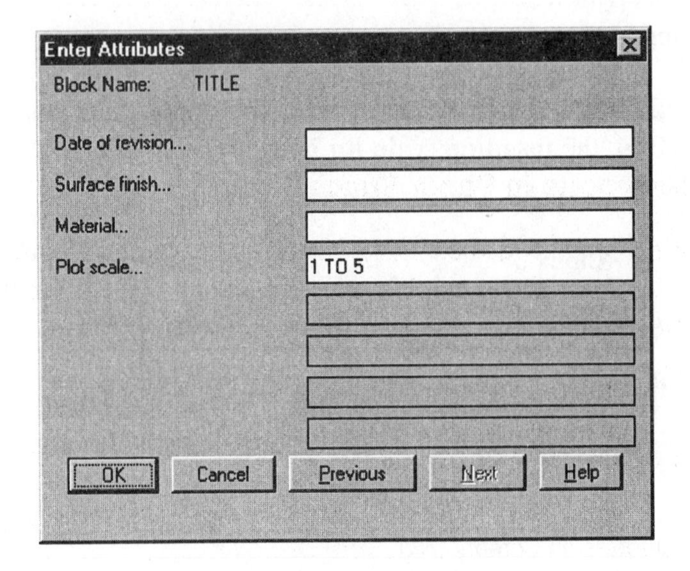

Figure 9.6 Second page

Select the [OK] button to exit.

[OK]

Title block insertion is complete. Apply the ZOOM command to zoom to the drawing extent. See Figure 9.7.

<View> **<Zoom>** **<All>**

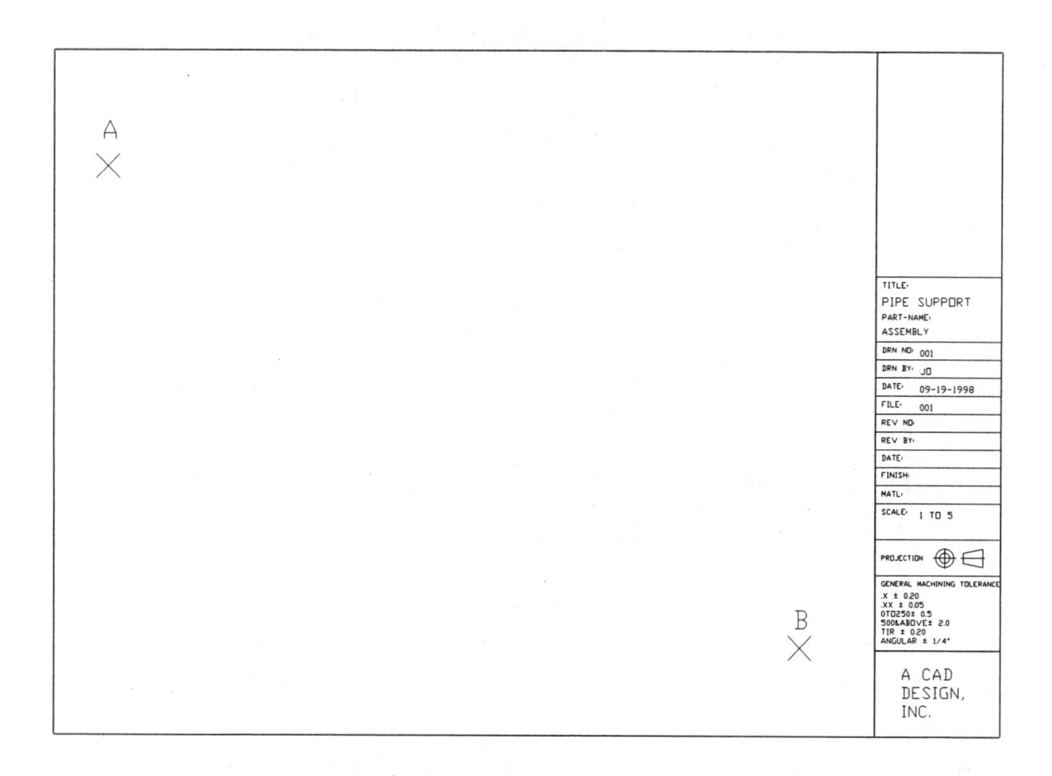

Figure 9.7 Title block inserted

Make Floating Viewports in Paper Space

After adding a title block, you then make floating viewports. Floating viewports are entities that you can manipulate. Apply the LAYER command to make a layer called Vports with color magenta for holding them. Set this layer current.

<Format> **<Layer...>**

Command: **LAYER**

[Layer

Name	Color	Linetype
Vports	Magenta	Continuous

Current layer: **Vports**

OK]

To create floating viewports, run the MVIEW command. See Figure 9.8.

<View> **<Floating Viewports>** **<1 Viewport>**

Command: **MVIEW**
ON/OFF/Hideplot/Fit/2/3/4/Restore/<First Point>: **[Select A (Figure 9.7).]**
Other corner: **[Select B (Figure 9.7).]**

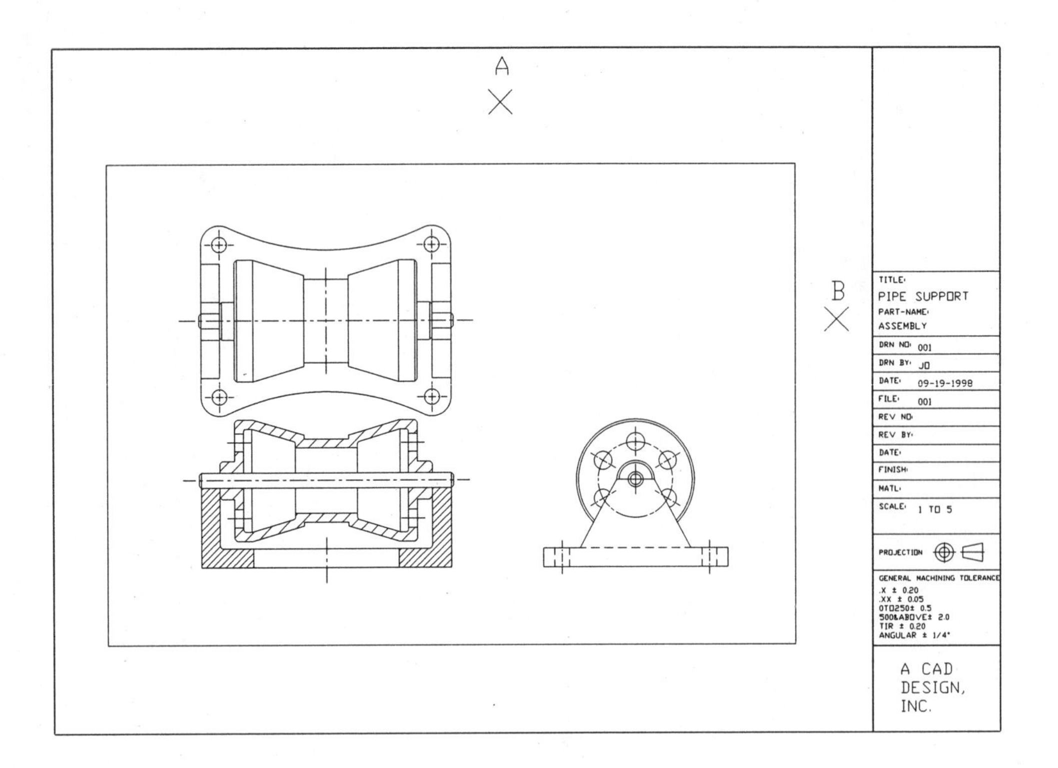

Figure 9.8 A floating viewport added

After you make a floating viewport, the entities in the model space reappear. You can regard the floating viewport as a window that cuts through the paper document. Through the window, you look into the model space environment from the paper space.

Apply the MVIEW command again to make another floating viewport. You can see that floating viewports can overlap each other. See Figure 9.9.

<View> **<Floating Viewports>** **<1 Viewport>**

Command: **MVIEW**
ON/OFF/Hideplot/Fit/2/3/4/Restore/<First Point>: **[Select A (Figure 9.8).]**
Other corner: **[Select B (Figure 9.8).]**

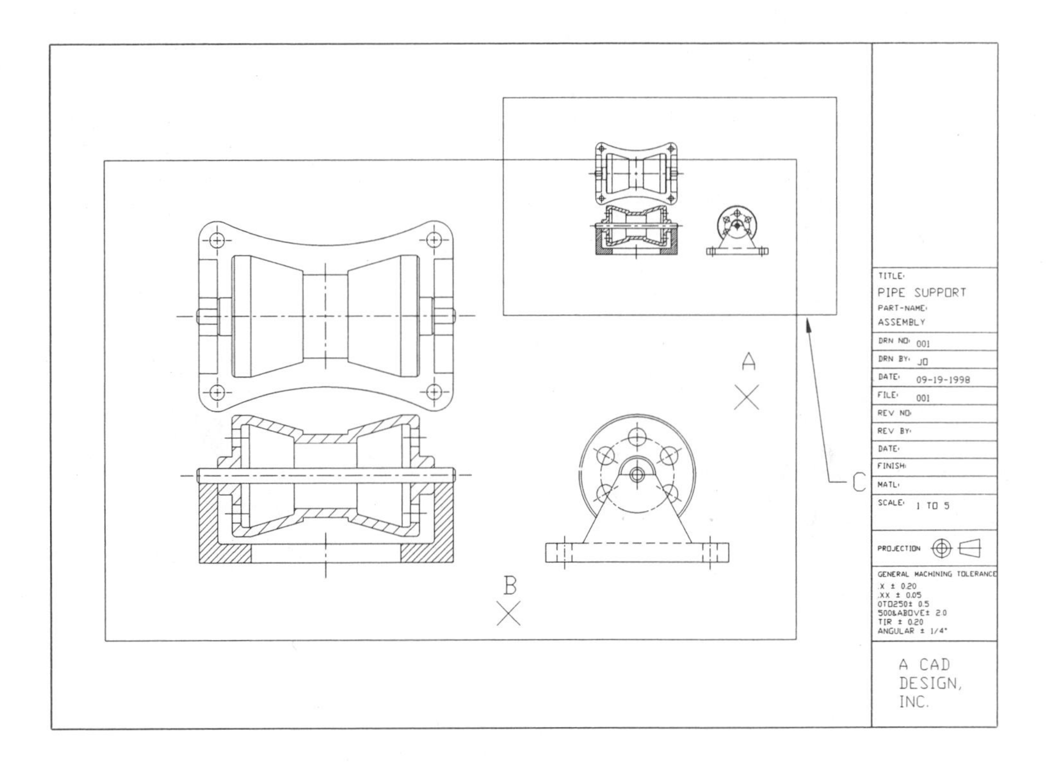

Figure 9.9 Second floating viewport added

You have created two floating viewports in paper space environment. In AutoCAD terms, there are two kinds of viewports: tiled viewports and floating viewports. In model space environment, you can use the VPORTS command to divide the screen display into a number of tiles; each is called a tiled viewport. In paper space environment, you can use the MVIEW command to make viewports that display objects constructed in model space. Because these kinds of viewports can be moved around and can overlap each other, they are called floating viewports.

Edit Objects in Paper Space

Now you are working in paper space environment. Through the floating viewports, you see the objects created in model space environment.

Issue the ERASE command and try to pick the entities inside the floating viewport.

<Modify> **<Erase>**

Command: **ERASE**
Select objects: **[Select A (Figure 9.9).]**
Other corner: **[Select B (Figure 9.9).]**
0 found
Select objects: **[Enter]**

As you see, you cannot manipulate the objects shown inside the floating viewports because they reside on another environment. Try the ERASE command on the edge of the second floating viewport. This time, the second viewport is erased. See Figure 9.10.

<Modify> <Erase>

Command: **ERASE**
Select objects: **[Select C (Figure 9.9).]**
Select objects: **[Enter]**

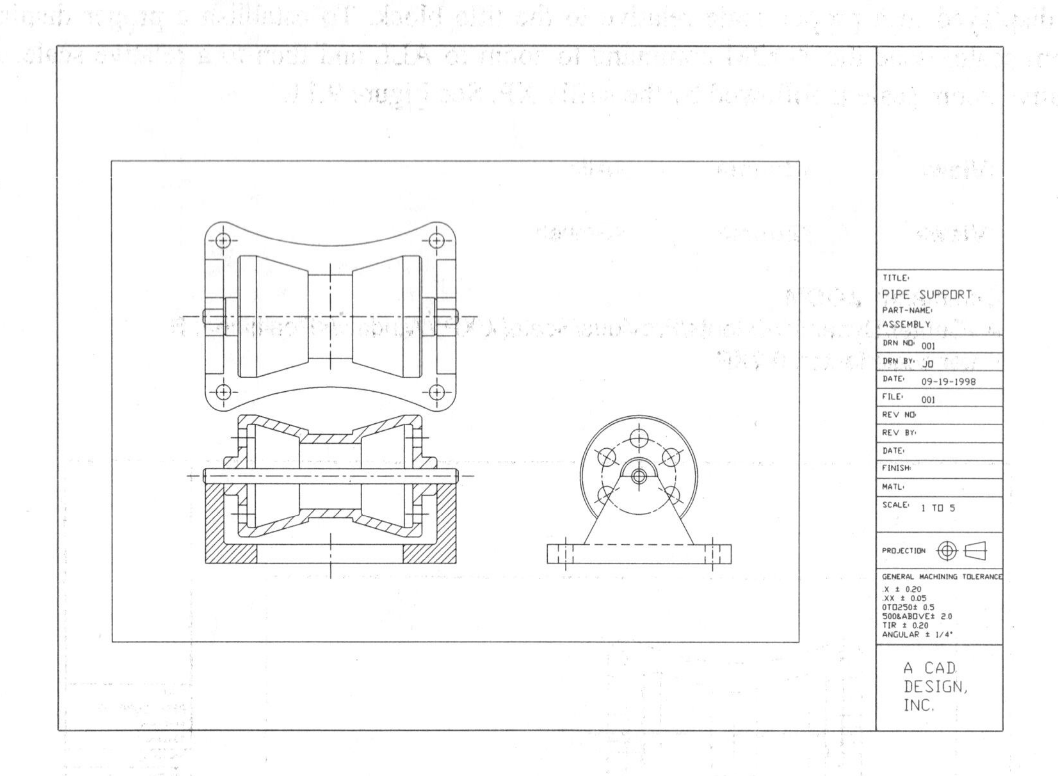

Figure 9.10 Second floating viewport erased

When you work in paper space environment, you can work only on objects created in paper space. Because the floating viewport is created in paper space, you can erase it. Besides the ERASE command, you can apply the COPY command, the MOVE command, the STRETCH command, and the SCALE command on the floating viewports.

Manipulate Entities Inside the Floating Viewports

In order to gain access to objects created in model space environment after you have set tilemode to zero, apply the MSPACE command. This command enables you to manipulate objects inside the floating viewports. Double select the [Paper] button of the status bar to change it to [Model].

[Status Bar] **[MODEL]**

Command: **MSPACE**

The special set square icon disappears, and the normal UCS icon reappears inside the floating viewport. Move the pointing device around. When it is inside the floating viewport, it changes to a pair of crosshairs. When it is outside the floating viewport, it changes to an arrowhead.

Now you can create and edit objects in model space in the normal way, but you cannot manipulate the objects created in paper space, although they appear on the screen.

In an engineering document, the objects shown inside the floating viewports need to be displayed in a proper scale relative to the title block. To establish a proper display zoom scale, issue the ZOOM command to zoom to ALL and then to a relative scale. A relative zoom scale is followed by the suffix XP. See Figure 9.11.

<View> **<Zoom>** **<All>**

<View> **<Zoom>** **<Scale>**

Command: **ZOOM**
All/Center/Dynamic/Extents/Previous/Scale(X/XP)/Window/<Realtime>: **S**
Enter scale factor: **0.2XP**

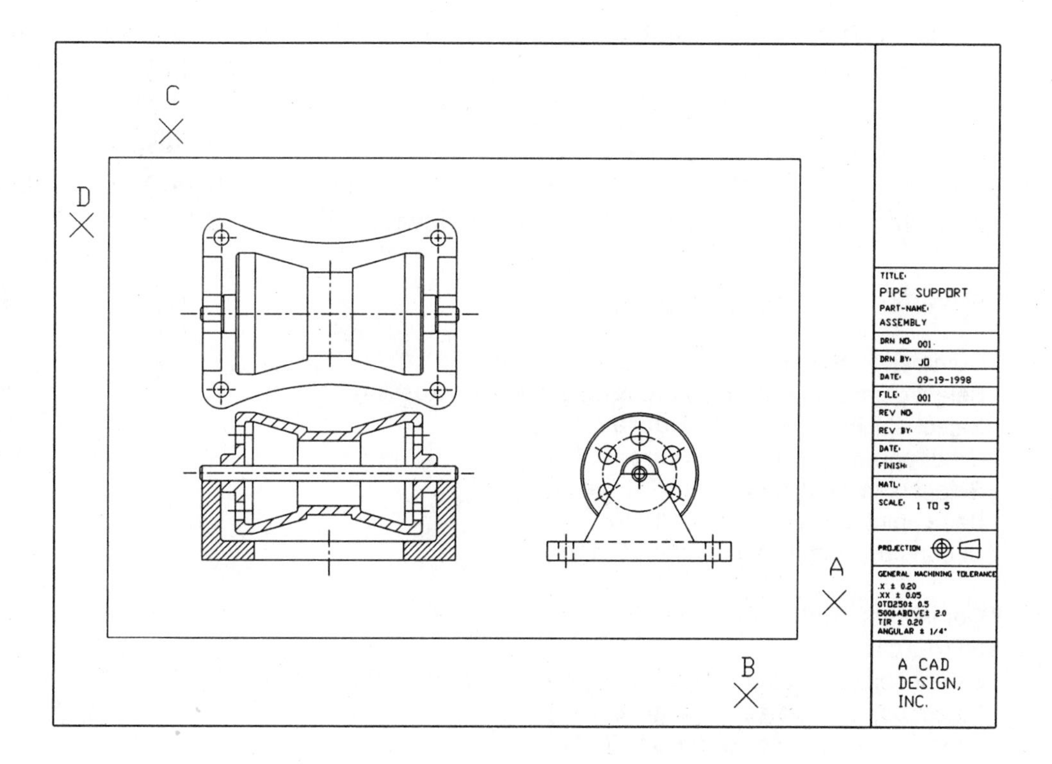

Figure 9.11 Zoom to a scale relative to scale in paper space

Here, the suffix XP indicates that the stated zoom scale is relative to the scale in paper space. A relative zoom scale of 0.2XP means that the entities in the floating viewport are zoomed to a scale of 0.2 relative to the entities in paper space. The scale of the objects as compared to the title block is 1 to 5.

Switch Back to Paper Space

After running the MSPACE command with the TILEMODE variable set to 0, you can perform any task that you can do when TILEMODE is set to 1. However, you should switch TILEMODE back to 1 if you are going to do a lot of entity creation and editing work, because the floating viewport is much smaller than the screen display.

Apply the PSPACE command to resume working in paper space environment by double selecting the [MODEL] button of the Status window to change it to [PAPER].

[Status Bar] **[PAPER]**

Command: **PSPACE**

Refer to your screen display. The special set square icon reappears, and the cursor changes to a pair of crosshairs.

To toggle between the two working spaces while TILEMODE is set to 0, use the MSPACE command and the PSPACE command.

Edit Floating Viewports

You may apply five editing commands on the floating viewports: ERASE, COPY, MOVE, SCALE, and STRETCH. Just previously, you erased a floating viewport. In order to become familiar with the effects of other editing commands on floating viewports, you will stretch, copy, scale, and move floating viewports in this drawing.

Run the STRETCH command to stretch the two opposite corners of the floating viewport to adjust the window size of the floating viewport. See Figure 9.12.

<Modify> **<Stretch>**

Command: **STRETCH**
Select objects to stretch by crossing-window or crossing-polygon...
Select objects: [**Select A (Figure 9.11).**]
Other corner: [**Select B (Figure 9.11).**]
Select objects: [**Enter**]
Base point or displacement: **10<0**
Second point of displacement: [**Enter**]

Command: [**Enter**]
STRETCH
Select objects to stretch by crossing-window or crossing-polygon...
Select objects: [**Select C (Figure 9.11).**]
Other corner: [**Select D (Figure 9.11).**]
Select objects: [**Enter**]
Base point or displacement: **10<180**
Second point of displacement: [**Enter**]

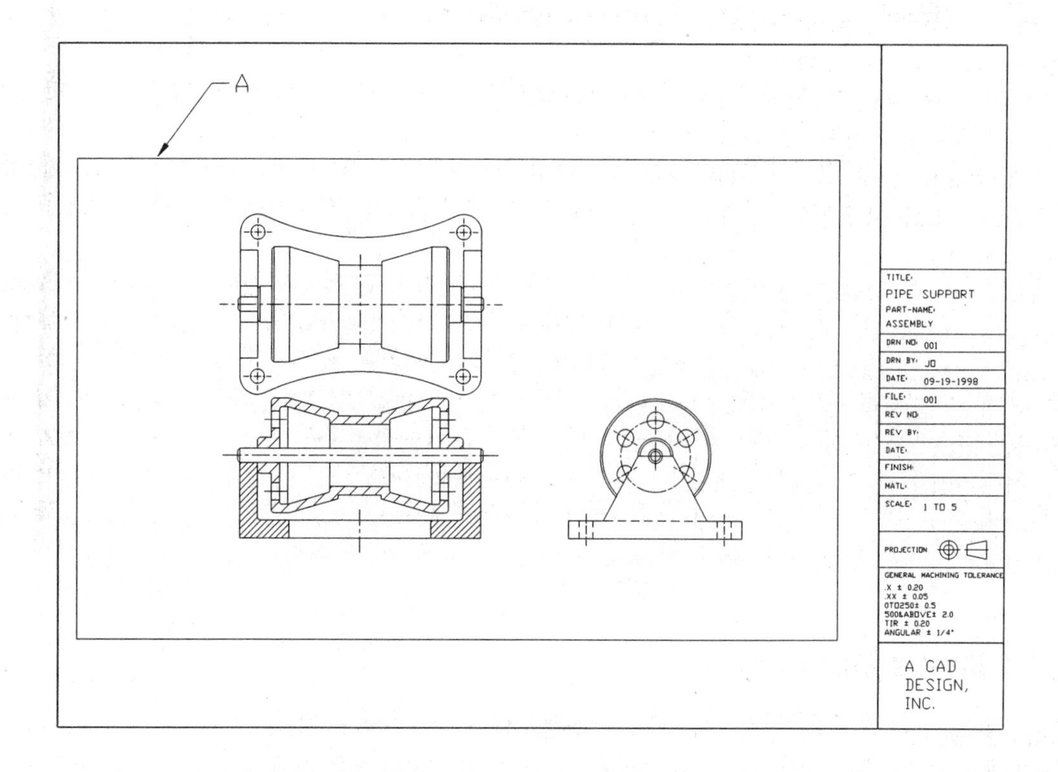

Figure 9.12 Floating viewport stretched

Refer to Figure 9.12 and compare it with Figure 9.11. You can see that the zoom scale of the entities inside the floating viewport is not affected by stretching the window size. By stretching a floating viewport, you simply adjust the window size.

You can copy a floating viewport. Use the COPY command. See Figure 9.13.

<Modify> **<Copy>**

Command: **COPY**
Select objects: **[Select A (Figure 9.12).]**
Select objects: **[Enter]**
<Base point or displacement>/Multiple: **30<45**
Second point of displacement: **[Enter]**

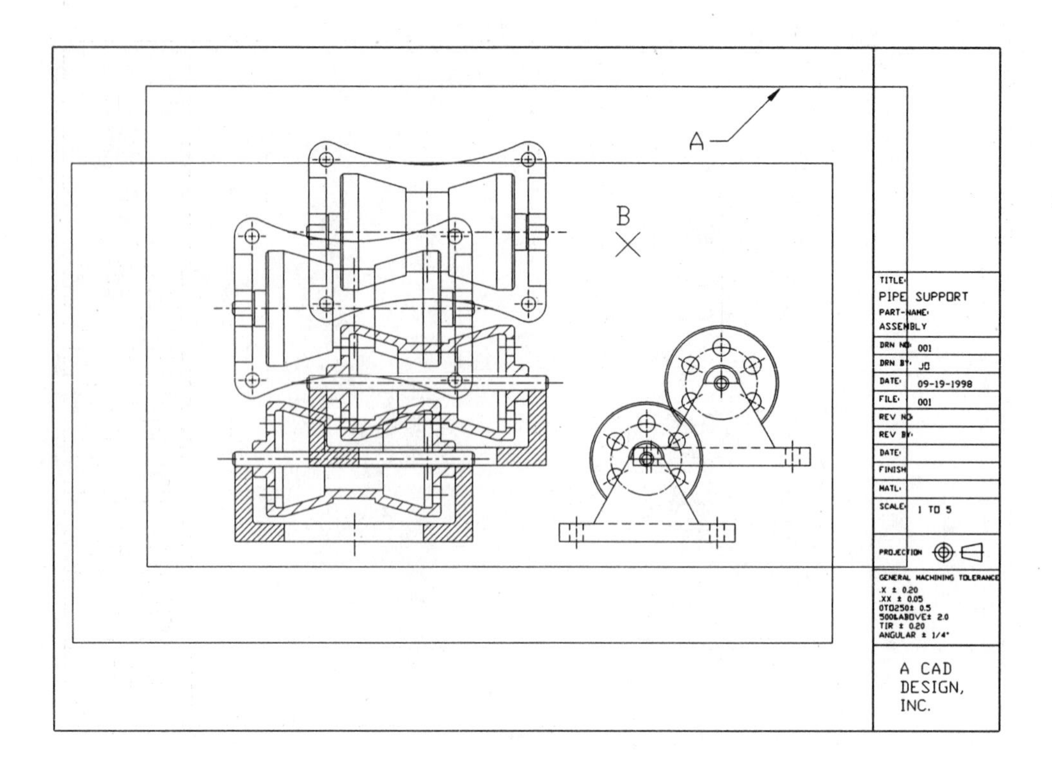

Figure 9.13 Floating viewport copied

Refer to Figure 9.13. Copying a floating viewport makes an exact replica of the original floating viewport, including the zoom scale inside the floating viewport.

To see the effects of scaling a floating viewport, run the SCALE command. See Figure 9.14.

<Modify> **<Scale>**

Command: **SCALE**
Select objects: [**Select A (Figure 9.13).**]
Select objects: [**Enter**]
Base point: [**Select B (Figure 9.13).**]
<Scale factor>/Reference: **0.6**

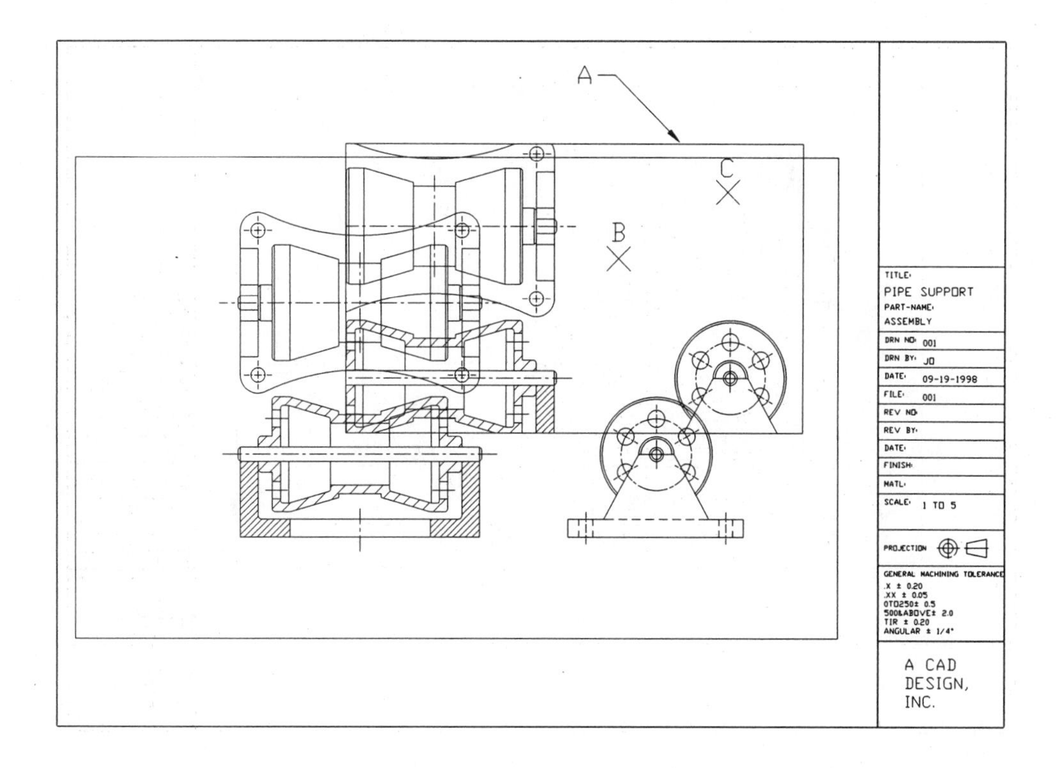

Figure 9.14 Second floating viewport scaled

Refer to Figure 9.14. You can see that scaling a floating viewport is scaling the window size. The zoom scale inside the floating viewport is not affected.

Apply the MOVE command to move the second floating viewport to the upper right corner of your drawing. See Figure 9.15.

<Modify> **<Move>**

Command: **MOVE**
Select objects: **[Select A (Figure 9.14).]**
Select objects: **[Enter]**
Base point or displacement: **[Select B (Figure 9.14).]**
Second point of displacement: **[Select C (Figure 9.14).]**

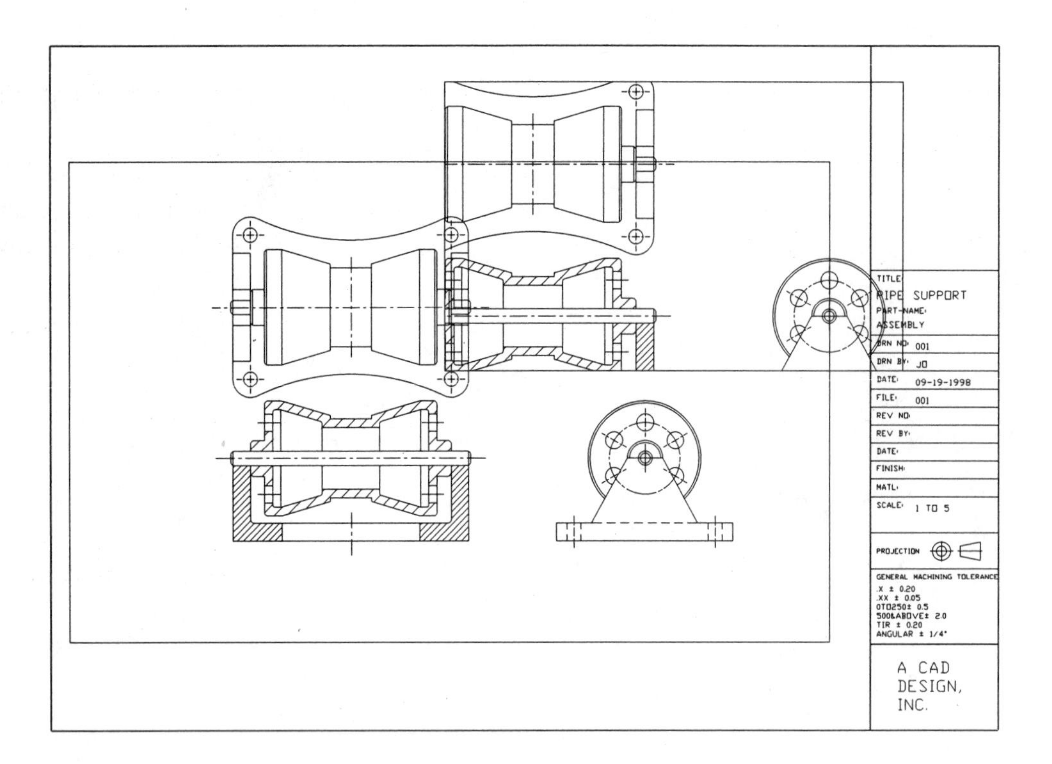

Figure 9.15 Second floating viewport moved

Stretch and move the second floating viewport according to Figure 9.16.

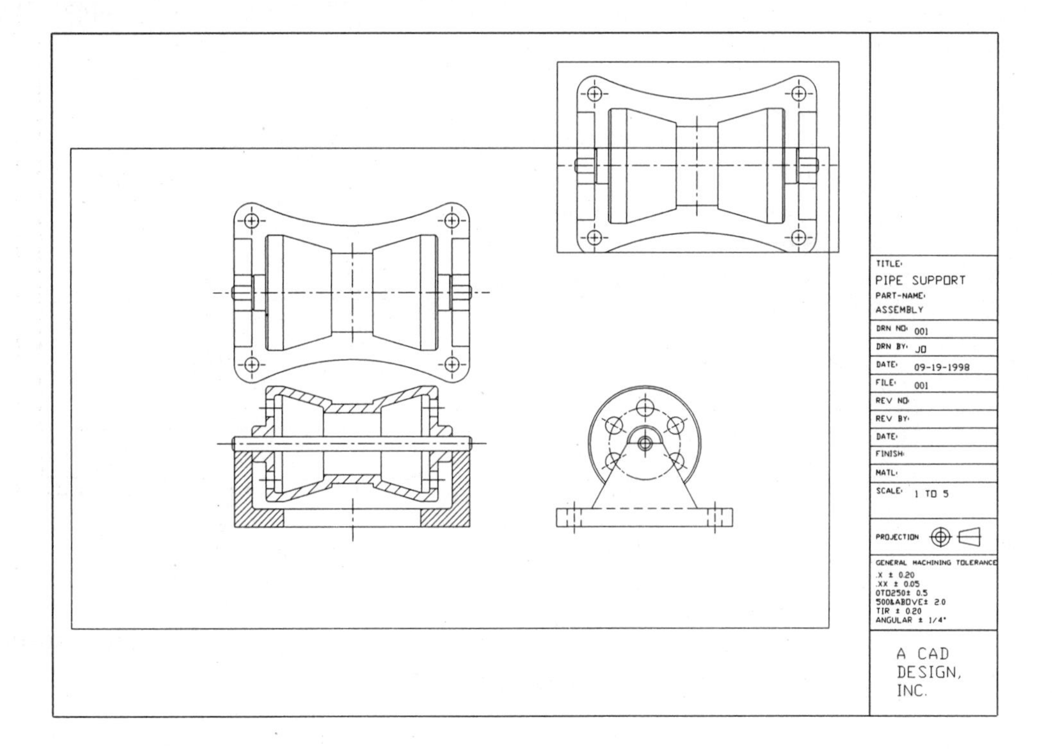

Figure 9.16 Second floating viewport stretched

Floating Viewport Visibility Control

Now you have two floating viewports: a large floating viewport showing three orthographic views and a small floating viewport showing only the top view.

Suppose you want to display all the entities in the large floating viewport and only those entities related to the second component of the pipe support assembly in the small floating viewport. To do so, you cannot simply turn off some layers, because that would turn them off in all the viewports.

Run the MSPACE command to gain access to model space environment and select the small floating viewport to make it the active floating viewport.

[Status Bar] **[MODEL]**

Command: **MSPACE**

To suppress visibility of layers in the selected viewports, use the LAYER command and select the icons under the Freeze in Current Viewport column.

<Format> **<Layer...>**

Command: **LAYER**

When the dialog box appears, select the icons related to the layers 0, Center, Hatch, and Hidden under the Freeze in Current Viewport column. See Figure 9.17.

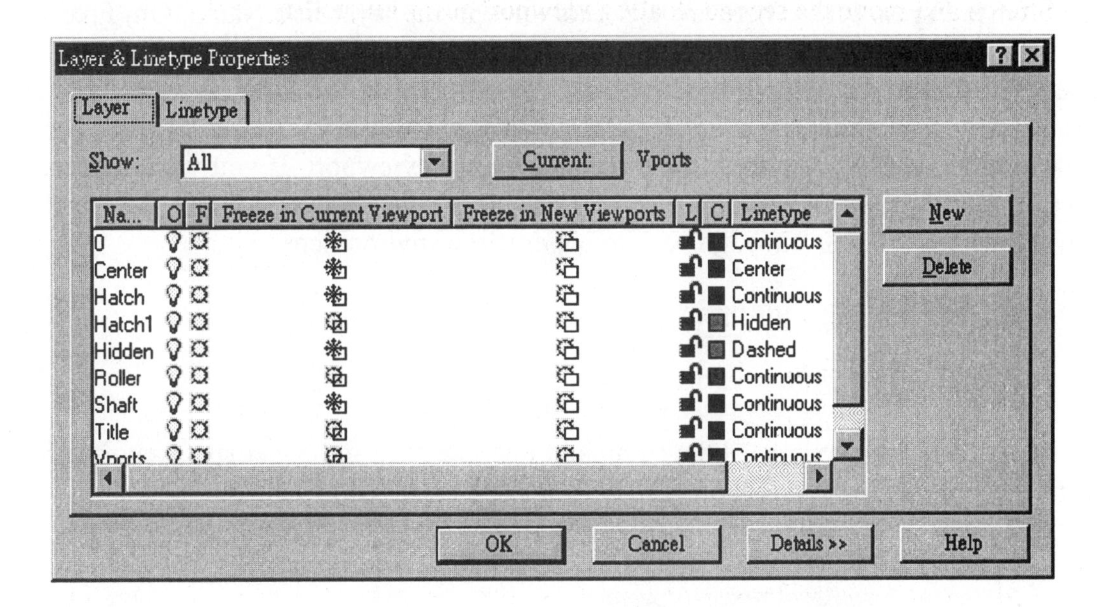

Figure 9.17 Selected layers frozen in the current viewport

Select the [OK] button to exit. The layers 0, Center, Hatch, and Hidden are frozen in the small floating viewport. See Figure 9.18.

[OK]

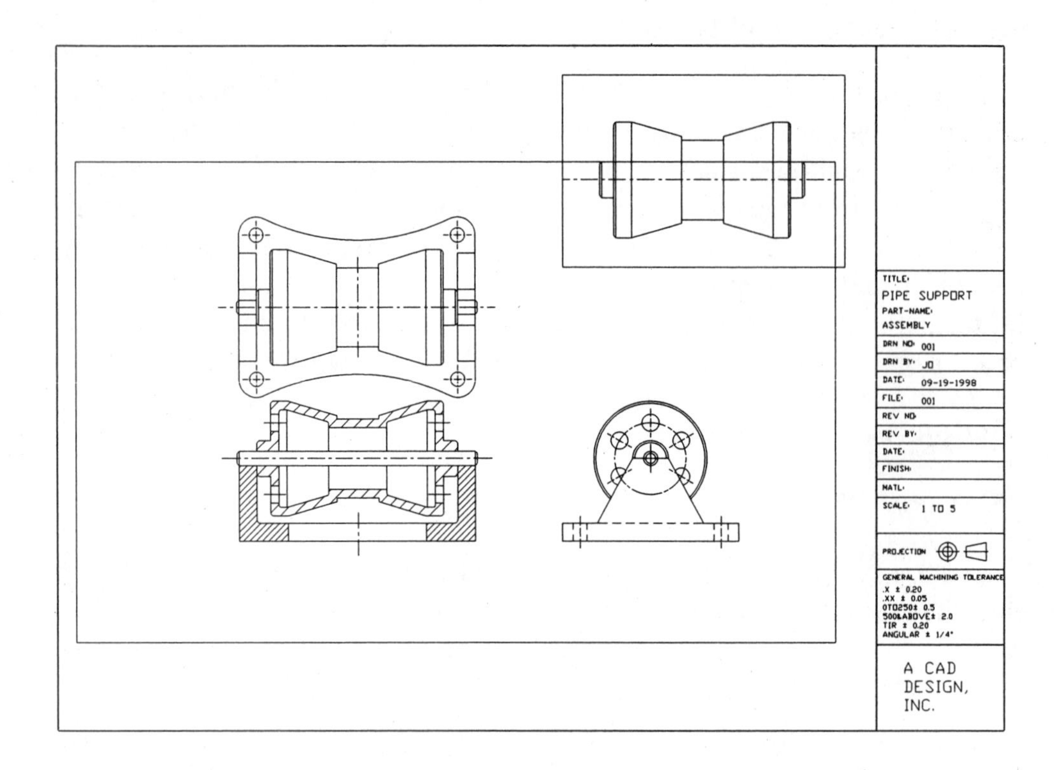

Figure 9.18 Some layers frozen in the second floating viewport

Refer to Figure 9.17. There are eight columns in the layer list: Name, On, Freeze in All Viewports, Freeze in Current Viewport, Freeze in New Viewports, Lock, Color, and Linetype. In Chapter 2 and other chapters, you worked on the LAYER command and learned how to control layers except Freeze in Current Viewport and Freeze in New Viewports. Indeed, you can freeze a layer in a selected viewport. If you want to freeze a layer in subsequently created new viewports, you can select the icon under the Freeze in New Viewports column. Now, go back to paper space environment.

[Status Bar] **[PAPER]**

Command: **PSPACE**

The command that works at the command line interface and paper space environment for freezing layers in selected floating viewports is VPLAYER.

Command: **VPLAYER**
?/Freeze/Thaw/Reset/Newfrz/Vpvisdflt:

Complete the Document

To complete the drawing, turn off the edges of the floating viewports. Apply the LAYER command. See Figure 9.19.

<Format> **<Layer...>**

Command: **LAYER**

[Layer

Name	On
Vports	Off

Current layer: **Title**

[OK]

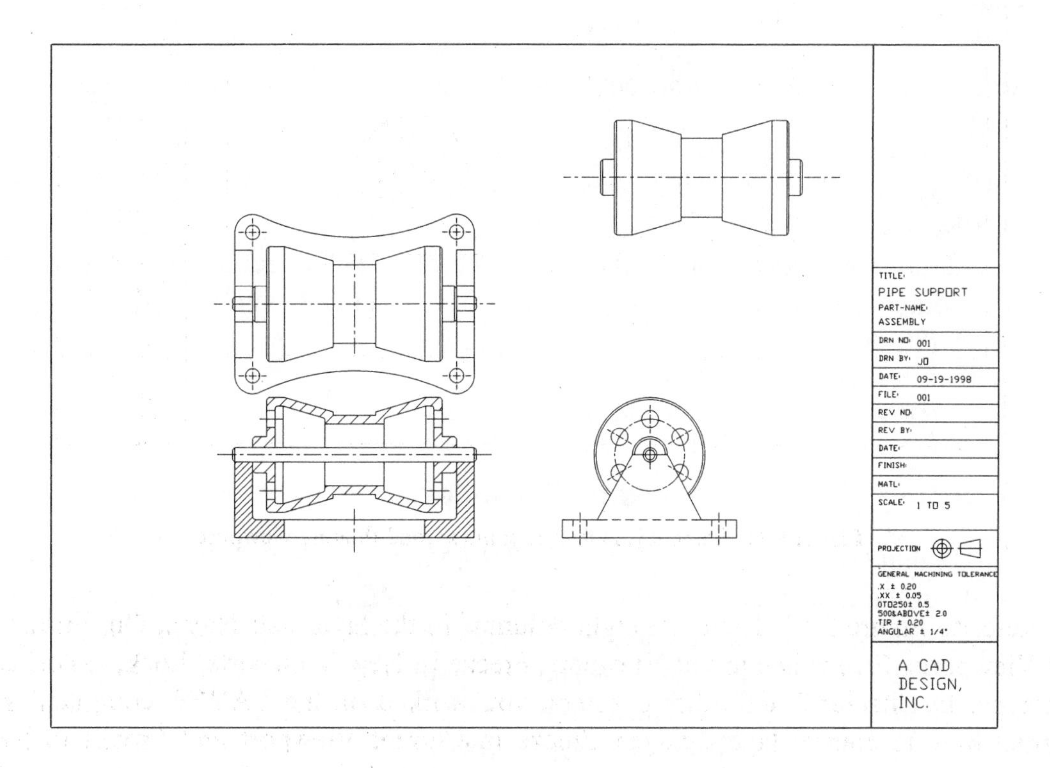

Figure 9.19 Completed document

Now you have a small floating viewport that shows only the top view of the second component and a large floating viewport that shows three orthographic views. Save the drawing.

<File> **<Save>**

The drawing document is complete. It is ready for plotting.

Dimensioning a Document

In Chapter 5, you learned how to add dimensions to a drawing. More recently, you learned that there are two working environments: paper space environment and model space environment. Depending on where you want to place the dimensions (tilemode=1, tilemode=0 and paper space, and tilemode=0 and model space), you have to make appropriate changes to the two dimension variables DIMSCALE and DIMLFAC.

Dimensioning with Tilemode=1

If TILEMODE is set to 1, you have to consider the relative zoom scale (?XP) of the floating viewport in paper space. Then you should use the reciprocal of this zoom scale as a factor to scale the size of the dimension components. For example, if the zoom scale is 0.2XP, the scale factor should be 1/0.2 = 5. In this case, set DIMSCALE to 5.

As for the DIMLFAC variable, set it to 1 so that the displayed dimension text is equal to the exact dimension value.

Dimensioning in Model Space with Tilemode=0

If you place dimensions in model space while TILEMODE is set to 0, you can simply let AutoCAD do the scaling of the size of the dimension components for you by setting DIMSCALE to 0.

When DIMSCALE is 0, AutoCAD uses the current relative zoom scale to adjust the size of the dimension components for you. Therefore, you must not do zoom in and zoom out during the dimensioning process if DIMSCALE is 0. If you do so, you may get unpredictable results.

Again, the DIMFLAC variable has to be set to 1.

Dimensioning in Paper Space with Tilemode=0

If you place dimensions in paper space while TILEMODE is set to 0, the dimensions will not appear in the model space after you switch TILEMODE back to 1. This can be an advantage, because it enables you to keep the model space free of any dimensions.

The main drawback of putting dimensions in paper space is that the definition points are not in model space. As a result, editing a feature in model space will not cause an automatic update of the dimensions in paper space.

Before you add dimensions, you need to know the relative zoom scale of the floating viewport and to make the necessary adjustment to the variable DIMLFAC, because measurement is taken on the paper instead of on the model. For example, if the zoom scale is 0.5XP, an actual length of 100 units will appear as 50 units in paper space. Therefore, you have to set the DIMLFAC variable to 2. As for the DIMSCALE variable, set it to 1.

DIMSCALE & DIMLFAC

The following is a summary of the settings explained in the previous paragraphs.

Tilemode	Working Environment	DIMSCALE	DIMLFACE
1	Model Space Only	Reciprocal of relative zoom scale	1
0	Model Space	0	1
0	Paper Space	1	Reciprocal of relative zoom scale

9.3 Document for a 3D Solid Model

A 3D solid model is ideal for design and manufacture. It has all the information required about the model. In addition to outputting a solid model in electronic data format for onward processing, you can prepare an engineering document from it.

Now you will prepare a document from the solid model of the gear box cover that you constructed in Chapter 6. Open the drawing of the gear box cover. See Figure 9.20.

<File> **<Open...>**

File name: **GBOX.DWG**

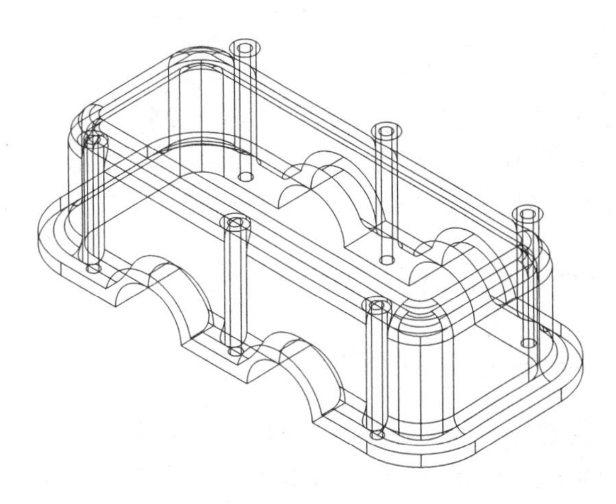

Figure 9.20 Solid model of angle block

By default, this model was created in the model space environment. To begin making an engineering document, switch the display to the paper space environment by selecting the [TILE] button of the status bar to set the system variable TILEMODE to zero.

[Status Bar] **[TILE]**

Command: **TILEMODE**
New value for TILEMODE: **0**

After you set the system variable TILEMODE to zero, the solid model in the model space disappears temporarily because you have switched to paper space environment. Now there should be nothing on your screen.

As in preparing a document for a 2D drawing, you will add a title block and then prepare floating viewports. Prior to inserting a drawing title block, make a new layer called TITLE. You will put the title block on this layer.

<Format> **<Layer...>**

Command: **LAYER**

Name	Color	Linetype
Title	White	Continuous

Current layer: **Title**

[OK]

Set the ATTDIA variable to 1 and then insert the title block.

Command: **ATTDIA**
New value for ATTDIA: **1**

<Insert> **<Block...>**

Command: **DDINSERT**

[Insert **File...**]

File name: **TITLE.DWG**

[Options
Specify Parameters on Screen **NO**
Insertion Point:
X: **1** Y: **1** Z: **1**
Scale:
X: **1** Y: **1** Z: **1**
Rotation:
Angle: **0**
OK]

When the attribute input dialog box appears, enter attribute values and then select the [OK] button. See Figure 9.21.

Enter Attribute	
PROJECT TITLE	GEAR BOX
Drawing number...	002
Your name...	JO
Drawing date...	12-1-1998
Drawing file number...	002
Material...	ALUMINUM
Plot scale...	1 TO 5

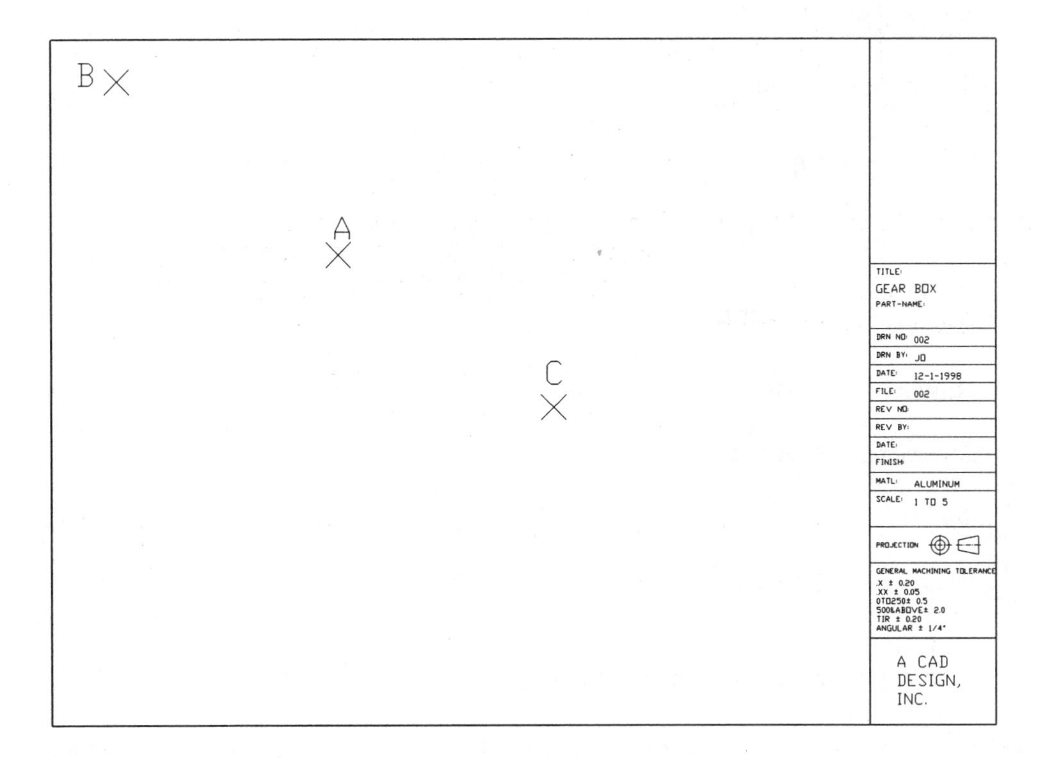

Figure 9.21 Title block inserted

Creating Floating Viewports

With a proper title block of known size in position, you can make the floating viewports for the solid model. To do so, you can use the SOLVIEW command. This command is tailor-made for 3D solid models. In addition to making the floating viewports like the MVIEW command does, the SOLVIEW command:

1. Creates a new layer called VPORTS for holding the viewport entities.
2. Prompts you to input a viewport name for each floating viewport. Then it creates three additional layers. The name of the first layer is the viewport name with the suffix "-dim," the second with "-hid," and the third with "-vis." The first layer, -dim, is reserved for placing the dimension entities. The layers -hid and -vis are used for placing the hidden lines and visible lines that are projected from the solid model by the SOLVIEW command. You should not put anything on these two layers. If you have already loaded the linetype Hidden, the layer -hid will use this linetype.
3. Allows you to create an orthographic viewport that aligns with a chosen drawing viewport.
4. Enables you to prepare a viewport for auxiliary view.
5. Allows you to prepare a viewport for projecting a sectional view by using the SOLDRAW command. If you specify a sectional view, a fourth layer for placing the hatching lines will be created for that viewport, with the suffix "-hat."
6. Freezes the additional layers in all other floating viewports so that they are visible only in the specific viewport.
7. Saves a display view using the name of the viewport. You can return to this saved view anytime after zooming and panning.

Use the LINETYPE command to load the linetype Hidden to the drawing file.

<Format> **<Linetype...>**

Command: **LINETYPE**

After loading the Hidden linetype, select the View item of the Setup item of the Solids cascading menu of the Draw pull-down menu to run the SOLVIEW command to prepare a floating viewport for the solid model.

<Draw> **<Solids>** **<Setup>** **<View>**

Command: **SOLVIEW**

The SOLVIEW command provides four options. To begin a new drawing, use the Ucs option because the other three options require a parent view. They project views from a selected view.

Ucs/Ortho/Auxiliary/Section/<eXit>: **UCS**

After choosing the Ucs option, you need to specify the orientation of the drawing view. Select the W option to prepare a floating viewport that corresponds to the plan view of WCS.

Named/World/?/<Current>: **W**

Unlike the MVIEW command that you need to zoom the objects in model space afterward, the SOLVIEW asks for a relative zoom scale.

Enter view scale: **1**

The next prompt asks for the location of the floating viewport. Select a point on the screen to locate the floating viewport.

View center: **[Select A (Figure 9.21).]**

After you select a point on the screen, a preview of the objects appears. Refer to your screen; the preview is too large when compared to the title block. Use the [ESC] key to cancel the SOLVIEW command. Then repeat the SOLVIEW command. This time, use a relative zoom scale of 0.2.

View center: ***Cancel***

Command: **[Enter]**
SOLVIEW
Ucs/Ortho/Auxiliary/Section/<eXit>: **UCS**
Named/World/?/<Current>: **W**
Enter view scale: **0.2**
View center: **[Select A (Figure 9.21).]**
View center: **[Enter]**

After confirming the relative zoom scale and the center location of the floating viewport, you need to specify the clipping size of the floating viewport.

Clip first corner: **[Select B (Figure 9.21).]**
Clip other corner: **[Select C (Figure 9.21).]**
View name: **TOP**
....

A top view with a relative zoom scale of 0.2 appears (Figure 9.22). If you run the LAYER command transparently, the layers Vports, Top-dim, Top-hid, and Top-vis are created.

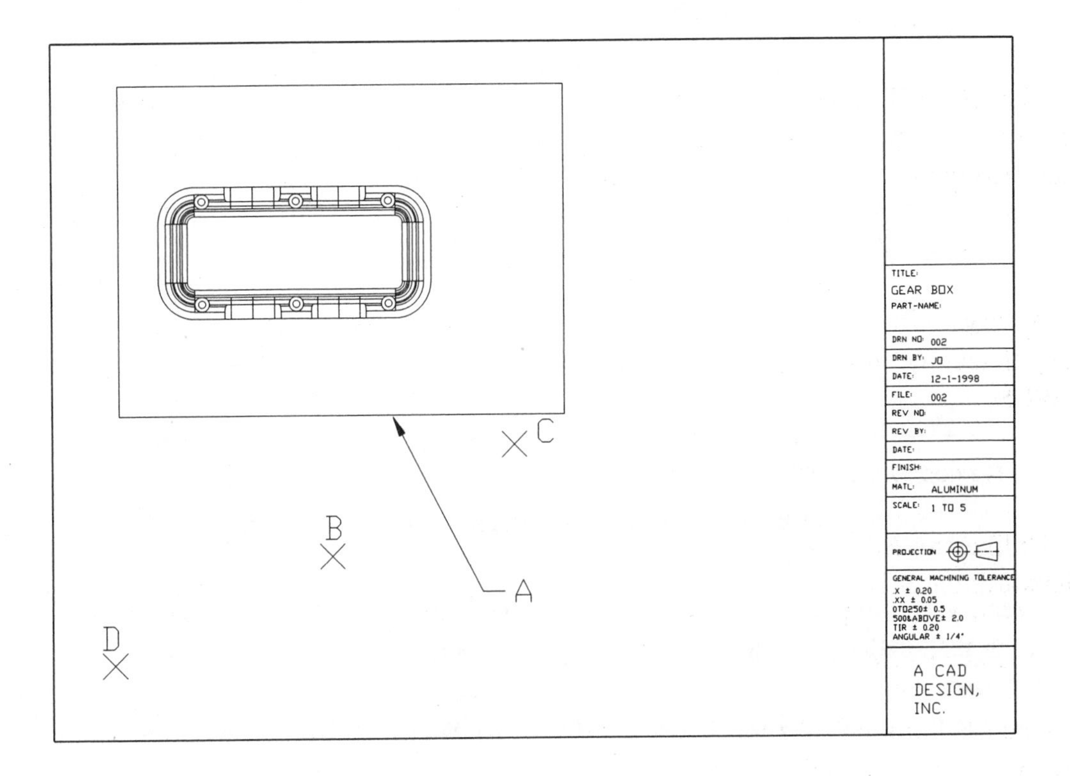

Figure 9.22 Floating viewport for the top view created

Continue the SOLVIEW command to create a front view, which is an orthographic view. Choose the O option. An orthographic view needs to be projected from an existing view; therefore, you have to select an edge of the top view to specify from which direction you would like to project. Select the lower edge of the top view, and then specify the location and the size of the new viewport. See Figure 9.23.

....
Ucs/Ortho/Auxiliary/Section/<eXit>: **O**
Pick side of viewport to project: **[Select A (Figure 9.22).]**
View center: **[Select B (Figure 9.22).]**
View center: **[Enter]**
Clip first corner: **[Select C (Figure 9.22).]**
Clip other corner: **[Select D (Figure 9.22).]**
View name: **FRONT**
....

Again, you will have created three additional layers: Front-dim, Front-hid, and Front-vis.

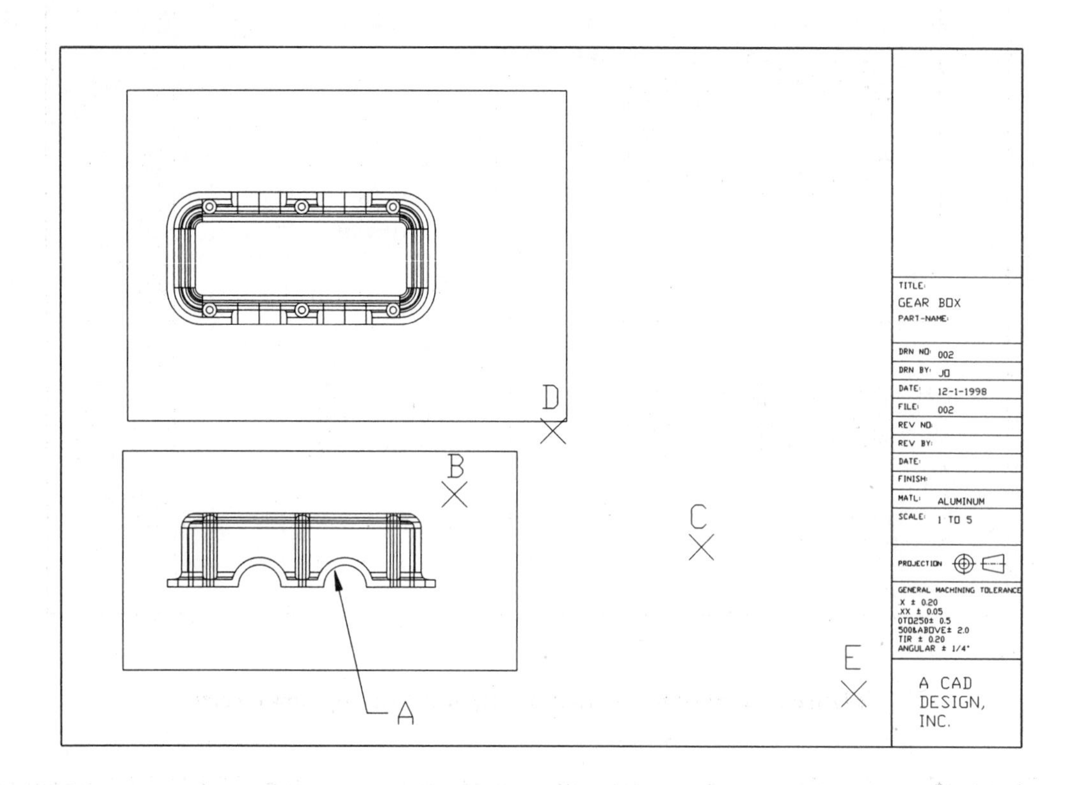

Figure 9.23 Floating viewport for the front view created

Next, you will prepare a viewport for projecting a sectional side view from the front view. A sectional view requires a cutting plane; therefore, you need to specify two points on the front view for this purpose. After defining a cutting plane, you have to indicate which side to project and where the viewport will be placed. See Figure 9.24.

....
Ucs/Ortho/Auxiliary/Section/<eXit>: **S**
Cutting Plane's 1st point: **CEN** of **[Select A (Figure 9.23.]**
Cutting Plane's 2nd point: **@1<90**
Side to view from: **[Select B (Figure 9.23).]**
Enter view scale: **0.2**
View center: **[Select C (Figure 9.23).]**
View center: **[Enter]**
Clip first corner: **[Select D (Figure 9.23).]**
Clip other corner: **[Select E (Figure 9.23).]**
View name: **SECT**
....

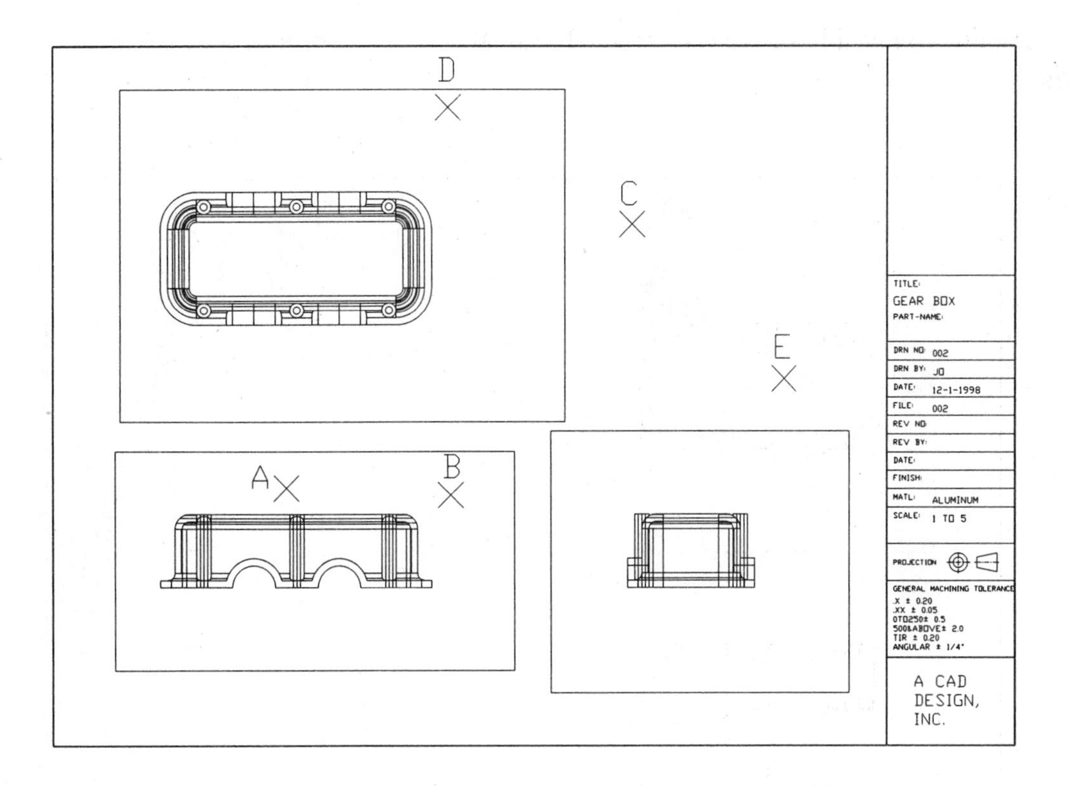

Figure 9.24 Floating viewport for the sectional side view created

In creating the floating viewport for a sectional view, the SOLVIEW command creates a layer Sect-hat for placing the hatching lines. Now you have prepared three aligned floating viewports: a view that bases on the plan view of the WCS, a front view projected from the plan view, and a sectional side view projected from the front view. At this moment, you can regard them as view windows through which you see the solid model in three directions. The additional layers created by the SOLVIEW command are, at this moment, empty. Later you will use the SOLDRAW command on these viewports to generate 2D orthographic views with outlines, hidden lines, and hatching lines on the additional layers.

Before you do that, you will add two more viewports: an auxiliary view and an isometric view. Continue the SOLVIEW command to produce an auxiliary viewport. Much as in making a viewport for a sectional view, you have to specify a plane by defining two points on a chosen viewport. See Figure 9.25.

....
Ucs/Ortho/Auxiliary/Section/<eXit>: **A**
Inclined Plane's 1st point: **[Select A (Figure 9.24) to select the front viewport and select A (Figure 9.24) again to select a point inside the floating viewport.]**
Inclined Plane's 2nd point: **@10<-45**
Side to view from: **[Select B (Figure 9.24).]**
View center: **[Select C (Figure 9.24).]**
View center: **[Enter]**
Clip first corner: **[Select D (Figure 9.24).]**
Clip other corner: **[Select E (Figure 9.24).]**
View name: **AU**
Ucs/Ortho/Auxiliary/Section/<eXit>: **[Enter]**

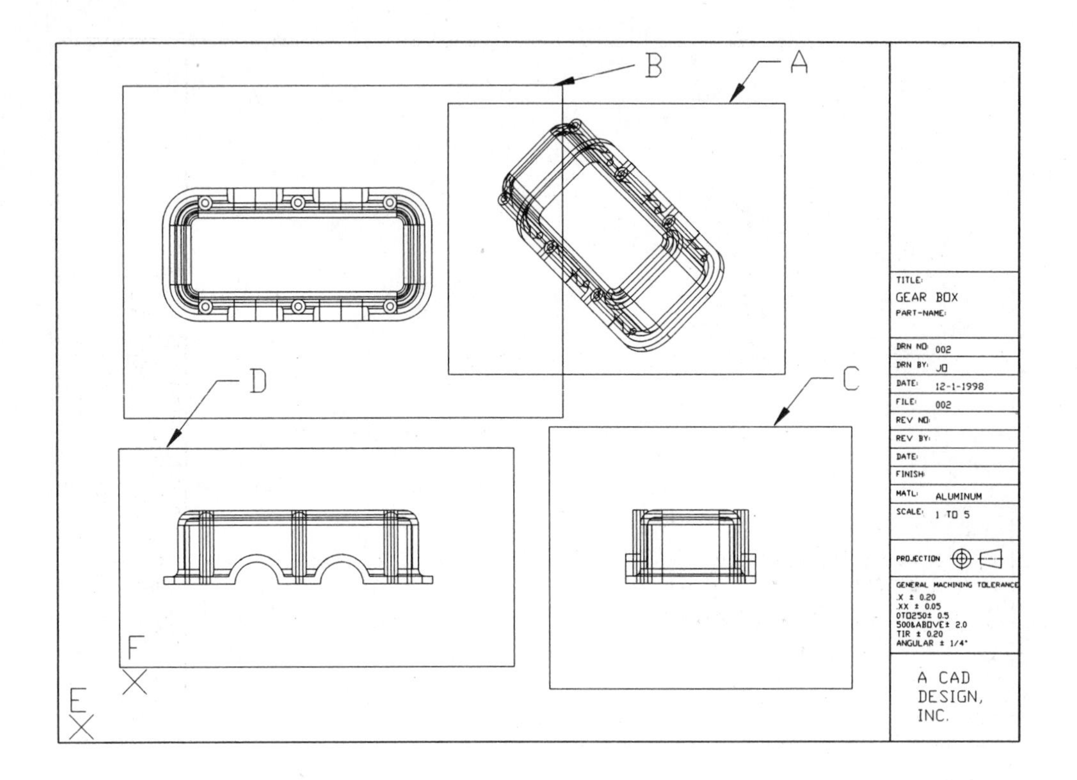

Figure 9.25 Floating viewport for auxiliary view created

Now you have four properly aligned floating viewports. As can be seen in Figure 9.25, there is not enough space for making an isometric view. To make space for the isometric view, you will relocate the floating viewports. In order not to jeopardize the alignment of these floating viewports, you need to move parent and child viewports together. The plan view is the parent of the front view. The front view is the parent of the auxiliary view and the sectional side view.

Run the MOVE command.

<Modify> **<Move>**

Command: **MOVE**
Select objects: **[Select A and B (Figure 9.25).]**
Select objects: **[Enter]**
Base point or displacement: **[Select C (Figure 9.25).]**
Second point of displacement: **[Select D (Figure 9.25).]**

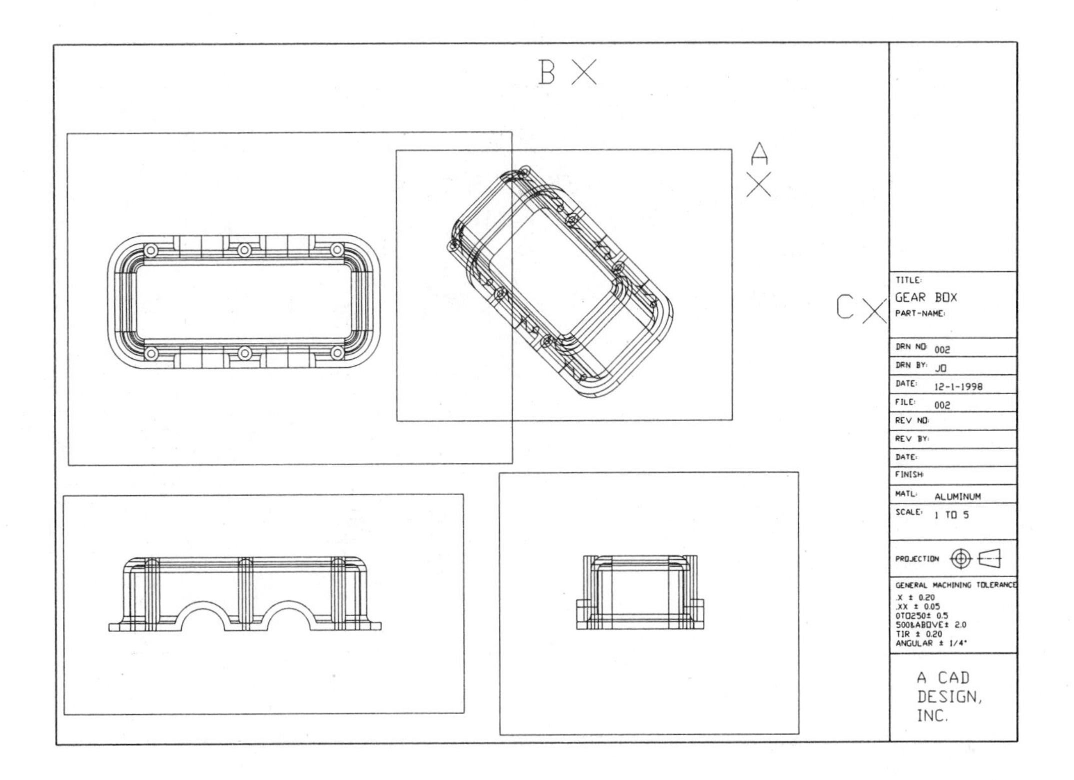

Figure 9.26 Four floating viewports moved

The next floating viewport is an isometric view. Because there is no option for making an isometric view, you have to make a UCS view, enter into model space, and change the viewing direction.

Run the SOLVIEW command to make a floating viewport that is independent of the three existing viewports. See Figure 9.27.

<Draw> **<Solids>** **<Setup>** **<View>**

Command: **SOLVIEW**
Ucs/Ortho/Auxiliary/Section/<eXit>: **UCS**
Named/World/?/<Current>: **W**
Enter view scale: **0.2**
View center: **[Select A (Figure 9.26).]**
View center: **[Enter]**
Clip first corner: **[Select B (Figure 9.26).]**
Clip other corner: **[Select C (Figure 9.26).]**
View name: **3D**
Ucs/Ortho/Auxiliary/Section/<eXit>: **[Enter]**

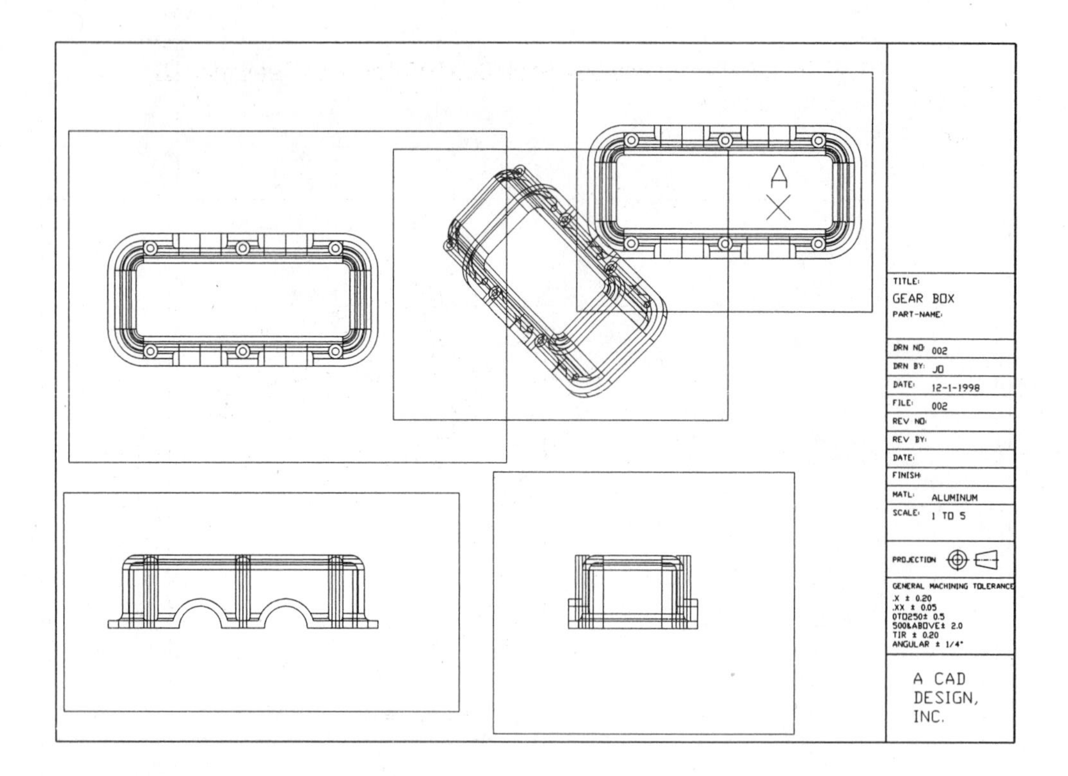

Figure 9.27 Fifth floating viewport created

For the time being, the new floating viewport displays the top view of the model. Before you can change it to a 3D view, you have to enter into the model space in the floating viewport.

Run the MSPACE command.

[Status Bar] **[MSPACE]**

Command: **MSPACE**

Because there are five floating viewports, select A (Figure 9.27) within the fifth viewport to choose it as the current viewport.

Run the VPOINT command to display a 3D view.

<View> **<3D Viewpoint>** **<SE Isometric>**

Command: **VPOINT**
Rotate/<View point>: ***1,-1,1**

Although the VPOINT command creates a display that occupies the entire area of the floating viewport, it does not allow you to specify a particular zoom scale relative to the paper space.

To zoom the view properly, use the ZOOM command to adjust the relative zoom scale to 0.2XP. See Figure 9.28.

<View> **<Zoom>** **<All>**

Command: **ZOOM**
All/Center/Dynamic/Extents/Previous/Scale(X/XP)/Window/<Realtime>: **S**
Enter scale factor: **0.2XP**

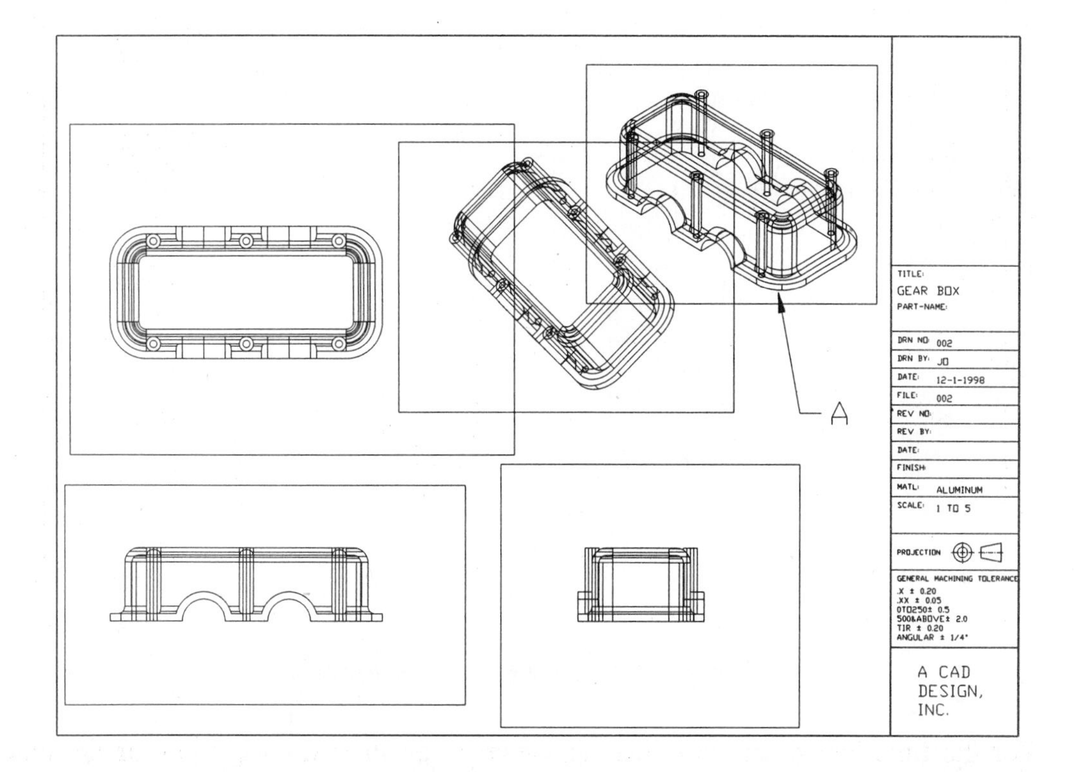

Figure 9.28 Fifth floating viewport changed to an isometric view

In addition to making an orthographic 3D view, you can prepare a perspective view. Run the DVIEW command. Set the camera to 35° from the XY plane and -65° from the X axis in the XY plane. Then choose the DISTANCE option to set camera distance and to turn on the perspective mode. See Figure 9.29.

<View> **<3D Dynamic View>**

Command: **DVIEW**
Select objects: **[Select A (Figure 9.28).]**
Select objects: **[Enter]**
CAmera/TArget/Distance/POints/PAn/Zoom/TWist/CLip/Hide/Off/Undo/<eXit>: **CA**
Toggle angle in/Enter angle from XY plane: **35**
Toggle angle from/Enter angle in XY plane from X axis: **-65**
CAmera/TArget/Distance/POints/PAn/Zoom/TWist/CLip/Hide/Off/Undo/<eXit>: **D**
New camera/target distance: **700**
CAmera/TArget/Distance/POints/PAn/Zoom/TWist/CLip/Hide/Off/Undo/<eXit>: **[Enter]**

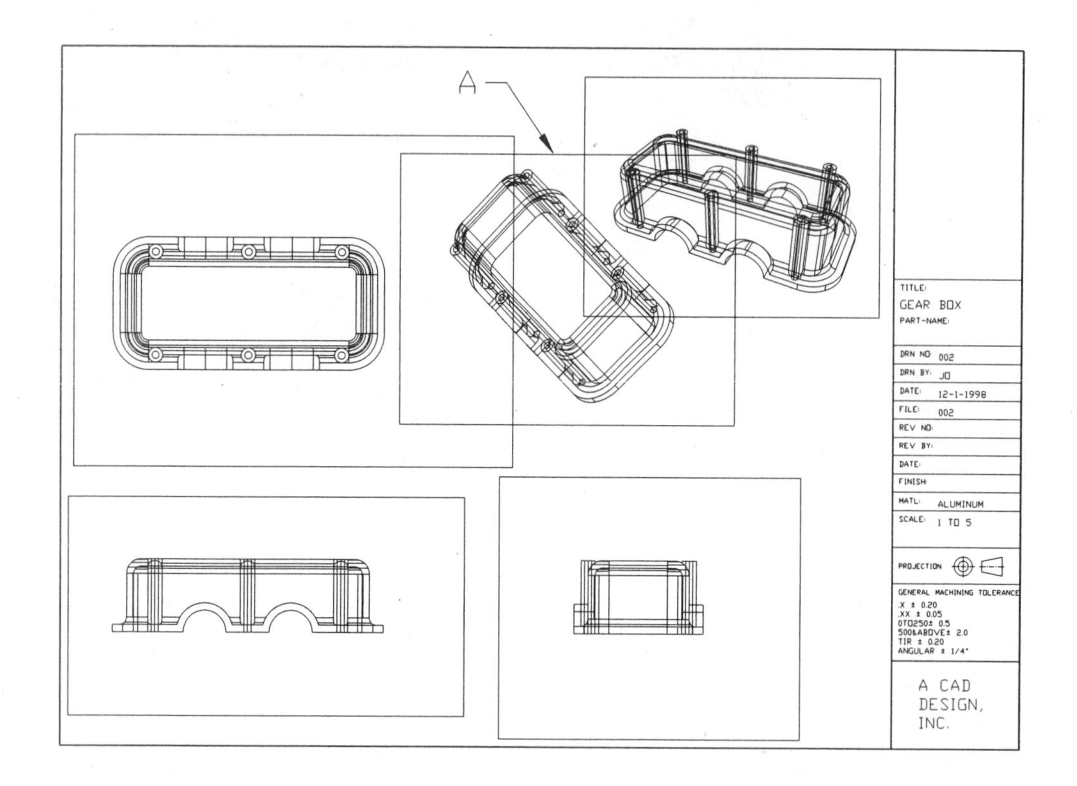

Figure 9.29 Fifth floating viewport changed to a perspective view

The perspective viewport is complete. You have learned how to create a UCS viewport, an orthographic viewport, an auxiliary viewport, a sectional viewport, and a 3D viewport.

For the time being, the auxiliary floating viewport is not required. To turn it off without deleting it, you can use the MVIEW command.

Run the MVIEW command to turn off the auxiliary viewport. See Figure 9.30.

<View> **<Floating Viewports>** **<Viewports Off>**

Command: **MVIEW**
ON/OFF/Hideplot/Fit/2/3/4/Restore/<First Point>: **OFF**
Select objects: **[Select A (Figure 9.29).]**
Select objects: **[Enter]**

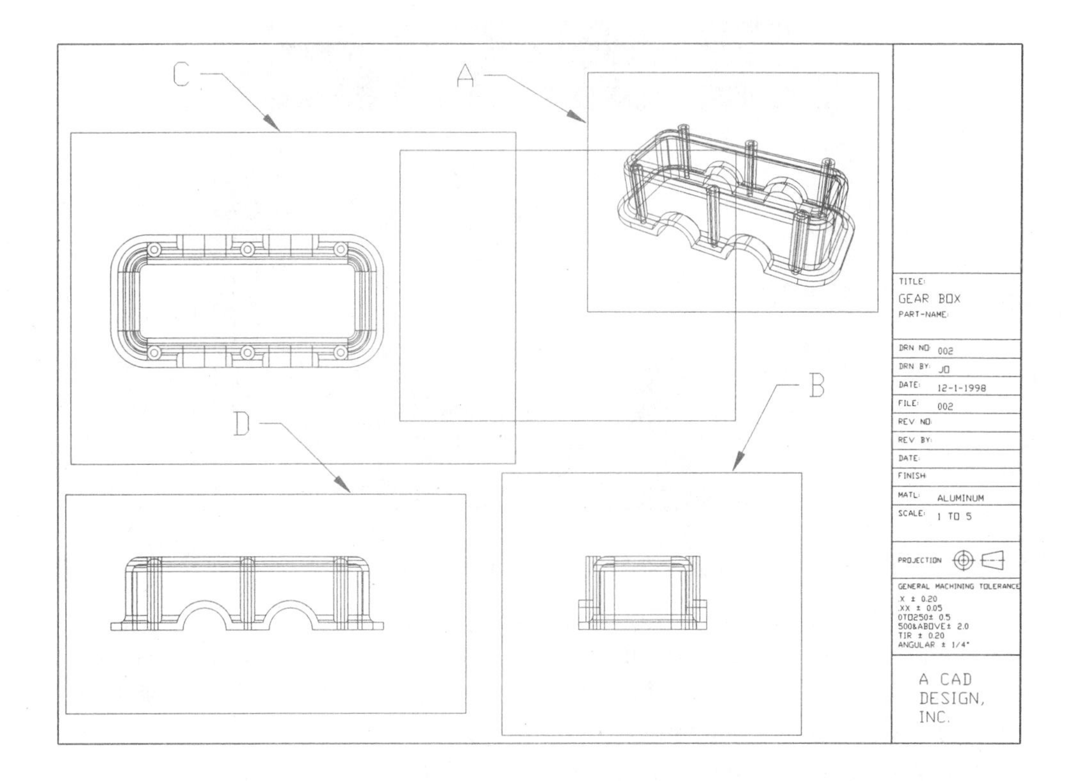

Figure 9.30 Auxiliary floating viewport turned off

The auxiliary viewport is turned off. If you want to turn it on again, repeat the MVIEW command with the ON option.

Saved Display View

As we have said, the SOLVIEW command saves the display view using the name of the floating viewport. To check these saved views, you can use the DDVIEW command or the VIEW command. The DDVIEW command brings out a dialog box. The VIEW command works at the command window. Select the Name Views... item of the View pull-down menu to run the DDVIEW command. See Figure 9.31.

<View> **<Named Views...>**

Command: **DDVIEW**

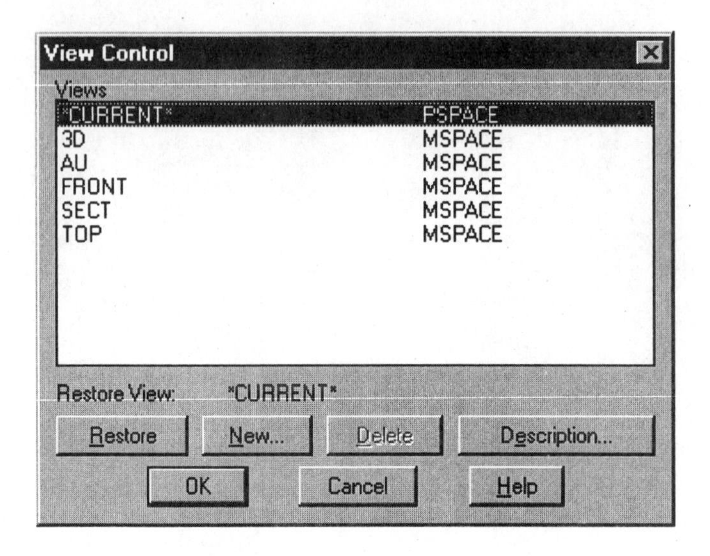

Figure 9.31 View Control dialog box

As you can see, apart from the current view in PSPACE, there are five saved display views in MSPACE. Each remembers the corresponding floating viewport. They are created by the SOLVIEW command.

Because you have changed the display of the 3D viewport to an isometric view and then to a perspective view, you have to overwrite the saved view 3D by running the DDVIEW command again.

Use the MSPACE command to change to model space environment.

[Status Bar] **[MODEL]**

Command: **MSPACE**

Select the perspective viewport to make it the current viewport, and run the DDVIEW command again.

<View> **<Named Views...>**

Command: **DDVIEW**

When the View Control dialog box appears, select the [New...] button to bring out the Define New View dialog box. See Figure 9.32.

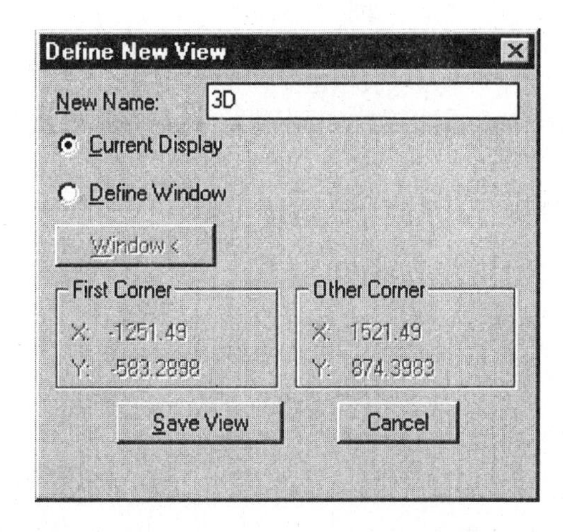

Figure 9.32 Define New View dialog box

In the Define New View dialog box, input 3D as the New Name, select the [Current Display] button, and then select the [Save View] button. Because there is a saved view called 3D, AutoCAD asks if you want to redefine it. Select the [Redefine] button to redefine the saved view, and then select the [OK] button to exit.

Return to the paper space by using the PSPACE command.

[Status Bar] **[PAPER]**

Command: **PSPACE**

Generate a 2D Drawing from a 3D Solid Model

At this moment, the floating viewports, including the sectional viewport, are simply viewing windows through which you can see the 3D solid model in several directions. To project 2D views from the 3D solid model, you have to run the SOLDRAW command, which does the following:

1. Projects the solid model onto a 2D plane to generate visible outlines. These lines reside on the layer "-vis."
2. Projects the solid model onto a 2D plane to generate hidden lines. These lines reside on the layer "-hid."
3. Freezes the layer that holds the solid model in all the floating viewports.
4. Projects a sectional view onto a 2D plane if the viewport is prepared as a sectional viewport. The hatching lines reside on the layer "-hat."

Select the Drawing item of the Setup item of the Solids cascading menu of the Draw pull-down menu to run the SOLDRAW command to project 2D orthographic drawing views from the 3D solid model. See Figure 9.33.

<Draw> **<Solids>** **<Setup>** **<Drawing>**

Command: **SOLDRAW**
Select viewports to draw:
Select objects: **[Select A, B, C, and D (Figure 9.30).]**
Select objects: **[Enter]**

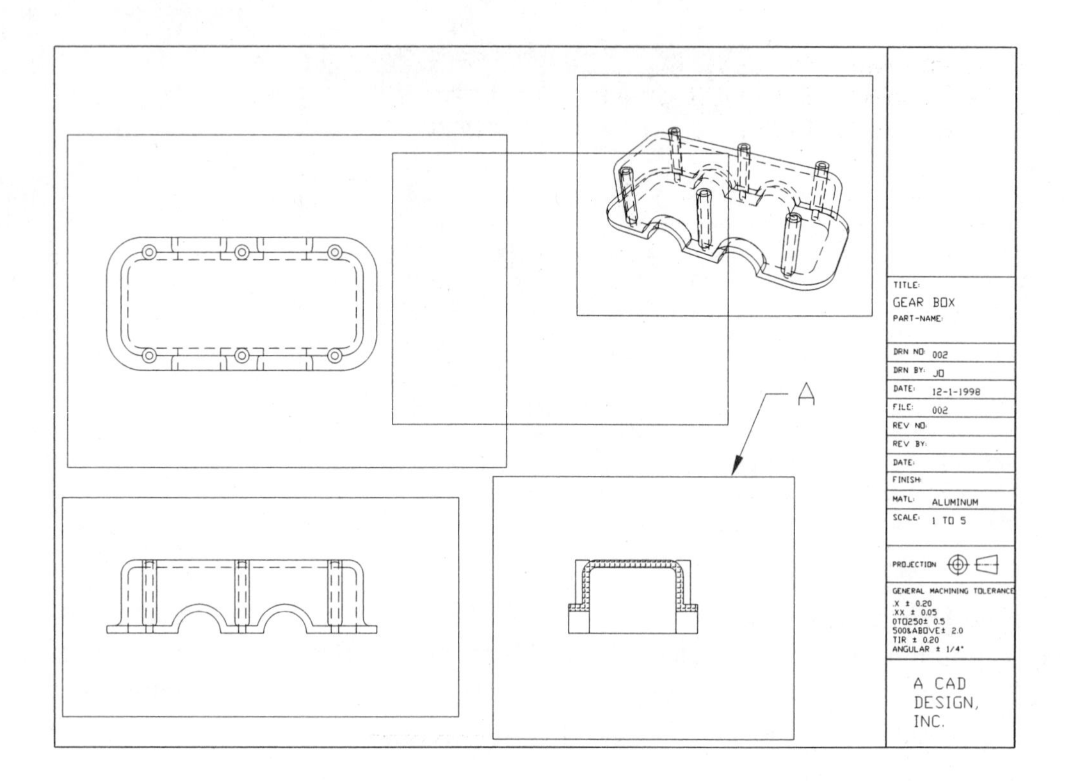

Figure 9.33 2D orthographic drawings projected from a 3D solid model

When projecting 2D orthographic views, AutoCAD uses default settings. To adjust the proportion of the hidden lines, set the linetype scale to 12.

<Format> **<Linetype...>**

Command: **LINETYPE**

[Global scale factor: **12**
OK]

In addition to adjusting the linetype proportion, you have to change the hatch pattern name and hatch pattern scale. Manipulate the HPNAME and HPSCALE variables.

Command: **HPNAME**
New value for HPNAME: **ANSI31**

Command: **HPSCALE**
New value for HPSCALE: **4**

As your screen display shows, the new linetype scale is reflected in the floating viewports, but not the hatch pattern and hatch pattern scale. To apply the new hatching variables to the sectional view, you must run the SOLDRAW command again. See Figure 9.34.

<Draw> **<Solids>** **<Setup>** **<Drawing>**

Command: **SOLDRAW**

Select viewports to draw:
Select objects: **[Select A (Figure 9.33).]**
Select objects: **[Enter]**

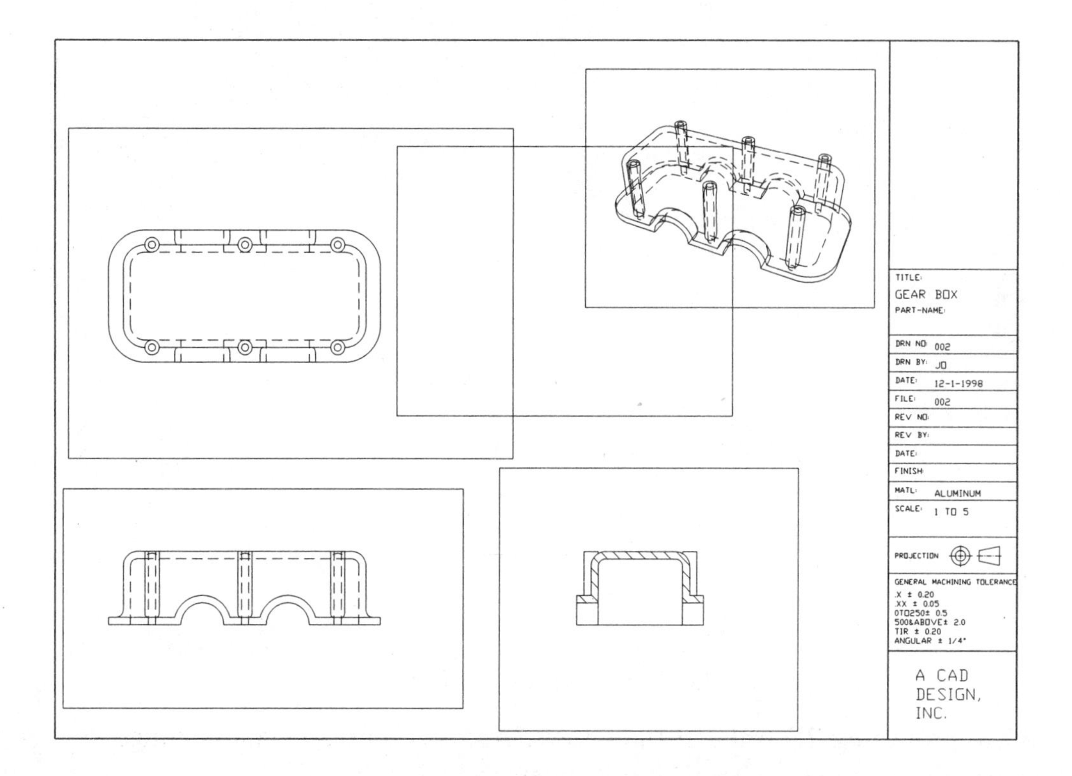

Figure 9.34 Sectional view regenerated

If you watch carefully, you will notice that the entities are erased and then regenerated on this viewport. When you use the SOLDRAW command on a viewport, it will first erase all the entities on the layers "-hid," "-vis," and "-hat." Then it generates the visible lines, hidden lines, and hatching lines again. Consequently, you should not use these layers for any other purpose because everything you put there might be erased.

To complete the drawing, you should freeze the layer VPORTS for two reasons: (1) The borders of the floating viewports reside on this layer, so you have to freeze this layer to hide the viewport borders, and (2) the viewports have been aligned properly by the SOLVIEW command, so freezing this layer prevents you from accidentally jeopardizing the alignment of the floating viewports.

Freeze the layer VPORTS. See Figure 9.35. The engineering drawing views are complete.

<Format> **<Layer...>**

Command: **LAYER**

Name	Freeze
Vports	Freeze

Current layer: **Title**

[OK]

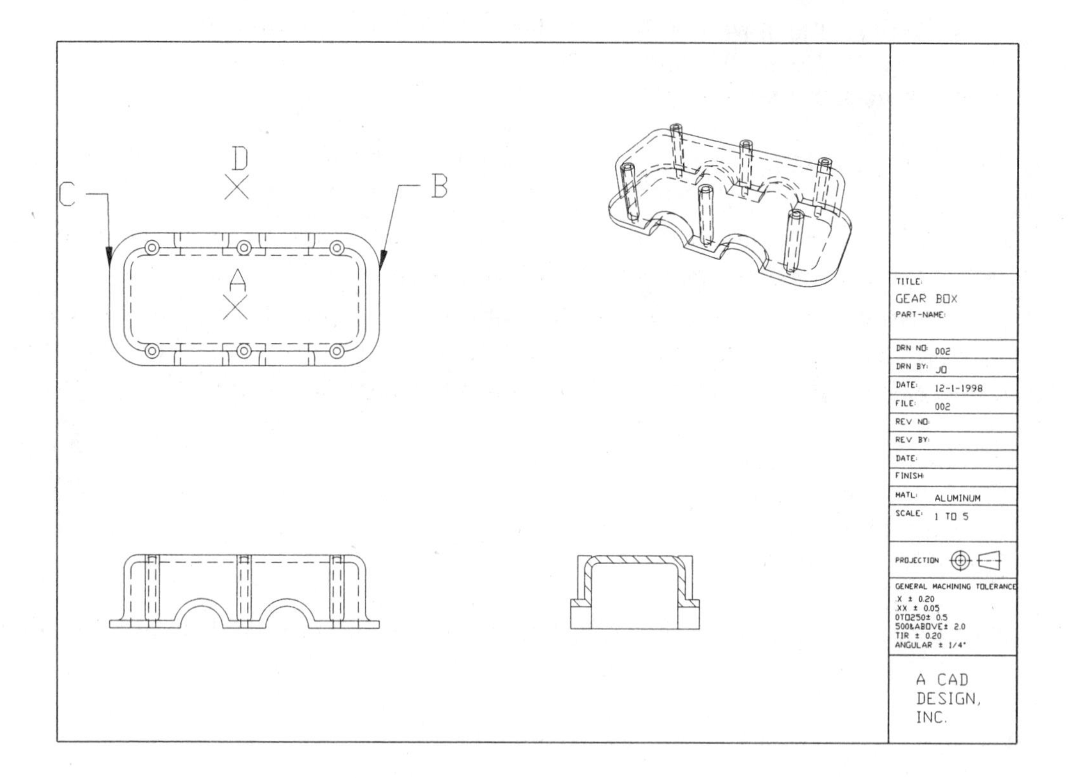

Figure 9.35 Layer Vports frozen

To summarize, the SOLVIEW command creates the floating viewports and the related layers and saves the display views. The SOLDRAW command generates the 2D views on the floating viewports that are created by the SOLVIEW command. This command does not work on the floating viewports that are created by the MVIEW command. If you have used the MVIEW command on solid models, you can use the SOLPROF command. The SOLDRAW command and the SOLPROF command work similarly; one is used for viewports created by the SOLVIEW command, and the other one is used for viewports created by the MVIEW command.

Dimensioning

As we have noted, each of the floating viewports created by the SOLVIEW command has a layer suffixed by the letters "-dim," which is reserved for dimensioning. These layers are frozen in all floating viewports except the relevant floating viewport.

When adding dimensions to the document of a solid model, there are a few points to observe:

1. Switch to model space by using the MSPACE command, not by setting the system variable TILEMODE. This enables you to place the dimensions in the orthographic views.
2. Place the dimension entities on the layer with the suffix "-dim" so that the dimensions will be displayed only in the relevant layer.
3. Set the UCS to VIEW such that the XY plane of the new UCS is parallel to the current floating viewport. Thus the dimension is placed on a plane parallel to the floating view.

To place the dimensions in model space, use the MSPACE command.

[Status Bar] **[MODEL]**

Command: **MSPACE**

You will add a linear dimension to the top view. Select the top view at A (Figure 9.35) to make this viewport the current floating viewport.

The dimensioning layer for this floating viewport is Top-dim; set it as the current layer. Because this layer is frozen in all the floating viewports except the front view viewport, entities created on this layer will only display in this viewport.

<Format> **<Layer...>**

Command: **LAYER**
Current layer: **Top-dim**

In order for the dimensions to be placed on a plane parallel to the display view, align the UCS to the viewport.

<Tools> **<UCS>** **<View>**

Command: **UCS**
Origin/ZAxis/3point/OBject/View/X/Y/Z/Prev/Restore/Save/Del/?/<World>: **V**

After you have switched to the model space, selected a floating viewport, made current a designated layer, and set the XY plane to align with the display view, you can create a dimension. See Figure 9.36.

<Dimension> **<Linear>**

Command: **DIMLINEAR**
First extension line origin or press ENTER to select: **END** of **[Select B (Figure 9.35).]**
Second extension line origin: **END** of **[Select C (Figure 9.35).]**
Dimension line location (Mtext/Text/Angle/Horizontal/Vertical/Rotated): **[Enter]**

Compare Figure 9.36 with your drawing. If you find the overall proportions of the dimension inappropriate, use the DDIM command to set the dimension style accordingly.

While you are dimensioning in model space, you can use the ZOOM command to zoom in and out whenever necessary. To reiterate, the SOLVIEW command creates a saved view for each floating viewport. The name of the saved display view is the name of the viewport. After zooming and panning, you can restore the display to the set view by running the DDVIEW command. If you have to adjust the size of any floating viewport that is created by the SOLVIEW command, you must update the saved view with the DDVIEW command before you can zoom or pan.

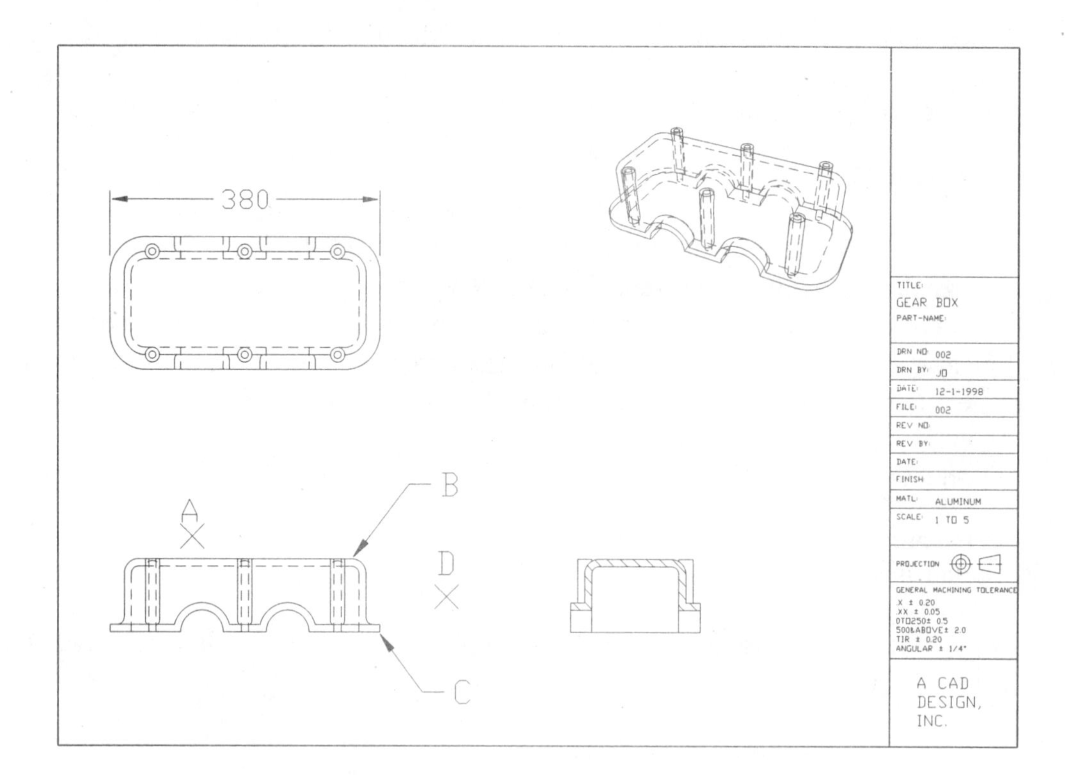

Figure 9.36 Linear dimension created in the top floating viewport

A dimension is created in the top viewport. To continue, you will make another linear dimension in the front view. Select A (Figure 9.36) to select the front view as the current floating viewport. Then set the current layer to Front-dim, align the UCS to the viewport, and create another linear dimension. See Figure 9.37.

<Format> **<Layer...>**

Command: **LAYER**
Current layer: **FRONT-DIM**

<Tools> **<UCS>** **<View>**

Command: **UCS**
Origin/ZAxis/3point/OBject/View/X/Y/Z/Prev/Restore/Save/Del/?/<World>: **V**

<Dimension> **<Linear>**

Command: **DIMLINEAR**
First extension line origin or press ENTER to select: **END** of [**Select B (Figure 9.36).**]
Second extension line origin: **END** of [**Select C (Figure 9.36).**]
Dimension line location (Mtext/Text/Angle/Horizontal/Vertical/Rotated): [**Enter**]

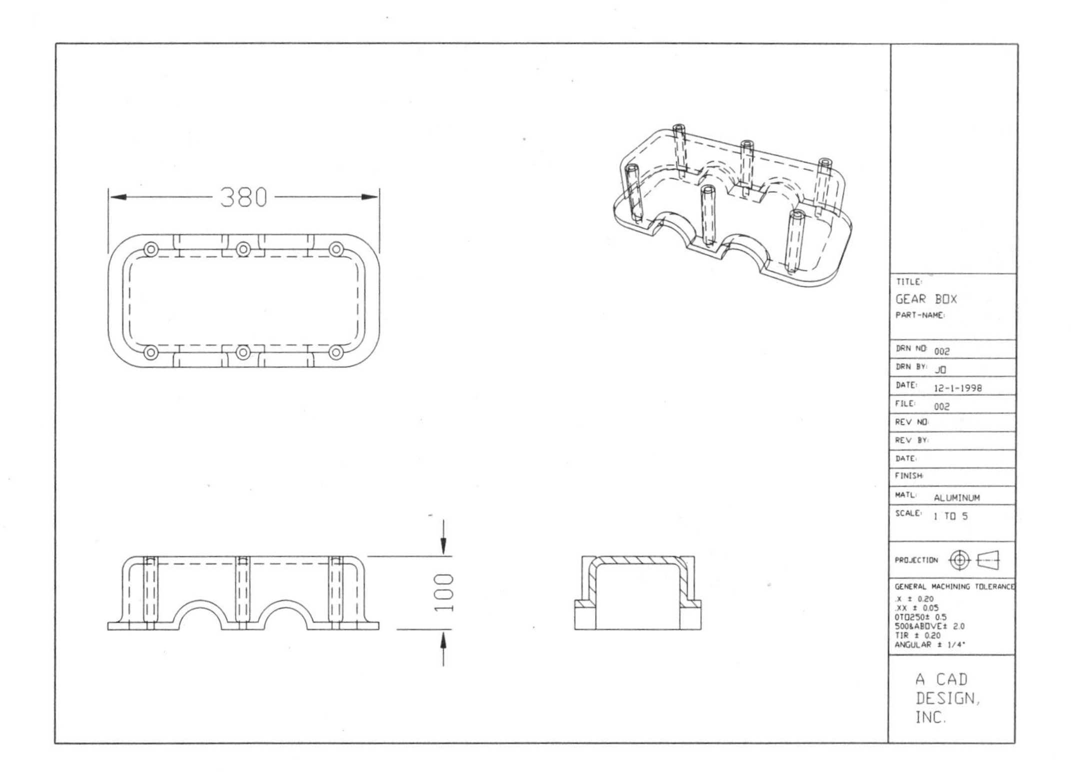

Figure 9.37 Dimension created in the front floating viewport

Now two dimensions are created. In accordance with Figure 9.38, continue to add other dimensions to the document. Remember to follow the steps: select the floating viewport in which you want to place a dimension, set the current layer to the layer that is reserved for dimensioning the selected floating viewport, set the UCS orientation to align with the selected floating view, and use the appropriate dimensioning command to create dimensions.

In the course of dimensioning, if you find the size of the floating viewport too small to display the dimensions, you should set to paper space environment and enlarge the floating viewport. Because there is a saved display view for each floating viewport, you should use the DDVIEW command to update the saved view after changing the size of the floating viewport.

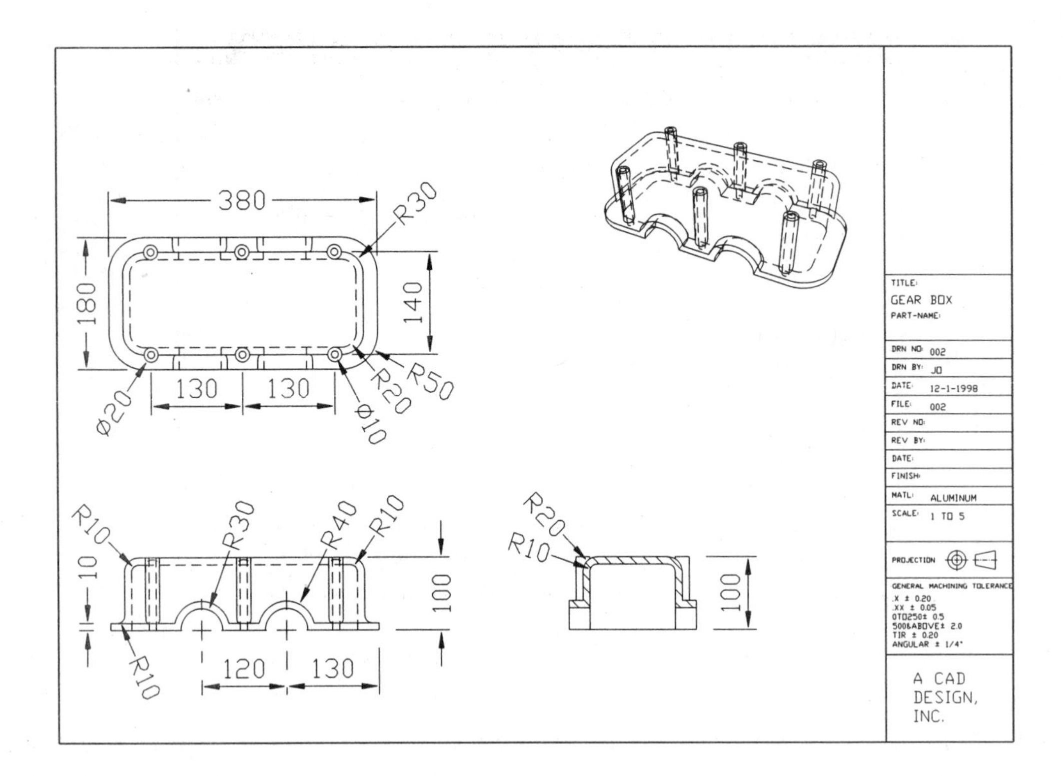

Figure 9.38 Fully dimensioned

Refer to Figure 9.38. Dimensioning is complete. Return to the paper space and save your drawing.

[Status Bar] **[PAPER]**

Command: **PSPACE**

<File> **<Save>**

The engineering document for the solid model of a gear box cover is complete.

9.4 Advanced Drawing Setup

In Chapter 2, you learned that the NEW command provides an advanced setup wizard. Now you will use this advanced wizard to set up a title block, together with a floating viewport, in paper space environment. Select the New... item of the File pull-down menu to run the NEW command and use the advanced setup wizard.

<File> **<New...>**

When the dialog box appears, select the Advanced Setup wizard and select Step 6. See Figure 9.39.

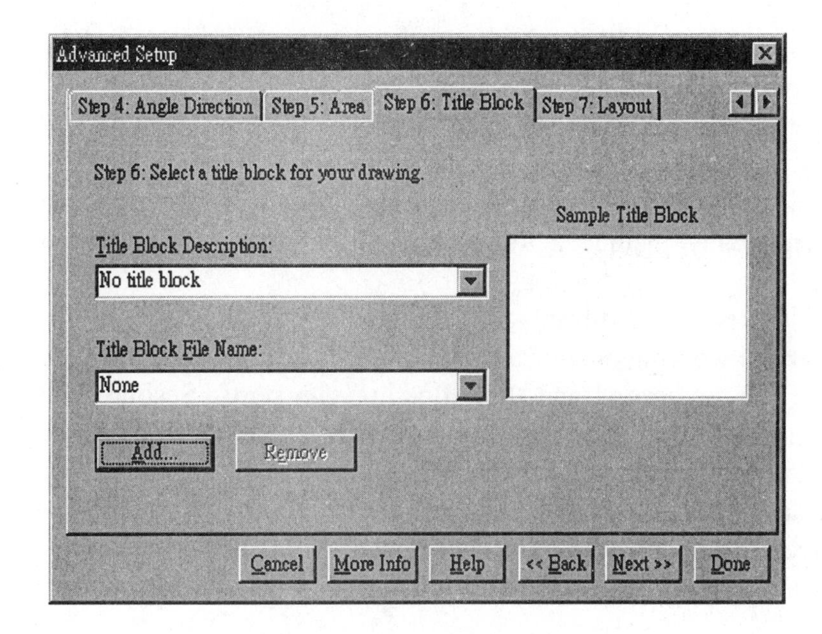

Figure 9.39 Advanced setup wizard

By default, there is no title block specified. To add a title block in your new drawing, you can select a file from the Title Block File Name file table. If you have prepared a title block similar to Figure 9.2 and want to include the title block in the title block list, select the [Add...] button and then select the file that you want to include.

After selecting the title block, select the Step 7 tab to indicate whether you want to place the title block in paper space or model space. See Figure 9.40.

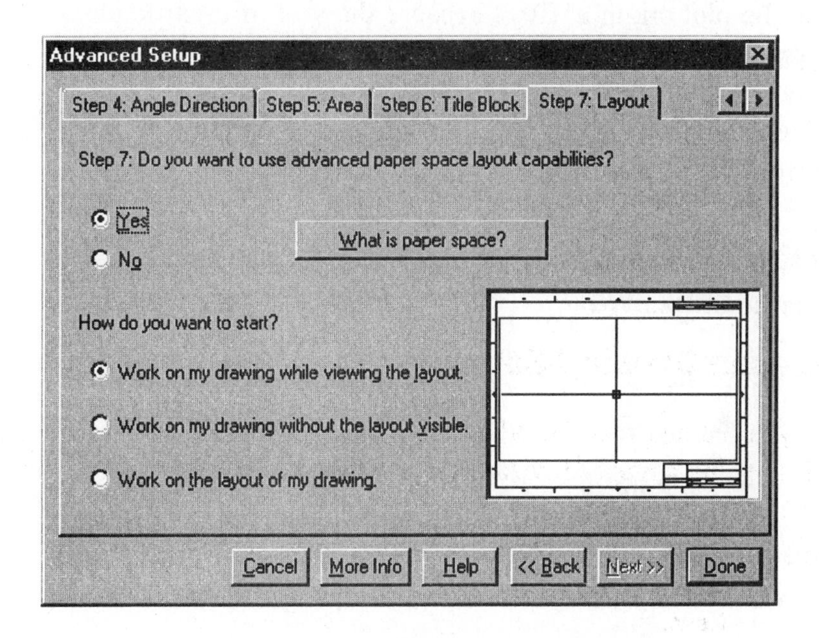

Figure 9.40 Step 7 of advanced setup wizard

Naturally, you should select the [Yes] button to place the title block in paper space environment. Regarding where you want to start your work, you have three options: work in model space with TILEMODE set to zero, work in model space with TILEMODE set to one, or work in paper space with TILEMODE set to zero. You can select any of these options to suit your own preference.

After selecting the [Done] button, you will get a new drawing with a title block and a floating viewport created in paper space environment.

The main advantage of setting up a new drawing by using the advanced wizard is that you can automatically include a title block and a floating viewport in paper space environment.

However, there are disadvantages. First, the default floating viewport created occupies the entire working area of the title block. You may need to resize the viewport to make room for the second and third floating viewports. Second, you may have to delete the default floating viewport and then use the SOLVIEW command to create orthographic viewports for a solid model. Third, there is no attribute value request, even though there are attribute templates defined in the title block drawing. Default attribute values are given. To change the default attribute values, you have to use the DDATTE command by selecting the Edit Attribute icon of the Modify II toolbar.

9.5 Plot a Document

You can plot a drawing in either model space or paper space. If you plot in model space, you have to set the plot scale according to the relationship between the actual drawing size and the paper size, and you have to set the plot origin according to the actual location of the entities.

You should avoid plotting in model space. To plot in paper space after you have created a document with title block and floating viewports, you can simply set the plot scale to 1 and the plot origin at (0,0), because the size of the title block is less than the actual plot size, and the insertion point for the title block is at (0,0). Plotting in paper space is highly recommended.

Open the document prepared for the solid model of the gear box cover. You will configure the plotting device for this drawing.

<File> **<Open...>**

File name: **GBOX.DWG**

Select the Print... item of the File pull-down menu to apply the PLOT command. See Figure 9.41.

<File> **<Print...>**

Command: **PLOT**

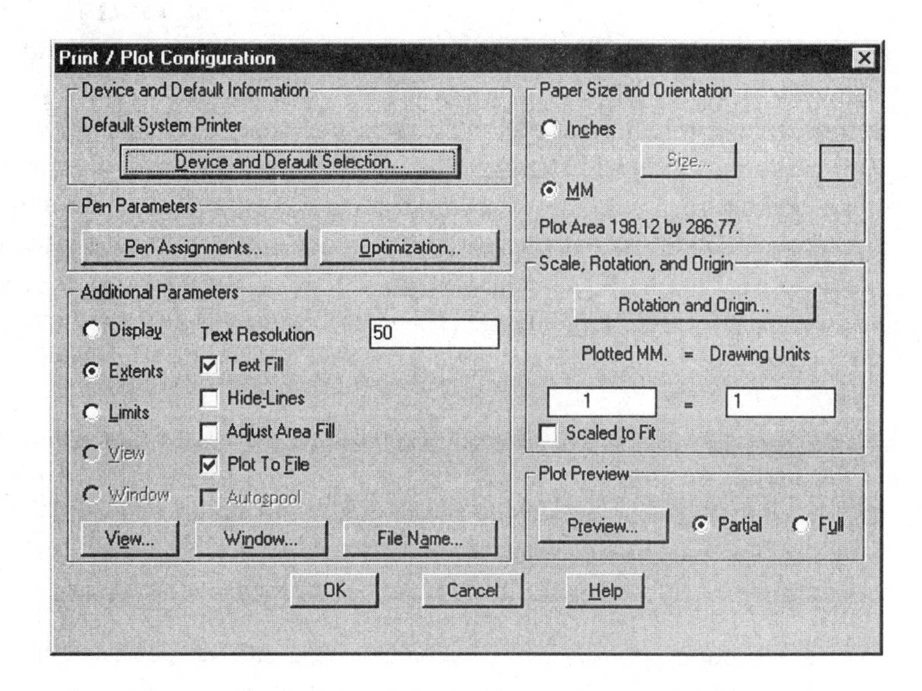

Figure 9.41 Print/Plot Configuration dialog box

Print/plot Configuration dialog box appears. There are six sections.

Device and Default Information

Select the [Device and Default Selection...] button to bring out a sub-dialog box. See Figure 9.42. The Device and Default information section concerns plotter information.

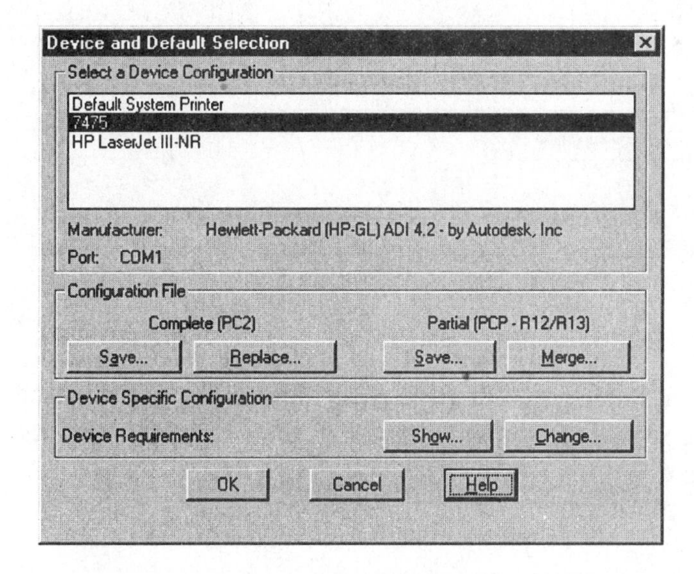

Figure 9.42 Device and Default Selection dialog box

In this dialog box, there are three sub-sections: Select a Device Configuration, Configuration File, and Device Specific Configuration.

The number of plotters available in the Select a Device Configuration section depends on how you configure AutoCAD. To add or remove plotters, you should cancel this command and select the Printer Setup... item of the File pull-down menu to use the CONFIG command. Now, select a plotter you want from this list.

You can store configuration information such as pen assignments, plot area, scale, paper size, and rotation in a configuration file. There are two kinds of configuration files: complete file and partial file. A complete configuration file stores all the configuration information. It has a .PC2 extension. A partial configuration file stores only the device-dependent information. It has a .PCP extension.

In the Configuration File section, you can save, replace, or merge configuration files. These files are useful when you plot a batch of drawings. See Batch Plotting in the next section.

In the Device Specific Configuration section, you can display and change the current device configuration information.

Select the [OK] button to exit the Device and Default Selection dialog box.

Pen Parameters

The Pen Parameters section controls pen parameters. You can set pen assignments and optimize pen motion. If you wish to set pen number, hardware linetype, speed, and pen width associated with each AutoCAD color, select the [Pen Assignments...] button. See Figure 9.43. In this dialog box, set pen assignments and select the [OK] button to exit.

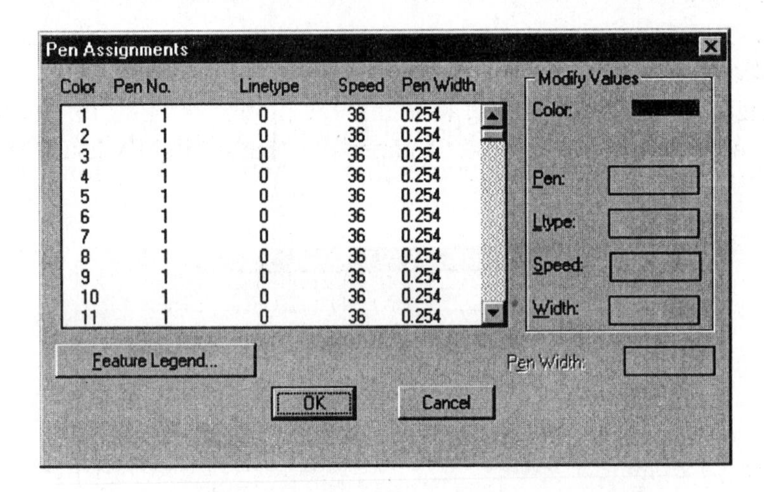

Figure 9.43 Pen Assignments dialog box

To optimize pen motion to minimize wasted pen motion, select the [Optimization...] button. See Figure 9.44. Select how you want to optimize the pen motion and then select the [OK] button to exit.

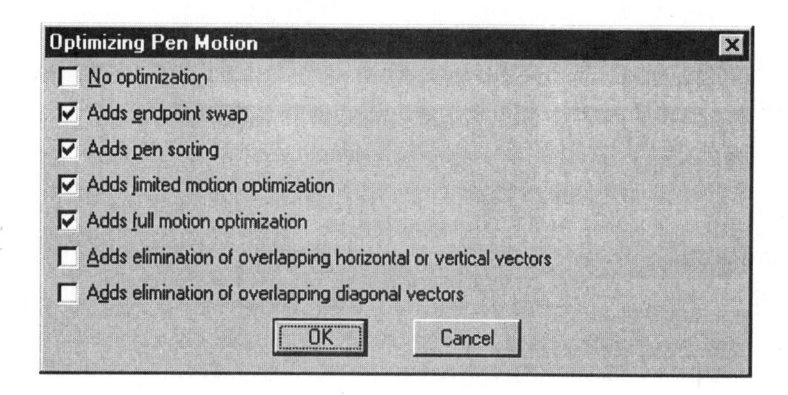

Figure 9.44 Optimizing Pen Motion dialog box

Additional Parameters

In the Additional Parameters section of the Print/Plot Configuration dialog box, you can specify which area of the drawing is to be plotted. You can remove hidden lines on a 3D object. You can even plot to a file. With a plot file, you can print a drawing without the need to use AutoCAD.

Paper Size and Orientation

The Paper Size and Orientation section sets the paper size and the orientation of the drawing in relation to the paper.

The paper size is the actual plotting size that the plotter can plot to, not the size of the paper. For a documented drawing, select a paper size that is slightly larger than the size of the chosen title block. See Figure 9.45.

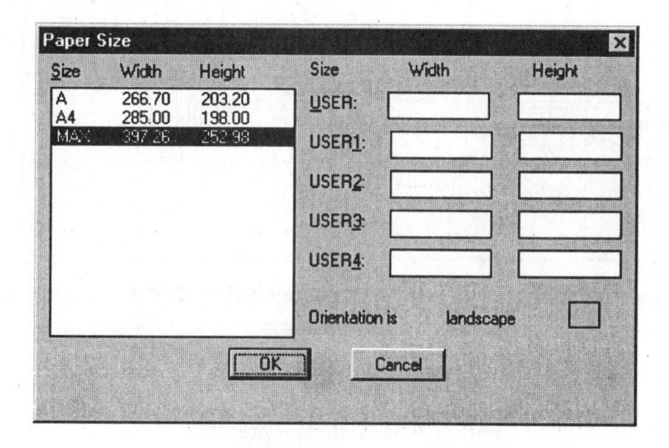

Figure 9.45 Paper Size dialog box

Scale, Rotation, and Origin

The Scale, Rotation, and Origin section sets the plot scale and the rotation of the drawing and the origin of the drawing relative to the plotted paper.

After you have prepared a document from a drawing and set tilemode to zero, you are plotting the title block with floating viewports on it. Naturally, the plot scale should be 1 to 1, and the plot origin should be at (0,0), which is the insertion point of the title block.

The rotation angle depends on the hardware configuration. If you are not sure, you can make use of the Plot Preview facility (see the next paragraph).

Plot Preview

The Plot Preview section is very useful because you can have a partial or full preview of the plotted drawing before you actually plot the drawing. This saves time and money.

Plot preview utility displays the outcome of plotting by using the configuration information you supplied in the previous sections. Therefore, you must ensure that the configuration information is correct.

Select the [Full] button and then the [Preview...] button to see a full preview. See Figure 9.46.

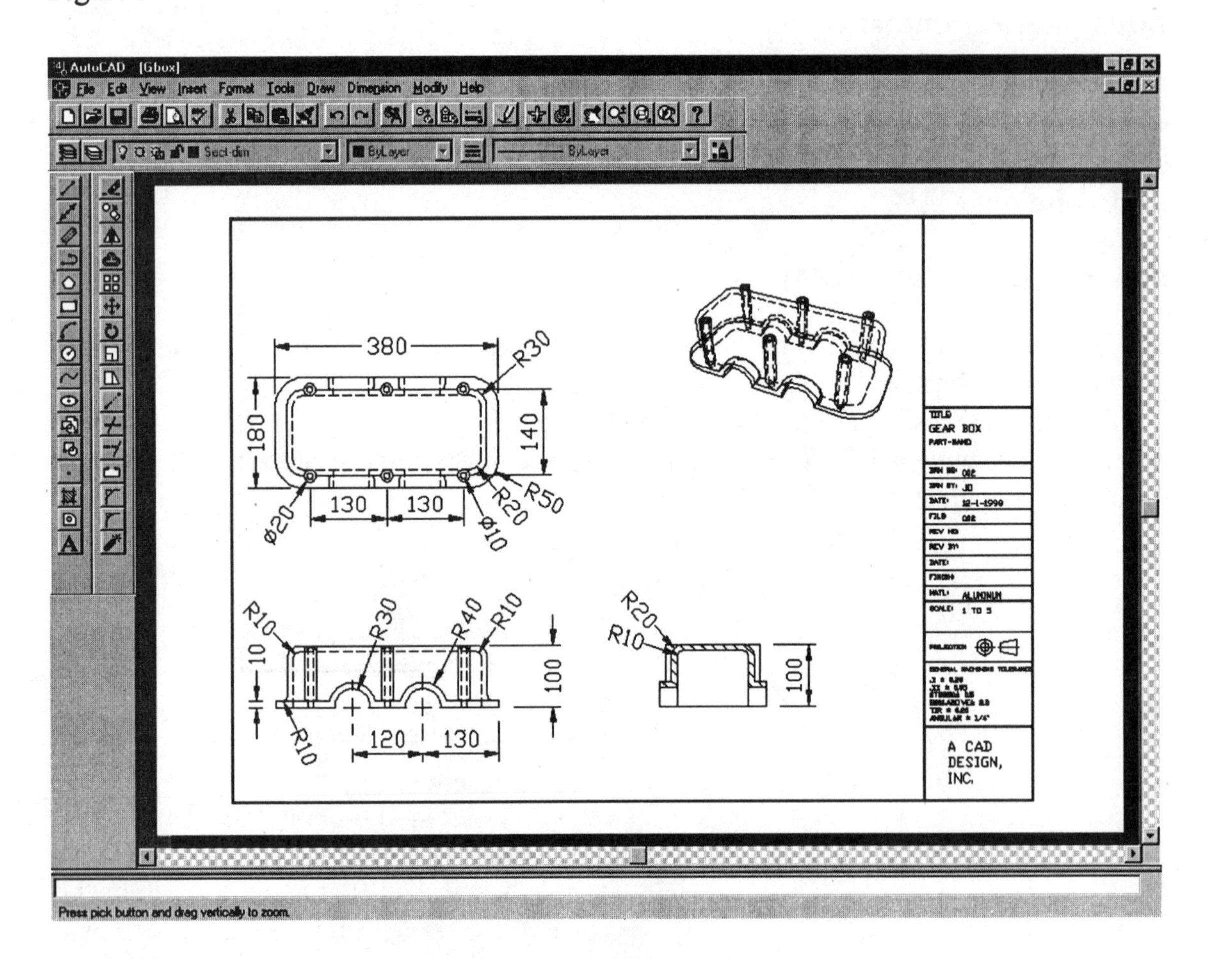

Figure 9.46 Full plot preview

If the plot preview is acceptable to you, select the [Exit] button of the preview dialog box to return to the Print/Plot Configuration dialog box.

Plot configuration is complete. Refer to Figure 9.42. Select the [Device and Default Selection...] button to bring up the Device and default selection dialog box. Then select the [Save...] button under the Complete (.PC2) heading to save the configuration information to a file named PRINT.PC2.

Now you can proceed in plotting your drawing by selecting the [OK] button of the

Print/Plot Configuration dialog box.

With the printer device properly configured, you can preview your drawing quickly by using the PREVIEW command.

<File> **<Preview>**

Command: **PREVIEW**

9.6 Batch Plotting

Batch plotting is a stand-alone application. It enables you to maintain a list of files and perform batch plotting of a set of files. To use this facility, you need to install Batch Plotting Utility to your computer. See Figure 9.47. If you have not installed this utility, you should add it to your computer by using the AutoCAD setup program.

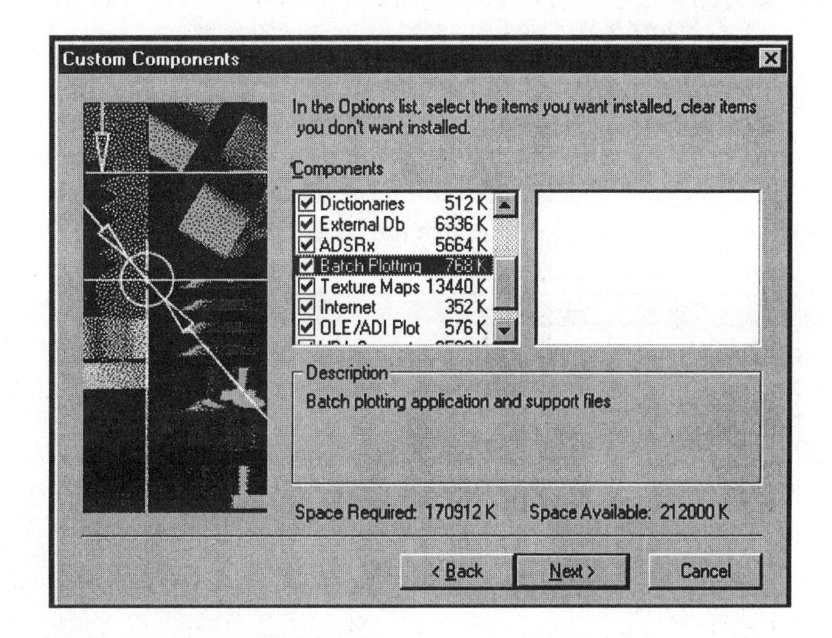

Figure 9.47 Include Batch Plotting Utility while installing AutoCAD

Batch plotting is a Visual Basic application. It starts AutoCAD and displays the AutoCAD Extended Batch Plot Utility dialog box. See Figure 9.48.

There are five tabs: File, Layers, Plot Area, Logging, and Plot Stamping. From the File tab, select the [Add Drawing...] button and select the files PIPESP.DWG and GBOX.DWG to add the two files to the List of Drawings to Plot. To remove a file from the list, select the file that you want to remove and then the [Delete Entry] button.

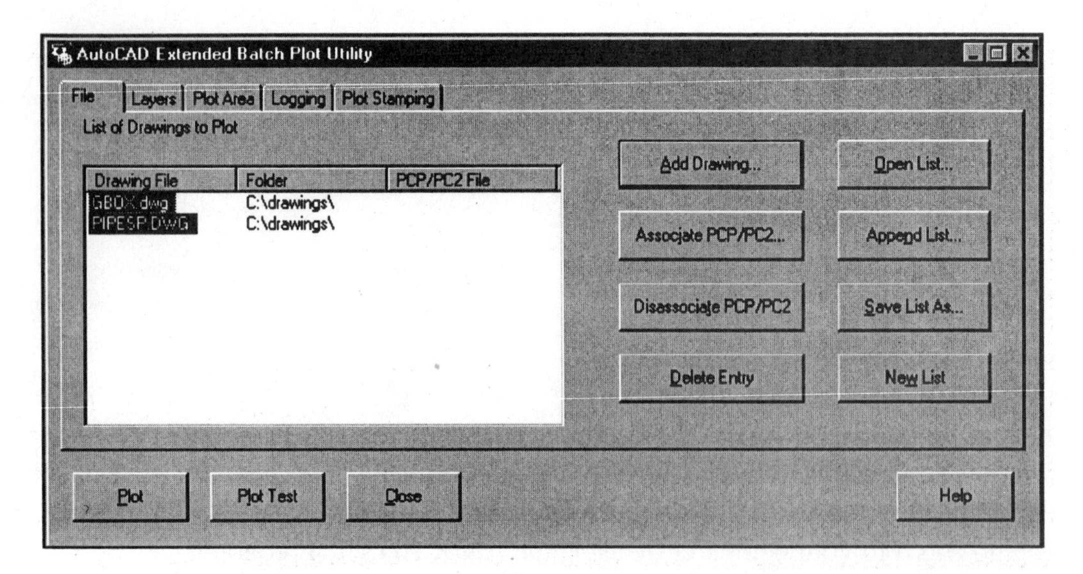

Figure 9.48 Two files added to the list and selected

Because plotting outcome depends on plotter configuration, you should associate a plot configuration file to the drawings to be plotted. Select the two files in the list, the [Associate PCP/PC2] button, and the configuration file PRINT.PC2. See Figure 9.49. If you want to remove the configuration file, use the [Disassociate PCP/PC2] button.

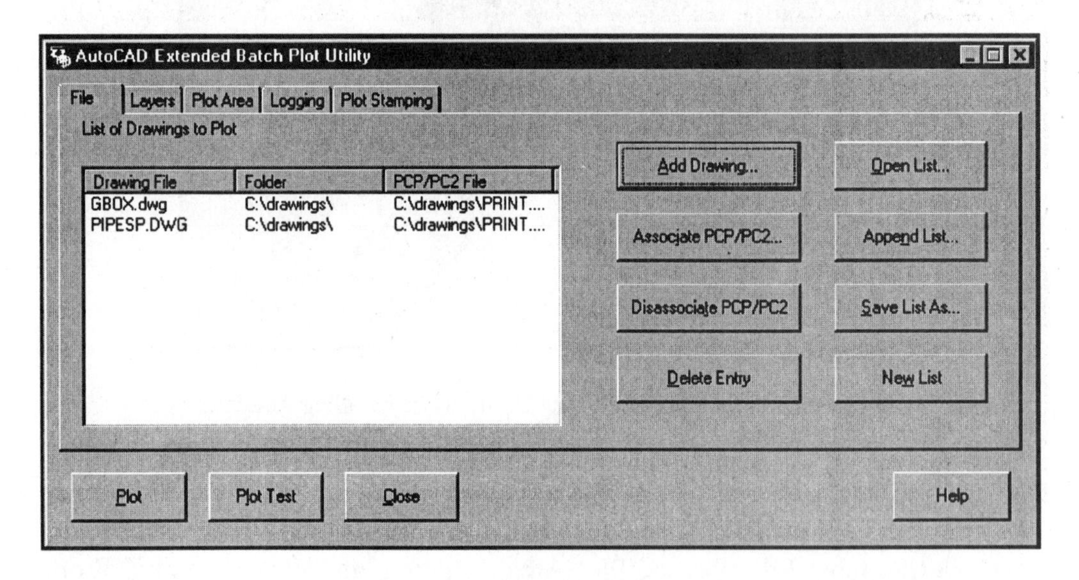

Figure 9.49 Configuration file selected

A list of drawings to be plotted is prepared. You can select the [Save List As...] button to save the list to a Batch Plot List file (with .BP2 extension) so that you can later use the [Open List...] button to retrieve the list. Alternatively, you can select the [Append List...] button to append this list to an existing Batch Plot List file. To start a new list, select the [New List] button.

After making a list, you can turn to the Layers tab. Here, you can turn on and off layers. See Figure 9.50.

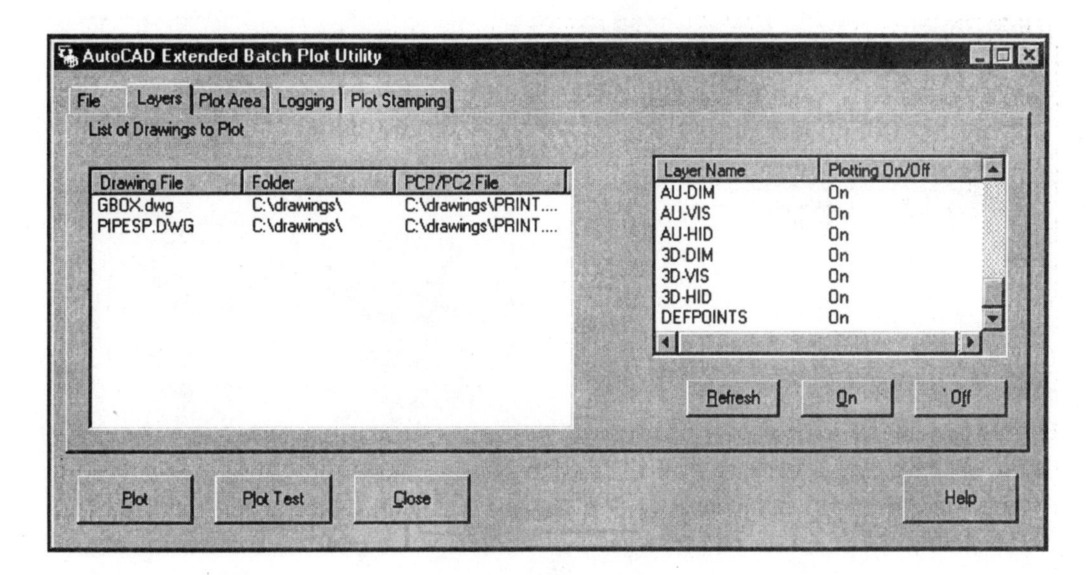

Figure 9.50 Layers tab

To control plot area, plot space, and plot scale, you can use the Plot Area tab. See Figure 9.51.

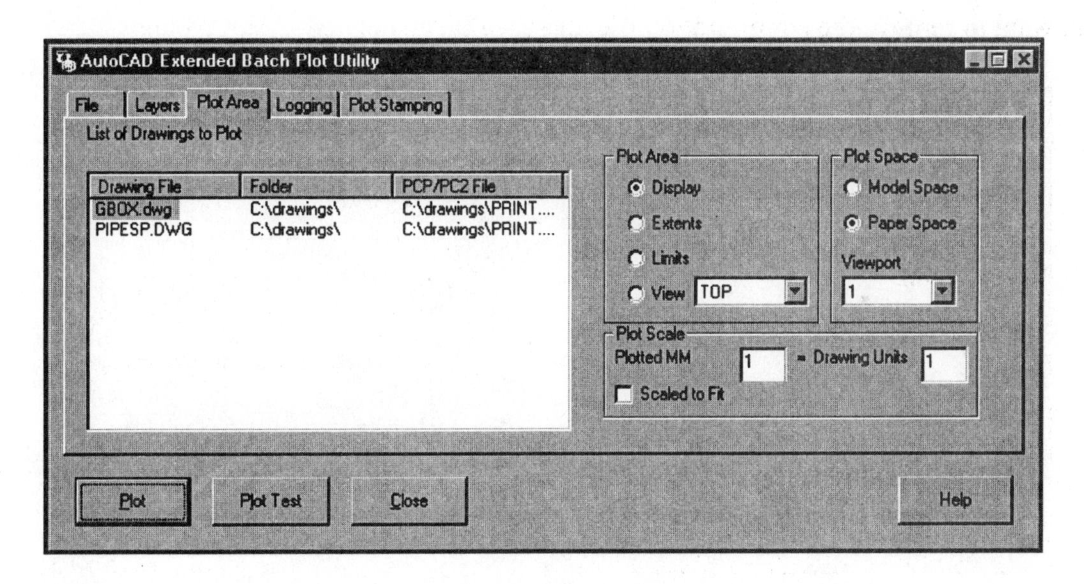

Figure 9.51 Plot Area tab

To log errors and plot-time events to a text file, you can use the Logging tab. See Figure 9.52.

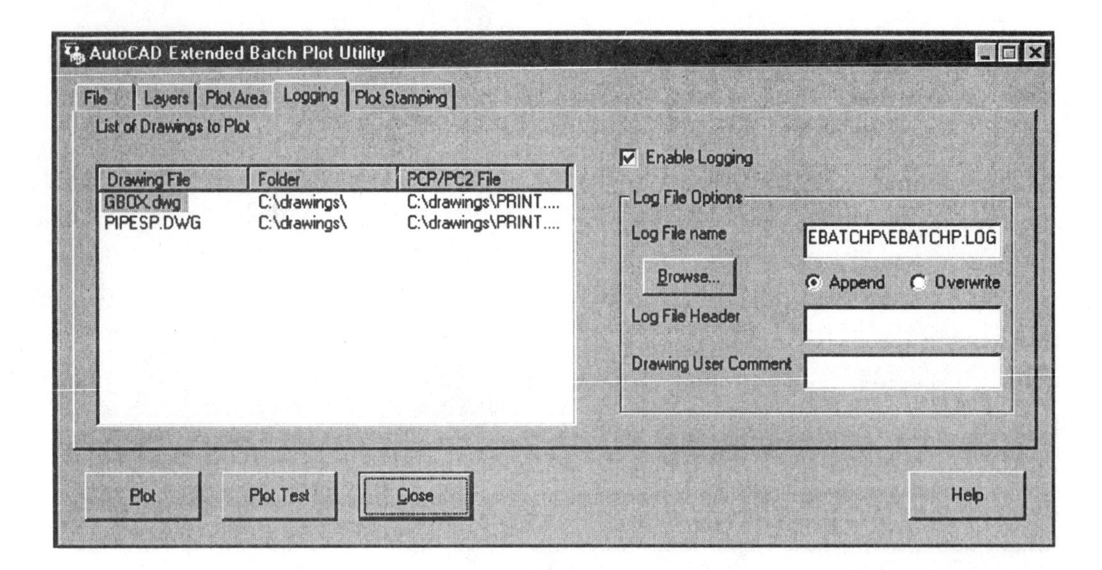

Figure 9.52 Logging tab

To place a text entity on the edge or the corner of the plotted drawing, select the Plot Stamping tab. See Figure 9.53.

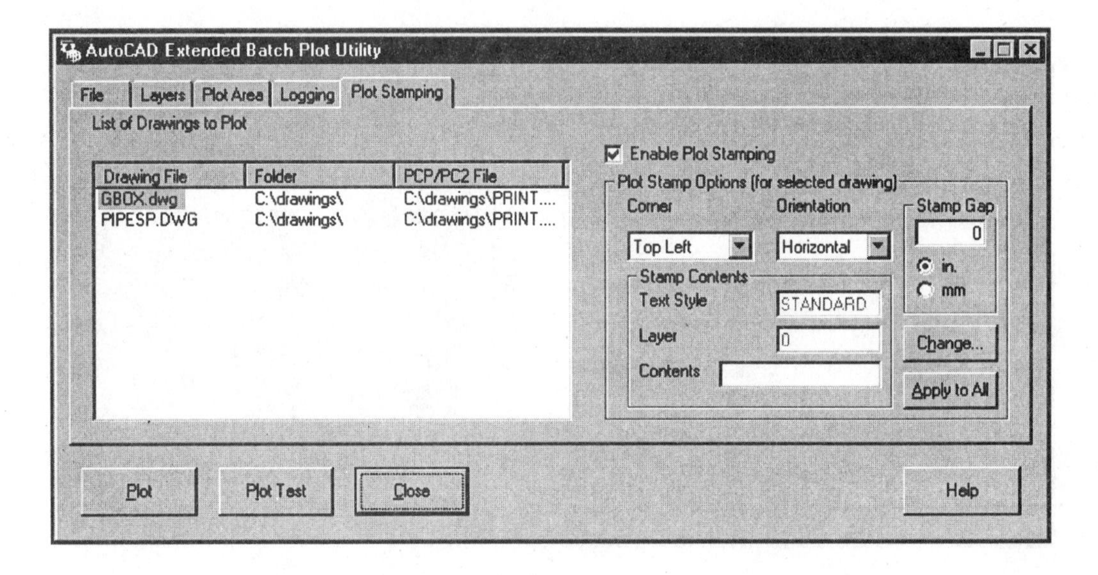

Figure 9.53 Plot Stamping tab

After setting up the Batch Plotting Utility, you should perform a virtual test by selecting the [Plot Test] button. Plot test of a batch of drawings determines if any part of a drawing is invalid or missing without actually plotting the drawings. If no error is encountered, you can select the [Plot] button to perform batch plotting. After batch plotting, you can close the utility by selecting the [Close] button. Note that selecting the [Close] button also closes AutoCAD.

9.7 Key Points and Exercises

There are two working environments in AutoCAD: model space environment and paper space environment. Model space environment is where you construct your model, 2D or 3D. Paper space environment is where you prepare an engineering document. To toggle between the two working environments, you can set the TILEMODE variable or select the TILE button of the status bar.

To prepare a document, there are two basic steps: insert a title block and create floating viewports in paper space environment.

A properly defined title block is useful in subsequent plotting of the drawing. Before inserting a title block, you can define attribute templates to the title block so that you can interactively include the attribute values as you insert the title block.

To display the model constructed in model space environment in an engineering document, you have to create floating viewports. There are two commands to make floating viewports: the MVIEW command and the SOLVIEW command. The MVIEW command can be used on both 2D models and 3D models. The SOLVIEW command is tailor-made for 3D solids. To generate the silhouette and profiles of solid models, you can use the SOLDRAW command or the SOLPROF command. The SOLDRAW command works on floating viewports created by using the SOLVIEW command. The SOLPROF command works on floating viewports created by using the MVIEW command.

With floating viewports created in paper space environment, you can work on the drawing in three ways. You can work on the model in model space environment with TILEMODE set to one, work on the model space environment with TILEMODE set to zero, or work on paper space environment with TILEMODE set to zero. With TILEMODE set to one, you can view the objects only in model space. With TILEMODE set to zero, you can view the objects in both model space and paper space, and you can choose to work on model space within a selected floating viewport by using the MSPACE command or to work on the paper space by using the PSPACE command.

Although you can add dimensions to a drawing in either paper or model space environment, you should put the dimensions in the model space. When floating viewports are prepared by using the SOLVIEW command, a layer with the suffix "-dim" is created automatically for each floating viewport. You should place the dimensions for the solid model in these layers.

With a properly prepared engineering document, you can proceed to plotting. The best result can be achieved by plotting in paper space environment with a plotting scale of 1 to 1. Plotting outcome depends on device configuration. Configuration may take some time, and you may need to configure the system to print on a number of devices. To save configuration settings, you can use complete or partial configuration files.

In this chapter, you made use of two working environments of AutoCAD to produce engineering documents from 2D and 3D drawings, and you learned how to do plotting. During preparation of the engineering document, you applied some techniques on attributes and block insertion.

Exercises 9.1

Why is it necessary to create an engineering title block before preparing an engineering document?

Exercises 9.2

AutoCAD has two working environments. What are they and what are their uses?

Exercises 9.3

Delineate the steps required to prepare an engineering document from a 2D drawing.

Exercises 9.4

Explain how to prepare an engineering document from a 3D solid model.

Exercises 9.5

Depict plotting procedure. Explain how to batch plot a set of drawings.

Exercises 9.6

Following the steps outlined in this chapter, create the engineering documents for the drawing files that you created in Chapters 3, 4, 5, 6, 7, and 8.

Chapter 10

Visualization and RP Application

10.1 Perspective View
10.2 Hide and Shade
10.3 Photo-realistic Rendering
10.4 Introduction to RP and STL
10.5 Key Points and Exercises

Aims and Objectives

The aim of this chapter is to familiarize you with the use of various visualization tools to produce 3D views, perspective views, hidden and shaded views, and photo-realistic rendered drawings; to introduce the basic principles of rapid prototyping; and to give you an opportunity to output a solid model to STL format. After studying this chapter, you should be able to:

- use various visualization tools,
- produce photo-realistic renderings, and
- output STL files from 3D solid models.

Overview

In Chapters 6, 7, and 8, you created 3D solid models. In this chapter, you will work on visualization of 3D objects and output 3D solids to rapid prototyping machines.

To make a realistic drawing, you can set the display to a perspective view. Normally, 3D solid objects are shown in wireframe mode. To see the 3D model better, you can remove hidden lines and apply shading. To get a photo-realistic rendering, you can add lights and assign materials. To achieve professional rendering and animation, you can export the 3D objects to 3D Studio format and use 3D Studio to do further work.

Rapid prototyping (RP) is becoming popular. You will learn the basic principles of rapid prototyping and export a 3D solid to stereolithography apparatus (SLA) file format required for making rapid prototypes.

10.1 Perspective View

Open the file ABLK.DWG that you created in Chapter 6. It is a model of an angle block. See Figure 10.1. You will set up a perspective view for this model.

<File> **<Open...>**

File name: **ABLK.DWG**

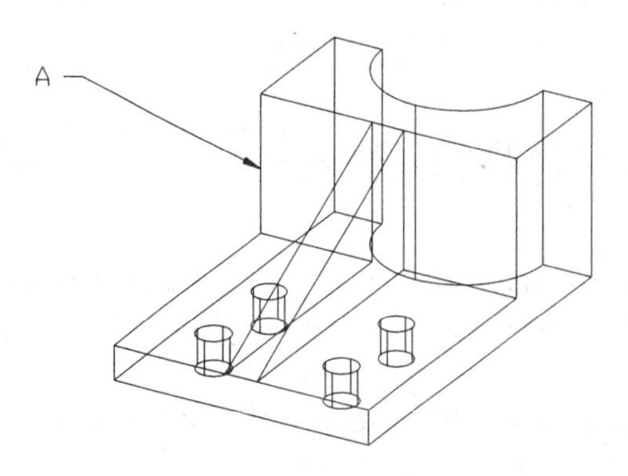

Figure 10.1 Solid model of an angle block

In previous chapters, you learned how to obtain a 3D view by using the VPOINT command and the DDVPOINT command. The 3D views obtained from these commands are based on orthographic projection techniques. The views are auxiliary views projected from a 3D direction. To get a perspective view, use the DVIEW command. From the View pull-down menu, select the 3D Dynamic View item to run the DVIEW command.

<View> **<3D Dynamic View>**

Command: **DVIEW**
Select objects: **[Enter]**

Depending on the complexity of the 3D object, it may take quite long time to generate a dynamic view. Therefore, this command lets you select objects for making previews. If you do not select anything, a 3D house will be displayed to give you an idea of how the final model will be positioned after you finished the command. See Figure 10.2.

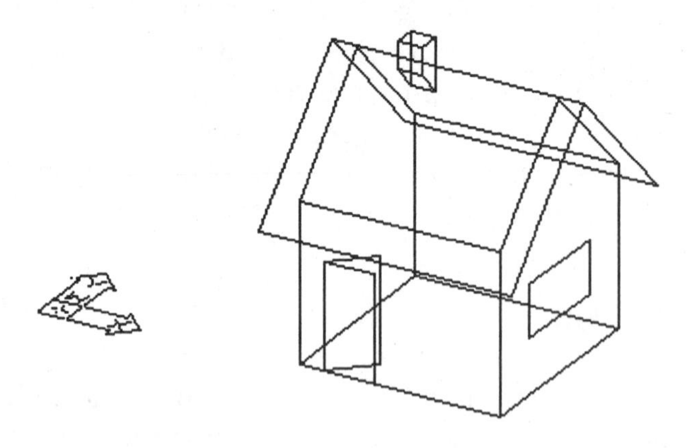

Figure 10.2 3D house displayed

If you want to display the 3D model instead of the 3D sample house, use the [ESC] key to cancel the command. Then repeat the command and select the model.

CAmera/TArget/Distance/POints/PAn/Zoom/TWist/CLip/Hide/Off/Undo/<eXit>: **[ESC]**

Command: **[Enter]**
DVIEW
Select objects: **[Select A (Figure 10.1).]**
Select objects: **[Enter]**

To trigger perspective display mode, use the Distance option. Then set the camera distance to 500 units. See Figure 10.3. Notice the shape of the UCS icon in your screen.

CAmera/TArget/Distance/POints/PAn/Zoom/TWist/CLip/Hide/Off/Undo/<eXit>: **D**
New camera/target distance: **500**

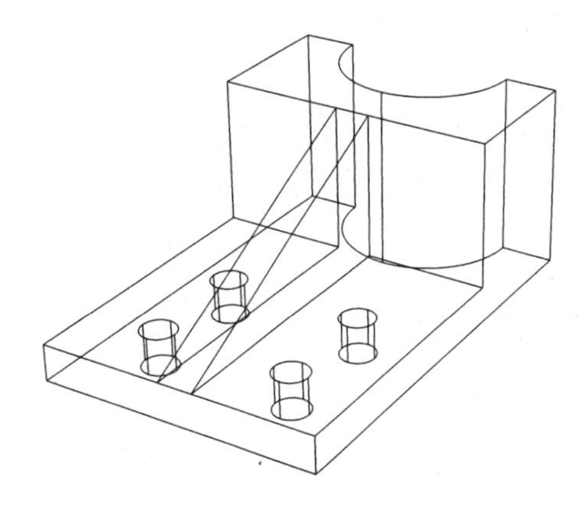

Figure 10.3 Perspective view

To refine the perspective view, use the Zoom option to set the camera lens size. Then set the distance again. See Figure 10.4.

CAmera/TArget/Distance/POints/PAn/Zoom/TWist/CLip/Hide/Off/Undo/<eXit>: **Z**
Adjust lenslength: **15**
CAmera/TArget/Distance/POints/PAn/Zoom/TWist/CLip/Hide/Off/Undo/<eXit>: **D**
New camera/target distance: **200**

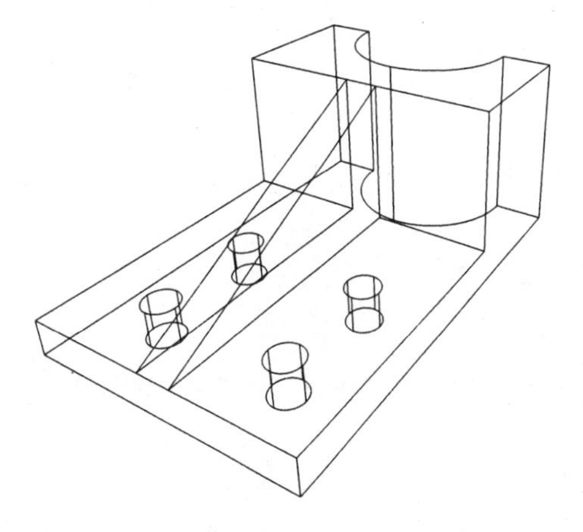

Figure 10.4 Camera lens and distance set

Compare Figure 10.3 with Figure 10.4 to see the effect of camera lens size and camera distance on the perspective view. To refine the view, set the camera to a new position. See Figure 10.5.

```
CAmera/TArget/Distance/POints/PAn/Zoom/TWist/CLip/Hide/Off/Undo/<eXit>: CA
Toggle angle in/Enter angle from XY plane: 20
Toggle angle from/Enter angle in XY plane from X axis: -150
```

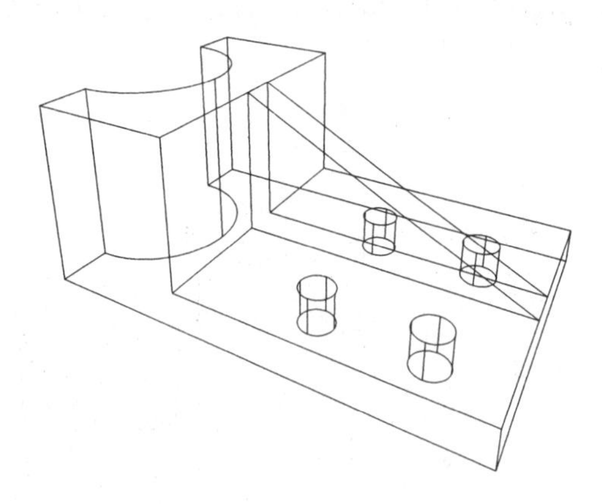

Figure 10.5 Camera set to a new position

In perspective mode, you cannot use the ZOOM command. Besides, display regeneration takes longer. If you want to turn off perspective mode, use the Off option. See Figure 10.6.

```
CAmera/TArget/Distance/POints/PAn/Zoom/TWist/CLip/Hide/Off/Undo/<eXit>: OFF
CAmera/TArget/Distance/POints/PAn/Zoom/TWist/CLip/Hide/Off/Undo/<eXit>: [Enter]
```

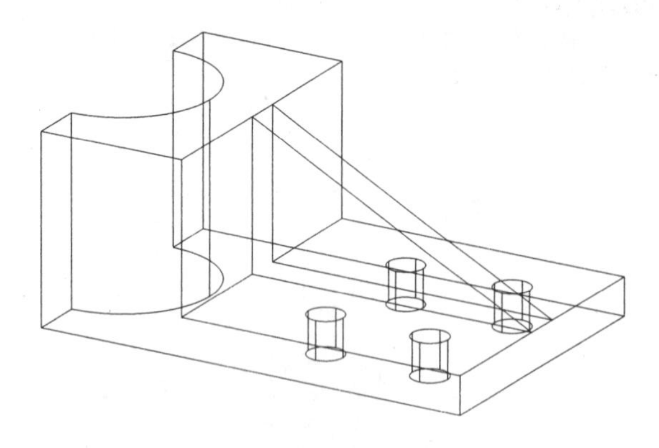

Figure 10.6 Perspective mode turned off

10.2 Hide and Shade

Having learned in previous chapters that there are wireframes, surfaces, and solids in a computer, you may notice that the display of a solid model shows only the wireframes. If you find it difficult to visualize the model, you can remove the hidden lines, shade the model, or render the model. Open the drawing GBOX.DWG that you saved in Chapter 6. It is a gear box cover. See Figure 10.7.

<File> **<Open...>**

File name: **GBOX.DWG**

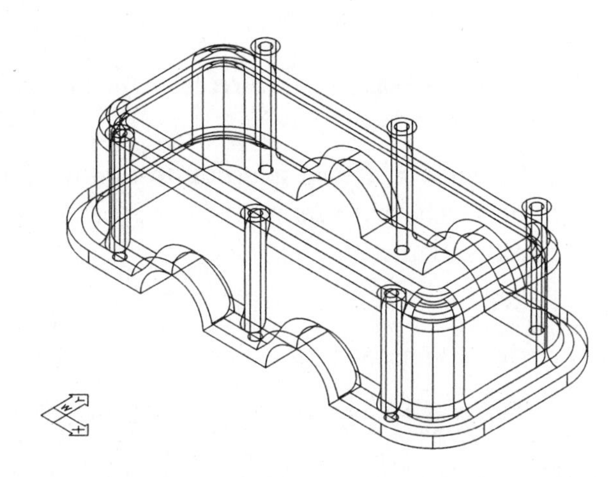

Figure 10.7 Solid model of a gear box cover displayed in wireframe mode

To obtain a view that removes hidden lines, select the Hide item from the View pull-down menu. See Figure 10.8.

<View> **<Hide>**

Command: **HIDE**

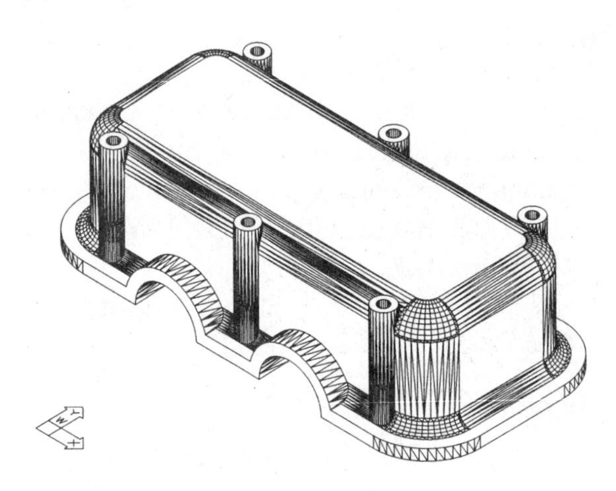

Figure 10.8 Hidden lines removed and curved faces displayed as facets

You may notice from Figure 10.8 that in addition to removing the hidden lines, the HIDE command displays the facets that represent the curved surfaces. To adjust facet resolution, set the FACETRES variable.

Command: **FACETRES**
New value for FACETRES <0.5000>: **0.01**

The FACETRES variable ranges from 0.01 to 10. Its default value is 0.5. A value of 10 gives the highest resolution. A value of 0.01 gives the lowest resolution. Use the HIDE command again. See Figure 10.9. Compare Figure 10.9 with Figure 10.8 to see the difference in facet resolution.

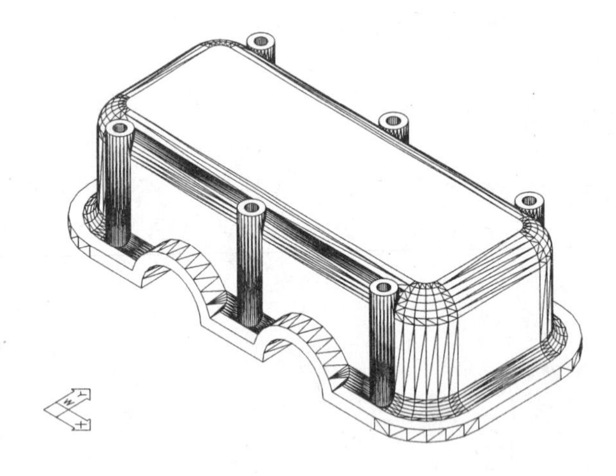

Figure 10.9 Facet resolution (FACETRES) set to 0.01

A command similar to hiding is the SHADE command. Select the 16 Color Filled item from the Shade cascading menu of the View pull-down menu. See Figure 10.10.

<View> **<Shade>** **<16 Color Filled>**

Command: **SHADE**

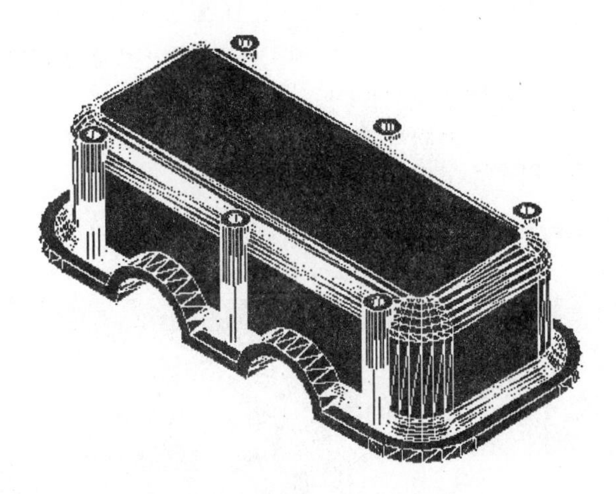

Figure 10.10 Shaded display

After shading, you have to regenerate the drawing before you can select the objects on the screen.

Command: **REGEN**

10.3 Photo-realistic Rendering

To render means to portray a scene or figure. Here you will continue to work on the drawing GBOX.DWG to produce photo-realistic renderings.

Edge and Surface Smoothness

Output quality of a rendered image is affected by the VIEWRES command and the FACETRES variable. They affect edge and surface smoothness, respectively.

In Chapter 3, you learned that the VIEWRES command controls the display accuracy of circles, arcs, and ellipses. In rendering, this command governs the smoothness of circular and elliptical edges. Run this command to set the zoom percent to 500. A zoom percent of 20000 gives the best result but takes longer.

Command: **VIEWRES**
Do you want fast zooms? **Y**
Enter circle zoom percent (1-20000): **500**

The FACETRES variable, besides affecting the facet resolution in hiding and shading, also sets the smoothness of the curved surfaces in a rendered drawing.

Command: **FACETRES**
New value for FACETRES: **0.01**

Render

After using the VIEWRES command and the FACETRES variable, select the Render... item of the Render cascading menu of the View pull-down menu to bring out the RENDER command dialog box. See Figure 10.11.

<View> **<Render>** **<Render...>**

Command: **RENDER**

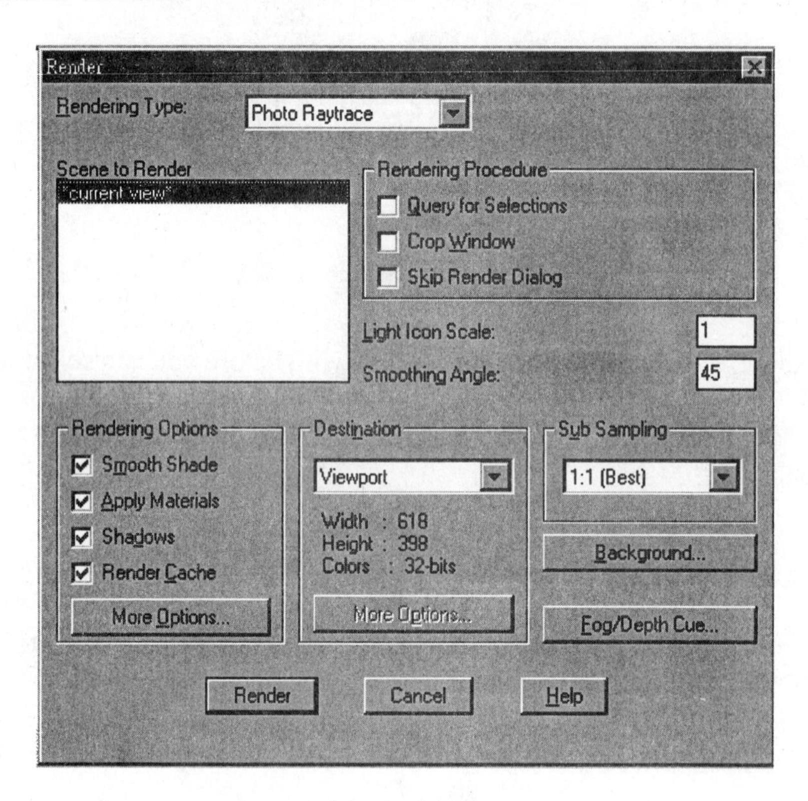

Figure 10.11 Render dialog box

Refer to Figure 10.11. There are three rendering types: Render, Photo Real, and Photo Raytrace. Render is the basic type of rendering provided. It takes the least time to generate a rendering. Photo Real displays bitmapped and transparent materials and generates volumetric and mapped shadows. It takes more time to generate. Photo Raytrace uses ray trace technique to generate reflections, refractions, and precise shadows. It gives the best result but takes the longest time to make a rendering. Select Photo Raytrace from the drop-down list.

The RENDER command renders a scene. In the Scene to Render box, you can select a saved scene or use the current view. Because you have not created a scene, use the current view and default light.

In the Rendering Options box, there are four check boxes and a button. Because the curved surfaces are represented by facets, select the [Smooth Shade] check box to smooth out the rough edges of the facets. To reiterate, the facet resolution is determined by the FACETRES variable. For the surface of the object to assume the material assigned by the

RMAT command or the ACI color of the drawing, select the [Apply Materials] check box.

The [Shadows] check box applies to Photo Real and Photo Raytracing rendering; it activates shadow casting. Check this box. In order to save time in subsequent rendering operations, select the [Render Cache] check box to write the rendering information to a cache file.

Depending on which kind of rendering you have chosen, the [More Options] button brings out different dialog boxes. Select this button. See Figures 10.12, 10.13, and 10.14.

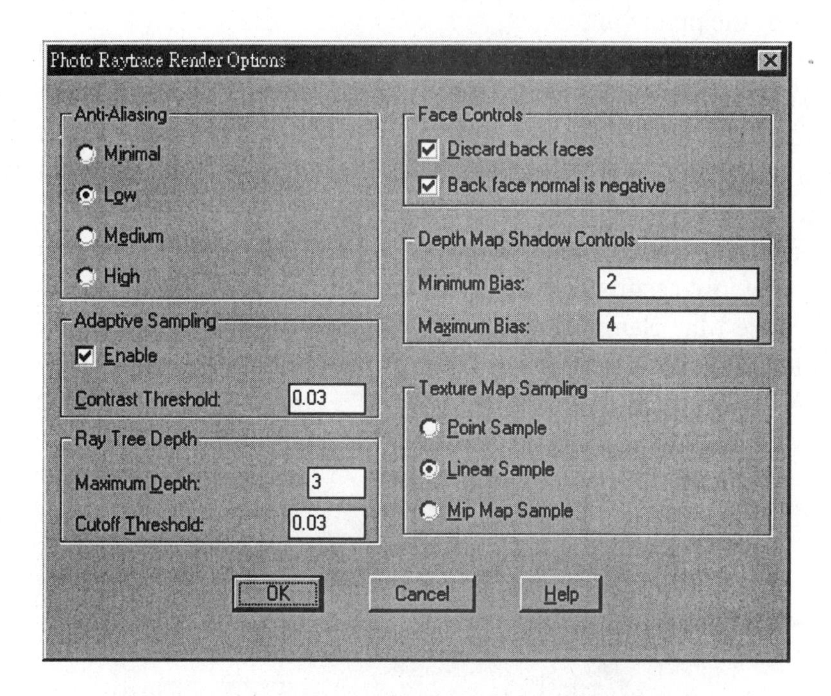

Figure 10.12 Photo Raytrace Render Options dialog box

Figure 10.12 shows the Photo Raytrace Render Options dialog box. In this dialog box, there are six boxes: Anti-Aliasing, Adaptive Sampling, Ray Tree Depth, Face Controls, Depth Map Shadow Controls, and Texture Map Sampling.

Computer-generated images are made of discrete pixels. Therefore, inclined straight lines and curves may appear to be jagged. This phenomenon is called aliasing. In order to offset the jagged effect, pixels in adjacent boundary objects are analyzed to determine the intermediate shading value; and lines are made thicker by adding shaded pixels of intermediate value. This method of minimizing the aliasing effect is called anti-aliasing. In the Anti-Aliasing box, you can determine how anti-aliasing is applied.

To apply anti-aliasing, rendering type has to be photo real or photo raytrace. There are four anti-alias options: Minimal, low, medium, and high. The minimal option computes the horizontally adjacent pixel. The low option also computes the horizontally adjacent pixels but uses four pixels as samples. The medium option takes nine pixels as samples. The high option takes 16 samples. The higher the value, the longer it takes to render a drawing.

Adaptive sampling is a technique to accelerate the anti-aliasing process. In the Adaptive Sampling box, you can enable adaptive sampling and determine the sensitivity of adaptive sampling by setting the value in the [Contrast threshold:] box. The value of contrast threshold ranges from 0 to 1. With a value of 1, the initial sample difference must be large to force further sampling. Therefore, the speed of rendering is increased.

In the Ray Tree Depth box, you can determine the maximum number of reflected and refracted rays to be considered by setting the value in the [Maximum Depth:] box, and you can determine how much the next bound contributes to the pixel value by setting the value in the [Cutoff Threshold:] box. A maximum depth of 3 sets the number of maximum bound rays to be three. A cutoff threshold of 0.03 sets the next bound ray to contribute 3% of the pixel value.

A surface has a front face and back face. You can speed up rendering by ignoring the back faces. To determine which is the back face, it is normal practice to consider as the back face a face with a vector pointing away from the display view. In the Face Controls box, you can choose to discard back faces in rendering and assign faces with negative vector direction as back faces.

In the Depth Map Shadow Controls box, you can adjust the movement of the shadow relative to the shadow-casting object to prevent self-shadowing and detached shadows.

In the Texture Map Sampling box, you can determine how to sample a texture map that is projected onto a surface that is smaller than the texture map.

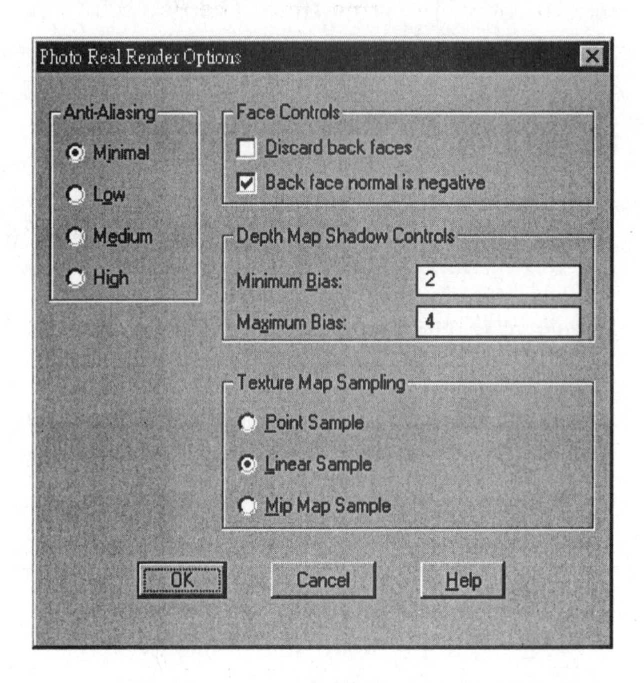

Figure 10.13 Photo Real Render Options dialog box

Figure 10.13 shows the Photo Real Render Options dialog box. There are four boxes: Anti-Aliasing, Face Controls, Depth Map Shadow Controls, and Texture Map Sampling boxes. Their functions are similar to those in the Photo Raytrace Render Options dialog box.

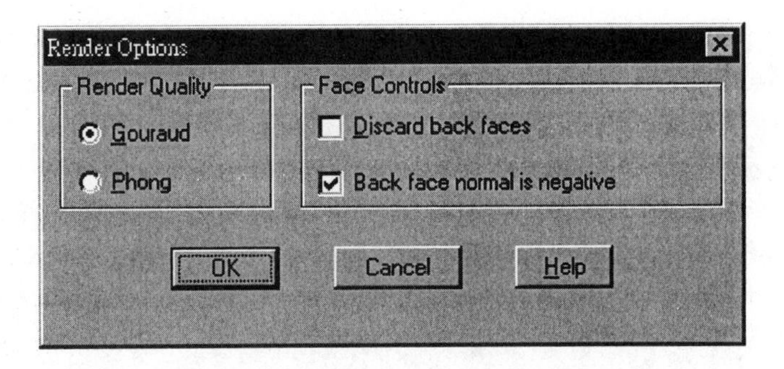

Figure 10.14 Render Options dialog box

Figure 10.14 shows the Render Options dialog box. The Render Quality box enables you to choose how a faceted surface is smoothened. There are two methods: Gouraud and Phong. Light intensity at a facet depends on the direction of the normal vector of the facet and the direction of the light source. Gouraud computes pixel intensity at each vertex of a facet and interpolates the intensity value along the entire facet. Phong averages the normal vector at each vertex of a facet, interpolates the vector value over the entire surface, and computes the intensity value of each pixel. Phong shading gives more accurate result but requires more rendering time. The Face Controls box enables you to discard back faces and to set faces with negative normals as back faces.

In the Rendering Procedure box (Figure 10.11), there are three check boxes: Query for Selections, Crop Window, and Skip Render Dialog.

By checking the Query for Selection box or the Crop Window box, you can render a portion of your drawing instead of rendering the entire scene. The Query for Selections box enables you to select objects to render. The Crop Window box enables you to describe a rectangular area to render. They are useful in fine-tuning the rendered image.

If you do not want to display the Render dialog box the next time you run the RENDER command, you can check the Skip Render Dialog box. To bring back the Render dialog box, you can select the Preferences item of the Render cascading menu of the View pull-down menu to bring up a Rendering Preferences dialog box. This dialog box is similar to the Render dialog box. See Figure 10.20. From the Rendering Preferences dialog box, you can un-check the Skip Render Dialog box.

Below the Rendering Procedure box there are two items. Whenever you include a light in your drawing, a light icon is displayed. The [Light Icon Scale:] box sets the scale of this icon. Curve surfaces are represented by facets and smoothened to give the appearance of a smooth surface. The [Smoothing Angle:] box determines the angle between adjacent facets for smoothing to be applied.

The Destination box enables you to output the rendered image to the viewport, rendered window, or file. If you select output to a file, you can select the [More Options...] button to configure file output. See Figure 10.15. Select Viewport as destination.

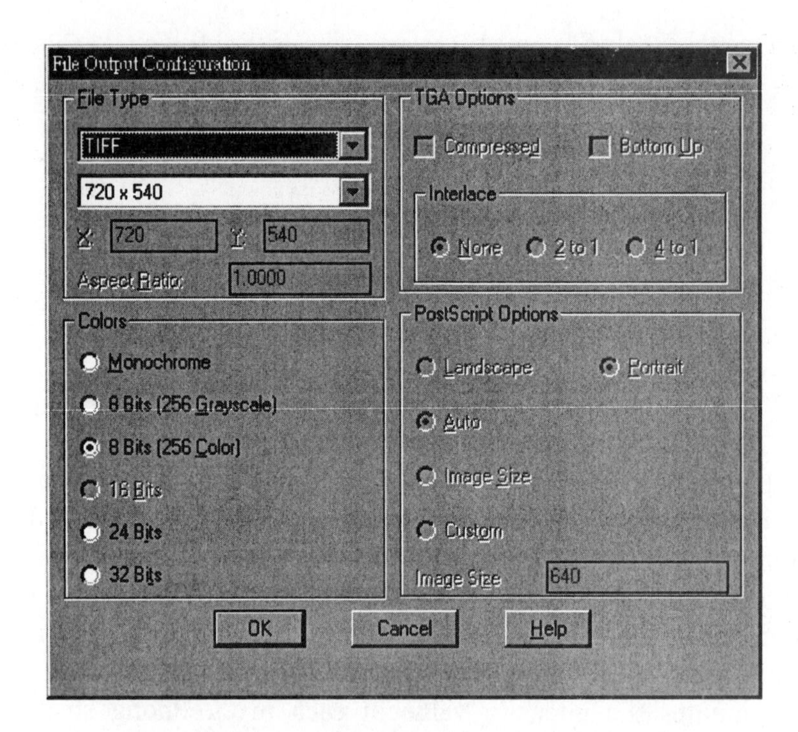

Figure 10.15 File Output Configuration dialog box

The Sub Sampling box of the main dialog box (Figure 10.11) enables you to reduce rendering time by rendering a fraction of all pixels.

The [Background...] button brings out the Background dialog box. Here you determine the background of the rendered image. See Figure 10.16.

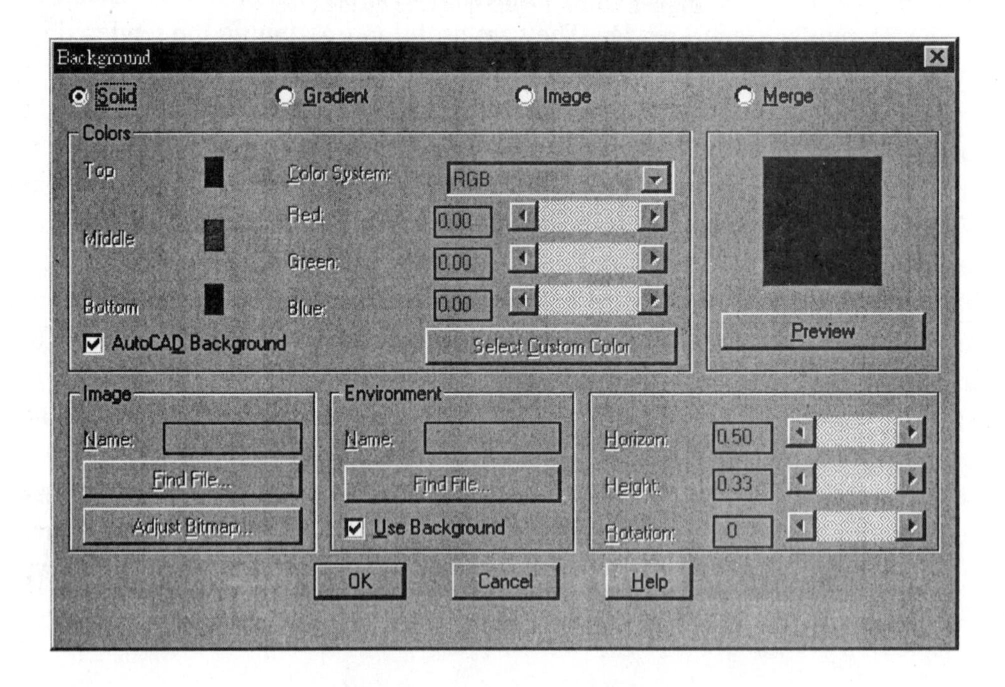

Figure 10.16 Background dialog box

The [Fog/Depth Cue...] button brings out the Fog/Depth Cue dialog box. Here you can enable fog and set the way the fog is to affect the rendered image. See Figure 10.17.

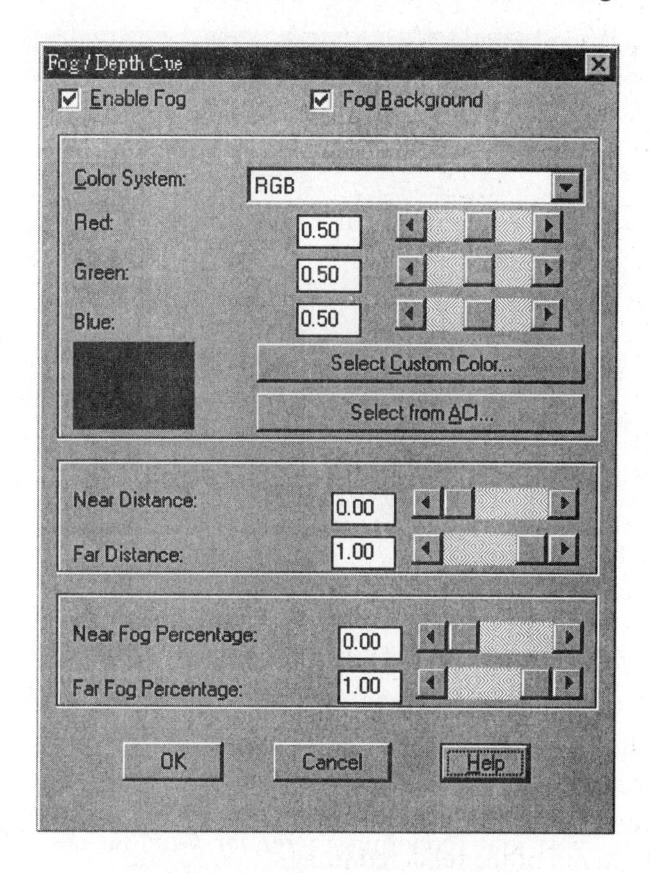

Figure 10.17 Fog/Depth Cue dialog box

After setting the options, select the [Render] button (Figure 10.11). See Figure 10.18.

Figure 10.18 Rendered drawing of the gear box cover

Repeat the RENDER command and select the Render Window in the Destination box. See Figure 10.19.

Command: **[Enter]**
RENDER

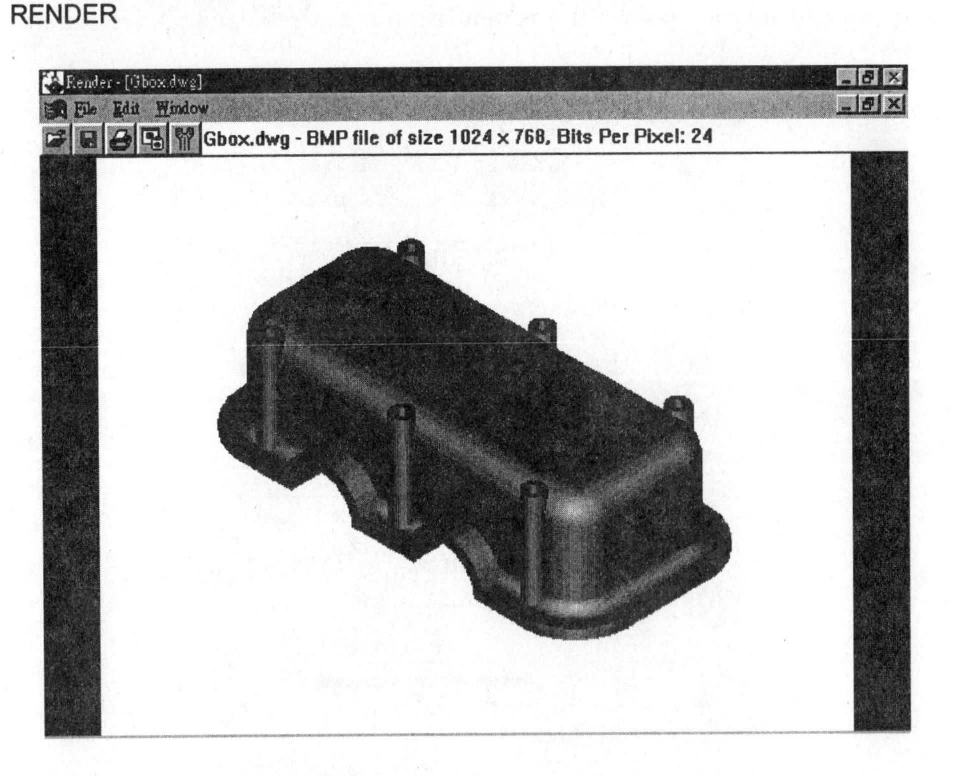

Figure 10.19 Render window

Rendering to viewport and rendering to render window are different. A rendered image that is rendered to viewport is restricted by the current display resolution. By rendering to render window, you can have a much higher resolution bipmap and use the render window to save the rendered image to a file.

To configure the RENDER command without rendering the objects, run the RPREF command. See Figure 10.20.

<View> **<Render>** **<Preferences...>**

Command: **RPREF**

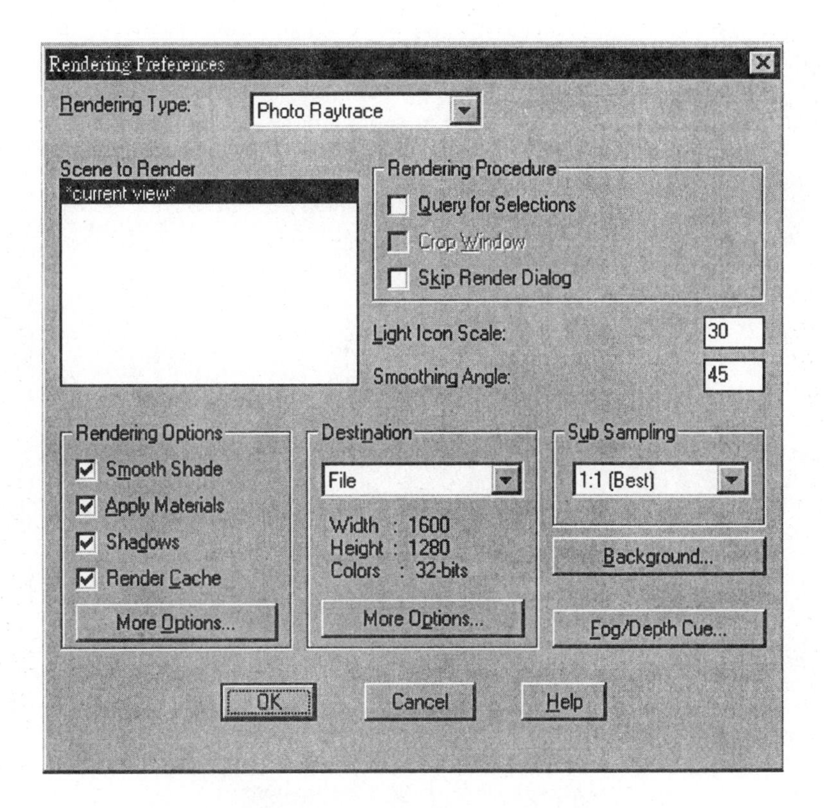

Figure 10.20 Rendering Preferences dialog box

Compare Figure 10.20 with Figure 10.11. They are identical except that one has an [OK] button and the other has a [Render] button. Set the Light Icon Scale to 30, and select the [OK] button to exit.

Light and Scene

To make a photo-realistic rendering, you need to set lights to the scene and assign material properties to the surfaces. A scene is a collection of lights and a display view. A scene has to be illuminated in order for you to obtain a rendered view. There are four kinds of lights: ambient, point light, distant light, and spotlight.

Ambient light comes from the surrounding environment. It gives a constant illumination of all the surfaces of an object.

A point light discharges light from a single point in all directions. You can set it to attenuate. Attenuation means that the intensity of a light diminishes as a factor of distance from the light source.

A distant light comes from very far way. It gives off uniform parallel light rays in a single direction. Its intensity does not diminish at any distance.

A spotlight delivers a cone of light from a single point. The cone of light has two zones: hotspot and falloff. Light is brightest within the hotspot cone and diminishes from the edge of the hotspot cone toward the outer edge of the falloff cone. You can set a spotlight to attenuate.

From the Render cascading menu of the View pull-down menu, select the Light... item to bring out the LIGHT command dialog box. See Figure 10.21.

<View> **<Render>** **<Light...>**

Command: **LIGHT**

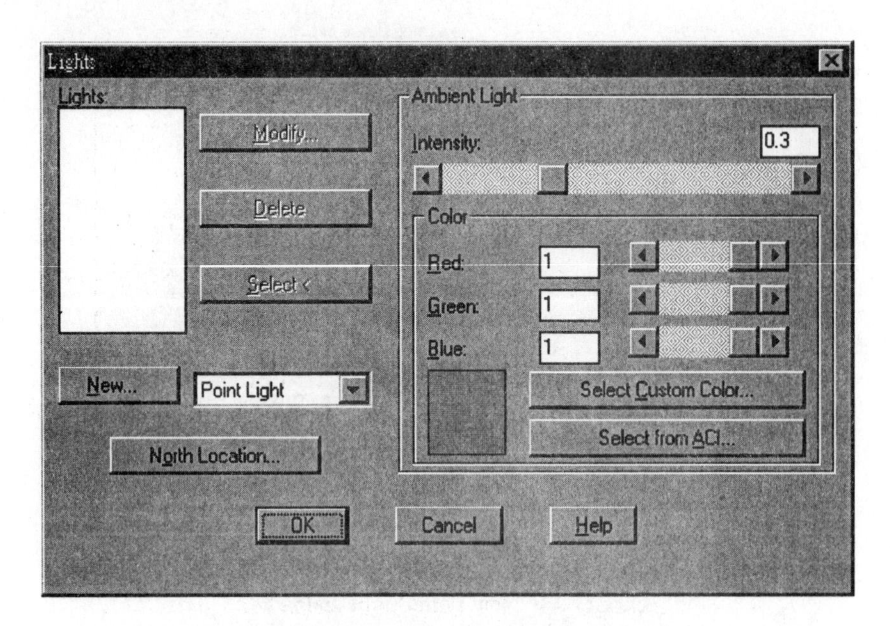

Figure 10.21 Lights dialog box

The Lights dialog box enables you to adjust the values of the ambient light and to create, modify, and delete point light, distant light, and spotlight.

Ambient light intensity ranges from zero to one. In order not to dull the image, keep the intensity value low. To adjust the color of ambient light, set the value of red, green, and blue colors. By selecting the [Select Custom Color...] button, you can set a color by specifying its HLS value. H is hue. It determines the color. L is luminous. It sets the luminosity of the color. S is saturation. It controls the saturation of the color. By selecting the [Select from ACI...] button, you can select a color from the AutoCAD Color Index (ACI). Use the default ambient color value of 0.3.

To create a point light, select the Point Light item from the light type drop-down list and select the [New...] button. Enter light name POINT1. See Figure 10.22.

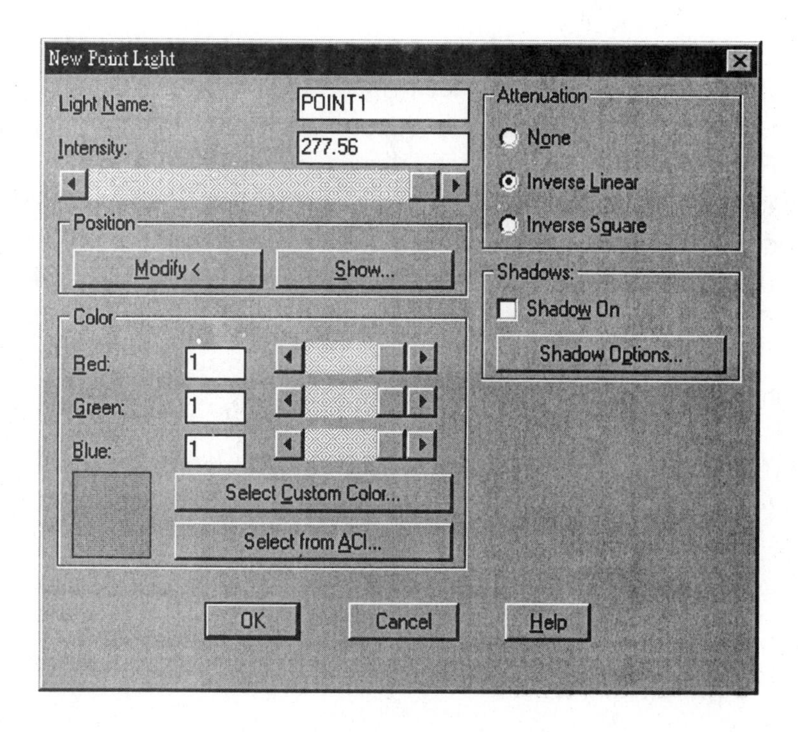

Figure 10.22 New Point Light dialog box

In the New Point Light dialog box, slide the Intensity slider bar to the right to set to maximum value. Then set the color value of Red, Green, and Blue to 1. This creates a white point light with maximum intensity.

A point light can attenuate. When you select inverse linear attenuation, light intensity diminishes in inverse proportion to the distance. When you select inverse square attenuation, light intensity diminishes in inverse proportion to the square of the distance. Select Inverse linear.

In addition to attenuation, you can decide whether shadow is to be cast and how shadow is applied. Set Shadow on and select the [Shadow Options...] button. See Figure 10.23.

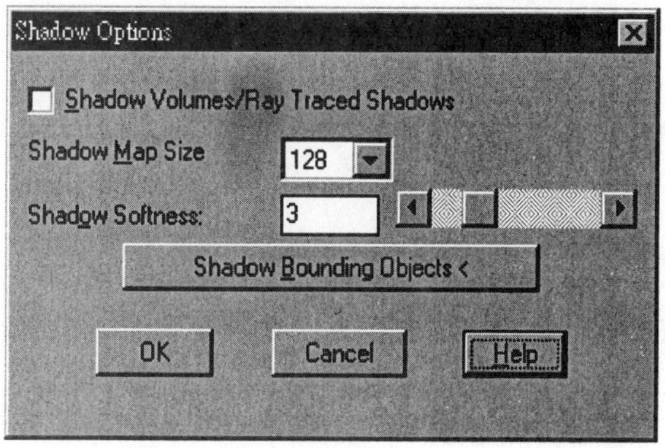

Figure 10.23 Shadow Options dialog box

As we have said, there are three kinds of renderings. If you use Photo Real render, checking the [Shadow Volumes/Ray Traced Shadows] produces volumetric shadows. If you use Photo Raytraced shadows, it produces raytraced shadows. If the [Shadow Volumes/Ray Traced Shadows] box is not selected, shadow maps will be used. You can set shadow map size, control shadow softness, and select shadow bounding objects. Select the [Shadow Volumes/Ray Traced Shadows] box and then the [OK] button.

To modify the point light position, select the [Modify<] button. The dialog box hides temporarily, and the default point light position is indicated.

Enter light location <current>: **@0,0,300**

Select the [OK] buttons to complete the LIGHT command. A point light is created. See Figure 10.24.

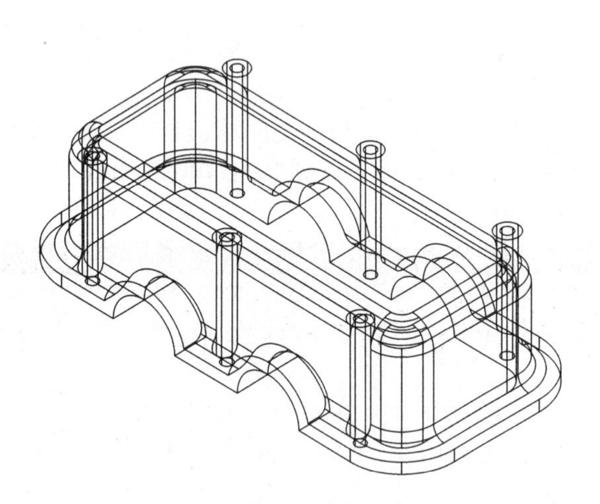

Figure 10.24 Point light added

To see the effect of this point light on rendering, use the RENDER command to render to the viewport. See Figure 10.25.

<View> **<Render>** **<Render...>**

Command: **RENDER**

Figure 10.25 Rendering with point light

Run the LIGHT command to create a distant light. The sun is a typical distant light. It comes from very far away. Light rays are parallel and do not attenuate.

<View> **<Render>** **<Light...>**

Command: **LIGHT**

Refer to the Lights dialog box shown in Figure 10.21. Select the Distant Light item from the light type drop-down box, and select the [New...] button. See Figure 10.26.

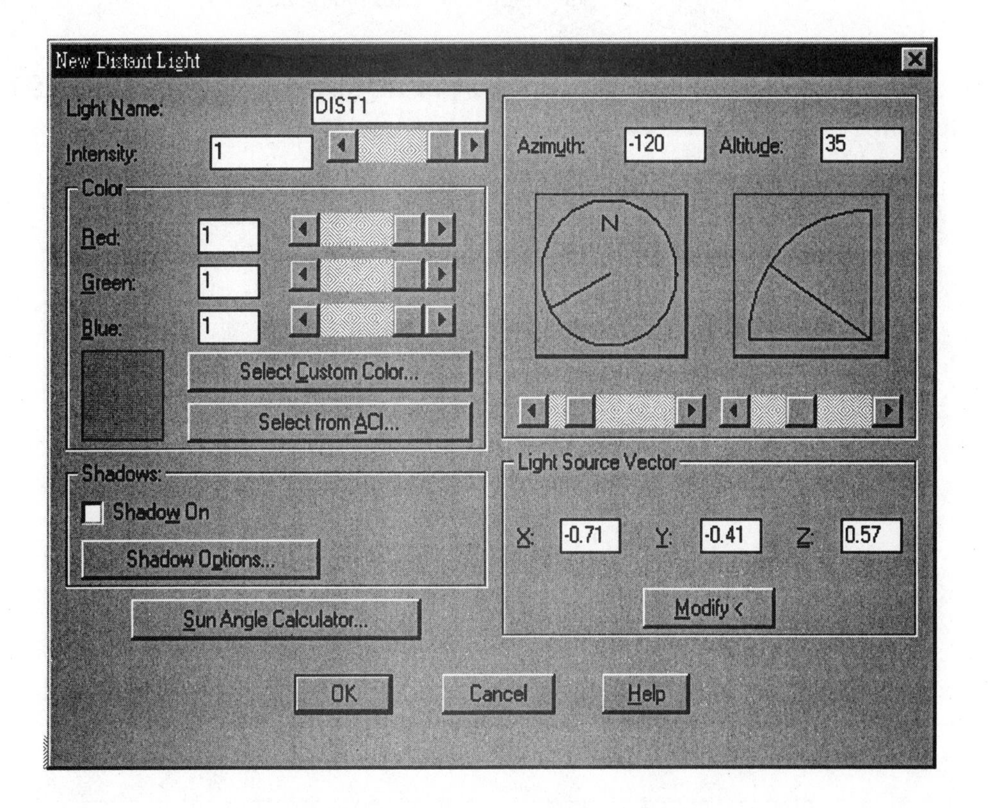

Figure 10.26 New Distant Light dialog box

Enter light name DIST1 as the light name. As for the point light, you can set intensity, color, and shadow. There are several ways to locate the distant light. Azimuth and altitude set the position of the distant light by using site-based coordinates. The light source vector enables you to specify a position in the drawing. The Sun Angle Calculator calculates the sun position by specifying time, date, and city from a map panel. Select the [Sun Angle Calculator...] button. See Figure 10.27.

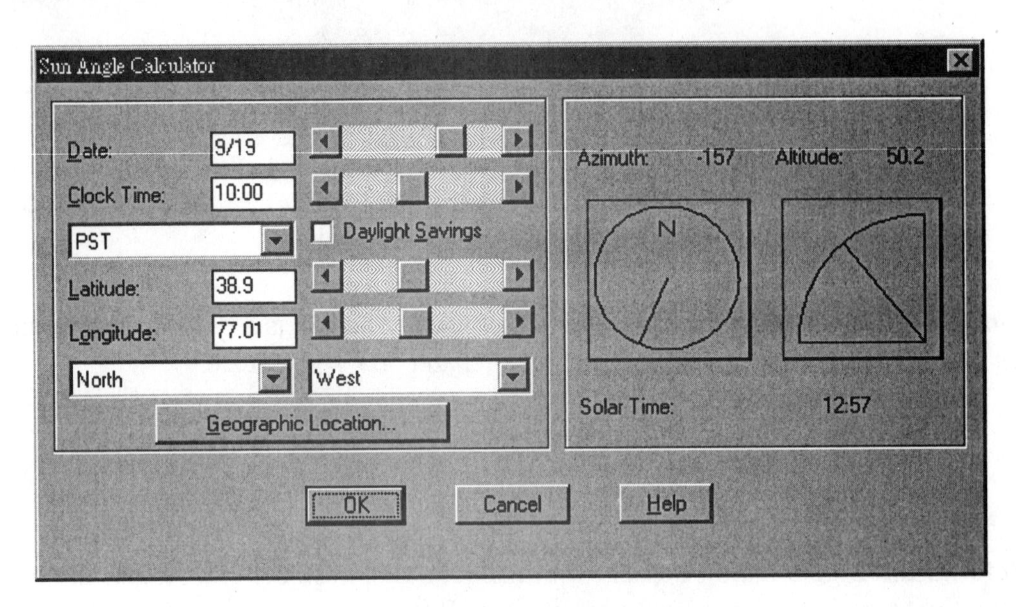

Figure 10.27 Sun Angle Calculator dialog box

In the Sun Angle Calculator dialog box, input the date of 9/19 and time of 11:00. Then select the [Geographic Location...] button. See Figure 10.28.

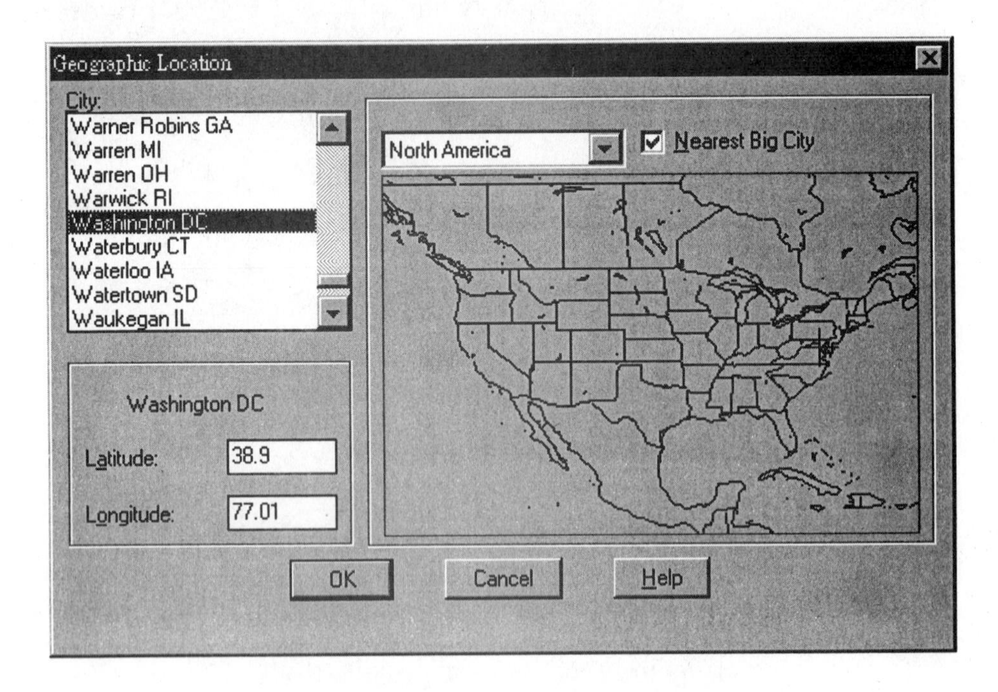

Figure 10.28 Geographic Location dialog box

The Geographic Location dialog box shows a map panel and the nearest big city that you can select. Suppose you are living in Washington DC, of North America. Select Washington DC. Then select the [OK] buttons to complete the command. A distant light is complete. See Figure 10.29.

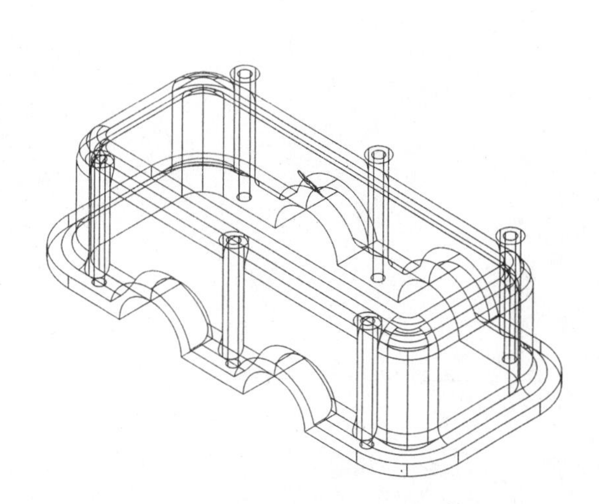

Figure 10.29 Distant light added

In addition to ambient light, you have created two kinds of light in your drawing: point and distant. In order to choose selectively the lights in a scene for rendering, you have to set up scenes. A scene enables you to remember the viewing position and select lights in making a rendered image. Before setting up scenes, save the current view by using the DDVIEW command.

<View> **<Named Views...>**

Command: **DDVIEW**

[View Control **New...**]

[Define New View
New Name: **ISO**
Save View]

[OK]

Now set the display to a perspective view. See Figure 10.30.

<View> **<3D Dynamic View>**

Command: **DVIEW**

Select objects: **[Select the model.]**
Select objects: **[Enter]**
CAmera/TArget/Distance/POints/PAn/Zoom/TWist/CLip/Hide/Off/Undo/<eXit>: **D**
New camera/target distance: **500**
CAmera/TArget/Distance/POints/PAn/Zoom/TWist/CLip/Hide/Off/Undo/<eXit>: **Z**
Adjust lenslength: **25**
CAmera/TArget/Distance/POints/PAn/Zoom/TWist/CLip/Hide/Off/Undo/<eXit>: **CA**
Toggle angle in/Enter angle from XY plane: **25**
Toggle angle from/Enter angle in XY plane from X axis: **-45**
CAmera/TArget/Distance/POints/PAn/Zoom/TWist/CLip/Hide/Off/Undo/<eXit>: **[Enter]**

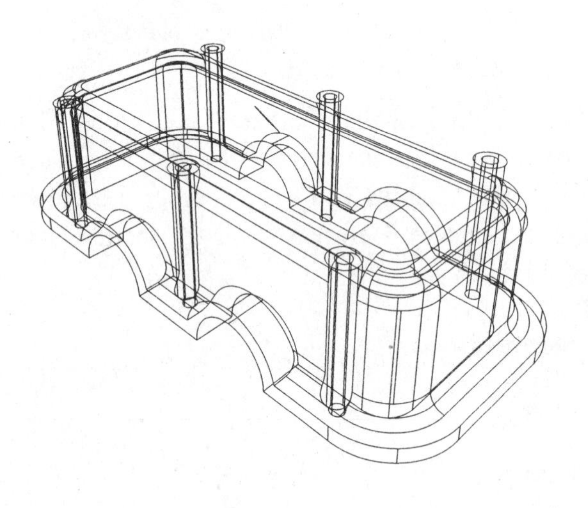

Figure 10.30 Perspective view

Save the perspective view named PERP by using the DDVIEW command.

<View> **<Named Views...>**

Command: **DDVIEW**

[View Control **New...**]

[Define New View
New Name: **PERP**
Save View]

[OK]

Now set up two scenes by using the SCENE command.

<View> **<Render>** **<Scene...>**

Command: **SCENE**

When the Scene dialog box appears, select the [New...] button. See Figure 10.31.

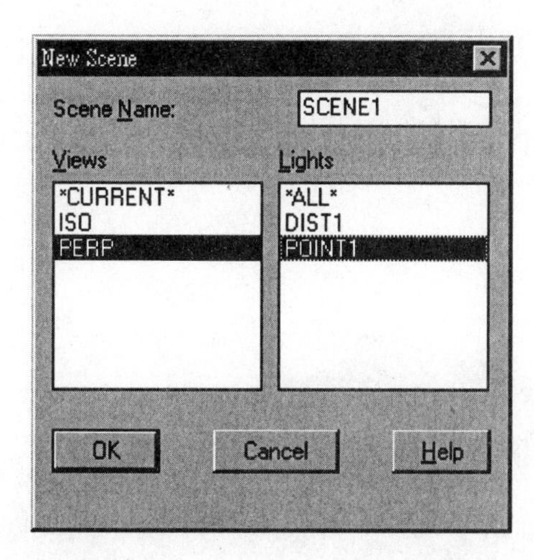

Figure 10.31 New Scene dialog box

In the New Scene dialog box, enter scene name as SCENE1. Then select PERP view and POINT1 light. Select the [OK] button.

On returning to the Scene dialog box, select the [New...] button and create a second scene named SCENE2 with PERP view and DIST1 light.

Now you have two scenes. Select the [OK] button to exit the SCENE command. Then run the RENDER command.

<View> **<Render>** **<Render...>**

Command: **RENDER**

In the Scene Name box, select SCENE1. Then select the [Render] button to render. See Figure 10.32.

Figure 10.32 Rendering perspective view with point light

Repeat the RENDER command to render SCENE2. See Figure 10.33.

Command: **[Enter]**
RENDER

Figure 10.33 Rendering perspective view with distant light

Compare Figure 10.33 with Figure 10.32 to see the difference between distant light and point light.

As we have noted, there are four kinds of light: ambient light, point light, distant light, and spotlight. Use the DDVIEW to restore the ISO view. Then use the LIGHT command to add a spotlight.

<View> **<Render>** **<Light...>**

Command: **LIGHT**

From the Lights dialog box, select the Spotlight item from the light type drop-down list, and select the [New...] button. Enter light name SPOT1. See Figure 10.34.

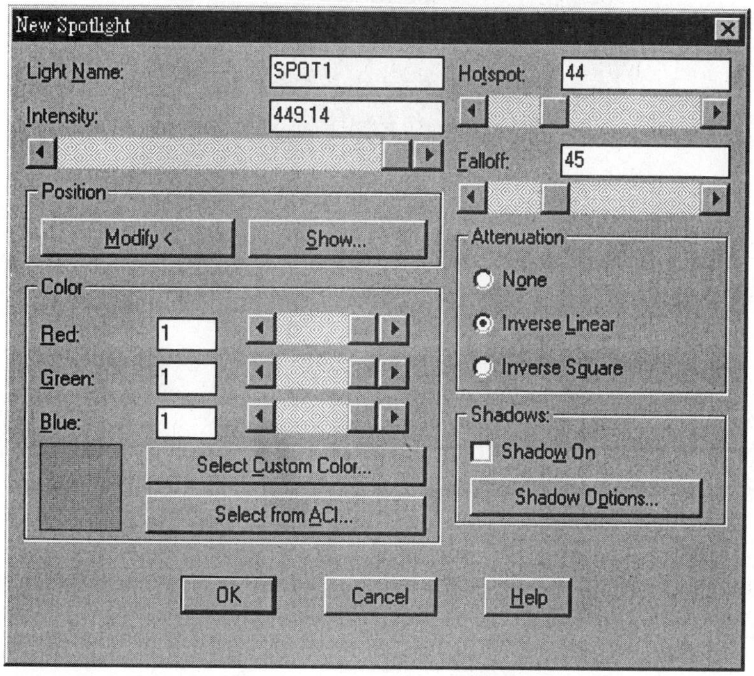

Figure 10.34 New Spotlight dialog box

Select the [Modify<] button to set the spotlight location.

Enter light target: **[Enter]**
Enter light location: **@-200,-200,200**

After locating the spotlight, select the [OK] button to complete the command. Now you have ambient light, point light, distant light, and spotlight in your drawing.

Run the SCENE command to create the third scene, SCENE3, that uses the perspective view and the spotlight.

<View> **<Render>** **<Scene...>**

Command: **SCENE**

Now use the RENDER command to render SCENE3. See Figure 10.35.

<View> **<Render>** **<Render...>**

Command: **RENDER**

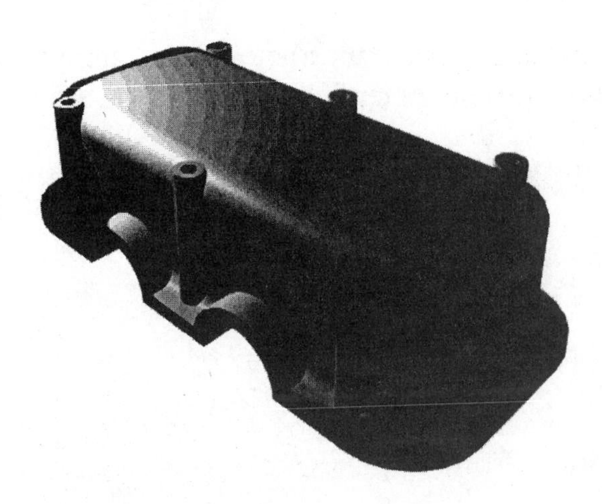

Figure 10.35 Rendering perspective view with spotlight

You can see in Figure 10.35 that the cone of light of the spotlight illuminates only a circular portion of the model. To add more light, set up the fourth scene, SCENE4.

<View> **<Render>** **<Scene...>**

Command: **SCENE**

Use the perspective view, the spotlight, and the point light. To select more than one light, hold down the [Control] key as you select objects.

Now render SCENE4. See Figure 10.36.

<View> **<Render>** **<Render...>**

Command: **RENDER**

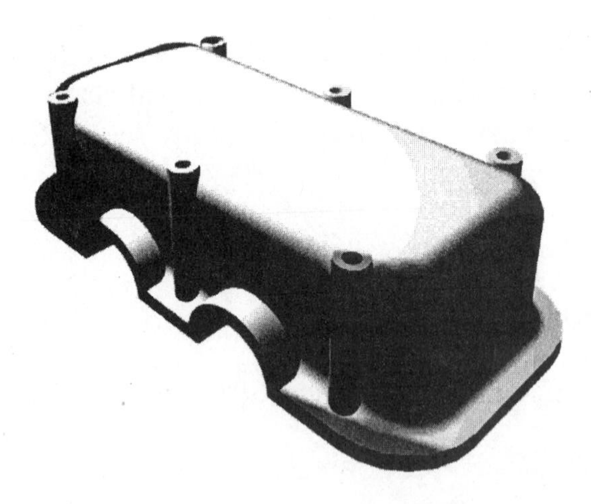

Figure 10.36 Rendering perspective view with spotlight and point light

As can be seen in Figure 10.36, some of the model is too bright and some is too dark. Try by yourself to adjust the intensity of the lights, to modify the size of the hotspot and falloff, and to render the model again. Figure 10.37 shows the rendering of SCENE4 with the intensity of the ambient light changed to 0.7 and the hotspot and falloff of the spotlight changed to 60 and 80, respectively.

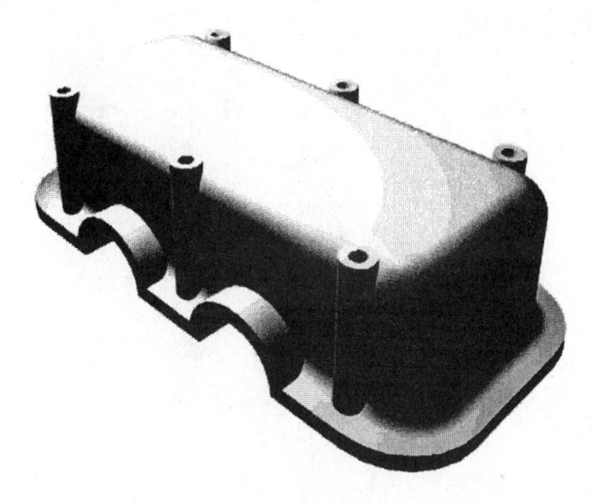

Figure 10.37 Ambient light increased and spotlight adjusted

Material and Mapping

To add more reality to the rendering, you can assign material properties to the objects in the scene. If the assigned material uses a bitmap, you have to select a mapping method and assign mapping coordinates. In order to use the texture maps and materials library provided by AutoCAD, you have to install the texture maps to your hard disk.

Select the Materials... item of the Render cascading menu of the View pull-down menu to run the RMAT command. See Figure 10.38.

<View> **<Render>** **<Materials...>**

Command: **RMAT**

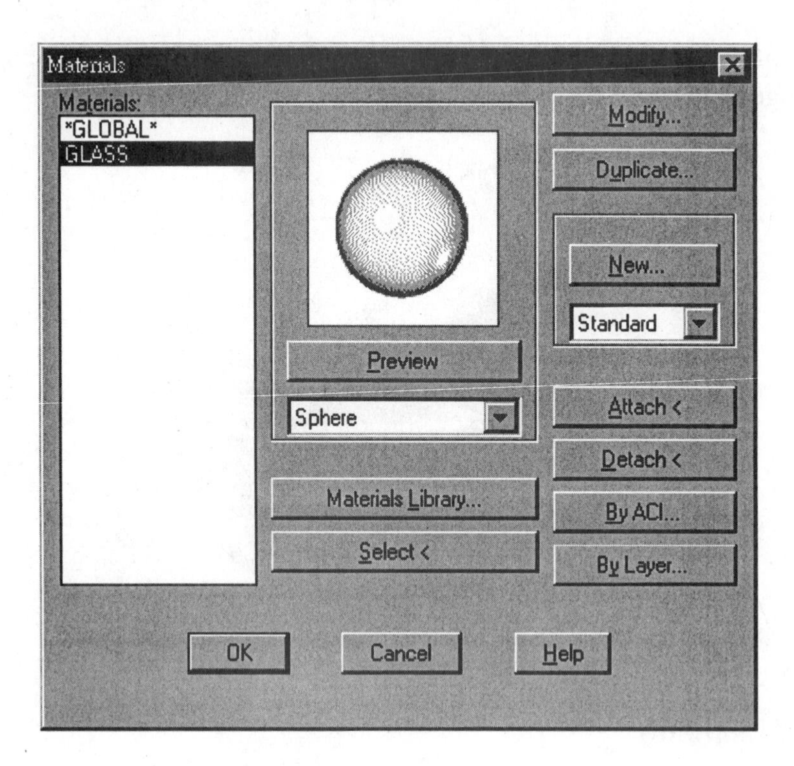

Figure 10.38 Materials dialog box

Refer to Figure 10.38. The Materials: box lists the materials available. The Preview window displays the selected material as a sphere or a cube. If you attach the material to a curved surface, you should use a sphere display. If you attach the material to a flat surface, you should use cube display. To select a material from the materials library, use the [Materials Library...] button to bring out the Materials Library dialog box. To get the material from an existing object, use the [Select<] button.

The [New...], [Modify...], and [Duplicate...] buttons bring out similar dialog boxes that enable you to create a new material and to modify and duplicate an existing material. Below the [New...] button, there is a drop-down list. Here you can choose to have the new material be standard, marble, granite, or wood.

After selecting a material, you can attach it to an object by using the [Attach<] button. To remove an attached material from an object, you can use the [Detach<] button. By using the [By ACI...] button, you can attach a material to an ACI. With the [By Layer...] button, you can attach a material to a layer.

Select the [Materials Library...] button. See Figure 10.39.

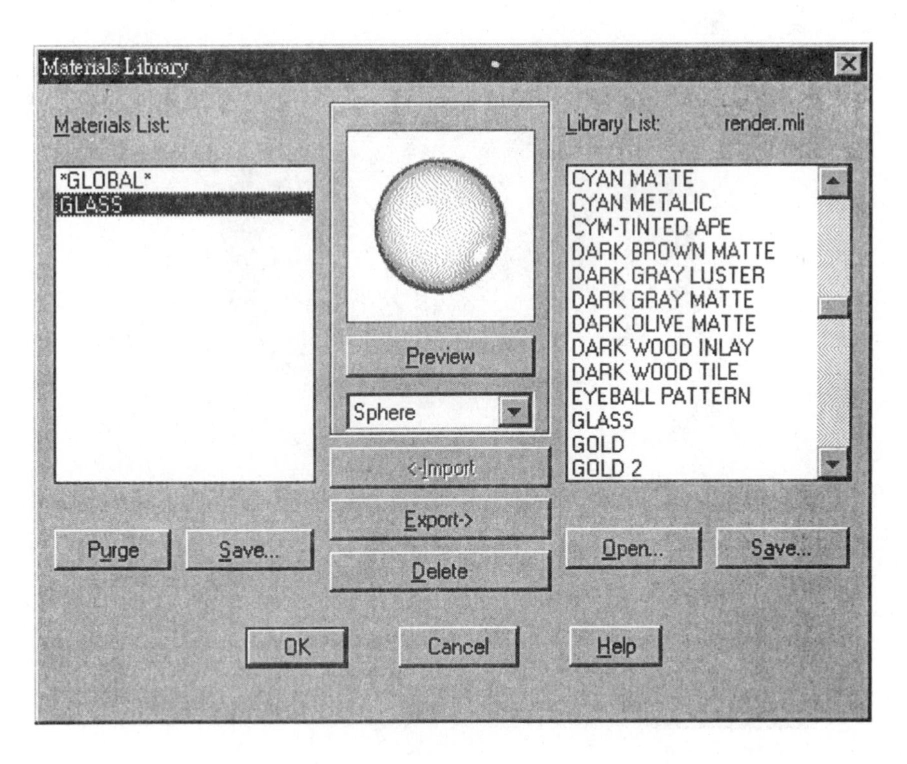

Figure 10.39 Materials Library dialog box

Refer to Figure 10.39. Select GLASS from the Library List, and then select the [<-Import] button to add to the Materials List. After that, select Sphere from the drop-down list and then the [Preview] button. If you want to remove the material from the Materials List, use the [Purge] button. Now, select the [OK] button to return to the main dialog box. See Figure 10.38 again.

On returning to the main dialog box, select the [Attach<] button to attach the GLASS material to the model.

Select objects to attach "GLASS" to: **[Select the model.]**
Select objects: **[Enter]**

Use the RENDER command to render SCENE4. See Figure 10.40.

<View> **<Render>** **<Render...>**

Command: **RENDER**

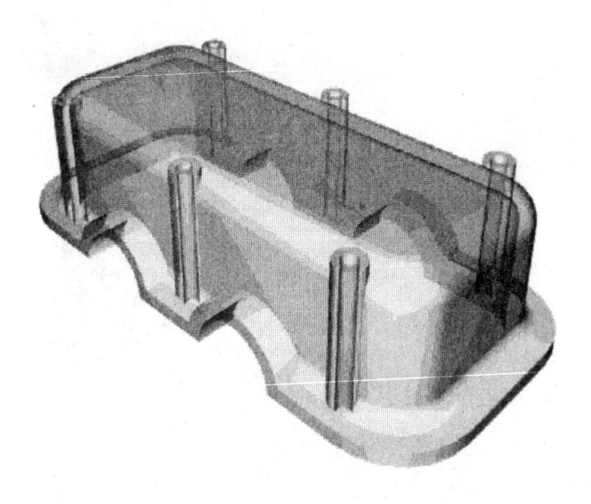

Figure 10.40 Model with GLASS material attached

Run the RMAT command again.

<View> **<Render>** **<Materials...>**

Command: **RMAT**

Select the [Materials Library...] button to bring up the Materials Library dialog box. Then import the material ZIGZAG PATTERN to the drawing. Next use the [Attach<] button to attach this material to the model. After that, render SCENE4. See Figure 10.41.

<View> **<Render>** **<Render...>**

Command: **RENDER**

Figure 10.41 Model with ZIGZAG PATTERN attached

Refer to Figure 10.41. This material has a bitmap. To map the material to the geometry, select the Mapping... item of the Render cascading menu of the View pull-down menu to use the SETUV command.

<View> **<Render>** **<Mapping...>**

Command: **SETUV**
Select objects: [**Select the model.**]
Select objects: [**Enter**]

You can choose one of the four methods to map the material to the geometry. Select Spherical and then the [Preview] button. See Figure 10.42.

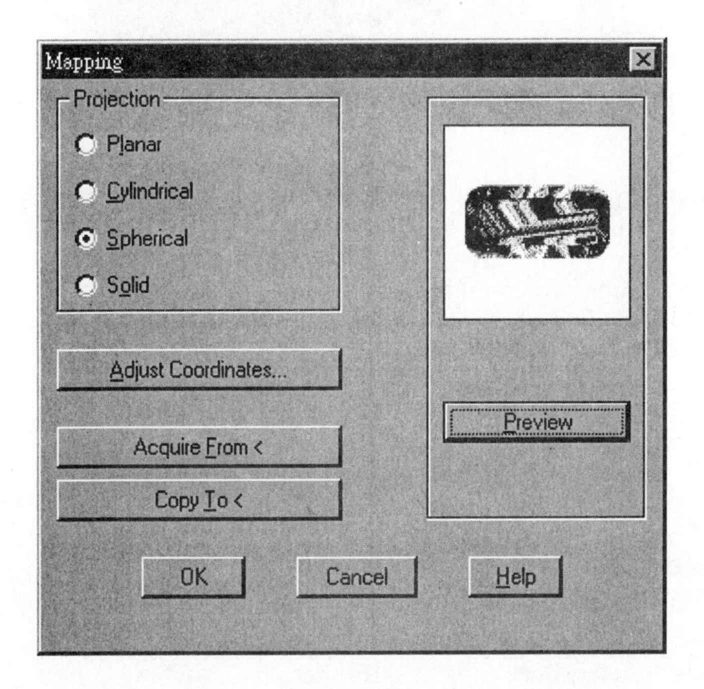

Figure 10.42 Mapping dialog box

If you want to adjust the positioning of the map relative to the model, select the [Adjust Coordinates...] button. Now, select the [OK] button to exit. After that, render SCENE4 again. See Figure 10.43.

<View> **<Render>** **<Render...>**

Command: **RENDER**

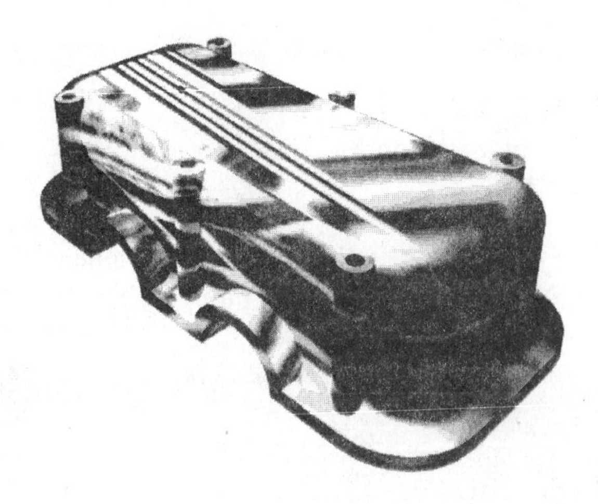

Figure 10.43 Rendering with ZIGZAG PATTERN mapped spherically

Run the RMAT command to create a new material.

<View> **<Render>** **<Materials...>**

Command: **RMAT**

Select the [New...] button to create a new material. See Figure 10.44.

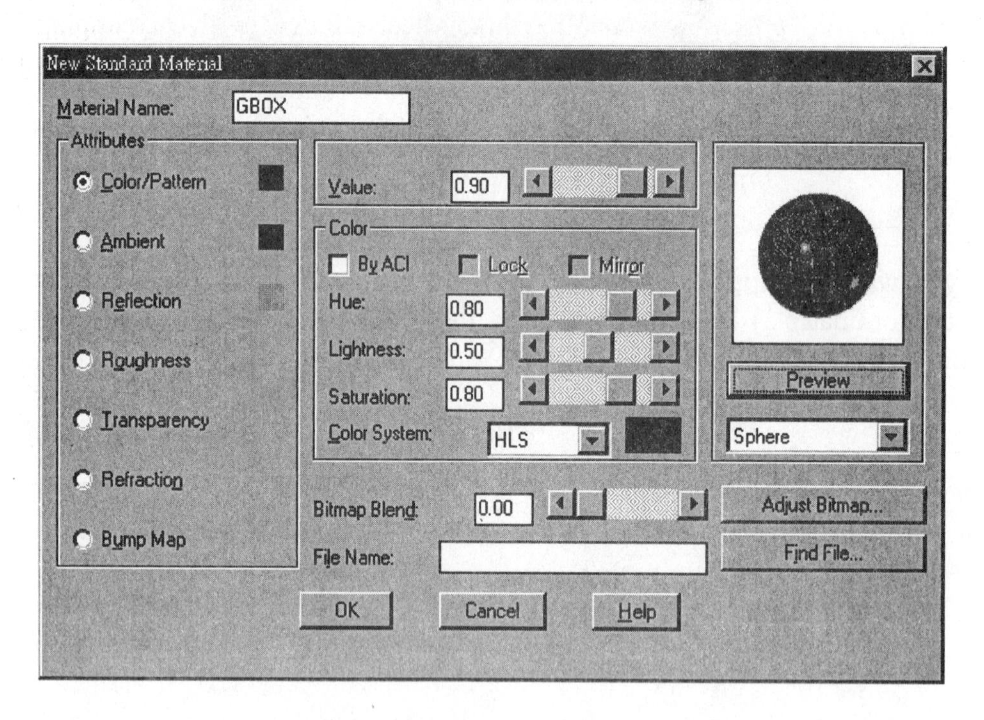

Figure 10.44 New Standard Material dialog box

A material has three colors: diffuse, ambient, and specular. Diffuse color is the main color of the material. It is the color of the material under diffuse light. Ambient color is the color of the material not under light but illuminated by ambient light. Specular color is the color of highlights under the direct reflection of light.

In the Attributes box, you can set diffuse, ambient, and specular color by selecting the items Color/Pattern, Ambient, and Reflection, respectively.

From the Color System drop-down list, select HLS to use Hue-Luminous-Saturation to set the color of the material.

From the Attributes box, select Color/Pattern to set the diffuse color of the material. Under the Color box, un-check the By ACI box. Then set the value of hue to 80, luminous to 50, and saturation to 80. To see the effect of the setting, select the [Preview] button. To adjust the color further, you can move the slider bar in the Value box. Set the value to 0.9.

Return to the Attributes box and select Ambient to set the shadow color of the material. Under the Color box, un-check both the By ACI and Lock boxes. Then set hue to 80, luminous to 30, and saturation to 30. Set the Value to 0.4

Now come to Reflection. Un-check By ACI, Lock, and Mirror. Then set hue to 80, luminous to 95, and saturation to 80. Set the Value to 1.

After setting the three colors of a material, select the Roughness item. Then adjust the Value to control the shininess of the material. Set the Value to 0.3.

To set the material to be transparent, use the Transparent item and set the Value. If you want to use a bitmap to control which area of the geometry is to be transparent, use a map by selecting the [Find File...] button.

If you use Photo-Raytraced rendering, you can use the Refraction item to determine how refractive the material is.

Finally, the Bump Map item enables you to select a bitmap to give a bumping effect.

Select the [OK] button to complete the new material. Then select the [Attach<] button to attach the material to the model.

Render SCENE4 again. See Figure 10.45.

<View> **<Render>** **<Render...>**

Command: **RENDER**

Figure 10.45 Rendered with new material attached

Export to and Import from 3D Studio

3D Studio is a professional tool for making photo-realistic renderings and animations. You can output 3D objects to 3D Studio format and use 3D Studio to do further work.

To output a 3D object to 3D Studio, use the 3DSOUT command.

```
Command: 3DSOUT
Select objects: [Select an object.]
Select objects: [Enter]
```

The command complementary to the 3DSOUT command is the 3DSIN command. This command inputs 3D Studio objects to become meshed objects in AutoCAD.

<Insert> **<3D Studio...>**

Command: **3DSIN**

10.4 Introduction to RP and STL

Rapid prototyping (RP) is a technology to create prototypes from 3D CAD models. Although there are different RP systems employing different techniques, they function similarly.

Open the file MOUSE.DWG that you saved in Chapter 7.

<File> **<Open...>**

File name: **MOUSE.DWG**

Set the current layer to S_u and turn off all other layers. See Figure 10.46.

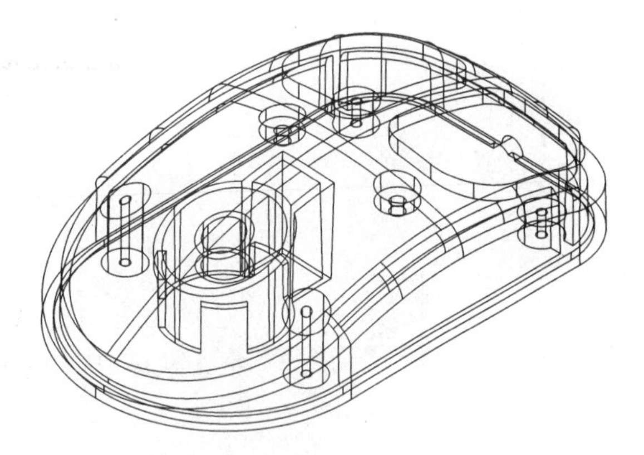

Figure 10.46 Upper casing of mouse assembly

Introduction to Rapid Prototyping (RP)

Unlike traditional machining processes that either remove material from a large piece of material or form material from molten or semi-molten material, RP systems use an additive process. The starting point of rapid prototyping is a 3D CAD model. From the 3D CAD model, the RP systems slice the electronic model into very thin layers. Then the RP systems make these thin layers one by one and compose the layers together to form the final model.

Figure 10.47 shows how the model of the mouse upper casing is sliced into thin layers by the RP systems.

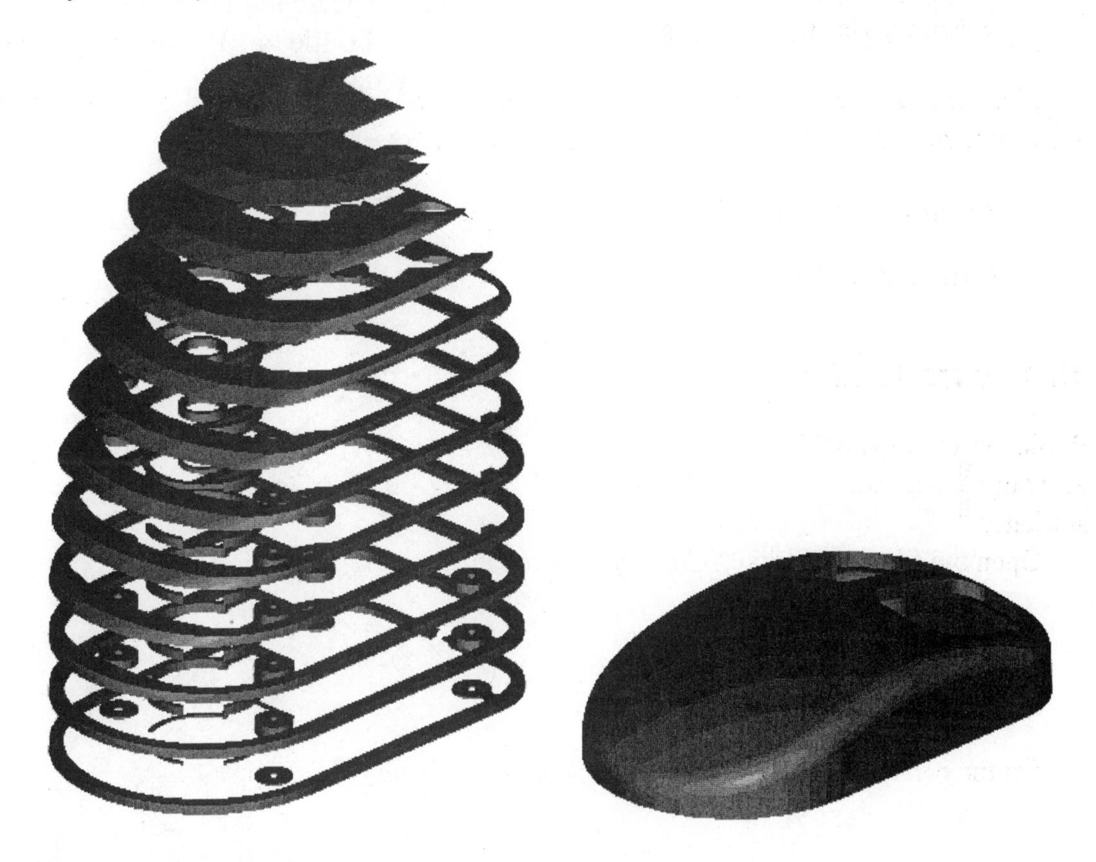

Figure 10.47 Solid model sliced into a number of thin layers

Stereolithography Format

Most RP systems require a file saved in Stereolithography Apparatus (SLA) format. AutoCAD solid models are readily exportable to RP systems.

Before exporting an AutoCAD solid model to SLA format, you have to translate the model in 3D so such that the entire model is lying on the positive side of the X, Y, and Z axes.

<Modify> **<Move>**

Select objects: **[Select the model.]**
Select objects: **[Enter]**
Base point or displacement: **100,100**
Second point of displacement: **[Enter]**

SLA format describes a solid model in facets. To control the facet resolution, you need to set the FACETRES variable. As we have said, the value of this variable ranges from 0.01 to 10.0. A value of 0.01 gives the lowest resolution.

Command: **FACETRES**
New value for FACETRES: **0.01**

The STL file describes a set of triangular facets by specifying their coordinates. To check the smoothness of the model before sending the STL file to RP machines, you can use the HIDE or SHADE command. Both commands display the edges of the faceted surfaces. Run the HIDE command. See Figure 10.47.

<View> **<Hide>**

Command: **HIDE**

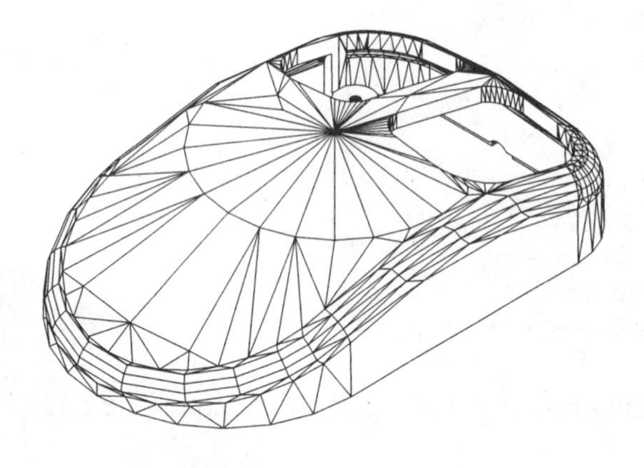

Figure 10.48 Facet resolution of 0.01

The facet resolution shown in Figure 10.48 is too coarse. Therefore, use the FACETRES variable to set facet resolution to 5. Now hide the model again. See Figure 10.49.

Command: **FACETRES**
New value for FACETRES: **5**

<View> **<Hide>**

Command: **HIDE**

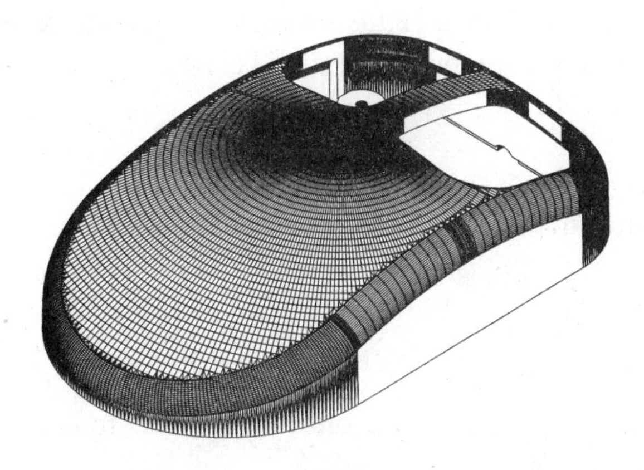

Figure 10.49 Facet resolution of 5

Compare Figure 10.49 with Figure 10.48 to see the effect of facet resolution on the smoothness of the model.

After translating the model to the positive side of the X, Y, and Z axes and setting the FACETRES variable, you can output the model. Run the STLOUT command.

```
Command: STLOUT
Select a single solid for STL output:
Select objects: [Select a model.]
Select objects: [Enter]
Create a binary STL file ? <Y>: Y
```

When the File dialog box appears, input a file name. The STL file is complete.

10.5 Key Points and Exercises

In addition to viewing the 3D object in a 3D direction, you can set to a perspective view. By default, 3D solids are displayed as wireframe objects in the screen display. To see the solid model better, you can remove the hidden lines and display a shaded view. Furthermore, you can display a photo-realistic rendering. In shaded and rendered views, curved surfaces are represented by facets. Therefore, you need to set the facet resolution. Rapid prototyping is an additive process to produce a physical model from a 3D computer model. To make an RP object, you export the 3D solid model to STL format.

In this chapter, you learned how to display a perspective view, to remove hidden lines, to apply shading, to produce photo-realistic renderings, and to output to stereolithography format. Perspective view gives a more realistic display than a 3D orthographic view. However, perspective view takes a longer time to regenerate, and you cannot zoom a perspective view. Hiding, shading, and rendering display the model in facets. To control facet resolution, set the FACETRES variable. The VIEWRES command affects the smoothness of circular and elliptical edges. Setting the zoom percentage to a higher value gives smoother edges. RP systems require STL file formats. AutoCAD 3D solids can be exported to STL format. Because the STL file describes a model in its faceted form, the FACETRES variable affects the resolution of the STL file. Setting to a higher value gives a more accurate STL file.

Exercises 10.1

Which system variable affects the display resolution and STL accuracy of solids with curved surfaces?

Exercises 10.2

Following the steps outlined in this chapter, create the rendered drawing for the 3D models that you created in Chapters 6, 7, and 8.

Appendix A

Application Window Configuration

A.1 Menu Bar Configuration

To load or unload the menu items onto the Windows menu bar from AutoCAD menu files, you can select the Customize Menus... item from the Tools pull-down menu to apply the MENULOAD command or type MENULOAD at the command line interface.

<Tools> **<Customize Menus...>**

Command: **MENULOAD**

The Menu Customization dialog box has two tabs, Menu Groups and Menu Bar. The first tab, the Menu Groups tab (Figure A.1), enables you to load or unload a menu file. Select the [Unload] button to unload all the menu items.

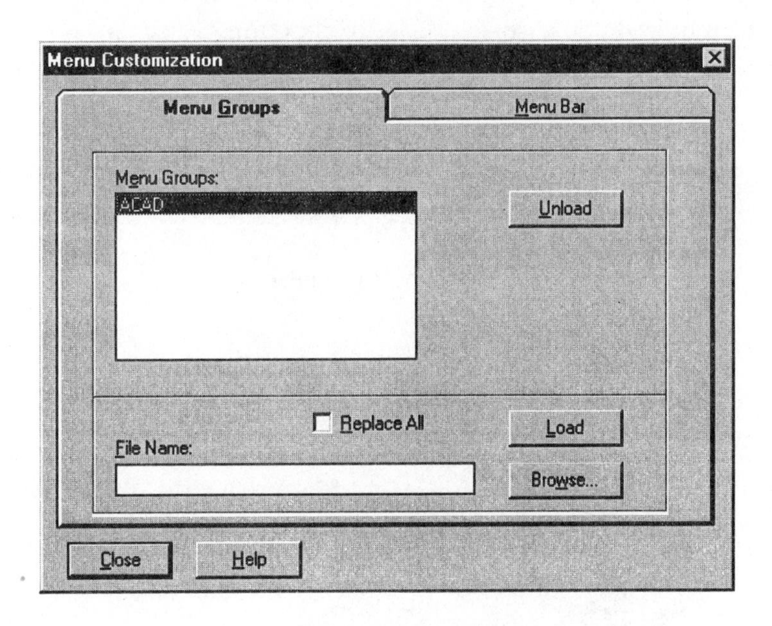

Figure A.1 Menu Groups tab of the MENULOAD command

All the menu items disappear. To load them back again, select the [Browse...] button to select a menu file. Then select the Replace All box and the [Load] button. The menu items come back again.

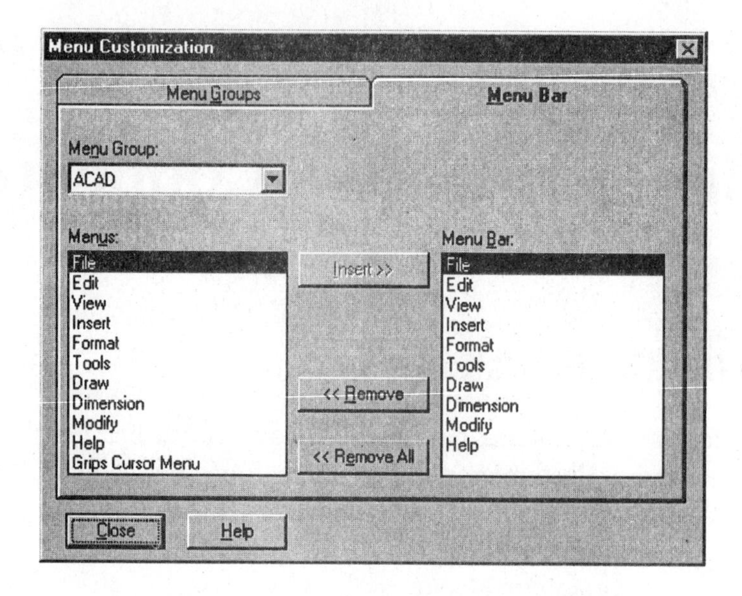

Figure A.2 Menu Bar tab of the MENULOAD command

The second tab, the Menu Bar tab (Figure A.2), enables you selectively to load or unload individual menu items to the menu bar. To the left of this page, there are menu items that are available in the selected menu file. To load these items to the Windows Menu Bar, select the items and the [Insert>>] button to put them to the right of the page. After loading the required items, select the [Close] button.

The 10 standard menu items in the pull-down menu are detailed in Figures A.3 - A.17.

Figure A.3 Menu bar

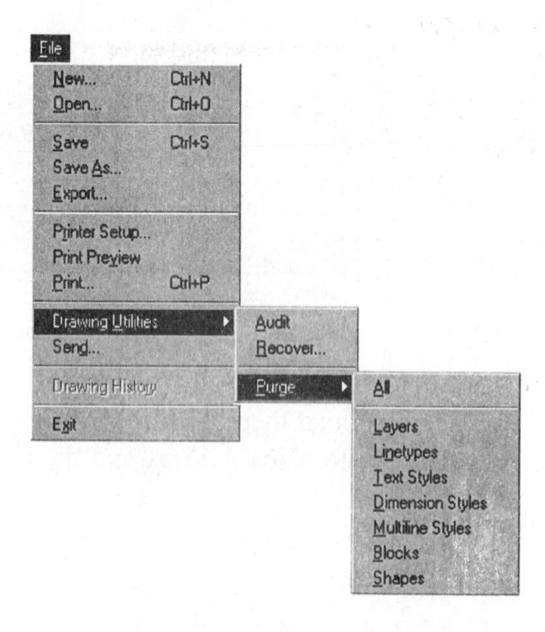

Figure A.4 Drawing Utilities cascading menu of File pull-down menu

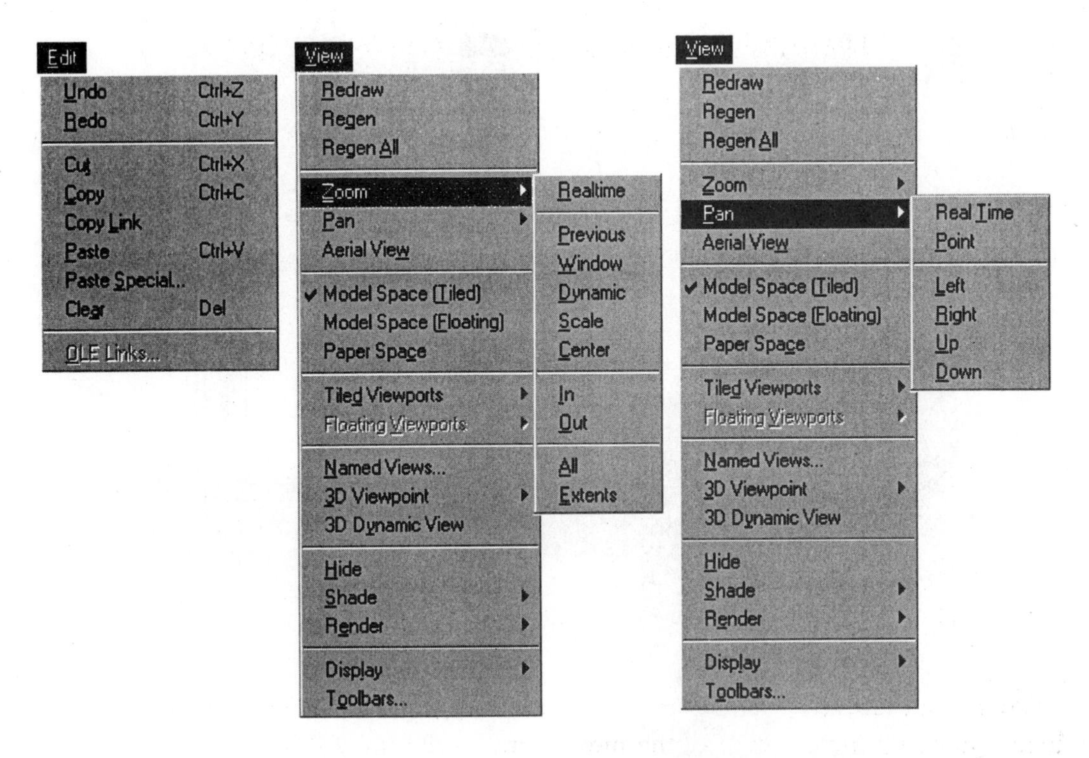

Figure A.5 Edit pull-down menu, and Zoom and Pan cascading menus of View pull-down menu

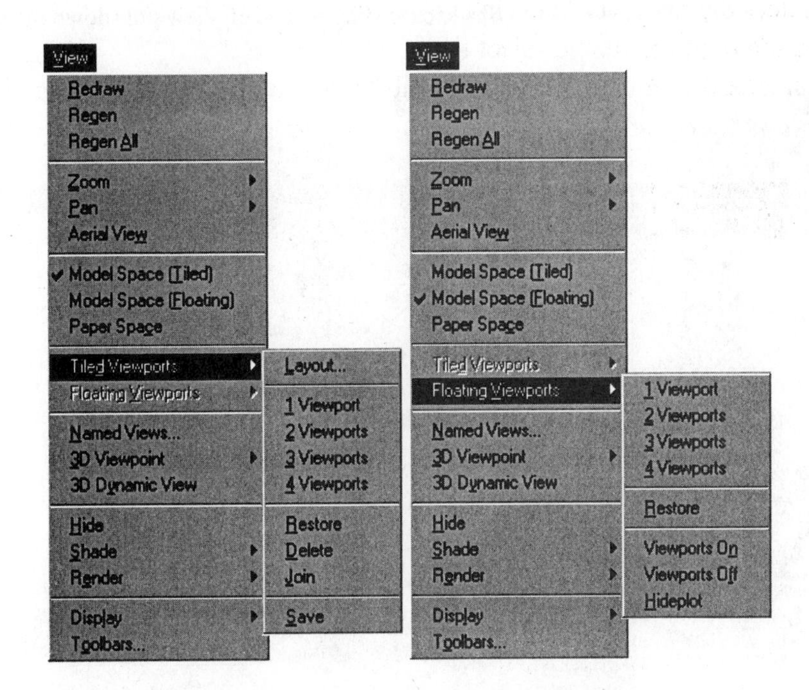

Figure A.6 Tiled Viewports and Floating Viewports cascading menus of View pull-down menu

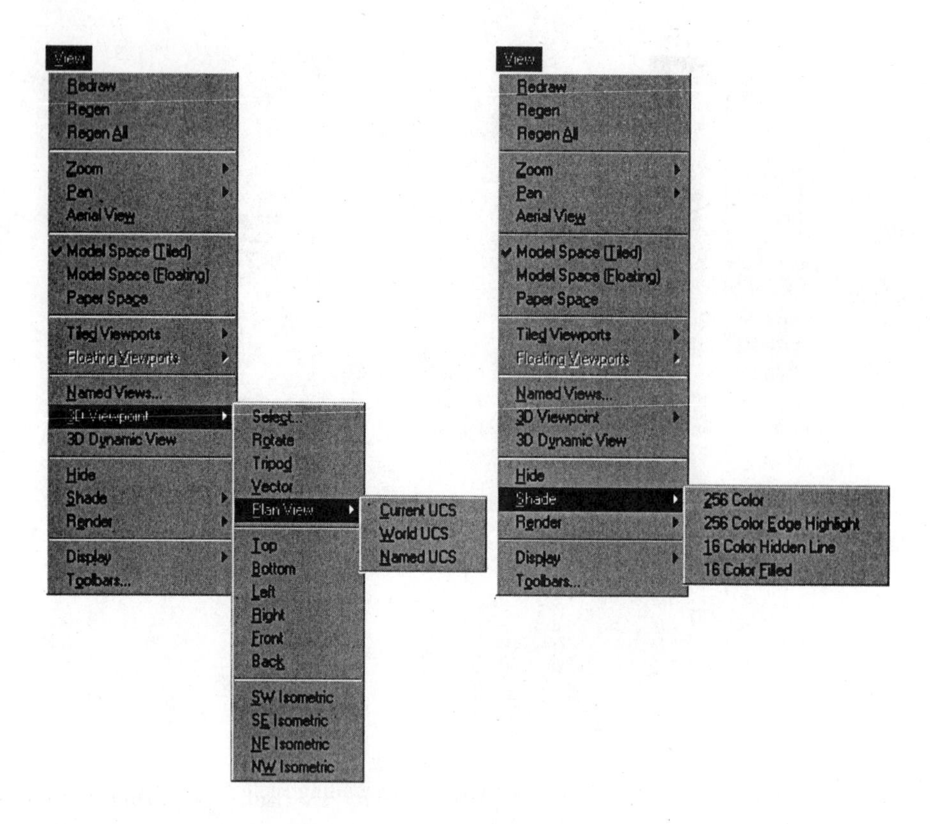

Figure A.7 3D Viewport and Shade cascading menus of View pull-down menu

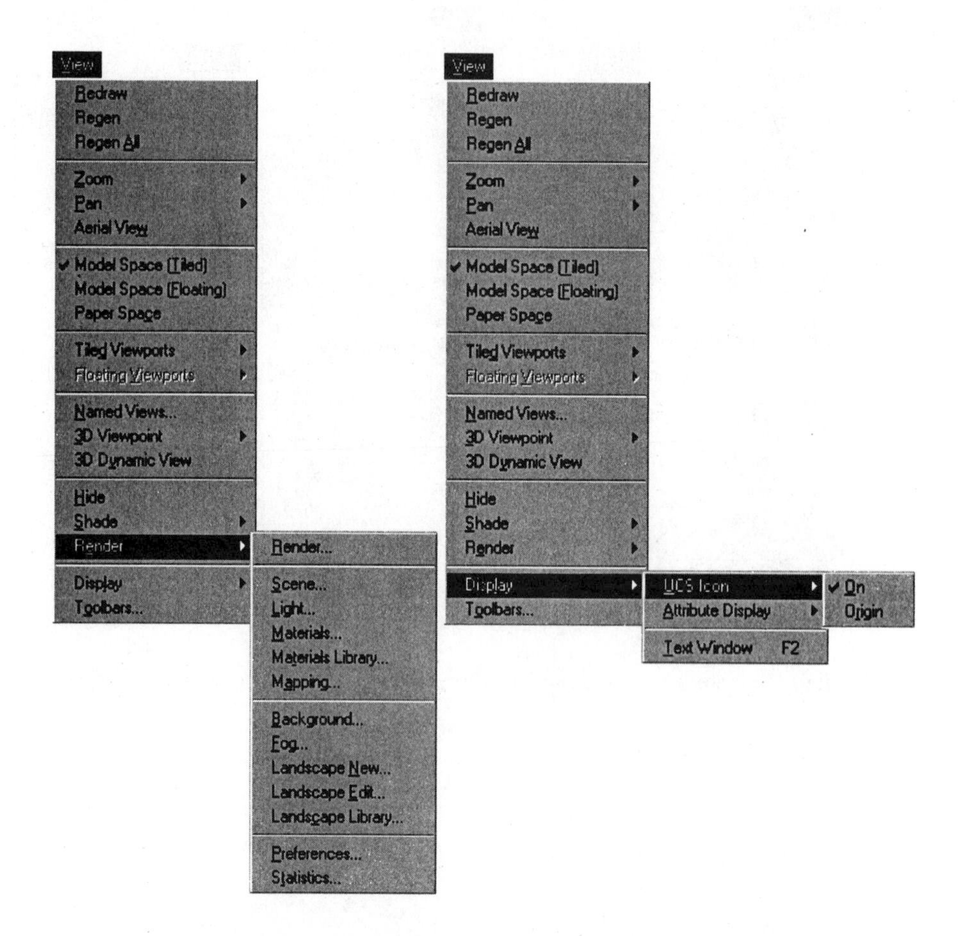

Figure A.8 Render and Display cascading menus of View pull-down menu

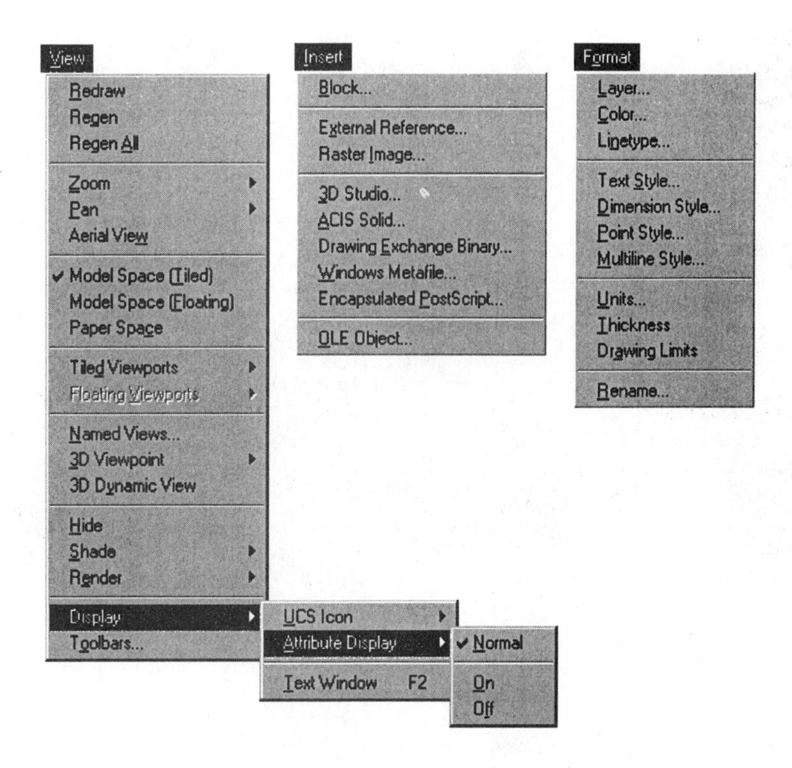

Figure A.9 Display cascading menu of View pull-down menu, Insert pull-down menu, and Format pull-down menu

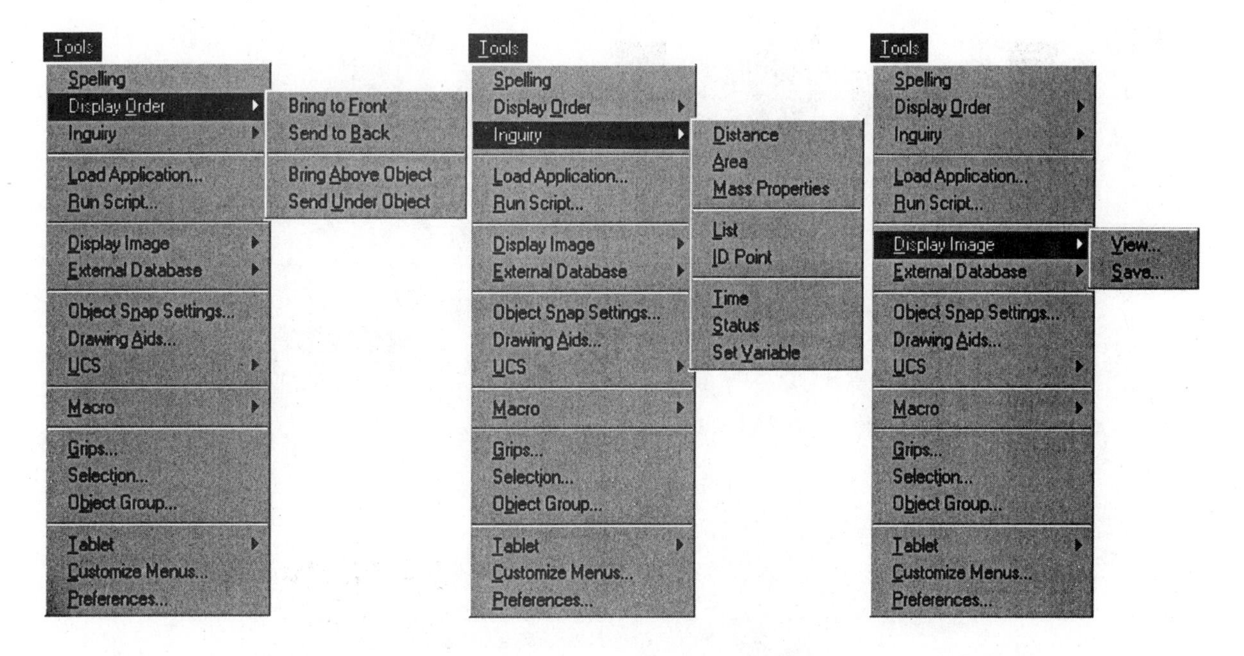

Figure A.10 Display Order, Inquiry, and Display Image cascading menus of Tools pull-down menu

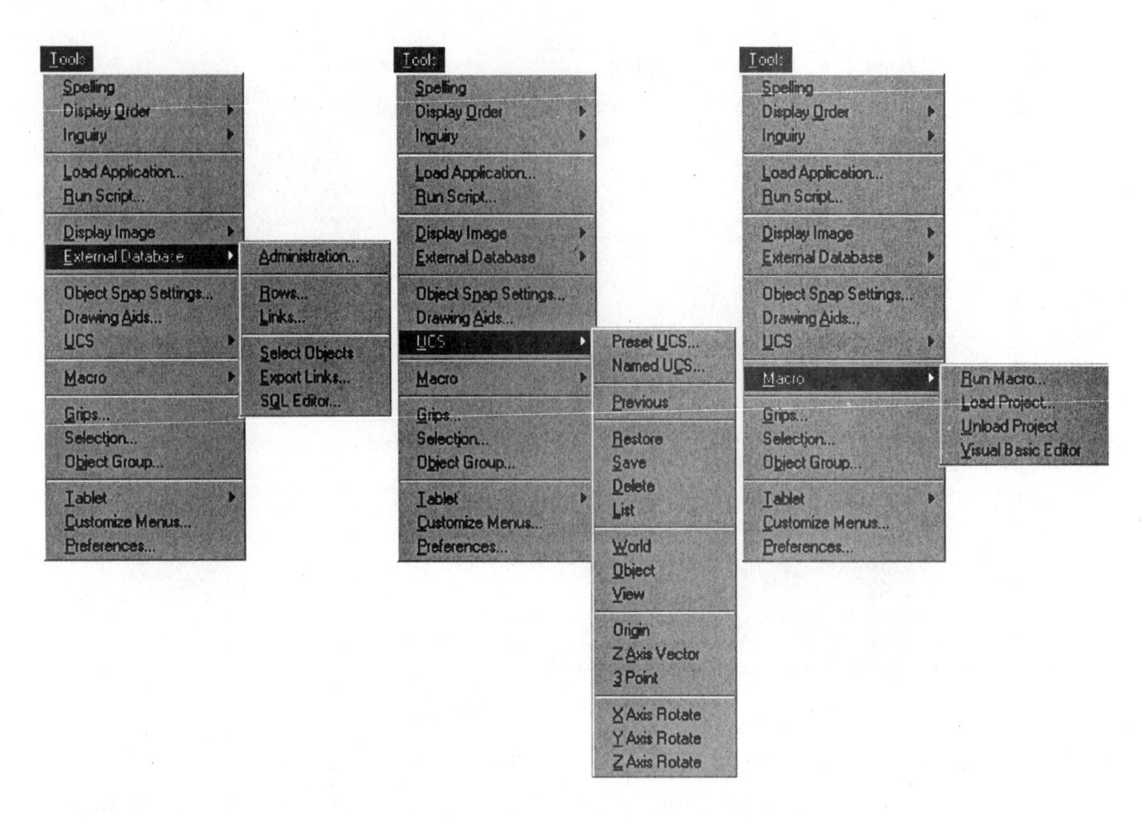

Figure A.11 External Database, UCS, and Macro cascading menus of Tools pull-down menu

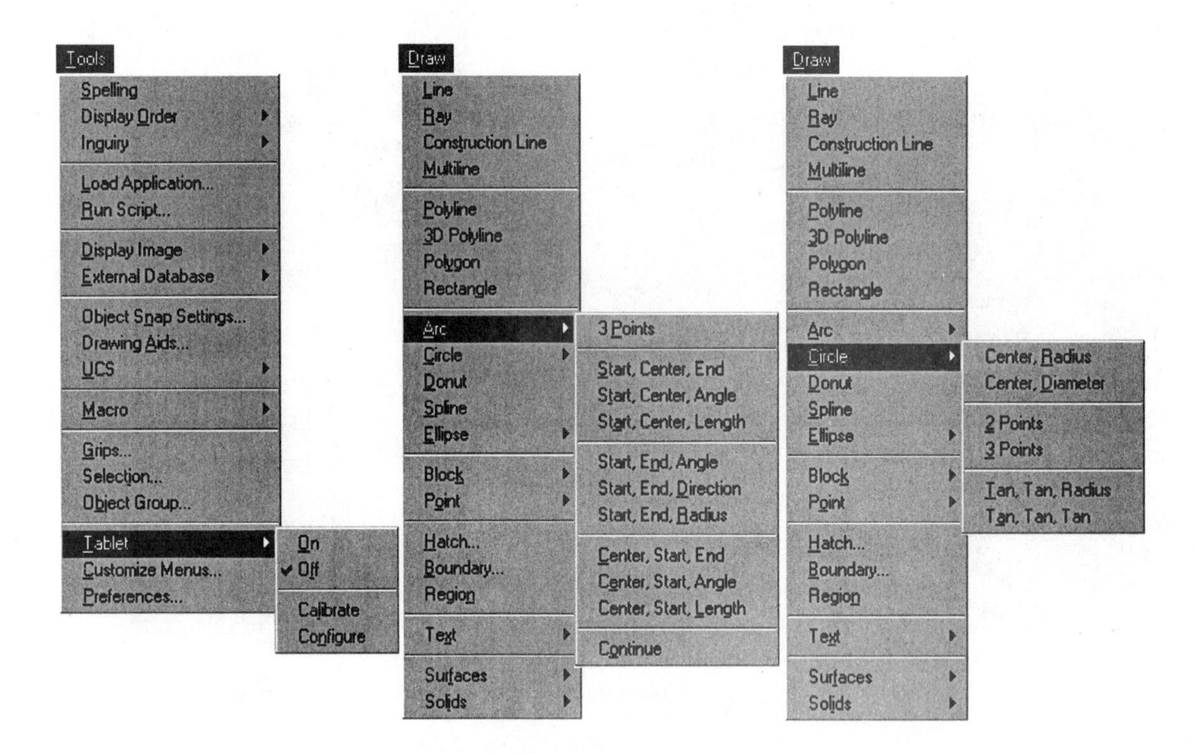

Figure A.12 Tablet cascading menu of Tools pull-down menu, and Arc and Circle cascading menus of Draw pull-down menu

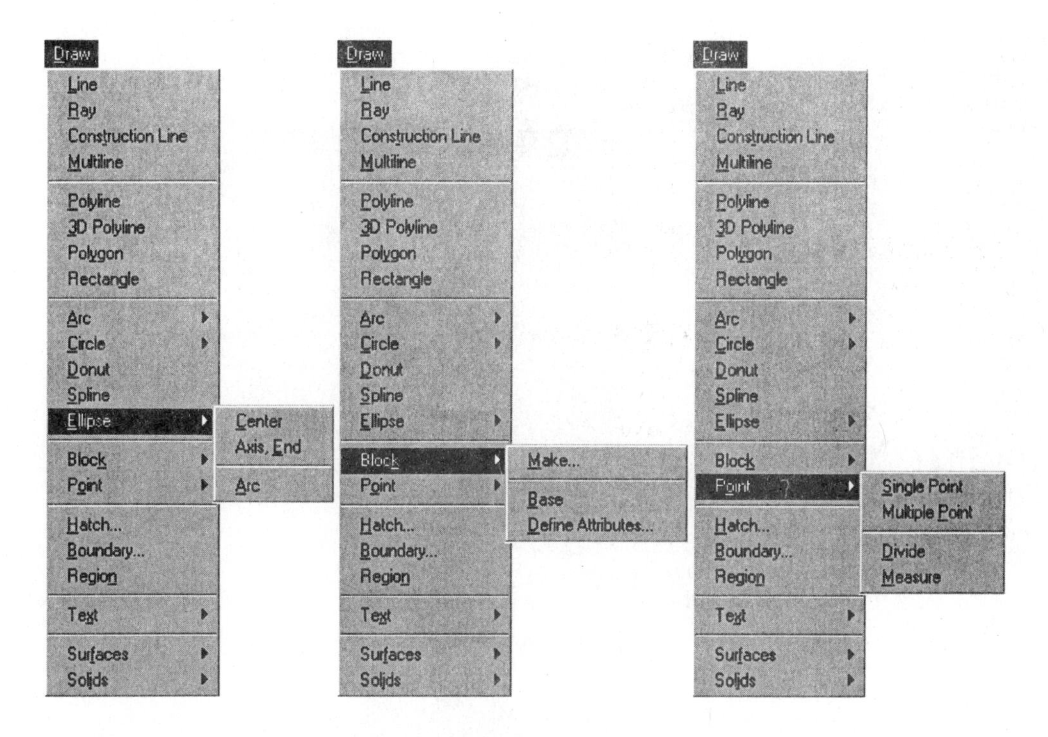

Figure A.13 Ellipse, Block, and Point cascading menus of Draw pull-down menu

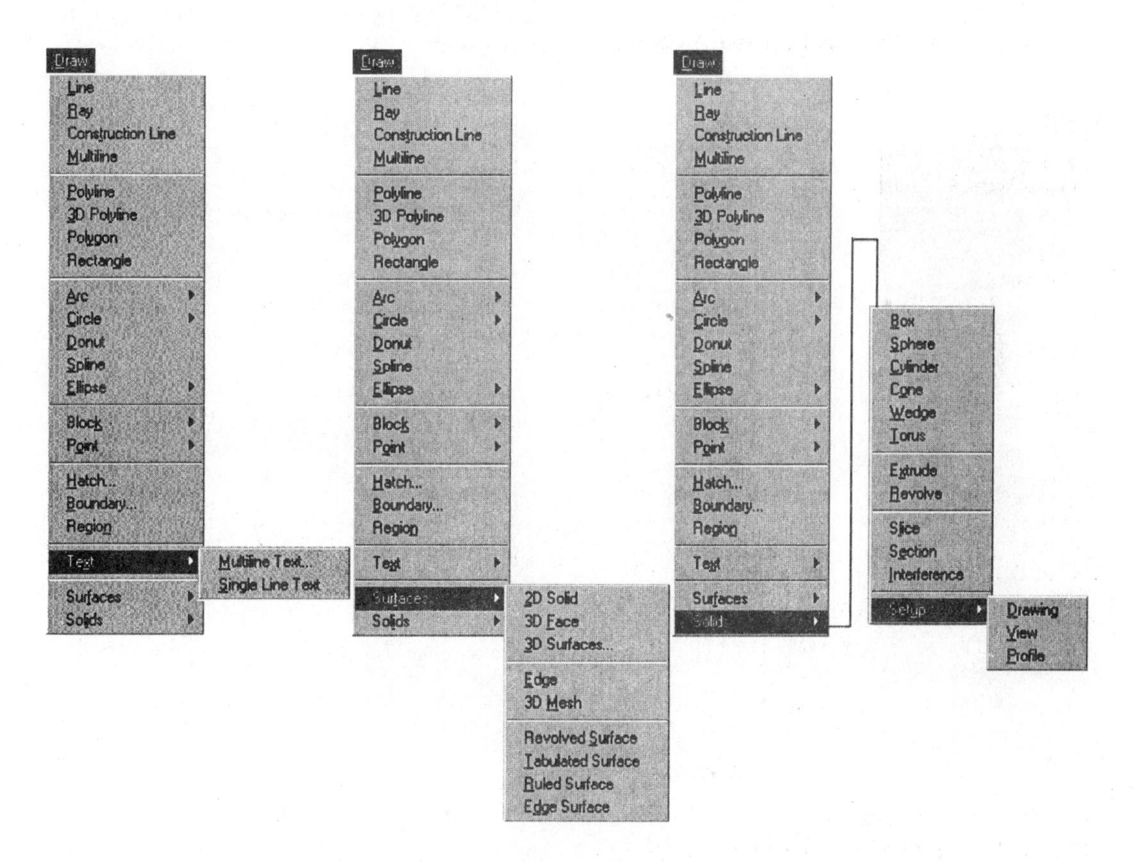

Figure A.14 Text, Surfaces, and Solids cascading menus of Draw pull-down menu

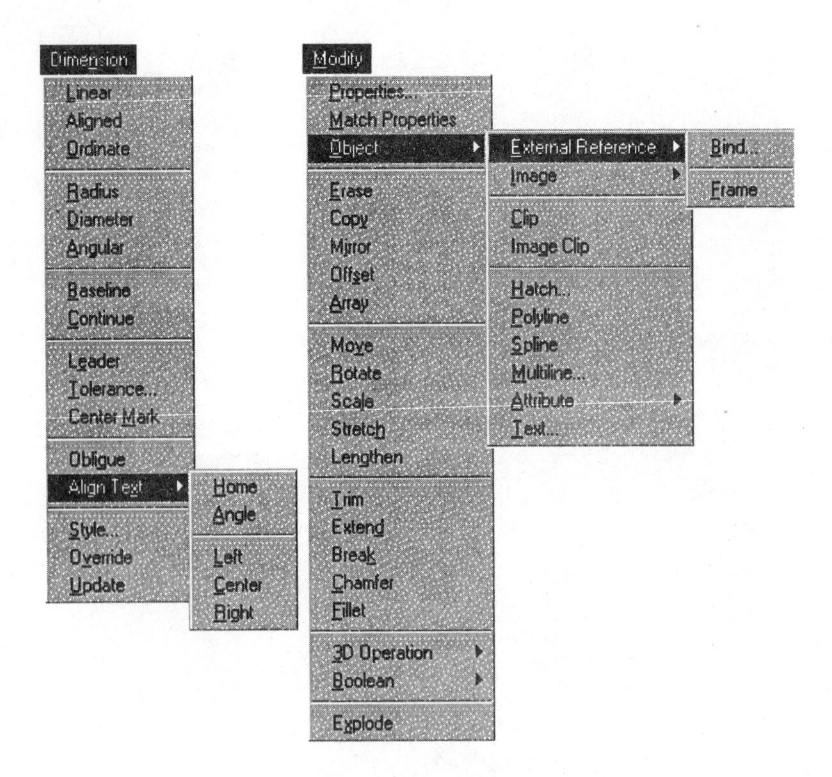

Figure A.15 Align Text cascading menu of Dimension pull-down menu and Object cascading menu of Modify pull-down menu

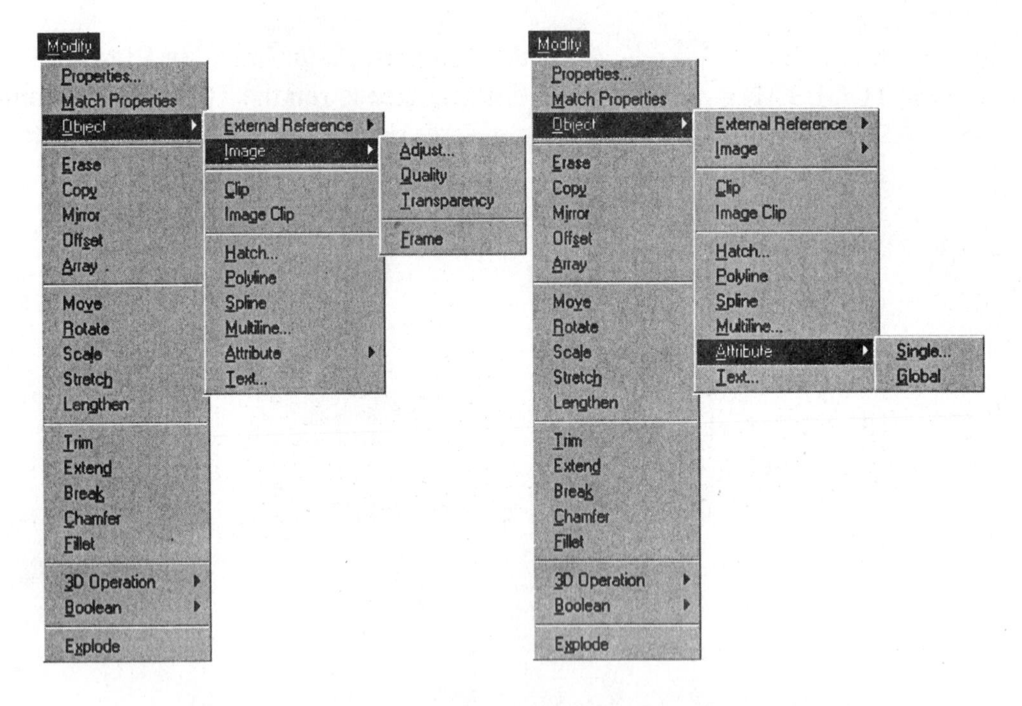

Figure A.16 Object cascading menus of Modify pull-down menu

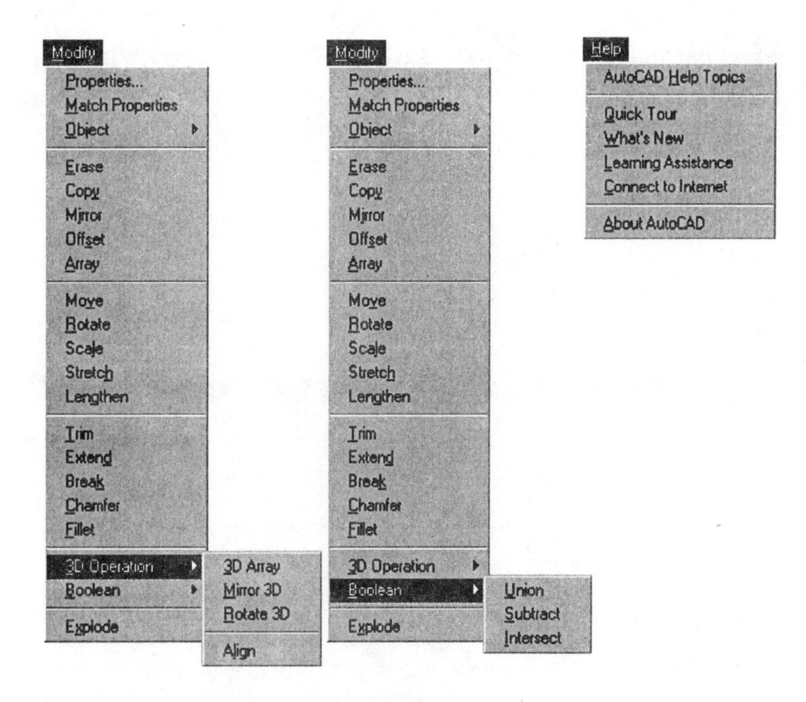

Figure A.17 3D Operation and Boolean cascading menus of Modify pull-down menu, and Help pull-down menu

A.2 Toolbar Configuration

To configure the toolbars, you can select the Toolbars... item from the View pull-down menu or type TOOLBAR at the command line interface to run the TOOLBAR command. See Figure A.18. (Another way of bringing up this dialog box is to place the cursor on any icon and then select the right button of your mouse.)

<View> **<Toolbars...>**

Command: **TOOLBAR**

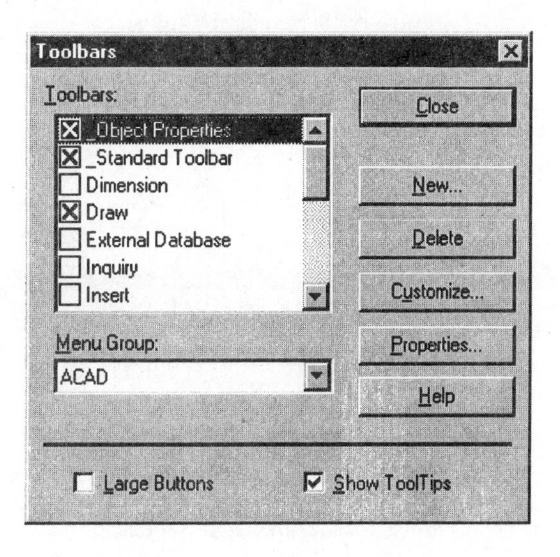

Figure A.18 Toolbar dialog box

Refer to the Toolbar dialog box. There is a list of available toolbars. Selecting an item displays or hides the respective toolbars. There are 17 standard tool bars. They are detailed in Figure A.19 - A.35.

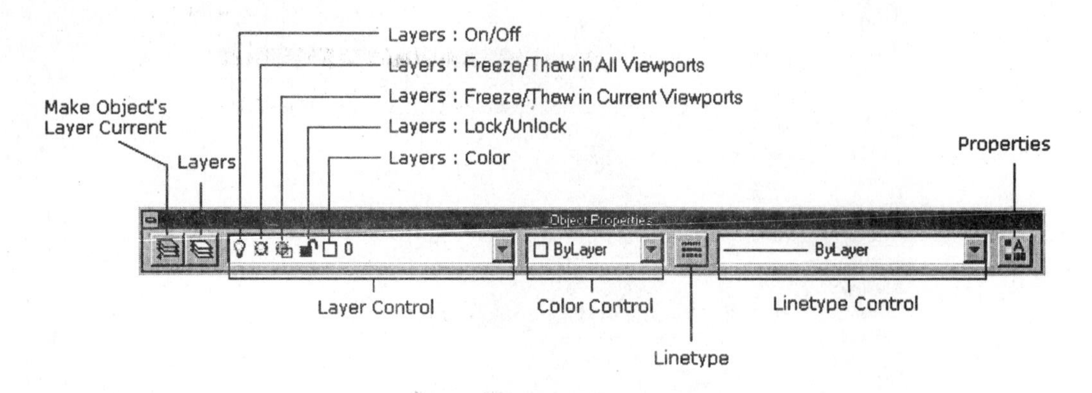

Figure A.19 Object Properties Toolbar

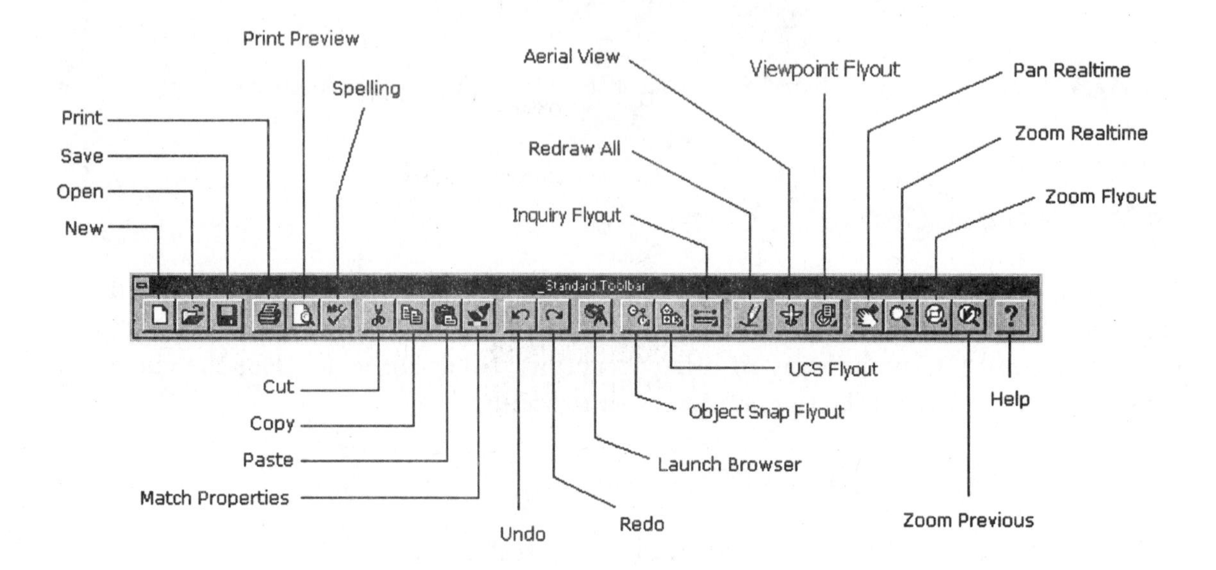

Figure A.20 Standard Toolbar

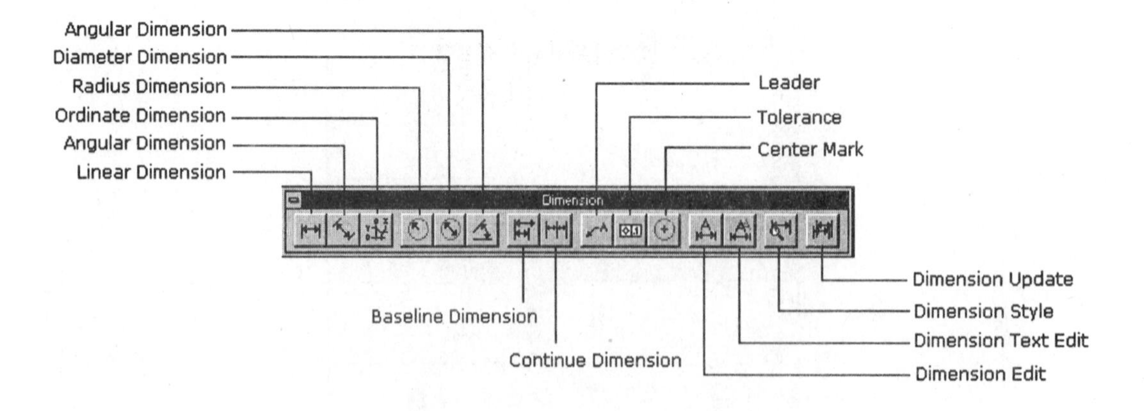

Figure A.21 Dimension Toolbar

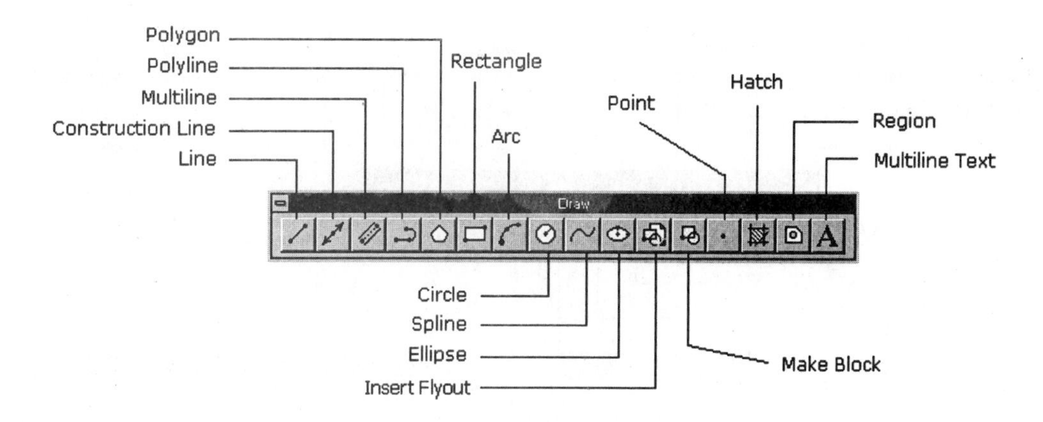

Figure A.22 Draw Toolbar

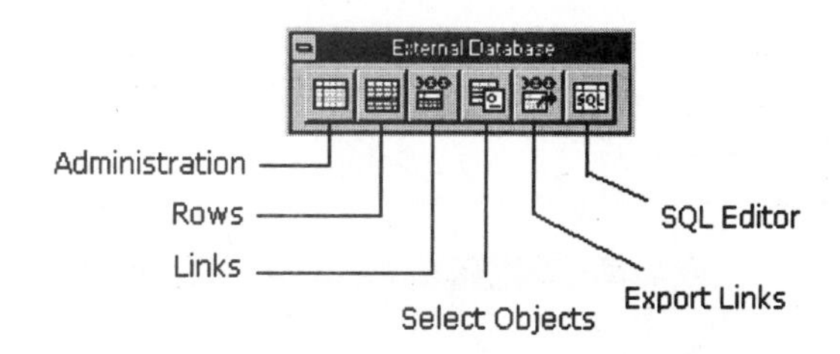

Figure A.23 External Database Toolbar

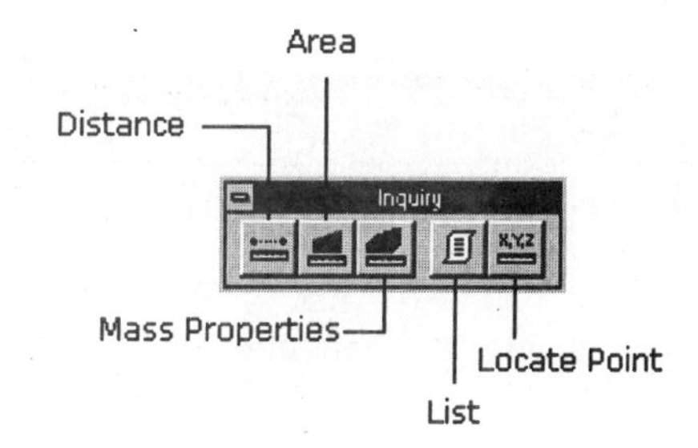

Figure A.24 Inquiry Toolbar

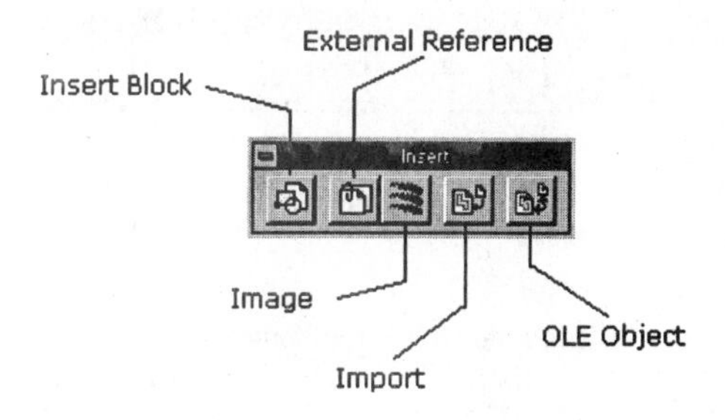

Figure A.25 Insert Toolbar

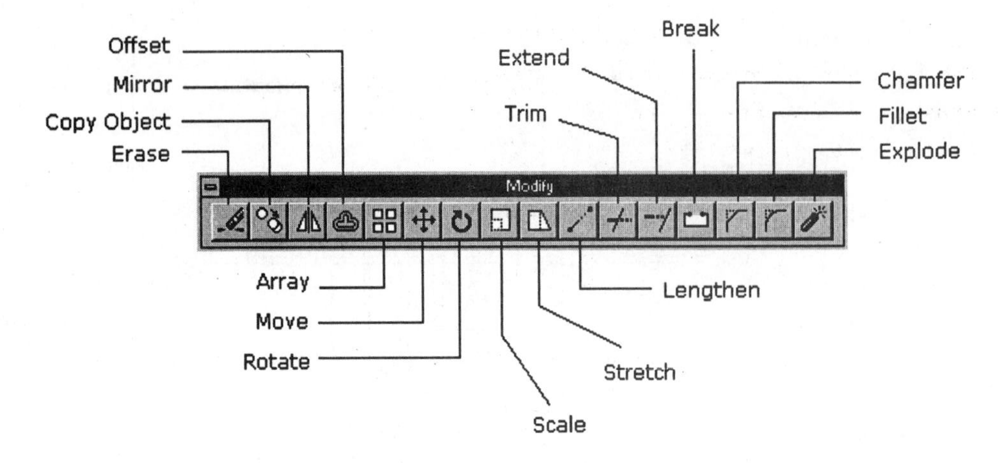

Figure A.26 Modify Toolbar

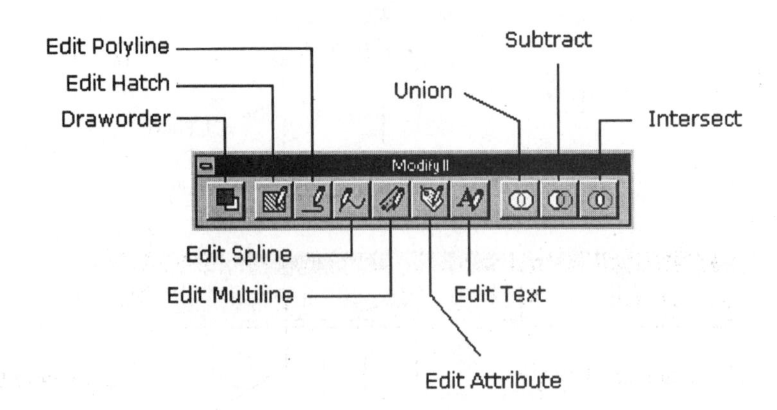

Figure A.27 Modify II Toolbar

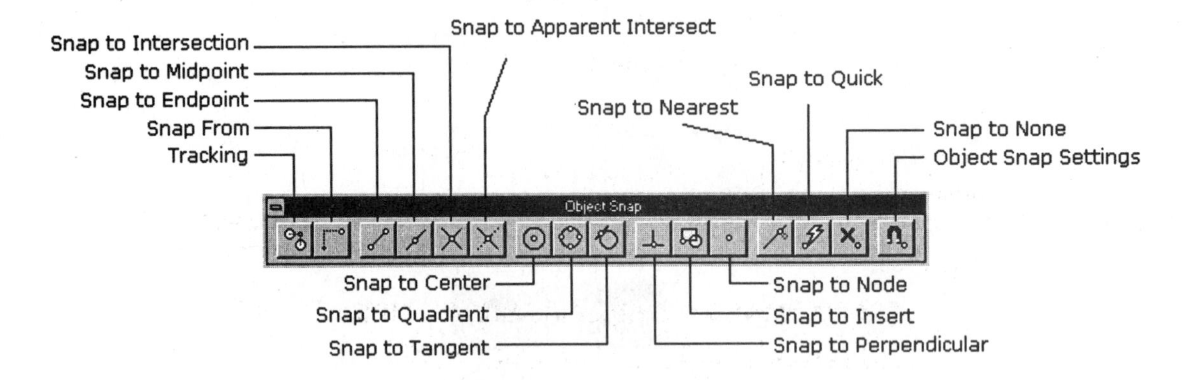

Figure A.28 Object Snap Toolbar

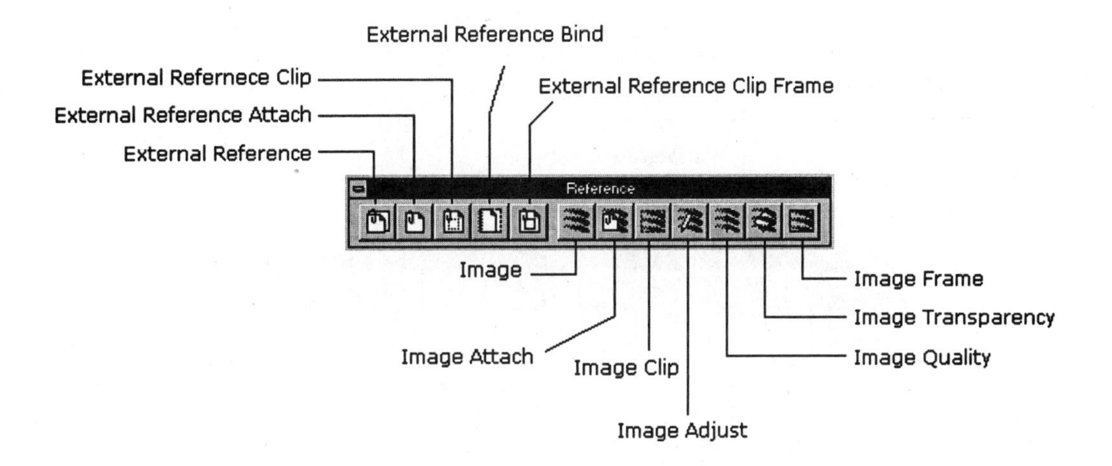

Figure A.29 Reference Toolbar

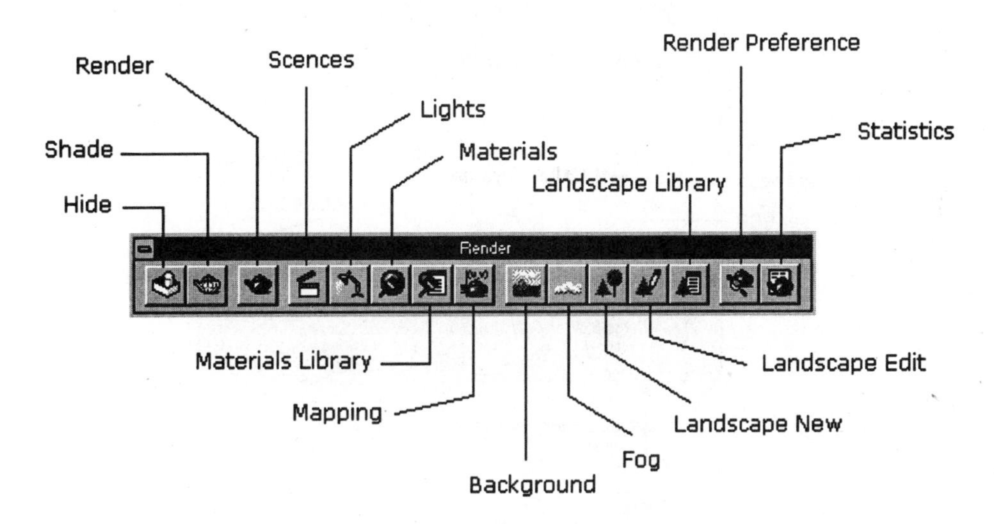

Figure A.30 Render Toolbar

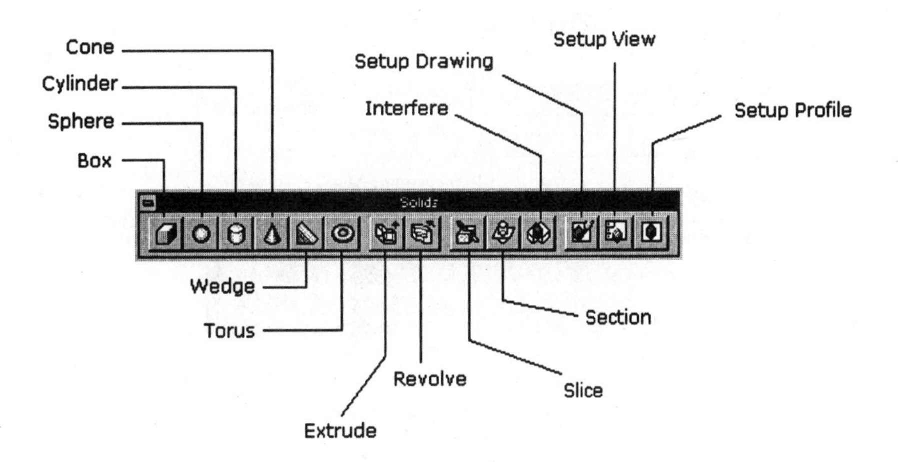

Figure A.31 Solids Toolbar

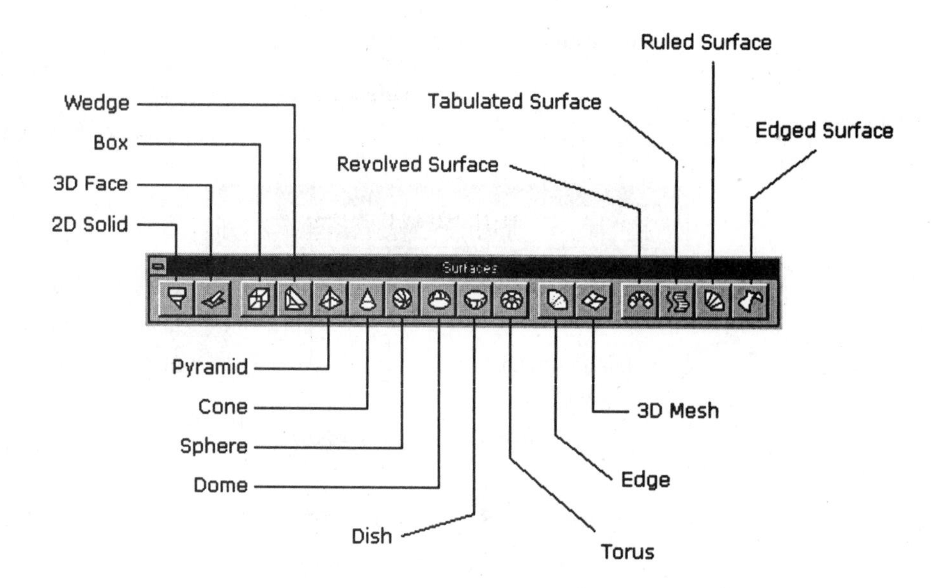

Figure A.32 Surfaces Toolbar

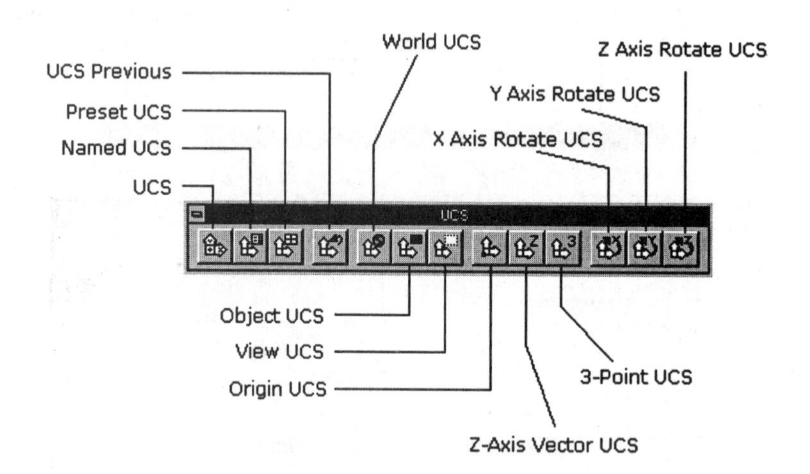

Figure A.33 UCS Toolbar

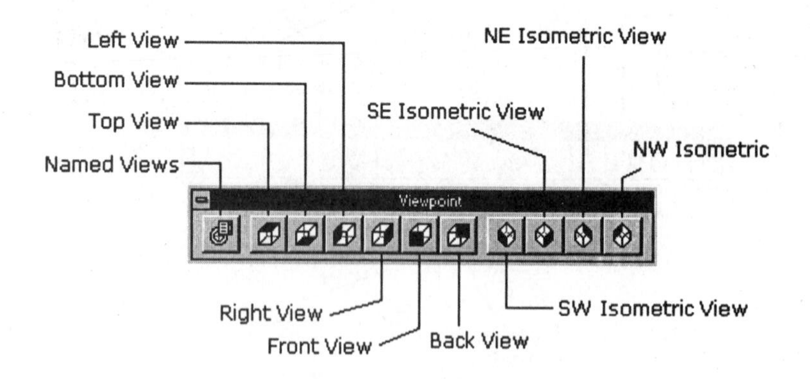

Figure A.34 Viewpoint Toolbar

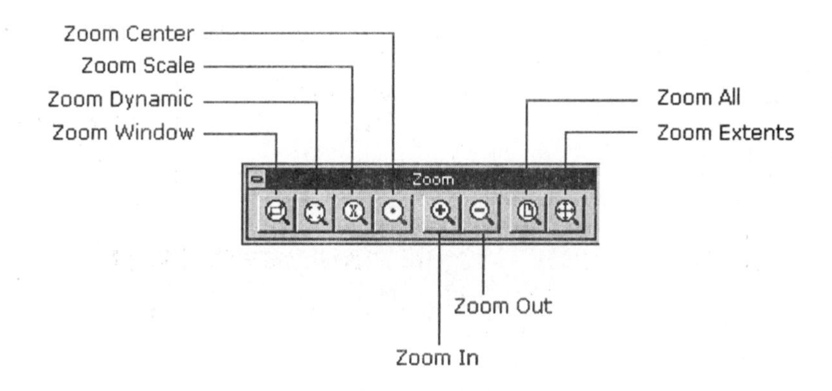

Figure A.35 Zoom Toolbar

A.3 Window Format Configuration

In general, the display format of AutoCAD application window depends on the Windows settings. However, you can use the PREFERENCES command to set the colors for the graphic window, menu areas, text windows, and commands and to set the fonts of the menus, text window, and command line. See Figure A.36.

<Tools> **<Preferences...>**

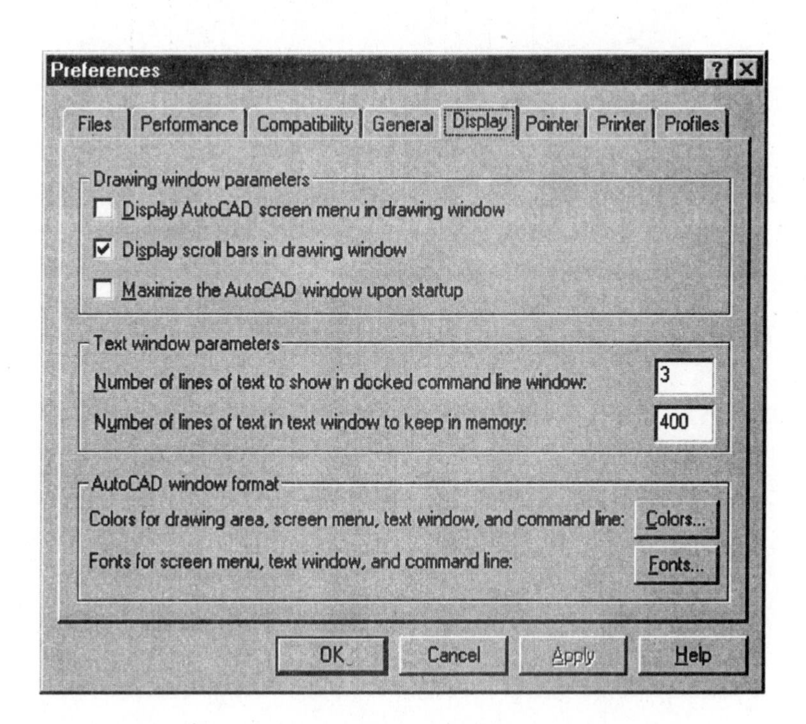

Figure A.36 Display tab of PREFERENCES command

Appendix B

Quick Command Reference

Brief descriptions of AutoCAD commands are given here for your quick reference. Commands prefixed by a ′ sign can be used transparently.

B.1 File and Plot Commands

AUDIT	enables you to evaluate the integrity of a drawing file.
NEW	enables you to start a new drawing file.
OPEN	enables you to open an existing drawing file.
PLOT	enables you to plot a drawing.
PREVIEW	enables you to display a plot preview.
QSAVE	enables you to save the current drawing.
QUIT	enables you to exit AutoCAD.
RECOVER	enables you to repair a damaged drawing file.
SAVE	enables you to save the current drawing to a new file name and continue working on the current drawing file.
SAVEAS	enables you to save the current drawing to a new file name and continue working on the new drawing file.

B.2 Export and Import Commands

3DSIN	enables you to import a 3D Studio file.
3DSOUT	enables you to export selected objects to a 3D Studio file.
ACISIN	enables you to import an ACIS file.
ACISOUT	enables you to export selected AutoCAD solid objects to an ACIS file.
AMECONVERT	enables you to convert AME solid models to AutoCAD solid objects.
BMPOUT	enables you to save selected objects to a bitmap file.
DWFOUT	enables you to export a drawing web format file.
DXBIN	enables you to import a specially coded binary file.
DXFIN	enables you to import a drawing interchange file.
DXFOUT	enables you to export a drawing interchange file.
EXPORT	enables you to export selected objects to other file formats.
IMPORT	enables you to import various file formats into AutoCAD.
PSIN	enables you to import a PostScript file.

PSOUT	enables you to export a PostScript file.
STLOUT	enables you to export a STL file from a 3D solid.
WMFIN	enables you to import a Windows metafile.
WMFOUT	enables you to export a Windows metafile.

B.3 Menu and Toolbar Commands

MENU	enables you to load a menu file.
MENULOAD	enables you to load partial menu files.
MENUUNLOAD	enables you to unload partial menu files.
TOOLBAR	enables you to manage toolbars.

B.4 Format Commands

'COLOR	enables you to set the color of new objects.
DDCOLOR	enables you to set the color of new objects by using dialog boxes.
'DDPTYPE	enables you to set point display mode by using dialog boxes.
'DDUNITS	enables you to set units of a drawing by using dialog boxes.
'LAYER	enables you to manage layers.
'LIMITS	enables you to control drawing limits.
'LINETYPE	enables you to manage linetypes.
'LTSCALE	enables you to set linetype scale factor.
MLSTYLE	enables you to define multiple parallel lines styles.
'UNITS	enables you to set units of a drawing.

B.5 Drawing Tool Commands

'APERTURE	enables you to set the size of the object snap target box.
'BLIPMODE	enables you to control the display of marker blips.
'CAL	enables you to evaluate mathematical and geometric expressions.
'DDGRIPS	enables you to manage grips by using dialog boxes.
'DDRMODES	enables you to set drawing aids.
'DDSELECT	enables you to set object selection modes.
'FILTER	enables you to manage object selection on the basis of properties.
'GRID	enables you to display a dot grid mesh.
GROUP	enables you to set and manage groups for selected objects.
'ISOPLANE	enables you to specify an isometric plane for constructing a 2D isometric drawing.
'ORTHO	enables you to constrain cursor movement to orthogonal directions.
'OSNAP	enables you to set running object snap modes and change the size of the object snap target box.
SELECT	enables you to place objects in the previous selection set.
'SNAP	enables you to restrict cursor movement to specific intervals.

B.6 UCS Commands

DDUCS	enables you to manage defined user coordinate systems (UCS).
DDUCSP	enables you to select a preset user coordinate system (UCS).
UCS	enables you to set and manage user coordinate systems (UCS).
UCSICON	enables you to manage the UCS icon.

B.7 Draw Commands

3D	enables you to construct 3D polygon mesh objects.
3DFACE	enables you to construct 3D face.
3DMESH	enables you to construct a set of free-form polygon meshes.
3DPOLY	enables you to construct a 3D polyline of straight line segments.
ARC	enables you to construct an arc.
BHATCH	enables you to add a hatch pattern to an enclosed area.
BOUNDARY	enables you to construct a region or a polyline from an enclosed area.
BOX	enables you to construct a 3D solid box.
CIRCLE	enables you to construct a circle.
CONE	enables you to construct a 3D solid cone.
CYLINDER	enables you to construct a 3D solid cylinder.
DIVIDE	enables you to place evenly spaced point objects or blocks along the length or perimeter of an object.
DONUT	enables you to construct filled circles and rings.
EDGESURF	enables you to construct a 3D polygon mesh from four edges.
ELLIPSE	enables you to construct an ellipse or an elliptical arc.
EXTRUDE	enables you to construct 3D solids by extruding 2D objects.
HATCH	enables you to add a hatch pattern to a specified boundary area.
LINE	enables you to construct straight line segments.
MEASURE	enables you to place point objects or blocks at measured intervals on an object.
MLINE	enables you to construct multiple parallel lines.
MTEXT	enables you to construct multiline text.
PFACE	enables you to construct a 3D polyface mesh vertex by vertex.
PLINE	enables you to construct 2D polylines.
POINT	enables you to construct point objects.
POLYGON	enables you to construct a regular polygon.
PSFILL	enables you to fill a 2D polyline outline with a PostScript pattern.
RAY	enables you to construct a semi-infinite line.
RECTANG	enables you to construct a rectangular polyline.
REGION	enables you to construct region objects from a selection set of existing objects.
REVOLVE	enables you to construct a 3D solid by revolving a 2D object.
REVSURF	enables you to construct a set of polygon meshes by rotating a curve.

RULESURF	enables you to construct a set of polygon meshes between two curves.
SKETCH	enables you to construct a series of freehand line segments.
SOLID	enables you to construct solid-filled polygons.
SPHERE	enables you to construct a 3D solid sphere.
SPLINE	enables you to construct a spline (NURBS) curve.
TORUS	enables you to construct a 3D donut-shaped solid.
TRACE	enables you to construct solid lines.
WEDGE	enables you to construct a 3D solid with a sloped face tapering along the X axis.
SECTION	enables you to construct a region across a selected section of a 3D solid.
TABSURF	enables you to construct a set of 3D polygon meshes from a path curve and a direction vector.
XLINE	enables you to construct an infinite line.

B.8 Edit and Modify Commands

3DARRAY	enables you to array objects in 3D space.
ALIGN	enables you to move and rotate objects to align with other objects.
ARRAY	enables you to array objects on a 2D plane.
BREAK	enables you to break an object in two or erase parts of objects.
CHAMFER	enables you to bevel the edges of objects.
CHANGE	enables you to change the properties or the point of existing objects.
CHPROP	enables you to change the color, layer, linetype, linetype scale, and thickness of an object.
COPY	enables you to copy objects.
DDCHPROP	enables you to change the color, layer, linetype, linetype scale, and thickness of an object by using dialog boxes.
DDEDIT	enables you to edit text and attribute definitions by using dialog boxes.
DDMODIFY	enables you to control the properties of existing objects.
ERASE	enables you to erase objects from a drawing.
EXPLODE	enables you to break a compound object into its component objects.
EXTEND	enables you to extend an object to meet another object.
FILLET	enables you to round off the edges of objects.
HATCHEDIT	enables you to edit a hatch object.
INTERSECT	enables you to form composite solids or regions by intersection.
LENGTHEN	enables you to lengthen an object.
'MATCHPROP	enables you to copy the properties from one object to one or more objects.
MIRROR	enables you to construct a mirror image of objects.
MIRROR3D	enables you to construct a mirror image of objects about a plane.
MLEDIT	enables you to edit multiple parallel lines.

MOVE	enables you to translate objects.
MULTIPLE	enables you to repeat the next command until it is canceled.
OFFSET	enables you to construct concentric circles, parallel lines, and parallel curves.
OOPS	enables you to restore erased objects.
PEDIT	enables you to edit polylines and 3D polygon meshes.
ROTATE	enables you to rotate objects about a base point.
ROTATE3D	enables you to rotate objects about a 3D axis.
SCALE	enables you to scale objects equally in the X-, Y-, and Z-directions.
SLICE	enables you to slice solids along a plane.
SPLINEDIT	enables you to edit a spline curve.
STRETCH	enables you to move or stretch objects.
SUBTRACT	enables you to form a composite solid or region by subtraction.
TRIM	enables you to trim objects at a cutting edge defined by other objects.
UNION	enables you to form composite region or solid by uniting.
XPLODE	enables you to break a compound object into its component objects.

B.9 Undo and Redo Commands

REDO	enables you to reverse the effects of the previous UNDO or U command.
U	enables you to reverse the most recent operation.
UNDO	enables you to reverse the effect of commands.

B.10 Block and External References Commands

ATTDEF	enables you to define an attribute.
'ATTDISP	enables you to control attribute visibility.
ATTEDIT	enables you to edit attribute values.
ATTEXT	enables you to extract attribute data.
ATTREDEF	enables you to redefine a block and update associated attributes.
'BASE	enables you to set the insertion base point.
BLOCK	enables you to create a block definition.
BMAKE	enables you to create a block definition by using dialog boxes.
DDATTDEF	enables you to define an attribute by using dialog boxes.
DDATTE	enables you to edit attribute values by using dialog boxes.
DDATTEXT	enables you to extract attribute data by using dialog boxes.
DDINSERT	enables you to insert a block or another drawing by using dialog boxes.
INSERT	enables you to insert a block or another drawing.
MINSERT	enables you to insert multiple instances of a block or another drawing in a rectangular array.

WBLOCK	enables you to write objects to a new drawing file.
XATTACH	enables you to attach another drawing to the current drawing.
XBIND	enables you to bind dependent symbols of an attached drawing to the current drawing.
XCLIP	enables you to clip the instance of an attached drawing.
XREF	enables you to manage attached drawings.

B.11 Annotation and Dimension Commands

DDIM	enables you to create and modify dimension styles.
DIM	enables you to access Dimensioning mode.
DIMALIGNED	enables you to construct an aligned linear dimension.
DIMANGULAR	enables you to construct an angular dimension.
DIMBASELINE	enables you to continue a linear, angular, or ordinate dimension from the baseline of the previous or selected dimension.
DIMCENTER	enables you to construct the center mark or the center lines of circles and arcs.
DIMCONTINUE	enables you to continue a linear, angular, or ordinate dimension from the second extension line of the previous or a selected dimension.
DIMDIAMETER	enables you to construct diameter dimensions for circles and arcs.
DIMEDIT	enables you to edit dimensions.
DIMLINEAR	enables you to construct linear dimensions.
DIMORDINATE	enables you to construct ordinate point dimensions.
DIMOVERRIDE	enables you to override dimension system variables.
DIMRADIUS	enables you to construct radial dimensions for circles and arcs.
DIMSTYLE	enables you to create and modify dimension styles on the command line interface.
DIMTEDIT	enables you to move and rotate dimension text.
DTEXT	enables you to add text and display it on screen while it is being typed.
LEADER	enables you to construct a line that connects annotation to a feature.
'STYLE	enables you to create named text styles.
TEXT	enables you to add a single line of text.
TOLERANCE	enables you to include geometric tolerances in the drawing.

B.12 Documentation Commands

MSPACE	enables you to switch from paper space to a model space viewport.
MVIEW	enables you to construct floating viewports.
MVSETUP	enables you to set up the specifications of a drawing.
PSPACE	enables you to switch from a model space viewport to paper space.
SOLDRAW	enables you to generate profiles and sections of 3D solids in viewports created with the SOLVIEW command.

SOLPROF	enables you to generate profile image of objects in viewports created with the MVIEW command.
SOLVIEW	enables you to construct floating viewports for 3D solids.
VPLAYER	enables you to control layer visibility within floating viewports.

B.13 View Commands

DDVIEW	enables you to manage named views by using dialog boxes.
DDVPOINT	enables you to set the display to a 3D view by using dialog boxes.
DRAWORDER	enables you to change the display order of objects.
DSVIEWER	enables you to open the Aerial View window.
DVIEW	enables you to set the display to parallel projection or perspective views.
EDGE	enables you to control the visibility of 3D face edges.
'FILL	enables you to control the filling of multilines, traces, solids, solid-fill hatches, and wide polylines.
HIDE	enables you to suppress the display of hidden lines of 3D objects.
'PAN	enables you to pan the display.
PLAN	enables you to set the display to a plan view.
'REDRAW	enables you to refresh the screen display of the current viewport.
'REDRAWALL	enables you to refresh the screen display of all viewports.
REGEN	enables you to regenerate the current viewport.
REGENALL	enables you to regenerate all viewports.
SHADE	enables you to display a flat-shaded image.
'VIEW	enables you to manage named views.
VIEWRES	enables you to set the resolution for object generation in the current viewport.
VPOINT	enables you to set the display to a 3D view.
VPORTS	enables you to divide the graphics area into multiple tiled viewports.
'ZOOM	enables you to zoom the viewport.

B.14 Image and Rendering Commands

BACKGROUND	enables you to set the background of scenes.
FOG	enables you to set visual cues for the apparent distance of objects.
IMAGE	enables you to insert images into an AutoCAD drawing file.
IMAGEADJUST	enables you to control the brightness, contrast, and fade values of the selected image.
IMAGEATTACH	enables you to attach an image object.
IMAGECLIP	enables you to create new clipping boundaries for an attached image object.
IMAGEFRAME	enables you to control the visibility of the frame of an attached image object.

IMAGEQUALITY	enables you to control the display quality of an attached image object.
LIGHT	enables you to manage lights and lighting effects.
MATLIB	enables you to import and export materials to and from a library of materials.
MSLIDE	enables you to construct a slide file of the current viewport.
RENDER	enables you to construct a realistically shaded image of a 3D wireframe or solid model.
REPLAY	enables you to display a BMP, TGA, or TIFF image.
RMAT	enables you to manage rendering materials.
SAVEIMG	enables you to save a rendered image.
SCENE	enables you to manage scenes.
SETUV	enables you to map materials onto geometry.
SHOWMAT	enables you to list the material type and attachment method for an object.
TRANSPARENCY	enables you to control the transparency of background pixels.
VSLIDE	enables you to display a slide file created with the MSLIDE command.

B.15 Inquiry Commands

'ABOUT	exhibits a dialog box to display information about AutoCAD.
AREA	enables you to calculate the area and perimeter of objects or of defined areas.
DBLIST	enables you to list database information for all the objects.
'DIST	enables you to measure the distance and angle between two points.
'HELP	enables you to display online help.
'ID	enables you to find out the coordinate values of a location.
INTERFERE	enables you to check the interference of two or more 3D solids.
LIST	enables you to display database information for selected objects.
MASSPROP	enables you to evaluate the mass properties of regions or solids.
STATS	enables you to display rendering statistics.
'STATUS	enables you to display drawing statistics, modes, and extents.
'TIME	enables you to display the date and time statistics of a drawing.
'TREESTAT	enables you to display information about the drawing's current spatial index.

B.16 Configuration Commands

'DRAGMODE	enables you to control the way dragged objects are displayed.
PREFERENCES	enables you to customize AutoCAD settings.
PSDRAG	enables you to control the appearance of a PostScript image as it is dragged into position with the PSIN command.
'QTEXT	enables you to control the display and plotting of text and attribute objects.

REDEFINE	enables you to restore AutoCAD internal commands undefined by the UNDEFINE command.
'REGENAUTO	enables you to control automatic regeneration of a drawing.
REINIT	enables you to reinitialize the digitizer, digitizer input/output port, and program parameters file.
RPREF	enables you to set rendering preferences.
'SETVAR	enables you to manage system variables.
SYSWINDOWS	enables you to arrange AutoCAD application windows.
TABLET	enables you to manage the digitizing tablet.
UNDEFINE	enables you to undefine an internal AutoCAD command.
WMFOPTS	enables you to set options for the WMFIN command.

B.17 Object Management Commands

CONVERT	enables you to convert 2D polylines and associative hatches to the optimized Release 14 format.
DDRENAME	enables you to rename objects by using dialog boxes.
'ELEV	enables you to set elevation and extrusion thickness properties of new objects.
PURGE	enables you to remove unused named objects.
RENAME	enables you to rename objects.

B.18 Utility Commands

BROWSER	enables you to launch the default Web browser.
'GRAPHSCR	enables you to switch from the text window to the graphics area.
LOGFILEOFF	enables you to close the log file opened by the LOGFILEON command.
LOGFILEON	enables you to open a log file that records the command line interface contents.
SHELL	enables you to access operating system commands.
'SPELL	enables you to check spelling.
'TEXTSCR	enables you to open the AutoCAD text window.

B.19 Windows Objects Commands

COPYCLIP	enables you to copy objects to the Windows clipboard.
COPYHIST	enables you to copy the command line history to the Windows clipboard.
COPYLINK	enables you to copy the current view to the Windows clipboard for linking to other OLE applications.
CUTCLIP	enables you to copy objects to the Windows clipboard and erases the objects from the drawing.
INSERTOBJ	enables you to insert a linked or embedded object.
OLELINKS	enables you to manage OLE links.

PASTECLIP	enables you to insert data from the Windows clipboard.
PASTESPEC	enables you to insert data from the Windows clipboard and control the format of the data.

B.20 Scripts Commands

'DELAY	enables you to add a timed pause to your script file.
'RESUME	enables you to resume an interrupted script file.
RSCRIPT	enables you to make a script file repeat continuously.
'SCRIPT	enables you to run a script file.

B.21 External Application Commands

'APPLOAD	enables you to load AutoLISP, ADS, and ARX applications.
ARX	enables you to load and unload ARX applications.
COMPILE	enables you to compile shape files and PostScript font files.
LOAD	enables you to load shape files for use by the SHAPE command.
SHAPE	enables you to insert a shape.

B.22 External Database Commands

ASEADMIN	enables you to manage external database commands.
ASEEXPORT	enables you to export link information for selected objects.
ASELINKS	enables you to manage links between objects and an external database.
ASEROWS	enables you to display and edit table data and create links and selection sets.
ASESELECT	enables you to create a selection set from rows that are linked to textual selection sets and graphic selection sets.
ASESQLED	enables you to execute Structured Query Language (SQL) statements.

B.23 Landscape Commands

LSEDIT	enables you to edit a landscape object.
LSLIB	enables you to maintain libraries of landscape objects.
LSNEW	enables you to add realistic landscape items, such as trees and bushes, to your drawings.

Appendix C

Export, Import, Windows Clipboard, and Raster Image

C.1 File Export

You can export an AutoCAD objects by using the EXPORT command. See Figure C.1.

<File> **<Export...>**

Command: **EXPORT**

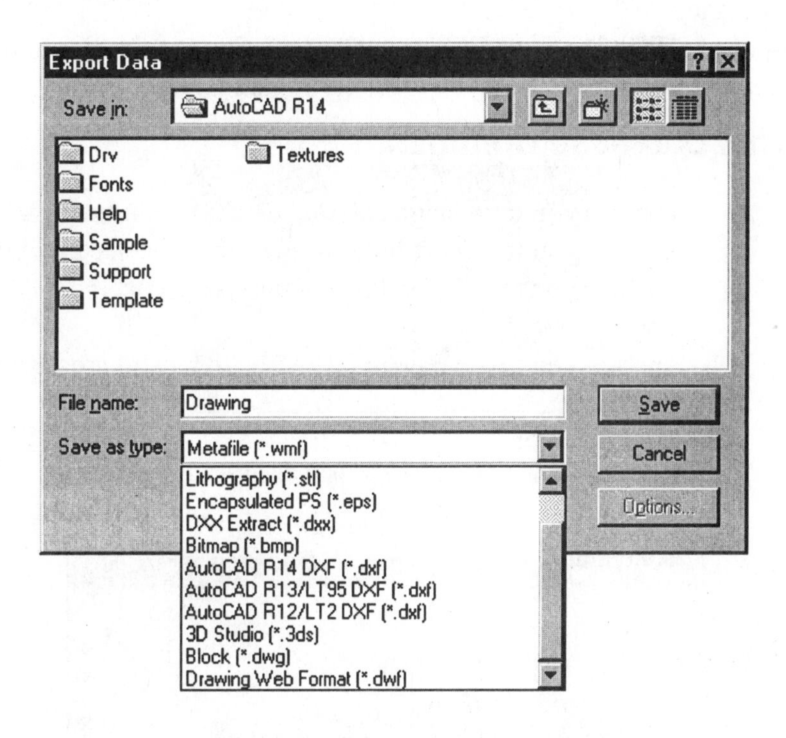

Figure C.1 Export Data dialog box

Using this command, you can output drawing objects in the following file types:

3DS exports objects to 3D Studio file. You can also use the 3DSOUT command.

BMP exports objects to Device-independent bitmap file. You can also use the BMPOUT command.

DWG exports objects to AutoCAD drawing file. You can also use the WBLOCK command.

DWF exports objects to AutoCAD drawing web format file. You can also use the DWFOUT command.

DXF exports objects to AutoCAD R14 drawing interchange file. You can also use the DXFOUT command.

DXF exports objects to AutoCAD R13/LT95 drawing interchange file. You can also use the DXFOUT command.

DXF exports objects to AutoCAD R12/LT2 drawing interchange file. You can also use the DXFOUT command.

DXX exports objects to Attribute extract DXF file. You can also use the ATTEXT command.

EPS exports objects to Encapsulated PostScript file. You can also use the PSOUT command.

SAT exports objects to ACIS solid object file. You can also use the ACISOUT command.

STL exports solid objects to stereo-lithography file. You can also use the STLOUT command.

WMF exports objects to Windows Metafile. You can also use the WMFOUT command.

If you output to DWF, DXF, or EPS formats, you can use the [Options...] button to set export options. See Figure C.2, Figure C.3, and Figure C.4.

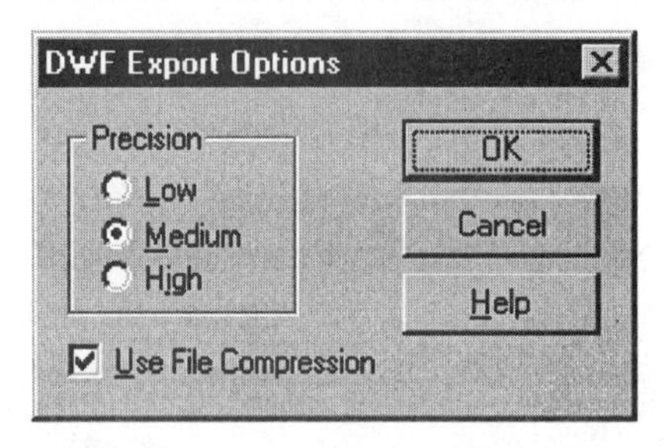

Figure C.2 DWF export options

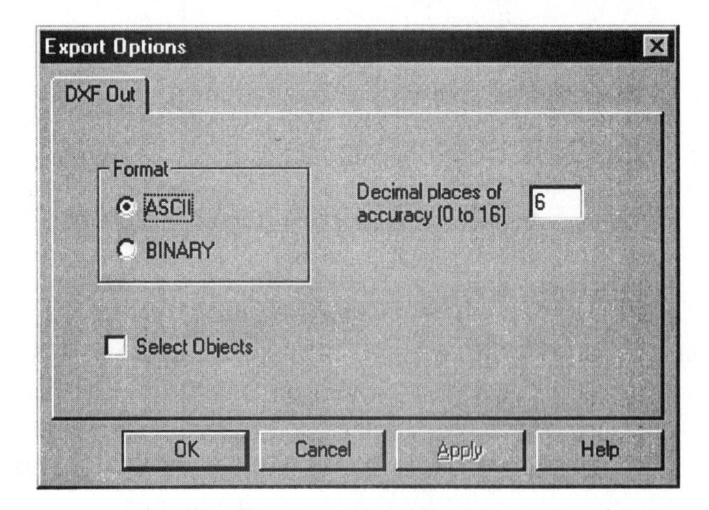

Figure C.3 DXF export options

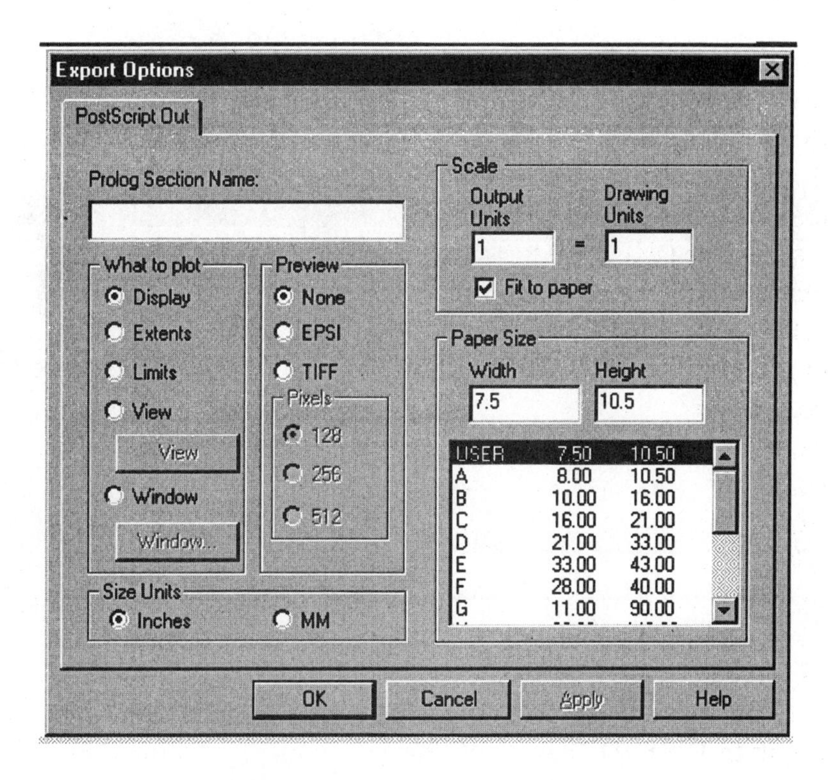

Figure C.4 EPS export options

C.2 File Import

To import various file formats into an AutoCAD drawing, you can use the IMPORT command. Select the Import icon of the Insert toolbar. If you cannot find the Insert toolbar, select the Toolbars... item from the View pull-down menu and then select the Insert item from the toolbars list. See Figure C.5.

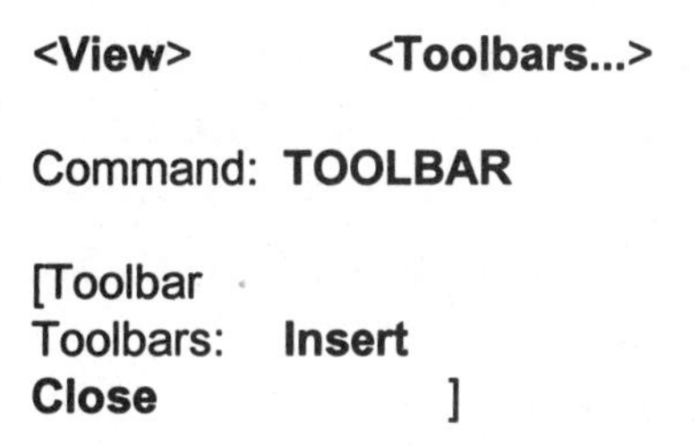

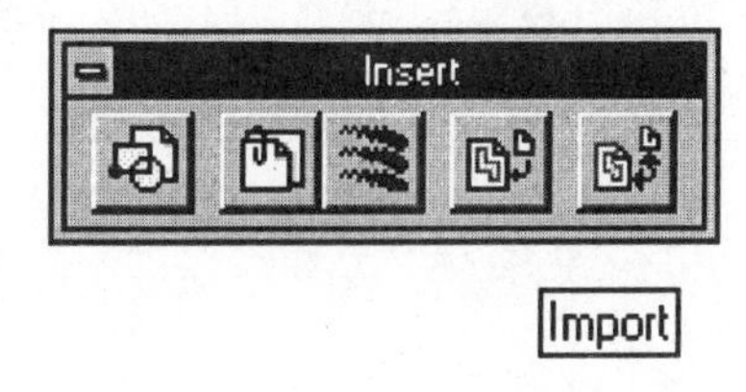

Figure C.5 Displaying the Insert toolbar

From the Insert toolbar, select the Import icon. See Figure C.6.

[Insert] **[Import]**

Command: **IMPORT**

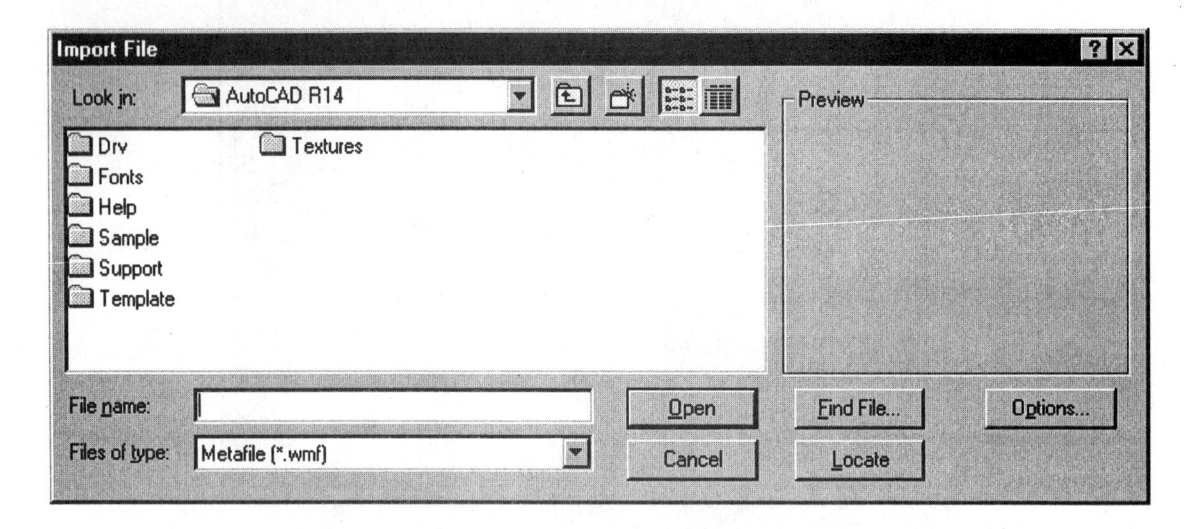

Figure C.6 Import File dialog box

Using the IMPORT command, you can import the following file types:

3DS imports a Studio file. You can also use the 3DSIN command by selecting the 3D Studio... item from the Insert pull-down menu.

DXF imports an AutoCAD drawing interchange file. You can also use the DXFIN command.

EPS imports an Encapsulated PostScript file. You can also use the PSIN command by selecting the Encapsulated PostScript... item from the Insert pull-down menu.

SAT imports an ACIS solid object file. You can also use the ACISIN command by selecting the ACIS Solids... item from the Insert pull-down menu.

WMF imports a Windows metafile. You can also use the WMFIN command by selecting the Windows Metafile... item from the Insert pull-down menu.

If you import a Metafile (WMF) file, you can use the [Options...] button. See Figure C.7.

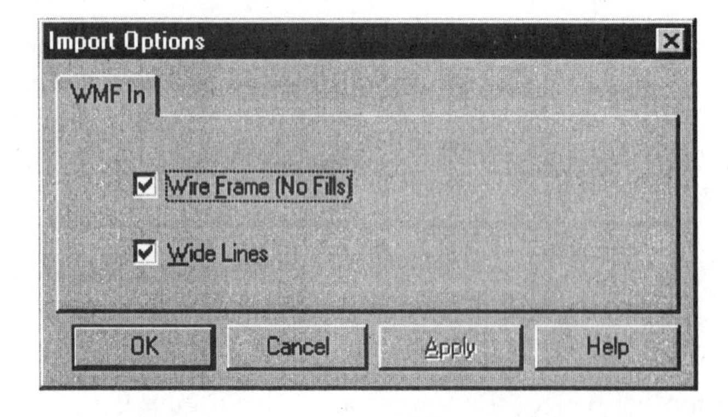

Figure C.7 WMFIN import options

C.3 Windows Clipboard

AutoCAD objects are compatible with Windows clipboard objects. You can copy selected AutoCAD objects to Windows clipboard by using the COPYCLIP command. The shortcut key is [^C]. [^C] means holding the [Control] key in conjunction with the [C] key of the keyboard.

[Standard toolbar] **[Copy to Clipboard]**

<Edit> **<Copy>**

Command: **COPYCLIP**

If you want to remove selected AutoCAD objects and place them to the clipboard, you can use the CUTCLIP command. The shortcut key is [^X].

[Standard toolbar] **[Cut to Clipboard]**

<Edit> **<Cut>**

Command: **CUTCLIP**

To insert data to AutoCAD drawing from the clipboard, you can use the PASTECLIP command. The shortcut key is [^V].

[Standard toolbar] **[Copy from Clipboard]**

<Edit> **<Paste>**

Command: **PASTECLIP**

To paste a linked or embedded object from the clipboard to an AutoCAD drawing, you can use the PASTESPEC command.

<Edit> **<Paste Special>**

Command: **PASTESPEC**

C.4 OLE Object

OLE means object linking and embedding. To insert a linked or embedded object, you can use the INSERTOBJ command by selecting the OLE Object... item from the Insert pull-down menu. See Figure C.8.

<Insert> **<OLE Object...>**

Command: **INSERTOBJ**

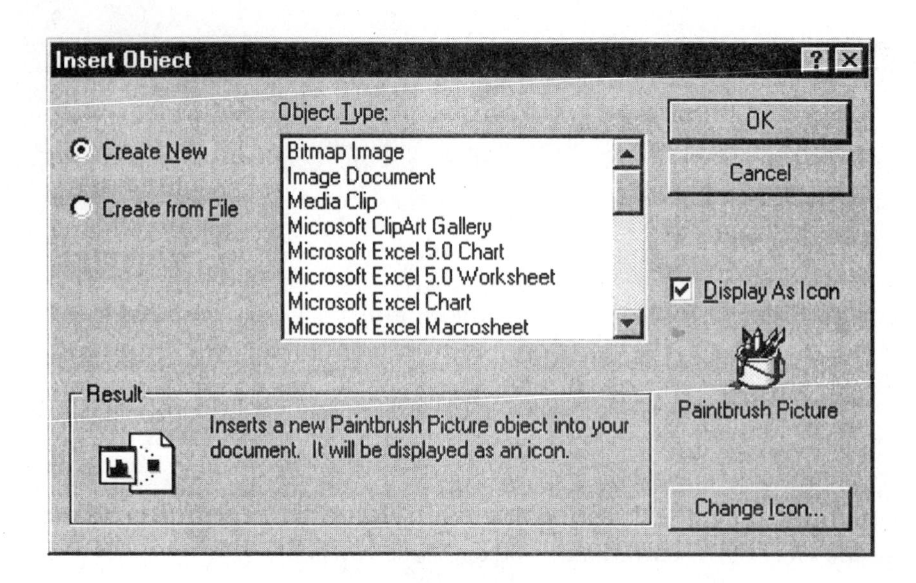

Figure C.8 Insert Object dialog box

Using the INSERTOBJ command to insert an object from an application that supports object linking and embedding to an AutoCAD drawing, you can either insert it by embedding or insert it by linking.

While working in AutoCAD environment, you can use the original application to edit the object. If you edit an embedded object, the embedded object changes but the original source object does not change. If you edit a linked object, the original object changes as well.

C.5 Raster Image Import

You can insert raster or bit-mapped bitonal, 8-bit gray, 8-bit color, or 24-bit color image file into an AutoCAD drawing. To insert and manage images, you can use the IMAGE command by selecting the Raster Image... item from the Insert pull-down menu. See Figure C.9. To insert images, you can use the IMAGEATTACH command by selecting the Image Attach icon of the Reference toolbar. See Figure C.10. These commands insert the images: BMP, TIF, RLE, JPG, GIF, and TGA.

<Insert> **<Raster Image...>**

Command: **IMAGE**

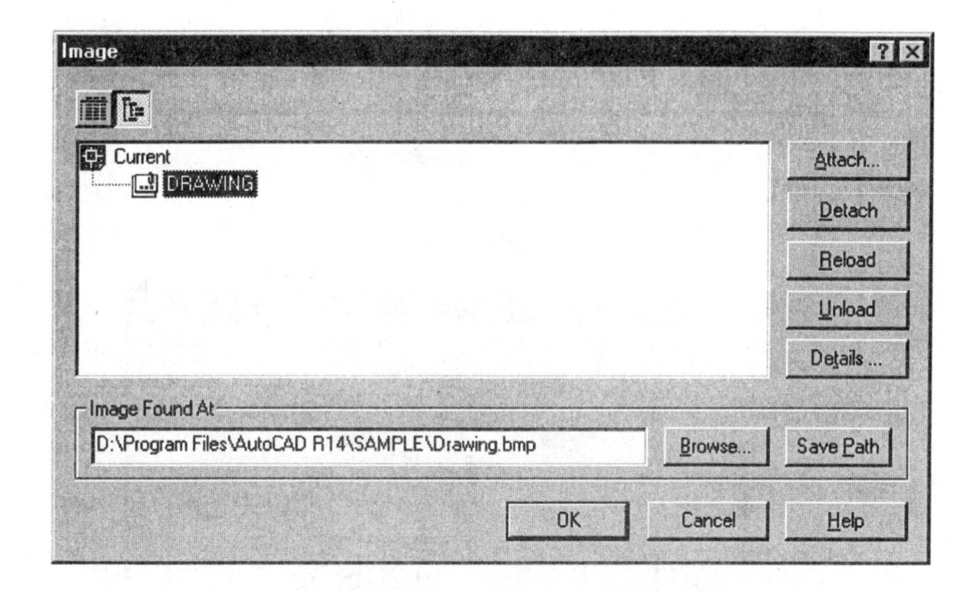

Figure C.9 Image dialog box

[Reference] **[Image Attach]**

Command: **IMAGEATTACH**

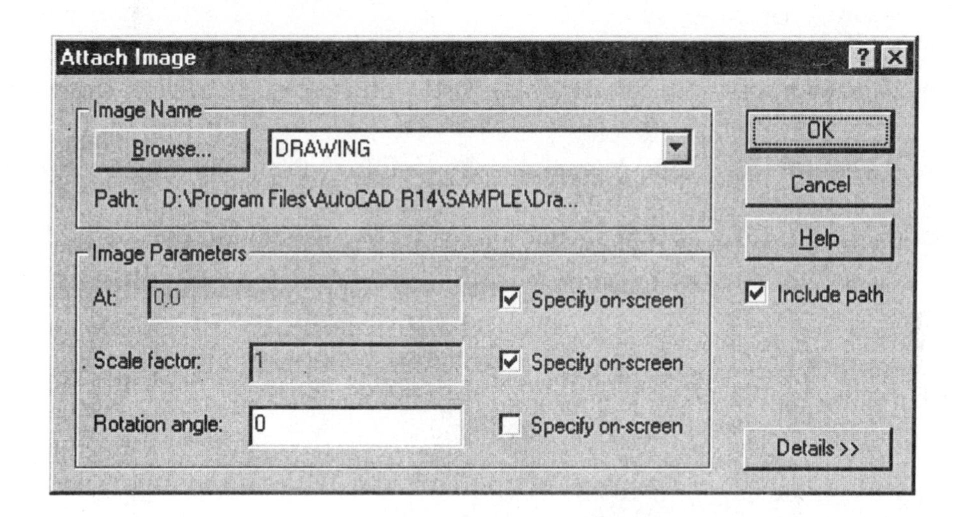

Figure C.10 Attach Image dialog box

C.6 Image Display and Saving

While you are working on an AutoCAD drawing, you can display an image in GIF, TGA, or TIFF format by using the REPLAY command. To apply this command, select View... from the Display Image cascading menu item of the Tools pull-down menu.

<Tools> **<Display Image>** **<View...>**

Command: **REPLAY**

To save the current display view as a BMP image, apply the SAVEIMG command. See Figure C.11.

<Tools> **<Display Image>** **<Save>**

Command: **SAVEIMG**

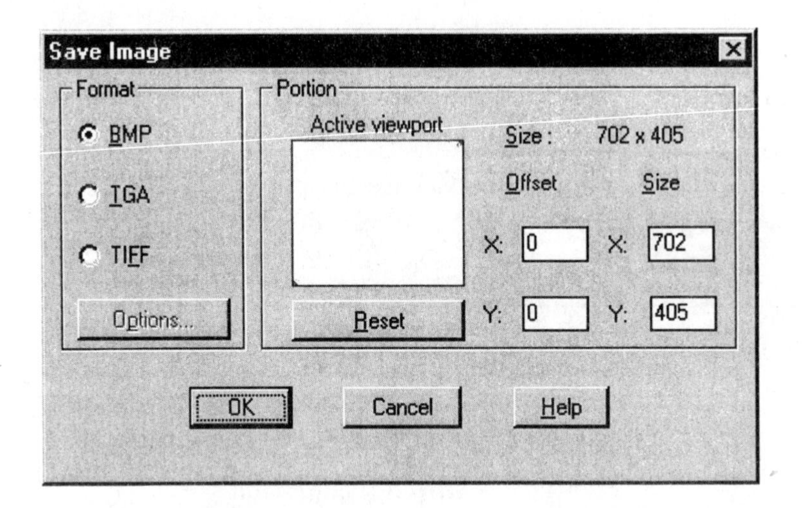

Figure C.11 Save Image dialog box

C.7 Slide Images

Slides are saved screen images. You can make a slide with the MSLIDE command and view a saved slide with the VSLIDE command.

Make a slide of the current screen display with the MSLIDE command.

Command: **MSLIDE**

When the dialog box appears, specify a slide file name.

Start a new drawing with the NEW command.

<File> **<New...>**

Command: **NEW**

Use the VSLIDE command to view the saved slide.

Command: **VSLIDE**

To restore the previous screen display after viewing a slide, apply the REDRAW command.

<View> **<Redraw>**

Command: **REDRAW**

Appendix D
Script File

A script file is a batch file running within the AutoCAD environment. It is a text file that can be created by any text editor. The file extension must be SCR for AutoCAD to recognize it as a script file. A script file contains lines of instructions that are valid command names and inputs to the commands. Script files are useful in presentation and demonstration. You can prepare a script file to demonstrate a sequence of operation. When writing a script file, take care not to include extra spaces in the script because a space is taken as a return key stroke.

Use a text editor to create a script file that contains the following lines. Do not include the <> symbols or the text enclosed within them. They are explanations for your reference.

```
CIRCLE          <Issues the CIRCLE command.>
120,150         <States the center position.>
25              <Specifies the radius.>
ZOOM            <Runs the ZOOM command.>
W               <Sets the W option.>
80,110          <Selects a first point.>
160,190         <Sets the other corner of the window.>
ERASE           <Executes the ERASE command.>
LAST            <Chooses the last object created.>
                <This blank line is equivalent to a return key stroke.>
REDRAW          <Calls for a redraw. Remember to add a return here.>
```

Save the foregoing lines to a file, and place the file in the current working directory. Run the script file with the SCRIPT command.

<Tools> **<Run Scripts...>**

Command: **SCRIPT**

When the File dialog box appears, select the saved script file and then the [OK] button. After that, AutoCAD draws on the screen a circle with a radius of 25 units at (120,150). Next, the display is zoomed to a window with corners at (80,110) and (160,190). The circle is then erased, and the screen display is redrawn.

Time Delay

The foregoing sequence would be too fast if you did not add a time delay in the script. Add a DELAY command in the script. Modify the script file as follows:

CIRCLE
120,150
25
ZOOM
W
80,110
160,190
DELAY <Adds a time delay in the script.>
3000 <Sets the time delay to 3000 milliseconds.>
ERASE
LAST
<This is a blank line.>
REDRAW

Run the script file again with the SCRIPT command.

<Tools> <Run Scripts...>

Command: **SCRIPT**

This time, the system halts for 3000 milliseconds after the circle is created. Then the displayed is zoomed. After that, it erases the circle and performs a redraw.

Replay

You can cause a script file to keep on repeating by adding the RSCRIPT command at the last line of the file.

CIRCLE
120,150
25
ZOOM
W
80,110
160,190
DELAY
3000
ERASE
LAST
<This is a blank line.>
DELAY <Adds another time delay.>
3000 <Sets the time delay to 3000 milliseconds.>
REDRAW

RSCRIPT <Causes the script to repeat endlessly.>

Run the script file again with the SCRIPT command.

<Tools> <Run Scripts...>

Command: **SCRIPT**

After you add the RSCRIPT command, the script file rewinds endlessly to draw a circle and then to erase it.

Interrupt and Replay

Sooner or later you have to terminate a script file that replays. You can press the [ESC] key. If you want to continue to play an interrupted script file, use the RESUME command.

Command: *Cancel*
Command: **RESUME**

The script continues to play afterwards.

Index

3D Cartesian coordinates, 411
3D coordinate systems, 411
3D cylindrical coordinates, 411
3D spherical coordinates, 411

A
ABOUT command, 252
Absolute coordinates, 177
Aerial view, 77
AM2SF command, 317
AMNEW command, 318
AMSOLCUT command, 317
Annotation, 242
APBOX variable, 85
APERTURE variable, 85
Aperture, 56
Application window, 4
ARC command, 57
AREA command, 254
ARRAY command, 106
ARRAY command, 54
Assembly of solid models, 489
Associative dimension, 257
Attach a file, 212
ATTDEF command, 228
ATTDEF command, 567
ATTDIA variable, 231
ATTDISP variable, 232
ATTEXT command, 234
ATTREDEF command, 236
Auditing, 23

B
BASE command, 195
Basic dimension, 304
Batch plotting, 615
BHATCH command, 119
Block attributes, 227
BLOCK command, 127
BMAKE command, 126
Boolean operations, 315
BOUNDARY command, 86
Boundary hatching, 118
BOX command, 320
BREAK command, 98

C
Calculation, 24
Cellular decomposition, 313
Center mark, 284
CHAMFER command, 105
CHAMFER command, 341
CHANGE command, 101
CHPROP command, 45
CIRCLE command, 54
Clipping external references, 221
Color, 47
Command history, 24
Command line interface, 7
Compatibility, 21
Complex region, 156
Components of a dimension, 256
CONE command, 335
Constructive solid geometry, 313
Coordinate display, 32
Coordinate systems, 177
COPY command, 79
COPYCLIP command, 689
Current layer, 43
Cursor menu, 5
Custom arrowheads, 299
CUTCLIP command, 689
CYLINDER command, 326

D
DBLIST command, 254
DDATTDEF command, 228
DDATTDEF command, 567
DDCHPROP command, 45
DDEDIT command, 248
DDGRIPS command, 169
DDIM command, 257
DDINSERT command, 185
DDMODIFY command, 246
DDMODIFY command, 68
DDPTYPE command, 97
DDRMODES command, 167
DDRMODES command, 37
DDSELECT command, 55
DDVIEW command, 75
DDVPOINT command, 324
Definition points, 257
DELAY command, 694
DELOBJ variable, 155
DIMALIGNED command, 278
DIMANGULAR command, 281
DIMBASELINE command, 279
DIMCENTER command, 284
DIMCONTINUE command, 280
DIMDIAMETER command, 282
DIMEDIT command, 295
Dimension style, 257
Dimension variables, 264
Dimensioning principles, 256
Dimensioning, 585
Dimensioning, 604
DIMLINEAR command, 275
DIMORDINATE command, 291
DIMOVERRIDE command, 297
DIMRADIUS command, 283
DIMTEDIT command, 297
Direct distance entry, 63
DIST command, 255
DIVIDE command, 127
DIVIDE command, 97
Documentation, 565
DONUT command, 148
Drawing limits, 32
DRAWORDER command, 86
DTEXT command, 243
DVIEW command, 622
Dynamic zoom, 76

E
Edit text, 246
ELLIPSE command, 167
ELLIPSE command, 63
Ending a drawing session, 23
Engineering title block, 565

Entity group, 174
Entity properties, 45
Entity selection, 173
ERASE command, 91
ESC key,
EXPLODE command, 160
EXPLODE command, 186
EXPLODE command, 197
EXPORT command, 685
EXTEND command, 107
External references, 207
EXTRUDE command, 353
EXTRUDE command, 361
Extruded solids, 314

F
FACETRES variable, 627
File preview, 22
FILL command, 143
FILLET command, 342
FILLET command, 82
FILTER command, 99
Filter points, 61
Floating viewport visibility control, 583
Floating viewport, 573
Format code, 250
Function keys, 16

G
Geometric reference frame, 304
Geometric snap tools, 175
Geometric tolerance symbol, 306
Geometric tolerance, 304
Global linetype scale, 72
Graphic window, 6
Grid mesh, 33
Grips, 169
GROUP command, 175
GROUP command, 360

H
HATCH command, 87
HATCHEDIT command, 121
Help dialog box, 13
HIDE command, 625

I
Icon menu, 15
ID command, 255
IMAGE command, 690
IMAGEATTACH command, 691
IMPORT command, 687
Inquiry, 252
Insertion base point, 195
INSERTOBJ command, 689
INTERFERE command, 463
Interference, 461
INTERSECT command, 156
INTERSECT command, 378
Isolines, 316
ISOPLANE command, 168

L
Lateral tolerance, 301
Layer creation, 39
Layer management, 39
Layer removal, 43
LEADER command, 287
Leader, 287
LENGTHEN command, 108
LENGTHEN command, 64
LIGHT command, 639
Light, 635
LIMITS command, 52
LINE command, 59
Linetype scale, 39
Linetype, 38
LIST command, 164
LIST command, 254
Load application, 25
Log file, 24

M
Marker blips, 56
Mass properties, 465
MASSPROP command, 465
Match properties, 46
Material and mapping, 647
MEASURE command, 99
Menu bar configuration, 659
MENULOAD command, 659
MINSERT command, 185
MIRROR command, 60
MIRROR3D command, 339
MLEDIT command, 144
MLEDIT command, 506
MLINE command, 135
MLINE command, 505
MLSTYLE command, 133
MLSTYLE command, 504
MOVE command, 103
MSLIDE command, 692
MSPACE command, 576
MTEXT command, 244
Multiline text, 244
Multiple lines, 133
MVIEW command, 573
MVIEW command, 598

N
Native solid, 314

O
Object properties toolbar, 48
OFFSET command, 64
OOPS command, 127
Open a file, 21
Ortho mode, 60
OSNAP command, 84
Overlay, 217

P
Paper space, 573
PASTECLIP command, 689
PASTESPEC command, 689
PDMODE variable, 97
Perspective view, 621
Pick first, 172
Pickbox size, 173
Pickbox size, 55
PLINE command, 146
PLINE command, 85
Plot a document, 610
PLOT command, 610
Plot preview, 614
POINT command, 96
Pointing device, 5
POLYGON command, 124
PREFERENCES command, 673
Preferences, 3
PREVIEW command, 615
Primitive solids, 314
PSPACE command, 578
PURGE command, 197

Q
Quick text mode, 251

R
Rapid prototype, 654
RAY command, 163
Re-initialization, 23
Real-time zoom, 77
Recalling a command, 10
Recovering a damaged file, 22
RECTANG command, 155
Redefining a block, 128
Redo, 34
REDRAW command, 56
REGEN command, 69
Regeneration speed, 70